Functional Materials for Electrocatalytic Energy Conversion

Functional Materials for Electrocatalytic Energy Conversion

Edited by Zhicheng Zhang, Meiting Zhao, and Yuchen Qin

WILEY-VCH

Editors

Prof. Zhicheng Zhang
Department of Chemistry, School of Science; Tianjin Key Laboratory of Molecular Optoelectronic Sciences
Tianjin University
China

Prof. Meiting Zhao
Department of Chemistry, Institute of Molecular Aggregation Science, Tianjin Key Laboratory of Molecular Optoelectronic Sciences
Tianjin University
Tianjin
China

Prof. Yuchen Qin
College of Science
Henan Agricultural University
Zhengzhou
China

Cover: © Andriy Onufriyenko/
Getty Images

Library of Congress Card No.: applied for

British Library Cataloguing-in-Publication Data
A catalogue record for this book is available from the British Library.

Bibliographic information published by the Deutsche Nationalbibliothek
The Deutsche Nationalbibliothek lists this publication in the Deutsche Nationalbibliografie; detailed bibliographic data are available on the Internet at <http://dnb.d-nb.de>.

Print ISBN: 978-3-527-35365-1
ePDF ISBN: 978-3-527-84635-1
ePub ISBN: 978-3-527-84637-5
oBook ISBN: 978-3-527-84636-8

Typesetting Straive, Chennai, India

Printed and bound by CPI Group (UK) Ltd, Croydon, CR0 4YY

C9783527353651_291124

Contents

Preface

With the continued rise in social energy demand and the decreasing availability of fossil fuels, there has been an increasing need for affordable, environmental friendly, and efficient renewable energy conversion and storage systems. This pursuit has become a critical challenge for modern society. Electrocatalysts play a crucial role in various energy storage and conversion systems, such as fuel cells, metal–air batteries, CO_2 conversion, and water splitting. This is accomplished by accelerating electrochemical reactions, which is essential for enhancing the efficiency of energy conversion.

This book aims to provide a comprehensive overview of functional materials in the electrocatalytic energy conversion process, including oxygen reduction, electrochemical CO_2 reduction, water splitting, nitrogen reduction, liquid fuel oxidation, and electrocatalytic biomass conversion. In the first section, the catalytic mechanisms and pathways of various electrocatalytic reactions are systematically discussed. The second section uncovers the design ideas, geometric/electronic structures, and structure–activity relationships of various state-of-the-art electrocatalysts in detail. Finally, the book proposes the challenges and opportunities of electrocatalytic functional materials for energy conversion in the future. Through the dissemination of knowledge and information in this book, we aim to stimulate increased interest and investment in further research and development in the field of electrocatalysis, ultimately leading to impactful advancements in the use of functional materials for practical applications.

30 January 2024

Zhicheng Zhang
Yuchen Qin
Meiting Zhao

Preface

Acknowledgments

We thank editors Ashok Ravi, Nandhini Karuppiah and Alice Qian from Wiley for their great efforts in editing this book.

About the Editors

Zhicheng Zhang is a professor of Tianjin University, China. He obtained his PhD degree from China University of Petroleum (Beijing) in 2012. Then he worked as a postdoctoral researcher in Tsinghua University. In 2014, he joined Nanyang Technological University (Singapore) as a research fellow. Since 2019, he joined Tianjin University as a full professor. His current research interests focus on the design and synthesis of functional metal-based nanomaterials and their applications in energy conversion and catalysis.

Meiting Zhao obtained his PhD degree from National Center for Nanoscience and Technology in 2014 under the guidance of Prof. Zhiyong Tang. Then he worked as a postdoctoral research fellow in Prof. Hua Zhang's group in Nanyang Technological University. In 2019, he joined Tianjin University as a professor. His research interests include the structure design, controlled synthesis and application of highly stable metal-organic frameworks (MOFs), covalent-organic frameworks (COFs), and their composites with nanomaterials for selective catalysis and energy conversion.

Yuchen Qin is currently a professor in College of Sciences, Henan Agricultural University. He received the PhD degree from China University of Petroleum (Beijing) in 2016. His current research interests are metal-based electrocatalysts, electrocatalysis, and photoelectrocatalysis.

1

Introduction

Ruoqing Xu[1], Jie Wang[2], Lilin Zhang[1], and Jingjie Ge[1]

[1] *The Hong Kong Polytechnic University, Department of Applied Biology and Chemical Technology, 11 YuK Choi Road, Hung Hom, Kowloon, HongKong*
[2] *Xihua University, Key Laboratory of Fluid and Power Machinery of Ministry of Education, Department of Materials Science and Engineering, School of Materials Science and Engineering, No. 9999, Hongguang Street, Pidu Area, Chengdu, 610039, China*

With the current rapid growth of energy required by society, the rapid increase in fossil fuel consumption has led to a series of environmental pollution problems, such as global warming, ecosystem destruction, and air pollution. Therefore, the exploration of low-cost, green, and clean renewable energy conversion and storage technologies has become one of the most serious challenges facing today's society [1]. Electrochemical energy storage and conversion systems such as fuel cells [2], metal–air batteries [3], water electrolysis devices [4], and carbon dioxide capture, storage, and reduction technologies [5] have emerged as important ways of dealing with environmental and energy crises. Electrocatalysis is a key technology for energy conversion, as it accelerates electrochemical reactions essential for improving conversion efficiency. It plays a central role in reactions such as water splitting, carbon dioxide reduction reaction (CO_2RR), nitrogen reduction reaction (N_2RR), and the production of liquid fuels, etc. For example, in water splitting, electrocatalysis facilitates the decomposition of water into hydrogen, a clean energy source essential for establishing a low-carbon economy. Catalysts enhance the energy conversion efficiency of electrocatalysis by reducing the activation energy of the reactants and optimizing the reaction pathways and rates. However, one of the most critical challenges is identifying suitable catalysts for various electrocatalytic reactions. These materials are expected to demonstrate efficient electrocatalytic activity, selectivity, and stability to ensure long-term, reliable energy conversion performance. Consequently, the properties of the electrode materials, such as composition, surface structure, and morphology, have to be carefully controlled according to the electrochemical conditions to achieve efficient and high-performance electrocatalysis.

Functional materials offer more attractive solutions for sustainable energy conversion due to their lower costs and wider resource availability compared to conventional noble metal electrocatalytic materials [6, 7]. In addition, research on functional materials has facilitated the development of novel energy conversion

Functional Materials for Electrocatalytic Energy Conversion, First Edition.
Edited by Zhicheng Zhang, Meiting Zhao, and Yuchen Qin.

technologies. For instance, CO_2RR enables the conversion of greenhouse gas CO_2 into valuable chemical fuels using renewable electricity, providing new opportunities to reduce resource extraction and achieve carbon neutrality [5, 8]. Similarly, N_2RR, which converts of N_2 to NH_3 by electrochemical reduction is considered a sustainable alternative process [9]. However, in electrocatalysis, the high dissociation energies of the C—O and N≡N bonds in the linear molecules of CO_2 and N_2 lead to their low chemical activity. The significant energy gap between the highest occupied molecular orbital (HOMO) and the lowest unoccupied molecular orbital (LUMO) of CO_2 and N_2 molecules further enhances their chemical stability. Besides, the low proton affinity of CO_2 and N_2 complicates direct protonation [10–12]. These properties make CO_2RR and N_2RR challenging. Consequently, scientists are focusing on developing innovative electrocatalysts to overcome the challenges posed by the high dissociation energies and significant energy gaps of CO_2 and N_2, while enhancing their chemical reactivity and protons affinity. Functional materials based on covalent organic frameworks (COFs) have emerged as promising catalysts for the efficient utilization of CO_2. In particular, COF-based functional materials with multiple active sites, such as single-metal sites, metal nanoparticles, and metal oxides, offer great potential for realizing CO_2 conversion and energy storage [13]. Besides, the researchers also focused on the important effects of vacancies, high-index facets, lattice strain, lattice disorder, and polymer–inorganic interface configurations on the enhancement of CO_2RR and N_2RR performance. They noted that defect engineering can enhance CO_2 and N_2 uptake and tune the electronic structure of the catalyst. In terms of interfacial engineering, polymers play an important role as supports, modifiers, or blenders of polymer–inorganic composites. The introduction of polymers can inhibit hydrogen evolution reaction (HER), enhance the concentrations of CO_2 and N_2, stabilize intermediates, and change the electronic structure of the catalyst. This modulation affects the binding energies of CO_2, N_2, and intermediates on the catalyst surface, leading to more efficient reactions [14].

Despite the significance and promising prospects of the aforementioned electrocatalytic process, a common challenge in these reactions is the relatively low energy conversion efficiency, which remains far from industrial viability. Conventional electrocatalysts and energy storage materials still face technical problems, including complicated preparation processes, high cost, and inadequate catalytic activity and stability. These limitations severely restrict their commercialization and further applications. To enhance the reaction rates of promising catalytic reactions, there is a need for low-cost, highly active, and durable electrochemical materials. In recent years, functional materials, particularly metal- and carbon-based nanomaterials, have garnered significant attention from researchers due to their large specific surface area and high surface activity. Moreover, the bonding and electronic states on the surface of these nanocatalysts are different from those inside the particles, and the incomplete coordination of the surface atoms increases the number of surface-active sites, thereby improving catalytic performance. As a result, functional materials with specific morphologies and architectures show great promise for energy conversion applications. This book focuses on four different types of functional materials, including metal-based materials, metallic compounds,

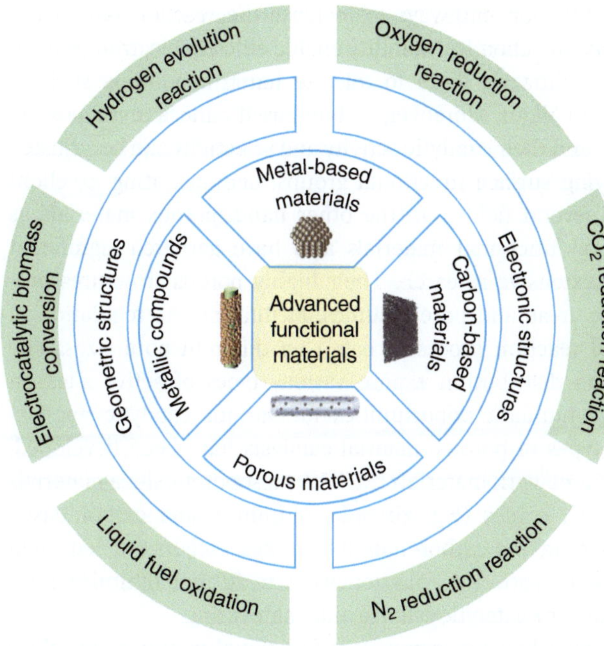

Figure 1.1 Illustration of advanced functional materials for electrochemical catalysis.

carbon-based materials, and porous materials. By regulating their geometric structures and electronic structures, these functional materials demonstrate specific catalytic effects on various electrochemical reactions in practical applications. The facilitating role of functional materials as catalysts in electrocatalytic energy conversion is explored from several perspectives (Figure 1.1).

The catalytic performance of metal-based nanocatalysts is influenced by their geometric and electronic properties, including size, morphology, phase, atomic distance, and composition. For example, the density of active sites and lattice strain on the surface of nanoparticles can be modified by adjusting their shape and size, thereby enhancing catalytic activity and selectivity. The most commonly employed metal electrocatalysts are transition metals, such as platinum, palladium, copper, iron, and nickel. In addition to metal-based catalysts, metallic compounds also exhibit promising catalytic performances for various electrocatalytic processes. These compounds are typically formed from metallic and nonmetallic elements. Similarly, the performance of metal compound catalysts is influenced by their composition, structure, and surface properties. Common metallic catalysts include transition metal oxides, transition metal sulfides, and transition metal nitrides. The metal elements in these compounds often possess a variety of oxidation states and coordination environments, which enable them to interact effectively with reactants and modulate the rate and selectivity of catalytic reactions.

In addition to metal-based functional materials, carbon-based materials exhibit excellent properties for electrocatalytic applications. These materials feature a highly specific surface area and abundant pore structure, providing abundant reactive active sites and diffusion pathways to increase the reaction rate. Their good electrical conductivity and chemical stability enable efficient electron transfer during catalytic reactions, allowing them to tolerate harsh conditions such as high temperatures, acid, and alkali. Moreover, carbon-based nanocatalysts possess tunable surface chemistry, and their catalytic activity and selectivity can be adjusted by introducing or modifying surface functional groups, demonstrating excellent catalytic performance in several fields. On the other hand, porous materials as a fundamental category of functional materials also have garnered significant attention and undergone extensive research. Their highly porous structures provide large specific surface areas and pore volumes, facilitating the regulation of adsorption, diffusion, and reaction processes based on different pore sizes. The tunable pore structure and distribution expose various types of active sites on their surfaces, providing abundant opportunities for various catalytic applications. Currently, various types of porous material catalysts have been developed, including metal–organic framework materials (MOFs), mesoporous silica materials (e.g. SBA-15 and MCM-41), oxides (e.g. zirconia, alumina), and carbon-based materials (e.g. activated carbon and carbon nanotubes). Researchers are dedicated to designing and synthesizing novel porous materials to further optimize their catalytic properties and advance catalytic science and engineering.

This book provides a comprehensive overview of functional materials and their specific performances in electrocatalytic energy conversions. We focus on metallic compounds, metal-based materials, carbon-based materials, and porous materials. By studying and optimizing the electrocatalytic properties of these functional materials, we aim to advance the development of electrocatalytic energy conversion technologies and contribute to the revolution toward sustainable energy. The goal of this book is to deepen readers' understanding of functional materials for electrocatalytic energy conversion while offering insights for future research and applications. We hope that through continuous innovation and collaboration, we can overcome the challenges we currently face and pave a way for a more sustainable and cleaner energy future.

Acknowledgment

The work was financially supported by the Hong Kong Polytechnic University (grant no. P0047728), the Key Project of Xihua University (grant no. Z222070), and Chengdu Vehicle Environmental Protection Technology Co., Ltd, (No. 2023-CDVEPTKF-04).

References

1 Zhang, J.N. (ed.) (2022). *Carbon-based Nanomaterials for Energy Conversion and Storage: Applications in Electrochemical Catalysis*, Springer Series in Materials Science, vol. 325. Springer.

2 Bailey, A., Andrews, L., Khot, A. et al. (2014). Hydrogen storage experiments for an undergraduate laboratory course—clean energy: hydrogen/fuel cells. *J. Chem. Educ.* 92 (4): 688–692.

3 Wang, H.-F. and Xu, Q. (2019). Materials design for rechargeable metal-air batteries. *Matter* 1 (3): 565–595.

4 Zhang, J., Zhang, Q., and Feng, X. (2019a). Support and interface effects in water-splitting electrocatalysts. *Adv. Mater.* 31 (31).

5 de Jesus Gálvez-Vázquez, M., Moreno-García, P., Xu, H. et al. (2020). Environment matters: CO_2RR electrocatalyst performance testing in a gas-fed zero-gap electrolyzer. *ACS Catal.* 10 (21): 13096–13108.

6 Kulkarni, A., Siahrostami, S., Patel, A., and Nørskov, J.K. (2018). Understanding catalytic activity trends in the oxygen reduction reaction. *Chem. Rev.* 118 (5): 2302–2312.

7 Wang, G., Jia, S., Gao, H. et al. (2023). The action mechanisms and structures designs of F-containing functional materials for high performance oxygen electrocatalysis. *J. Energy Chem.* 76: 377–397.

8 Xue, D., Xia, H., Yan, W. et al. (2020). Defect engineering on carbon-based catalysts for electrocatalytic CO_2 reduction. *Nanomicro Lett.* 13 (1): 5.

9 Tong, W., Huang, B., Wang, P. et al. (2020). Crystal-phase-engineered PDCU electrocatalyst for enhanced ammonia synthesis. *Angew. Chem.* 132 (7): 2671–2675.

10 Li, L., Li, X., Sun, Y., and Xie, Y. (2022). Rational design of electrocatalytic carbon dioxide reduction for a zero-carbon network. *Chem. Soc. Rev.* 51 (4): 1234–1252.

11 Shen, H., Choi, C., Masa, J. et al. (2021). Electrochemical ammonia synthesis: mechanistic understanding and catalyst design. *Chem* 7 (7): 1708–1754.

12 Pan, F. and Yang, Y. (2020). Designing CO_2 reduction electrode materials by morphology and interface engineering. *Energy Environ. Sci.* 13 (8): 2275–2309.

13 Lu, M., Zhang, M., Liu, J. et al. (2022). Covalent organic framework based functional materials: important catalysts for efficient CO_2 utilization. *Angew. Chem.* 134 (15): e200200003.

14 Chen, Y., Chen, C., Cao, X. et al. (2023). Recent advances in defect and interface engineering for electroreduction of CO_2 and N_2. *Acta Phys. -Chim. Sin.* 0 (0): 2210053.

Part I

Advanced Functional Materials for Electrocatalytic Energy Conversion

2

Density Functional Theory for Electrocatalytic Energy Conversion

Wei Li and Ding Yi

Beijing Jiaotong University, School of Physical Science and Engineering, Department of Physics, No.3 Shangyuancun, Beijing 100044, China

2.1 Introduction

Currently, power generation from renewable energy sources, such as solar and wind, has become a mature technology, but the generated electricity is intermittent [1, 2]. Developing electrocatalytic technology is an effective strategy to achieve energy conversion to a more sustainable and efficient form [3, 4]. On the one hand, electrocatalysis can convert abundant molecules into high-value chemicals, such as hydrogen, ammonia and hydrocarbons, under mild conditions. On the other hand, it can convert chemical energy into electricity without carbon emissions, such as fuel cells [5–9]. This involves many important electrocatalytic reactions, including the hydrogen evolution reaction (HER), oxygen evolution reaction (OER), oxygen reduction reaction (ORR), nitrogen reduction reaction (NRR), carbon dioxide reduction reaction (CO_2RR), and so on. Numerous scientific studies aim to enhance electrocatalytic activity and improve energy conversion efficiency. However, these complex electrocatalytic reactions are often limited by slow kinetic processes and high reaction overpotentials [6, 10, 11]. Therefore, the design and synthesis of catalysts with high catalytic activity and selectivity have become increasingly crucial.

Generally, there are two main strategies to enhance catalyst activity. First, designing the geometric structure of a catalyst to expose more active sites. Second, regulating the electronic structure of a catalyst to improve its intrinsic activity [12]. In recent years, a wide range of strategies for preparing electrocatalysts have been proposed, and the mechanisms for many electrocatalytic reactions have also been improved [13–19]. However, the current catalytic performance still lags far behind practical application standards, and in-depth studies through experiments have become more costly and time-consuming, severely limiting the development of highly efficient electrocatalysts. Therefore, the first-principles calculations based on density functional theory (DFT) have gained more and more attention.

Functional Materials for Electrocatalytic Energy Conversion, First Edition.
Edited by Zhicheng Zhang, Meiting Zhao, and Yuchen Qin.

With the rapid development of computer technology, DFT calculations have become an effective tool to study reaction processes and explain reaction mechanisms. In the field of electrocatalysis, DFT calculations have successfully explained and predicted numerous experimental phenomena [20, 21]. First, DFT calculations can provide insights into the changes in Gibbs free energy and transition state energy barriers during reaction processes, offering theoretical support for investigating elementary steps and identifying the potential-determining step (PDS). Second, DFT calculations can intuitively provide adsorption models of various reaction intermediates on the catalyst surface, offering cognition for the reaction process from atomic and molecular levels. Last but not least, DFT calculations can provide information on the electronic structures of catalytic active sites as well as reaction intermediates, including charge transfer, orbital coupling, and even spin states, which are powerful methods to analyze the structure–activity relationships of catalysts. Based on the above, combining DFT with experiments allows for a deeper understanding of the mechanisms of electrocatalytic reactions to reasonably predict and design efficient electrocatalysts. It also enables rapid screening of catalysts, significantly reducing the trial-and-error costs in catalyst development and providing convenience for the preparation.

This chapter aims to illustrate how DFT, as a practical research method, is applied in electrocatalytic energy conversion. From a computational perspective, several common electrocatalytic reactions are introduced, including HER, ORR and NRR, and the reaction mechanisms, DFT calculation methods, representative descriptors, and catalyst structure–activity relationships of these electrocatalytic reactions are described in detail, respectively.

2.2 Computational Methods

2.2.1 Reaction Free Energy

Nørskov et al. [22] proposed the computational hydrogen electrode (CHE) model as early as 2004, utilizing the characteristics of the standard hydrogen electrode (SHE), where the reaction described by Eq. (2.1) is in equilibrium.

$$H^+ + e^- \rightarrow 1/2H_2(g) \tag{2.1}$$

$$G_{(H^+)} + G_{(e^-)} = 0.5G_{(H_2)} \tag{2.2}$$

This established a numerical relationship between the Gibbs free energy of proton–electron pairs and the Gibbs free energy of hydrogen gas (Eq. (2.2)). When pH = 0 and $U = 0$ V vs. SHE, the Gibbs free energy of the proton–electron pair is equal to half of the Gibbs free energy of hydrogen in standard conditions (1 bar in the gas phase at 298 K). This greatly simplifies the computational process for complex electrocatalytic reactions. Combining a portion of thermodynamic data with calculations achievable by DFT itself, one can compare the stability of intermediates, plot the free energy diagrams, and determine the reaction PDS.

As is known, the energy provided by DFT calculations is under 0 K, and the simulated atoms are completely static, excluding the effect of zero-point energy, temperature, and entropy. Therefore, to calculate the system's free energy, it is necessary to make corrections to the energy obtained from DFT calculations, as shown in Eq. (2.3).

$$G = E + E_{ZPE} + TS \tag{2.3}$$

where E is the energy obtained from the DFT calculations directly, E_{ZPE} is the correction of zero-point energy, TS is the contribution of temperature and entropy, which can be obtained by calculating the vibrational frequency through DFT or looking up the standard tables. For each elementary step, the free energy change at pH = 0 and $U = 0$ V vs. SHE can be written as

$$\Delta G_0 = G_{(\text{final})} - G_{(\text{initial})} = \Delta E + \Delta E_{ZPE} - T\Delta S \tag{2.4}$$

If we further consider the effect of U and pH, the free energy change can be written as

$$\Delta G = \Delta G_0 + \Delta G_{pH} + \Delta G_U = \Delta E + \Delta E_{ZPE} - T\Delta S + \Delta G_{pH} + \Delta G_U \tag{2.5}$$

where $\Delta G_{pH} = -k_B T\ln(H^+) = pH \times k_B T\ln 10$ (k_B is the Boltzmann constant), and $\Delta G_U = -neU$ (U is the applied potential relative to SHE, e is the elementary charge, and n is the number of electrons transferred).

The above content describes the main computational method of DFT for calculating the change of reaction-free energy. Additionally, depending on specific requirements, DFT can also be used to calculate adsorption energy, formation energy, and so on. The energy analysis of different electrocatalytic reactions will be discussed in detail in subsequent sections.

2.2.2 Electronic Structures

The study of electronic structures of catalysts and various adsorbates is also the core of DFT calculations in electrocatalysis. Here, we briefly introduce several analysis tools that will be mentioned later. First, for the charge (electron) transfer during the reaction process, the Bader charge analysis is widely used, which accurately quantifies the charge of each atom. Charge density difference is also a good method that can visualize the charge transfer in coordinate space. The combination of them enables an accurate analysis of charge distribution. Second, the orbital coupling between catalytic active sites and reaction intermediates can also be well analyzed. By calculating the projected density of states (PDOS), information such as orbital overlap, energy distribution, and electron occupation can be obtained. Moreover, the calculation of crystal orbital Hamilton population (COHP) provides data on the interaction form and strength between any two orbitals of the interacting atoms.

The above analysis methods can comprehensively reveal the structure–activity relationship between catalysts and electrocatalytic reactions, providing theoretical guidance for the design of efficient catalysts.

2.3 Application of DFT in Electrocatalysis

2.3.1 Hydrogen Evolution Reaction

Hydrogen energy is a secondary energy source with great development potential and has attracted much attention because of its advantages such as high energy efficiency and cleanliness [5, 23]. The utilization of renewable energy for hydrogen production through electrocatalytic HER is a promising technology. Currently, noble metal catalysts such as Pt, Ir and Pd are generally used as catalysts in HER [24–26]. However, their large-scale application is impeded by many shortcomings, such as high cost, low stability and susceptibility to poisoning [27, 28]. Therefore, the study and design of electrocatalysts are the key to solving these problems. Using DFT calculations to investigate reaction mechanisms and the electronic structures of catalysts has become a powerful tool for exploring HER.

2.3.1.1 Reaction Mechanism

HER is the cathodic half-reaction of electrochemical water splitting. During the reaction process, proton-coupled electron transfer (PCET) plays an important role [29, 30]. In acidic solutions, the reaction can be represented as: $2H^+ + 2e^- \rightarrow H_2$. Specifically, the reaction can be divided into two steps: the Volmer step and the Tafel or Heyrovsky step. In the Volmer step, because of the high concentration of H^+ in acidic electrolytes, H^+ first strongly interacts with the surface of the electrocatalyst via a PCET pathway, leading to the generation of an adsorbed H*. The second step of the reaction has two possible pathways: (i) If the concentration of H* is high, and they are adsorbed on adjacent active sites, two H* can directly combine to generate one molecule of H_2, which is then released into the environment (Tafel step). (ii) The H* generated in the Volmer step will generate a H_2 molecule via the second PCET pathway (Heyrovsky step). The three steps can be represented by the following reaction equations:

$$\text{Volmer step: } H^+ + e^- + {}^* \rightarrow H^* \tag{2.6}$$

$$\text{Tafel step: } 2H^* \rightarrow H_2 + 2^* \tag{2.7}$$

$$\text{Heyrovsky step: } H^+ + e^- + H^* \rightarrow H_2 + {}^* \tag{2.8}$$

In alkaline or neutral solutions, water splitting is the primary source of H^+, and the reaction can be represented as: $2H_2O + 2e^- \rightarrow H_2 + 2OH^-$, and these three steps are modified as:

$$\text{Volmer step: } H_2O + e^- + {}^* \rightarrow H^* + OH^- \tag{2.9}$$

$$\text{Tafel step: } 2H^* \rightarrow H_2 + 2^* \tag{2.10}$$

$$\text{Heyrovsky step: } H_2O + e^- + H^* \rightarrow H_2 + OH^- + {}^* \tag{2.11}$$

Generally speaking, in acidic conditions, appropriate adsorption strength of H* is conducive to the reaction, but in alkaline or neutral conditions, the production of H^+ requires the splitting of H_2O. Since the breaking of the H–OH bond entails a high dissociation barrier, the Volmer step is typically the rate-determining step (RDS). Compared with acidic HER, alkaline HER is more difficult [31].

2.3.1.2 Detailed Computational Methods for HER

The adsorption of H* is involved in both acidic and alkaline HER, so the adsorption free energy of hydrogen (ΔG_{H^*}) can be used to describe the performance of HER and is represented as follows:

$$\Delta G_{H^*} = G_{H^*} - G_* - 0.5G_{H_2} \tag{2.12}$$

During the reaction, both the adsorption and desorption of H* are involved, and the two processes compete with each other. As with the Sabatier principle, catalysts with high activity should ensure that the reaction intermediates have an optimal adsorption strength to balance the adsorption and desorption. If the adsorption of H* is too weak, the reaction cannot proceed, but if the adsorption is too strong, desorption becomes difficult, limiting the subsequent generation of H_2 and occupying active sites [32]. Therefore, the ΔG_{H^*} is closely related to the activity of the electrocatalyst. The ideal value of ΔG_{H^*} is close to 0 eV [7, 33], as shown by the red line in Figure 2.1, and the performance of catalysts with weaker H adsorption (blue line), such as Au catalyst and catalysts with stronger H adsorption (green line), such as Pt, Ni, Mo are not perfect yet [19]. This hydrogen adsorption free energy diagram visually demonstrates the catalytic activity of catalysts.

For alkaline HER, exploring the free energy change of H_2O dissociation is often required [34]. A lower dissociation barrier as well as an appropriate H* adsorption often corresponds to higher activity of alkaline HER.

2.3.1.3 Descriptors

As mentioned above, the adsorption and desorption of H* on the catalyst surface have a competitive relationship, too strong and too weak adsorption strength

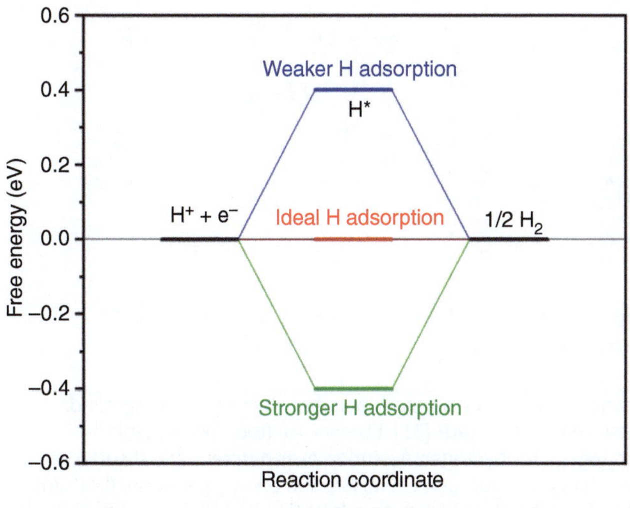

Figure 2.1 Diagram of adsorption free energy of hydrogen.

can affect the activity of HER. Therefore, ΔG_{H^*} can provide a semi-quantitative prediction of catalyst activity. Trasatti [35] first proposed a clear linear relationship between the logarithm of HER current density and the strength of metal–hydrogen (M–H) bonds. This work summarized the acidic HER performance on various transition metal (TM) surfaces and observed a volcano-shaped relationship as shown in Figure 2.2a. It can be clearly seen that metal surfaces with both strong and weak H binding correspond to lower HER current density, while metals located at the peak of the volcano plot, such as Pt and Pd, have optimal adsorption strength and thus have the best HER activity. Nørskov et al. [19] calculated the ΔG_{H^*} on several TM surfaces by DFT calculations and compared it with the experiment results. The volcano-shaped relationship is also observed as shown in Figure 2.2b. ΔG_{H^*} of metals close to 0 eV are near the peak of the volcano plot, corresponding to the highest acidic HER activity.

Another useful descriptor is the d-band center, which was first proposed by Nørskov and Hammer in the early 2000s [36], suggesting that the d-band center (relative to Fermi level) of TMs plays an important role in determining the binding strength between the metal and adsorbates (Figure 2.3a). The interaction of an adsorbate state with the d-electrons of a metal surface often gives rise to bonding and antibonding states. Generally, higher (lower) d-band center corresponds to less (more) electron filling into the antibonding state and further corresponds to strong (weak) binding. Figure 2.3b shows the calculated ΔG_{H^*} with respect to the d-band centers of different catalysts, and a linear relationship was obtained [37]. Therefore, the d-band center can also be used as an independent descriptor to predict catalytic activity.

2.3.1.4 Structure–Activity Relationship

As mentioned above, regulating the adsorption strength of H* to an appropriate range is key to increasing the activity of HER electrocatalysts. Yi et al. [38] used DFT calculations to predict the acidic HER performance of various single-atom doped

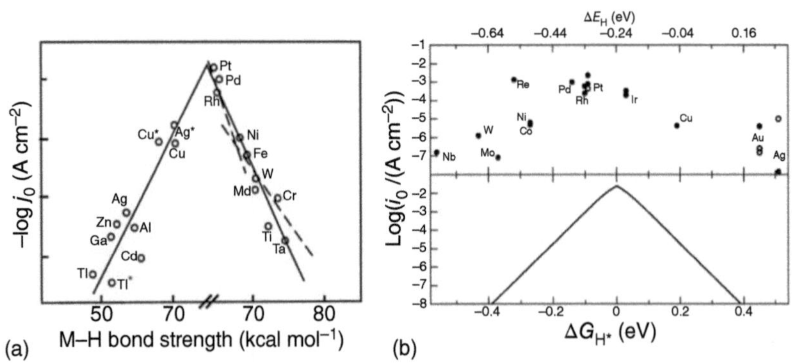

Figure 2.2 (a) Exchange current for HER vs. strength of metal–hydrogen (M–H) bond. Source: Reproduced with permission of Trasatti [35], Elsevier. (b) (top) Experimentally measured exchange current, log(i_0), for hydrogen evolution over different metal surfaces plotted as a function of the calculated hydrogen chemisorption energy per atom. (bottom) The relationship between the free energy of hydrogen adsorption and the exchange current density. Source: Reproduced with permission of Nørskov et al. [19], The Electrochemical Society.

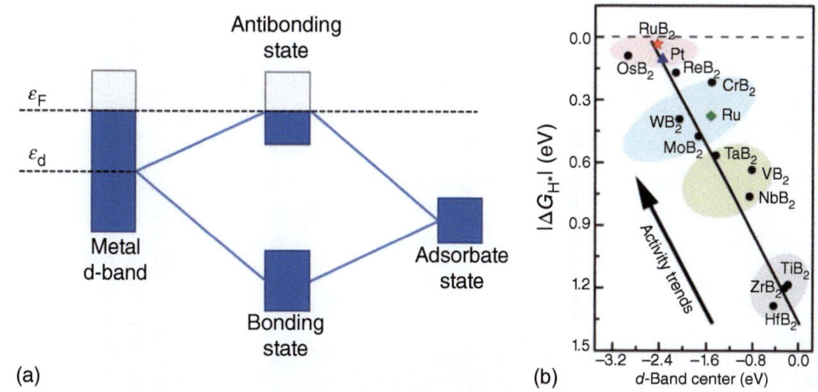

Figure 2.3 (a) Schematic illustration of d band center model. (b) Fitted linear relationship between hydrogen adsorption free energy (ΔG_{H*}) and d-band center of MB_2, Pt, and Ru. Source: Reproduced with permission of Li et al. [37], Wiley-VHC.

titania nanosheets and found that by changing the doped atoms, a continuous influence on the charge of active site O atoms was achieved, thereby affecting the electron occupation of the antibonding orbitals formed by the overlap of O-2p and H-1s orbitals (Figure 2.4a, b). This enabled continuous regulation of HER activity (Figure 2.4c) and has been confirmed by experiments. Peng et al. [39] loaded Ru nanoparticles onto WO nanowires and achieved excellent HER performance under acidic conditions. Theoretical simulations revealed that an obvious charge transfer occurred at the Ru–WO interface, enhancing the local electron density of the active site Ru, thereby increasing the adsorption energy of H and optimizing the HER catalytic performance.

For alkaline HER, since the Volmer step involves the breaking of the H–OH bond, enhancing the adsorption of H_2O has a positive impact on HER activity. Zhou et al. [40] designed a NiPt alloy electrocatalyst with a large number of edge active sites. DFT calculations showed that when the Ni atom is moved from inner to surface, the d_{z^2}, d_{xz} and d_{yz} orbitals are significantly activated, and when the Ni atom continues to be moved to the edge position, the d_{xy}, d_{yz}, and $d_{x^2-y^2}$ orbitals are further activated (Figure 2.4d, e). The final effect is that the activation of the d_{yz} orbital is maximized, and the coupling between O-2p and Ni-d_{yz} orbital is the strongest (Figure 2.4f), which weakens the energy barrier of H_2O dissociation obviously. Zhu et al. [41] synthesized RuCo nanoscrew catalysts with step and edge sites, exhibiting excellent water electrosplitting performance. DFT calculations further demonstrated that the d-band center of the Ru active site shifts upward with the decrease of coordination number. This indicates that Ru active sites located at the edge positions have stronger adsorption effects on H_2O, which can reduce the dissociation barrier of H–OH bond and promote alkaline HER process.

2.3.2 Oxygen Reduction Reaction

The ORR is the cathodic reaction in fuel cells, playing a crucial role in the field of electrocatalysis. However, due to the existence of multiple electron transfer steps

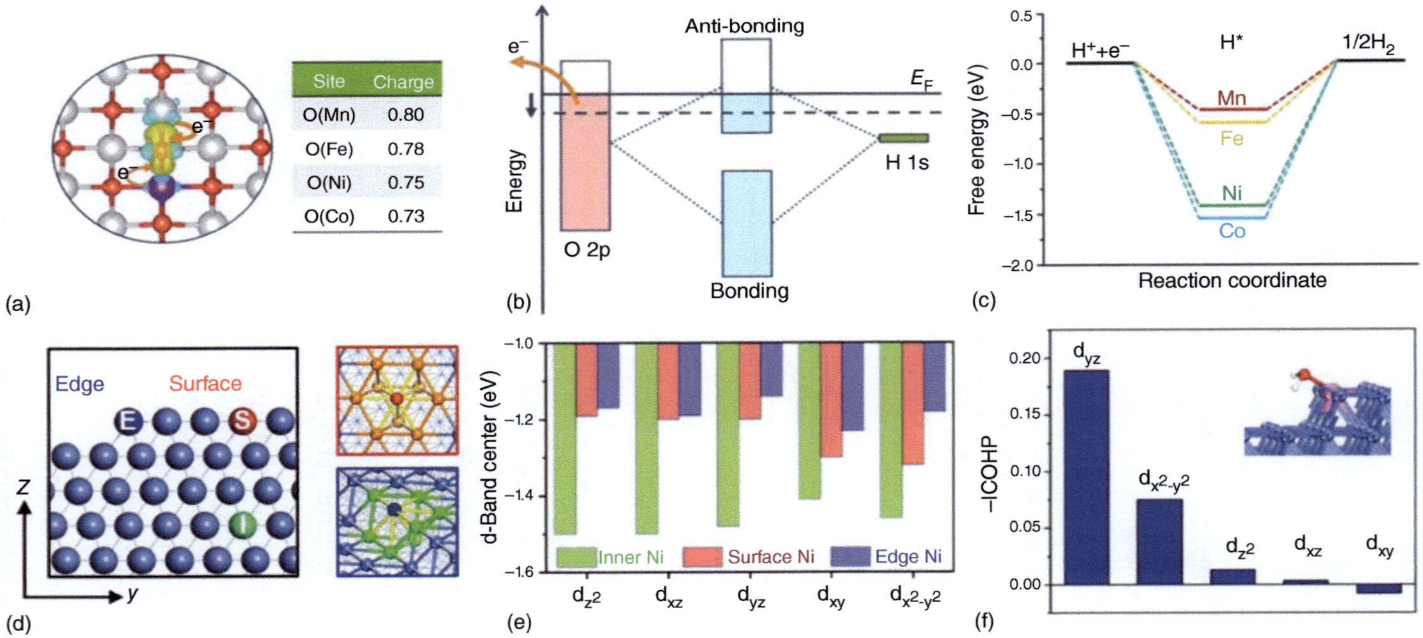

Figure 2.4 (a) Simulated charge transfer from Ti and TM atoms to O atom. (b) Illustration of overlapping between O-2p orbital and H-1s orbital. (c) Adsorption free energy of hydrogen with different TM doping. Source: Panels a–c reproduced with permission of Yi et al. [38], Wiley-VCH. (d) NiPt alloy model. Positions of inner (I), surface (S), edge (E) Ni atoms, and the detailed coordination environment of surface and edge Ni. (e) The d-band center of the five d orbitals for inner, surface, and edge Ni atoms, respectively. (f) Integrated COHP values of five d orbitals of edge Ni with the orbitals of O atoms. Source: Panels d–f reproduced with permission of Zhou et al. [40], Wiley-VCH.

and slow reaction kinetics, the efficiency of energy conversion is limited [9, 11]. In recent years, the ORR has been extensively studied from various aspects, including electrode materials, electrochemical interfaces, and electrode potentials [42–44]. Deep understanding of the reaction mechanism is beneficial for accelerating the advancement of ORR catalysts. This section will introduce the significance of DFT calculations in exploring and designing high-activity ORR electrocatalysts.

2.3.2.1 Reaction Mechanism

There are two possible pathways for ORR: one is the two-electron ($2e^-$) pathway that generates hydrogen peroxide ($O_2 + 2H^+ + 2e^- \rightarrow H_2O_2$), and the other is a complex four-electron ($4e^-$) pathway that ultimately generates water ($O_2 + 4H^+ + 4e^- \rightarrow 2H_2O$). Ideally, the $4e^-$ ORR pathway can provide a larger electrical potential of 1.23 V than the $2e^-$ pathway (0.7 V), which is favorable for fuel cells [45]. Thereby, we will mainly introduce the $4e^-$ ORR pathway.

For the $4e^-$ ORR, the whole reaction involves four elementary steps and multiple reaction intermediates, such as OOH*, O*, and OH*, as shown in Figure 2.5.

The ORR pathways under acidic conditions are summarized as follows:

$$O_2 + H^+ + e^- + {}^* \rightarrow OOH^* \tag{2.13}$$

$$OOH^* + H^+ + e^- \rightarrow O^* + H_2O \tag{2.14}$$

$$O^* + H^+ + e^- \rightarrow OH^* \tag{2.15}$$

$$OH^* + H^+ + e^- \rightarrow H_2O + {}^* \tag{2.16}$$

$$\text{The overall reaction is: } O_2 + 4H^+ + 4e^- \rightarrow 2H_2O \tag{2.17}$$

The ORR pathways under alkaline/neutral conditions are summarized as follows:

$$O_2 + H_2O + e^- + {}^* \rightarrow OOH^* + OH^- \tag{2.18}$$

$$OOH^* + e^- \rightarrow O^* + OH^- \tag{2.19}$$

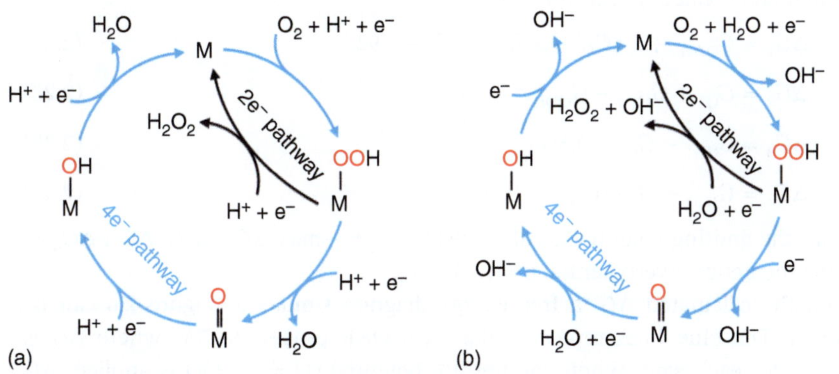

Figure 2.5 ORR process of two-electron pathway and four-electron pathway under (a) acidic and (b) alkaline conditions, respectively.

$$O^* + H_2O + e^- \rightarrow OH^* + OH^- \tag{2.20}$$

$$OH^* + e^- \rightarrow OH^- + ^* \tag{2.21}$$

$$\text{The overall reaction is: } O_2 + 2H_2O + 4e^- \rightarrow 4OH^- \tag{2.22}$$

2.3.2.2 Detailed Computational Methods for ORR

Because the intermediates (OOH*, O*, and OH*) in the ORR process are thermodynamically unstable, their free energies are obtained from the free energy of stoichiometrically appropriate amounts of H_2O and H_2. Besides, the Gibbs free energy of $O_2(g)$ is obtained through the equation: $G_{O_2(g)} = 2G_{H_2O(l)} - 2G_{H_2} + 4 \times 1.23$ eV. Furthermore, the free energy of $H_2O(l)$ can be approximated by calculating the free energy of $H_2O(g)$ at 298.15 K and 0.035 bar (saturation vapor pressure), due to the inaccuracy of DFT calculations when dealing with the hydrogen bond.

The free energy change (ΔG_1, ΔG_2, ΔG_3, ΔG_4) for ORR under acidic conditions, which corresponds to Eqs. (2.13)–(2.16), respectively, can be calculated using the following equations:

$$\begin{aligned}\Delta G_1 &= G_{OOH*} - \left(G_{O_2} + G_* + G_{H^+} + G_{e^-}\right)\\ &= G_{OOH*} - (2G_{H_2O} - 2G_{H_2} + 4 \times 1.23 \text{ eV}) - G_* - \left(G_{H^+} + G_{e^-}\right)\\ &= G_{OOH*} + 1.5G_{H_2} - 2G_{H_2O} - G_* - 4.92\end{aligned} \tag{2.23}$$

$$\begin{aligned}\Delta G_2 &= (G_{O^*} + G_{H_2O}) - \left(G_{OOH^*} + G_{H^+} + G_{e^-}\right)\\ &= G_{O^*} + G_{H_2O} - G_{OOH^*} - 0.5G_{H_2}\end{aligned} \tag{2.24}$$

$$\begin{aligned}\Delta G_3 &= G_{OH^*} - (G_{O^*} + G_{H^+} + G_{e^-})\\ &= G_{OH^*} - G_{O^*} - 0.5G_{H_2}\end{aligned} \tag{2.25}$$

$$\begin{aligned}\Delta G_4 &= (G_{H_2O} + G_*) - \left(G_{OH^*} + G_{H^+} + G_{e^-}\right)\\ &= G_{H_2O} + G_* - G_{OH^*} - 0.5G_{H_2}\end{aligned} \tag{2.26}$$

The derivation of the free energy change for each elementary step in alkaline ORR is similar and is calculated as follows:

$$\Delta G_1 = G_{OOH^*} + 1.5G_{H_2} - 2G_{H_2O} - G_* - 4.92 \tag{2.27}$$

$$\Delta G_2 = G_{O^*} + G_{H_2O} - G_{OOH^*} - 0.5G_{H_2} \tag{2.28}$$

$$\Delta G_3 = G_{OH^*} - G_{O^*} - 0.5G_{H_2} \tag{2.29}$$

$$\Delta G_4 = G_{H_2O} + G_* - G_{OH^*} - 0.5G_{H_2} \tag{2.30}$$

Thus, the limiting potential is calculated by $U_L = -\max(\Delta G_1, \Delta G_2, \Delta G_3, \Delta G_4)/e$ and the theoretical overpotential $\eta = 1.23 - U_L$.

With the calculated ΔG, a free energy diagram similar to Figure 2.6 can be obtained. The blue lines represent the ideal ORR process at 0 V, where ΔG is −1.23 eV for each step. When equilibrium potential (U = 1.23 V) is applied, ΔG becomes 0 eV, indicating that the reaction is in an equilibrium state (red lines). Further increasing the applied potential will cause the reaction to stop. However, in

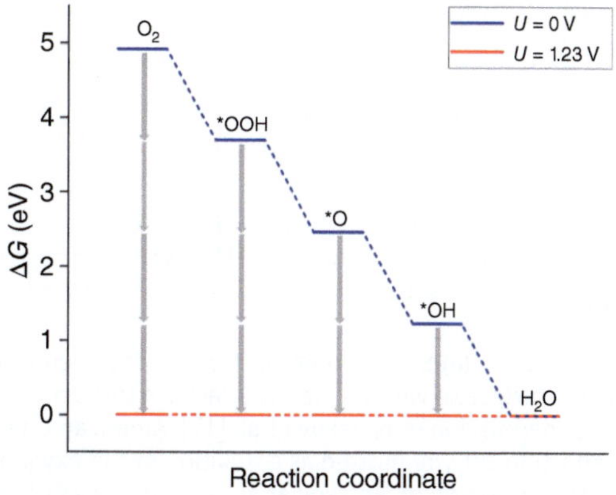

Figure 2.6 Free energy diagram for ideal ORR at 0 and 1.23 V.

reality, ΔG will deviate from the ideal state due to the varying binding strength of different reaction intermediates on the electrocatalyst surfaces. Nørskov et al. [17] carried out a theoretical study of the ORR catalytic activity on the Pt(111) surface. It shows that the last step (i.e. Eqs. (2.16) and (2.26)) is the PDS, corresponding to the limiting potential of 0.75 V, which agrees well with the experimentally observed maximum potential for large ORR current densities [46–48].

2.3.2.3 Descriptors

The ORR process involves various reaction intermediates such as OOH*, O*, and OH*, and their different binding strengths on catalyst surfaces affect the limiting potential. Since all these intermediates are bonded to the catalyst surface through O atoms, there must be a strong correlation between their adsorption energies. A distinct linear relationship can be observed between ΔG_{OH^*} and ΔG_{O^*} or ΔG_{OOH^*}, characterized by different slopes, as shown in Figure 2.7a [17]. Based on such a linear relationship, it becomes feasible to use one of the intermediates as a descriptor for predicting the ORR activity of catalysts. Nørskov et al. [22] used adsorption free energy of O* (ΔG_{O^*}) as a descriptor to establish a volcano-shaped relationship, as shown in Figure 2.7b. On the left leg of the volcano plot, the O atom forms a strong bond with these metals, which means that the ORR is limited by O* or OH* protonation. Conversely, for metals that have weaker binding with O, which corresponds to the right leg of the volcano plot, the PDS is typically the reduction of O_2* to OOH*. Moreover, O exhibits optimal adsorption on the Pt surface, resulting in the highest ORR activity.

According to the volcano plot in Figure 2.7b, it can be known that reducing the adsorption energy of O on the Pt surface by approximately 0.2 eV can lead to the top catalytic activity. Inspired by this rule, the construction of Pt-based alloy catalysts doped with different metal atoms has been extensively investigated [49, 50]. Data on the ORR catalytic performance of some Pt-based alloys are presented in Figure 2.7c, where Pt_3M alloys show remarkable activity [51]. Markovic et al. [52] prepared a series of Pt_3M (M = Ni, Co, Fe, Ti, V) alloys, and their ORR activity was all improved. Duan et al. [53] conducted theoretical studies on the origin of the improved ORR activity in these catalysts. The results revealed that the d-band centers of these alloys

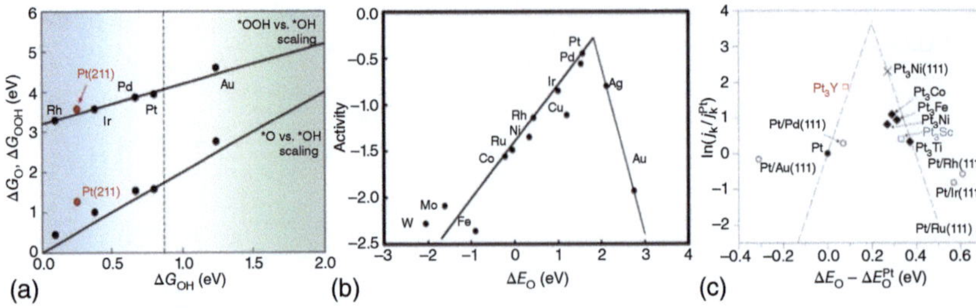

Figure 2.7 (a) Scaling relationships for the chemisorption energies of *OOH and *O for the (111) surface of various metals (black circles) using *OH as a descriptor. Source: Reproduced with permission of Nørskov et al. [17], American Chemical Society. (b) Trends in oxygen reduction activity plotted as a function of the oxygen binding energy. Source: Reproduced with permission of Nørskov et al. [22], American Chemical Society. (c) Volcano plots for the ORR on Pt-based TM alloy. Source: Reproduced with permission of Greely et al. [51], Springer Nature.

all shifted downward, indicating a weakened adsorption. However, the regulation of O adsorption should be moderate. The ORR performance of Pt/Ru, Pt/Ir, and Pt/Rh alloys [54, 55] is reduced because of the weak adsorption (Figure 2.7c).

2.3.2.4 Structure–Activity Relationship

Based on DFT calculations, studying the electronic structures of a catalytic system is a powerful means to reveal the structure–activity relationship. Jiang et al. [56] devised a two-dimensional heterostructure composed of single-TM-doped graphene and electride (Y_2C or Ca_2N) substrate (TM-Gr-Y_2C/Ca_2N). DFT calculations indicate an increase in electron transfer from Fe-Gr-Y_2C to O_2 molecules (0.40–0.62e^-), as well as Co-Gr-Ca_2N (0.38–0.49e^-), compared to TM-Gr, which enhanced the adsorption of O_2 on the metal active site. Further calculations on doping different single-TM atoms revealed a linear relationship between the charge transfer and the adsorption energy of O_2. This indicates that by controlling the amount of transferred charge to O atoms, the adsorption energy of O_2 can be regulated to an appropriate state, and the ORR activity can be improved. Peng et al. [57] prepared a series of metal nanoparticles deposited on black phosphorus and achieved excellent ORR activity. The DFT calculations found that there was a significant charge redistribution between the substrate and the metal nanoparticles, which led to an increase in the local electron density on the metal nanoparticles, thus enhancing the adsorption of O_2.

Moreover, the structure–activity relationship can also be revealed by exploring the orbital interactions between the reaction intermediates and the active sites. Wen et al. [58] successfully explained the reason why graphene oxide (GO) embedded with Li atom has remarkable ORR activity. As shown in Figure 2.8, the weak adsorption of OOH* limits the overpotential of ORR on GO. When COOLi groups are formed by the embedded Li at the graphene edges, the average O–C–C bond angle is 109.1° (Figure 2.8a), which is closer to the perfect sp^3 hybridization (109.5°) compared with that of 105.5° on graphene plane (Figure 2.8b). This greatly promotes the adsorption of OOH* during the ORR process and improves the

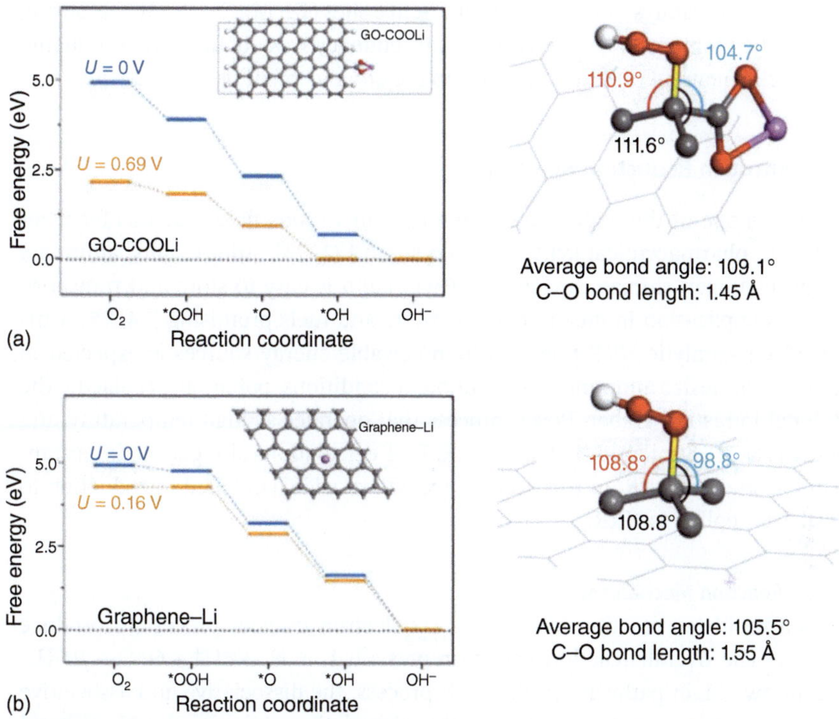

Figure 2.8 Free energy diagram for ORR on (a) GO-COOLi and (b) graphene-Li at $U = 0$ V (vs. RHE, blue curves) and onset potential (vs. RHE, orange curves), as well as the corresponding adsorption configuration for *OOH. Black, white, red, and violet balls represent C, H, O, and Li atoms, respectively. Source: Reproduced with permission of Wen et al. [58], Wiley-VCH.

activity of ORR. Lu et al. [59] simulated the different adsorption configurations of O_2 under the external magnetic field by changing the angle between O_2 molecules and the catalyst surface. By analyzing the electronic structures, it was found that the coupling between the O_2-π^* and Pt-d_{yz} orbitals becomes weakened as the angle increases. The weakened coupling corresponds to the decrease in ORR activity, which is consistent with the experimental results.

In recent years, several studies have shown that the spin state of catalysts can also affect ORR activity [60, 61]. Taking the alkaline ORR process as an example, the reactant O_2 is paramagnetic, while the product OH^- is diamagnetic, and there is a change in spin state during the reaction. Currently, research related to spin regulation is mainly focused on catalysts such as perovskite oxides [62–64], spinel oxides [65–67], and metal nitrogen complex carbon [68–71]. Liu et al. [70] established a model of Fe–N–C catalyst with FeN_4 active site. By constructing different defects on the carbon substrate, the spin magnetic moments of Fe could be continuously regulated. Further calculations showed a clear linear relationship between the magnetic moment of Fe and ΔG_{max}. Specifically, a smaller magnetic moment of Fe corresponds to better catalytic performance for ORR. Yang et al. [71] implanted Mn–N moieties into the Fe–N–C catalyst forming diatomic active sites (Fe, Mn/N–C). By measuring the magnetic moment of Fe atoms, a similar tendency was obtained, that reducing the magnetic moment of Fe is conducive to the enhancement of ORR activity.

2.3.3 Nitrogen Reduction Reaction

Ammonia is one of the widely used chemicals in various fields such as fertilizer production, pharmaceutical synthesis, and so on [72, 73]. Additionally, ammonia has high hydrogen storage capacity (17.6 wt%) and is easy to store and transport, making its application in green energy carriers and fuels promising [74, 75]. Currently, electrocatalytic NRR powered by renewable energy sources is expected to efficiently synthesize ammonia under ambient conditions, potentially replacing the traditional industrial Haber–Bosch process that operates at high temperature and pressure [76, 77]. Similar to HER and ORR, DFT calculations also play an important role in the study of NRR. In this section, we will provide a detailed introduction to how DFT is applied to NRR.

2.3.3.1 Reaction Mechanism

NRR is a multistep reaction involving a six-electron transfer process to produce ammonia. The overall reaction can be represented as $N_2 + 6H^+ + 6e^- \rightarrow 2NH_3$. There are two main pathways in the NRR process: the dissociative and associative pathways, as shown in Figure 2.9. In the dissociative pathway, the N≡N bond breaks first, and then each N atom is hydrogenated until two molecules of NH_3 are generated. In the associative pathway, the two N atoms remain bonded to each other during hydrogenation. Depending on the adsorption configuration of N_2 and the sequence of N atom hydrogenation, associative pathway can be further divided into three pathways: distal pathway, alternating pathway, and enzymatic pathway. Due to the mild conditions in electrocatalysis compared to thermal catalysis, the associative pathway is more favorable. In addition, a new Mars–van Krevelen pathway involving lattice N atom has been extensively studied [78–80], particularly in TM nitride catalysts. In this pathway, lattice N atoms participate in the reaction, and N-vacancies are refilled by N_2 molecules.

Here, we take the associative distal pathway as an example to introduce the NRR process. The NRR pathways and reaction intermediates involved are summarized as follows:

$$N_2(g) + {}^* \rightarrow N_2{}^* \tag{2.31}$$

$$N_2{}^* + H^+ + e^- \rightarrow N_2H^* \tag{2.32}$$

$$N_2H^* + H^+ + e^- \rightarrow N_2H_2{}^* \tag{2.33}$$

$$N_2H_2{}^* + H^+ + e^- \rightarrow N^* + NH_3 \tag{2.34}$$

$$N^* + H^+ + e^- \rightarrow NH^* \tag{2.35}$$

$$NH^* + H^+ + e^- \rightarrow NH_2{}^* \tag{2.36}$$

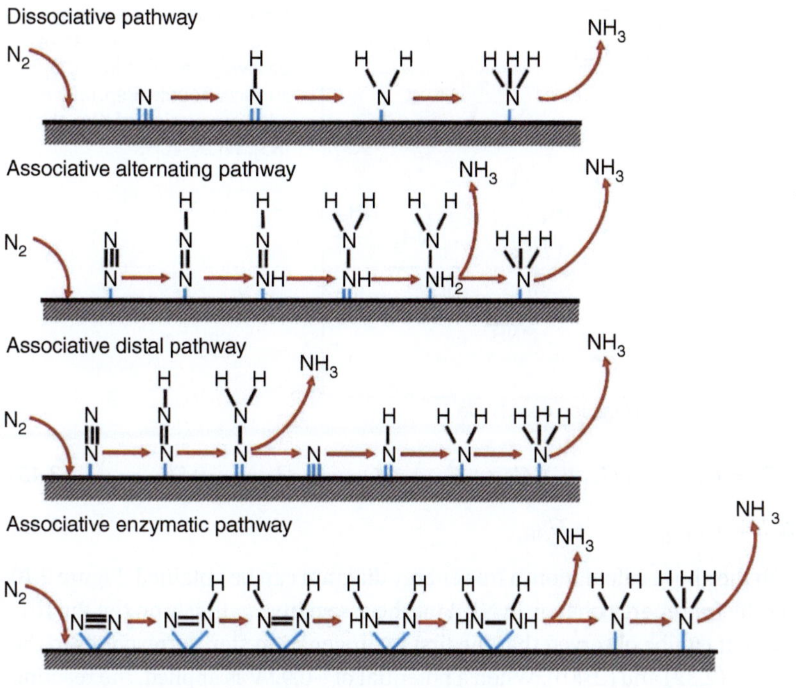

Figure 2.9 Schematic diagram of pathways of NRR.

$$NH_2{}^* + H^+ + e^- \rightarrow NH_3{}^* \tag{2.37}$$

$$NH_3{}^* \rightarrow NH_3(g) + {}^* \tag{2.38}$$

It is worth noting that HER is the main competing reaction for NRR, which only involves a 2 e$^-$ transfer process, and has more advantages in reaction kinetics [81, 82].

2.3.3.2 Detailed Computational Methods for NRR

In DFT calculations, the free energy of each elementary step in NRR can be calculated by using the following equations (taking the associative distal pathway as an example):

$$\Delta G_1 = G_{N_2{}^*} - G_{N_2(g)} - G_* \tag{2.39}$$

$$\Delta G_2 = G_{N_2H^*} - \left(G_{N_2{}^*} + G_{H^+} + G_{e^-}\right) = G_{N_2H^*} - G_{N_2{}^*} - 0.5G_{H_2} \tag{2.40}$$

$$\Delta G_3 = G_{N_2H_2*} - \left(G_{N_2H*} + G_{H^+} + G_{e^-}\right) = G_{N_2H_2*} - G_{N_2H^*} - 0.5G_{H_2} \tag{2.41}$$

$$\Delta G_4 = (G_{N^*} + G_{NH_3}) - \left(G_{N_2H_2{}^*} + G_{H^+} + G_{e^-}\right) = G_{N^*} + G_{NH_3} - G_{N_2H_2{}^*} - 0.5G_{H_2} \tag{2.42}$$

$$\Delta G_5 = G_{NH^*} - (G_{N^*} + G_{H^+} + G_{e^-}) = G_{NH^*} - G_{N^*} - 0.5G_{H_2} \tag{2.43}$$

$$\Delta G_6 = G_{NH_2{}^*} - (G_{NH^*} + G_{H^+} + G_{e^-}) = G_{NH_2{}^*} - G_{NH^*} - 0.5G_{H_2} \tag{2.44}$$

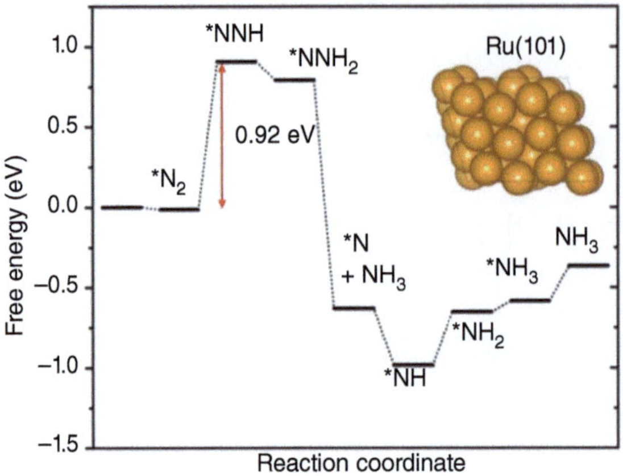

Figure 2.10 Free energy diagram for the associative pathway on a flat Ru(101) surface. Source: Reproduced with permission of Kim et al. [83], Wiley-VCH.

$$\Delta G_7 = G_{NH_3^*} - \left(G_{NH_2^*} + G_{H^+} + G_{e^-}\right) = G_{NH_3^*} - G_{NH_2^*} - 0.5G_{H_2} \tag{2.45}$$

$$\Delta G_8 = G_{NH_3(g)} + G_* - G_{NH_3^*} \tag{2.46}$$

Through the above calculation, a free energy diagram can be obtained. Figure 2.10 illustrates the free energy change in NRR for the associative pathway on the Ru(101) surface [83]. It can be observed that the first hydrogenation step corresponds to the PDS (i.e. Eqs. (2.32) and (2.40)). When a potential of −0.92 V is applied, the reaction begins to occur.

2.3.3.3 Descriptors

To identify suitable descriptors for NRR, Nørskov et al. [81] conducted detailed DFT calculations of the reaction process of NRR on various TM surfaces. The results showed that the limiting potential exhibits a volcano-shaped relationship with the adsorption free energy of nitrogen atom (ΔG_{N*}) on the catalyst surface, as shown in Figure 2.11a. It can be observed that on the right leg of the volcano plot (marked by the magenta lines), the PDS corresponds to the first hydrogenation step ($N_2^* + H^+ + e^- \rightarrow N_2H^*$), while on the left leg of the volcano plot (marked by the red lines), the PDS becomes the last hydrogenation step ($NH_2^* + H^+ + e^- \rightarrow NH_3^*$). The change of the PDS indicates the existence of a scaling relationship, where the stronger adsorption of N atoms on the active sites, the stronger adsorption of N_2 molecules, and the higher activation of N≡N bonds. On the contrary, it also results in strong adsorption of NH_x, which makes the subsequent hydrogenation steps more difficult to proceed. Therefore, to achieve high NRR performance, it is important to regulate the adsorption strength of nitrogen on catalyst surfaces to an optimal state, which is closer to the peak of volcano plot. Liu et al. [84] also observed the presence of the volcano-shaped relationship in graphitic carbon nitride (g-C_3N_4) supported transition metal (TM) single-atom. Inspired by biological nitrogen fixation, Huang et al. [85] explored the NRR activity of metal dimers supported on 2D expanded phthalocyanine by DFT calculations. In this work, they used the adsorption energy of N_2H^* (ΔE_{N_2H*}) as a descriptor and constructed a similar volcano plot to screen for optimal catalysts. Furthermore, they considered the presence of the competing

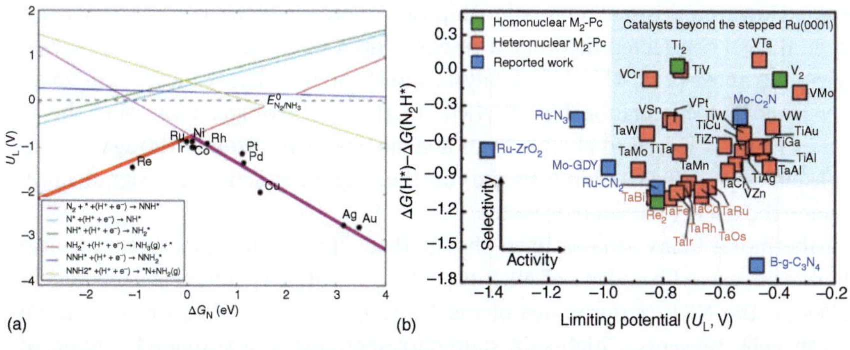

Figure 2.11 (a) Volcano-shaped relationship of limiting potential with adsorption of N atom at (211) surface of fcc metals in NRR. Source: Reproduced with permission of Nørskov et al. [81], Royal Society of Chemistry. (b) Limiting potential (U_L) vs. $\Delta G(H^*) - \Delta G(N_2H^*)$. Source: Panels c,d reproduced with permission of Huang et al. [85], American Chemical Society.

HER reaction and utilized the adsorption free energy difference between H* and N_2H^* ($\Delta G_{H*} - \Delta G_{N_2H*}$) as a descriptor to identify catalysts with high selectivity (Figure 2.11b).

2.3.3.4 Structure–Activity Relationship

It is known that the two N atoms in N_2 are sp-hybridized, forming a strong N≡N triple bond (941 kJ mol^{-1}) composed of one σ bond and two π bonds. Légaré et al. [16] first pointed out that the activation of N_2 is attributed to the coupling of empty TM-d orbital and occupied N_2-σ orbital, as well as the coupling of occupied TM-d orbital and empty N_2-π* orbital, which enables electron transfer between TM and N_2 through an "acceptance and donation" mechanism. The enhanced π* orbital and the weakened σ orbital together lead to the activation of N≡N triple bond. This mechanism also holds for N_2 adsorption on some nonmetallic active sites (with partially occupied p orbitals) [86, 87]. Ling et al. [87] designed a catalyst of boron single-atom-doped g-C_3N_4 (B/g-C_3N_4), where B atom is sp^3 hybridized resulting in three occupied orbitals and one empty orbital. The empty sp^3 orbital accepts electrons from the σ orbital of N_2, while the occupied sp^3 orbitals donate electrons to the π* orbital of N_2, which leads to a superb limiting potential of −0.2 V.

According to the activation mechanism, transferring more electrons to N_2-π* orbitals is an effective strategy. Zafari et al. [88] conducted a theoretical study on various transition metals implanted on boron–arsenide (BAs) and boron–antimony (BSb). Among them, W-BSb exhibited the optimal catalytic activity. Compared with W-BAs, W-BSb transfers more electrons to the N_2 molecule. Further PDOS analysis revealed that the N_2-π* orbital is partially filled with the d-electrons of W, shifting the N_2-π* orbital below the Fermi level. Similarly, Niu et al. [89] improved the NRR performance of MoS_2 by doping Fe or V. DFT calculations demonstrated that more electrons transfer to N_2 after doping, which significantly reduces the energy barrier for the first hydrogenation step. As mentioned, the strong adsorption of N (or N_2) can lead to the change of PDS from the first hydrogenation step to the last. However, this scaling relationship can be broken by designing active sites. Gao et al. [90] conducted theoretical research on a range of TM-doped Cu-based single-atom alloys (TM_1-Cu(111)) catalysts and found that these catalysts can break the volcano-shaped relationship in NRR. In the last hydrogenation step, the N atom in NH_2* is forced to change from a stable sp^3 hybridization (bridge site) to an unstable sp^2 hybridization (top site) on the single-TM active site, significantly reducing the energy barrier of this step.

Furthermore, many studies have shown that altering the spin state can also enhance the catalytic activity of NRR [91, 92]. Li et al. [93] conducted theoretical studies on the NRR performance of the Fe–N–C catalysts. In this work, the Fe atom in FeN_3 presents a high-spin state corresponding to a strong adsorption of N_2 molecules on it. When the N_2 molecules are adsorbed on the FeN_3 site, spin polarization occurred obviously on N atoms, indicating that a high-spin state Fe can significantly activate the N≡N bond. Similar phenomena in small Fe clusters supported on N-doped graphene were also observed [94]. Gao et al. [95] compared the NRR performance between the Mo–N–C and FeMo–N–C catalysts where the doped Fe significantly reduced the magnetic moment of Mo (from 2.63 to $0.74\mu_B$). By analyzing the PDOS, it was found that the Mo–N–C catalyst exhibits four unpaired d orbitals (d_{z^2}, d_{yz}, $d_{x^2-y^2}$ and d_{xz}) below the Fermi level that can overlap with the orbitals of N_xH_y intermediates, leading to a strong NH_2* adsorption. After Fe doping, the Mo active site has only two unpaired d orbitals ($d_{x^2-y^2}$ and d_{z^2}), weakening its binding with NH_2* and facilitating the last hydrogenation step.

2.4 Conclusion

In summary, DFT has shown great application prospects in the field of electrocatalysis, primarily in the following three aspects. First, analyze each elementary step in electrocatalytic reactions at the atomic and molecular levels to obtain the free energy change, so that the most optimal reaction pathway and PDS can be identified. Second, establish simple descriptors to accelerate the screening of efficient electrocatalysts, greatly reducing trial-and-error costs and providing guidance for experimental synthesis. Third, unveil the electronic structures of catalyst active sites and reaction intermediates, which helps to explore the structure–activity relationship.

Although DFT calculations have been widely applied in electrocatalysis, this method still faces a series of challenges. The accuracy of calculations for complex electrocatalytic reactions is limited. First, DFT calculations rely on constructing models that are approximate to experiments, but there may still be significant differences between the models and the experimentally synthesized catalysts. Second, DFT calculations provide results in ideal conditions, while the interactions between electrodes and electrolytes in real environments are more complex, involving factors such as pH, solvation effect, surface coverage, and applied potential. Therefore, there are usually deviations between the results of DFT calculations and experimental observations. It is a challenge to take all the factors into account, which is also a current research hotspot. In the future, with further advancements in theoretical simulation methods, DFT calculations will play a more significant role in the field of electrocatalysis.

Acknowledgment

This work was supported by the National Natural Science Foundation of China (grant no. 21905019).

References

1 Lewis, N.S. and Nocera, D.G. (2006). Powering the planet: chemical challenges in solar energy utilization. *Proc. Natl. Acad. Sci.* 103: 15729–15735.
2 Chu, S. and Majumdar, A. (2012). Opportunities and challenges for a sustainable energy future. *Nature* 488: 294–303.
3 Zhang, Y.L., Luo, M., Yang, Y. et al. (2019). Advanced multifunctional electrocatalysts for energy conversion. *ACS Energy Lett.* 4: 1672–1680.
4 O'Mullane, A.P., Escudero-Escribano, M., Stephens, I.E.L., and Krischer, K. (2019). The role of electrocatalysis in a sustainable future: from renewable energy conversion and storage to emerging reactions. *ChemPhysChem* 20: 2900–2903.
5 Turner, J.A. (2004). Sustainable hydrogen production. *Science* 305: 972–974.
6 Lan, R., Irvine, J.T., and Tao, S. (2013). Synthesis of ammonia directly form air and water at ambient temperature and pressure. *Sci. Rep.* 3: 1145.
7 Jiao, Y., Zheng, Y., Jaroniec, M., and Qiao, S.Z. (2015). Design of electrocatalysts for oxygen and hydrogen involving energy conversion reactions. *Chem. Soc. Rev.* 44: 2060–2068.
8 Whipple, D.T. and Kenis, P.J.A. (2010). Prospects of CO_2 utilization via direct heterogeneous electrochemical reduction. *J. Phys. Chem. Lett.* 1: 3451–3458.
9 Debe, M.K. (2012). Electrocatalyst approaches and challenges for automotive fuel cells. *Nature* 486: 43–51.
10 Govindarajan, N., Kastlunger, G., Heenen, H.H., and Chan, K. (2022). Improving the intrinsic activity of electrocatalysts for sustainable energy conversion: where are we and where can we go? *Chem. Sci.* 13: 14–26.
11 Yang, C., Costamagna, P., Srinivasan, S. et al. (2001). Approaches and technical challenges to high temperature operation of proton exchange membrane fuel cells. *J. Power Sources* 103: 1–9.
12 Benck, J.D., Hellstern, T.R., Kibsgaard, J. et al. (2014). Catalyzing the hydrogen evolution reaction (HER) with molybdenum sulfide nanomaterials. *ACS Catal.* 4: 3957–3971.
13 Zhu, Q., Qu, Y., Liu, D. et al. (2020). Two-dimensional layered materials: high-efficient electrocatalysts for hydrogen evolution reaction. *ACS Appl. Nano Mater.* 3: 6270–6292.

14 Shi, L., Yin, Y., Wang, S., and Sun, H. (2020). Rational catalyst design for N_2 reduction under ambient conditions: strategies toward enhanced conversion efficiency. *ACS Catal.* 10: 6870–6899.

15 Huang, X.X., Shen, T., Sun, S., and Hou, Y. (2021). Synergistic modulation of carbon-based, precious-metal-free electrocatalysts for oxygen reduction reaction. *ACS Appl. Mater. Interfaces* 13: 6989–7003.

16 Légaré, M.A., Bélanger-Chabot, G., Dewhurst, R.D. et al. (2018). Nitrogen fixation and reduction at boron. *Science* 359: 896–900.

17 Kulkarni, A., Siahrostami, S., Patel, S., and Nørskov, J.K. (2018). Understanding catalytic activity trends in the oxygen reduction reaction. *Chem. Rev.* 118: 2302–2312.

18 Fang, Z., Zhao, W., Shen, T. et al. (2023). Spin-modulated oxygen electrocatalysis. *Precis. Chem.* 1: 395–417.

19 Nørskov, J.K., Bligaard, T., Logadottir, A. et al. (2005). Trends in the exchange current for hydrogen evolution. *J. Electrochem. Soc.* 152: J23–J26.

20 Nørskov, J.K., Bligaard, T., Rossmeisl, J., and Christensen, C.H. (2009). Towards the computational design of solid catalysts. *Nat. Chem.* 1: 37–46.

21 Nørskov, J.K., Kibsgaard, J., Dickens, C.F. et al. (2017). Combining theory and experiment in electrocatalysis: insights into material design. *Science* 355: eaad4998.

22 Nørskov, J.K., Rossmeisl, J., Logadottir, A. et al. (2004). Origin of the overpotential for oxygen reduction at a fuel-cell cathode. *J. Phys. Chem. B* 108: 17886–17892.

23 Subbaraman, R., Rossmeisl, J., Logadottir, A. et al. (2011). Enhancing hydrogen evolution activity in water splitting by tailoring Li^+-$Ni(OH)_2$-Pt interfaces. *Science* 334: 1256–1260.

24 Li, C.Q. and Baek, J.B. (2020). Recent advances in noble metal (Pt, Ru, and Ir)-based electrocatalysts for efficient hydrogen evolution reaction. *ACS Omega* 5: 31–40.

25 Sarkar, S. and Peter, S.C. (2018). An overview on Pd-based electrocatalysts for the hydrogen evolution reaction. *Inorg. Chem. Front.* 5: 2060–2080.

26 Esposito, D.V., Hunt, S.T., Stottlemyer, A.L. et al. (2010). Low-cost hydrogen-evolution catalysts based on monolayer platinum on tungsten monocarbide substrates. *Angew. Chem. Int. Ed.* 49: 9859–9862.

27 Bhatt, M.D. and Lee, J.Y. (2020). Advancement of platinum (Pt)-free (non-Pt precious metals) and/or metal-free (non-precious-metals) electrocatalysts in energy applications: a review and perspectives. *Energy Fuel* 34: 6634–6695.

28 Xiong, B.Y., Chen, L., and Shi, J. (2018). Anion-containing noble-metal-free bifunctional electrocatalysts for overall water splitting. *ACS Catal.* 8: 3688–3707.

29 Costentin, C. and Saveant, J.M. (2017). Theoretical and mechanistic aspects of proton-coupled electron transfer in electrochemistry. *Curr. Opin. Electrochem.* 1: 104–109.

30 Warburton, R. E., Soudackov, A. V. and Hammes-Schiffer, S. (2022). Theoretical modeling of electrochemical proton-coupled electron transfer. Chem. Rev. 122: 10599.

31 Mahmood, N., Yao, Y., Zhang, J.-W. et al. (2018). Electrocatalysts for hydrogen evolution in alkaline electrolytes: mechanisms, challenges, and prospective solutions. *Adv. Sci.* 5: 1700464.

32 Sabatier, P. (1920). *La Catalyse en Chimie Organique*. Paris et Liege: Librairie Polytechnique.

33 Zheng, Y., Jiao, Y., Jaroniec, M., and Qiao, S.Z. (2014). Advancing the electrochemistry of the hydrogen-evolution reaction through combining experiment and theory. *Angew. Chem. Int. Ed.* 52: 2–16.

34 Mao, J.J., He, C.-T., Pei, J. et al. (2018). Accelerating water dissociation kinetics by isolating cobalt atoms into ruthenium lattice. *Nat. Commun.* 9: 4958.

35 Trasatti S. (1972). Work function, electronegativity, and electrochemical behaviour of metals: III. Electrolytic hydrogen evolution in acid solutions J. Electroanal. Chem. Interfacial Electrochem. 39: 163–184.

36 Nørskov, J.K. and Hammer, B. (2000). Theoretical surface science and catalysis-calculations and concepts. *Adv. Catal.* 45: 71–129.

37 Li, Q.J., Zou, X., Ai, X. et al. (2018). Revealing activity trends of metal diborides toward pH-universal hydrogen evolution electrocatalysts with Pt-like activity. *Adv. Energy Mater.* 9: 201803369.

38 Yi, D., Lu, F., Zhang, F. et al. (2020). Regulating charge transfer of lattice oxygen in single-atom doped titania for hydrogen evolution. *Angew. Chem. Int. Ed.* 59: 15855–15859.

39 Peng, L.X., Su, L., Yu, X. et al. (2022). Electron redistribution of ruthenium-tungsten oxides Mott-Schottky heterojunction for enhanced hydrogen evolution. *Appl. Catal. B Environ.* 308: 121229.

40 Zhou M., Li, H., Long, A. et al. (2021). Modulating 3d orbitals of Ni atoms on Ni-Pt edge sites enables highly-efficient alkaline hydrogen evolution. Adv. Energy Mater. 11: 2101789.

41 Zhu, T., Huang, J., Huang, B. et al. (2020). High-index faceted RuCo nanoscrews for water electrosplitting. *Adv. Energy Mater.* 47: 16146840.

42 Xu, Y., Shao, M., Mavrikakis, M., and Adzic, R. (2009). Recent development in the electrocatalysis of the O_2 reduction reaction. In: *Fuel Cell Catalysis: A Surface Science Approach* (ed. M.T.M. Koper). Hoboken: John Wiley & Sons Inc.

43 Bagger, A., Castelli, I.E., Hansen, H., and Rossmeisl, J. (2018). Fundamental atomic insight in electrocatalysis. In: *Handbooks of Materials Modeling* (ed. W. Andreoni and S. Yip), 1–31. Springer Nature.

44 Gottesfeld, S. (2009). Electrocatalysis of oxygen reduction in polymer electrolyte fuel cells: a brief history and a critical examination of present theory and diagnostics. In: *Fuel Cell Catalysis: A Surface Science Approach* (ed. M.T.M. Koper). Hoboken: John Wiley & Sons Inc.

45 Zhao, Y.G., Saseendran, D.P.A., Huang, C. et al. (2023). Oxygen evolution/reduction reaction catalysts: from *in situ* monitoring and reaction mechanisms to rational design. *Chem. Rev.* 123: 6257–6358.

46 Markovic, N. (2002). Surface science studies of model fuel cell electrocatalysts. *Surf. Sci. Rep.* 45: 117–229.

47 Yeager, E.B., Adžić, R.R., Cahan, B.D., and Yeager, E.B. (1944). Structural effects in electrocatalysis: oxygen reduction on platinum low index single-crystal surfaces in perchloric acid solutions. *J. Electroanal. Chem.* 377: 249–259.

48 Gasteiger, H.A., Kocha, S.S., Sompalli, B., and Wagner, F.T. (2005). Activity benchmarks and requirements for Pt, Pt-alloy, and non-Pt oxygen reduction catalysts for PEMFCs. *Appl. Catal. B* 56: 9–35.

49 Nørskov, J.K., Mun, B.S., Mayrhofer, K.J. .J. et al. (2006). Changing the activity of electrocatalysts for oxygen reduction by tuning the surface electronic structure. *Angew. Chem. Int. Ed.* 45: 2897–2901.

50 Chorkendorff, I., Malacrida, P., Hansen, M.H. et al. (2016). Tuning the activity of Pt alloy electrocatalysts by means of the lanthanide contraction. *Science* 352: 73–76.

51 Greeley, J., Stephens, I.E.L., Bondarenko, A.S. et al. (2009). Alloys of platinum and early transition metals as oxygen reduction electrocatalysts. *Nat. Chem.* 1: 552–556.

52 Markovic, M., Mun, B.S., Arenz, M. et al. (2007). Trends in electrocatalysis on extended and nanoscale Pt-bimetallic alloy surfaces. *Nat. Mater.* 6: 241–247.

53 Duan, Z.Y., Wang, G.F. et al. (2011). A first principles study of oxygen reduction reaction on a Pt(111) surface modified by a subsurface transition metal M (M = Ni, Co, or Fe). *Phys. Chem. Chem. Phys.* 13: 20178–20187.

54 Mavrikakis, M., Xu, Y., Zhang, J. et al. (2007). Bimetallic and ternary alloys for improved oxygen reduction catalysis. *Top. Catal.* 46: 276–284.

55 Adzic, R.R. (2013). Tuning the catalytic activity of Ru@Pt core–shell nanoparticles for the oxygen reduction reaction by varying the shell thickness. *J. Phys. Chem. C* 117: 1748–1753.

56 Jiang, J., Liu, C., Gu, C. et al. (2023). Interlayer charge transfer regulates single-atom catalytic activity on electride/graphene 2D heterojunctions. *J. Am. Chem. Soc.* 145: 4774–4783.

57 Peng, Y., Lu, B., Wang, N. et al. (2019). Oxygen reduction reaction catalyzed by black-phosphorus-supported metal nanoparticles: impacts of interfacial charge transfer. *ACS Appl. Mater. Interfaces* 11: 24707–24714.

58 Wen, S.D., Liu, B., Li, W. et al. (2022). A battery process activated highly efficient carbon catalyst toward oxygen reduction by stabilizing lithium oxygen bonding. *Adv. Funct. Mater.* 32: 2203960.

59 Lu, F., Wang, J., Li, J. et al. (2020). Regulation of oxygen reduction reaction by the magnetic effect of $L1_0$-PtFe alloy. *Appl. Catal. B Environ.* 278: 119332.

60 Gracia, J. (2017). Photosystem II acts as a spin-controlled electron gate during oxygen formation and evolution. *J. Am. Chem. Soc.* 139: 16604–16608.

61 Gracia, J., Sharpe, R., and Munarriz, J. (2018). Principles determining the activity of magnetic oxides for electron transfer reactions. *J. Catal.* 361: 331–338.

62 Gao, D., Wang, T., Zhang, Z. et al. (2020). Engineered spin state in Ce doped $LaCoO_3$ with enhanced electrocatalytic activity for rechargeable Zn-air batteries. *Nano Energy* 74: 104948.

63 Pena, M.A. and Fierro, J.L. (2001). Chemical structures and performance of perovskite oxides. *Chem. Rev.* 101: 1981–2017.

64 Beall, C.E., Fabbri, E., and Schmidt, T.J. (2021). Perovskite oxides based electrodes for the oxygen reduction and evolution reactions: the underlying mechanism. *ACS Catal.* 11: 3094–3114.

65 Zhou, Y., Sun, S., Xi, S. et al. (2018). Super exchange effects on oxygen reduction activity of edge sharing [$Co_xMn_{1-x}O_6$] octahedra in spinel oxide. *Adv. Mater.* 30: 1705407.

66 Sun, Y., Ren, X., Sun, S. et al. (2021). Engineering high-spin state cobalt cations in spinel zinc cobalt oxide for spin channel propagation and active site enhancement in water oxidation. *Angew. Chem. Int. Ed.* 60: 14536–14544.

67 Sun, Y., Liao, H., Wang, J. et al. (2020). Covalency competition dominates the water oxidation structure activity relationship on spinel oxides. *Nat. Catal.* 3: 554–563.

68 Zhai, Y., Niu, H., Ding, J. et al. (2021). Unraveling the origin of sulfur-doped Fe-N-C single-atom catalyst for enhanced oxygen reduction activity: effect of iron spin-state tuning. *Angew. Chem. Int. Ed.* 60: 25404–25410.

69 Xue, D.P., Yuan, P., Jiang, S. et al. (2023). Altering the spin state of Fe-N-C through ligand field modulation of single-atom sites boosts the oxygen reduction reaction. *Nano Energy* 105: 108020.

70 Liu, M., Fu, J., Lin, Y. et al. (2022). Insights into the activity of single-atom Fe-N-C catalysts for oxygen reduction reaction. *Nat. Commun.* 13: 2075.

71 Yang, G., Zhu, J., Yuan, P. et al. (2021). Regulating Fe-spin state by atomically dispersed Mn-N in Fe-N-C catalysts with high oxygen reduction activity. *Nat. Commun.* 12: 1734.

72 Zheng, J., Liao, F., Wu, S. et al. (2019). Efficient non-dissociative activation of dinitrogen to ammonia over lithium promoted ruthenium nanoparticles at low pressure. *Angew. Chem. Int. Ed.* 58: 17335–17341.

73 Chen, J.G., Crooks, R.M., Seefeldt, L.C. et al. (2018). Beyond fossil fuel-driven nitrogen transformations. *Science* 80: eaar6611.

74 Ren, Y., Yu, C., Tan, X. et al. (2021). Strategies to suppress hydrogen evolution for highly selective electrocatalytic nitrogen reduction: challenges and perspectives. *Energy Environ. Sci.* 14: 1176–1193.

75 Demirci, U.B. and Miele, P. (2011). Chemical hydrogen storage: 'material' gravimetric capacity versus 'system' gravimetric capacity. *Energy Environ. Sci.* 4: 3334–3341.

76 Martín, A.J., Shinagawa, T., and Pérez-Ramírez, J. (2019). Electrocatalytic reduction of nitrogen: from Haber–Bosch to ammonia artificial leaf. *Chem* 5: 263–283.

77 van der Ham, C.J.M., Koper, M.T.M., and Hetterscheid, D.G.H. (2014). Challenges in reduction of dinitrogen by proton and electron transfer. *Chem. Soc. Rev.* 43: 5183–5191.

78 Skulason, E., Garden, A.L., Howalt, J.G. et al. (2016). Electroreduction of N_2 to ammonia at ambient conditions on mononitrides of Zr, Nb, Cr, and V: a DFT guide for experiments. *ACS Catal.* 6: 635–646.

79 Abghoui, Y. and Skúlason, E. (2017). Electrochemical synthesis of ammonia via Mars-Van Krevelen mechanism on the (111) facets of Group III-VII transition metal mononitrides. *Catal. Today* 286: 78–84.

80 Abghoui, Y. and Skúlason, E. (2017). Computational predictions of catalytic activity of zincblende (110) surfaces of metal nitrides for electrochemical ammonia synthesis. *J. Phys. Chem. C* 121: 6141–6151.

81 Montoya, J.H., Tsai, C., Vojvodic, A., and Nørskov, J.K. (2015). The challenge of electrochemical ammonia synthesis: a new perspective on the role of nitrogen scaling relations. *ChemSusChem* 8: 2180–2192.

82 Choi, C., Back, S., Kim, N.-Y. et al. (2018). Suppression of hydrogen evolution reaction in electrochemical N_2 reduction using single-atom catalysts: a computational guideline. *ACS Catal.* 8: 7517–7525.

83 Kim, C., Song, J.-Y., Choi, C. et al. (2022). Atomic-scale homogeneous Ru-Cu alloy nanoparticles for highly efficient electrocatalytic nitrogen reduction. *Adv. Mater.* 34: 2205270.

84 Liu, X., Jiao, Y., Zheng, Y. et al. (2019). Building up a picture of the electrocatalytic nitrogen reduction activity of transition metal single-atom catalysts. *J. Am. Chem. Soc.* 141: 9664–9672.

85 Huang, S.P., Gu, J., Lin, S. et al. (2020). Tackling the activity and selectivity challenges of electrocatalysts toward the nitrogen reduction reaction via atomically dispersed diatom catalysts. *J. Am. Chem. Soc.* 142: 5709–5721.

86 Zhang, L., Ding, L.-X., Chen, G.-F. et al. (2019). Ammonia synthesis under ambient conditions: selective electroreduction of dinitrogen to ammonia on black phosphorus nanosheets. *Angew. Chem. Int. Ed.* 58: 2612–2616.

87 Ling, C.Y., Niu, X., Li, Q. et al. (2018). Metal-free single atom catalyst for N_2 fixation driven by visible light. *J. Am. Chem. Soc.* 140: 14161–14168.

88 Zafari, M., Umer, M., Nissimagoudar, A.S. et al. (2022). Unveiling the role of charge transfer in enhanced electrochemical nitrogen fixation at single-atom catalysts on BX sheets (X = As, P, Sb). *J. Phys. Chem. Lett.* 13: 4530–4537.

89 Niu, L.J., Wang, D., Xu, K. et al. (2021). Tuning the performance of nitrogen reduction reaction by balancing the reactivity of N_2 and the desorption of NH_3. *Nano Res.* 14: 4093–4099.

90 Gao, D.L., Yi, D., Sun, C. et al. (2022). Breaking the volcano-shaped relationship for highly efficient electrocatalytic nitrogen reduction: a computational guideline. *ACS Appl. Mater. Interfaces* 14: 52806–52814.

91 Nørskov, J.K., Bukas, V.J., Shadravan, V. et al. (2022). A spin promotion effect in catalytic ammonia synthesis. *Nat. Commun.* 13: 2382.

92 Zhang, J.N., Cheng, W., Yuan, P. et al. (2021). Boosting nitrogen reduction to ammonia on FeN_4 sites by atomic spin regulation. *Adv. Sci.* 8: 2102945.

93 Li, X.F., Li, Q.-K., Cheng, J. et al. (2016). Conversion of dinitrogen to ammonia by FeN_3-embeddded graphene. *J. Am. Chem. Soc.* 138: 8706–8709.

94 Cui, C.N., Zhang, H., and Luo, Z. (2020). Nitrogen reduction reaction on small iron clusters supported by N-doped graphene: a theoretical study of the atomically precise active-site mechanism. *Nano Res.* 13: 2280–2288.

95 Gao, S., Liu, X., Wang, Z. et al. (2023). Spin regulation for efficient electrocatalytic N_2 reduction over diatomic Fe-Mo catalyst. *J. Colloid Interface Sci.* 630: 215–223.

3

Electrocatalytic Reaction Mechanism for Energy Conversion

Hao Ma and Yuanmiao Sun

Chinese Academy of Sciences, Shenzhen Institute of Advanced Technology, Institute of Technology for Carbon Neutrality, Shenzhen 1068 Xueyuan Avenue, Shenzhen University Town, Shenzhen, 518055, Guangdong, China

3.1 Introduction

Electrocatalysis refers to an electrochemical reaction process wherein catalysts are utilized to lower the activation energy, thereby promoting reaction rates [1–3]. This process is pivotal for enhancing energy conversion efficiency and driving the advancement of sustainable energy technologies, including fuel cells, water electrolyzers, and batteries. Understanding the mechanism of electrocatalytic reactions is essential due to their complex influence on the kinetics of the reaction. However, the presence of various species on the electrode surface, such as reactants, intermediates, and products, can lead to intricate reaction pathways that are challenging to interpret. Besides, multiple mechanisms may occur simultaneously with equal thermodynamical likelihood, which makes it difficult to identify the most feasible one. Consequently, a comprehensive understanding of different electrocatalytic mechanisms in chemical reactions is essential to unravel the complex interplay of species on the surface of the catalyst and their impact on the electrocatalytic process.

As the development of electrocatalytic technology continues, several important aspects of electrocatalytic mechanisms deserve special attention. In detail, firstly, intermediates are considered to be temporary species formed and consumed within the electrocatalytic process, which can be used to identify the pathway of a reaction and explain how a reaction evolves at a molecular level. By studying the intermediates and the steps in which they are involved, chemists can better understand the reaction mechanisms and design strategies to control or manipulate the reaction. By mastering the key reaction intermediates that determine the reaction pathway, it is possible to optimize the structure of the catalysts and therefore gain enhanced catalytic activity. Secondly, different catalytic sites on the catalyst surface may exhibit distinguished catalytic activity, as they may possess distinct structural characteristics and adsorption affinity, leading to differentiated reaction performance in catalytic reactions. The above phenomena can be attributed to the electronic structure, surface defects, and interactions between reaction intermediates and the

Functional Materials for Electrocatalytic Energy Conversion, First Edition.
Edited by Zhicheng Zhang, Meiting Zhao, and Yuchen Qin.

catalytic sites. Besides, unfavorable sites could lead to side reactions, energy losses, or an overall reduction in catalytic efficiency. Therefore, a more precise mastery of the surface sites on the catalyst is beneficial for optimizing catalyst design and enhancing catalytic activity and selectivity. Thirdly, the surface coverage of adsorbents on the catalyst surface is also a noticeable point for optimizing the catalyst. Elaborating on the competitive relationships and effects among different coverages to help one maximize beneficial coverages or adjusting catalytic conditions to minimize the effect of harmful coverages on the reaction are effective strategies to achieve enhanced catalytic activity. Lastly, the electron transfer process is a crucial and inevitable step in catalytic reactions, which influences the competition and selection of reaction pathways. By analyzing the electron transfer process, it is possible to gain a deeper understanding of the kinetic nature of the reaction, including reaction rates and activation energies, so that the electronic structure of the catalyst can be further optimized to be more suitable for a particular reaction selection. In brief, an in-depth awareness of electrocatalytic mechanisms is essential for the advancement of various fields, from the conversion and storage of energy to medicine, environmental protection, and materials science. It can not only provide us with new insight for the development of new highly efficient electrocatalytic materials but also establish a theoretical foundation to advance new technologies with improved performance, sustainability, and applicability.

This chapter aims to introduce the electrocatalytic mechanisms of various electrochemical reactions to the readers, especially different feasible reaction pathways and some basic concepts. Examples of some widely researched electrocatalytic reactions are introduced, i.e. hydrogen evolution reaction (HER) [4], oxygen evolution reaction (OER) [5], oxygen reduction reaction (ORR) [6], CO_2 reduction reaction (CO_2RR) [7], and nitrogen reduction reaction (NRR) [8]. By analyzing the fundamental principles of electrocatalysis, we will introduce the underlying ideas governing electrocatalysis to the reader and pique their interest in exploring more about the enormous variety of electrocatalysts and their uses.

3.2 Electrochemical Parameters of Electrocatalysts

To explain the electrocatalytic process of electrochemical reactions, some important parameters need to be evaluated, including the overpotential/onset potential, Faradic efficiency (FE), Gibb's free energy, Tafel slope, turnover frequency (TOF), and exchange current density. Understanding and optimizing these parameters collectively contributes to creating high-efficiency and practical electrocatalysts for various applications, including fuel cells, electrolysis, and other electrochemical devices. Here is a brief explanation of some of the important electrochemical parameters.

3.2.1 Overpotential

Overpotential is defined as the extra potential required to overcome the kinetic barriers and drive a desired electrochemical reaction [9]. In general, the required overpotential to reach a 10 mA cm^{-2} current density serves as a valuable benchmark for comparing the catalytic activities of different electrocatalysts. Electrocatalysts that can achieve this current density with lower overpotentials are generally more efficient and desirable for practical applications, as they signify better kinetics and reduced energy losses in the electrochemical reaction. This approach simplifies the evaluation process and aids in identifying promising candidates for further development in energy conversion and storage technologies.

3.2.2 Faradic Efficiency

In electrochemistry, the ratio of the experimentally detected amount of objective gas to the amount of theoretically generated gas is known as FE, which serves as a critical parameter in assessing the effectiveness of electrocatalysts and the overall efficiency of electrochemical processes [10]. The FE is calculated by:

$$\varepsilon_{\text{Faradaic}} = \frac{Q_{\text{prod}}}{Q_{\text{total}}} = \frac{\alpha n F}{Q} \times 100\% \tag{3.1}$$

Here, Q_{total} is the total charge passing through the reaction process and Q_{prod} presents the charges that form a particular product, α is the transferred number of electrons (e.g. $\alpha = 6$ for methanol), F and n are Faraday's constant (96,485 C mol^{-1}) and the number of moles of a desired product, respectively, while Q is the total charge passed. For example, in HER, by comparing the actual amount of hydrogen gas produced with the theoretically expected amount, researchers can gain insights into the extent to which electrons are utilized for the intended reaction versus being lost to undesired side reactions. This parameter quantifies the utilization efficiency of transferred electrons for the desired reaction and plays a pivotal role in designing and optimizing electrochemical systems, especially for applications like hydrogen production, where efficiency is crucial for sustainable energy technologies.

3.2.3 Gibbs Free Energy

One thermodynamic potential that can be employed to determine the maximum amount of work in thermodynamics is known as the Gibbs free energy (ΔG). The adsorbate adsorption or desorption on an electrocatalyst's surface during the electrocatalytic process can be well evaluated by ΔG. The ΔG indicates whether the adsorbate adsorption/desorption process is thermodynamically favorable or not, and it is evaluated by:

$$\Delta G = \Delta E + \Delta E_{\text{ZPE}} - T\Delta S \tag{3.2}$$

where ΔE indicates the adsorption binding energy, ΔE_{ZPE}, and ΔS are the variation of zero-point energy and reaction entropy between the reactant and product, respectively. T is the absolute temperature. As is known to all, the Sabatier principle highlights the significance of proper charge transfer between the adsorbate and the electrocatalyst surface. When this charge transfer occurs effectively, it facilitates the formation of catalytic production. However, there is a fine balance [11, 12].

An overly strong interaction between the adsorbed reactant and electrocatalyst might hinder the desorption process, limiting the rate of catalytic production. Conversely, if the bonding is too weak, it can adversely affect the adsorption process. Therefore, based on the Sabatier principle, the calculation of ΔG can aid in understanding the thermodynamic and kinetic characteristics of electrocatalytic reactions and provide guidance for designing more efficient catalysts.

3.2.4 Tafel Slope

A Tafel plot is a graphical representation used in electrochemistry to analyze the relation between the overpotential (η) and the steady-state current density (j) [13]. Tafel plots are commonly used to study the kinetics of electrochemical reactions and to understand the rate-limiting steps involved in the reaction mechanism. In a Tafel plot, the logarithm of the current density ($\log j$) is plotted against the η.

$$\eta = a + b\log(j) \tag{3.3}$$

The slope of this linear section of the Tafel plot is known as the Tafel slope (b). The Tafel slope represents the change in overpotential required to increase the current density by a factor of 10. The smaller Tafel slope demonstrates that the electrocatalysts are able to significantly improve the reaction rate (current density) with a relatively small change in overpotential.

3.2.5 Turnover Frequency

TOF is a crucial parameter in catalysis that quantifies the catalytic activity of specific active sites within electrocatalytic material [14]. In the electrocatalytic reaction, TOF represents the rate at which individual catalytic centers facilitate the catalytic production from reactants. TOF is expressed as the number of reaction cycles or molecular reactions that occur at the catalytic site per unit of time.

$$\text{TOF} = \frac{\text{volumetric rate of reaction}}{\text{number of catalytic centers per volume}} \tag{3.4}$$

A higher TOF value indicates a more efficient catalyst because it signifies that more reaction events are occurring at each catalytic center per unit time. This parameter is particularly useful for comparing different catalysts' intrinsic activity in catalytic reactions, irrespective of the loading mass or electrode area.

3.2.6 Exchange Current Density

The exchange current density (i_0) is a fundamental metric in electrochemistry used to quantify the rate of an electrochemical reaction at equilibrium, specifically at the electrode–electrolyte interface. It represents the current density at which the rate of the forward reaction (oxidation or reduction) is equal to the rate of the reverse reaction [15]. A typical example is the HER process. The theoretical i_0 is calculated based on the averaged hydrogen adsorption free energies (ΔG_{H*}) for a given electrochemical reaction. If $\Delta G_{H*} \leq 0$, the i_0 is expressed at pH = 0 by:

$$i_0 = -ek_0 \frac{1}{1 + \exp\left(-\Delta G_{H^*}/k_B T\right)} \tag{3.5}$$

If $\Delta G_{H*} > 0$, the i_0 is calculated by:

$$i_0 = -ek_0 \frac{1}{1 + \exp\left(-\Delta G_{H^*}/k_B T\right)} \tag{3.6}$$

Here, k_0 is the rate constant. In fact, i_0 is an important parameter for many electrochemical applications, including batteries and corrosion studies. It provides insights into the activity and efficiency of electrocatalysts and helps in the design and optimization of electrochemical devices.

3.3 Fundamentals of Electrocatalytic HER

Hydrogen can be considered one of the most promising resources to address the global energy challenges. Electrochemical water splitting holds the potential as a sustainable strategy for hydrogen production. Despite the exceptional performance of Pt-based materials in HER, the low abundance and high cost constrain their widespread industrial applications [16, 17]. The advancement of electrocatalysts for HER is of paramount significance for the clean energy industry in the future. The acceleration of HER reaction kinetics through efficient electrocatalysts can facilitate the production of hydrogen gas (H_2), subsequently bolstering the efficiency of green hydrogen generation. As a two-electron transfer process in a cathode, the HER involves the reduction of protons from water molecules to produce H_2. Active catalysts play a role by reducing the energy barriers associated with each step of the process. The generally accepted HER pathways in acidic/alkaline environments can be described by the Volmer–Heyrovsky and Volmer–Tafel mechanisms [16–19] (see Figure 3.1).

For acidic media,

$$H^+ + e^- \rightarrow H^* \quad (\text{Volmer step}) \tag{3.7}$$

$$H^* + H^+ + e^- \rightarrow H_2 \quad (\text{Heyrovsky step}) \tag{3.8}$$

$$2H^* \rightarrow H_2 \quad (\text{Tafel step}) \tag{3.9}$$

(I) In the first step of the HER under acidic conditions, hydrogen atoms are adsorbed onto the electrode's surface. This adsorption is facilitated by the reaction between electrons and protons, which leads to the formation of adsorbed hydrogen species (i.e. electrochemical hydrogen adsorption). **(II)** Following the adsorption of hydrogen, the subsequent step involves a desorption reaction. In this step, H_2 is produced as a result of the interaction between the adsorbed hydrogen species, electrons, and protons. **(III)** The third step further contributes to H_2 generation. Here, two adsorbed hydrogen atoms (H*) react together, resulting in the formation

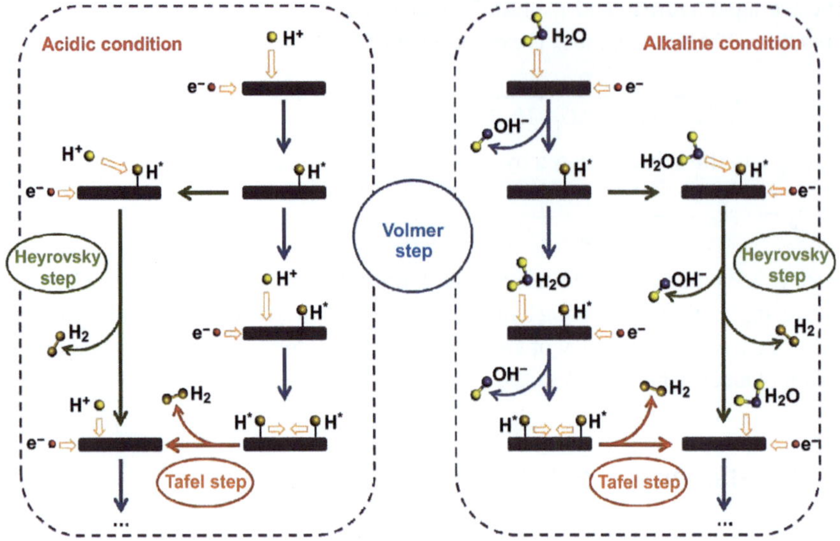

Figure 3.1 Schematic pathways for the HER mechanism on the surface of the catalyst under acidic and alkaline conditions. Source: [19]/Springer Nature/CC BY 4.0.

of an H_2 molecule. This combination of adsorbed hydrogen species is also a crucial mechanism for H_2 production.

The Volmer and Heyrovsky steps of the HER in alkaline media involve splitting apart water molecules to produce protons, while the Tafel step remains unchanged in comparison to that in the acid media. The reaction formula for HER in an alkaline medium is represented as:

$$H_2O + e^- \rightarrow H^* + OH^- \text{ (Volmer step)} \tag{3.10}$$

$$H^* + H_2O + e^- \rightarrow OH^- + H_2 \text{ (Heyrovsky step)} \tag{3.11}$$

$$2H_2O + 2e^- \rightarrow 2OH^- + H_2 \text{ (Tafel step)} \tag{3.12}$$

3.4 Fundamentals of Electrocatalytic OER

OER is a familiar half-reaction in many aqueous catalytic processes, particularly in the electrolysis of water [12, 20–22]. Because the anodic water oxidation to produce oxygen has a higher energy barrier compared to the cathodic hydrogen evolution, OER often becomes the rate-limiting step in water splitting, which means that the kinetics of OER can significantly impact the speed and overall efficiency of hydrogen production for water electrolysis.

The conventional adsorption evolution mechanism (AEM) is one of the widely acknowledged OER mechanisms and is distinguished by four consecutive proton-coupled electron transfer steps [23]. In general, the OER pathway tends to

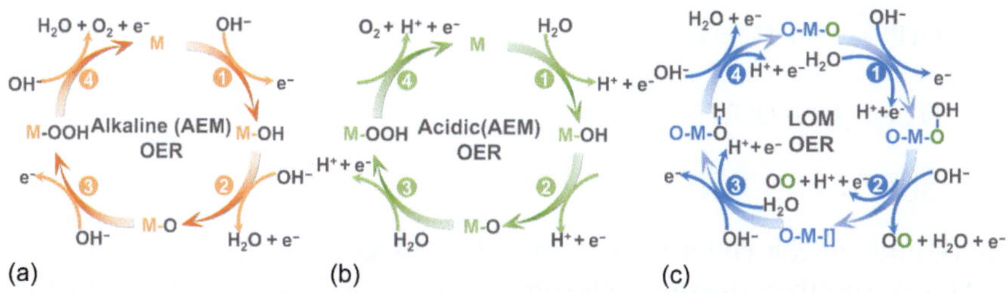

Figure 3.2 Schematic pathways for the OER mechanism, namely, (a–b) AEM (alkaline/acidic) and (c) LOM (acidic-inner ring/alkaline-outer ring). Source: Reproduced with permission from Zhang et al. [23]. Copyright 2022 American Chemical Society.

be sensitive to the solution's pH, the reaction process can be described as follows in alkaline media (see Figure 3.2a):

$$^* + OH^- \rightarrow OH^* + e^- \tag{3.13}$$

$$OH^* + OH^- \rightarrow O^* + H_2O + e^- \tag{3.14}$$

$$O^* + OH^- \rightarrow OOH^* + e^- \tag{3.15}$$

$$OOH^* + OH^- \rightarrow * + O_2 + H_2O + e^- \tag{3.16}$$

Initially, OH^* is adsorbed onto the catalytic metal sites, typically transition metal cations, and then the adsorbed OH^* loses a proton (H^+) and an electron (e^-) to become an adsorbed oxygen atom (O^*). The O—O bond is formed via the intermediate OOH^*. Eventually, OOH^* further releases O_2 through a single electron transfer process, recovering the metal's active center. The overall reaction pathway can be represented as $OH^* \rightarrow O^* \rightarrow OOH^* \rightarrow O_2$. A significant characteristic of the AEM is its linear scaling relationship with the adsorption energy for the adsorbed intermediates, which is expressed as $\Delta G_{OOH^*} = \Delta G_{OH^*} + 3.2 \pm 0.2\,\text{eV}$ [24]. However, it is crucial to note that the actual difference between ΔG_{*OH} and ΔG_{*OOH} should exceed the ideal model value of 2.46 eV. As a result, the minimum overpotential required for the AEM mechanism is 0.37 eV [21, 25]. This value serves as a fundamental guideline for constructing an activity volcano plot from the binding strength of O^*. Thus, in the framework of the AEM, breaking or weakening the linear scaling relation between ΔG_{OH^*} and ΔG_{OOH^*} is key to reducing the overpotential of OER. Similarly, in acidic media, the OER process also involves three different intermediates, i.e. OH^*, O^*, and OOH^* (see Figure 3.2b), but the initial step of breaking the H—OH bond is the most crucial and often the slowest in acidic conditions. It should be mentioned that non-precious metal catalysts can be highly active for OER in alkaline media, but they often face stability issues in acidic environments. Due to the above limitation, substantial research efforts aimed at improving the durability of non-noble metal catalysts, especially in acidic conditions:

$$^* + H_2O \rightarrow OH^* + H^+ + e^- \tag{3.17}$$

$$OH^* \rightarrow O^* + H^+ + e^- \quad (3.18)$$

$$O^* + H_2O \rightarrow OOH^* + H^+ + e^- \quad (3.19)$$

$$OOH^* \rightarrow * + O_2 + H^+ + e^- \quad (3.20)$$

In addition to conventional AEM, the lattice oxygen participation mechanism (LOM) has recently been extensively considered as an alternative reaction pathway in (hydro)oxides (see Figure 3.2c) [23]. For LOM, **(I)** the catalyst contains lattice oxygen atoms within its crystal structure, and these lattice oxygen atoms are involved in the oxygen evolution process as active sites. **(II)** OH^- ions directly interact with the activated lattice oxygen through a nucleophilic attack, forming an intermediate species OOH*. **(III)** The OOH* species eventually undergoes further reactions that lead to the release of O_2 as one of the products. Simultaneously, this step is accompanied by the creation of a lattice oxygen vacancy (V_O) within the catalyst. **(IV)** The V_O can be refilled with either OH^- ions (in alkaline conditions) or H_2O (in acidic conditions) from the surrounding electrolyte, recovering the lattice oxygen sites. In 2019, Hu et al. conducted in situ and operando spectroscopic investigations that directly confirmed the involvement of lattice oxygen during the OER reaction [26]. Up to now, researchers continue to explore and refine their understanding of this mechanism [27, 28].

In fact, the LOM mechanism includes two other possible pathways to produce O_2 [29]. Taking alkaline media as an example, as shown in Figure 3.3 (left panel), the single-metal-site mechanism (SMSM) is the process of adsorption of OH^- and deprotonation by using single metal sites as the catalytic center. The catalyst surface reconstruction can cause direct interaction between activated lattice oxygen and O* intermediates to form O_2 molecules. Another possible pathway for the LOM mechanism is regarded as the dual-metal-site mechanism (DMSM) (see the right panel in Figure 3.3), i.e. if the catalyst contains bimetallic sites that are coupled by two bridge oxygen atoms (oxos), the relatively short distance between

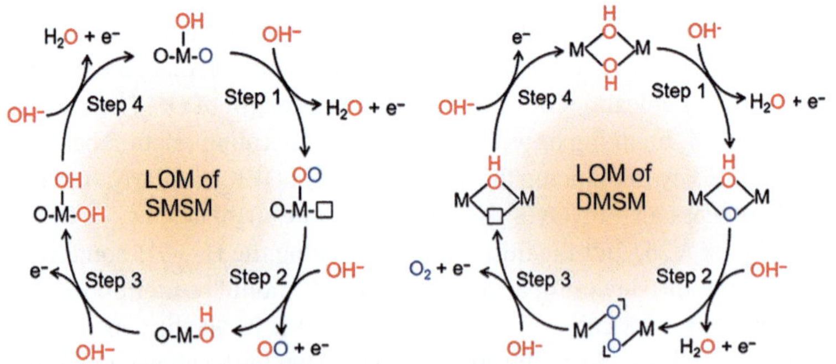

Figure 3.3 Schematic illustrations of two possible LOM pathways in alkaline conditions, including SMSM and DMSM. Here, SMSM and DMSM represent the single- and dual-metal-site mechanisms, respectively. Source: Reproduced with permission from Zhang et al. [29]. Copyright 2021 Royal Society of Chemistry.

the two adjacent metal sites allows for efficient coupling of O—O. Initially, the metal–OH–metal motifs on the catalyst surface deprotonate to form a metal-oxo species, and then two adjacent metal-oxo groups are directly coupled to form an O—O bond instead of forming an OOH* intermediate. Once the O—O bond is formed, O_2 is generated and released. Simultaneously, vacant metal sites are filled by OH^-, repairing the catalyst for the next OER cycles. These pathways involve the direct coupling of oxygen atoms to form O_2, which can be of significant interest in the exploration of OER catalysts with high efficiency and stability.

3.5 Fundamentals of Electrocatalytic ORR

ORR refers to the reduction half-reaction in which O_2 is reduced to water (H_2O) or hydrogen peroxide (H_2O_2). This reaction typically involves the gain of electrons by oxygen atoms, resulting in the formation of H_2O or H_2O_2. ORR is an essential reaction in metal–air batteries, fuel cells, supercapacitors, and the synthesis of specific chemicals [30]. The stoichiometries of the ORR vary according to the surrounding medium, which includes $4e^-$ and $2e^-$ pathways:

$4e^-$ pathway as follows:

$$\text{Acid medium:}\quad O_2 + 4e^- + 4H^+ \rightarrow 2H_2O \tag{3.21}$$

$$\text{Alkaline medium:}\ O_2 + 4e^- + 2H_2O \rightarrow 4OH^- \tag{3.22}$$

$$\text{Solid oxide:}\ O_2 + 4e^- \rightarrow 2O^{2-} \tag{3.23}$$

$2e^-$ pathway as follows:

$$\text{Acid medium:}\ O_2 + 2e^- + 2H^+ \rightarrow H_2O_2 \tag{3.24}$$

$$\text{Alkaline medium:}\ O_2 + 2e^- + H_2O \rightarrow HO_2^- + OH^- \tag{3.25}$$

The cathode reaction in fuel cells, particularly in alkaline fuel cells, proton-exchange membrane fuel cells (PEMFCs), and solid oxide fuel cells, typically follows the $4e^-$ pathway. In contrast, the $2e^-$ pathway reaction often occurs as a secondary or side reaction alongside the $4e^-$ pathway. Additionally, the $2e^-$ pathway can be employed for the synthesis of H_2O_2.

Taking the ORR in acid media as an example, the $4e^-$ ORR dissociation mechanism proceeds through the following basic steps [31]:

$$O_2 + * + H^+ + e^- \rightarrow OOH^* \tag{3.26}$$

$$OOH^* + H^+ + e^- \rightarrow O^* + H_2O \tag{3.27}$$

$$O^* + H^+ + e^- \rightarrow OH^* \tag{3.28}$$

$$OH^* + H^+ + e^- \rightarrow * + H_2O \tag{3.29}$$

whereas the dissociation process of the 2e⁻ ORR can be summed up as follows:

$$O_2 + * + H^+ + e^- \rightarrow OOH^* \tag{3.30}$$

$$OOH^* + H^+ + e^- \rightarrow * + H_2O_2 \tag{3.31}$$

Moreover, another pathway by direct O_2 dissociation of oxygen atoms along the following steps, and is followed by steps (3.28) and (3.29) in an acidic solution.

$$O_2 + * + * \rightarrow O^* + O^* \tag{3.32}$$

To attain elevated catalytic activity, the adsorption of OOH* species must be boosted, whereas to achieve superior selectivity, it is crucial to diminish the adsorption of O* species, which results from the dissociation of OOH*. Specifically, following the hydrogenation of the O_2 molecule by the H^+/e^- pair, two possible reaction pathways might further lead to the protonation of OOH*: (a) $OOH^* + H^+ + e^- \rightarrow O^* + H_2O$ (R1) or (b) $OOH^* + H^+ + e^- \rightarrow H_2O_2$ (R2). The energy barrier for H_2O_2 formation (R2) should be less than that for the hydrogenation of OOH* to O* (R1) to guarantee that the catalyst thermodynamically favors the formation of H_2O_2 over H_2O. Therefore, for a catalyst that is highly selective for producing H_2O_2, the Gibbs free energy of O* ($\Delta G(O^*)$) should be greater than 3.52 eV. Overall, the exploration and optimization of ORR have far-reaching implications for both clean energy technologies and environmentally friendly chemical processes, contributing to the development of more sustainable and efficient approaches to energy storage, conversion, and chemical synthesis.

3.6 Fundamentals of Electrocatalytic CO_2RR

As global temperature rises and the impacts of climate change intensify, it is now critically important to reduce greenhouse gas emissions, especially carbon dioxide (CO_2), and has become an urgent global imperative [32]. The unsustainable use of fossil fuels and emissions from industrial processes are major sources of CO_2 emission, making it crucial to transform and mitigate CO_2.

The research into electrocatalytic CO_2 reduction (ECR) has garnered significant attention as a groundbreaking technology. It involves the use of electrocatalysts to convert CO_2 molecules into valuable carbon-based compounds such as carbon monoxide (CO), methane (CH_4), and ethanol (C_2H_5OH). This process not only holds the potential to reduce CO_2 emissions but also to convert CO_2 into high-value chemicals and fuels, enabling the efficient reuse of carbon resources. The ECR on a cathodic catalyst is a complex multistep process, which encompasses the adsorption of CO_2 molecules onto the surface of catalyst, interactions occurring at the catalyst–electron–proton–CO_2, and the subsequent desorption of the products generated in earlier stages. Figure 3.4 shows the formation mechanism of the five primary C1 products throughout the ECR process [33]. The reaction process can involve various electron transfer steps, including 2, 4, 6, and 8-electron processes,

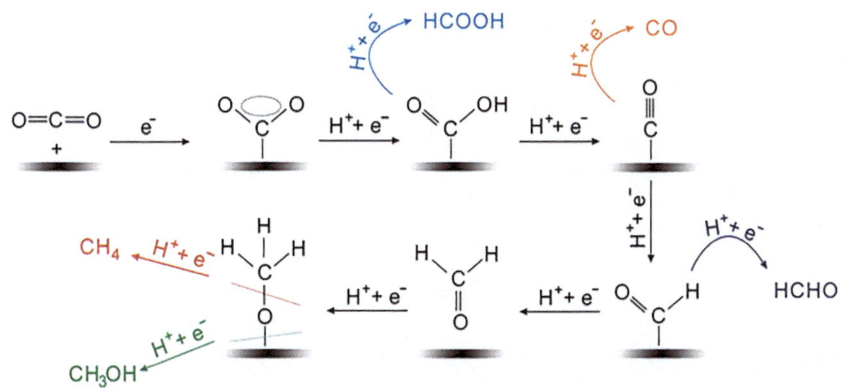

Figure 3.4 Reaction mechanism and pathways of the five major C1 products in the ECR reaction. Source: Reproduced with permission from Zhang et al. [33]. Copyright 2017 WILEY-VCH.

which lead to the production of different products, namely CO or HCOOH, HCHO, CH_3OH, and CH_4, as illustrated by the following Reactions (3.33)–(3.37):

$$CO_2 + 2H^+ + 2e^- \rightarrow CO + H_2O, \quad E^0 = -0.52\,V \tag{3.33}$$

$$CO_2 + 2H^+ + 2e^- \rightarrow HCOOH, \quad E^0 = -0.61\,V \tag{3.34}$$

$$CO_2 + 4H^+ + 4e^- \rightarrow HCHO + H_2O, \quad E^0 = -0.51\,V \tag{3.35}$$

$$CO_2 + 6H^+ + 6e^- \rightarrow CH_3OH + H_2O, \quad E^0 = -0.38\,V \tag{3.36}$$

$$CO_2 + 8H^+ + 8e^- \rightarrow CH_4 + 2H_2O, \quad E^0 = -0.24\,V \tag{3.37}$$

The process of CO_2 adsorption on the catalyst's surface is crucial in the overall ECR reaction. The interaction between the CO_2 molecule and the surface can perturb the C=O bond. Meanwhile, the catalyst and CO_2 molecule share the electrons, which effectively weakens the bond and makes it more susceptible to the subsequent reaction. Then, the generated $CO_2^{\bullet-}$ intermediate will form either HCOOH or CO by the proton transfer process. For instance, a high product selectivity of CO can be induced by the large adsorption energy of $CO_2^{\bullet-}$ and the comparatively moderate adsorption energy of CO on Au, while Cu-based catalysts have been reported to yield many kinds of carbon compounds (alcohols and hydrocarbon species) [34]. Next, in the process of synthesizing hydrocarbons, hydrogenation of adsorbed CO to generate adsorbed CHO* is a crucial step. Due to the lower thermodynamic stability of the CHOH, the formed CHO* would not convert into CHOH but would be reduced to CH_3OH or CH_4 with further protonation process. When the proton adds to the oxygen atom rather than the carbon atom in the stable CH_3O* intermediate for the surface of Cu_2O, CH_3OH can be produced [34]. On the contrary, Cu is in favor of the formation of CH_4 as a result of the addition of a proton to the carbon atom [35].

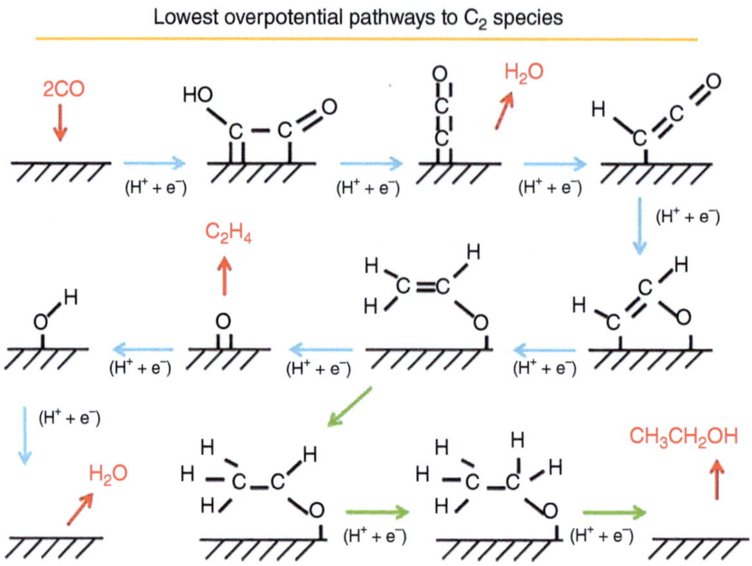

Figure 3.5 Reaction mechanism and pathways for the two C_2 products in the ECR reaction, including CH_3CH_2OH (green) and C_2H_4 (blue). Source: Reproduced with permission from Calle-Vallejo and Koper [36]. Copyright 2013 WILEY-VCH.

By contrast, the formation of C2 products necessitates a high coverage of C1 intermediates and stable C2 intermediates. The production of C_2H_4 and C_2H_5OH mainly results from the C–C coupling process (Reactions (3.38) and (3.39)),

$$2CO_2 + 8H_2O + 12e^- \rightarrow C_2H_4 + 12OH^-, \quad E^0 = -0.34\,V \tag{3.38}$$

$$2CO_2 + 9H_2O + 12e^- \rightarrow C_2H_5OH + 12OH^-, \quad E^0 = -0.33\,V \tag{3.39}$$

The formation pathways of C_2H_4 and C_2H_5OH are shown in Figure 3.5 [36]. For the products of C_2H_4 and C_2H_5OH, the coupling of two CO molecules to form C_2O_2* commonly serves as the rate-determining step. Generally, OH groups are more reactive than O atoms to interact with carbonyl groups due to their propensity for protonation and subsequent dehydration reactions. This explains why water from the surface is removed and left behind with reactive *CCO groups (see Figure 3.5). Hereafter, the C=C double bond in the CCO* groups is preserved in the subsequent proton–electron additions for the formation of CHCO*, CHCHO*, and CH_2CHO* intermediates. Next, the reaction pathway for C_2H_4 continues by protonating the C atom bonded to the O atom, leading to the transient formation of CH_2CH_2O*. The C—O bond can be quickly cleaved, resulting in the desorption of C_2H_4 and a remaining O* adsorbate. The remaining O* undergoes hydrogenation to form OH* and subsequently converts to H_2O. Alternatively, the CH_3CH_2OH pathway can also be triggered via an ethoxy intermediate (CH_3CH_2O*). For example, the surfaces of Cu_3Ni, Cu_3Co, and Cu_3Rh have been demonstrated to be excellent catalysts for producing the formate, whereas Cu_3Pd and Cu_3Pt surfaces exhibit a preference for methanol production [37]. Additionally, it has been noted that obtaining a high methanol selectivity is highly dependent on these surfaces' inclination to produce the CH_2OH^* intermediate.

Currently, ECR reaction still faces numerous challenges, including the enhancement of reaction efficiency, the manipulation of product selectivity, the decrease of reaction overpotentials, and the identification of efficient and stable electrocatalysts. Therefore, in-depth research and technological innovations in CO_2 catalytic reduction are paramount to realizing its potential as a sustainable solution for CO_2 reduction and energy conversion.

3.7 Fundamentals of Electrocatalytic NRR

Ammonia is widely used in various sectors, including pharmaceuticals, chemical industry, agriculture, and energy [38–41]. The current industrial production of ammonia entails substantial consumption of fossil fuels, posing environmental hazards [42]. Electrocatalytic N_2 reduction (NRR) into ammonia presents an appealing alternative, with the potential to facilitate ammonia synthesis using more moderate conditions within small-scale, decentralized, and on-site electrolysis cells [43, 44]. The mechanism of NRR to synthesize ammonia can be broadly categorized into two main types [45]: dissociative and associative pathways, as shown in Figure 3.6.

As shown in Figure 3.6a, for the dissociative pathway of ammonia synthesis, the catalyst captures N_2 molecules and directly opens the triple bond of N≡N, leading to the formation of two adsorbed N atoms. Subsequently, protonation of these two N atoms leads to the formation of NH_3. The industrial Haber–Bosch method is based

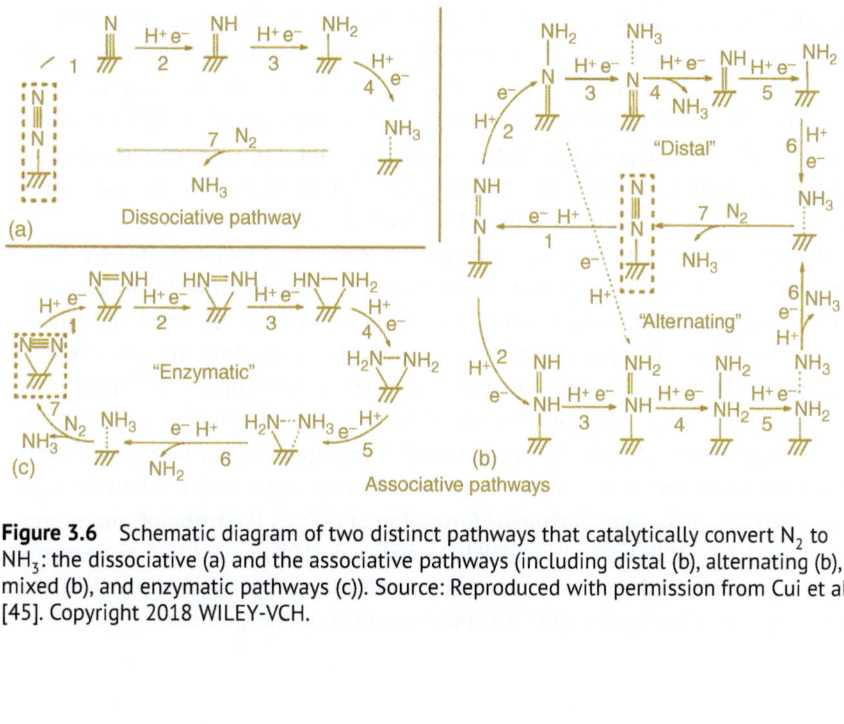

Figure 3.6 Schematic diagram of two distinct pathways that catalytically convert N_2 to NH_3: the dissociative (a) and the associative pathways (including distal (b), alternating (b), mixed (b), and enzymatic pathways (c)). Source: Reproduced with permission from Cui et al. [45]. Copyright 2018 WILEY-VCH.

on this mechanism. In the associative pathway synthesis of ammonia, there are three possible pathways: distal, alternating, and enzymatic pathways. The adsorption mode of the N_2 molecule determines the specific pathway for the associative pathway of ammonia synthesis. When N_2 molecules are adsorbed at the active catalytic site via an end-on pattern (i.e. the N_2 molecule is vertically oriented), the synthesis of ammonia can proceed through the distal, alternating, or mixed pathways.

In the distal pathway, as depicted in Figure 3.6b, after an N_2 molecule is adsorbed through an end-on pattern, the formation and release of the first NH_3 molecule are the result of hydrogenation, which occurs first on the outer N atom furthest from the active site. Subsequently, the second NH_3 molecule is subsequently produced as a result of hydrogenation of the adsorbed N atoms. The exposed active site then captures another N_2 molecule for further ammonia synthesis. In the alternating pathway, both N atoms within the adsorbed N_2 molecule undergo sequential hydrogenation, leading to the successive generation and release of two NH_3 molecules. Additionally, a mixed pathway exists, as indicated by the dashed lines in Figure 3.6b. In this pathway, the outer N atom within the N_2* molecule initially interacts with two hydrogen atoms, followed by the hydrogenation of the N atom closer to the active site, resulting in the consecutive generation and release of two NH_3 molecules. In addition, when N_2 molecules adopt a side-on adsorption configuration at the catalytic site (i.e. the active site forms bonds with both N atoms in N_2), ammonia synthesis will follow the enzymatic pathway. In this specific mechanism, two N atoms undergo hydrogenation alternately, culminating in the successive generation and release of two NH_3 molecules, as shown in Figure 3.6c.

Compared with the dissociative/associative pathways with direct dissociation and cooperative dissociation, a Mars-van Krevelen (MvK) mechanism can occur on metal nitrides (MNs) [46]. During this ammonia synthesis process, hydrogenation reactions initially take place on the N atoms located at the lattice sites of MNs, resulting in the generation and release of NH_3. Subsequently, a vacant N site is left on the catalyst, which is able to capture an N_2 molecule with one N atom filling the vacant site while the other remains exposed on the surface. At this point, the exposed surface N atom exhibits higher reactivity, facilitating protonation reactions and the production of NH_3. The above cycle repeats as further hydrogenation reactions occur on the N atoms at the lattice sites of MNs.

Furthermore, the synthesis of ammonia on transition metal surfaces can be explained by a surface hydrogenation process [47]. In this process, H^+ ions adsorb onto the metal surface, which undergoes reduction to form adsorbed *H. Subsequently, the surface H* species can reduce the N_2 molecules from the surrounding environment, leading to the formation of N_2H_2*. Then, the formed N_2H_2* undergoes a series of protonation reactions that ultimately result in the production of NH_3. It should be mentioned that the adsorption of H^+ ions onto the metal surface and their subsequent reduction represent the rate-determining steps, and a relatively high energy barrier is necessary during the formation of N_2H_2*. For instance, the energy barrier on the Pd(100) surface and Pd(101) surface is 1.98 and 1.82 eV, respectively [48]. The surface hydrogenation mechanism elucidates the phenomenon of the transformation from N_2 molecules to ammonia on Pd metal surfaces.

3.8 Summary

In conclusion, understanding the catalytic mechanism involves a detailed comprehension of the molecular-level events and reaction pathways occurring in catalytic reactions, including how the catalyst influences the reaction, how molecules interact on the catalyst surface, and the intricate steps of the reaction. Identifying the catalyst's active sites is the initial step to understanding the catalytic mechanism. Meanwhile, investigate the formation of reaction intermediates and reaction pathways, thereby guiding optimization measures to increase the overall reaction rate and catalytic activity. Next, learning about the preferred reaction pathways and how to inhibit side reactions also helps in the design of highly selective catalytic systems as well as in the interpretation of observed experimental phenomena, such as product distributions and reaction rates. This provides a more in-depth and comprehensive interpretation of experimental results and increases the acquaintance of catalytic reactions. Scientists can target the design and synthesis of new catalysts based on corresponding catalytic mechanisms, driving innovation in catalysis. In summary, understanding catalytic mechanisms is a key process to gain insight into intermolecular interactions and reaction pathways by advancing catalytic reactions from the macroscopic to the microscopic level. It provides the basis for advances in catalytic science and engineering and contributes to solving major challenges in energy, environment, and sustainable development.

References

1 Seh, Z.W., Kibsgaard, J., Dickens, C.F. et al. (2017). Combining theory and experiment in electrocatalysis: insights into materials design. *Science* 355 (6321): eaad4998.

2 Ju, H., Badwal, S., and Giddey, S.A. (2018). Comprehensive review of carbon and hydrocarbon assisted water electrolysis for hydrogen production. *Appl. Energy* 231: 502–533.

3 Tomczak, S.K. (2019). Comparison of the financial standing of companies generating electricity from renewable sources and fossil fuels: a new hybrid approach. *Energies* 12 (20): 3856.

4 Zou, X. and Zhang, Y. (2015). Noble metal-free hydrogen evolution catalysts for water splitting. *Chem. Soc. Rev.* 44 (15): 5148–5180.

5 Suen, N.-T., Hung, S.-F., Quan, Q. et al. (2017). Electrocatalysis for the oxygen evolution reaction: recent development and future perspectives. *Chem. Soc. Rev.* 46 (2): 337–365.

6 Dai, L., Xue, Y., Qu, L. et al. (2015). Metal-free catalysts for oxygen reduction reaction. *Chem. Rev.* 115 (11): 4823–4892.

7 Song, R., Zhu, W., Fu, J. et al. (2019). Electrode materials engineering in electrocatalytic CO_2 reduction: energy input and conversion efficiency. *Adv. Mater.* 2019: 1903796.

8 Qing, G., Ghazfar, R., Jackowski, S.T. et al. (2020). Recent advances and challenges of electrocatalytic N_2 reduction to ammonia. *Chem. Rev.* 120 (12): 5437–5516.

9 Nørskov, J.K., Rossmeisl, J., Logadottir, A. et al. (2004). Origin of the overpotential for oxygen reduction at a fuel-cell cathode. *J. Phys. Chem. B* 108 (46): 17886–17892.

10 Filimonenkov, I.S., Bouillet, C., Kéranguéven, G. et al. (2019). Carbon materials as additives to the OER catalysts: RRDE study of carbon corrosion at high anodic potentials. *Electrochim. Acta* 321: 134657.

11 Nørskov, J.K., Bligaard, T., Logadottir, A. et al. (2005). Trends in the exchange current for hydrogen evolution. *J. Electrochem. Soc.* 152 (3): J23.

12 Rossmeisl, J., Logadottir, A., and Nørskov, J.K. (2005). Electrolysis of water on (oxidized) metal surfaces. *Chem. Phys.* 319 (1–3): 178–184.

13 Fletcher, S. (2009). Tafel slopes from first principles. *J. Solid State Electrochem.* 13 (4): 537–549.

14 Voiry, D., Chhowalla, M., Gogotsi, Y. et al. (2018). Best practices for reporting electrocatalytic performance of nanomaterials. *ACS Nano* 12 (10): 9635–9638.

15 Dickinson, E.J.F. and Wain, A.J. (2020). The Butler–Volmer equation in electrochemical theory: origins, value, and practical application. *J. Electroanal. Chem.* 872: 114145.

16 Zheng, Y., Jiao, Y., Vasileff, A. et al. (2018). The hydrogen evolution reaction in alkaline solution: from theory, single crystal models, to practical electrocatalysts. *Angew. Chem. Int. Ed.* 57 (26): 7568–7579.

17 Mahmood, N., Yao, Y., Zhang, J. et al. (2018). Electrocatalysts for hydrogen evolution in alkaline electrolytes: mechanisms, challenges, and prospective solutions. *Adv. Sci.* 5 (2): 1700464.

18 Gong, M., Wang, D.-Y., Chen, C.-C. et al. (2016). A mini review on nickel-based electrocatalysts for alkaline hydrogen evolution reaction. *Nano Res.* 9 (1): 28–46.

19 Wei, J., Zhou, M., Long, A. et al. (2018). Heterostructured electrocatalysts for hydrogen evolution reaction under alkaline conditions. *Nano-Micro Lett.* 10 (4): 75.

20 Chatenet, M., Pollet, B.G., Dekel, D.R. et al. (2022). Water electrolysis: from textbook knowledge to the latest scientific strategies and industrial developments. *Chem. Soc. Rev.* 51 (11): 4583–4762.

21 Song, J., Wei, C., Huang, Z.-F. et al. (2020). A review on fundamentals for designing oxygen evolution electrocatalysts. *Chem. Soc. Rev.* 49 (7): 2196–2214.

22 Chen, F.-Y., Wu, Z.-Y., Adler, Z. et al. (2021). Stability challenges of electrocatalytic oxygen evolution reaction: from mechanistic understanding to reactor design. *Joule* 5 (7): 1704–1731.

23 Zhang, Y., Fu, Q., Song, B. et al. (2022). Regulation strategy of transition metal oxide-based electrocatalysts for enhanced oxygen evolution reaction. *Acc. Mater. Res.* 3 (10): 1088–1100.

24 Man, I.C., Su, H., Calle-Vallejo, F. et al. (2011). Universality in oxygen evolution electrocatalysis on oxide surfaces. *ChemCatChem* 3 (7): 1159–1165.

25 Montoya, J.H., Seitz, L.C., Chakthranont, P. et al. (2017). Materials for solar fuels and chemicals. *Nat. Mater.* 16 (1): 70–81.

26 Lee, S., Banjac, K., Lingenfelder, M. et al. (2019). Oxygen isotope labeling experiments reveal different reaction sites for the oxygen evolution reaction on nickel and nickel iron oxides. *Angew. Chem. Int. Ed.* 131 (30): 10401–10405.

27 Zhou, Y., Sun, S., Wei, C. et al. (2019). Significance of engineering the octahedral units to promote the oxygen evolution reaction of spinel oxides. *Adv. Mater.* 31 (41): 1902509.

28 Sun, Y., Liao, H., Wang, J. et al. (2020). Covalency competition dominates the water oxidation structure–activity relationship on spinel oxides. *Nat. Catal.* 3 (7): 554–563.

29 Zhang, N. and Chai, Y. (2021). Lattice oxygen redox chemistry in solid-state electrocatalysts for water oxidation. *Energy Environ. Sci.* 14 (9): 4647–4671.

30 Ma, R., Lin, G., Zhou, Y. et al. (2019). A review of oxygen reduction mechanisms for metal-free carbon-based electrocatalysts. *npj Comput. Mater.* 5 (1): 78.

31 Guo, X., Lin, S., Gu, J. et al. (2019). Simultaneously achieving high activity and selectivity toward two-electron O_2 electroreduction: the power of single-atom catalysts. *ACS Catal.* 9 (12): 11042–11054.

32 Francke, R., Schille, B., and Roemelt, M. (2018). Homogeneously catalyzed electroreduction of carbon dioxide—methods, mechanisms, and catalysts. *Chem. Rev.* 118 (9): 4631–4701.

33 Zhang, L., Zhao, Z.-J., and Gong, J. Nanostructured materials for heterogeneous electrocatalytic CO_2 reduction and their related reaction mechanisms. *Angew. Chem. Int. Ed.* 56 (38): 11326–11353.

34 Xie, H., Wang, T., Liang, J. et al. (2018). Cu-based nanocatalysts for electrochemical reduction of CO_2. *Nano Today* 21: 41–54.

35 Manthiram, K., Beberwyck, B.J., and Alivisatos, A.P. (2014). Enhanced electrochemical methanation of carbon dioxide with a dispersible nanoscale copper catalyst. *J. Am. Chem. Soc.* 136 (38): 13319–13325.

36 Calle-Vallejo, F. and Koper, M.T.M. (2013). Theoretical considerations on the electroreduction of CO to C_2 species on Cu(100) electrodes. *Angew. Chem. Int. Ed.* 125 (28): 7423–7426.

37 Hirunsit, P., Soodsawang, W., and Limtrakul, J. (2015). CO_2 electrochemical reduction to methane and methanol on copper-based alloys: theoretical insight. *J. Phys. Chem. C* 119 (15): 8238–8249.

38 Légaré, M.-A., Bélanger-Chabot, G., Dewhurst, R.D. et al. (2018). Nitrogen fixation and reduction at boron. *Science* 359 (6378): 896–900.

39 Li, H., Shang, J., Ai, Z. et al. (2015). Efficient visible light nitrogen fixation with BiOBr nanosheets of oxygen vacancies on the exposed {001} facets. *J. Am. Chem. Soc.* 137 (19): 6393–6399.

40 Li, X.-F., Li, Q.-K., Cheng, J. et al. (2016). Conversion of dinitrogen to ammonia by FeN_3-embedded graphene. *J. Am. Chem. Soc.* 138 (28): 8706–8709.

41 Zhao, J. and Chen, Z. (2017). Single Mo atom supported on defective boron nitride monolayer as an efficient electrocatalyst for nitrogen fixation: a computational study. *J. Am. Chem. Soc.* 139 (36): 12480–12487.

42 Duinen, G.V. (2008). How a century of ammonia synthesis changed the world. *Nat. Geosci.* 1: 636–639.

43 Nishibayashi, Y. (2015). Recent progress in transition-metal-catalyzed reduction of molecular dinitrogen under ambient reaction conditions. *Inorg. Chem.* 54 (19): 9234–9247.

44 Van Der Ham, C.J.M., Koper, M.T.M., and Hetterscheid, D.G.H. (2014). Challenges in reduction of dinitrogen by proton and electron transfer. *Chem. Soc. Rev.* 43 (15): 5183–5191.

45 Cui, X., Tang, C., and Zhang, Q. (2018). A review of electrocatalytic reduction of dinitrogen to ammonia under ambient conditions. *Adv. Energy Mater.* 8 (22): 1800369.

46 Ye, T.-N., Park, S.-W., Lu, Y. et al. (2020). Contribution of nitrogen vacancies to ammonia synthesis over metal nitride catalysts. *J. Am. Chem. Soc.* 142 (33): 14374–14383.

47 Ling, C., Zhang, Y., Li, Q. et al. (2019). New mechanism for N_2 reduction: the essential role of surface hydrogenation. *J. Am. Chem. Soc.* 141 (45): 18264–18270.

48 Wang, J., Yu, L., Hu, L. et al. (2018). Ambient ammonia synthesis via palladium-catalyzed electrohydrogenation of dinitrogen at low overpotential. *Nat. Commun.* 9 (1): 1795.

Part II

Advanced Functional Materials for Electrocatalytic Hydrogen Evolution Reaction

4

Metal-Based Materials for Electrocatalytic Hydrogen Evolution Reaction

Rongbo Sun[1,*], *Xiaoqian Wang*[2,*], *Gang Xie*[3], *Hongwen Huang*[4], *and Jingjie Ge*[5]

[1] *Sinochem Holdings Co Ltd, Sinochem Energy High-tech Co Ltd, 2 West Tieying Middle Rd., Fengtai District, Beijing, 100031, China*
[2] *King Abdullah University of Science and Technology (KAUST), Advanced Membranes and Porous Materials Center, Physical Sciences and Engineering Division, 4700 King Abdullah University of Science and Technology, Thuwal, 23955-6900, Saudi Arabia*
[3] *Southwest Petroleum University, State Key Laboratory of Oil & Gas Reservoir Geology and Exploitation, 8 XinDu Rd., XinDu District, Chengdu, Sichuan, 610500, China*
[4] *Hunan University, College of Materials Science and Engineering, 2 Lushan Rd., Yuelu District, Changsha, Hunan, 410082, China*
[5] *The Hong Kong Polytechnic University, Department of Applied Biology and Chemical Technology, Faculty of Science, 11 Yucai Rd. Hung Hom, Kowloon, PR China*

4.1 Introduction

To overcome the shackles of conventional fossil energy and focus on solving environmental problems, the exploration of new clean energy is further launched. In the context of renewable energy sources, hydrogen (H_2) energy has garnered significant attention due to its superior caloric value, sustainability, abundant availability, and negligible environmental impact [1, 2]. Currently, the cost-effective industrial production of hydrogen relies heavily on steam reformation of hydrocarbons, a process plagued by significant CO_2 emissions. The proportion of renewable energy-generated H_2 is <4% of the total hydrogen product worldwide. In response, the scientific community has invested substantial effort in developing sustainable and eco-friendly hydrogen production methods. These include electrochemical water splitting, photochemical water splitting, and acid hydrolysis of metallic hydrides [3, 4]. H_2 is receiving increasing attention as a zero-carbon energy carrier: it is expected that by 2050, more than 20% of the world's CO_2 emissions will be diminished through H_2 energy substitution, and hydrogen energy consumption will account for 18% of the world's energy market [5].

Among the various ways to produce H_2, water electrolysis technology integrated with renewable energy resources has the advantage of zero carbon emissions

*These authors contributed equally to this work.

Functional Materials for Electrocatalytic Energy Conversion, First Edition.
Edited by Zhicheng Zhang, Meiting Zhao, and Yuchen Qin.

in the whole process to produce green H_2. The electrochemical water-splitting process includes anodic oxygen evolution reactions (OER) and cathodic hydrogen evolution reactions (HER) [6, 7]. HER is one of the essential steps in electrochemical hydrolysis. This process usually needs a large driving overpotential, which can lead to high energy consumption, thus requiring high-performance catalysts to achieve rapid kinetics [8]. Currently, platinum-based materials are considered the most efficient HER catalysts. Due to its optimal environment for adsorption and desorption of H atoms, it typically exhibits relatively low overpotential and high exchange current density. According to the DFT calculation criteria, precious metals, such as Pt, Ir, and Ru, show higher catalytic activity [9, 10]. However, due to the scarcity and high price of precious metals, the promotion of water electrolysis technology is obstructed [11, 12]. Besides, its industrial applicability is further limited by its poor durability, propensity for dissolving and aggregation, and low tolerance to poisoning during electrocatalysis [13].

On the other hand, numerous nonprecious metal-based electrocatalysts are inexpensive, environmentally benign, and have high electrical conductivity and electrocatalytic efficacy on par with noble metals. With continued technological advancement, these electrocatalysts could replace precious materials in the water electrolysis process. Nevertheless, creating these affordable and effective non-precious electrocatalysts requires significant study, and there is currently a long way to go toward industrialization. In summary, the ideal electrocatalyst should meet the following four conditions: (i) large reserves of raw materials, low price, and pollution; (ii) easy preparation and mass production; (iii) high catalytic activity, low starting potential, and overpotential; (iv) strong durability and long working time. Therefore, to fully evaluate the activity of electrocatalysts, it is necessary to test and calculate related indicators, including overpotential, current exchange density, resistance, activation energy, Tafel slope, and stability.

4.1.1 Mechanism of the Electrocatalytic HER

Paets van Troostwijk and Deiman first reported in 1789 that water plus electricity produces hydrogen and oxygen [14]. Electrochemical water decomposition consists of a two-part reaction, that is, HER at the cathode and OER at the anode as shown in Figure 4.1a [15]. HER occurs at the cathode surface, regardless of whether it is in an acidic, alkaline, or neutral solution, with the theoretically required potential of 0 V. The large number of experimental measurements, advanced characterizations, and theoretical analyses indicate that the HER process consists of three main basic steps and that the types of reactions are varied under different electrolyte environments. The HER mechanism can be expressed as the following equation when the electrolyte is an acidic system (Eqs. 4.1–4.3).

$$\text{Volmer reaction: } H^+ + e^- + {}^* \rightarrow H^* \tag{4.1}$$

$$\text{Heyrovsky reaction: } H^* + H^+ + e^- \rightarrow H_2 \tag{4.2}$$

$$\text{Tafel reaction: } H^* + H^* \rightarrow H_2 + 2^* \tag{4.3}$$

where the * represents the active site. The Volmer reaction constitutes the initial phase of the HER, which involves the acquisition of an electron by a hydrogen ion and culminating in its attachment to the catalyst's surface [18, 19]. In the subsequent stage, the generated H^* undergoes a reaction either with another H^* (pathway 1) or with H^+ (pathway 2), leading to the release of an H_2 molecule. Concurrently, to ascertain the specific type of reaction, numerous researchers have employed the Tafel slope as a critical indicator. When the reaction adheres to the Volmer mechanism, the corresponding Tafel slope is approximately 120 mV dec^{-1}. And in the Heyrovsky and Tafel mechanisms, the respective Tafel slopes are approximately 40 and 30 mV dec^{-1} [20, 21]. In scenarios where the electrolytic environment is alkaline or neutral, modifications to the above formulas become necessary, with the pertinent equations delineated as (4.4), (4.5), and (4.6) [22].

$$\text{Volmer reaction: } H_2O + e^- + {}^* \rightarrow H^* + OH^- \tag{4.4}$$

$$\text{Heyrovsky reaction: } H_2O + e^- + H^* \rightarrow H_2 + OH^- \tag{4.5}$$

$$\text{Tafel reaction: } H^* + H^* \rightarrow H_2 + 2^* \tag{4.6}$$

There, the Tafel values continue to match those observed in an acidic solution, with the notable distinction being the involvement of H_2O as the reactant instead of H^+. In summary, the electrocatalytic HER involves two fundamental reaction mechanisms: the Volmer–Heyrovsky and Volmer–Tafel pathways [23]. It is well-established that in an acidic solution, the rate-determining step for Pt is the Tafel process, which significantly enhances the kinetics of the HER. Conversely, Pt exhibits a moderate electrocatalytic performance associated with the Volmer–Heyrovsky step under alkaline and neutral conditions [24]. Figure 4.1b illustrates the mechanisms underlying the HER over a range of electrocatalysts in varying solvent environments [16].

4.1.2 Theoretical Method for Describing the Efficient HER Catalyst

The prevailing consensus in the scientific community is that the inherent catalytic activity of HER electrocatalysts is intricately linked with the Gibbs free energy of hydrogen adsorption (ΔG_{H^*}) [25, 26]. The H_2 desorption and H^* adsorption in electrocatalytic processes are quantifiable through the catalyst's inherent ΔG_{H^*} value. Grounded in the Sabatier principle, the catalytic surface's ability to adsorb H^* and desorb H_2 must be optimal, and the binding force of H^* on this surface sufficiently robust to expedite the initiation of the reaction and the products' quick release. Consequently, enhancing the HER efficacy of these catalysts necessitates that ΔG_{H^*} approximates as closely as possible to the thermoneutral value ($\Delta G_{H^*} \approx 0$) [27]. Nørskov and coworkers delineated the volcano curve correlation between ΔG_{H^*} and the exchange current value (j_0, an activity index of catalysts), which was utilized to intuitively assess or forecast the HER activity of electrocatalysts [28]. As evidenced in Figure 4.1c, metals positioned proximal to the pinnacle of the volcano plot demonstrate optimal H^* adsorption interactions. Typically, metal Pt was regarded as the gold standard of HER performance due to its proximity to the peak

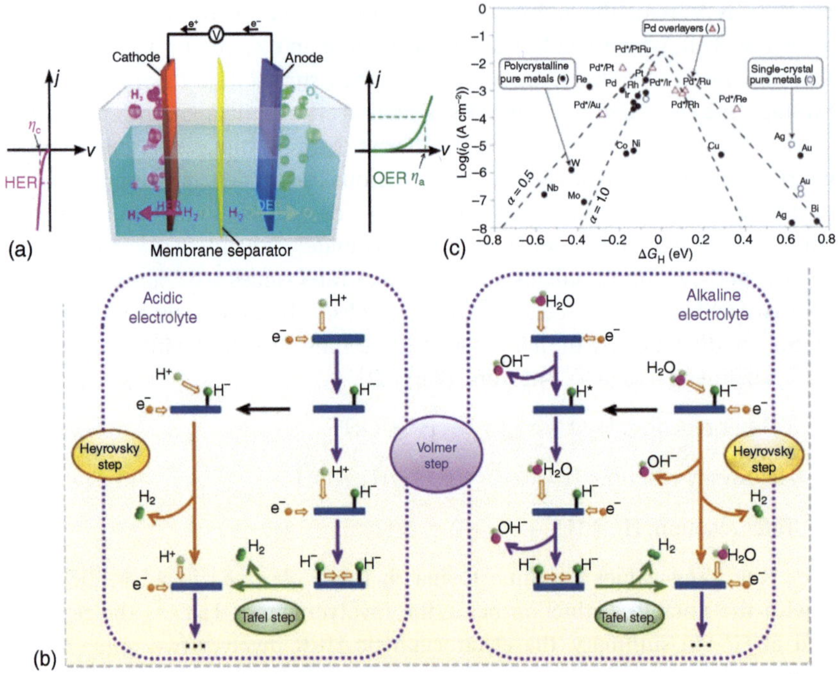

Figure 4.1 (a) Schematic diagram of HER and OER in the electrolyzer. Source: Reproduced with permission from Zhu et al. [15], American Chemical Society. (b) Schematic illustration of the HER in different media. Source: Reproduced with permission from Pang et al. [16], Royal Society of Chemistry. (c) Volcano plot for the HER for various pure metals and metal overlayers. ΔG_H values are calculated at 1 bar of H_2 (298 K) and a surface hydrogen coverage of either 1/4 or 1/3 ML. The two curved lines correspond to the activity predictions of simple mean-field, microkinetic models, assuming transfer coefficients (α) of 0.5 and 1.0, respectively. Source: Reproduced with permission from Greeley et al. [17], Springer Nature.

of the volcano plot, characterized by the highest exchange current and a minimal ΔG_{H^*} value nearing zero. When ΔG_{H^*} is less than zero, H^* adsorption rises as ΔG_{H^*} falls; conversely, when ΔG_{H^*} is greater than zero, H^* adsorption falls as ΔG_{H^*} rises, and the exchange current density falls exponentially as $|\Delta G_{H^*}|$ rises. Furthermore, among the various noble metals, Pd and Ru are acknowledged as promising alternatives to Pt in this context. Additionally, the volcano plot analysis reveals that Ir and Rh also exhibit optimal ΔG_{H^*} values conducive to HER performance [17].

4.2 Electrocatalytic HER Activity on Metal-Based Materials

The HER ranks among the most extensively researched electrocatalytic processes within the domain of electrochemistry [29]. Over recent years, substantial research endeavors have been dedicated to the design and synthesis of high-efficiency HER catalysts, culminating in the creation and experimental evaluation of numerous catalysts with varied elemental compositions and structural frameworks. In summary, platinum continues to be recognized as the most active catalyst for HER, primarily attributed to its hydrogen-binding energy for hydrogen adsorption and desorption nearing the ideal value of zero. Furthermore, additional noble metals including Pd, Ru, Ir, and Rh have also been explored as potential catalysts for their applicability in water electrolysis [30–33]. Although the platinum group metal (PGM)-based catalysts exhibit superior catalytic activities toward HER, their high cost and limited availability pose significant barriers to widespread commercialization. Consequently, a primary objective in research has been to reduce the loading without hampering their catalytic performance [34]. Meanwhile, considering the relatively low cost, non-PGM-based materials have also been developed intensively [35–38]. In the following sections, typical examples of metal-based HER electrocatalysts will be introduced by element composition classification, and their unique characteristics will be emphasized.

4.2.1 PGM-Based Materials

4.2.1.1 Pt-Based Materials

Numerous studies have concentrated on enhancing the reaction kinetics of the HER through the development of innovative, highly active, and cost-effective Pt-based catalysts, facilitating the economic generation of green hydrogen. In recent years, research efforts have been directed toward augmenting the atomic utilization efficiency of Pt catalysts by engineering a diversity of nanostructures, encompassing designs of various sizes, shapes, heterostructures, alloys, clusters, and single-atom (SA) configurations [39–41].

The fabrication of stable platinum alloys exhibiting high HER performance not only reduces the platinum loading density but also enhances the HER efficiency of platinum [42, 43]. This approach was employed to optimize the bond strength between platinum and hydrogen by adjusting the d-band center of Pt. Hu and coworkers have successfully developed a Pt_3V alloy electrocatalyst, characterized by its high efficiency for HER [44]. The multiple valence states of vanadium enable greater flexibility in adjusting the electronic structure of Pt, thereby fine-tuning the catalytic activity for HER. The HER performance of the alloy was evaluated in both 0.5 M H_2SO_4 and 1 M KOH. At a current density of 10 mA cm^{-2}, the measured overpotential was 20 and 25 mV in acidic and basic conditions, respectively. The computed Gibbs free energy of hydrogen adsorption for the Pt_3V alloy was determined to be 0.26 eV, in contrast to −0.3 eV for Pt(111). It is reported that the presence of V in the alloy results in a negative shift of platinum's d-band center, thereby reducing the hydrogen adsorption energy. Moreover, the less electronegative subsurface V atoms donate electrons to adjacent Pt atoms, which further contributes to the downward shift of platinum's d-band center. The calculated d-band center of the alloy was found to exhibit a correlation with the increase in the ΔG_{H*} value. Li and coworkers pioneered the synthesis of PtRu bimetallic nanoparticles supported on carbon cloth ($PtRu/CC_{1500}$), utilizing an innovative and straightforward electrochemical deposition technique [45]. The $PtRu/CC_{1500}$

electrocatalyst demonstrated superior mass activity (MA) compared to Pt/C in catalyzing the HER across a broad pH range. Intriguingly, the PtRu/CC_{1500} catalyst exhibited optimal water dissociation energy, a result of the synergistic combination of the chemical properties of both metals at a finely tuned atomic ratio.

Relative to Pt-based binary alloys, the multimetallic Pt-based nanostructures have proven to be more efficacious in augmenting the HER catalytic performance. Scientists formulate the multimetallic nanomaterials through the amalgamation of platinum with two or more metals, which results in enhanced surface area, altered electronic structures, and strategically designed morphologies. Furthermore, the incorporation of nonprecious metals offers the advantage of diminishing the reliance on scarce and costly Pt metal. Chen et al. adeptly synthesized a trimetallic PtNiCo hollow alloy 3D multipod (PtNiCo HAMPs) catalyst utilizing a one-pot solvothermal approach, synergistically combined with an acid etching technique [46]. Acid etching is employed to enhance the exposure of surface area for intermediates. Within a 1.0 M KOH solution, the PtNiCo alloy multipod catalyst demonstrated outstanding performance in the HER. Chen and coworkers engineered a PtCuFe one-dimensional nanochain structure composed of numerous interconnected nanoparticles. High-resolution transmission electron microscopy (HRTEM) disclosed that these nanoparticles exhibited a notable shrinkage in lattice fringe, displaying spacings of 0.226 nm compared to 0.239 nm in pristine Pt (111) nanoparticles [47]. The nanochain structure was evaluated for its HER efficacy in 1 M KOH, benchmarked against Pt/C. This assessment revealed that the nanochain structure exhibited superior HER performance, evidenced by an overpotential of 19 mV for the alloy, in contrast to 40 mV for Pt/C, at a current density of 10 mA cm^{-2}.

Typically, nonmetallic atoms exhibit a propensity to lose electrons, and this phenomenon is attributed to the negative charge effect. The incorporation of nonmetallic atoms into the lattice structure can effectively modify the d-band center, thereby regulating the adsorption behavior of H^* at active sites [48]. Consequently, nonmetallic atoms (such as N, P, and S) are frequently employed to fine-tune ΔG_{H^*} at platinum active sites, thereby enhancing the overall catalytic behavior [49, 50]. Furthermore, Wang and coworkers developed a boron–phosphide-activated Pt/graphene (BPed–Pt/GR) catalyst, illustrating the potent activation effect of phosphorus on the platinum surface [51]. Electrochemical analyses revealed that the specific activity of the Pt/GR and BPed–Pt/GR catalyst at 100 mV in a 1.0 M KOH solution was 44.2 and 154.70 mA cm^{-2}, respectively. The marked differences in activity were observed in BPed–Pt/GR catalysts with varying Pt–P compositions (Figure 4.2a,b). DFT calculations indicated that the doping of phosphorus atoms alters the position of the d-band center (Figure 4.2c). As shown in Figure 4.2d, the BPed–Pt/GR catalyst containing 18–25% Pt—P bonds exhibited the highest current densities and the most favorable ΔG_{H^*} value, demonstrating the influential role of Pt—P bond content in dictating the catalytic efficiency of hydrogen evolution.

Given the established correlation between MA, defined as the catalytic current per unit mass of active metal, and the number of exposed active sites, significant research endeavors have been invested in reducing the size of Pt-based nanoparticles to cluster or SA dimensions [52, 53]. Yang et al. documented Pt SAs

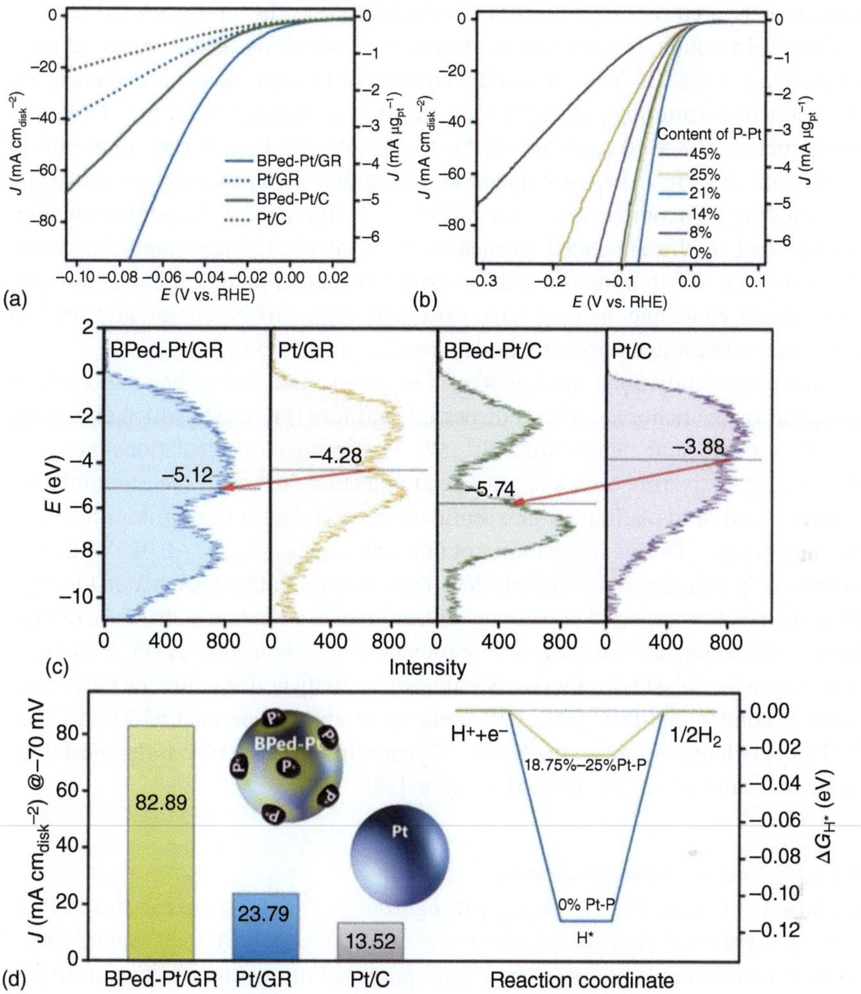

Figure 4.2 (a) HER polarization curves of BPed–Pt/GR, and (b) BPed–Pt/GR with different contents of Pt–P bonds. After iR correction in 1.0 M KOH at 25 °C normalized by the geometrical area of the disk. (c) High-resolution valence-band Pt VB XPS spectra of Pt relative to the Fermi level as an analog of the density of states with the black lines indicating the position of the d-band centers, and (d) current densities at 70 mV vs. RHE and free-energy diagram for the HER at the potential $U = 0$ relative to the standard hydrogen electrode with different Pt–P contents. Source: Reproduced with permission from Wang et al. [51] Wiley-VCH.

and PtCo alloy loaded on a N-doped porous carbon framework electrocatalyst (CoPt–PtSA/NDPCF) [54]. The CoPt–PtSA/NDPCF catalyst, featuring a platinum loading of 0.4 wt%, demonstrated an ultrahigh MA of $7.5\,A\,mg_{Pt}^{-1}$ in a 1.0 M KOH solution, attributed to the advantageous electron effect, thereby significantly enhancing the utilization efficiency of platinum atoms. Wang and coworkers utilized a straightforward in situ strategy to synthesize platinum single atoms loaded on layered α-$Ni_{2/3}Fe_{1/3}(OH)_2$(Pt–Ni_2Fe_1) [55]. The Pt–Ni_2Fe_1-24 catalyst, synthesized through 24-hour in-situ oxidation, exhibited markedly improved HER kinetics in 1.0 M KOH. The successful synthesis of Pt single atoms is evidenced by the distinctive scattering path of Pt–O in the Pt L_3-edge extended X-ray absorption fine structure (EXAFS) spectrum of Pt–Ni_2Fe_1-24. The HER polarization curves reveal that the Pt–Ni_2Fe_1-24 catalyst demonstrated a significantly elevated MA (~7.5 mA mg_{Pt}^{-1}) compared to 20 wt% Pt/C (~1.2 mA mg_{Pt}^{-1}). The phenomenon is attributed to the enhanced intrinsic activity at each single-atomic platinum site. However, maintaining the stable separation state of platinum atoms remains a significant challenge, as they have extremely high surface energy and tend to aggregate, with limited durability under specific voltages [56].

In a separate study, Song and coworkers employed onion-like carbon nanospheres as a platform to anchor atomically dispersed platinum (Pt_1/OLC), which effectively served as an efficient catalyst for HER [57]. First-principles calculations indicated that the high curvature of this onion-like architecture facilitated the accumulation of electrons around platinum atoms, emulating metal sites at the corners and edges of nanoparticles. This configuration not only achieved a ΔG_{H*} of −0.01 eV but also induced a tip-enhanced local electric field, thereby accelerating the reaction kinetics. Yu and coworkers reported on the customization of the electronic structure of platinum single atoms on N-doped mesoporous hollow carbon spheres (Pt_1/NMHCS), resulting in a catalyst with markedly enhanced activity and stability for HER compared to commercial Pt/C (MA, 2.07 A $mg^{-1}{}_{Pt}$ at an overpotential of 50 mV). The DFT calculations pinpointed the N_1–Pt_1–C_2 coordination structure as the most thermodynamically advantageous configuration [58].

4.2.1.2 Other PGM-Based Materials

Ru-based materials are increasingly recognized as promising alternatives to Pt in serving as potential electrocatalysts for water electrolysis. With the rapid advancement of nanotechnology, researchers have proposed numerous effective strategies to prepare efficient ruthenium-based HER electrocatalysts [59, 60]. Kong and coworkers introduced a trapping–bonding strategy to achieve the super-assembly of surface-enriched ruthenium nanoclusters on a phytic acid-modified nitrogen-doped carbon framework, termed NCPO–Ru NCs [61]. The as-synthesized NCPO–Ru NCs, benefiting from the extensive exposure of active sites and the dense distribution of ultra-small ruthenium nanoclusters, exhibit exceptional electrocatalytic performance for HER (9.6 and 14.3 times higher than commercial Pt/C and Ru/C, respectively). In a recent study, Wang and coworkers synthesized ruthenium nanoparticles on oxygen-deficient WO_{3-x}, a process that involved the insertion of protons into WO_3 [62]. The proton transfer process from WO_{3-x} to Ru nanoparticles is identified as a rate-determining step, which significantly contributes to the enhanced performance in HER. Electrochemical measurements demonstrated a remarkable 24-fold enhancement in the HER activity of Ru–WO_{3-x} compared to the commercial Ru/C in a 1 M phosphate-buffered saline (PBS) solution. Theoretical calculations unveiled that WO_{3-x} possesses the capability to catalyze the dissociation of water, yielding protons that subsequently migrate to Ru nanoparticles. The step involving the transfer of protons from WO_{3-x} to Ru is identified as the rate-determining stage, thus explaining the significant improvement in HER.

Furthermore, it is worth noting that the morphology of metals can exert a substantial influence on the performance. Among the diverse forms of noble metal structures, the three-dimensional (3D) nanostructure stands out for its abundant catalytic sites and inherent resistance to aggregation. Huang and coworkers introduced snowflake-like RuCu nanosheets onto a carbon black substrate (RuCu NSs/C) consisting of crystalline hexagonal-close-packed (*hcp*) ruthenium and amorphous copper with a profusion of accessible channels [63]. Specifically, the RuCu NSs/C prepared in this manner displayed remarkably low overpotentials of only 19 and 20 mV in 0.5 M H_2SO_4 and 1 M KOH, both at a current density of 10 mA cm^{-2}. XPS and DFT investigations unveiled that the surface electronic activation of RuCu NSs can be attributed to their channel-rich structure. Furthermore, it was observed that a noticeable short-range lattice distortion occurred in the vicinity of the channel region, leading to lattice relaxation, and consequently reducing the energy expenditure associated with bond dissociation intermediates. In a recent work, the researchers synthesized Ru multilayered nanosheets (Ru NSs) characterized by a superlattice structure supported on carbon. These Ru NSs exhibited a low overpotential of 24 mV at a current density of 10 mA cm^{-2} in the alkaline solution [64]. According to HRTEM, Ru layers were stacked at twisted angles ranging from 2° to 30°, which helped construct the superlattice. DFT calculations provided insight into the exceptional HER performance, indicating that it stemmed from the strain effect within the Ru interlayers, which resulted in lattice contraction and a decrease in the adsorption energy of H^*. The research team led by Lou undertook the fabrication of SA Ru catalysts to further augment their catalytic activity [65]. Their study introduced the incorporation of individual Ru atoms into a carbon matrix (ECM@Ru). Remarkably, the SAC catalyst demonstrated exceptional catalytic performance with a minimal overpotential.

Rh-based materials, characterized by their rarity, and possessing a suitable density of states (DOS) for HER, stand out in the field. The current literature on Rh-based catalysts converges on a shared objective: enhancing catalytic activity while minimizing the Rh content [66]. In a recent study, Li et al. delved into the RhO_x system for alkaline HER, unveiling that RhO_2 exhibits superior activity and stability [67]. Zhang et al. pioneered the development of Rh nanocrystals encapsulated in single-walled carbon nanotubes (SWNTs) [68]. Their findings indicated that this unique combination polarizes electron clouds surrounding the Rh nanocrystals toward the SWNT, rendering it efficacious for HER, OER, and ORR. This multifunctionality is attributed to the synergistic effect of the supporting material, which alters the hydrogen adsorption and charge transfer characteristics of the composite. Furthermore, Rh is frequently employed alongside synergistic supports such as graphene, nitrogen-doped graphene (NG), graphene oxides, graphitic carbon nitride, and sulfur-doped graphenes (SGs), enhancing its catalytic efficacy. A notable investigation in this domain is the work by Liu et al., who engineered Rh nanoparticles supported on nitrogen and sulfur-doped graphene (NSG) [69]. This catalyst underwent rigorous testing for HER in a diverse array of environments, encompassing alkaline, acidic, and seawater solutions, and demonstrated activity on par with that of Pt/C. Significantly, the research indicates an elevated electron concentration at the interface when Rh particles are in interaction with nitrogen (N) and nitrogen, sulfur (N,S)-doped graphene supports. In instances where Rh is present independently, its pronounced hydrogen adsorption performance predisposes it to significant hydride poisoning. Within this framework, SACs appear to have gained a competitive advantage over commercial Pt/C catalysts, attributable to the disparity in Rh loading in SACs compared to Pt in Pt/C, typically spanning two to three orders of magnitude. Recently, Guan and associates reported on the HER performance of an authentic Rh SAC supported on NG in various pH environments, including acidic, neutral, and alkaline, observing remarkable activity across the board [70]. The first-principles investigation elucidated that the catalytically active sites for HER comprised an Rh atom intricately coordinated to three nitrogen (N) sites on the NG.

Pd-based materials have been the subject of extensive research as potential alternatives to Pt-based catalysts across a wide range of applications. A comparative analysis of Pt and Pd reveals that these two noble metals exhibit both shared and distinct characteristics. Pd and Pt both belong to the PGMs, sharing similar physical and chemical attributes such as stability in air, the capacity to form alloys, and the potential for nanostructure design [71]. The aforementioned comparable properties of Pd make it attractive for high-performance electrocatalysis.

Aiming to augment the number of active sites and reduce costs, Ohm and Domke employed a nano-confinement strategy to cultivate atomically dispersed Pd/Ir islands on Au(111), thereby enhancing the HER performance through meticulous control over the morphology, density, and size of these islands [72]. Wang and associates successfully synthesized palladium nanosheets (Pd NSs) with an average thickness ranging from three to eight monolayers, specifically to modulate the strain. The development of ultrathin Pd NSs with adjustable intrinsic strain has led to a significant enhancement in hydrogen adsorption energy and optimization of HER catalytic reactivity, underscoring the intrinsic structural advantages of low-dimensional Pd-based nanostructures [73]. Pei et al. utilized nickel hydroxide as a support for palladium nanoclusters and ruthenium single atoms, employing this catalyst effectively in both HER and OER [74]. Subsequent to the adjustment of the Pd/Ru ratio in the palladium nanoclusters and ruthenium single atoms, the $Pd_{12}Ru_3/Ni(OH)_2/C$ catalyst demonstrated superior performance in both HER and OER processes, a result attributed to the synergistic effect of the proximal palladium and ruthenium atoms.

To further reduce the content of noble metals in the pursuit of cost-effective electrocatalysts, the fabrication of bimetallic catalysts incorporating a substantial proportion of non-noble metals emerges as a viable approach. Nonetheless, achieving a balance between the stability of such catalysts and the exceedingly low loading of noble metals presents a considerable challenge. To identify the most effective solution to this challenge, Liu et al. synthesized nanoporous NiPd/metallic glass and strategically minimized the palladium loading in the alloys by employing the dealloying method and galvanic replacement to deposit palladium on the surface [75]. Furthermore, the promotion of sulfur species could be an additional factor contributing to its heightened activity. As an illustration, Xu et al. incorporated phosphorus into Pd_4S nanowires to enhance their performance in acidic HER [76]. Following the integration of the two nonmetal atoms into the palladium crystal lattice, the structure transitioned from a cubic to a P-Pd_4S tetragonal phase, characterized by the substitution of some S atoms with P atoms. Within this distinct structural framework, P and S function as proton attraction centers, thereby inducing polarization and facilitating the formation of Pd–H_{ads} in the Volmer step. The optimization of hydrogen desorption led to an overpotential of 47 mV for P-Pd_4S NWs to attain a current density of 10 mA cm^{-2} in 0.5 M H_2SO_4. Additionally, the Tafel slope of P-Pd_4S NWs was measured at 32.7 mV dec^{-1}, closely mirroring that of Pt/C.

Ir-based material is among the most promising candidates for water splitting, attributed to their inherent catalytic activity, broad pH tolerance, and exceptional stability. Typically, the compositional optimization of Ir-based materials is directed toward minimizing the use of iridium while simultaneously enhancing the intrinsic catalytic efficiency of the active sites. A key strategy involves downsizing the active metals to the nanocluster scale or the level of atomic dispersion. This approach proves effective in maximizing atomic efficiency and optimizing electronic and geometric properties of the metals [77]. Additionally, the strategic introduction of foreign atoms, be they noble or base metals, into Ir-based materials for the modification of active sites is noteworthy. This includes optimizing electronic and geometric structures, thereby inducing electronic/geometric and synergistic effects between iridium and alloyed metals or carriers, often culminating in a marked enhancement in catalytic activities.

The robust interactions between Ir and its supporting material are expected to result in the precise adjustment of the electronic structure, thereby enhancing the electrocatalytic performance of Ir in the HER. Li et al. discovered that the vacant d orbitals of Ir atoms could be effectively diminished through interactions with electronegative carbon and nitrogen species in the surrounding environment. These interactions serve to equilibrate the thermodynamics of hydrogen adsorption and desorption at Ir sites, consequently expediting the HER [78]. Moreover, to broaden the scope and utility of SA catalysts, Chou and coworkers introduced a comprehensive approach for fabricating a diverse array of single noble metal atoms (including Ru, Rh, Pd, Ir, Pt, among others) immobilized on nitrogen-doped carbon (referred to as M_1@NC) to facilitate the HER [79]. Notably, Ir_1@NC exhibited the most auspicious electrocatalytic performance in the context of the HER, characterized by an exceptionally low overpotential of 174 mV at a current density of 10 mA cm^{-2} and an exceptionally shallow Tafel slope of 146 mV per decade in a 0.5 M H_2SO_4 electrolyte.

Concerning Ir-based alloys, the enhancement of their inherent HER activity is intricately associated with the fine-tuning of hydrogen adsorption and desorption characteristics. This tuning is a consequence of the profound influence exerted by the incorporation of foreign metals on the electronic structure of Ir. As an illustrative example, considering the IrFe alloy, it exhibited superior performance in the HER when contrasted with its individual constituents and even outperformed the commercial 40% Pt/C catalyst [80]. The introduction of Fe into the alloy resulted in a substantial adjustment of Ir's Fermi level through electron transfer, driven by the disparity in electronegativity between the two elements. This adjustment led to a ΔG_{H*} (Gibbs free energy change for hydrogen adsorption) closer to zero (−0.25 vs. −0.33 eV), consequently enhancing the catalytic activities for the HER. In contrast to metallic iridium, Ir oxides exhibit lower catalytic activity in the HER due to their higher d-band center energy (−.77 eV) when compared to that of Ir (−2.57 eV) and platinum (Pt, −2.47 eV). Consequently, IrO_2 exhibits a more robust H–Ir binding affinity compared to pure Ir (−0.96 vs. −0.36 eV), impeding the desorption of hydrogen and consequently leading to a decline in its HER activity [81]. Guo and coworkers presented their findings on IrNi alloy nanoflowers (IrNi NFs) as an exceptional electrocatalyst for the HER under acidic conditions, showcasing an impressively low overpotential of 25 mV at a current density of 10 mA cm^{-2} and a notably shallow Tafel slope of 29.7 mV per decade. These performance metrics significantly outperform commercial Pt/C and Ir/C catalysts [82]. Furthermore, a self-assembled configuration of IrNi alloy nanoparticles (IrNi NPs) supported on a carbon nanotube sponge (referred to as DSIrNi@CNTS) was synthesized via a controlled thermal shock process under extreme conditions [83]. HRTEM in conjunction with geometric phase analysis provided visual evidence that the rapid cooling process from elevated temperatures, combined with variations in atomic radii, effectively immobilized a multitude of dislocations within the IrNi NPs, yielding consequential strain effects. This catalyst demonstrated outstanding electrocatalytic performance for the alkaline HER, characterized by an impressively low overpotential of a mere 17 mV at a current density of 10 mA cm^{-2}, and this high performance was sustained for an extended duration exceeding 45 hours.

4.2.2 Non-PGM-Based Materials

In recent years, there has been a considerable surge in interest regarding non-noble metal-based materials. Current endeavors are focused on the creation of compounds utilizing diverse transition metals (including Fe, Co, Ni, and Cu) to replace noble metal HER electrolysis catalysts, and these alternative catalysts exhibit commendable catalytic activity. Researchers are employing a range of engineering strategies, including the incorporation of nonmetallic elements, to develop hydrogen evolution electrocatalysts that are not only highly active but also cost-effective [84].

Wang and coworkers conducted a comprehensive investigation by merging iron oxide with nickel-based compounds via intricate interface engineering. Additionally, they introduced sulfur and selenium anions into the system to scrutinize the impacts of interface engineering and anionic defects on the electronic structure and catalytic sites of the compound, as illustrated in Figure 4.3a–f [85]. DFT calculations unveiled that the S site serves as the most optimal active site for the HER when employing the Ni_3S_2 catalyst without the introduction of the Fe_2O_3 compound, as depicted in Figure 4.3d,e. Conversely, in the presence of the Fe_2O_3 compound, the Ni site emerges as the superior active site for HER. The surface structure of the catalyst is visually depicted in Figure 4.3f. These computational results corroborate that the catalysts $Fe_2O_3/NiSe_2$ and Fe_2O_3/Ni_3S_2 were synthesized through interfacial engineering while concurrently introducing anions (Se, S). This

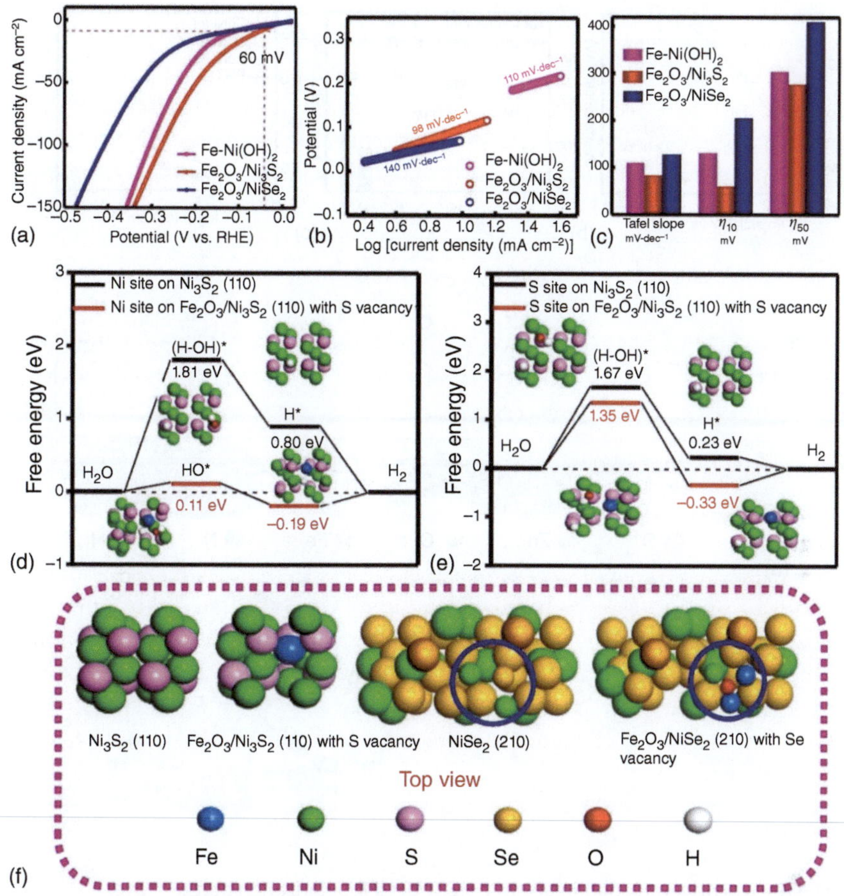

Figure 4.3 (a) LSV curves, (b) Tafel slopes for HER, (c) comparison of overpotential and Tafel slopes for HER. HER energetic pathways comparison between pristine Ni_3S_2 and Fe_2O_3/Ni_3S_2 with S vacancy on (d) Ni sites, and (e) S sites. (f) Top view of catalyst surface structure. Source: Reproduced with permission from Wang et al. [85] Elsevier.

intricate construction of heterostructures and the incorporation of anionic dopants precipitated the migration of active sites.

Apart from heteroatom doping, bimetallic catalysts showcase a synergistic effect, which holds the potential to significantly enhance the performance of the HER. To illustrate, Co single atoms derived from MOFs effectively segregate Pt nanoparticles through an isolating effect and establish robust Co–Pt interactions, as detailed in reference [86]. In an alternate scenario, the incorporation of either Co or Fe into Ni single atoms facilitated the adsorption and desorption processes of intermediate products, as elucidated in reference [87]. Shao and coworkers conducted a synthesis of polymorphic interfacial hydrogen evolution catalysts, specifically CoNi alloys, which comprise a steadfast *hcp* phase alongside its stable counterpart, the face-centered cubic (*fcc*) phase. These alloys give rise to a stable *hcp/fcc* interfacial structure known as an in-situ polymorphic interface (*inf*).

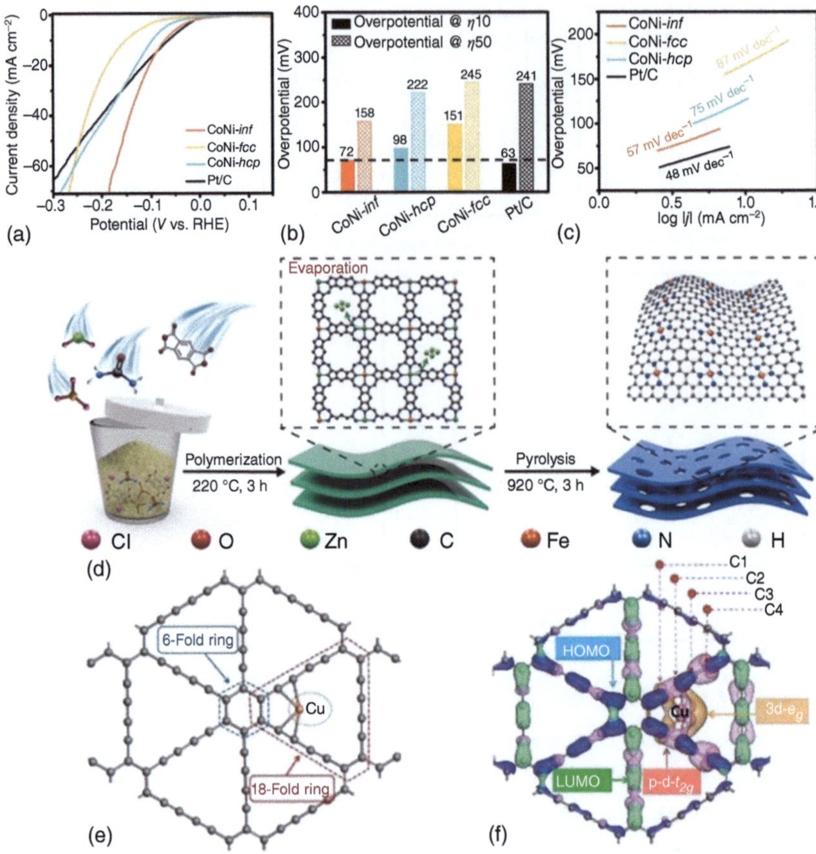

Figure 4.4 Activity evaluation in alkaline HER. (a) HER polarization plots, (b) overpotentials at current densities of 10 and 50 mA cm^{-2}, and (c) Tafel plots of CoNi-inf, CoNi-*hcp*, CoNi-*fcc*, and the commercial Pt/C. Source: Reproduced with permission from Huang et al. [88], Wiley-VCH. (d) Synthesis illustration of Fe single atoms/N-doped porous carbon. Source: Reproduced with permission from Li et al. [89] Wiley-VCH. (e) Schematic illustration of the unit structure of Cu^0/GDY. (f) Real-space HOMO (blue isosurface) and LUMO (green isosurface) contour plots, and the charge densities for the coupled p and the d-t_{2g} component (pink isosurface) and the d-e_g component (brown isosurface) of Cu of Cu^0/GDY. (C1, C2, C3, and C4 indicate the active H-adsorbed C-sites). Source: Reproduced with permission from Li et al. [90] Wiley-VCH.

As depicted in Figure 4.4a–c, the collective HER activity of CoNi-*inf* is notably enhanced, underscoring the role of the phase interface in promoting the activation of molecules and, consequently, enhancing the HER properties of the alloy [88].

Due to their cost-effectiveness, efficient atomic utilization, and the abundance of exposed active sites, non-noble SACs have found extensive application in the field of electrocatalytic HER [91–93]. Through the utilization of a polymerization–pyrolysis–evaporation approach, Fe single atoms were skillfully dispersed onto the surface of nitrogen-doped porous carbon. In-depth examination reveals that following the high-temperature pyrolysis of polymers in conjunction with Zn atom evaporation (as illustrated in Figure 4.4d), the Fe atoms were adeptly isolated and firmly anchored by four nitrogen atoms [89]. The electrocatalyst exhibited a minimal overpotential of merely 202 mV at a current density of 10 mA cm^{-2}, and a sustained current density was maintained throughout the extensive 70,000-second duration of the long-term stability test. In a separate study, researchers employed an in situ reduction method to precisely anchor copper atoms onto graphdiyne (GDY). Notably, the Cu atoms were primarily situated at the acetylenic ring corner (18-fold ring) in proximity to the 6-fold ring edge, as illustrated in Figure 4.4e [90]. Arising from the overlap of C-2p and Cu-3d-t_{2g} orbitals, the robust p–d coupling instigated a closed-shell effect while optimizing the charge exchange characteristics. This phenomenon holds the potential to significantly enhance the HER performance, as depicted in Figure 4.4f.

4.3 Conclusion and Outlook

In this chapter, we initially explored the active sites of common metal-based materials and the associated mechanism for HER. Simultaneously, considering the conditions required for practical applications, we summarized typical examples of metal-based materials composed of various elements for HERs. While some characterization and synthesis challenges persist in harnessing the potential of metal-based materials, this endeavor undoubtedly enriches the field and deepens our understanding of catalytic processes.

(1) Narrowing the gap between industrial applications and small-scale experimental performance is imperative. Researchers have developed a range of advanced metal-based electrocatalysts with remarkable electrocatalytic HER performance. However, most studies have focused on reducing the overpotential of electrocatalysts, which are accompanied by negligible current densities such as 10 and 100 mA cm^{-2}. Moreover, electrochemical cycling experiments have not exceeded 200 hours. Unlike experimental studies, industrial applications necessitate electrocatalysts to deliver current densities exceeding 500 mA cm^{-2}, maintained for over a month. Most experimental electrocatalysts fail to meet the growing demands of industrial applications, posing a pressing challenge in narrowing the gap between empirical studies and industrial requirements for current metal-based electrocatalysts.
(2) During water electrolysis, catalytical active sites may experience localized collapses, oxidation, and other issues leading to deactivation. Accurately identifying these deactivation sites enables a more thorough analysis of the catalyst.
(3) The active sites' utilization efficiency should be reported to accurately depict the catalyst's actual hydrogen production. Numerous sites may experience localized collapses, oxidation, and other issues during the electrolysis of water, which eventually causes the sites to become inactive. A more comprehensive catalyst evaluation can be made if these deactivation sites can be accurately identified.
(4) Advanced characterization techniques and comprehensive theoretical calculations are essential. While numerous advanced characterization techniques have

been developed to analyze surface and structural features, interactions between active sites and intermediates play a pivotal role in catalytic activity. However, an in-depth investigation of these interactions remains unexplored. Therefore, employing advanced in situ characterization techniques like in situ Raman, in situ FTIR, and in situ XAS to monitor changes in active sites and the binding states between intermediates and active sites is imperative for metal-based electrocatalysts.

(5) In addition to in situ techniques, DFT calculations are essential for studying metal-based catalyst reaction mechanisms. A close integration of DFT calculations with in-situ characterization will contribute to a deeper understanding of the reaction mechanism. It is worth noting that, in some theoretical simulations of catalysts, the models constructed could be relatively simple, and many real-world situations are easily disregarded. In widely published literature, many structural models of crystalline materials are built in perfect conditions. However, in the actual situation, the produced samples frequently contain amorphous substances in addition to the crystalline phase region, and these amorphous substances often exhibit excellent catalytic activity during the HER reaction process.

Acknowledgment

This work was supported by the Hong Kong Polytechnic University (grant no. P0047728).

References

1 Lin, F., Li, M., Zeng, L. et al. (2023). Intermetallic nanocrystals for fuel-cells-based electrocatalysis. *Chem. Rev.* 123 (22): 12507.

2 Jin, M., Zhang, X., Niu, S. et al. (2022). Strategies for designing high-performance hydrogen evolution reaction electrocatalysts at large current densities above 1000 mA cm^{-2}. *ACS Nano* 16 (8): 11577.

3 Gao, R., Zhu, J., and Yan, D. (2021). Transition metal-based layered double hydroxides for photo(electro)chemical water splitting: a mini review. *Nanoscale* 13 (32): 13593.

4 Krishnan, A., Ajith, A., Krishnan, A.V. et al. (2023). Ni-based electro/photo-catalysts in HER – a review. *Surf. Interfaces* 36: 102619.

5 Van Renssen, S. (2020). The hydrogen solution? *Nat. Clim. Chang.* 10 (9): 799.

6 Zhu, Y., Qian, L., Zhong, Y. et al. (2020). Metal oxide-based materials as an emerging family of hydrogen evolution electrocatalysts. *Energy Environ. Sci.* 13 (10): 3361.

7 Beyene, B.B. and Hung, C.H. (2020). Recent progress on metalloporphyrin-based hydrogen evolution catalysis. *Coord. Chem. Rev.* 410: 213234.

8 Chen, X., Wan, J., Zheng, M. et al. (2023). Engineering single atomic ruthenium on defective nickel vanadium layered double hydroxide for highly efficient hydrogen evolution. *Nano Res.* 16 (4): 4612.

9 Gao, G., Zhu, G., Chen, X. et al. (2023). Optimizing Pt-based alloy electrocatalysts for improved hydrogen evolution performance in alkaline electrolytes: a comprehensive review. *ACS Nano* 17 (21): 20804.

10 Li, H. and Li, G. (2023). Novel palladium-based nanomaterials for multifunctional ORR/OER/HER electrocatalysis. *J. Mater. Chem. A* 11 (17): 9383.

11 Hansen, J.N., Prats, H., Toudahl, K.K. et al. (2021). Is there anything better than Pt for HER? *ACS Energy Lett.* 6 (4): 1175.

12 Shi, Z., Zhang, X., Lin, X. et al. (2023). Phase-dependent growth of Pt on MoS_2 for highly efficient H_2 evolution. *Nature* 621 (7978): 300.

13 Li, Y., Sun, Y., Qin, Y. et al. (2020). Recent advances on water-splitting electrocatalysis mediated by noble-metal-based nanostructured materials. *Adv. Energy Mater.* 10 (11): 1903120.

14 De Levie, R. (1999). The electrolysis of water. *J. Electroanal. Chem.* 467 (1): 92.

15 Zhu, J., Hu, L., Zhao, P. et al. (2020). Recent advances in electrocatalytic hydrogen evolution using nanoparticles. *Chem. Rev.* 120 (2): 851.

16 Pang, Q., Fan, X., Sun, K. et al. (2023). Advanced progress of rhenium (Re)-based electrode materials in electrocatalytic hydrogen evolution: a review. *J. Mater. Chem. A* 11 (27): 14451.

17 Greeley, J., Jaramillo, T.F., Bonde, J. et al. (2006). Computational high-throughput screening of electrocatalytic materials for hydrogen evolution. *Nat. Mater.* 5 (11): 909.

18 Yang, M., Zhou, Y.N., Cao, Y.N. et al. (2020). Advances and challenges of Fe-MOFs based materials as electrocatalysts for water splitting. *Appl. Mater. Today* 20: 100692.

19 Li, J., Zhang, J., Shen, J. et al. (2023). Self-supported electrocatalysts for the hydrogen evolution reaction. *Mater. Chem. Front.* 7 (4): 567.

20 Anantharaj, S., Ede, S.R., Karthick, K. et al. (2018). Precision and correctness in the evaluation of electrocatalytic water splitting: revisiting activity parameters with a critical assessment. *Energy Environ. Sci.* 11 (4): 744.

21 Zhu, C., Gao, D., Ding, J. et al. (2018). TMD-based highly efficient electrocatalysts developed by combined computational and experimental approaches. *Chem. Soc. Rev.* 47 (12): 4332.

22 Dastafkan, K. and Zhao, C. (2020). Recent trends in alkaline hydrogen evolution using nonprecious multi-metallic electrocatalysts. *Curr. Opin. Green Sustainable Chem.* 25: 100342.

23 Bodhankar, P.M., Sarawade, P.B., Kumar, P. et al. (2022). Nanostructured metal phosphide based catalysts for electrochemical water splitting: a review. *Small* 18 (21): 2107572.

24 Anantharaj, S., Noda, S., Jothi, V.R. et al. (2021). Strategies and perspectives to catch the missing pieces in energy-efficient hydrogen evolution reaction in alkaline media. *Angew. Chem. Int. Ed.* 60 (35): 18981.

25 Tian, X., Zhao, P., and Sheng, W. (2019). Hydrogen evolution and oxidation: mechanistic studies and material advances. *Adv. Mater.* 31 (31): 1808066.
26 Zhou, Z., Pei, Z., Wei, L. et al. (2020). Electrocatalytic hydrogen evolution under neutral pH conditions: current understandings, recent advances, and future prospects. *Energy Environ. Sci.* 13 (10): 3185.
27 El-Refaei, S.M., Russo, P.A., and Pinna, N. (2021). Recent advances in multimetal and doped transition-metal phosphides for the hydrogen evolution reaction at different pH values. *ACS Appl. Mater. Interfaces* 13 (19): 22077.
28 Seh, Z.W., Kibsgaard, J., Dickens, C.F. et al. (2017). Combining theory and experiment in electrocatalysis: Insights into materials design. *Science* 355 (6321): eaad4998.
29 Wang, J., Xu, F., Jin, H. et al. (2017). Non-noble metal-based carbon composites in hydrogen evolution reaction: fundamentals to applications. *Adv. Mater.* 29 (14): 1605838.
30 Liu, Y., Wang, Q., Zhang, J. et al. (2022). Recent advances in carbon-supported noble-metal electrocatalysts for hydrogen evolution reaction: syntheses, structures, and properties. *Adv. Energy Mater.* 12 (28): 2200928.
31 Liang, J., Ge, Y., He, Z. et al. (2021). Wet-chemical synthesis and applications of amorphous metal-containing nanomaterials. *Nano Res.* 16 (4): 4289.
32 Ding, J., Yang, H., Zhang, S. et al. (2022). Advances in the electrocatalytic hydrogen evolution reaction by metal nanoclusters-based materials. *Small* 18 (52): 2204524.
33 Wang, Y., Huang, X., and Wei, Z. (2021). Recent developments in the use of single-atom catalysts for water splitting. *Chin. J. Catal.* 42 (8): 1269.
34 Li, C. and Baek, J.B. (2020). Recent advances in noble metal (Pt, Ru, and Ir)-based electrocatalysts for efficient hydrogen evolution reaction. *ACS Omega* 5 (1): 31.
35 Zhao, Y., Ling, T., Chen, S. et al. (2019). Non-metal single-iodine-atom electrocatalysts for the hydrogen evolution reaction. *Angew. Chem. Int. Ed.* 58 (35): 12252.
36 Li, X., Zhao, L., Yu, J. et al. (2020). Water splitting: from electrode to green energy system. *Nano Micro Lett.* 12 (131): 1.
37 Hao, J., Wu, K., Lyu, C. et al. (2023). Recent advances in interface engineering of Fe/Co/Ni-based heterostructure electrocatalysts for water splitting. *Mater. Horizons* 10 (7): 2312.
38 Gao, X., Dai, Q., Lu, X., and Kawi, S. (2023). Carbon-supported non-noble metal single-atom catalysts for electro-catalytic hydrogen evolution reaction. *Int. J. Hydrog. Energy* 48 (45): 17106.
39 Zheng, D., Yu, L., Liu, W. et al. (2021). Structural advantages and enhancement strategies of heterostructure water-splitting electrocatalysts. *Cell Rep. Phys. Sci.* 2 (6): 100443.
40 Xu, Y., Zhang, X., Liu, Y. et al. (2023). A critical review of research progress for metal alloy materials in hydrogen evolution and oxygen evolution reaction. *Environ. Sci. Pollut. Res.* 30 (5): 11302.
41 Zhang, Q. and Guan, J. (2020). Single-atom catalysts for electrocatalytic applications. *Adv. Funct. Mater.* 30 (31): 2000768.

42 Cui, Z., Jiao, W., Huang, Z.Y. et al. (2023). Design and synthesis of noble metal-based alloy electrocatalysts and their application in hydrogen evolution reaction. *Small* 19 (35): 2301465.

43 Wang, H., Fu, W., Yang, X. et al. (2020). Recent advancements in heterostructured interface engineering for hydrogen evolution reaction electrocatalysis. *J. Mater. Chem. A* 8 (15): 6926.

44 Da, Y., Jiang, R., Tian, Z. et al. (2023). Development of a novel Pt_3V alloy electrocatalyst for highly efficient and durable industrial hydrogen evolution reaction in acid environment. *Adv. Energy Mater.* 13 (16): 2300127.

45 Li, L., Zhang, G., Wang, B. et al. (2020). Electrochemical formation of PtRu bimetallic nanoparticles for highly efficient and pH-universal hydrogen evolution reaction. *J. Mater. Chem. A* 8 (4): 2090.

46 Chen, H.Y., Niu, H.J., Han, Z. et al. (2020). Simple fabrication of trimetallic platinum-nickel-cobalt hollow alloyed 3D multipods for highly boosted hydrogen evolution reaction. *J. Colloid Interface Sci.* 570: 205.

47 Fan, F., Chen, D.H., Yang, L. et al. (2022). PtCuFe alloy nanochains: synthesis and composition-performance relationship in methanol oxidation and hydrogen evolution reactions. *J. Colloid Interface Sci.* 628: 153.

48 Zhang, C., Liang, X., Xu, R. et al. (2021). H_2 in situ inducing strategy on Pt surface segregation over low Pt doped $PtNi_5$ nanoalloy with superhigh alkaline HER activity. *Adv. Funct. Mater.* 31 (14): 2008298.

49 Zhang, J., Chen, X., Lei, Y. et al. (2022). Highly rechargeable lithium oxygen batteries cathode based on boron and nitrogen Co-doped holey graphene. *Chem. Eng. J.* 428: 131025.

50 Rao, Q., Hu, F.X., Gan, L.-Y. et al. (2022). Boron-nitrogen-Co-doping nanocarbons to create rich electroactive defects toward simultaneous sensing hydroquinone and catechol. *Electrochim. Acta* 402: 139427.

51 Wang, X., Bai, L., Lu, J. et al. (2019). Rapid activation of platinum with black phosphorus for efficient hydrogen evolution. *Angew. Chem. Int. Ed.* 131 (52): 19236.

52 Cheng, Q., Hu, C., Wang, G. et al. (2020). Carbon-defect-driven electroless deposition of Pt atomic clusters for highly efficient hydrogen evolution. *J. Am. Chem. Soc.* 142 (12): 5594.

53 Speck, F.D., Paul, M.T.Y., Ruiz-Zepeda, F. et al. (2020). Atomistic insights into the stability of Pt single-atom electrocatalysts. *J. Am. Chem. Soc.* 142 (36): 15496.

54 Yang, W., Cheng, P., Li, Z. et al. (2022). Tuning the cobalt-platinum alloy regulating single-atom platinum for highly efficient hydrogen evolution reaction. *Adv. Funct. Mater.* 32 (39): 2205920.

55 Wang, L., Zhang, L., Ma, W. et al. (2022). In situ anchoring massive isolated Pt atoms at cationic vacancies of α-$Ni_xFe_{1-x}(OH)_2$ to regulate the electronic structure for overall water splitting. *Adv. Funct. Mater.* 32 (31): 2203342.

56 Zhang, H., An, P., Zhou, W. et al. (2018). Dynamic traction of lattice-confined platinum atoms into mesoporous carbon matrix for hydrogen evolution reaction. *Sci. Adv.* 4 (1): eaao6657.

57 Liu, D., Li, X., Chen, S. et al. (2019). Atomically dispersed platinum supported on curved carbon supports for efficient electrocatalytic hydrogen evolution. *Nat. Energy* 4 (6): 512.

58 Kuang, P., Wang, Y., Zhu, B. et al. (2021). Pt single atoms supported on N-doped mesoporous hollow carbon spheres with enhanced electrocatalytic H_2-evolution activity. *Adv. Mater.* 33 (18): 2008599.

59 Yuan, M., Wang, C., Wang, Y. et al. (2021). General fabrication of RuM (M = Ni and Co) nanoclusters for boosting hydrogen evolution reaction electrocatalysis. *Nanoscale* 13 (30): 13042.

60 Xu, J., Wang, S., Yamg, C. et al. (2021). Free-standing two-dimensional ruthenium-beryllium nanosheets for alkaline hydrogen evolution. *Chem. Eng. J.* 421 (1): 129741.

61 Liang, Q., Li, Q., Xie, L. et al. (2022). Superassembly of surface-enriched Ru nanoclusters from trapping-bonding strategy for efficient hydrogen evolution. *ACS Nano* 16 (5): 7993.

62 Chen, J., Chen, C., Qin, M. et al. (2022). Reversible hydrogen spillover in Ru-WO_{3-x} enhances hydrogen evolution activity in neutral pH water splitting. *Nat. Commun.* 13 (14): 5382.

63 Yao, Q., Huang, B., Zhang, N. et al. (2019). Channel-rich RuCu nanosheets for pH-universal overall water splitting electrocatalysis. *Angew. Chem. Int. Ed.* 58 (39): 13983.

64 Zhang, J., Mao, X., Wang, S. et al. (2022). Superlattice in a Ru superstructure for enhancing hydrogen evolution. *Angew. Chem. Int. Ed.* 61 (14): e202116867.

65 Zhang, H., Zhou, W., Lu, X. et al. (2020). Implanting isolated Ru atoms into edge-rich carbon matrix for efficient electrocatalytic hydrogen evolution. *Adv. Energy Mater.* 10 (23): 2000882.

66 Arulraj, R., Prabu, R., and Sara Fidha, M. (2023). Rh for HER electrocatalysis? A critical analysis of recent studies and thoughts on the same! *J. Mater. Chem. A* 11 (46): 25216.

67 Li, Z., Feng, Y., Liang, Y. et al. (2020). Stable rhodium (IV) oxide for alkaline hydrogen evolution reaction. *Adv. Mater.* 32 (25): 1908521.

68 Zhang, W., Zhang, X., Chen, L. et al. (2018). Single-walled carbon nanotube induced optimized electron polarization of rhodium nanocrystals to develop an interface catalyst for highly efficient electrocatalysis. *ACS Catal.* 8 (9): 8092.

69 Liu, Y., Hu, X., Huang, B., and Xie, Z. (2019). Surface engineering of Rh catalysts with N/S-codoped carbon nanosheets toward high-performance hydrogen evolution from seawater. *ACS Sustain. Chem. Eng.* 7 (23): 18835.

70 Guan, J., Wen, X., Zhang, Q., and Duan, Z. (2020). Atomic rhodium catalysts for hydrogen evolution and oxygen reduction reactions. *Carbon* 164 (30): 121.

71 Wang, T., Chutia, A., Brett, D.J. et al. (2021). Palladium alloys used as electrocatalysts for the oxygen reduction reaction. *Energy Environ. Sci.* 14 (5): 2639.

72 Ohm, D. and Domke, K.F. (2021). Controlled deposition of 2D-confined Pd or Ir nano-islands on Au(111) following Cu UPD, and their HER activity. *J. Electroanal.* 896 (1): 115285.

73 Wang, L., Zeng, Z., Gao, W. et al. (2019). Tunable intrinsic strain in two-dimensional transition metal electrocatalysts. *Science* 363 (6429): 870.

74 Pei, A., Li, G., Zhu, L. et al. (2022). Nickel hydroxide-supported Ru single atoms and Pd nanoclusters for enhanced electrocatalytic hydrogen evolution and ethanol oxidation. *Adv. Funct. Mater.* 32 (51): 2208587.

75 Liu, X., Zhu, S., Liang, Y. et al. (2022). Self-standing nanoporous NiPd bimetallic electrocatalysts with ultra-low Pd loading for efficient hydrogen evolution reaction. *Electrochim. Acta* 411 (10): 140077.

76 Xu, Y., Ren, K., Ren, T. et al. (2020). Phosphorus-triggered modification of the electronic structure and surface properties of Pd_4S nanowires for robust hydrogen evolution electrocatalysis. *J. Mater. Chem. A* 8 (38): 19873.

77 Yang, J., Li, W., Wang, D., and Li, Y. (2021). Single-atom materials: small structures determine macroproperties. *Small Struct.* 2 (2): 2000051.

78 Li, F., Han, G.F., Noh, H.J. et al. (2019). Balancing hydrogen adsorption/desorption by orbital modulation for efficient hydrogen evolution catalysis. *Nat. Commun.* 10 (6): 4060.

79 Lai, W.H., Wang, H., Zheng, L. et al. (2020). General synthesis of single-atom catalysts for hydrogen evolution reactions and room-temperature Na-S batteries. *Angew. Chem. Int. Ed.* 59 (49): 22171.

80 Jiang, P., Huang, H., Diao, J. et al. (2019). Improving electrocatalytic activity of iridium for hydrogen evolution at high current densities above $1000\,mA\,cm^{-2}$. *Appl. Catal. B* 258 (5): 117965.

81 Kim, S.J., Jung, H., Lee, C. et al. (2019). Comparative study on hydrogen evolution reaction activity of electrospun nanofibers with diverse metallic Ir and IrO_2 composition ratios. *ACS Sustain. Chem. Eng.* 7 (9): 8613.

82 Lv, F., Zhang, W., Yang, W. et al. (2020). Ir-based alloy nanoflowers with optimized hydrogen binding energy as bifunctional electrocatalysts for overall water splitting. *Small Methods* 4 (6): 1900129.

83 Liu, S., Hu, Z., Wu, Y. et al. (2020). Dislocation-strained IrNi alloy nanoparticles driven by thermal shock for the hydrogen evolution reaction. *Adv. Mater.* 32 (48): 2006034.

84 Liu, J., Zhang, H., Qiu, M. et al. (2020). A review of non-precious metal single atom confined nanomaterials in different structural dimensions (1D-3D) as highly active oxygen redox reaction electrocatalysts. *J. Mater. Chem. A* 8 (5): 2222.

85 Shao, Z., Liu, R., Xue, H. et al. (2021). Regulating the electronic structure of ultrathin Ni-based chalcogenide nanosheets through iron modification towards high electrocatalytic activities. *Chem. Eng. J.* 416: 129098.

86 Liang, L., Jin, H., Zhou, H. et al. (2021). Cobalt single atom site isolated Pt nanoparticles for efficient ORR and HER in acid media. *Nano Energy* 88: 106221.

87 Luo, W., Wang, Y., Luo, L. et al. (2022). Single-atom and bimetallic nanoalloy supported on nanotubes as a bifunctional electrocatalyst for ultrahigh-current-density overall water splitting. *ACS Catal.* 12 (2): 1167.

88 Tan, X., Geng, S., Ji, Y. et al. (2020). Closest packing polymorphism interfaced metastable transition metal for efficient hydrogen evolution. *Adv. Mater.* 32 (40): 2002857.

89 Pan, Y., Liu, S., Sun, K. et al. (2018). A bimetallic Zn/Fe polyphthalocyanine-derived single-atom Fe-N_4 catalytic site: a superior trifunctional catalyst for overall water splitting and Zn-air batteries. *Angew. Chem. Int. Ed.* 57 (28): 8614.

90 Hui, L., Xue, Y., Yu, H. et al. (2020). Loading copper atoms on graphdiyne for highly efficient hydrogen production. *ChemPhysChem* 21 (19): 2145.

91 Chen, W., Pei, J., He, C.T. et al. (2018). Single tungsten atoms supported on MOF-derived N-doped carbon for robust electrochemical hydrogen evolution. *Adv. Mater.* 30 (30): 1800396.

92 Yang, X., Song, W., Zhang, T. et al. (2023). Review of emerging atomically precise composite site-based electrocatalysts. *Adv. Energy Mater.* 13 (37): 2301737.

93 Peng, X., Hou, J., Mi, Y. et al. (2021). Bifunctional single-atomic Mn sites for energy-efficient hydrogen production. *Nanoscale* 13 (9): 4767.

5

Metal Compounds for Electrocatalytic Hydrogen Evolution Reaction

Tingting Yin, Wei Jiang, and Guigao Liu

Nanjing University of Science and Technology, School of Chemistry and Chemical Engineering, National Special Superfine Powder Engineering Research Center, 200 Xiaolingwei, Nanjing, 210094, Jiangsu, China

5.1 Introduction

With growing awareness of issues related to climate change, pollution, and energy security associated with dwindling supplies of fossil fuels, the discovery of clean and renewable sources of energy has become one of the greatest challenges to the sustainable development of societies [1–4]. Because of its high energy density, hydrogen has been regarded as an ideal energy carrier [5, 6]. Hydrogen production is therefore essential for a clean, low-carbon economy [7]. In recent years, efficient hydrogen production has become increasingly popular globally with the announcement of hydrogen and fuel cell vehicle programs in all parts of the world [8, 9]. Up to now, the main conventional method to produce hydrogen has been coal gasification, producing carbon dioxide pollution, and still relying on fossil fuels [10].

Hydrogen (H_2), as a new energy carrier with high energy density and zero carbon emissions, has been recognized as a credible alternative to traditional fossil fuels. Electrocatalytic water splitting for hydrogen production is a clean and sustainable technology for converting renewable energy sources into green hydrogen [11]. Hydrogen production through electrocatalytic water splitting consists of two half-cell reactions: the hydrogen evolution reaction (HER) at the cathode and the oxygen evolution reaction (OER) at the anode. The anodic OER involves four electron transfers, while the cathodic HER involves two electron transfers, suggesting that the HER is kinetically favorable compared to the OER [12]. Thus, the development of cost-effective, efficient, and excellently durable electrocatalysts for the HER plays an important role in electrolytic water separation. The most effective HER catalysts are known to be based on platinum group metals (PGMs), and Pt is commonly used as a reference catalyst in research. However, their scarcity and limited availability for industrial use, in addition to their high cost, make them unsuitable for large-scale industrial use [13, 14]. Hence, to overcome this challenge while ensuring activity and stability, low-cost transition metal-based compound catalysts for use in alkaline media have been investigated in recent years, and these catalysts

Functional Materials for Electrocatalytic Energy Conversion, First Edition.
Edited by Zhicheng Zhang, Meiting Zhao, and Yuchen Qin.

have satisfactory intrinsic activity and stability for HER, which is important for realizing the hydrogen economy. Virtually, all studies on the catalysis of HER in alkaline media by metal-based chalcogenides, oxides or hydroxides, carbides, nitrides, and phosphides date back more than half a century ago, demonstrating the potential of metal-based catalysts [15–18]. Consequently, the design of advanced HER electrocatalysts using metal substrates has attracted sufficient attention from researchers around the world.

A number of review articles have summarized the development of metal compounds (MCs) for HER from a materials and properties perspective [19, 20]. For example, MCs doped with heteroatoms, introducing defects into MCs, as well as strain and phase engineering, and interfacial engineering are all ways to deepen the understanding of HER MCs. Nevertheless, the HER kinetics of MCs still lack comprehensive knowledge and understanding; in particular, there are gaps in designing and optimizing the mass transport of MCs based on their atomic structure features and larger-scale electrode aggregation to further develop their applications in industrial water electrolyzers [21].

In recent years, metal-based compound catalysts have shown excellent performance in HER electrocatalysis, but there are few systematic and comprehensive summaries on this subject. First, we give a brief overview of all electrocatalysts based on MCs and their derivatives. A discussion of their physicochemical properties and the potential for electrocatalytic applications is presented. Then, transition metal chalcogenides, transition metal oxygen (hydroxide) oxides, transition metal carbides (TMCs)/nitrides (TMNs), and transition metal phosphides (TMPs) are reviewed, with a focus on the structure effect of MCs on the performance. Finally, we provide some insights into the current progress of MC electrocatalysts as well as the challenges that remain, and future directions for this promising field.

5.2 Metal Compounds as HER Electrocatalysts

5.2.1 Metal Chalcogenides

Because of their intriguing structural and electronic properties, metal chalcogenides represent the largest class of materials in electrocatalysis and have been used in HER [22–24]. Metal chalcogenides can be categorized into layered transition metal dichalcogenides (TMDs) and non-layered metal chalcogenides (NMCs), with TMD-based electrocatalysts having the greatest potential for low-cost, high-performance HER [25–27].

5.2.1.1 Transition Metal Dichalcogenides

Layered TMDs are important two-dimensional (2D) nanomaterials with a unique thickness dependence, a rich phase diagram encompassing a wide range of compositions and crystalline structures, and a variety of different chemical and physical properties that make them attractive for many potential applications. As shown in Figure 5.1, layered TMDs can be described by the general chemical formula

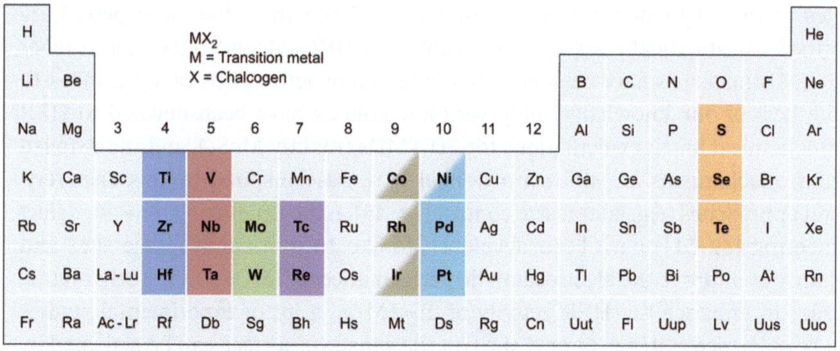

Figure 5.1 Structure of monolayered TMDs. About 40 different layered TMD compounds exist. The transition metals and the three chalcogen elements that predominantly crystallize in those layered structures are highlighted in the periodic table. Partial highlights for Co, Rh, Ir, and Ni indicate that only some of the dichalcogenides form layered structures. Source: Reproduced with permission of Chhowalla et al. [28], Springer Nature.

MX_2 (M is a transition metal element and X is an acetal, such as S or Se) [28]. Once these properties are controlled, 2D TMDs will also be a powerful platform for fundamental studies of interlayer coupling, phase engineering, charge transfer, and interfacial phenomena. Moreover, the uniqueness of 2D TMDs lies in their polycrystalline structures. For example, the three common crystal structures of MoS_2, 2H (hexagonal), 1T (tetragonal), and 3R (rhombohedral), depend on the different coordination modes between Mo and S atoms and the stacking order of the layers [29, 30].

Up until now, there have been two main methods for the synthesis of TMDs. They are the top-down liquid exfoliating method and the bottom-up growth method [31]. For example, TMD nanostructures are usually synthesized by gas-phase chemical deposition, which consists of reacting a substrate-confined metal precursor with a synthetic aldehyde reagent at elevated temperatures [32–35]. While gas-phase methods produce high-quality TMD monolayers that are ideal for exploring new quantum phenomena and demonstrating high-performance devices, their low yields, substrate confinement, and harsh synthesis conditions limit their scalability and range of applications. Liquid exfoliating is widely used to produce high-quality 2D TMD suspensions by direct sonication [36–38]. However, the nanosheets in the resulting dispersion are usually highly polydisperse, which is not ideal for practical applications. It has been found that solution routes to TMD nanostructures have emerged as an alternative synthesis method that addresses several of these drawbacks. These methods typically result in the preparation of high-quality MC nanomaterials with a wide range of applications in optics, biomedicine, and catalysis. Based on density functional theory (DFT) calculations, Hinnemann et al. first suggested that the Mo edge sites of MoS_2 would be hydrogen evolution active [39]. Since then, a large number of studies have explored not only MoS_2 but also other 2D-TMDs, phosphides, and carbons as HER catalysts.

As a HER descriptor, ΔG_H was developed to guide the design of HER electrocatalysts. In electrocatalytic applications, by DFT calculations and experiments,

X sites at the edges of the layers (rather than sites within the basal plane) are identified as catalytically active sites. Employing DFT-calculated ΔG_H, a number of 2D-TMD materials have been explored in search of active HER catalysts [40–42]. To the best of our knowledge, only very few studies have been devoted to HER electrochemical barrier calculations for 2D TMDs (mainly MoS_2), and the detailed reaction mechanisms are still controversial [43]. Also inspired by this discovery, several approaches (e.g. lateral size control [44, 45], nanostructuring [46–48], defect engineering [49–51]) have been developed to expose more active edge sites and thus further improve the electrocatalytic performance of TMDs. On the other hand, in order to enhance the HER activity of 2D-TMDs, a lot of experimental studies have been carried out, e.g. M or X sites on the substrate surface can be activated by doping with heteroatoms. Experiments and DFT calculations show that both metal atom (Fe, Co, Ni) and nonmetal atom (B, N, O) doping can effectively improve the HER activity of TMDs [52–54].

***MoS*$_2$** In this section, we will briefly describe several methods (e.g. heteroatom doping [55, 56], defect engineering, interaction engineering, phase transition engineering) used to tweak the ΔG_{H*} and energy band structure of MoS_2, which is one of the most classic materials among TMDs, to enhance its activity.

MoS_2 represents a representative TMD compound with an S-Mo-S triatomic layer structure. It was discovered in 2005 by Hinnemann et al., and later Jaramillo et al. demonstrated that the active site of MoS_2 is an edge site rather than a basal site [39, 57]. Following the identification of the active site of MoS_2 materials, a variety of chemical and/or physical strategies have been developed to expose and/or extend their active edge sites.

The low catalytic performance of pristine MoS_2 is attributed to the low concentration of exposed catalytic sites and semiconductivity. One possible way to improve the electrocatalytic performance is to increase the active site density, activate the intrinsic activity of the original inert sites, and tune the electronic structure of the catalyst through structural engineering. Structural engineering, including crystalline modulation and nanostructure fabrication, has been used to improve the conductivity or activity of pristine catalysts. The edges, vacancies, and boundaries have been identified as intrinsic catalytic sites. One successful strategy for constructing a large number of catalytic sites has been dopants and heterojunctions.

1T, 2H, and 3R are the main crystalline phases of MoS_2. The 1T phase one was reported to exhibit better performance in catalyzing hydrogen precipitation reactions compared to the corresponding semiconducting prismatic 2H phase one, which could be attributed to the significant decrease in the basal plane charge transfer resistance and the increase in the active sites. The electrocatalytic properties of metallic 1T-MoS_2 were summarized by Zhu and co-workers [58]. The synthesis of 1T-MoS_2 is more challenging due to the thermodynamic instability of the structure and the easy phase transition to the more stable 2H-phase under external disturbances.

Flower-like 2H-MoS_2 nanostructures with high-density exposed edges were synthesized directly on graphite substrates by a simple chemical vapor deposition method starting from $MoCl_5$, and elemental sulfur precursors, and the hydrogen evolution process was experimentally documented in detail to improve the HER catalytic efficiency [59]. Exfoliated MoS_2 (1T-MoS_2) nanostructural morphology was preserved, but the exfoliated nanosheets were more disordered and compressed onto the substrate surface compared to the grown 3D MoS_2 nanoflowers. Catalytic activity can be easily assessed by directly growing and exfoliating MoS_2 nanosheets on a conductive graphite substrate, which can be attached to a standard rotating disk electrode device during three-electrode electrochemical measurements using a 0.5 M H_2SO_4(aq) electrolyte, and continuously rinsed with H_2. After lithium intercalation at 60 °C for 48 hours, the peeled 1T-MoS_2 nanosheets significantly improved HER activity. Early onset potential and smaller Tafel slopes indicate that hydrogen adsorption is closer to equilibrium. A Tafel slope of 43 mV dec^{-1} and an earlier onset of significant H_2 evolution confirm that 1T-MoS_2 is one of the most catalytically active HER electrocatalysts.

Because most of the marginal catalytic sites in bulk MoS_2 are situated next to each other, reaction kinetics and mass transfer are severely limited. Improved electrochemical performance of MoS_2 can be achieved by modifying mass transfer, improving electron transport and diffusion, and avoiding restacking. Reconstructing various nanostructures is a simple but effective way to improve electrochemical performance. For example, to maximize diffusivity, Kibsgaard et al. synthesized an extended 3D mesoporous MoS_2 network structure with a bimetallic morphology [60]. The HER electrochemical activity of the bilobate MoS_2 was then evaluated. The HER activity data of bilobate MoS_2 in the form of cyclic voltammograms at 50 mV s^{-1}. By comparing the difference in capacitance current between −0.1 V and +0.4 V regions with the reversible hydrogen electrode (RHE), it is evident that samples with longer deposition time exhibit increased surface area. The current density versus RHE measured at −0.3 V voltage reveals a clear trend of increasing activity with increasing thickness. However, the normalized data relative to the surface area of each sample show that the increase in HER current at higher current densities is not linearly proportional to surface area. The activity trend is reversed, with thinner films obtaining higher turnover frequencies than thicker films.

Other Transition Metal Dichalcogenides Previously, it was known that the edge sites of TMDs dominated the electrochemical reaction, while the basal surface proved to be inert [61–63]. One of the best techniques for tuning HER activity is substitutional doping in TMDs [64, 65]. By utilizing this technique, the activity of HER can be significantly increased and easily used for efficient energy generation. In this context, various studies have been devoted to the modulation of the electronic structure by various means, such as doping, to enhance the activity of HER. It has been proved that doping of metallic elements in TMDs can change the electronic configuration of the catalyst and optimize the hydrogen adsorption energy [66, 67]. Therefore, it is considered an effective method to obtain excellent HER activity for TMDs.

For example, various approaches have been employed to improve the HER properties of WS_2 electrocatalysts, including doping with certain heteroatoms and controlling the morphology or structure to increase available active sites. Among them, appropriate metal doping strategies can effectively promote the HER activity of transition MCs by simultaneously increasing active sites and electrical conductivity. With the two critical points of conductivity and active sites, metal-doped WS_2 will be a promising system to further explore how to effectively promote the HER process. In particular, Das et al. were able to activate the catalytic activity of S atoms in the basal plane and make them active sites themselves by doping with Zn [68]. They successfully prepared a Zn-doped WS2 catalyst by hydrothermal method at ambient temperature. The electrical conductivity of pure WS_2 can also be enhanced by Zn doping, leading to excellent HER, and it has been experimentally and theoretically demonstrated that the synthesized Zn-doped WS_2 catalysts have higher electrochemical activities compared to pure WS_2. The electrochemical HER activity of all Zn-doped WS_2 catalysts was measured in 0.5 M H_2SO_4 by employing a typical three-electrode setup. The HER performance was found to be superior to that of pure WS_2. The HER performance was gradually increased with the increase of Zn^{2+} concentration up to WS_2-40 and then decreased with the concentration of Zn further increasing, which implies that the doping concentration of Zn plays a key role in the enhancement of the HER performance.

5.2.1.2 Non-Layered Metal Chalcogenides

Especially in the past few years, non-layered nanomaterials with strong chemical bonds in three dimensions have also received increasing attention. The study of non-layered nanomaterials not only greatly enriches the family of 2D nanomaterials, but also provides some novel electronic properties due to their abundant surface dangling bonds and obvious structural distortions [69–71]. NMCs are a class of metal chalcogenides with three-dimensional crystal structures. To overcome the problem of difficulty in synthesizing 2D NMCs by conventional methods (e.g. liquid exfoliation), new approaches have been developed, such as directed attachment growth, 2D template synthesis, and selective extraction-assisted exfoliation. To date, a variety of non-layered nanomaterials have been successfully developed, including metal chalcogenides (M_aX_b, where M = Co, Ni, Zn, Cu, In, etc. and X = S, Se, etc.) [72]. The unique and peculiar structural and electronic properties of non-layered nanomaterials allow them to be characterized by short ion diffusion paths, multiple surface active sites, tunable band gaps, and enhanced electrical conductivity and carrier mobility, which greatly facilitates their application in the field of advanced energy storage and conversion, such as supercapacitors, rechargeable batteries, catalysis, and photodetectors. Among them, 2D NMCs containing bimetallic cations (e.g. Fe-NiS_2) usually show better HER performance due to the increased density and accessibility of catalytically active sites [73].

$CoSe_2$ $CoSe_2$ as a typical metallic material among these pyrite dichalcogenides has an inherent high electrical conductivity that ensures efficient electron transfer during electrocatalysis, which is superior to that of other semiconducting pyrites. The research group of Xie and co-workers has recently realized atomically thin $CoSe_2$ by conventional liquid-phase exfoliation, which provides an existing platform for the realization of the above-established model [74]. For the first time, a subtle deformation of the atomic arrangement was conceptualized and induced into the ultrathin nanosheets by the incorporation of heterogeneous Mn^{2+}, and the strategy of exposing the active edge sites was rationally designed to optimize the HER activity.

NiS Low-cost nickel sulfide (NiS_x) has been extensively investigated for electrocatalytic hydrogen production [75, 76]. Qin et al. identified that NiS/Ni_3S_4 composites with different compositions were synthesized by a simple hydrothermal method, and the obtained NiS_x samples showed the highest performance in both electrocatalytic and photocatalytic hydrogen production. The basic mechanism was revealed by theoretical calculations and experimental studies [77]. SEM and TEM images of various NiS_x-m samples, in which the hydrothermally prepared NiS_x-m samples all present hollow spheres composed of NiS_x nanoparticles, indicate polycrystalline properties. The hollow structure is considered to be very favorable for the HER reaction due to its large surface area and tight interpenetrating network. Across all five samples, electrocatalytic H_2 evolution was achieved under negative bias, where it was found that the overpotential steadily decreases and then increases as the molar ratio of S^{2-}/Ni^{2+} increases.

Fe_xS Metallo-rich transition metal chalcogenides have recently emerged as alternatives to conventional noble metal-based hydrogen evolution catalysts. In particular, pentlandite material (Pn) with the general formula M_9X_8 (where M is a transition metal and X is a chalcogenide) has proved to be an effective candidate for replacing platinum due to its high hydrogen evolution activity. The elemental composition of pentlandite is flexible enough to allow targeted synthesis of Pn materials with precise stoichiometric control through various routes [78]. Along these lines, varying the composition of the catalyst can initially optimize the HER activity of the catalyst. For example, Smialkowski et al. found an efficient catalyst among the trimetallic pentacene $(Fe, Co, Ni)_9S_8$ with different metal ratios, which resulted in a patent application for acidic HER [79]. To probe electrochemical HER behavior, overpotentials at $-50\ mA\ cm^{-2}$ (η_{50}) were assessed, which showed a significant decrease in HER performance from high to low iron content. Considering the complex phase composition of the materials, the different reactivities may be attributed to phase-dependent interactions. Further focused observation of the phase-pure compounds revealed that the overpotential decreases when the Fe or Ni content is high and the Fe/Ni distribution is more balanced.

5.2.2 Transition Metal Oxides and Hydroxides

Transition metal oxides (TMOs) and transition metal hydroxides (TMHs) are an important family of functional materials with diverse compositions and structures, but they are traditionally considered to be inactive for HER and have rarely been studied for HER due to their poor electrical conductivity [80–82]. Many recent studies have shown that the electrocatalytic activity of TMOs and TMHs can be improved by reducing their thickness [83–85]. More catalytically active sites and higher electrical conductivity are necessary for improving their HER performance [86, 87].

5.2.2.1 Transition Metal Oxides

The high efficiency of HER catalysts depends on their electronic structure, conductivity, and intrinsic catalytic activity. TMOs have great potential to replace platinum-based catalysts, but further improvements in intrinsic catalytic activity are limited to the conductivity of the material. There are two major classes of 2D TMOs: layered TMOs and non-layered TMOs. The layered class includes metal trioxides (e.g. MoO_3 [88]), which have graphene-like layered structures and can be converted into 2D nanosheets by typical exfoliation methods (e.g. liquid exfoliation) [89]. Most 2D TMOs can be directly used as electrocatalysts without further modification.

TMOs have great potential for the evolution of high-purity hydrogen and oxygen, with the added value of process simplicity, zero CO_2 emissions, and low pollution [90, 91]. The pursuit of efficient, cost-effective, and non-precious-metal electrocatalysts in the advancement of HER has emerged as a pivotal research focus in the recent drive toward large-scale industrialization [92, 93]. Among various metal oxide electrocatalysts for HER, molybdenum oxides [94–96] and tungsten oxides [97, 98] have received increasing attention due to their excellent HER performance. However, most TMOs exhibit inherently low electrical conductivity and unfavorable hydrogen adsorption–desorption capabilities, which hinder their use as selective electrocatalysts for HER in alkaline media. Therefore, it is important to improve the electron mobility of these oxides during the reaction. For example. Shen et al. prepared nanoflowering MoO_2 and porous MoO_2 nanosheets using nickel foam substrates and reported good HER activity in alkaline solutions [99].

TiO_2 TiO_2 nanomaterials are well-known for their numerous applications, which range from common products (e.g. sunscreens) to advanced devices (e.g. photovoltaic cells) and also include a range of environmental and biomedical applications. Over the past decades, the importance and diversity of these applications have stimulated great interest and substantial progress in the preparation, characterization, and fundamental understanding of TiO_2 nanomaterials. Pandey et al. presented the first systematic study of the geometric and electronic properties and HER catalytic activity of ionic liquid (IL) C_2mim TfO and $(TiO_2)_n$ nanocluster hybrid structures, with $n = 2–12$, using state-of-the-art DFT first-principles calculations [100]. The results in the η values at $n = 4$, 5, and 7 display significant minima. These small η values indicate high catalyst activity. With an overpotential of about 0 V, the IL/$(TiO_2)_5$ hybrid system is again an ideal catalyst for HER, competing with the conventional efficient HER catalyst Pt. The excellent η values for $n = 5$ may be related to the charge transfer between the hydrogen molecules and the oxygen molecules nearest to the $(TiO_2)_5$ nanoclusters.

MoO_x Molybdenum oxide is regarded as a promising candidate HER electrocatalyst because of its excellent electronic conductivity and fine stability in acidic solutions. For example, Sun et al. prepared the MoO_x-Pd/CC catalysts by a simple two-step method of dip-coating and electrochemical reduction, and demonstrated their excellent catalytic efficiency in HER under acidic environment [101]. As shown in Figure 5.2, the electrocatalytic activity of the bare Pd/CC and MoO_x/CC electrodes

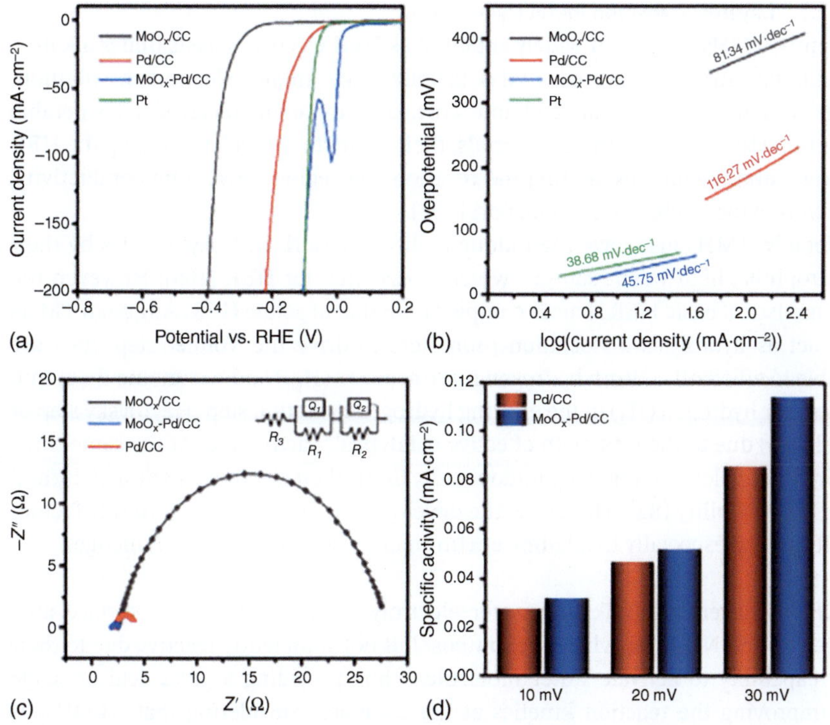

Figure 5.2 (a) Polarization curves of MoO_x/CC, Pd/CC, MoO_x-Pd/CC and commercial Pt mesh after iR-correction. (b) Corresponding Tafel plots for MoO_x/CC, Pd/CC, MoO_x-Pd/CC, and commercial Pt mesh. (c) Nyquist plots of MoO_x/CC, Pd/CC, and MoO_x-Pd/CC. (d) Specific activities of Pd/CC and MoO_x-Pd/CC at overpotentials of 10, 20, and 30 mV in N_2 saturated 0.5 M H_2SO_4. Source: Reproduced with permission of Sun et al. [101], Springer Nature.

for HER was relatively lower than that of MoO_x-Pd/CC, which showed an ultra-low overpotential of 25 mV at a current density of 10 mA cm^{-2}. Furthermore, the MoO_x-Pd/CC electrode showed the smallest Tafel slope value, indicating that it had the highest HER kinetic activity. The high HER activity of MoO_x-Pd/CC could be attributed to its self-supported configuration which could accelerate the gas diffusion and electron transport.

WO_3 Unlike most TMOs, tungsten oxides have high thermodynamic stability over a wide pH range, especially in acidic media. However, tungsten oxides have rarely been used as efficient HER electrocatalysts due to their poor activity. To enhance the HER, Liu et al. optimized the adsorption energy of H on WO_3 by doping with Co heteroatoms [102]. In their study, the HER electrocatalytic activity of Co-doped WO_3/CF was investigated at different pH values. Particularly, at pH = 14, Co-doped WO_3/CF showed much superior electrocatalytic activity for H_2 evolution as compared with one without Co doping.

5.2.2.2 Layered Transition Metal Hydroxides

Recently, TMHs have been widely explored as HER electrocatalysts under alkaline conditions [103]. 2D layered TMHs usually have unique electron distribution, large specific surface area, and low cost, and therefore received considerable attention [104–106]. For these materials, their basic strategies for reducing the HER overpotential include expanding the reaction site/region, increasing conductivity, and improving bubble or gas transport [107].

Notably, TMHs also have the unique ability to adsorb hydroxyl species by their electrophilic, high-valent cations, which is essential for HER [108]. However, the mechanism of basic HER is more complex than that of acidic HER. Although TMHs can act as hydrolysis dissociation promoters to drive the Volmer step, they are unable to efficiently adsorb hydrogen intermediates (H_{ad}) and recombine them into molecular hydrogen (H_2) in the ensuing hydrogen evolution step (Heyrovsky step or Tafel step) due to the lack of an effective catalytically active site. At the same time, their inherent low conductivity undoubtedly limits their electrocatalytic activity and long-term stability [82]. Therefore, the development of TMHs cxv with satisfactory HER activity, especially in alkaline environments, remains a major challenge.

$Ni(OH)_2$ Conventional alkaline water electrolyzers have used high surface area Raney Ni and Ni alloys, which are low cost but not sufficiently reactive due to their low capability to activate water molecules, thus providing a great deal of scope for improving the reaction kinetics at the cathode. Considering that $Ni(OH)_2$ is effective for cleaving the HO—H bond, the hybrid of $Ni(OH)_2$ with Ni could lead to much enhanced HER under alkaline conditions. For example, the HER on Ni electrodes modified by $Ni(OH)_2$ nanoclusters was shown by Danilovic et al. to be much faster than on bare Ni surfaces (Figure 5.3) [109].

$Co(OH)_2$ The studies on the performance of cobalt-based catalysts in recent years have mainly focused on heteroatom doping, morphology modulation, and heterostructure construction. Although transition metal-modified cobalt-based catalysts have been widely reported, the effect of noble metal modification on the performance of cobalt-based catalysts has not been studied deeply enough. Han et al. successfully prepared Ru anchored on $Co(OH)_2$ nanowire arrays (Ru/$Co(OH)_2$ NWAs) on nickel foams using a hydrothermal method [110]. The Ru/$Co(OH)_2$ NWAs catalysts exhibited excellent electrochemical catalytic performances for HER with a low overpotential of 96 mV at 10 mA cm^{-2} and a small Tafel slope of 63 mV dec^{-1}.

NiFeMo LDH Guo et al. employed a convenient and friendly hydrothermal strategy to support a unique molybdenum-doped nickel–iron alloy–LDH nanosheets on nickel foam ($Ni_xFe_1Mo_1$-LDH/NF), which was used as an HER electrode material to explore its electrocatalytic properties [111]. In order to further improve the electroconductivity, the composition-optimized nanosheets were grown in situ on rGO homogeneously modified nickel foam ($Ni_6Fe_1Mo_1$-LDH/rGO/NF), which showed excellent HER performance and stability. The HER performance of all catalysts was tested by linear scanning voltammetry (LSV) in 1.0 M KOH solution (Figure 5.4). Among all the catalysts, $Ni_6Fe_1Mo_1$-LDH/NF showed the best HER activity, indicating that molybdenum doping in NiFe–LDH accelerated the reaction kinetics and improved the HER performance.

5.2.3 Transition Metal Carbides and Nitrides

In the last few years, TMCs and TMNs have been most extensively studied for HERs [112–114]. TMCs and TMNs have been interesting as electrode-active materials due to their superior physical properties, like high conductivity and good chemical stability [115, 116]. Electrochemical characterization of Group IV–VI TMCs and the analysis of the band structure of these carbides indicate that not only tungsten carbide but also molybdenum, chromium, and titanium carbides are likely to be highly active in HER [117]. The discovery of the platinum-like catalytic behavior of tungsten carbide (WC) since 1973 (reported by Levy and Boudart) has advanced the development of the potential electrochemical properties of tungsten carbides [118].

TMCs based on IVB–VIB group elements, often referred to as "interstitial alloys," are pre-prepared by incorporating carbon or nitrogen atoms into the interstitial sites of the parent metal [119]. Transition metal nitrides are analogous to carbides and are of interest as low cost alternatives to Group VIII noble metals in a variety of reactions [120]. For example, molybdenum nitride has a catalytic activity close to that of platinum in hydrocarbon hydrogenolysis. In addition, the preparation of metal nitrides is similar to that of carbides. As summarized by Chen et al., the structural differences between carbides and nitrides can be explained by electronic effects [121]. In reactions involving the donation of d electrons, such as HER, the catalytic activity of metal nitrides will be higher than that of the parent metal [122].

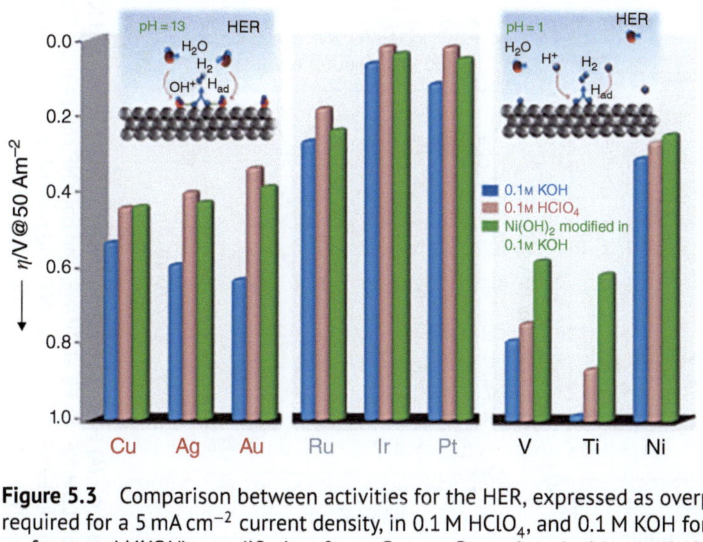

Figure 5.3 Comparison between activities for the HER, expressed as overpotential (h) required for a 5 mA cm^{-2} current density, in 0.1 M $HClO_4$, and 0.1 M KOH for both bare metal surfaces and $Ni(OH)_2$-modified surfaces. Source: Reproduced with permission of Danilovic et al. [109], Wiley-VCH.

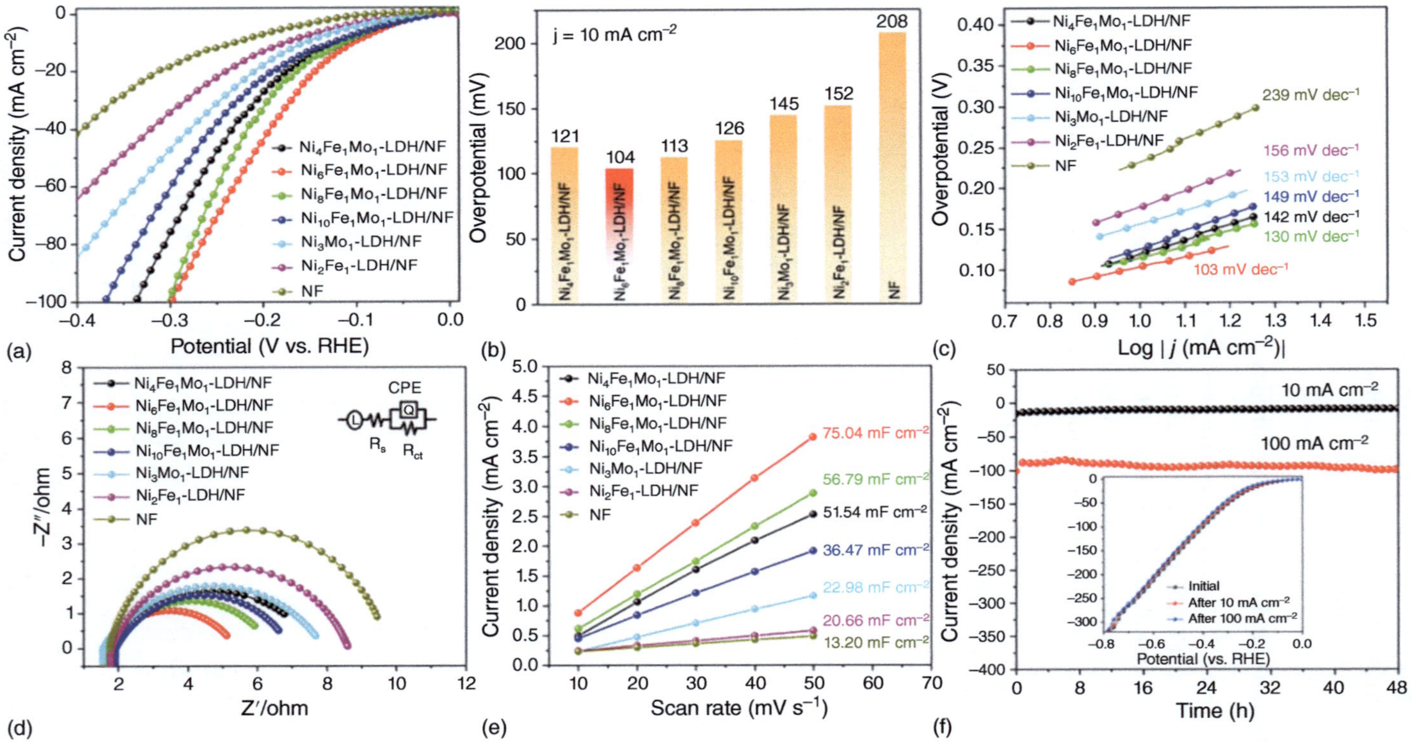

Figure 5.4 (A) HER polarization curves, (B) Tafel plots, and (C) the HER performance comparison chart of $Ni_xFe_1Mo_1$-LDH/NF ($x = 4, 6, 8, 10$), Ni_3Mo_1-LDH/NF, Ni_2Fe_1-LDH/NF, and bare NF composites. (D) Nyquist plots, and (E) capacitive current densities of $Ni_xFe_1Mo_1$-LDH/NF ($x = 4, 6, 8, 10$), Ni_3Mo_1-LDH/NF, Ni_2Fe_1-LDH/NF, and bare NF plotted against different scan rates. (F) Chronoamperometry measurements of the $Ni_6Fe_1Mo_1$-LDH/NF composite for 48 h at 10 and 100 mA cm^{-2} (inset: comparison of HER polarization curves before and after the test). Source: Reproduced with permission of Guo et al. [111], American Chemical Society.

5.2.3.1 W_xC

Tungsten carbides have recently been investigated as potential candidates for HER catalysts due to their high thermal stability and their unique tungsten–carbon chemical bonding and noble metal-like d-states. However, the poor electrical conductivity of tungsten carbides still hinders their catalytic role in HER. It is noted that metal-N active centers are effective in enhancing electrocatalytic performance because of their Pt-like electrical structure and enhanced electronic conductivity. Introducing nitrogen into tungsten carbides might contribute to the improvement of the catalytic performance for HER. For example, Xu et al. reported on the synthesis of metal carbide and metal nitride nanocomposites to improve HER performance [114]. The irregular product consists of small cores encapsulated in a carbon shell with an average diameter of 100 nm. The formation of WN and W_2C nanocrystals encapsulated in a nitrogen-doped carbon layer was further confirmed by HRTEM and EDS mapping. The polarization curves of WN–W_2C core-shell nanocomposites in 0.5 M H_2SO_4 electrolyte at different carbonation heating rates. WN–W_2C–2R has the best catalytic activity with an overpotential of only 242 mV at 10 mA cm^{-2}, which is much lower than other reported nitrogen-doped carbon materials. It is also shown that the synergistic interaction between the three different material phases can provide abundant active sites, while nitrogen doping can reduce the Gibbs free energy of the adsorbed hydrogen protons of the neighboring carbon atoms. The high HER activity can be attributed to the synergistic effect of the unique three-phase configuration of the active sites consisting of interfacial W–N coupling centers, indicating that the metal-N coupling centers are effective catalytic sites. The low Tafel slope of the WN–W_2C–2R nanocomposite catalyst suggests that the HER process may be carried out through the Volmer–Heyrovsky mechanism.

5.2.3.2 Mo_xC

Molybdenum is much more abundant in nature than precious metals such as platinum. Molybdenum-based compounds are highly active for HER and are expected to replace the precious platinum. Among these molybdenum-based compounds, molybdenum carbide is considered to be a good alternative to platinum due to its excellent catalytic activity and high stability in acidic and alkaline media. The Mo_2C phase is more active than the other phases towards HER due to its special electronic structure (similar to the electronic state of platinum). Upadhyay et al. reported a cost-effective and environmentally friendly method for the one-step synthesis of pure phases Mo_2C and MoC embedded in graphitic carbon via an in situ carburization reduction route [123]. From the experimental studies, it was found that the pure phase Mo_2C exhibited high performance in HER with a Tafel slope of 129.7 mV dec^{-1}, while MoC exhibited low activity in HER with a Tafel slope of 266 mV dec^{-1}.

5.2.3.3 Co_xC

Cobalt carbide is also one of the most active HER electrocatalysts in acidic and alkaline aqueous solutions. Recently, Kawashima et al. reported the successful enhancement of the electrochemical HER activity of cobalt carbide in acidic media by hybridizing cobalt metal (Co) with cobalt carbide (Co_xC: Co_3C and Co_2C) via a polyol process [124]. Moreover, computational and experimental results revealed that Co_2C has a higher intrinsic activity for hydrogen evolution compared to Co_3C.

5.2.3.4 MXenes

From the discovery of $Ti_3C_2T_x$ by Gogotsi's group in 2011, MXenes received increasing attention as HER electrocatalysts due to their unique properties, including outstanding structural and chemical stability, high electrical conductivity, large active surface area, and good hydrophilicity [125, 126]. In the review by Bai et al., the preparation of MXenes is described, followed by a focus on recent research advances in MXenes-based HER electrocatalysts, and finally the discussion on strategies to optimize the performance of these electrocatalysts [127]. In recent years, great advances in the design of MXenes-based HER electrocatalysts have been made. For example, Yuan et al. reported a series of Pt-modified $Ti_3C_2T_x$ (T = O, OH, F; MXene) catalysts prepared by a wet impregnation method followed by photo-induced reduction [128]. The obtained Pt-$Ti_3C_2T_x$ catalyst has an overpotential of 55 mV at a current density of 10 mA cm^{-2}. This high activity is mainly dependent on the electronic state of the platinum, which in turn is influenced by the modified $Ti_3C_2T_x$ surface terminals (Ti–OH or Ti–O). The enhanced charge transfer between the O terminals and the platinum accelerates the HER kinetics and thus improves the overall catalytic performance. In order to improve the electron transport and the intrinsic chemical activity of MXenes for electrocatalytic applications in hydrogen production, Yoon et al. synthesized the P-doped V_2CT_x MXenes that showed a low onset overpotential of −28 mV, a Tafel slope of 74 mV dec^{-1}, and a low overpotential of −163 mV in 0.5 M H_2SO_4 at 10 mA cm^{-2} [129]. Computational and experimental results suggest that the high activity of the catalysts is because the P—C bond in P–V_2CT_x could act as an active site, thus weakening the binding strength of H and facilitating the desorption of H_{ads} in HER kinetics. In another study, Wang et al. elaborated on a novel CoC_2O_4@MXene tubular catalyst with rapid and complete remodeling properties for enhanced HER performance [130]. High electronic accessibility and abundant electrolyte diffusion channels induced spontaneous and transient reconfiguration of CoC_2O_4@MXene to form $Co(OH)_2$@MXene. In situ characterization and DFT calculations showed that the reconfigured $Co(OH)_2$@MXene was a truly active species, which could adjust the electronic structure and optimize the energy barriers for water dissociation and H* adsorption/desorption, consequently leading to a high performance in HER with 10 and 1000 mA cm^{-2} obtained at low overpotentials of 28 and 216 mV, respectively, under alkaline conditions.

5.2.4 Transition Metal Phosphide

While Pt is the benchmark catalyst for HER, its rarity, high cost, and durability issues have hindered its wide application. Fortunately, TMPs, such as Ni_2P, CoP, and Fe_2P, have recently been shown to be promising as alternatives to precious Pt catalysts [131, 132]. However, the currently available TMPs still suffer from two major drawbacks: the small surface area and poor electrical conductivity due to the lack of suitable charge carriers and insufficient catalytically active metal components in the compositions, making them unsuitable for large-scale hydrogen production. Their variable chemical and structural composition, electronic structure, and electrical conductivity, however, make them still attractive for HER. In 2013, Lewis et al. experimentally demonstrated the ability of Ni_2P to promote HER in acidic media [133]. In addition, the metal phosphides were earlier prepared using yellow phosphorus, but this material is highly toxic and the existing synthesis methods suffer from complex operation and extreme conditions [134]. Therefore, it is important and challenging to find cost-effective ways to develop nanostructures of TMPs with high efficiency and stability [135].

5.2.4.1 FeP_x

TMPs have the advantage of low manufacturing costs. Amongst the reported studies on TMP-based electrocatalysts, iron phosphide has received a lot of attention because iron is the second most common metallic element [136]. Compared with sulfides and nitrides, TMPs show clear superiority in catalytic applications [137]. For example, Wang et al. developed iron phosphides supported on iron foam (FeP_x@Fe) as an efficient HER electrocatalyst [138]. The electrocatalytic activity of FeP_x@Fe for HER is shown in Figure 5.5. FeP_x@Fe requires an overpotential of 124 mV to approach 10 mA cm^{-2}. The smaller overpotential compared to other samples suggests that FeP_x@Fe is more efficient in hydrogen production. It also has a smaller Tafel slope (78 mV dec^{-1}). The long-term durability of FeP_x@Fe was examined by chronopotentiometry at 10 mA cm^{-2}. The change in potential was very small throughout the chronopotentiometric test. The LSV curves after cycling showed only a small decay of activity, indicating the long-term durability of FeP_x@Fe.

5.2.4.2 MoP

Few earth-rich catalysts have shown efficacy for hydrogen reduction reactions in strong acids. Molybdenum phosphide (MoP) is one of the best materials that has been reported to be highly active for the hydrogen reduction reaction with excellent stability under acidic conditions. For example, Kibsgaard et al. reported the MoP catalysts prepared via a wet chemical method and demonstrated that such catalysts had a very high catalytic activity for HER with the turnover frequency (TOF) of 0.19 H_2 s^{-1} as the overpotentials of 150 mV [139]. Furthermore, after treating the MoP catalysts in an H_2S atmosphere, sulfur could be further introduced into their surface, resulting in the formation of the molybdenum phosphosulfide (MoP|S) catalysts. As shown in Figure 5.6a and b, the MoP|S catalysts exhibited the much-increased activity as compared with the original MoP catalysts. Meanwhile, they also maintained high stability in accelerated durability tests under the acidic environment (Figure 5.6c). The superior activity and stability of MoP|S could be attributed to the synergistic effect between sulfur and phosphorus.

5.2.4.3 CoP

Cobalt phosphide (CoP) possesses metal-like properties and high resistance to acid and alkali corrosion which render it with favorable activities and stabilities for

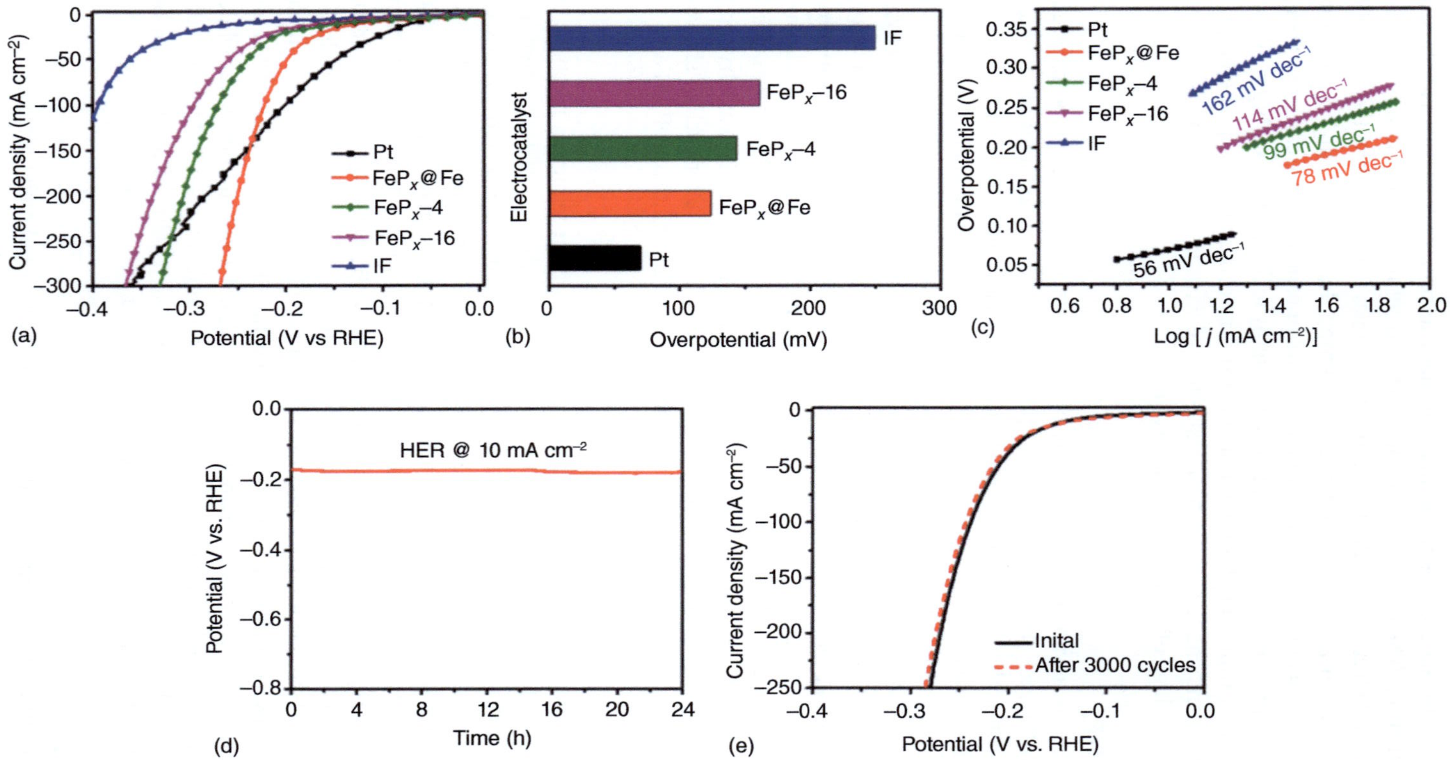

Figure 5.5 (a) LSV curves, (b) overpotentials at 10 mA cm^{-2}, and (c) Tafel plots of Pt, FeP_x@Fe, FeP_x-4, FeP_x-16 and iron foam (IF). (d) The chronopotentiometric curve of FeP_x@Fe at 10 mA cm^{-2}. (e) LSV curves of FeP_x@Fe before and after 3000 cycles of CV scanning. Source: Reproduced with permission of Wang et al. [138], Springer Nature.

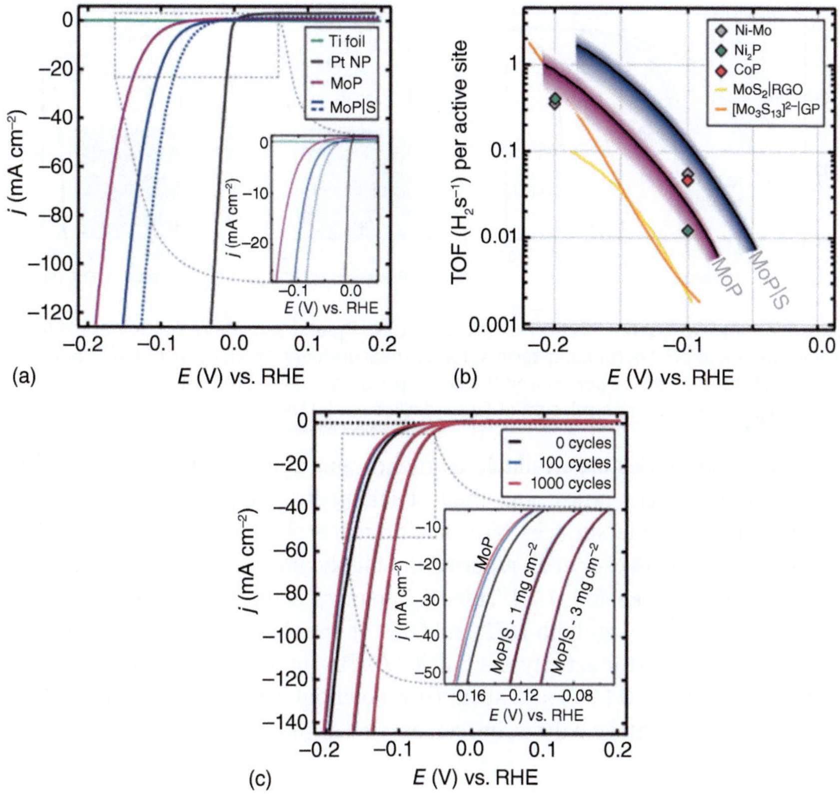

Figure 5.6 HER activity of MoP and MoP|S. Source: Reproduced with permission of Kibsgaard et al. [139], Wiley-VCH.

HER. As one of classic non-platinum group (NPG) HER catalysts with low cost, CoP shows great promise to replace platinum group catalysts in commercial-grade polymer electrolyte membrane (PEM) electrolyzer applications. In this regard, King et al. reported the demonstration of the application of CoP from lab-scale experiments to a commercial-scale 86 cm^2 PEM electrolyzer which first bridges the gap between decades of laboratory-scale, solution-based NPG catalyst development and industrial-scale electrolyzer operation [140]. In their commercial-scale electrolyzer system, the CoP catalysts exhibited high activity and stability to produce hydrogen continuously for more than 1700 hours at 1.86 A cm^{-2}, which is comparable to those of platinum catalysts but with significant material cost savings. This work highlights the great potential of CoP-based NPG catalysts for industrial PEM electrolyzer applications.

5.2.4.4 NiP

Nickel phosphide has been widely used for electrocatalytic HER due to its unique electronic properties, efficient electrocatalytic performance, and excellent corrosion resistance. However, the HER activity of nickel phosphide electrocatalysts is still very low in electrolyzers and requires further investigation. It is worth noting

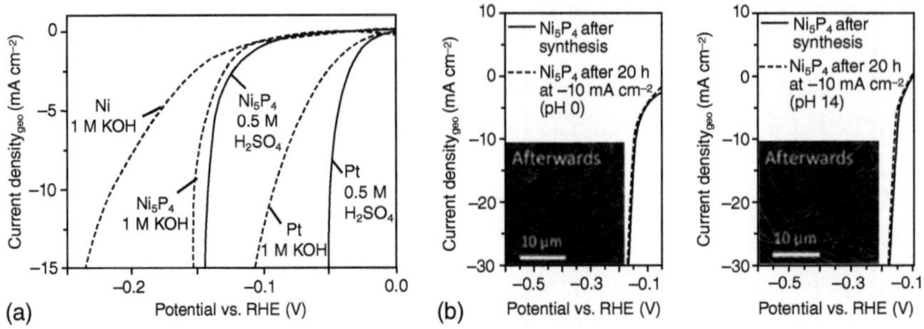

Figure 5.7 (a) Hydrogen evolution reaction of Pt, Ni_5P_4, and pristine nickel foil at pH 0 (straight line) and pH 14. (b) Long-term stability measurement of Ni_5P_4 in 1 M KOH and 0.5 M H_2SO_4 for a 20 hours period and the corresponding SEM images after testing. Source: Reproduced with permission of Ledendecker et al. [142], Wiley-VCH.

that nickel phosphides are available as mono- and poly-phosphides with different nickel/phosphorus ratios, and their stoichiometric compositions, crystalline phases, and crystalline surfaces have an influence on the electrocatalytic HER activity [141]. Ledendecker et al. reported the highly ordered Ni_5P_4 nanostructures grown directly on nickel foil [142]. This Ni_5P_4 nanostructure exhibited bifunctional electrocatalytic activity for both HER and OER. It was shown that Ni_5P_4 performed considerably better HER activity than nickel (0.21 V in alkali, unstable in acid) in acid (overpotential c. 0.14 V) and alkali (overpotential c. 0.15 V) and is one of the highest non-noble metal catalysts, but not as good as platinum foils (Figure 5.7).

5.3 Conclusion and Outlook

MCs have received a great deal of attention for HER from the research community. Here, we summarize the recent progress of emerging MC nanomaterials for HER electrocatalysis. The catalyst design principles and advanced methods for determining active sites as well as various engineering strategies have been covered with many examples. To serve as an excellent catalyst, an MC needs to be able to effectively reduce the overpotential of the reaction, increase the current density, and improve the efficiency. In addition, it is required to reduce the cost of electrocatalysis. Using various physical, chemical, or electrochemical nanotechnologies, MC catalysts with various morphologies have been developed. By the rational design of the structure, the catalysts can achieve increased specific surface area with abundant active centers exposed and enhanced electrical conductivity, consequently delivering improved HER activity.

However, so far, the application of HER technology is mostly limited to laboratory-scale studies, and actual industrial application devices still face significant technical barriers. In order to commercialize these technologies, a range of high-performance electrocatalysts with fast reaction kinetics and high yields are required. Therefore, a comprehensive comprehension of the fundamental principles, concepts, and knowledge underlying HER electrocatalysis is imperative for the rational design of advanced electrocatalysts. This encompasses not only the advancement of techniques for material preparation and characterization but also the promotion of interdisciplinary thinking and further exploration of electrocatalytic technologies in conjunction with diverse disciplines. Additionally, it necessitates the investigation into highly efficient, stable, safe, cost-effective, and environmentally friendly catalysts suitable for large-scale hydrogen production in industrial settings. Consequently, greater emphasis should be placed on harnessing the potential application of these MC-based electrocatalysts in HER.

Acknowledgments

This research received financial support from the National Natural Science Foundation of China (Project No. 22275088, 52101260), the Project of Shuangchuang Scholar of Jiangsu Province (Project No. JSSCBS20210212), the Fundamental Research Funds for the Central Universities (Project No. 30921011203), and the Start-Up Grant (Project No. AE89991/340) from Nanjing University of Science and Technology.

References

1 Jiao, Y., Zheng, Y., Jaroniec, M., and Qiao, S.Z. (2015). Design of electrocatalysts for oxygen- and hydrogen-involving energy conversion reactions. *Chem. Soc. Rev.* 44 (8): 2060.

2 Seh, Z.W., Kibsgaard, J., Dickens, C.F. et al. (2017). Combining theory and experiment in electrocatalysis: Insights into materials design. *Science* 355 (6321): eaad4998.

3 Jin, H., Guo, C., Liu, X. et al. (2018). Emerging two-dimensional nanomaterials for electrocatalysis. *Chem. Rev.* 118 (13): 6337.

4 Avani, A.V. and Anila, E.I. (2022). Recent advances of MoO_3 based materials in energy catalysis: applications in hydrogen evolution and oxygen evolution reactions. *Int. J. Hydrogen Energy* 47 (47): 20475.

5 Yu, Y., Liu, R., Sun, Y. et al. (2023). Sub-nanometric materials for hydrogen evolution reaction. *Mater. Chem. Front.* 8: 159.

6 Buathet, S., Simalaotao, K., Reunchan, P. et al. (2020). Electrochemical performance of Bi_2Te_3 heterostructure thin film and Cu_7Te_4 nanocrystals on undoped and In^{3+}-doped WO_3 films for energy storage applications. *Electrochim. Acta* 341: 136049.

7 Brandon, N.P. and Kurban, Z. (2017). Clean energy and the hydrogen economy. *Philos. Trans. R. Soc. Lond. A* 375 (2098): 20160400.

8 Hao, H., Mu, Z., Liu, Z., and Zhao, F. (2018). Abating transport GHG emissions by hydrogen fuel cell vehicles: chances for the developing world. *Front. Energy* 12 (3): 466.

9 Ambrose, A.F., Rasiah, R., Al-Amin, A.Q., and Chen, Z. (2018). Introducing hydrogen fuel cell vehicles in Malaysia. *J. Asia Pac. Econ.* 23 (2): 263.

10 Boukhvalov, D.W., Cheng, J., D'Olimpio, G. et al. (2021). Unveiling the mechanisms ruling the efficient hydrogen evolution reaction with mitrofanovite Pt_3Te_4. *J. Phys. Chem. Lett.* 12 (35): 8627.

11 Wu, H., Huang, Q., Shi, Y. et al. (2023). Electrocatalytic water splitting: mechanism and electrocatalyst design. *Nano Res.* 16 (7): 9142.

12 Zou, X. and Zhang, Y. (2015). Noble metal-free hydrogen evolution catalysts for water splitting. *Chem. Soc. Rev.* 44 (15): 5148.

13 Ming, M., Zhang, Y., He, C. et al. (2019). Room-temperature sustainable synthesis of selected platinum group metal (PGM = Ir, Rh, and Ru) nanocatalysts well-dispersed on porous carbon for efficient hydrogen evolution and oxidation. *Small* 15 (49): 1903057.

14 Krstajić Pajić, M.N., Dobrota, A.S., Mazare, A. et al. (2023). Activation of osmium by the surface effects of hydrogenated TiO_2 nanotube arrays for enhanced hydrogen evolution reaction performance. *ACS Appl. Mater. Interfaces* 15 (26): 31459.

15 Chen, Z., Duan, X., Wei, W. et al. (2019). Recent advances in transition metal-based electrocatalysts for alkaline hydrogen evolution. *J. Mater. Chem. A* 7 (25): 14971.

16 Peng, X., Pi, C., Zhang, X. et al. (2019). Recent progress of transition metal nitrides for efficient electrocatalytic water splitting. *Sustain. Energ. Fuels* 3 (2): 366.

17 Kumaravel, S., Karthick, K., Sankar, S.S. et al. (2021). Current progressions in transition metal based hydroxides as bi-functional catalysts towards electrocatalytic total water splitting. *Sustain. Energy Fuels* 5 (24): 6215.

18 Xiong, L., Qiu, Y., Peng, X. et al. (2022). Electronic structural engineering of transition metal-based electrocatalysts for the hydrogen evolution reaction. *Nano Energy* 104: 107882.

19 Yu, P., Wang, F., Shifa, T.A. et al. (2019). Earth abundant materials beyond transition metal dichalcogenides: a focus on electrocatalyzing hydrogen evolution reaction. *Nano Energy* 58: 244.

20 Solanki, R., Patra, I., Ahmad, N. et al. (2022). Investigation of recent progress in metal-based materials as catalysts toward electrochemical water splitting. *J. Environ. Chem. Eng.* 10 (4): 108207.

21 Venkateshwaran, S., Ajith, A., Duraisamy, V. et al. (2023). Tailoring of 1T phase-domain MoS_2 active sites with bridging $S2^{2-}$/apical S^{2-} phase-selective precursor modulation for enriched HER kinetics. *Inorg. Chem.* 62 (2): 841.

22 Kim, Y., Tiwari, A.P., Prakash, O., and Lee, H. (2017). Activation of ternary transition metal chalcogenide basal planes through chemical strain for the hydrogen evolution reaction. *ChemPlusChem* 82 (5): 785.

23 Siegmund, D., Blanc, N., Smialkowski, M. et al. (2020). Metal-rich chalcogenides for electrocatalytic hydrogen evolution: activity of electrodes and bulk materials. *ChemElectroChem* 7 (7): 1514.

24 Vega-Granados, K., Gochi-Ponce, Y., and Alonso-Vante, N. (2022). Electrochemical interfaces on chalcogenides: some structural perspectives and synergistic effects of single-surface active sites. *Curr. Opin. Electrochem.* 33: 100955.

25 Chia, X. and Pumera, M. (2018). Layered transition metal dichalcogenide electrochemistry: journey across the periodic table. *Chem. Soc. Rev.* 47 (15): 5602.

26 Zhang, X., Lai, Z., Ma, Q., and Zhang, H. (2018). Novel structured transition metal dichalcogenide nanosheets. *Chem. Soc. Rev.* 47 (9): 3301.

27 Zhou, N., Yang, R., and Zhai, T. (2019). Two-dimensional non-layered materials. *Mater. Today Nano* 8: 100051.

28 Chhowalla, M., Shin, H.S., Eda, G. et al. (2013). The chemistry of two-dimensional layered transition metal dichalcogenide nanosheets. *Nat. Chem.* 5 (4): 263.

29 Wilson, J.A. and Yoffe, A.D. (1969). The transition metal dichalcogenides discussion and interpretation of the observed optical, electrical and structural properties. *Adv. Phys.* 18 (73): 193.

30 Sun, Y., Terrones, M., and Schaak, R.E. (2021). Colloidal nanostructures of transition-metal dichalcogenides. *Acc. Chem. Res.* 54 (6): 1517.

31 Rani, S., Sharma, M., Verma, D. et al. (2022). Two-dimensional transition metal dichalcogenides and their heterostructures: role of process parameters in top-down and bottom-up synthesis approaches. *Mater. Sci. Semicond. Process.* 139: 106313.

32 Wong, S.L., Liu, H., and Chi, D. (2016). Recent progress in chemical vapor deposition growth of two-dimensional transition metal dichalcogenides. *Prog. Cryst. Growth Charact. Mater.* 62 (3): 9.

33 Shinde, N.B., Francis, B., Ramachandra Rao, M.S. et al. (2019). Rapid wafer-scale fabrication with layer-by-layer thickness control of atomically thin MoS_2 films using gas-phase chemical vapor deposition. *APL Mater.* 7 (8): 081113.

34 Lei, J., Xie, Y., and Yakobson, B.I. (2021). Gas-phase "prehistory" and molecular precursors in monolayer metal dichalcogenides synthesis: the case of MoS_2. *ACS Nano* 15 (6): 10525.

35 Zeng, H., Wen, Y., Yin, L. et al. (2023). Recent developments in CVD growth and applications of 2D transition metal dichalcogenides. *Front. Phys.* 18 (5): 53603.

36 Gholamvand, Z., McAteer, D., Backes, C. et al. (2016). Comparison of liquid exfoliated transition metal dichalcogenides reveals $MoSe_2$ to be the most effective hydrogen evolution catalyst. *Nanoscale* 8 (10): 5737.

37 Lee, H., Han, H., Park, C. et al. (2021). Halide perovskite nanocrystal-enabled stabilization of transition metal dichalcogenide nanosheets. *Small* 18 (6): 2106035.

38 Chen, I.W.P., Li, Y.-H., Su, Y.-L., and Huang, G.-X. (2022). Layer-by-layer exfoliation of transition-metal dichalcogenides by amino acid in water for promoting hydrogen evolution reaction. *J. Phys. Chem. C* 126 (14): 6207.

39 Hinnemann, B., Moses, P.G., Bonde, J. et al. (2005). Biornimetic hydrogen evolution: MoS_2 nanoparticles as catalyst for hydrogen evolution. *J. Am. Chem. Soc.* 127 (15): 5308.

40 Tsai, C., Chan, K., Abild-Pedersen, F., and Nørskov, J.K. (2014). Active edge sites in $MoSe_2$ and WSe2 catalysts for the hydrogen evolution reaction: a density functional study. *Phys. Chem. Chem. Phys.* 16 (26): 13156.

41 Tsai, C., Chan, K., Nørskov, J.K., and Abild-Pedersen, F. (2015). Theoretical insights into the hydrogen evolution activity of layered transition metal dichalcogenides. *Surf. Sci.* 640: 133.

42 Noh, S.H., Hwang, J., Kang, J. et al. (2018). Tuning the catalytic activity of heterogeneous two-dimensional transition metal dichalcogenides for hydrogen evolution. *J. Mater. Chem. A* 6 (41): 20005.

43 Huang, Y., Nielsen, R.J., Goddard, W.A., and Soriaga, M.P. (2015). The reaction mechanism with free energy barriers for electrochemical dihydrogen evolution on MoS_2. *J. Am. Chem. Soc.* 137 (20): 6692.

44 Jin, H., Michaud-Rioux, V., Gong, Z.-R. et al. (2019). Size dependence in two-dimensional lateral heterostructures of transition metal dichalcogenides. *J. Mater. Chem. C* 7 (13): 3837.

45 Bruix, A., Lauritsen, J.V., and Hammer, B. (2021). Size-dependent phase stability in transition metal dichalcogenide nanoparticles controlled by metal substrates. *Nanoscale* 13 (22): 10167.

46 Zou, Y.C., Chen, Z.G., Liu, S. et al. (2018). Atomic insights into phase evolution in ternary transition-metal dichalcogenides nanostructures. *Small* 14 (22): 1800780.

47 Xiao, Y., Zhou, M., Liu, J. et al. (2019). Phase engineering of two-dimensional transition metal dichalcogenides. *Sci. China Mater.* 62 (6): 759.

48 Ullah, N., Guziejewski, D., Yuan, A., and Shah, S.A. (2023). Recent advancement and structural engineering in transition metal dichalcogenides for alkali metal ions batteries. *Materials* 16 (7): 2559.

49 Lin, Z., Carvalho, B.R., Kahn, E. et al. (2016). Defect engineering of two-dimensional transition metal dichalcogenides. *2D Mater.* 3 (2): 022002.

50 Hu, Z., Wu, Z., Han, C. et al. (2018). Two-dimensional transition metal dichalcogenides: interface and defect engineering. *Chem. Soc. Rev.* 47 (9): 3100.

51 Su, H., Pan, X., Li, S. et al. (2023). Defect-engineered two-dimensional transition metal dichalcogenides towards electrocatalytic hydrogen evolution reaction. *Carbon Energy* 5 (6): e296.

52 Zhao, Y., Xu, K., Pan, F. et al. (2016). Doping, contact and interface engineering of two-dimensional layered transition metal dichalcogenides transistors. *Adv. Funct. Mater.* 27 (19): 1603484.

53 Ruqia, B., Kanti Kabiraz, M., Wook Hong, J., and Choi, S.-I. (2022). Catalyst activation: surface doping effects of group VI transition metal dichalcogenides towards hydrogen evolution reaction in acidic media. *J. Energy Chem.* 72: 217.

54 Sun, L., Gao, M., Jing, Z. et al. (2022). 1 T-phase enriched P doped WS_2 nanosphere for highly efficient electrochemical hydrogen evolution reaction. *Chem. Eng. J.* 429: 132187.

55 Gao, X., Zhou, Y., Cheng, Z. et al. (2021). Distance synergy of single Ag atoms doped MoS_2 for hydrogen evolution electrocatalysis. *Appl. Surf. Sci.* 547: 149113.

56 Gao, G., Sun, Q., and Du, A. (2016). Activating catalytic inert basal plane of molybdenum disulfide to optimize hydrogen evolution activity via defect doping and strain engineering. *J. Phys. Chem. C* 120 (30): 16761.

57 Jaramillo, T.F., Jorgensen, K.P., Bonde, J. et al. (2007). Identification of active edge sites for electrochemical H_2 evolution from MoS_2 nanocatalysts. *Science* 317 (5834): 100.

58 Jiao, Y., Hafez, A.M., Cao, D. et al. (2018). Metallic MoS_2 for high performance energy storage and energy conversion. *Small* 14 (36): 1800640.

59 Lukowski, M.A., Daniel, A.S., Meng, F. et al. (2013). Enhanced hydrogen evolution catalysis from chemically exfoliated metallic MoS_2 nanosheets. *J. Am. Chem. Soc.* 135 (28): 10274.

60 Kibsgaard, J., Chen, Z., Reinecke, B.N., and Jaramillo, T.F. (2012). Engineering the surface structure of MoS_2 to preferentially expose active edge sites for electrocatalysis. *Nat. Mater.* 11 (11): 963.

61 Tan, C., Luo, Z., Chaturvedi, A. et al. (2018). Preparation of high-percentage 1T-phase transition metal dichalcogenide nanodots for electrochemical hydrogen evolution. *Adv. Mater.* 30 (9): 1705509.

62 Zhuang, M., Xu, G.-L., Gan, L.-Y. et al. (2019). Sub-5 nm edge-rich 1T′-$ReSe_2$ as bifunctional materials for hydrogen evolution and sodium-ion storage. *Nano Energy* 58: 660.

63 Hwang, Y. and Shin, N. (2021). Colloidal synthesis of $MoSe_2/WSe_2$ heterostructure nanoflowers via two-step growth. *Materials* 14 (23): 7294.

64 Chia, X., Sutrisnoh, N.A.A., Sofer, Z. et al. (2018). Morphological effects and stabilization of the metallic 1T phase in layered V-, Nb-, and Ta-doped WSe_2 for electrocatalysis. *Chem. Eur. J.* 24 (13): 3199.

65 Chia, H.L., Mayorga-Martinez, C.C., Sofer, Z. et al. (2020). Vanadium dopants: a boon or a bane for molybdenum dichalcogenides-based electrocatalysis applications. *Adv. Funct. Mater.* 31 (8): 2009083.

66 Chen, L., Zhu, Y., Li, J. et al. (2020). Phase modulation and chemical zctivation of $MoSe_2$ by phosphorus for electrocatalytic hydrogen evolution reaction. *Energy Technol.* 8 (5): 1901503–1901501.

67 Guo, F.-b., Zhao, X.-y., Lei, H.-y. et al. (2022). Bimetallic doping-derived heterostructures in NiCo-WSe2 to promote hydrogen evolution reaction. *J. Alloys Compd.* 924: 166538.

68 Das, A., Roy, D., Kumar Das, B. et al. (2023). Zinc doping induced WS_2 accelerating the HER and ORR kinetics: a theoretical and experimental validation. *Catal. Today* 423: 113921.

69 Wang, F., Wang, Z., Shifa, T.A. et al. (2016). Two-dimensional non-layered materials: synthesis, properties and applications. *Adv. Funct. Mater.* 27 (19): 1603254.

70 Dou, Y., Zhang, L., Xu, X. et al. (2017). Atomically thin non-layered nanomaterials for energy storage and conversion. *Chem. Soc. Rev.* 46 (23): 7338.

71 Tan, J., Zhang, Z., Zeng, S. et al. (2022). Dual-metal precursors for the universal growth of non-layered 2D transition metal chalcogenides with ordered cation vacancies. *Sci. Bull.* 67 (16): 1649.

72 Tan, C., Cao, X., Wu, X.-J. et al. (2017). Recent advances in ultrathin two-dimensional nanomaterials. *Chem. Rev.* 117 (9): 6225.

73 Long, X., Li, G., Wang, Z. et al. (2015). Metallic iron–nickel sulfide ultrathin nanosheets as a highly active electrocatalyst for hydrogen evolution reaction in acidic media. *J. Am. Chem. Soc.* 137 (37): 11900.

74 Liu, Y., Hua, X., Xiao, C. et al. (2016). Heterogeneous spin states in ultrathin nanosheets induce subtle lattice distortion to trigger efficient hydrogen evolution. *J. Am. Chem. Soc.* 138 (15): 5087.

75 Yu, X.Y., Yu, L., Wu, H.B., and Lou, X.W. (2015). Formation of nickel sulfide nanoframes from metal–organic frameworks with enhanced pseudocapacitive and electrocatalytic properties. *Angew. Chem. Int. Edit.* 54 (18): 5331.

76 Yang, L., Wu, X., Zhu, X. et al. (2015). Amorphous nickel/cobalt tungsten sulfide electrocatalysts for high-efficiency hydrogen evolution reaction. *Appl. Surf. Sci.* 341: 149.

77 Qin, Z., Chen, Y., Huang, Z. et al. (2016). Composition-dependent catalytic activities of noble-metal-free NiS/Ni_3S_4 for hydrogen evolution reaction. *J. Phys. Chem. C* 120 (27): 14581.

78 Tetzlaff, D., Pellumbi, K., Baier, D.M. et al. (2020). Sustainable and rapid preparation of nanosized Fe/Ni-pentlandite particles by mechanochemistry. *Chem. Sci.* 11 (47): 12835.

79 Smialkowski, M., Siegmund, D., Stier, K. et al. (2022). Trimetallic pentlandites $(Fe,Co,Ni)_9S_8$ for the electrocatalytical HER in acidic media. *ACS Mater. Au* 2 (4): 474.

80 Feng, H., Xu, Z., Ren, L. et al. (2018). Activating titania for efficient electrocatalysis by vacancy engineering. *ACS Catal.* 8 (5): 4288.

81 Li, S., Yang, N., Liao, L. et al. (2018). Doping β-$CoMoO_4$ nanoplates with phosphorus for efficient hydrogen evolution reaction in alkaline media. *ACS Appl. Mater. Interfaces* 10 (43): 37038.

82 Liu, Q., Wang, E., and Sun, G. (2020). Layered transition-metal hydroxides for alkaline hydrogen evolution reaction. *Chin. J. Catal.* 41 (4): 574.

83 Yin, H. and Tang, Z. (2016). Ultrathin two-dimensional layered metal hydroxides: an emerging platform for advanced catalysis, energy conversion and storage. *Chem. Soc. Rev.* 45 (18): 4873.

84 Spencer, M.A., Fortunato, J., and Augustyn, V. (2022). Electrochemical proton insertion modulates the hydrogen evolution reaction on tungsten oxides. *J. Chem. Phys.* 156 (6): 064704.

85 Wang, L., Lin, C., Huang, D. et al. (2015). Optimizing the volmer step by single-layer nickel hydroxide nanosheets in hydrogen evolution reaction of platinum. *ACS Catal.* 5 (6): 3801.

86 Wu, Y., Sun, R., and Cen, J. (2020). Facile synthesis of cobalt oxide as an efficient electrocatalyst for hydrogen evolution reaction. *Front. Chem.* 8: 386.

87 Zhu, L., Li, Z., Yang, M. et al. (2023). An effective approach to enhance hydrogen evolution reaction and hydrogen oxidation reaction by Ni doping to MoO_3. *Small.*

88 Balendhran, S., Walia, S., Nili, H. et al. (2013). Two-dimensional molybdenum trioxide and dichalcogenides. *Adv. Funct. Mater.* 23 (32): 3952.

89 Kalantar-zadeh, K., Ou, J.Z., Daeneke, T. et al. (2016). Two dimensional and layered transition metal oxides. *Appl. Mater. Today* 5: 73.

90 Shang, T.-T., Liu, X.-Y., and Gu, L. (2016). Interface of transition metal oxides at the atomic scale. *Sci. China Phys. Mech. Astron.* 59 (9): 697001.

91 Xiong, W., Yin, H., Wu, T., and Li, H. (2023). Challenges and opportunities of transition metal oxides as electrocatalysts. *Chem. Eur. J.* 29 (5): e202202872.

92 Eftekhari, A. (2017). Electrocatalysts for hydrogen evolution reaction. *Int. J. Hydrogen Energy* 42 (16): 11053.

93 Yuan, J., Liu, Y., Bo, T., and Zhou, W. (2020). Activated HER performance of defected single layered TiO_2 nanosheet via transition metal doping. *Int. J. Hydrogen Energy* 45 (4): 2681.

94 Jin, Y., Wang, H., Li, J. et al. (2016). Porous MoO_2 nanosheets as non-noble bifunctional electrocatalysts for overall water splitting. *Adv. Mater.* 28 (19): 3785.

95 Datta, R.S., Haque, F., Mohiuddin, M. et al. (2017). Highly active two dimensional α-MoO_{3-x} for the electrocatalytic hydrogen evolution reaction. *J. Mater. Chem. A* 5 (46): 24223.

96 Haque, F., Zavabeti, A., Zhang, B.Y. et al. (2019). Ordered intracrystalline pores in planar molybdenum oxide for enhanced alkaline hydrogen evolution. *J. Mater. Chem. A* 7 (1): 257.

97 Zhang, X., Jin, G., Wang, D. et al. (2021). Crystallographic phase and morphology dependent hydrothermal synthesis of tungsten oxide for robust hydrogen evolution reaction. *J. Alloys Compd.* 875: 160054.

98 Xu, W., Ni, X.M., Zhang, L.J. et al. (2022). Tuning the electronic structure of tungsten oxide for enhanced hydrogen evolution reaction in alkaline electrolyte. *ChemElectroChem* 9 (5): e202101300.

99 Jin, Y. and Shen, P.K. (2015). Nanoflower-like metallic conductive MoO_2 as a high-performance non-precious metal electrocatalyst for the hydrogen evolution reaction. *J. Mater. Chem. A* 3 (40): 20080.

100 Pandey, D.K., Kagdada, H.L., Materny, A., and Singh, D.K. (2021). Hybrid structure of ionic liquid and TiO_2 nanoclusters for efficient hydrogen evolution reaction. *J. Phys. Chem. A* 125 (12): 2653.

101 Sun, J., Zhang, X., Jin, M. et al. (2020). Robust enhanced hydrogen production at acidic conditions over molybdenum oxides-stabilized ultrafine palladium electrocatalysts. *Nano Res.* 14 (1): 268.

102 Liu, L., Zhao, Y., Wang, Y. et al. (2021). Single-atom Co doped in ultrathin WO_3 arrays for the enhanced hydrogen evolution reaction in a wide pH range. *ACS Appl. Mater. Interfaces* 13 (45): 53915.

103 Anantharaj, S., Noda, S., Jothi, V.R. et al. (2021). Strategies and perspectives to catch the missing pieces in energy-efficient hydrogen evolution reaction in alkaline media. *Angew. Chem. Int. Edit.* 60 (35): 18981.

104 Bhavanari, M., Lee, K.-R., Tseng, C.-J. et al. (2022). New insights into interface charge-transfer mechanism of copper-iron layered double hydroxide cathodic electrocatalyst in alkaline electrolysis. *J. Environ. Chem. Eng.* 10 (2): 107287.

105 Sahoo, D.P., Das, K.K., Mansingh, S. et al. (2022). Recent progress in first row transition metal layered double hydroxide (LDH) based electrocatalysts towards water splitting: a review with insights on synthesis. *Coord. Chem. Rev.* 469: 214666.

106 Yuan, H.Y., Liu, P.F., and Yang, H.G. (2022). Peculiar double-layered transition metal hydroxide nanosheets. *Matter* 5 (4): 1063.

107 Chen, T., Zhang, R., Chen, G. et al. (2019). Plasma-doping-enhanced overall water splitting: case study of NiCo hydroxide electrocatalyst. *Catal. Today* 337: 147.

108 Long, X., Wang, Z., Xiao, S. et al. (2016). Transition metal based layered double hydroxides tailored for energy conversion and storage. *Mater. Today* 19 (4): 213.

109 Danilovic, N., Subbaraman, R., Strmcnik, D. et al. (2012). Enhancing the alkaline hydrogen evolution reaction activity through the bifunctionality of $Ni(OH)_2$/metal catalysts. *Angew. Chem. Int. Edit.* 51 (50): 12495.

110 Han, X., Li, Y., Wang, X. et al. (2024). Ru anchored on $Co(OH)_2$ nanowire arrays as highly effective electrocatalyst for full water splitting. *Int. J. Hydrogen Energy* 51: 769.

111 Guo, J., Wang, K., Zhang, H., and Zhang, H. (2022). Enhanced electrocatalytic activity of Mo-doped NiFe layered double hydroxide nanosheet arrays for the hydrogen evolution reaction. *ACS Appl. Nano Mater.* 6 (1): 379.

112 Kimura, Y., Inoue, T., Yin, F., and Tsuzaki, K. (2008). Inverse temperature dependence of toughness in an ultrafine grain-structure steel. *Science* 320 (5879): 1057.

113 Chen, W.-F., Muckerman, J.T., and Fujita, E. (2013). Recent developments in transition metal carbides and nitrides as hydrogen evolution electrocatalysts. *Chem. Commun.* 49 (79): 8896.

114 Xu, S., Chu, S., Yang, L. et al. (2018). Tungsten nitride/carbide nanocomposite encapsulated in nitrogen-doped carbon shell as an effective and durable catalyst for hydrogen evolution reaction. *New J. Chem.* 42 (24): 19557.

115 Guler, U., Boltasseva, A., and Shalaev, V.M. (2014). Refractory plasmonics. *Science* 344 (6181): 263.

116 Liu, T., Hu, J., Luo, Y., and Li, X. (2022). Hot-carrier physics of transition-metal carbides and nitrides: insight from electronic structure. *Phys. Rev. Appl.* 18 (1): 014058.

117 Alhajri, N.S., Anjum, D.H., Hedhili, M.N., and Takanabe, K. (2016). Generation and characteristics of IV-VI transition metal nitride and carbide nanoparticles using a reactive mesoporous carbon nitride. *ChemistrySelect* 1 (2): 290.

118 Levy, R.B. and Boudart, M. (1973). Platinum-like behavior of tungsten carbide in surface catalysis. *Science* 181 (4099): 547.

119 Meng, T. and Cao, M. (2018). Transition metal carbide complex architectures for energy-related applications. *Chem. Eur. J.* 24 (63): 16716.

120 Theerthagiri, J., Lee, S.J., Murthy, A.P. et al. (2020). Fundamental aspects and recent advances in transition metal nitrides as electrocatalysts for hydrogen evolution reaction: a review. *Curr. Opin. Solid State Mater. Sci.* 24 (1): 100805.

121 Oyama, S. (1992). Preparation and catalytic properties of transition metal carbides and nitrides. *Catal. Today* 15 (2): 179.

122 Chen, J.G. (1996). Carbide and nitride overlayers on early transition metal surfaces: preparation, characterization, and reactivities. *Chem. Rev.* 96 (4): 1477.

123 Upadhyay, S. and Pandey, O.P. (2020). One-pot synthesis of pure phase molybdenum carbide (Mo2C and MoC) nanoparticles for hydrogen evolution reaction. *Int. J. Hydrogen Energy* 45 (51): 27114.

124 Kawashima, K., Shin, K., Wygant, B.R. et al. (2020). Cobalt metal–cobalt carbide composite microspheres for water reduction electrocatalysis. *ACS Appl. Energy Mater.* 3 (4): 3909.

125 Naguib, M., Mochalin, V.N., Barsoum, M.W., and Gogotsi, Y. (2013). 25th anniversary article: MXenes: a new family of two-dimensional materials. *Adv. Mater.* 26 (7): 992.

126 Kang, Z., Khan, M.A., Gong, Y. et al. (2021). Recent progress of MXenes and MXene-based nanomaterials for the electrocatalytic hydrogen evolution reaction. *J. Mater. Chem. A* 9 (10): 6089.

127 Bai, S., Yang, M., Jiang, J. et al. (2021). Recent advances of MXenes as electrocatalysts for hydrogen evolution reaction. *npj 2D Mater. Appl.* 5 (1): 78.

128 Yuan, Y., Li, H., Wang, L. et al. (2019). Achieving highly efficient catalysts for hydrogen evolution reaction by electronic state modification of platinum on versatile $Ti_3C_2T_x$ (MXene). *ACS Sustain. Chem. Eng.* 7 (4): 4266.

129 Yoon, Y., Tiwari, A.P., Choi, M. et al. (2019). Precious-metal-free electrocatalysts for activation of hydrogen evolution with nonmetallic electron donor: chemical composition controllable phosphorous doped vanadium carbide MXene. *Adv. Funct. Mater.* 29 (30): 1903443.

130 Wang, L., Hao, Y., Deng, L. et al. (2022). Rapid complete reconfiguration induced actual active species for industrial hydrogen evolution reaction. *Nat. Commun.* 13 (1): 5785.

131 Shao, Y., Shi, X., and Pan, H. (2017). Electronic, magnetic, and catalytic properties of thermodynamically stable two-dimensional transition-metal phosphides. *Chem. Mater.* 29 (20): 8892.

132 Yoon, S.J., Lee, S.J., Kim, M.H. et al. (2023). Recent tendency on transition-metal phosphide electrocatalysts for the hydrogen evolution reaction in alkaline media. *Nanomaterials* 13 (18): 2613.

133 Popczun, E.J., McKone, J.R., Read, C.G. et al. (2013). Nanostructured nickel phosphide as an electrocatalyst for the hydrogen evolution reaction. *J. Am. Chem. Soc.* 135 (25): 9267.

134 Reddy, D.A., Kim, H.K., Kim, Y. et al. (2016). Multicomponent transition metal phosphides derived from layered double hydroxide double-shelled nanocages as an efficient non-precious co-catalyst for hydrogen production. *J. Mater. Chem. A* 4 (36): 13890.

135 Hou, M., Teng, X., Wang, J. et al. (2018). Multiscale porous molybdenum phosphide of honeycomb structure for highly efficient hydrogen evolution. *Nanoscale* 10 (30): 14594.

136 Parra-Puerto, A., Ng, K.L., Fahy, K. et al. (2019). Supported transition metal phosphides: activity survey for HER, ORR, OER, and corrosion resistance in acid and alkaline electrolytes. *ACS Catal.* 9 (12): 11515.

137 Su, J., Zhou, J., Wang, L. et al. (2017). Synthesis and application of transition metal phosphides as electrocatalyst for water splitting. *Sci. Bull.* 62 (9): 633.

138 Wang, Y., Ma, B., and Chen, Y. (2019). Iron phosphides supported on three-dimensional iron foam as an efficient electrocatalyst for water splitting reactions. *J. Mater. Sci.* 54 (24): 14872.

139 Kibsgaard, J. and Jaramillo, T.F. (2014). Molybdenum phosphosulfide: an active, acid-stable, earth-abundant catalyst for the hydrogen evolution reaction. *Angew. Chem. Int. Edit.* 53 (52): 14433.

140 King, L.A., Hubert, M.A., Capuano, C. et al. (2019). A non-precious metal hydrogen catalyst in a commercial polymer electrolyte membrane electrolyser. *Nat. Nanotechnol.* 14 (11): 1071.

141 Hu, C., Lv, C., Liu, S. et al. (2020). Nickel phosphide electrocatalysts for hydrogen evolution reaction. *Catalysis* 10 (2): 188.

142 Ledendecker, M., Krick Calderón, S., Papp, C. et al. (2015). The synthesis of nanostructured Ni_5P_4 films and their use as a non-noble bifunctional electrocatalyst for full water splitting. *Angew. Chem. Int. Edit.* 54 (42): 12361.

6

Carbon-Based Materials for Electrocatalytic Hydrogen Evolution Reaction

Zhangyou Wang[1,], Sundus Umer[2,*], Lu Hao[2,*], Fangge Cheng[2], Jiawen Li[2], and Jia Liu[2]*

[1] *Tianjin University, School of Chemical Engineering and Technology, No. 92 Weijin Road, Nankai District, Tianjin, 300072, China*
[2] *Tianjin University, School of Science, Tianjin Key Laboratory of Molecular Optoelectronic Sciences, Department of Chemistry, No. 92 Weijin Road, Nankai District, Tianjin, 300072, China*

6.1 Introduction

Carbon, as the most important component on Earth, can form diverse structures. Exploring the properties and potential applications of carbon-based materials is a timeless topic. For chemical processes, carbon-based materials often play an important role in the reaction process, catalyzing the occurrence and controlling the reaction pathway. Especially in the catalytic conversion of energy materials, the research and development of carbon-based catalysts have gained attention. It is widely known that carbon-based fossil fuels are of great importance to daily life. Nevertheless, fossil fuels are a nonrenewable energy source with limited reserves. Therefore, it is urgent to exploit eco-friendly and cost-effective renewable energy sources. Hydrogen is a clean energy with vast sources and high combustion heat, and the combustion product is only water. Electrolysis of water to produce hydrogen is a high-quality energy substitution strategy. In theory, the required voltage for electrolysis of water is only 1.23 V, but the reaction kinetics is sluggish [1], thus limiting the application of electrocatalytic hydrogen production [2, 3]. Therefore, highly efficient catalysts are required to reduce the energy consumption in hydrogen evolution reaction (HER) process.

To promote HER, metals and metal oxides have been discovered, especially noble metals, which display better properties than transition metals, such as Pt, RuO_2, and IrO_2. However, their applications are limited owing to their cost, scarcity, and poor stability [4, 5]. Therefore, carbon-based materials have become the preferred choice for HER catalysts due to their affordability and stability [6]. Additionally, carbon-based materials exhibit less heavy metal pollution and are environmentally friendly to our environment [7]. As for how to further improve the catalytic activity

* These authors contributed equally to this work.

Functional Materials for Electrocatalytic Energy Conversion, First Edition.
Edited by Zhicheng Zhang, Meiting Zhao, and Yuchen Qin.

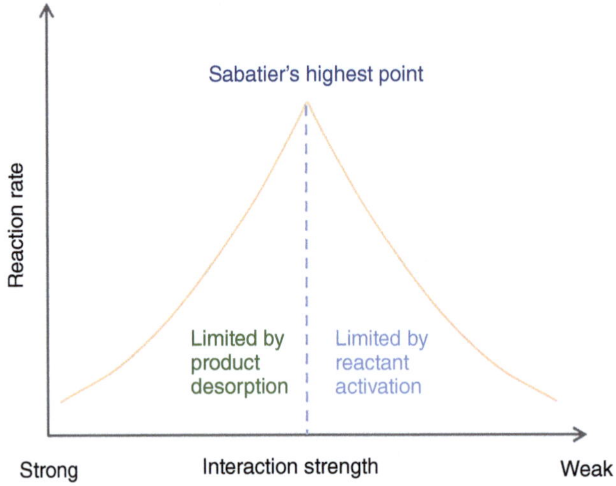

Figure 6.1 "Volcanic" diagram of Sabatier's principle. Source: Adapted from Li et al. [12].

of carbon-based materials, researchers developed a series of strategies to increase the active surface area and enhance the intrinsic catalytic activity [8]. In order to enhance the surface area, porous carbon, zero-dimensional (0D) [9–11], and two-dimensional (2D) carbon material were explored. Besides carbon materials, metal–organic framework (MOF) is another effective catalyst for HER. MOFs are formed by connecting metal ions with organic ligands through coordination bonds and have adjustable composition and structure. Due to good dispersion and stable structure, an efficient carbon-based catalyst can be obtained for highly efficient HER. Therefore, in this chapter, we will focus on discussing the applications of porous carbon, low-dimensional carbon materials, MOF materials, and atomic-level metal carbon-based materials in the field of HER.

Besides increasing the material's active surface area, enhancing the intrinsic catalytic activity is necessary to explore the reaction mechanism of the material. HER reaction kinetics is sluggish, so it is necessary to involve catalysts that may improve the reaction process, and what is the principle of catalyst selection? In chemical reactions, some metal-based catalysts, follow Sabatier's principle (Figure 6.1). This principle states that the interaction intensity between the catalyst with the best performance and the reaction intermediate should be moderate; that is, the catalyst can promote the activation of reactants on the one hand and facilitate the desorption of products on the other hand. If the binding force is too strong, it is not conducive to desorption. On the contrary, the weak binding force is not conducive to the activation of reactants. In short, the occurrence of both cases will slow down the reaction rate. The following "volcanic map" can directly reflect Sabatier's principle [12]. The interaction intensity at the highest point of the "volcanic" pattern curve is the largest, and this maximum value is the limit value that the reaction rate can reach. Therefore, understanding and applying Sabatier's principle is of great significance.

Since hydrogen cannot be directly generated from hydronium ions and electrons, the reaction must be carried out under catalytic conditions. Sabatier's principle well explains why there are significant differences in activity between catalysts and why Pt-based catalysts have the best catalytic activity [13]. However, "volcanic maps" cannot explain the different properties of protons at different pH conditions. For example, in an acidic environment, the proton donor is a hydronium ion, whereas, in an alkaline environment, the proton donor is a water molecule. The mechanism in acidic and alkaline environments is different for HER because of the different proton donors.

HER reaction equations are shown below in different pH conditions:

$$\text{Alkaline medium: } 2H_2O + 2e^- \rightarrow H_2 + 2OH^-$$

$$\text{Acidic medium: } H_3O^+ + 2e^- \rightarrow H_2 + 2H_2O$$

Sabatier's principle cannot explain why the reactivity is higher under acidic conditions and lower under alkaline conditions. Overall, the water decomposition reaction needs a voltage of 1.23 V to produce hydrogen at the cathode. Moreover, HER reaction is a two-electron reversible reaction, and the process of hydrogen generation is not carried out in one step. This phenomenon is elaborated in the next section.

6.2 The Fundamentals of HER

6.2.1 Mechanistic of HER

The HER is a simple half-cell reaction in which water is reduced at cathode to generate H_2, as it entails a basic proton–electron transfer reaction (Figure 6.2) [14]. Alkaline electrolysis, proton exchange membrane electrolysis/diaphragm (PEM), and solid oxide water electrolysis in acidic media are the main methods of electrolysis. HER measurement can be characterized by three electrode systems: reference electrode, counter, and working electrode. The factors to consider during production are electrolytic medium, rate determination steps (RDS), and reaction kinetics [15].

The initiation of this process is facilitated by the abundance of hydrogen ions (protons) in the acid. First, protons couple with electrons and attach to the catalyst surface, known as the Volmer reaction (step 1). The adsorbed hydrogen H* binds with the hydrated proton and then connects with electrons to produce H_2, which is called the Heyrovsky step (2). The generation of H_2 is directly achieved by binding two adsorbed hydrogen ions [8], which is called the Tafel step (3).

$$\text{Volmer step: } H^+ + e^- \rightarrow H^* \qquad \text{(Electrochemical adsorption)}$$

$$\text{Heyrovsky step: } H^* + e^- + H^+ \rightarrow H_2 \qquad \text{(Electrochemical desorption)}$$

$$\text{Tafel step: } H^* + H^* \rightarrow H_2 \qquad \text{(Chemical desorption)}$$

In alkaline media, the reduction of water generates more hydroxide ions, and the Volmer step is accompanied by the formation of H* and OH^-. In acids, the Tafel step

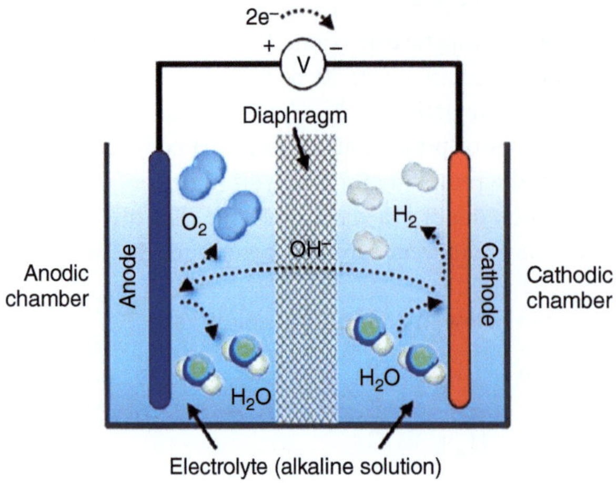

Figure 6.2 Schematic illustration of electrolyzer for HER. Source: Rodríguez, et al., [14]/MDPI/CC BY 4.0.

exhibits similarity, but in the Heyrovsky stage, the generation of OH ions and H_2 is simultaneous.

$$\text{Volmer step}: H_2O + e^- \rightarrow H^* + OH^- \qquad \text{(Electrochemical adsorption)}$$

$$\text{Heyrovsky step}: H_2O + e^- + H^* \rightarrow H_2 + OH^- \qquad \text{(Electrochemical desorption)}$$

$$\text{Tafel step}: H^* + H^* \rightarrow H_2 \qquad \text{(Chemical desorption)}$$

6.2.2 Kinetics and Rate-Determining Steps in HER

The HER mechanism adsorbs/desorbs intermediate hydrogen through the Volmer–Heyrovsky or Volmer–Tafel steps [16] (Figure 6.3):

1. ***Volmer–Tafel.*** Initially, the discharge of protons occurs (Volmer step), resulting in an increase in hydrogen coverage on the catalyst surface. The probability of two adsorbed H atoms forming molecules increases with the hydrogen coverage. Subsequently, the discharge rate equals the recombination rate of hydrogen and the formation rate of H_2.
2. ***Volmer–Heyrovsky.*** The initial proton release is similar in both mechanisms. In this pathway, as the coverage of the adsorbed H increases, so does the probability that a proton will collide with the adsorbed H itself (assuming that the Volmer–Tafel step is negligible). The discharge reaction equilibrates with the electrochemical desorption and finally reaches a steady state.

Both mechanisms mentioned above lead to three possible RDSs, namely Volmer, Heyrovsky, and Tafel [17]. The onset potential, overpotential, Tafel slope, electrochemical impedance, stability, turnover frequency (TOF), electrochemical active

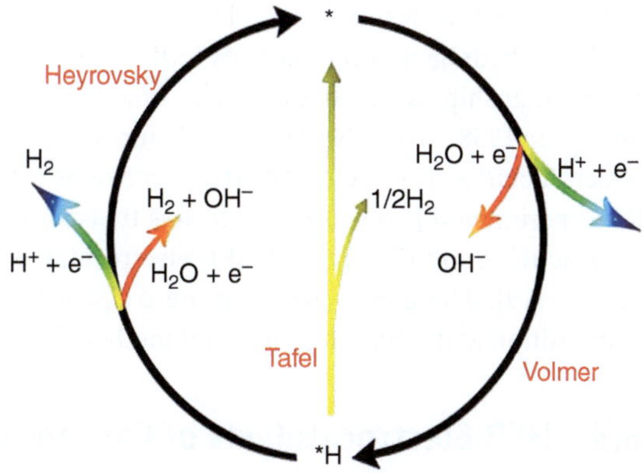

Figure 6.3 Schematic illustration of HER mechanisms on the surface of an electrocatalyst under acidic and alkaline conditions. Blue: alkaline, Red: acidic. Source: Reproduced with permission of Wang et al. [16], Elsevier.

surface area, and Faraday efficiency are parameters used for analyzing electrochemical processes [18]. Overpotential and Tafel slope are the most significant parameters for evaluating and estimating HER reaction activity [8]. In all cases, the catalytic current rises exponentially with η. However, the rate varies among different RDSs. The Tafel equation is an important analytical technique that helps describe RDSs and mechanisms:

$$\eta = b \log(j) + a$$

where "a" represents the Tafel constant, and "b" represents the Tafel slope. The mechanism analysis of HER on the catalyst surface can be carried out using the above equation. When the Volmer step becomes rate-determining, a Tafel slope of $120\,\text{mV dec}^{-1}$ arises, while for Heyrovsky or Tafel RDS, 40 and $30\,\text{mV dec}^{-1}$ are observed, respectively. The Tafel slope is independent of the surface area of the catalyst (i.e. the magnitude of the catalytic current) and presents specific values based on the RDS, as described above. A lower Tafel slope is always preferred in electrocatalysts, as higher catalytic currents can be achieved at relatively low overpotentials [17].

In the experiment, the Tafel slope is used as an indicator to evaluate reaction kinetics and reveal possible reaction mechanisms. The cyclic transformation from * to H* can be explained by thermodynamics in theory. DFT can calculate the Gibbs free energy (ΔG_{H*}) of adsorption/desorption on the catalyst crystal surface, where ΔE_{H*} is the adsorption energy, ΔS_H is the entropy change during the adsorption/desorption process, ΔE_{ZPE} is the difference in zero-point energy between the adsorbed H and the desorbed H atoms, and T is the temperature.

$$\Delta G_{H*} = \Delta E_{H*} + \Delta E_{ZPE} + T\Delta S_{H*}$$

It is evident that the absorption and desorption processes of hydrogen are key steps based on the two reaction mechanisms of hydrogen evolution. Therefore, the hydrogen bond strength on the catalyst surface can predict and determine HER activity [19]. Specifically, the weak binding energy between active sites and hydrogen atoms is favorable for hydrogen desorption, but hydrogen absorption is not effective. Alternatively, strong binding energy can effectively absorb hydrogen but reduces the desorption of H_2. Therefore, an ideal electrocatalyst possesses moderate hydrogen bond energies [20]. In addition, Parson et al. first pointed out the relationship between the Gibbs free energy (ΔG_{H*}) of hydrogen adsorption and j_0 in 1958, while Norskov et al. further plotted the famous volcanic curve, which includes the calculated ΔG_{H*} on X-axis for different transition metals, and the experimental j_0 on Y-axis indicates that ΔG_{H*} is a representative descriptor of HER activity, and the optimal HER electrocatalyst should have a ΔG_{H*} value close to zero [19]. Therefore, changing the ΔG_{H*} value of materials by changing their composition and structure is a useful method for optimizing catalytic activity.

6.3 HER Electrocatalysts of Carbon-Based Materials

Carbon-based electrocatalysts have also been extensively studied, but their catalytic performance in alkaline and acidic electrolytes needs to be further improved, especially in HER. Therefore, researchers have developed strategies to enhance the activity of carbon-based electrocatalytic materials [21]. These strategies include increasing specific activity and the density of active sites. In addition to heteroatom doping, methods such as edge functionalization and internal defect engineering have been reported to modify the electronic structure of carbon and its intrinsic activity [22]. Strategies aimed at increasing or exposing more active sites include nanostructures, porous structures, and 3D morphology construction. In addition to being used individually, the above methods are also used in combination to improve performance. Furthermore, various other active species have also been introduced to hybridize with carbon, which improves HER electrocatalysis [16]. This approach expands the potential applications of carbon-based materials. Finally, in order to develop efficient carbon materials, it is important to identify the active mechanisms and sites.

There are various carbon-based electrocatalysts for HER, including but not limited to porous carbon [23], graphene [1], carbon nanotubes (CNTs) [24], carbon nanofibers [25], carbon nanosheets [26], carbon shells [27], carbon quantum dots [28], carbon N doping, MOF-derived carbon materials [29], heteroatom-doped carbon [30], and atomic metal doping [31] of carbon materials. The preparation methods of carbon-based electrocatalysts usually include pyrolysis, hydrothermal carbonization, chemical vapor phase reaction, ball milling, and arc discharge [32]. The following section will provide a detailed introduction to some carbon-based HER electrocatalysts and various preparation methods.

6.3.1 Carbon-Based Metal-Free Electrocatalysts

Efforts have been devoted worldwide to developing carbon-based metal-free electrocatalysts (C-MFEC) to replace or minimize the use of precious metals.

Compared to metal catalysts, C-MFEC has many advantages, such as Earth abundance, low price, controllable structure at molecular level, and compatibility with multiple catalytic active sites [33]. The catalytic activity of most metal-based catalysts strictly depends on the properties of their metal elements. In contrast, C-MFEC introduces various heteroatoms (heteroatom doping), defects (defect doping) [34], and adsorbents (charge transfer doping) into various 0D to three-dimensional (3D) carbon structures with specific reactive active sites [35]. Among them, carbon catalysts based on Co-doped and multidoped porous nanofibers have shown great prospects for improving electrochemical performance due to the synergistic effect of different dopants [34].

Yanan et al. developed N, P, and O Co-doped porous carbon nanosheets (defined as NPOC) through one-pot pyrolysis. NPOC has a cross-linked 3D porous nanostructure. Due to the electronic modulation effect of heteroatoms on the carbon structure and the 3D porous nature of crosslinking, NPOC exhibits excellent electrocatalytic activity for HER. At a current density of $10\,mA\,cm^{-2}$, NPOC provides high HER activity at an overpotential of 232 mV, which is lower than Pt/C and IrO_2, and other catalysts obtained from different amounts of phosphorus sources. In the low overpotential range of 0–250 mV, the Tafel slope is much lower than the slope of NPOC, indicating that Pt/C exhibits excellent HER kinetics at relatively low current densities. On the contrary, at high overpotential, the Tafel slope of NPOC is much lower than the slope of Pt/C, indicating its superior activity [36].

In another study by He et al., a novel nitrogen-doped porous carbon fiber was synthesized from silk fibroin under acidic and alkaline conditions through chemical activation, serving as a metal-free catalyst for HER. Research has found that the morphology of the prepared catalyst is related to the concentration of KCl. Electrospun fiber membranes with high KCl concentration are fibrous in shape because after carbonization β-sheet structures are damaged by KCl. This work reveals the potential of porous carbon nanofibers as HER electrocatalysts in alkaline and acidic media [37].

6.3.1.1 Acidic HER Performance

The LSV curves of commercial 20% Pt/C catalyst, SCNF, and SPCNF are compared. Among all the samples synthesized, 4% SPCNF exhibits the best HER catalytic activity in an acid, as its LSV curve has a small overpotential turning point. A current density of $10\,mA\,cm^{-2}$ (expressed as $\eta 10$) is used to characterize the electrocatalytic performance of the catalyst. The $\eta 10$ value of 4%-SPCNF was the smallest (310.86 ± 12.93 mV). The excellent catalytic performance of 4% SPCNF can be attributed to the large specific surface area generated by the porous structure generated by chemical activation, as well as the large through-holes between carbon fibers, which may expose more HER active sites and this structure can enhance electrolyte diffusion, thus promoting the hydrogen adsorption process, which is considered a key step in HER. In addition, the Gibbs free energy of hydrogen adsorption ΔG_{H*} in DFT calculation shows that in N-doped carbon, the ratio of pyrrole N and graphite N have smaller ΔG_{H*}. Graphite N is more active as ΔG_{H*} is only 0.81 eV [38]. Among all the prepared samples, 4%-SPCNF exhibits the smallest Tafel slope, indicating faster reaction kinetics. Furthermore, 4%-SPCNF has the smallest R_{ct}. The excellent stability can be attributed to the nanoporous carbon of 4%-SPCNF, which is almost free from electrolyte corrosion. Additionally, in case of 4%-SPCNF, $\eta 10$ only increased to 3.40 mV after a 1500CV cycle, indicating significant durability.

6.3.1.2 Alkaline HER Performance

Among all LSV curves of samples, 4%-SPCNF also exhibited excellent HER catalytic activity in an alkaline. The results of the histogram of $\eta 10$ values show that all electrocatalysts have higher $\eta 10$ values than those under acidic conditions. This is because free protons are difficult to find in alkaline electrolytes [39]. Compared with other synthetic catalysts, the 4%-SPCNF sample has the smallest $\eta 10$ value, indicating excellent catalytic activity. The effect of catalyst morphology on HER catalytic activity under alkaline conditions is similar to that under acidic conditions, where porous structure and fiber morphology can synergistically improve catalytic performance. The Tafel slope of 4%-SPCNF is the smallest, indicating faster reaction kinetics. Due to the lack of free protons, the Tafel slope under alkaline conditions is greater, and free protons rely on water dissociation reactions that are believed to accompany or precede the Volmer reaction [40]. The R_{ct} of the active sample (SPCNF) is smaller than that of the SCNF with a smaller semicircular diameter, indicating that its porous structure is conducive to the diffusion of electrolytes and thus improves the electron mobility during the catalytic reaction. In addition, the stability and durability of 4%-SPCNF in 1M KOH were also tested. After 1500 cycles, the η value of 4%-SPCNF increased to 21.38 mV, which showed good durability. After operating at a current density of 10 mA cm^{-2} for 10 hours, the overpotential increased to 39 mV, indicating excellent stability of the catalyst, which can be attributed to the corrosion resistance of the carbon matrix [37].

According to the results of Tafel slope in an acidic medium, the HER mechanism of SCNF and SPCNF may be the Volmer–Heyrovsky' process. The reactions can be described as follows:

$$^* + H^+ + e^- \rightarrow H^* \text{ (Volmer step)}$$

and

$$^* + H^+ + H^* + e^- \rightarrow H_2 \text{ (Heyrovsky' step)}$$

Here, * refers to the active site of the catalyst.

In an alkaline medium, it can be inferred that the reaction mechanism of the catalyst synthesized may be Volmer–Heyrovsky'. The process can be described as:

$$^* + H_2O + e^- \rightarrow H^* + OH^- \text{ (Volmer step)}$$

and

$$H_2O + H^* + e^- \rightarrow H_2 + OH^- + ^* \text{ (Heyrovsky' step)}$$

In addition, the Volmer step is considered a rate-determining step.

Carbon fiber has unique physical and chemical properties and is a sustainable material used as a catalytic carrier for HER. Heckova et al. developed porous carbon microfibers using needle-free electrospinning technology [41].

The process involves the removal of PVP by washing, which resulted in the formation of porosity in polyacrylonitrile/polyvinylpirrolidone (PAN/PVP) precursor fibers, and further, it was followed by carbonization of PAN. The most effective method was use of sodium dodecyl sulfate (SDS) as an anionic surfactant.

The interaction between SDS and PVP chains leads to the formation of micelles, regular porous structures in a fibrous carbon matrix are observed as the carbonization process completes.

The electrocatalytic activity of unstable carbon fibers PAN/PVP/SDS carbonized in an H_2 reducing atmosphere is the highest. The overpotential value of −288 is required for the porous carbon sample at a cathode current density of 10 mA cm^{-2}, and the overpotential value of −483 is required for a current density of 20 mA cm^{-2}, indicating good HER activity, which can be related to the porous structure. In contrast, the Tafel slopes of the catalyst were 185, 183, 133, and 104 [mV dec^{-1}] respectively, indicating that the HER reaction was carried out through rapid Heyrovsky steps and Volmer RDSs. The high value of b indicates that carbon fibers are inert in water electrocatalysis [42]. However, the introduction of pores has the possibility of electrolytes penetrating the interstitial pores of fibers. The smaller diameter of the semicircle (PAN/PVP/SDS non-stab.H_2) represents enhanced electron transfer, which accelerates HER dynamics. The polarization curves of all samples after 500 cycles were analyzed. The polarization curves before and after the cycle were similar, suggesting good stability. A higher pore content in carbon fibers after H_2 reduction and heat treatment yields fast reaction kinetics for HER.

6.3.2 Low-Dimensional Carbon Material and Heteroatom-Doped Carbon

6.3.2.1 Carbon Quantum Dots

For the HER, using materials with large active surface area and high electron transfer efficiency as catalysts can significantly enhance the efficiency of H_2 generation. Low-dimensional carbon catalysts, such as 0D carbon quantum dots (CQDs), are composed of ultrafine, dispersed, quasi-spherical carbon nanoparticles (NPs) with sizes below 10 nm [43]. The unique structure of CQDs enables them to possess chemical stability, large specific surface area, good dispersibility, as well as excellent conductivity. In addition, they are easy to synthesize, cost-effective, safe, and nontoxic, which makes them feasible as highly efficient HER catalysts. The surface of CQDs has abundant functional groups, such as –COOH, –OH, and –NH_2, which endow them with good hydrophilicity and enhance their interaction with water molecules in the solution. These functional groups can also coordinate with metal ions, providing numerous favorable sites for the preparation of highly active catalytic materials. CQDs are perfect electron donors and acceptors, promoting more efficient electron transfer. Their microscale size can be adjusted, and their electron configuration can be altered by simple modifications, such as doping with heteroatoms, which increases their electron-capturing capacity and enhances electron transfer efficiency [44]. This can also lower the dissociation energy of reactant intermediates to improve the reaction kinetics. Additionally, preparing CDQ-metal composites can further generate HER catalysts with superior performance, which are currently the mainstream CQD catalysts for HER. Metals easily interact with H ions, forming M—H bonds and further promoting H_2 generation. The large specific surface area of CQDs, when loaded with metal, can provide abundant active sites, facilitating the adsorption, desorption, generation, and diffusion of H_2, thus promoting the HER process [8]. However, in practical applications, CQDs tend to undergo severe aggregation due to their high surface free energy, leading to a decrease in catalytic activity [45]. Therefore, modifying CDQ materials to prevent the aggregation or clustering caused by the interaction between CQDs themselves or metal particles to maintain their excellent dispersibility and large active surface area, is the key to preparing excellent CDQ catalysts.

Liu et al. [46] achieved large-scale synthesis of N-CQDs using a convenient *N*-methyl pyrrolidone (NMP) reflux-assisted polymerization method, which can be scaled up under atmospheric pressure and has practical application significance. The N-CQDs obtained from a reflux time of 17 hours exhibited a more uniform shape and size distribution, maximizing the HER performance. The N/C molar ratio of these N-CQDs was as high as 46%, and the nitrogen atoms were mainly in the form of pyrrolidine-N. Nitrogen atom doping can improve electrocatalytic performance by modulating the electronic properties of adjacent carbon atoms through intramolecular charge transfer. What is more, the doped nitrogen atoms in N-CQDs also participate in the electrocatalytic hydrogen evolution process as active sites in the form of pyrrolic-N, which is electron-withdrawing compared to graphitic-N, thus favoring H adsorption. These synergistic effects determine the remarkable catalytic performance of N-CQDs for HER.

Ruthenium (Ru) is a widely used catalyst with bonding strength similar to hydrogen and has a much lower cost than Pt, making it an ideal HER catalyst material [47]. However, their overpotential and electrochemical stability need to be improved, while CQDs have good stability, high conductivity, and a large active surface area, making them ideal conductive carriers for Ru composites to prepare catalysts with both performance and stability. Li et al. [48] used three carbon-based materials (biomass porphyria, porphyria-CQDs, and porphyria activated carbon) as carriers for hybrid materials and loaded highly dispersed Ru NPs, comparing their HER catalytic performance. Among them, Ru NPs catalyst supported on porphyria-CQDs (Ru@CQDs) prepared by solvent thermal method had the highest catalyst activity. The overpotential of Ru@CQDs was 65 mV at 10 mA cm^{-2} in 1.0 M KOH, and Ru@CQDs prepared at 800 C had the best HER activity due to their high crystallinity. Compared with the other two materials, Ru@CQDs had better dispersion and higher catalytic activity. This can be attributed to the strong coordination between the Ru's d orbitals and the surface functional groups of CQD. The excellent HER electrocatalytic performance is the result of the synergistic interaction between CQDs and Ru: proper crystallinity promotes electron transfer between the metal and the carrier. CQDs in the hybrid inhibit the aggregation of Ru NPs, resulting in improved catalytic performance and stability. Liu et al. [49] proposed and developed a universal route for producing ultrafine Ru-based bimetallic electrocatalysts using CQDs as carbon and nitrogen sources – transition metal atoms were uniformly doped into the Ru lattice via a solvent thermal method, forming RuM NPs attached to CQDs. The generality of this method was demonstrated by synthesizing three types of bimetallic RuM/CQDs (RuNi, RuMn, RuCu). The test results showed that RuNi/CQDs catalyst exhibited excellent HER performance at all pH values.

In particular, RuNi/CQDs only required overpotentials of 13 mV (1 M KOH), 58 mV (0.5 M H_2SO_4), and 18 mV (1 M PBS) to achieve a current density of 10 mA cm^{-2} for HER, and maintained its catalytic activity after 10,000 cycles, outperforming most reported materials for HER.

Chen et al. [50] used a one-step CVD technique to synthesize Ni_5P_4 and $Mn_xNi_{5-x}P_4$ nanosheets precursors, which were then immersed in a CQDs solution and annealed to obtain CQDs/$Mn_xNi_{5-x}P_4$ catalysts. In a 0.5 M H_2SO_4 electrolyte, this catalyst achieved a current density of 10 mA cm^{-2} with only 31 mV overpotential and exhibited a low Tafel slope of 41.0 mV dec^{-1} and a large exchange current density of 1.753 mA cm^{-2}. In addition, CQDs/$Mn_xNi_{5-x}P_4$ also showed high catalytic activity and stability in alkaline conditions. Mn doping effectively reduced the charge transfer resistance during the HER process. Javan et al. [51] deposited Ni–Cu bimetallic nanocomposites on a glassy carbon electrode (GCE) using a simple electrochemical method and modified them with CQDs to prepare Ni–Cu/RCQD/GCE. In an acidic medium, the Ni–Cu/RCQDs/GCE exhibited an overpotential of 230 mV at a current density of 10 mA cm^{-2} and a small Tafel slope of 83.5 mV dec^{-1}. By comparing the performance of Ni–Cu/RCQDs/GCE, Ni–Cu/GCE, and pure Ni electrodes, it was found that the presence of RCQDs provided a large active surface area and enhanced the synergistic effect between Ni and Cu. Ni–Cu/RCQDs/GCE achieved efficient HER catalysis by increasing the cathodic current density, reducing overpotential, and lowering the charge transfer resistance. Yang et al. [52] prepared a hybrid material of nickel nanoparticles/carbon quantum dots (Ni/CQDs) using a simple pyrolysis method with $Ni(AC)_2$ and CQDs as raw materials. The test results showed that Ni/CQDs prepared with 20 mg CQDs at 500 °C exhibited the best HER catalytic performance. Under visible light irradiation, the HER performance was further improved, showing higher current density and lower Tafel slope (77 mV dec^{-1}) after illumination. The catalytic performance of Ni/CQDs was mainly determined by the synergistic effect of Ni NPs, CQDs, and the Ni–O–C interface. In summary, CQDs can be combined with other nonprecious metal catalysts, such as Ni and Mn, to prepare superior HER catalysts. In this case, CQDs primarily act as good electron transport carriers and provide a large active surface area to assist the transfer of electrons for the metal elements.

Transition metal carbides (TMCs), in particular, 2D layered TMCs, also known as MXenes, due to their platinum-like activity achieved through surface modification, have become popular catalysts for the HER [53]. Theoretically, the edge sites of MXenes are considered highly active starting points for the HER reaction due to their weak hydrogen bonding energy and low hydrolysis barrier [54]. Reducing the dimensionality of MXenes to CQDs is an effective way to expose their edge sites. Jiang et al. [55] employed a Li intercalation strategy to cut 2D MXene nanosheets into 0D CQDs, enriching the active edge sites and optimizing the density of states of active Ti atoms to lower the energy barrier. The optimal monolayer Ti_3C_2 quantum dots exhibited an activity nearly 13.7 times higher than their nanosheets. MXene QDs also demonstrated excellent stability, experiencing only about 27 mV of overpotential decay after 2000 cycles (Figure 6.4). Combining DFT and XPS analysis, metal atoms at the edges of the quantum dots were more catalytically active for the HER, particularly those

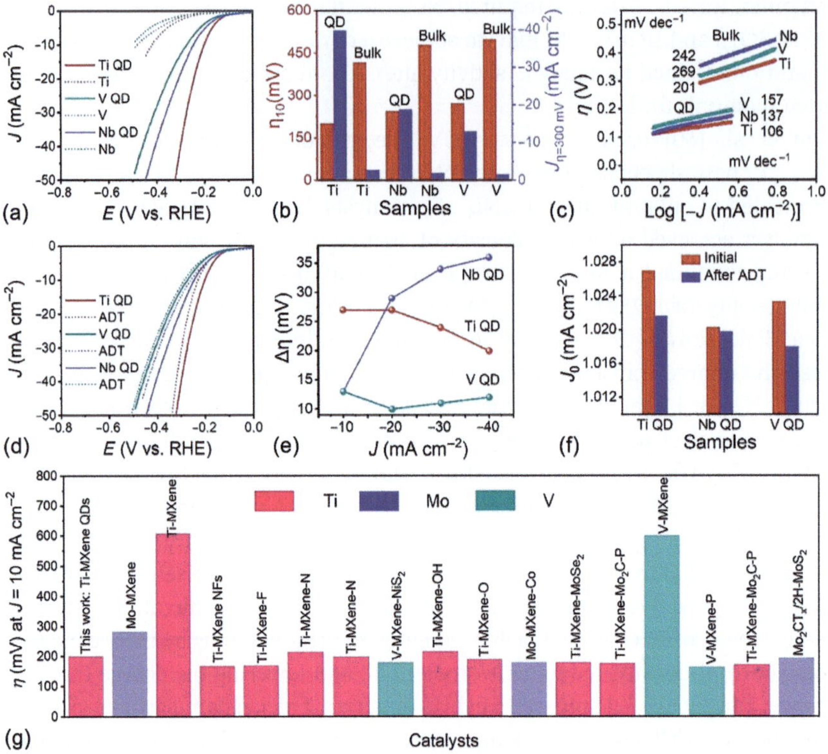

Figure 6.4 Electrochemical evaluation of Ti, V, Nb base MXene QDs toward HER. Source: Reproduced with permission of Jiang et al. [55], Elsevier.

bonded to hydroxyl-terminated groups. This strategy provides insight into the effective reduction dimension of MXene-based HER catalysts. Xue et al. [56] constructed 0D/2D graphene quantum dots (GQD)/$Ti_3C_2T_x$ MXene heterojunctions through a tunable assembly process – thorough mixing of $Ti_3C_2T_x$ nanosheets with a certain amount of graphene oxide quantum dots (GOQD) solution, followed by a hydrothermal process to connect the oxygen-containing species of GOQD, such as carboxyl and hydroxyl groups, with the functional groups on the surface of $Ti_3C_2T_x$, providing a stable 0D/2D heterojunction structure. The resulting GQD/$Ti_3C_2T_x$ heterojunction with an appropriate GQD content (5 wt%) showed a low onset potential of 130 mV, a small Tafel slope of 89 mV dec^{-1}, and excellent long-term durability for the HER.

In summary, for CQD catalysts used in HER, improving electrode stability, corrosion resistance, and maintaining catalyst dispersion are the key aspects of CQD catalyst development. Developing more CQD-metal composite catalysts and utilizing the synergistic effect between CQDs and metal NPs are crucial for the development of CQD catalysts.

6.3.2.2 Carbon Nanotube Catalysts for HER

In the context of catalysts made from low-dimensional carbon materials for the HER, one-dimensional (1D) materials typically possess large surface areas and high aspect ratios, beneficial for the distribution of active sites. Among them, CNTs consist of carbon atoms arranged in a hexagonal pattern, forming multiple coaxial layers ranging from a few to tens of layers. CNTs have nanoscale dimensions, conjugated all-carbon structures, and strong molecular symmetry, exhibiting excellent stability, optoelectronic properties, thermodynamic properties, and catalytic activity [57]. CNTs possess substantial surface area, stability, and high conductivity, and their surface can be easily modified due to the abundance of active functional groups. However, the application of CNT catalysts faces similar challenges as CDQs – issues such as agglomeration and poor intrinsic chemical reactivity have seriously affected the development of CNT catalysts. Modification techniques such as defect engineering, heterostructure construction, and surface modification are currently the mainstream methods for preparing CNT catalysts [58]. Doping CNTs with heteroatoms (such as N, P, and B) allows for controlled modulation of the electronic structure and charge density distribution, leading to the generation or exposure of more active sites on the CNT surface, thereby enhancing its chemical reactivity [59]. Similar to CDQ catalysts, the combination of CNTs with metal materials, such as metal monomers or metal oxides, to prepare CNT-metal composites results in improved HER catalytic performance. The introduction of heteroatoms can effectively prevent the agglomeration of metal NPs and produce more active sites, thereby enabling the fabrication of efficient and stable HER catalysts [60]. It is worth noting that in this case, CNTs not only provide active sites but also serve as conductive carriers, accelerating electron transport and exerting synergistic effects with the loaded metals and heteroatoms to enhance catalytic activity. CNTs exhibit good electron transferability, and ionic liquids (ILs) often contain specific functional groups (such as amino or sulfonic acid groups) that have a strong affinity for protons in solution. Combining their characteristics, the preparation of ionic liquid-functionalized carbon nanotubes (IL-CNTs) also holds great potential as an HER electrocatalyst.

Lu et al. [61] successfully synthesized nitrogen-doped carbon nanotubes (NCNTs) containing abundant graphitic nitrogen (g-N) on nickel foam (NF) using acetonitrile (ACN) as a precursor through atmospheric pressure chemical vapor deposition (APCVD) at low temperatures. The synthesized NCNTs@NF exhibited enhanced HER catalytic activity with a current density of $-20\,\text{mA}\,\text{cm}^{-2}$ at an overpotential of 171 mV and a Tafel slope of $119\,\text{mV}\,\text{dec}^{-1}$. The results and DFT calculations revealed that the synthesis of NCNTs on NF at lower temperatures ensured the presence of abundant g-N, which facilitated the HER process. Additionally, the bamboo-like morphology of NCNTs with numerous edge defects provided a favorable catalytic environment. The nitrogen doping altered the electronic structure, further promoting reversible hydrogen adsorption and desorption, thereby providing numerous active sites for the HER process. Inspired by polydopamine (PDA), Qu et al. [62] reported a clever "grafting-pyrolysis" approach to prepare nitrogen and sulfur Co-doped carbon materials. Firstly, PDA was deposited on the surface of multiwalled CNTs, followed by chemical grafting with a sulfur precursor at room temperature, and then high-temperature pyrolysis was conducted. This method allowed for the in situ and uniform incorporation of nitrogen and sulfur Co-doping into the carbon framework, with a sulfur doping efficiency of up to 5.6%.

The prepared N, S–CNTs exhibited excellent bifunctional electrocatalytic activity for oxygen evolution reaction OER and HER in alkaline solution. Experiment results revealed that N, S–CNTs possessed layered pores, high electrochemical double-layer capacitance (C_{dl}), and abundant void spaces, facilitating mass transfer. Furthermore, the high conductivity of CNTs and strong coupling with PDA promoted charge transfer. The multicomponent doping process generated many active sites, contributing to the remarkable HER catalytic performance of N, S–CNTs.

By utilizing the unique structure and excellent conductivity of CNTs, they can be combined with metal materials to create catalysts that exhibit improved HER catalytic performance. Tavakkoli et al. [63] first dissolved Pt ions from Pt foil into sulfuric acid and then used a simple electrodeposition method to achieve pseudoatomic scale dispersion of platinum on the sidewalls of single-walled carbon nanotubes (SWNTs) as single atoms or sub-nanometer clusters. Under acidic conditions, SWNT-Pt consists primarily of a small number to tens of Pt atoms dispersed on the sidewalls of the carbon nanotubes, with a Pt loading on the electrode surface of only 0.19–0.75 at%. The dispersed Pt atoms greatly enhance the active sites on the surface. Experimental results and DFT calculations indicate that the curved structure of SWNTs plays a crucial role in stabilizing the Pt atoms, making it an important support for achieving stable dispersion of Pt atoms. Furthermore, due to the activation effect of Pt atoms, SWNTs also exhibit high HER activity comparable to bulk Pt. Nguyen et al. [64] heat-treated multiwalled carbon nanotubes (MWCNTs) at an extremely high temperature of 2600 °C to obtain heat-treated multiwalled carbon nanotubes (hMWCNTs) and studied their HER performance. MWCNTs possess excellent electrochemical charge transfer kinetics and lower electrochemical charge transfer resistance, making hMWCNTs exhibit good HER performance. When hMWCNTs are further combined with Pt, hMWCNTs-Pt composite materials demonstrate even better HER performance, with a Tafel slope of only 10.9 mV dec^{-1} at an overpotential of 0.04 V vs. RHE. Sun et al. [65] synthesized MWCNTs/$Co_{0.85}Se$ magnetic nanoparticles supported on carbon nanotubes using a one-pot hydrothermal method. After the composite formation, the size of $Co_{0.85}Se$ nanoparticles decreases, resulting in increased distance between $Co_{0.85}Se$ nanoparticles and the hybrid material, thereby enhancing the saturation magnetization and coercivity. Experimental results show that the current density of the 30 wt% MWCNTs/Co0.85Se catalyst is 10 mA cm^{-2} at an overpotential of 266 mV/RHE in an acidic solution, with a Tafel slope of 60.5 mV dec^{-1}. Furthermore, it exhibits good HER catalytic performance and stability with minimal decay after 1500 cycles. The high HER activity of MWCNTs/$Co_{0.85}Se$ can be attributed to several factors based on material characterization – the inherent semimetallic properties of $Co_{0.85}Se$ formed by the overlapping of Co 3d and Se 4p spin electrons, the high surface area of $Co_{0.85}Se$ synthesized using a polar organic solvent ($N_2H_4{\cdot}H_2O$), and the presence of numerous surface defects providing ample reaction sites.

Compared to metal NPs, metal carbides, phosphides, and nitrides exhibit better HER catalytic performance. Combining them with CNTs results in composite material HER catalysts with more advantages. Tabassum et al. [66] synthesized a CoP@BCN structure (CoP encapsulated in BCN nanotubes) through pyrolysis and phosphidation reactions. The doping of heteroatoms effectively prevents the aggregation of CoP NPs in BCN nanotubes, and the synergistic effect generated by B and N Co-doping provides abundant active sites for CoP@BCN-1 nanotubes. The current density of CoP@BCN-1 nanotube electrocatalyst can reach 100 mA cm^{-2} in acidic, alkaline, and PBS solutions, with overpotentials of 87, 215, and 122 mV at −10 mA cm^{-2}, respectively, demonstrating excellent electrochemical stability. Based on testing and characterization results, metal-based hydrogenase with metal centers is introduced as the active site in BCN nanotubes. In addition, the 1D carbon nanotubes have smaller charge transfer resistance, faster electrochemical kinetics, and larger active surface area, while the P sites can be used as proton carriers, accelerating the kinetics of protons (H^+), significantly enhancing the HER activity. Zhang et al. [67] reported a simple method to synthesize MoP/CNTs hybrid structure through the solution method by first synthesizing an intermediate $MoO_x(OH)_y$/CNTs in ethanol, followed by phosphidation transformation to obtain MoP/CNTs. In MoP/CNTs, well-dispersed and crystalline MoP nanoparticles with small volume, are uniformly attached to the sidewalls of CNTs. In pH-universal electrolyte, MoP/CNTs hybrid exhibits excellent catalytic activity for HER; at cathodic current densities of 10 mA cm^{-2} in acidic 0.5 M H_2SO_4, neutral 1 M phosphate buffer solution, and alkaline 1 M KOH electrolyte, the overpotentials are 83, 102, and 86 mV, respectively, showing superior performance in nonprecious metal HER catalyst systems. The addition of CNTs improves the conductivity and allows for the uniform distribution of MoP NPs due to their special structure. Additionally, the interaction between metal ions and carbon nanotubes reduces the aggregation and sintering of MoP nanoparticles, increasing the active sites. Liu et al. [68] first prepared Co/N-doped carbon nanotubes through thermally induced self-catalysis and then prepared Ru@Co/NCNTS catalysts by electrodepositing ruthenium nanoclusters using the displacement reaction. Characterization results show that the crystalline state of Ru nanoclusters is intact, and they are uniformly dispersed on CNTs, exposing a large number of active sites of Ru. Moreover, Ru itself exhibits a strong electron effect, which endows Ru@Co/NCNTs with high activity for HER in both acidic and alkaline electrolytes. In alkaline and acidic media, the overpotentials are only 48 and 92 mV at a working current density of 10 mA cm^{-2}, respectively, and they show excellent durability, with a hydrogen production current efficiency close to 100%, outperforming most reported ruthenium-based and cobalt-based catalysts.

Due to the unique structure and abundant functional groups on the surface of CNTs, they can be combined with ILs to prepare high-performance HER catalysts. Li et al. [69] prepared IL-functionalized multiwalled carbon nanotubes (IL-MWCNTs) composite materials and used them as nonmetallic electrocatalysts for HER. Among them, 1-aminopropyl-3-methylimidazolium bromide functionalized carbon nanotubes (AMIM-Br-MWCNTs) showed the best HER performance. The experimental results showed that the electrocatalytic performance of AMIM-Br-MWCNTs was attributed to the synergistic effect of the covalent bond structure. At the same time, IL could replace Nafion solution as a binder, which not only reduced the cost but also improved the conductivity of the electrode. Through improvement, Li et al. [70] proposed an in situ functionalization strategy of carbon nanotubes with ILs to enhance the HER performance. The imidazolium ILs were directly synthesized and Immobilized on the CNT skeleton through a simple and fast reflux method, and various CNT-IM-X species were obtained through anion exchange. Experimental results and DFT calculations showed that this direct IL functionalization was beneficial to the electron transfer process, as ILs could directly act as electron acceptors and had strong hydrogen adsorption capacity. Among various CNT-IM-X species, 1-butyl-3-methylimidazolium chloride (CNT-IM-Cl) showed excellent electrocatalytic activity in acidic media, achieving a current density of $10\,mA\,cm^{-2}$ with only 135 mV overpotential, and exhibited low onset potential and Tafel slope of 80 mV and $38\,mV\,dec^{-1}$, respectively, with good stability.

In general, CNTs as 1D carbon materials play an important role in HER catalysts. After simple modification, CNTs themselves can act as the catalytic active centers for the HER process. When combined with metal materials, ILs, and other catalytic active substances, CNTs primarily appear as carriers, utilizing their excellent conductivity, unique structure, and large specific surface area to further enhance the catalytic performance of the composite materials, resulting in highly efficient and stable HER catalysts.

6.3.2.3 Graphene, N-Doped Carbon, and g-C_3N_4

The consumption of traditional fossil fuels such as coal, oil, and natural gas not only leads to resource depletion, but also causes serious environmental problems such as acid rain, greenhouse effect, and biodiversity loss [71]. Electrocatalytic HER provides the possibility for the mass production and storage of hydrogen energy. In this section, we will focus on introducing efficient electrocatalysts for HER, including graphene, N-doped carbon, and g-C_3N_4.

Graphene-Based Electrocatalysts for HER In recent years, carbon materials, including porous carbon, CNTs [72], graphene [73], and graphdiyne [74], have been studied and developed. HER electrocatalysts have shown a broad and fascinating prospect in this field. Graphene and its derivatives are ideal materials for HER electrocatalysts due to their (i) large specific surface area, (ii) strong electron transfer ability, (iii) good stability, and (iv) easy adjustment of structure and surface properties. Therefore, significant progress has been made in the research and development of advanced materials, leading to the emergence of various advanced synthetic strategies to fabricate graphene-based HER catalysts with different nanostructures [75] (Figure 6.5).

We will discuss three categories of graphene-based HER electrocatalysts: (i) 2D, (ii) 3D porous, and (iii) heteroatom-doped graphene-based electrocatalysts.

2D graphene is a single-layer carbon atom arranged in a hexagonal lattice. Although the graphene layer is chemically inert, bending of the graphene layer changes the charge distribution and thus improves the electrochemical activity [76]. 2D graphene exhibits relatively low activity in HER. However, by chemical modification strategies, the electronic structure and chemical reactivity of graphene sheets can be adjusted [75].

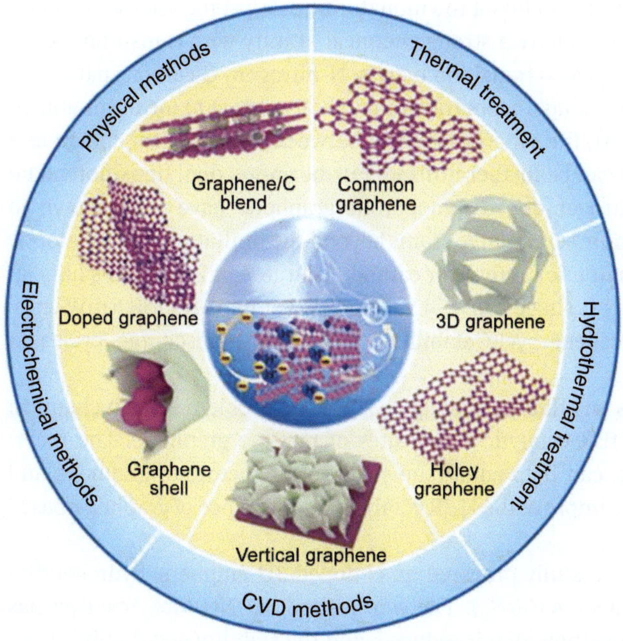

Figure 6.5 Schematic illustration for graphene-based HER electrocatalyst synthesized by different methods. Source: Reproduced with permission of Huang et al. [75], Wiley-VCH.

Theoretical studies [77] have demonstrated that graphene is able to interact with different metals to effectively modulate its working function. The graphene-metal interaction is strongly dependent on the metal surface. In addition, 2D graphene can serve as a good platform to disperse other HER active materials; graphene not only controls the morphology and microstructure of the active material, thereby exposing more electroactive sites but also promotes strong synergistic coupling effects that enhance the dynamics of the HER process.

The use of graphene-based nanomaterials often results in reaggregation or restacking phenomena, leading to the loss of the unique structure of the original graphene sheet and significant coverage of electrochemical active sites. Consequently, a substantial decline in catalytic activity is observed [78]. To solve this problem, the research direction of designing and constructing a 3D graphene architecture with interconnected porous frameworks has attracted a lot of attention, which can not only effectively avoid the reaggregation and restacking of graphene layers but also provide a multidimensional growth platform for other HER active materials [79]. 3D porous graphene has attracted wide attention due to its unique porosity, excellent electron transfer ability, and multifunction [80].

Chen [81] reported progress in addressing the challenge of developing single-atom catalysts using 3D nanoporous graphene. In the curved structure of 3D graphene, the presence of topological and geometric defects provides significant advantages. They provide additional chemical active sites, thereby improving the catalytic loading capacity, performance, and stability of the monatomic catalyst together with chemical dopants. These catalysts have a strong chemical affinity with transition metals and form active M–N_x–C_y (M = transition metals, N-nitrogen, C-carbon) parts.

By introducing a second single-atom transition metal catalyst to form a double or multiactive M–N_x–C_y part, the synergistic effect between the chemical dopant and the single-atom catalyst can be further utilized. Furthermore, dual transition metal atoms supported by nitrogen-doped nanoporous graphene have been shown to effectively increase the number of active atoms beyond the limitations of monatomic catalysts. The dual monatomic catalyst for cathode charging of Zn-air cells has significant energy efficiency (voltage gap 0.74 V), high power density (150.6 mW cm^{-2} 250 mA cm^{-2}), and excellent cycle stability (>500 charge–discharge cycles at 10 mA cm^{-2}) [82].

Heteroatom doping is an efficient strategy to solve the problem of high chemical inertness and lack of HER catalytic active sites of pristine graphene. The incorporation of heteroatoms can help adjust the electronic structure of graphene and increase the active site to improve the HER catalytic performance of graphene-based materials [83].

Chen's group [84] successfully prepared an N-doped 3D graphene hydrogel film (NG) with molecular clusters (MoS_x). The synthesized 3D NG-MoS_x catalyst has good HER activity, the overpotential is reduced, and the stability and durability are improved due to the increase of MoS_x defect sites on NG. Experimental studies have found that NG–Mo has many functional groups, such as COOH, C–O, and C–N, and has a high water absorption affinity, which is suitable for electrocatalysis. Further study of the HER mechanism shows that the double active site process on the NG–Mo catalyst is very likely. MoS_x clusters are particularly active in catalyzing HER because they interact with edge water molecules and defect sites to promote water dissociation. In addition, pyridine and pyridine C–N formed on NG–Mo are other active centers of HER. Interestingly, the activity of NG without MoS_x alone was only 12.9% of that of NG–Mo, highlighting the critical role of MoS_x as the main active species in this hybrid catalyst.

Deng's group [85] prepared highly torn amine-functionalized N-doped graphene (HT-AFNG) by a simple hydrothermal method. They used HT–AFNG as an efficient electrocatalyst for HER. The excellent HER performance of HT–AFNG can be attributed to the functionalization of amines, the synergistic interaction between nitrogen and graphene resulting in more active sites, and the high electron transport properties of HT–AFNG.

N-Doped Carbon Electrocatalysts for HER The doping of heteroatoms into carbon can regulate the physical and chemical properties of carbon, thus providing more active sites. It is worth noting that this strategy can produce more carbon-based materials and promote the development of efficient electrocatalysts for HER. N-doped carbon is a common HER catalyst. The incorporation of nitrogen atoms improves the catalytic performance of carbon without affecting the conductivity of carbon. When nitrogen atoms are introduced into the carbon structure, the amount of nitrogen and its local structure determine the performance of such carbon-based catalysts [18].

Jia et al. [86] reported a convenient, environmentally friendly, and low-cost synthesis strategy. They synthesized and constructed Co_2P/NPPC composites in one step at 800 °C. The good pore structure and high specific surface area of NPPC make it an ideal carrier for loading Co_2P particles, providing an abundance of active sites. NeP Co-doping and Co_2P act together on Co_2P/NPPC-800, showing good HER performance and stability in acidic and alkaline environments.

Yang [87] prepared Zn, Mn, S, and P (ZnMnSP/NDC) loaded on N-doped carbon by pyrolysis, adsorption, vulcanization, and phosphorylation using zeolite imidazolate zinc framework (ZIF-8) as a precursor, which has excellent HER catalytic performance and long-term stability. This work proposes a new strategy for the preparation of heteroatom-loaded nitrogen-doped carbon as a highly stable and efficient catalyst.

Recently, N-doped carbon materials have been mostly used as bifunctional or multifunctional catalysts for water-splitting. Jiang et al. [88] reported a defect-rich and ultrathin N-doped carbon nanosheet material that can be used as an advanced nonfunctional metal electrocatalyst for ORR, OER, and HER. Through DFT calculations, they found that the carbon atom near the graphite nitrogen dopant is the most active site, and the graphite nitrogen dopant is more important than the pyridine and pyrrole N dopants. The theoretical calculation results are consistent with the experimental results. The sample prepared at high temperature (1000 °C) has a higher percentage of graphite nitrogen, abundant micropores, and edge defects than that at low temperature (800 °C) and has stronger catalytic performance.

***g-C_3N_4* Electrocatalysts for HER** A semiconductor polymer, g-C_3N_4, with the advantages of high natural abundance, low cost, and easy synthesis, is often used to construct electrocatalysts with transition metals or transition metal oxides to promote the occurrence of HER so as to solve the problem of energy shortage [89, 90].

Liu et al. [91] reported that the hierarchical structure of graphitic carbon nitride (g-C_3N_4) nanobelts-graphene provides a large accessible surface area, multiple electron transfer channels, and a short diffusion distance, achieving good charge separation and transfer, effectively accelerating the electrochemical process of HER.

Geng et al. [92] optimized the HER catalytic activity of molybdenum disulfide by contacting S-doped g-C_3N_4 to form a 2D layered heterojunction. The MoS_2-C_3N_4-60 catalyst has a high charge transfer rate, which can adjust the electronic structure and provide more active sites to improve HER activity. Experiments show that MoS_2-C_3N_4-40 has the highest charge transfer resistance, which is due to the less coupling between S-doped g-C_3N_4 and molybdenum disulfide, which is not enough to enhance the electron transfer between molybdenum disulfide and S-doped g-C_3N_4. MoS_2-C_3N_4-80 has a larger charge transfer resistance than MoS_2-C_3N_4-60, which is due to the greater binding of S-doped g-C_3N_4 to molybdenum disulfide. Doping g-C_3N_4 with more S in MoS_2-C_3N_4-60 will reduce the active sites exposed by molybdenum disulfide.

Silva et al. [93] reported that a hybrid electrocatalyst based on non-noble metals can promote the occurrence of efficient HER. They found that the interface effect of the hybrid heterostructure can lead to a significant enhancement of the catalytic activity of g-C_3N_4. In particular, they focused on the structure of g-C_3N_4/X

(X = WTe_2, MoS_2 hybrid electrocatalyst, and graphene). Based on the research results, they concluded that the N—H bond has strong s-p hybridization characteristics, and they confirmed that g-C_3N_4/WTe_2 can show the best performance for the properties studied.

Dong et al. [94] reported an efficient HER electrocatalyst consisting of bimetallic cobalt–iron sulfide ultrathin nanosheets supported by a g-C_3N_4 catalyst (CoS_2/FeS_2/CN). Electrochemical experiments show that ultrathin nanosheets can expose more active sites, and the addition of g-C_3N_4 and Fe to CoS_2 can improve its catalytic stability and activity. In addition, the Co-doping of g-C_3N_4 and Fe with CoS_2 can adjust the electronic structure at the interface. CoS_2/FeS_2/CN has excellent HER performance, reaching $10\,mA\,cm^{-2}$ in acidic solution, with an overpotential of only 76.5 and 175.6 mV in alkaline solution.

Carbon-based materials are promising catalytic materials for replacing precious metal-based (such as Pt) materials for HER. Promoting the development of carbon-based materials can help solve the problem of high demand for energy but insufficient energy supply in the current society. Graphene, N-doped carbon, and g-C_3N_4 have been developed rapidly in recent years, which are increasingly known for their low-cost and high-activity characteristics. With their own strengths, a new type of catalyst is constructed with other elements, which can reduce the overpotential required for HER, and show good environmental friendliness while saving energy. Therefore, the carbon-based catalyst has high economic benefits and broad development prospects. Although few studies have been carried out so far, it is still possible to design and develop carbon-based catalysts more innovatively to achieve higher activity and longer stability.

6.3.3 MOF-Derived Electrocatalysts

MOFs, with their unique crystalline porous structures, high specific surface area, and tunable compositions, offer broad prospects for applications in electrocatalysis [95]. Recent studies have focused on deriving advanced materials from MOFs for applications in HER, critical processes in sustainable energy production.

A significant aspect of recent research has been the innovative synthesis of MOF-derived materials. Techniques like vapor phase growth, controlled vulcanization [96], and topotactic transformation [97] have been employed to create materials with enhanced physical and chemical properties. These methods lead to the formation of heterostructures, bimetallic composites, and nanostructures with improved conductivity, stability, and catalytic activity.

Bimetallic and polymetallic sulfides derived from MOFs have demonstrated enhanced HER activities compared to their single metal counterparts, highlighting the benefits of alloying strategies [98]. The introduction of transition metal chalcogenides [99], such as sulfides, selenides [100], and tellurides, into MOFs has also been explored [101]. These compounds exhibit strong ionic bonds and effective hydrogen adsorption, making them suitable for HER.

Recent advancements have also seen the development of 2D MOFs and their derivatives, focusing on single metal and bimetallic/polymetallic MOFs for enhanced HER performance [98]. He et al. presented bifunctional MOFs with dual active sites in the Co–catecholate (Co–CAT) created through Ru, Ir, or Rh doping for overall water-splitting [102]. Among these catalysts, RuCo–CAT showed outstanding bifunctional activity, superior to Pt/C and RuO_2 in both the HER and OER reactions. Theoretical calculations show that the introduction of Ru atoms provides favorable adsorption energy for the hydrogen intermediate, while the Co center lowers the energy barrier for the rate-determining step and thus becomes the active site for HER and OER, respectively. The integration of Ru atoms also enhances the electrical conductivity and water adsorption capacity of Co–CAT, synergistically enhancing its bifunctional catalytic activity.

Sun et al. focus on enhancing HER efficiency in layered double hydroxides (LDHs) [97]. They reported a novel MOF-mediated topotactic transformation tactic to create holey ternary CoFeNi LDHs on NF. These LDHs exhibit polygonal mesopores with atomistic edge steps and lattice defects. The optimized catalyst with a Co/Fe/Ni ratio of c. 1 : 1 : 1 requires only an external voltage of 1.49 V to afford a water-splitting current density of $10\,mA\,cm^{-2}$. Advanced spectroscopies and DFT reveal that the synergetic effect among under-coordinated metal centers lowers the energy barrier for water dissociation. Furthermore, the doping of Fe modulates the density of states (DOS) of Co and Ni, enhancing the intrinsic activity of HER.

The catalytic efficiencies of most MOFs are limited by low electrical conductivity and inappropriate adsorption/desorption energy of intermediates. Geng et al. introduced cobalt (Co) into conductive Cu–catecholate (Cu–CAT) nanorod arrays grown directly on flexible carbon cloth (CuCo–CAT/CC) to enhance the catalytic efficiency [103]. These Co-incorporated CuCAT nanorods arrays exhibit significantly enhanced HER performance with low overpotentials of 52 and 143 mV at $10\,mA\,cm^{-2}$ in alkaline and neutral media, respectively. The synergistic effect of the bimetallic Co and Cu nodes is crucial for enhanced performance. Co-doping optimizes the adsorption energy of hydrogen on Cu sites and enhances water molecule adsorption. The Cu sites, with an optimized Gibbs free energy (ΔG_{H*}), accelerate the Heyrovsky step, making them the primary active sites, while Co sites assist in water adsorption (ΔE_{H_2O}) and hydrogen generation. In addition, the high conductivity of Cu–CAT nanorods, further enhanced by Co-doping, facilitates charge transfer during HER.

Furthermore, the development of freestanding 3D heterostructure films combining MOF-derived components with other materials, such as reduced graphene oxide, has opened new avenues in electrocatalysis. MOF-derived materials possess well-defined 3D morphology, which facilitates efficient mass transport and promotes the exposure of active sites in electrocatalytic processes [104]. Yan et al. developed a freestanding 3D heterostructure film, which is a synergistic combination of nickel–metal–organic framework (Ni–MOF) derived carbon nanotube microspheres and reduced graphene oxide (rGO) sheets [105]. When N-doped CNTs with hollow-microsphere (HM) Ni–MOF was used, the film (called Ni@N-HCGHF) demonstrated a low overpotential of 95 mV at $10\,mA\,cm^{-2}$ for HER. This performance is attributed to the effective synergy between the N-doped carbon shell and Ni nanoparticles. The film's innovative design ensures a controlled dopant, high surface area, hierarchical pores, and efficient electron and ion transfer pathways, contributing to its superior electrocatalytic activity and stability.

The advancement in MOF-derived electrocatalysts represents a significant step toward efficient and sustainable hydrogen production. By leveraging the unique properties of MOFs and employing innovative synthesis techniques, researchers have developed materials that exhibit superior catalytic performance in crucial energy-related reactions.

6.3.4 Atomic Metal Doping of Carbon Materials

The exploration of sustainable energy solutions has become a cornerstone of modern scientific endeavors [106]. Hydrogen, as a renewable and environmentally friendly energy carrier, stands out as the optimal substitute for fossil fuels [107]. The HER is a cornerstone reaction in electrochemical water-splitting, a process integral to the production of green hydrogen [9]. Traditionally, platinum and its alloys have been the benchmark catalysts for HER due to their low overpotential and high catalytic activity [108]. However, the scarcity and high costs of platinum have led to the exploration of alternative catalysts. Single-atom catalysts (SACs) have been used for the preparation of HER due to their cost-effectiveness [109, 110].

Presently, an optimal electrocatalyst for HER should exhibit the following key attributes: (i) robust electrocatalytic activity across a broad pH range in electrolyte solutions; (ii) exceptional durability, maintaining stable activity in complex electrolytes; (iii) a straightforward synthesis process conducive to industrial production; (iv) minimal usage of noble metals to reduce costs; and (v) easy recyclability and reusability [111]. Consequently, the strategic design of electrocatalysts based on their structure and morphology stands out as an effective approach to augment the electrocatalytic performance of the HER. Various strategies can be employed in synthesizing catalysts to achieve high performance. Resulting catalysts may possess several desirable properties, including (i) a large specific surface area and numerous exposed active sites; (ii) improved electrical contact; (iii) increased porosity between adjacent nanostructures within the catalyst, facilitating efficient charge transport deep into the catalyst surface; and (iv) enhanced crystal and topological defects, elevating the material's electron density and conductivity. Notably, the size of metal particles emerges as a crucial factor influencing the properties and performance of an electrocatalyst. Both theoretical and experimental studies have shown that sub-nanometer catalysts have better electrocatalytic activity and selectivity compared with nanoscale catalysts [112, 113]. Additionally, low-coordination and unsaturated atoms often play a pivotal role as active sites in catalysis [114]. Consequently, reducing particle size to nano and atomic levels, such as nanoclusters and single atoms, is a promising strategy to enhance HER.

Reducing metal nanoparticles to nanoclusters or even single atoms can significantly enhance the catalytic activity and selectivity toward electrochemical reactions [115]. One example is the synthesis of single-atom Pt catalysts by atomic layer deposition (ALD), which allows precise control of particle size from nanoparticles to single atoms [116]. These Pt-based catalysts show high activity and notable stability in acidic environments [82].

In addition, a study conducted by Kuang et al. showed that the electrocatalytic activity of Pt (Pt) single atoms in the HER was significantly increased compared to Pt nanoparticles and commercially available 20% Pt/C catalysts when Pt (Pt) single atoms were loaded on nitrogen-doped mesoporous hollow carbon spheres (Pt_1/NMHCS) catalysts. [117]. This superior performance is attributed to a strong electronic metal–support interaction (EMSI) effect, which tailors the electronic structure of Pt atoms. The HAADF–STEM analysis confirms the even dispersion of Pt_1 single atoms across the NMHCS support. Experimentally, the Pt_1/NMHCS catalysts show a low overpotential of 40 mV at a current density of 10 mA cm^{-2}, a large TOF, and a high mass activity of 2.07 A $mg^{-1}{}_{Pt}$ at 50 mV overpotential, indicating their high efficiency. Specifically, Pt_1/NMHCS displays outstanding durability in acidic electrolytes, maintaining their catalytic activity over prolonged usage. The results show that the unique N_1–Pt_1–C_2 coordination structure in Pt_1/NMHCS plays a crucial role in this performance, as it facilitates the reduction of adsorbed protons and enhances H–H coupling. This study unveils the path to developing Pt_1-single-atom electrocatalysts, combining high efficiency and durability with a specific coordination environment tailored for the desired sustainable activity in HER.

Apart from Pt SACs, Ru single atoms (Ru SACs) were also introduced in the conductive substrate and investigated as HER electrocatalysts. For example, Yao and coworkers synthesized Ru SACs anchored on hierarchical porous nitrogen-doped carbon (NMC-Ru_{SA+NC}), offering a cost-effective and efficient solution for HER [118]. NMC-Ru_{SA+NC} demonstrates an ultralow overpotential for achieving a high current density of 500 mA cm^{-2} in 1.0 M KOH (185 mV) and 0.5 M H_2SO_4 (161 mV). Experimental and DFT results reveal that the coupling of Ru single atoms and nanoclusters can synergistically produce strong electronic coupling effects. These effects prove beneficial for achieving rapid water dissociation and optimizing the strength of metal-adsorbate H*. Additionally, the hierarchical porous structure of NMC-Ru_{SA+NC} facilitates the swift transport of reactants to active sites and the removal of produced H_2, thereby enhancing the performance of HER at high current densities.

Besides, single-atom non-noble metal materials [119] have been developed for HER. For example, Qiu and his colleagues utilized CVD to anchor single nickel atoms onto nanoporous graphene (np-Ni), demonstrating superior HER performance with low overpotential of ≈50 mV at 10 mA cm^{-2} and low Tafel slope of 45 mV dec^{-1} in 0.5 m H_2SO_4 [120]. The study also delves into the catalytic mechanism of these nickel-doped graphene catalysts. The results show that the high HER activity is attributed to the unique electronic structure of the single-atom Ni dopant interacting with the graphene substrate. The Ni atoms are theorized to occupy carbon sites in the graphene lattice, creating an efficient catalytic site due to the sp–d orbital charge transfer. This results in the formation of a hybrid orbital near the Fermi level, which becomes unoccupied. This local structural unit, composed of the Ni dopant and the surrounding carbon atoms, becomes an active site for catalysis. The presence of the H* atom forms a chemical bond with this hybrid orbital, effectively reducing the energy required for H* adsorption. Furthermore, the charge transfer between the Ni dopants and the surrounding carbon atoms also serves to stabilize the Ni species in acidic solutions. This chemical bonding, resulting from the charge transfer, plays a significant role in maintaining the stability of the Ni-doped graphene. Therefore, the synergistic interaction between the substitutional single-atom Ni dopant and the surrounding carbon atoms not only enhances the HER catalysis but also ensures the long-term stability of the Ni-doped graphene material.

In addition, many high-performance electrocatalysts toward HER have been explored [121]. For instance, Su et al. showed that individual metal atoms dispersed on a carbon substrate can establish remote communication with the metal nanoparticles supported on it. They supported ruthenium (Ru) nanoparticles by dispersing single metal atoms (e.g. Fe, Co, and Ni) on an oxygen-doped graphene substrate (Ru/M@OG) (Figure 6.6a–c) [122]. This interaction leads to a synergistic electronic coupling with the nanoparticles, allowing for precise control over their electrocatalytic activity. The hybrid catalysts display superior HER activity, boasting an overpotential of 13 mV at a current density of 10 mA cm^{-2} and ultralow Tafel slopes of 22.8 mV dec^{-1}, outperforming the commercial Pt/C electrocatalyst (Figure 6.6d and e). DFT calculations suggest that modifying an O-doped graphene substrate with single metal atoms effectively enhances the EMSI with Ru nanoparticles. This enhancement leads to the redistribution of electrons in the nanoparticles, resulting in an optimized adsorption free energy of H* species on Ru particles and ultimately boosting the HER activity (Figure 6.6f). Additionally, the presence of metal atoms on the O-doped graphene strengthens the binding of the Ru nanoparticles to the substrate, preventing their agglomeration and improving their stability. Our findings highlight the effectiveness of this approach in developing highly active catalysts for HER without the need for platinum.

In summary, SACs, particularly those comprising transition metals, have demonstrated excellent HER activity. Their high surface area and unique electronic properties enable efficient catalytic reactions, often surpassing the performance of traditional platinum-based catalysts. Furthermore, the ability to tune the electronic structure of SACs by manipulating the metal–support interactions open new avenues for optimizing HER activity. In addition, SACs can also provide simplified ideal model systems to evaluate experimental results directly through DFT. These developments underscore the versatility and adaptability of SACs in various environments.

Despite these advancements, challenges persist in the field. The stability of SACs under long-term operational conditions, particularly at high current densities, remains a concern. Additionally, scaling up the synthesis of SACs for industrial applications poses significant challenges. There is also a need to further optimize the electronic and structural properties of SACs to enhance their catalytic performance. Addressing these challenges requires a deeper understanding of the fundamental aspects of SACs, including the nature of metal–support interactions and the dynamics of the catalytic process. Besides, advanced in situ characterization techniques are needed, such as operando atomic resolution imaging that can directly observe the reaction process at the active site and highly sensitive surface

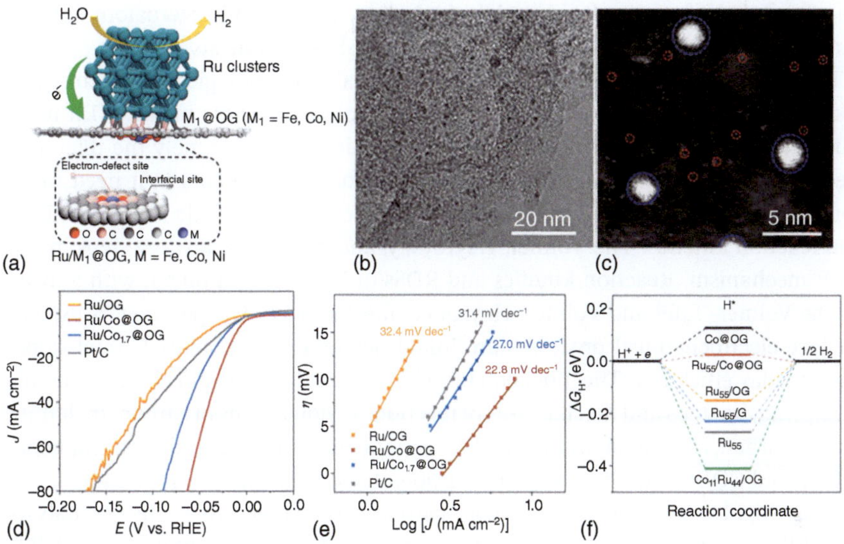

Figure 6.6 (a–c) Schematic representation and characterization of Ru-based catalysts, (d–e) HER performance of Ru-based catalysts and commercial Pt/C, (f) free energy diagrams for hydrogen evolution at zero potential and pH = 0 on various catalysts by DFT calculation. Source: Su et al. [122]/Wiley-VCH/CC BY 4.0.

spectroscopy techniques that can address the molecular structure and catalytic behavior of the SACs during HER.

6.4 Summary

The chapter emphasizes the importance of renewable energy as well as the potential of carbon-based electrochemical catalysts for HER. Sluggish reaction kinetics can reduce electrocatalytic hydrogen production efficiency of HER process. As a result, great efforts are made to develop catalysts that may improve the hydrogen energy efficiency of HER.

In an in-depth discussion of several carbon-based compounds with promising catalytic activity as well as strategies for improving their activity, it is noteworthy that metals and metal oxides perform exceptionally in HER, mainly precious metals like Pt, IrO_2, and RuO_2. However, high cost and limited access prevent these metals from being widely used at a commercial scale. On the other hand, carbon-based materials like GQDs, carbon dots (CQD), and polymer dots (PD) have shown exceptional performance owing to their adjustable structures and physicochemical characteristics. These carbon-based composite materials are also sustainable and have lower rate of heavy metal pollutants as compared to conventional materials.

Furthermore, the catalyst's HER performance is improved by altering the electronic structure and particle size of catalysts via heteroatom doping approaches. Low-dimensional carbon materials, particularly graphene-based catalysts, demonstrate exceptional conductivity and stability during electrocatalytic activity for HER.

The formation of composite materials, including sulfides and heteroatom doping, can significantly enhance the activity of graphene-based catalysts.

The mechanism of HER highlights three main technologies in electrolysis: alkaline electrolysis, PEM electrolysis, and electrolysis in acidic media. HER measurement is characterized by using three electrode systems, namely the reference electrode, counter electrode, and working electrode. The factors that need to be considered in the hydrogen production process include electrolytic medium, RDSs, and reaction kinetics. The Volmer, Heyrovsky, and Tafel steps are key steps in the HER mechanism. Reaction kinetics and RDSs of HER are examined, with a focus on the Volmer–Tafel and Volmer–Heyrovsky mechanisms. The parameters used to analyze electrochemical processes are initial potential, overpotential, Tafel slope, and Faraday efficiency. The Tafel equation is an important analytical technique for describing the steps and mechanisms of reaction. Various forms of carbon, including porous carbon, carbon nanotubes, carbon nanofibers, carbon nanosheets, graphene, MOFs, and CQDs, have been explored as potential electrocatalytic materials.

It is highly desirable to develop carbon-based metal-free electrocatalysts (C-MFECs) as an alternative to precious metal catalysts. The advantages of C-MFEC include its affordability, availability, and compatibility with a variety of catalytic active sites. Strategies to further increase catalytic activity include introducing heteroatoms, defects, and adsorbents into different carbon architectures.

In addition, CQDs, described as 0D carbon nanoparticles less than 10 nm in size, offer a number of benefits when used as catalysts. They possess exceptional conductivity, good dispersibility, a large specific surface area, and chemical stability. Besides, they are nontoxic, simple to synthesize, and cost-effective. Abundant functional groups on the surface of CQD improve its interactions with metal ions and water molecules, providing suitable conditions for the formation of highly active catalytic materials. Electrons can be donated and accepted by CQD, facilitating efficient electron transfer.

Their size and electronic configuration can be adjusted through simple modifications, such as doping with heteroatoms or combining CQDs with metals, which increases their electron capture ability and improves electron transfer efficiency. These properties improve reaction kinetics by reducing the dissociation energy of intermediate reactants. On the other hand, carbon nanotubes are 1D carbon-based materials that can also serve as catalysts for HER.

In addition, graphene and its derivatives are ideal materials for HER electrocatalysts due to their large specific surface area, strong electron transfer ability, good stability, and adjustable surface properties. 2D graphene-based electrocatalysts can be chemically modified to enhance their electrochemical activity. The interaction between graphene and metal can be regulated, leading to hybridization of metal and graphene states, even for monolayer graphene which results in band splitting. In addition, 2D graphene can serve as a carrier for dispersing other HER active materials, promoting the kinetics of HER. The drawback of reaggregation and restacking of graphene-based nanomaterials has attracted much attention, and 3D porous graphene structures have been developed to address this drawback. These structures prevent the loss of electrochemical active sites and provide a growth platform for other HER active materials. Doping nitrogen and other heteroatoms into graphene can improve its catalytic performance by adjusting its electronic structure and increasing the number of active sites.

In addition, MOFs have a unique structure and adjustable composition, making them a demanding candidate for HER electrocatalysis. The incorporation of transition metal chalcogenides (such as sulfides, selenides, and tellurides) into MOFs demonstrates the potential for HER due to their strong ionic bonds and effective hydrogen adsorption. Compared with single metal sulfides, the catalytic activity of bimetallic and polymetallic sulfides derived from MOFs has been improved, highlighting the advantages of alloying strategies. Further progress includes the development of 2D MOFs and their derivatives, including single metal and bimetallic/multimetal MOFs, to improve HER performance by incorporating dopants such as Ru, Ir, or Rh, which outperform benchmark materials. In addition, the introduction of MOF-derived components and other materials, such as reduced graphene oxide into 3D heterostructure films has shown promising prospects in improving mass transfer and exposing active sites, thus exhibiting excellent electrocatalytic activity and stability.

The optimal electrocatalyst for HER should have strong activity, extraordinary durability, a simple synthesis process, minimal use of precious metals, and easy recovery over a wide pH range. Reducing metal nanoparticles to nanoclusters and single atoms can improve catalytic activity and selectivity. The reduction in size increases the surface volume ratio, geometric effects, and electronic properties of the catalyst. The performance of Pt and Ru SACs loaded on nitrogen-doped carbon materials indicates that their electrocatalytic activity and stability are improved compared to nanoparticles. In addition, due to the unique electronic structure of doped atoms, the use of nonprecious metal materials, such as single nickel atoms on nanoporous graphene, exhibits excellent HER performance.

Overall, this chapter provides valuable insights into the development of efficient and sustainable hydrogen production catalysts, focusing on carbon-based materials and their modifications. We hope these material synthesis methods and reaction mechanisms can be helpful and inspiring for readers engaged in the field of HER.

References

1 Nemiwal, M., Zhang, T.C., and Kumar, D. (2021). Graphene-based electrocatalysts: hydrogen evolution reactions and overall water splitting. *Int. J. Hydrog. Energy* 46 (41): 21401.

2 Shah, S.S.A., Najam, T., Aslam, M.K. et al. (2020). Recent advances on oxygen reduction electrocatalysis: correlating the characteristic properties of metal organic frameworks and the derived nanomaterials. *Appl. Catal. B Environ.* 268: 118570.

3 Wu, H., Wang, J., Jin, W. et al. (2020). Recent development of two-dimensional metal-organic framework derived electrocatalysts for hydrogen and oxygen electrocatalysis. *Nanoscale* 12 (36): 18497.

4 Lai, W.H., Zhang, L.F., Hua, W.B. et al. (2019). General π-electron-assisted strategy for ir, Pt, Ru, Pd, Fe, Ni single-atom electrocatalysts with bifunctional active sites for highly efficient water splitting. *Angew. Chem. Int. Ed.* 58 (34): 11868.
5 Zheng, T., Shang, C., He, Z. et al. (2019). Intercalated iridium diselenide electrocatalysts for efficient pH-universal water splitting. *Angew. Chem. Int. Ed.* 58 (41): 14764.
6 Luo, M., Yang, Y., Sun, Y. et al. (2019). Ultrathin two-dimensional metallic nanocrystals for renewable energy electrocatalysis. *Mater. Today* 23: 45.
7 Nemiwal, M. and Kumar, D. (2021). Recent progress on electrochemical sensing strategies as comprehensive point-care method. *Monatsh. Chem.* 152 (1): 1.
8 Selvaraju, N., Ravichandran, K., and Venugopal, G. (2023). A short review of the kinetic parameters of carbon quantum dots for electrocatalytic hydrogen evolution reaction. *Int. J. Hydrog. Energy* 48 (10): 3807.
9 Zhu, J., Hu, L., Zhao, P. et al. (2020). Recent advances in electrocatalytic hydrogen evolution using nanoparticles. *Chem. Rev.* 120 (2): 851.
10 Wu, J., Liu, J., Yuan, S. et al. (2015). Sulfur transformation during hydrothermal dewatering of low rank coal. *Energy Fuel* 29 (10): 6586.
11 Wu, H., Lu, S., and Yang, B. (2022). Carbon-dot-enhanced electrocatalytic hydrogen evolution. *Acc. Mater. Res.* 3 (3): 319.
12 Li, J., Ji, Y., Chen, J., and Wen, Z. (2021). Marking electrocatalysts on the "volcanic belt" of hydrogen electrode reactions. *J. Phys. Chem. C* 125 (10): 5587.
13 Kibsgaard, J., Jaramillo, T.F., and Besenbacher, F. (2014). Building an appropriate active-site motif into a hydrogen-evolution catalyst with thiomolybdate $[Mo_3S_{13}]^{2-}$ clusters. *Nat. Chem.* 6 (3): 248.
14 Rodríguez, J. and Amores, E. (2020). CFD modeling and experimental validation of an alkaline water electrolysis cell for hydrogen production. *Processes* 8 (12): 1634.
15 Li, Z., Yue, Y., Peng, J., and Luo, Z. (2023). Phase engineering two-dimensional nanostructures for electrocatalytic hydrogen evolution reaction. *Chin. Chem. Lett.* 34 (1): 107119.
16 Wang, J., Kong, H., Zhang, J. et al. (2021). Carbon-based electrocatalysts for sustainable energy applications. *Prog. Mater. Sci.* 116: 100717.
17 Murthy, A.P., Madhavan, J., and Murugan, K. (2018). Recent advances in hydrogen evolution reaction catalysts on carbon/carbon-based supports in acid media. *J. Power Sources* 398: 9.
18 Zhou, W., Jia, J., Lu, J. et al. (2016). Recent developments of carbon-based electrocatalysts for hydrogen evolution reaction. *Nano Energy* 28: 29.
19 Lu, J., Yin, S., and Shen, P.K. (2018). Carbon-encapsulated electrocatalysts for the hydrogen evolution reaction. *Electrochem. Energy Rev.* 2 (1): 105.
20 Wang, B. and Zhang, F. (2022). Main descriptors to correlate structures with the performances of electrocatalysts. *Angew. Chem. Int. Ed.* 61 (4): e202111026.
21 Zhang, L., Xiao, J., Wang, H., and Shao, M. (2017). Carbon-based electrocatalysts for hydrogen and oxygen evolution reactions. *ACS Catal.* 7 (11): 7855.

22 Han, Y., Yan, X., Wu, Q. et al. (2023). Defect-derived catalysis mechanism of electrochemical reactions in two-dimensional carbon materials. *Small Struct.* 4 (10): 2300036.

23 Xiao, X., Li, Z., Xiong, Y., and Yang, Y.-W. (2023). Irmo nanocluster-doped porous carbon electrocatalysts derived from cucurbit[6]uril boost efficient alkaline hydrogen evolution. *J. Am. Chem. Soc.* 145 (30): 16548.

24 Nangan, S., Natesan, T., Sukmas, W. et al. (2023). Waste plastics derived nickel-palladium alloy filled carbon nanotubes for hydrogen evolution reaction. *Chemosphere* 341: 139982.

25 Yu, S., Zou, Y., Wang, Q. et al. (2022). Self-supported Co–Mo sulfide in electrospun carbon nanofibers as electrocatalysts for hydrogen evolution reaction in alkaline medium. *J. Alloys Compd.* 911: 165094.

26 Huang, Y., Hu, Z., Huang, L.A. et al. (2023). Phosphorus-modified cobalt single-atom catalysts loaded on crosslinked carbon nanosheets for efficient alkaline hydrogen evolution reaction. *Nanoscale* 15 (7): 3550.

27 Di, F., Wang, X., Farid, S., and Ren, S. (2023). CoS_2 with carbon shell for efficient hydrogen evolution reaction. *Int. J. Hydrog. Energy* 48 (47): 17758.

28 Rimal, V., Mahapatra, S.S., and Srivastava, P.K. (2022). Metal-free oleic acid-derived carbon dots as efficient catalysts for hydrogen evolution reaction. *J. Appl. Electrochem.* 53 (2): 285.

29 Gunaseelan, H., Munde, A.V., Patel, R., and Sathe, B.R. (2023). Metal-organic framework derived carbon-based electrocatalysis for hydrogen evolution reactions: a review. *Mater. Today Sustainability* 22: 100371.

30 Yang, C., Tao, S., Huang, N. et al. (2020). Heteroatom-doped carbon electrocatalysts derived from nanoporous two-dimensional covalent organic frameworks for oxygen reduction and hydrogen evolution. *ACS Appl. Nano Mater.* 3 (6): 5481.

31 Zhang, Y., Yan, Z., Gao, Y. et al. (2022). Electron modulation by atomic ir site decoration in porous Co/N Co-doped carbon for electrocatalytic hydrogen evolution. *Nano Res.* 16 (2): 2011.

32 Sahoo, M.R., Ray, A., Ahuja, R., and Singh, N. (2023). Activation of metal-free porous basal plane of biphenylene through defects engineering for hydrogen evolution reaction. *Int. J. Hydrog. Energy* 48 (28): 10545.

33 Feng, L., Fu, C., Li, D. et al. (2023). High-density low-coordination Ni single atoms anchored on Ni-embedded nanoporous carbon nanotubes for boosted alkaline hydrogen evolution. *Dalton Trans.* 52 (28): 9684.

34 Hu, C., Gao, Y., Zhao, L., and Dai, L. (2022). Carbon-based metal-free electrocatalysts: recent progress and forward looking. *Chem. Catal.* 2 (9): 2150.

35 Walter, C., Beltrán-Suito, R., Schwarze, M. et al. (2023). Elemental chalcogens acting as metal-free electrocatalysts for effective alkaline and acidic hydrogen evolution reaction. *Catal. Today* 423: 113917.

36 Zhang, Y., Chen, L., Hou, L., and Liu, X. (2023). N-, p-, and o-doped porous carbon as advanced trifunctional metal-free electrocatalysts. *Catal. Sci. Technol.* 13 (3): 910.

37 He, H., Zhang, Y., Zhang, W. et al. (2022). Porous carbon nanofibers derived from silk fibroin through electrospinning as n-doped metal-free catalysts for hydrogen evolution reaction in acidic and alkaline solutions. *ACS Appl. Mater. Interfaces* 14 (1): 834.

38 Jiao, Y., Zheng, Y., Davey, K., and Qiao, S.-Z. (2016). Activity origin and catalyst design principles for electrocatalytic hydrogen evolution on heteroatom-doped graphene. *Nat. Energy* 1 (10): 16130.

39 Xu, J., Zhong, M., Song, N. et al. (2023). General synthesis of Pt and Ni Co-doped porous carbon nanofibers to boost her performance in both acidic and alkaline solutions. *Chin. Chem. Lett.* 34 (2): 107359.

40 Zheng, Y., Jiao, Y., Vasileff, A., and Qiao, S.-Z. (2018). The hydrogen evolution reaction in alkaline solution: from theory, single crystal models, to practical electrocatalysts. *Angew. Chem. Int. Ed.* 57 (26): 7568.

41 Heckova, M., Streckova, M., Orinakova, R. et al. (2020). Porous carbon fibers for effective hydrogen evolution. *Appl. Surf. Sci.* 506: 144955.

42 Liu, Y., Yue, C., Sun, F. et al. (2023). Superhydrophilic molybdenum phosphide quantum dots on porous carbon matrix for boosting hydrogen evolution reaction. *Chem. Eng. J.* 454: 140105.

43 Guo, Y.G., Hu, J.S., and Wan, L.J. (2008). Nanostructured materials for electrochemical energy conversion and storage devices. *Adv. Mater.* 20 (15): 2878.

44 Wu, H., Feng, C., Zhang, L. et al. (2021). Non-noble metal electrocatalysts for the hydrogen evolution reaction in water electrolysis. *Electrochem. Energy Rev.* 4 (3): 473.

45 Tian, J., Chen, J., Liu, J. et al. (2018). Graphene quantum dot engineered nickel-cobalt phosphide as highly efficient bifunctional catalyst for overall water splitting. *Nano Energy* 48: 284.

46 Liu, Y., Ye, N., Li, X. et al. (2021). Nitrogen-doped carbon quantum dots via a facile reflux assisted polymerization of *N*-methyl-pyrrolidone for hydrogen evolution reaction. *J. Solid State Chem.* 293: 121781.

47 Liu, Y., Yong, X., Liu, Z. et al. (2019). Unified catalyst for efficient and stable hydrogen production by both the electrolysis of water and the hydrolysis of ammonia borane. *Adv. Sustainable Syst.* 3 (5): 1800161.

48 Li, W., Wei, Z., Wang, B. et al. (2020). Carbon quantum dots enhanced the activity for the hydrogen evolution reaction in ruthenium-based electrocatalysts. *Mater. Chem. Front.* 4 (1): 277.

49 Liu, Y., Li, X., Zhang, Q. et al. (2020). A general route to prepare low-ruthenium-content bimetallic electrocatalysts for pH-universal hydrogen evolution reaction by using carbon quantum dots. *Angew. Chem. Int. Ed.* 59 (4): 1718.

50 Chen, W., Qin, Z., McElhenny, B. et al. (2019). The effect of carbon quantum dots on the electrocatalytic hydrogen evolution reaction of manganese–nickel phosphide nanosheets. *J. Mater. Chem. A* 7 (37): 21488.

51 Javan, H., Asghari, E., Ashassi-Sorkhabi, H., and Haghighi, M.M. (2020). A low-cost platinum-free electrocatalyst based on carbon quantum dots decorated

Ni–Cu hierarchical nanocomposites for hydrogen evolution reaction. *Int. J. Hydrog. Energy* 45 (38): 19324.

52 Yang, Y., Liu, J., Guo, S. et al. (2015). A nickel nanoparticle/carbon quantum dot hybrid as an efficient electrocatalyst for hydrogen evolution under alkaline conditions. *J. Mater. Chem. A* 3 (36): 18598.

53 Qin, T., Wang, Z., Wang, Y. et al. (2021). Recent progress in emerging two-dimensional transition metal carbides. *Nanomicro Lett.* 13 (1): 183.

54 Yang, X., Gao, N., Zhou, S., and Zhao, J. (2018). MXene nanoribbons as electrocatalysts for the hydrogen evolution reaction with fast kinetics. *Phys. Chem. Chem. Phys.* 20 (29): 19390.

55 Jiang, B., Yang, T., Wang, T. et al. (2022). Edge stimulated hydrogen evolution reaction on monodispersed MXene quantum dots. *Chem. Eng. J.* 442: 136119.

56 Xue, Y., Xie, Y., Xu, C. et al. (2022). 0D/2D heterojunction of graphene quantum dots/MXene nanosheets for boosted hydrogen evolution reaction. *Surf. Interfaces* 30: 101907.

57 Wu, Q., Yang, L., Wang, X., and Hu, Z. (2017). From carbon-based nanotubes to nanocages for advanced energy conversion and storage. *Acc. Chem. Res.* 50 (2): 435.

58 He, S., Wu, M., Li, S. et al. (2022). Research progress on graphite-derived materials for electrocatalysis in energy conversion and storage. *Molecules* 27 (24): 8644.

59 Jiao, S., Fu, X., Wang, S., and Zhao, Y. (2021). Perfecting electrocatalystsviaimperfections: towards the large-scale deployment of water electrolysis technology. *Energy Environ. Sci.* 14 (4): 1722.

60 Yang, L., Jiang, S., Zhao, Y. et al. (2011). Boron-doped carbon nanotubes as metal-free electrocatalysts for the oxygen reduction reaction. *Angew. Chem. Int. Ed.* 50 (31): 7132.

61 Lu, X., Yang, X., Wang, L. et al. (2023). N-doped carbon nanotubes with high amount of graphitic nitrogen as an excellent electrocatalyst for water splitting in alkaline solution. *J. Electroanal. Chem.* 931: 117160.

62 Qu, K., Zheng, Y., Jiao, Y. et al. (2016). Polydopamine-inspired, dual heteroatom-doped carbon nanotubes for highly efficient overall water splitting. *Adv. Energy Mater.* 7 (9): 1602068.

63 Tavakkoli, M., Holmberg, N., Kronberg, R. et al. (2017). Electrochemical activation of single-walled carbon nanotubes with pseudo-atomic-scale platinum for the hydrogen evolution reaction. *ACS Catal.* 7 (5): 3121.

64 Nguyen, T.K., Bannov, A.G., Popov, M.V. et al. (2018). High-temperature-treated multiwall carbon nanotubes for hydrogen evolution reaction. *Int. J. Hydrog. Energy* 43 (13): 6526.

65 Sun, X., Habibul, N., and Du, H. (2021). $Co_{0.85}Se$ magnetic nanoparticles supported on carbon nanotubes as catalyst for hydrogen evolution reaction. *Chin. J. Catal.* 42 (1): 235.

66 Tabassum, H., Guo, W., Meng, W. et al. (2017). Metal-organic frameworks derived cobalt phosphide architecture encapsulated into B/N Co-doped graphene nanotubes for all pH value electrochemical hydrogen evolution. *Adv. Energy Mater.* 7 (9): 1601671.

67 Zhang, X., Yu, X., Zhang, L. et al. (2018). Molybdenum phosphide/carbon nanotube hybrids as pH-universal electrocatalysts for hydrogen evolution reaction. *Adv. Funct. Mater.* 28 (16): 1706523.

68 Liu, Z., Yang, X., Hu, G., and Feng, L. (2020). Ru nanoclusters coupled on Co/N-doped carbon nanotubes efficiently catalyzed the hydrogen evolution reaction. *ACS Sustain. Chem. Eng.* 8 (24): 9136.

69 Li, T., Cui, Z., Yuan, W. et al. (2016). Ionic liquid functionalized carbon nanotubes: metal-free electrocatalyst for hydrogen evolution reaction. *RSC Adv.* 6 (16): 12792.

70 Li, T., Chen, Y., Hu, W. et al. (2021). Ionic liquid in situ functionalized carbon nanotubes as metal-free catalyst for efficient electrocatalytic hydrogen evolution reaction. *Nanoscale* 13 (8): 4444.

71 Chu, S. and Majumdar, A. (2012). Opportunities and challenges for a sustainable energy future. *Nature* 488 (7411): 294.

72 Nadolska, M., Przesniak-Welenc, M., Lapinski, M., and Sadowska, K. (2021). Synthesis of phosphonated carbon nanotubes: new insight into carbon nanotubes functionalization. *Materials* 14 (11): 2726.

73 Garcia-Miranda Ferrari, A., Brownson, D.A.C., and Banks, C.E. (2019). Investigating the integrity of graphene towards the electrochemical hydrogen evolution reaction (HER). *Sci. Rep.* 9 (1): 15961.

74 Xue, Y., Huang, B., Yi, Y. et al. (2018). Anchoring zero valence single atoms of nickel and iron on graphdiyne for hydrogen evolution. *Nat. Commun.* 9 (1): 1460.

75 Huang, H., Yan, M., Yang, C. et al. (2019). Graphene nanoarchitectonics: recent advances in graphene-based electrocatalysts for hydrogen evolution reaction. *Adv. Mater.* 31 (48): e1903415.

76 Jaimes-Paez, C.D., Vences-Alvarez, E., Salinas-Torres, D. et al. (2023). Microwave-assisted synthesis of carbon-supported Pt nanoparticles for their use as electrocatalysts in the oxygen reduction reaction and hydrogen evolution reaction. *Electrochim. Acta* 464: 142871.

77 Fukushima, T. and Murakoshi, K. (2019). Molecularly defined graphitic interface toward proton manipulation. *Curr. Opin. Electrochem.* 17: 158.

78 Mao, J., Iocozzia, J., Huang, J. et al. (2018). Graphene aerogels for efficient energy storage and conversion. *Energy Environ. Sci.* 11 (4): 772.

79 Liu, Z., He, T., Jiang, Q. et al. (2022). A review of heteroatomic doped two-dimensional materials as electrocatalysts for hydrogen evolution reaction. *Int. J. Hydrog. Energy* 47 (69): 29698.

80 Liao, Y., Zhu, R., Zhang, W. et al. (2023). Ultrafast synthesis of 3D porous flash graphene and its adsorption properties. *Colloids Surf. A* 676: 132178.

81 Johnson, I., Han, J., and Chen, M. (2022). 3d nanoporous graphene based single-atom electrocatalysts for energy conversion and storage. *Acc. Mater. Res.* 3 (10): 1011.

82 Zhang, Q. and Guan, J. (2020). Single-atom catalysts for electrocatalytic applications. *Adv. Funct. Mater.* 30 (31): 2000768.

83 Cencerrero, J., Romero, A., de Lucas-Consuegra, A. et al. (2023). Towards metal-free nitrogen-doped graphene aerogels as efficient electrocatalysts in hydrogen evolution reaction. *FlatChem.* 42: 100554.

84 Chen, S., Duan, J., Tang, Y. et al. (2015). Molybdenum sulfide clusters-nitrogen-doped graphene hybrid hydrogel film as an efficient three-dimensional hydrogen evolution electrocatalyst. *Nano Energy* 11: 11.

85 Deng, B., Wang, D., Jiang, Z. et al. (2018). Amine group induced high activity of highly torn amine functionalized nitrogen-doped graphene as the metal-free catalyst for hydrogen evolution reaction. *Carbon* 138: 169.

86 Jia, S., Cheng, Y., Huang, Q. et al. (2022). One-step synthesis of Co_2P/N-P Co-doped porous carbon composites derived from soybean derivatives as acidic and alkaline her electrocatalysts. *Int. J. Hydrog. Energy* 47 (59): 24796.

87 Yang, X. (2022). ZiF-8 derived Zn, Mn, S and P Co-loaded on N doped carbon as efficient electrocatalyst for hydrogen evolution reaction (HER). *Int. J. Hydrog. Energy* 47 (42): 18314.

88 Jiang, H., Gu, J., Zheng, X. et al. (2019). Defect-rich and ultrathin N doped carbon nanosheets as advanced trifunctional metal-free electrocatalysts for the ORR, OER and HER. *Energy Environ. Sci.* 12 (1): 322.

89 Niu, W., Li, Z., Marcus, K. et al. (2017). Surface-modified porous carbon nitride composites as highly efficient electrocatalyst for Zn-air batteries. *Adv. Energy Mater.* 8 (1): 1701642.

90 Volokh, M., Peng, G., Barrio, J. et al. (2019). Carbon nitride materials for water splitting photoelectrochemical cells. *Angew. Chem. Int. Ed.* 58 (19): 6138.

91 Liu, Y., Yu, H., Quan, X., and Shalom, M. (2014). Efficient and durable hydrogen evolution electrocatalyst based on nonmetallic nitrogen doped hexagonal carbon. *Sci. Rep.* 4: 6843.

92 Geng, S., Yang, W., and Yu, Y.S. (2019). Building MoS_2/S-doped g-C_3N_4 layered heterojunction electrocatalysts for efficient hydrogen evolution reaction. *J. Catal.* 375: 441.

93 Silva, J.L., Brena, B., and Araujo, C.M. (2020). G-C_3N_4/WTe_2 hybrid electrocatalyst for efficient hydrogen evolution reaction. *J. Phys. Chem. C* 124 (16): 8726.

94 Dong, Y., Liu, S., Deng, W. et al. (2024). Modulating electronic structures of bimetallic Co-Fe sulfide ultrathin nanosheet supported on g-C_3N_4 promoting electrocatalytic hydrogen evolution performance. *J. Colloid Interface Sci.* 653 (Pt B): 1557.

95 Furukawa, H., Cordova, K.E., O'Keeffe, M., and Yaghi, O.M. (2013). The chemistry and applications of metal-organic frameworks. *Science* 341 (6149): 1230444.

96 Liu, T., Li, P., Yao, N. et al. (2019). Self-sacrificial template-directed vapor-phase growth of mof assemblies and surface vulcanization for efficient water splitting. *Adv. Mater.* 31 (21): e1806672.

97 Sun, H., Chen, L., Lian, Y. et al. (2020). Topotactically transformed polygonal mesopores on ternary layered double hydroxides exposing under-coordinated metal centers for accelerated water dissociation. *Adv. Mater.* 32 (52): e2006784.

98 Zhou, Y., Abazari, R., Chen, J. et al. (2022). Bimetallic metal-organic frameworks and MOF-derived composites: recent progress on electro- and photoelectrocatalytic applications. *Coord. Chem. Rev.* 451: 214264.

99 Chen, Z., Qing, H., Zhou, K. et al. (2020). Metal-organic framework-derived nanocomposites for electrocatalytic hydrogen evolution reaction. *Prog. Mater. Sci.* 108: 100618.

100 Thangasamy, P., He, R., Randriamahazaka, H. et al. (2023). Collectively exhaustive electrochemical hydrogen evolution reaction of polymorphic cobalt selenides derived from organic surfactants modified Co-MOFs. *Appl. Catal. B Environ.* 325: 122367.

101 Sanati, S., Morsali, A., and García, H. (2022). First-row transition metal-based materials derived from bimetallic metal-organic frameworks as highly efficient electrocatalysts for electrochemical water splitting. *Energy Environ. Sci.* 15 (8): 3119.

102 He, Y., Yan, F., Zhang, X. et al. (2023). Creating dual active sites in conductive metal-organic frameworks for efficient water splitting. *Adv. Energy Mater.* 13 (20): 2204177.

103 Geng, B., Yan, F., Zhang, X. et al. (2021). Conductive CuCo-based bimetal organic framework for efficient hydrogen evolution. *Adv. Mater.* 33 (49): e2106781.

104 Wang, H.F., Chen, L., Pang, H. et al. (2020). Mof-derived electrocatalysts for oxygen reduction, oxygen evolution and hydrogen evolution reactions. *Chem. Soc. Rev.* 49 (5): 1414.

105 Yan, L., Xu, Y., Chen, P. et al. (2020). A freestanding 3D heterostructure film stitched by MOF-derived carbon nanotube microsphere superstructure and reduced graphene oxide sheets: a superior multifunctional electrode for overall water splitting and Zn-air batteries. *Adv. Mater.* 32 (48): e2003313.

106 Liu, Y., Liu, S., Wang, Y. et al. (2018). Ru modulation effects in the synthesis of unique rod-like Ni@Ni_2P-Ru heterostructures and their remarkable electrocatalytic hydrogen evolution performance. *J. Am. Chem. Soc.* 140 (8): 2731.

107 Staffell, I., Scamman, D., Velazquez Abad, A. et al. (2019). The role of hydrogen and fuel cells in the global energy system. *Energy Environ. Sci.* 12 (2): 463.

108 He, T., Wang, W., Shi, F. et al. (2021). Mastering the surface strain of platinum catalysts for efficient electrocatalysis. *Nature* 598 (7879): 76.

109 Swain, S., Altaee, A., Saxena, M. et al. (2022). A comprehensive study on heterogeneous single atom catalysis: current progress, and challenges. *Coord. Chem. Rev.* 470: 214710.

110 Cai, J., Javed, R., Ye, D. et al. (2020). Recent progress in noble metal nanocluster and single atom electrocatalysts for the hydrogen evolution reaction. *J. Mater. Chem. A* 8 (43): 22467.

111 Qin, X., Ola, O., Zhao, J. et al. (2022). Recent progress in graphene-based electrocatalysts for hydrogen evolution reaction. *Nanomaterials* 12 (11): 1806.

112 Vajda, S., Pellin, M.J., Greeley, J.P. et al. (2009). Subnanometre platinum clusters as highly active and selective catalysts for the oxidative dehydrogenation of propane. *Nat. Mater.* 8 (3): 213.

113 Lei, Y., Mehmood, F., Lee, S. et al. (2010). Increased silver activity for direct propylene epoxidation via subnanometer size effects. *Science* 328 (5975): 224.

114 Remediakis, I.N., Lopez, N., and Norskov, J.K. (2005). Co oxidation on rutile-supported au nanoparticles. *Angew. Chem. Int. Ed.* 44 (12): 1824.

115 Zang, W., Kou, Z., Pennycook, S.J., and Wang, J. (2020). Heterogeneous single atom electrocatalysis, where "singles" are "married". *Adv. Energy Mater.* 10 (9): 1903181.

116 Cheng, N., Stambula, S., Wang, D. et al. (2016). Platinum single-atom and cluster catalysis of the hydrogen evolution reaction. *Nat. Commun.* 7: 13638.

117 Kuang, P., Wang, Y., Zhu, B. et al. (2021). Pt single atoms supported on N-doped mesoporous hollow carbon spheres with enhanced electrocatalytic H_2-evolution activity. *Adv. Mater.* 33 (18): e2008599.

118 Yao, H., Wang, X., Li, K. et al. (2022). Strong electronic coupling between ruthenium single atoms and ultrafine nanoclusters enables economical and effective hydrogen production. *Appl. Catal. B Environ.* 312: 121378.

119 Hossain, M.D., Liu, Z., Zhuang, M. et al. (2019). Rational design of graphene-supported single atom catalysts for hydrogen evolution reaction. *Adv. Energy Mater.* 9 (10): 1803689.

120 Qiu, H.J., Ito, Y., Cong, W. et al. (2015). Nanoporous graphene with single-atom nickel dopants: an efficient and stable catalyst for electrochemical hydrogen production. *Angew. Chem. Int. Ed.* 54 (47): 14031.

121 Mo, F., Zhou, Q., Xue, W. et al. (2023). The optimized catalytic performance of single-atom catalysts by incorporating atomic clusters or nanoparticles: in-depth understanding on their synergisms. *Adv. Energy Mater.* 13 (38): 2301711.

122 Su, P., Pei, W., Wang, X. et al. (2021). Exceptional electrochemical HER performance with enhanced electron transfer between Ru nanoparticles and single atoms dispersed on a carbon substrate. *Angew. Chem. Int. Ed.* 60 (29): 16044.

7

Porous Materials for Electrocatalytic Hydrogen Evolution Reaction

Lixin Yi[1,*], *Xiaoyan Qu*[2,*], *and Zhengqing Liu*[1]

[1] *Northwestern Polytechnical University, Institute of Flexible Electronics (IFE), Xi'an Institute of Biomedical Materials and Engineering, Frontiers Science Center for Flexible Electronics, No.1 Dongxiang Road, Chang'an District, Xi'an, Shaanxi 710129, China*
[2] *Xi'an Jiaotong University, Frontier Institute of Science and Technology, Instrument Analysis Center, State Key Laboratory for Manufacturing Systems Engineering, No.28 Xianning West Road, Xi'an, Shaanxi 710000, China*

7.1 Introduction

Nowadays, the sustainable development of human society has been profoundly impacted by the critical issues of energy and the environment. Fossil fuels, namely oil, coal, and natural gas, continue to dominate the energy demand by a significant margin, accounting for over 80% [1]. However, the excessive reliance on fossil fuels, surging energy requirements, and escalating global environmental concerns have spurred extensive research into renewable energy technologies as a solution to the pressing energy crisis [2–5]. Hydrogen (H_2) has emerged as an ideal clean energy source that has captured worldwide attention [6–9]. Not only is H_2 a promising clean energy option, but it also holds significant importance as a chemical raw material. Among the various methods of H_2 production, electrolysis of water stands out as a reliable and cost-effective approach to achieve industrial scale and affordable H_2 production. This method becomes even more advantageous when paired with electricity derived from green energy sources like wind, geothermal, and solar energy [10–13]. Consequently, the production of H_2 through water electrolysis plays an important role in the advancement of hydrogen energy [14]. During the electrolysis process, water molecules are decomposed into H_2 and oxygen (O_2) by utilizing an electric current. This method of H_2 production not only eliminates the need for fossil fuels but also emits water vapor, making it exceptionally eco-friendly. Furthermore, the electrolysis of water offers several advantages beyond its environmental friendliness. One notable advantage is the capability to produce high-purity H_2, making it suitable for a wide range of applications. As we move forward, H_2 power is anticipated to play a crucial role in various domains, and the method of

[*] These authors contributed equally to this section.

Functional Materials for Electrocatalytic Energy Conversion, First Edition.
Edited by Zhicheng Zhang, Meiting Zhao, and Yuchen Qin.

creating green H_2 through water electrolysis exhibits tremendous potential for its generation and utilization.

During the process of electrolysis of water, the reduction reaction occurs at the cathode where protons or hydrated hydrogen ions accept electrons. This reaction, known as the hydrogen evolution reaction (HER), leads to the production of H_2. The HER can proceed through either the Volmer–Heyrovsky mechanism or the Volmer–Tafel mechanism, with similar steps for both alkaline and acidic conditions [15].

Under alkaline conditions, the steps involved in the HER are as follows:

$$H_2O + e^- \rightarrow OH^- + H_{ads}\ \text{(Volmer)} \tag{7.1}$$

$$H_{ads} + H_2O + e^- \rightarrow OH^- + H_2\ \text{(Heyrovsky)} \tag{7.2}$$

$$\text{or } 2H_{ads} \rightarrow H_2\ \text{(Tafel)} \tag{7.3}$$

Under acidic conditions, the steps of the HER are as follows:

$$H^+ + e^- + * \rightarrow H_{ads}\ \text{(Volmer)} \tag{7.4}$$

$$H_{ads} + H^+ + e^- \rightarrow H_2\ \text{(Heyrovsky)} \tag{7.5}$$

$$\text{or } 2H_{ads} \rightarrow H_2\ \text{(Tafel)} \tag{7.6}$$

In these equations, * represents the active site on the catalyst surface, and "ads" indicates the adsorption state of the intermediate (H_{ads}).

To enhance the rate of the electrochemical reaction under the same conditions, the first step is to identify the rate-determining step (RDS) of the HER reaction, which refers to the slowest step in the overall process. Once identified, appropriate techniques can be employed to expedite the electrochemical reaction rate. The determination of the RDS can be achieved by analyzing the Tafel slope of the reaction during the HER [16]. Specifically, when the Tafel slope is 40 mV dec^{-1}, the RDS is the Heyrovsky process, and when the Tafel slope is 30 mV dec^{-1}, the RDS is the Tafel process. On the other hand, when the Tafel slope is 120 mV dec^{-1}, the RDS is the Volmer process. This process involves the participation of a crucial species, *H. The rate of the intrinsic chemical reaction depends on the binding energy of the *H catalyst surface, which can be derived from the Langmuir adsorption model [17].

It is crucial to adhere to Sabatier's law, which highlights the significance of maintaining a moderate adsorption power of the active intermediate species on the catalyst surface [18]. If adsorption is weak, there will be insufficient intermediate product present on the catalyst surface, causing a considerable reduction in the overall reaction rate. On the other hand, if the adsorption is too strong, the reaction product will bond firmly to the catalyst's active site, preventing it from being desorbed from the surface, ultimately resulting in the reaction coming to an end (as observed in the Langmuir adsorption model). From a physicochemical perspective, the adsorption of H* and the desorption of H_2 in the HER process can be assessed by computing the free energy of hydrogen adsorption (ΔG_{H*}) using density functional theory (DFT). The optimal value of ΔG_H is ideally zero, as it theoretically maximizes the reaction rate of the HER process, leading to a maximal exchange current density (j_0). Thus, the efficiency of the HER relies heavily on the chosen catalyst. Achieving significant efficiency in electrocatalytic hydrogen production necessitates the use of superior electrocatalysts [19–23]. Currently, platinum, ruthenium, palladium, and other valuable metals are deemed optimal catalysts due to their ability to effectively reduce overpotential [24]. However, their high costs and limited resources impose restrictions on their widespread application, further driving the pursuit of low-cost, efficient, long-lasting, and environmentally friendly catalysts, which have become a prominent area of research.

Porous nanomaterials are versatile electrocatalytic materials due to their multilevel porosity, diverse pore size distributions, extensive surface area, excellent electrical and thermal conductivity, and strong chemical resistance [25–30]. These attributes impart rich physicochemical properties, rendering porous nanomaterials economically practical and environmentally benign for electrocatalysis [31–34]. The pore structure of porous nanomaterials has the potential to significantly increase the number of active catalytic centers [35–37]. Furthermore, the aforementioned pore structures can serve as conduits for improving mass transfer during the HER process and facilitate access to surface-active sites during catalysis. Additionally, the catalytic capability is closely linked to the mass transfer of electrolytes and H_2 bubbles. In cases where conductor materials are used, pore channels bolster the electron transfer performance considerably. Furthermore, the collaborative effect between the electrocatalyst and the host material boosts the innate activity of the active sites. This leads to an optimization of the hydrogen adsorption and desorption capacity. Research has demonstrated that pore geometry is a vital parameter that crucially influences ion transfer via the pore channels [38–41]. This ultimately leads to excellent electrochemical performance. Thus, it is possible to categorize porous nanomaterials based on their pore channel dimensions. This section classifies the channels in porous materials into one-dimensional (1D), two-dimensional (2D), and three-dimensional (3D) channels to provide an objective classification. Accordingly, materials possessing these channels are defined as 1D porous materials, 2D porous materials, and 3D porous materials, respectively. Table 7.1 summarizes the HER performance of some typical porous nanomaterials from recent years.

7.2 Porous 1D Nanomaterials

1D nanostructures, like nanowires, nanofibers, nanoribbons, nanorods, and nanotubes, are recognized as highly promising materials for energy-related applications. Porous nanomaterials offer several advantages in their application areas by enhancing controllable nanoscale porosity. Therefore, porous materials have a wide range of applications in energy storage, gas storage, adsorption, catalysis, separation, and sensing. Especially in the field of electrocatalysis, porous materials have unique properties [70–74]. Electrocatalytic technologies have benefited greatly from porous materials. These materials provide larger specific surface areas, more active sites, and faster mass transfer rates [75–78]. Furthermore, porous 1D

Table 7.1 The HER performance of some typical porous catalysts.

Electrocatalysts	Channel type	Overpotential at 10 mA cm^{-2} (mV)	Tafel slope (mV dec^{-1})	Electrolyte	References
PtPdRhIrNi nanowires	1D	22	21.6	0.5 M H_2SO_4	[42]
PtPdRhIrNi nanowires	1D	55	55.7	1.0 M KOH	[42]
RhTe nanowires	1D	28	29.6	0.5 M H_2SO_4	[43]
RhTe nanowires	1D	29	30.2	1.0 M KOH	[43]
RhTe nanowires	1D	49	47.5	1.0 M PBS	[43]
Pt_3Co@nitrogen-doped carbon nanotubes (NCNT)	1D	42	27.2	0.5 M H_2SO_4	[44]
Pt_3Co@NCNT	1D	36	34.8	1.0 M KOH	[44]
Co@NCNT	1D	200	82	0.5 M H_2SO_4	[45]
$MoSe_2$/porous NCNT	1D	102	53	0.5 M H_2SO_4	[46]
CoMoP-600 nanotubes	1D	220	136	0.5 M H_2SO_4	[47]
MoS_x flat boxes	1D	206	58	0.5 M H_2SO_4	[48]
MoS_x nanotubes	1D	210	65	0.5 M H_2SO_4	[48]
β-Mo_2C nanotubes	1D	172	62	0.5 M H_2SO_4	[49]
β-Mo_2C nanotubes	1D	112	55	1.0 M KOH	[49]
SnO_2 nanotubes	1D	147.6	104	0.5 M H_2SO_4	[50]
SnO_2 microplates	1D	181.3	134	0.5 M H_2SO_4	[50]
CoFeO@black phosphazene (BP) nanosheets	2D	88	51	1.0 M KOH	[51]
BPIr@carbon cloth (CC)_be nanosheets	2D	25	30.9	0.5 M H_2SO_4	[52]

Table 7.1 (Continued)

Electrocatalysts	Channel type	Overpotential at 10 mA cm^{-2} (mV)	Tafel slope (mV dec^{-1})	Electrolyte	References
BPIr@CC_be nanosheets	2D	1.98	91	1.0 M KOH	[52]
BPIr@CC_be nanosheets	2D	121	160	1.0 M PBS	[52]
Co-metal–organic framework (MOF) /H_2	2D	30	88	1.0 M KOH	[53]
Ni-MOF@Pt	2D	43	30	0.5 M H_2SO_4	[54]
Ni-MOF@Pt	2D	102	88	1.0 M KOH	[54]
Pt/Ni-MOF	2D	25	42.1	1.0 M KOH	[55]
Mo_2CT_x	2D	189	70	0.5 M H_2SO_4	[56]
N-Ti_2CT_x	2D	215	67	0.5 M H_2SO_4	[57]
Pt SASs/AG	2D	12	29.3	0.5 M H_2SO_4	[58]
IrCo@nitrogen-doped graphene layer (NC)	2D	24	23	0.5 M H_2SO_4	[59]
$Co_{0.85}Se$@nitrogen-doped graphene nanosheets (NG)	2D	209	36.1	0.5 M H_2SO_4	[60]
$Co_{0.85}Se$@NG	2D	227	75.9	1.0 M KOH	[60]
P, Mo-Ru@phosphorus-doped porous carbon matrix (PC)	3D	21	21.7	1.0 M KOH	[61]
A-FeCoP/CPN	3D	110	101	1.0 M KOH	[62]
Co-Mo_2C-g-C3N4 nanosheets (CN)	3D	112	105	0.5 M H_2SO_4	[63]
Co-Mo_2C-CN	3D	92	89	1.0 M KOH	[63]
MoC-Mo_2C/nitrogen-doped carbon dodecahedra (PNCDs)	3D	121	60	1.0 M KOH	[64]
WC-W_2C/PNCDs	3D	101	90	1.0 M KOH	[64]

(Continued)

Table 7.1 (Continued)

Electrocatalysts	Channel type	Overpotential at 10 mA cm^{-2} (mV)	Tafel slope (mV dec^{-1})	Electrolyte	References
CoN_x/C	3D	133	57	0.5 M H_2SO_4	[65]
CoN_x/C	3D	170	75	1.0 M KOH	[65]
CoN_x/C	3D	247	/	1.0 M PBS	[65]
MX/CN/RGO	3D	382 (100 mA cm^{-2})	76	0.5 M H_2SO_4	[66]
Ni–Co–Fe–P NBs	3D	65	32	1.0 M KOH	[67]
Co_3Mo/Cu	3D	12	40	1.0 M KOH	[68]
CoS_2	3D	196	113	1.0 M KOH	[69]

nanomaterials unite the advantages of 1D nanostructures and porous structures, which include grain size, surface area, ion diffusion length, and pore structure [79]. By exploiting the benefits of both the 1D structure and porous nature, it is possible to further improve the performance of electrocatalysis. This section will introduce some typical porous 1D nanomaterials for HER.

Porous nanowires represent a typical example of 1D porous nanomaterials. For instance, Wang et al. [42] developed a self-templating strategy based on eutectic orientation, employing a bottom-up synthesis approach. They successfully prepared nanoporous nanowires (NPNWs) of PtPdRhIrNi and NiPtPdRhIrAl with finely layered nanostructures and ultrafine ligament sizes of approximately 2 nm, using a combination of fast-solidification and de-alloying techniques (see Figure 7.1). The PtPdRhIrNi NPNWs display superior electrocatalytic HER activity and stability. In 0.5 M H_2SO_4 and 1.0 M KOH electrolytes, they demonstrate low overpotentials of 22 and 55 mV at a current density of 10 mA cm^{-2}, with Tafel slopes of 21.6 and 55.7 mV dec^{-1}, respectively. This remarkable electrocatalytic performance can be attributed to the unique structure of the PtPdRhIrNi NPNWs and the high entropy effect. The high entropy effect facilitates the optimization of the electronic structure of the active surface, while the distinctive porous nanowire structure enhances electron transfer and electrolyte diffusion. Besides, Zhang et al. [43] conducted a study in which they synthesized 1D mesoporous RhTe nanowires (m-RhTe NWs) using a two-template approach. The m-RhTe NWs exhibited remarkable HER activity over a wide pH range. Achieving a current density of 10 mA cm^{-2} in 0.5 M H_2SO_4, 1.0 M KOH, and 1.0 M phosphate-buffered saline (PBS) required only overpotentials of 28, 29, and 49 mV, respectively. The Tafel slopes were 29.6, 30.2, and 47.5 mV dec^{-1} for the respective electrolytes. The m-RhTe NWs also demonstrated good stability, with a Faraday efficiency of 97%. The 1D channeling of the m-RhTe NWs plays a crucial

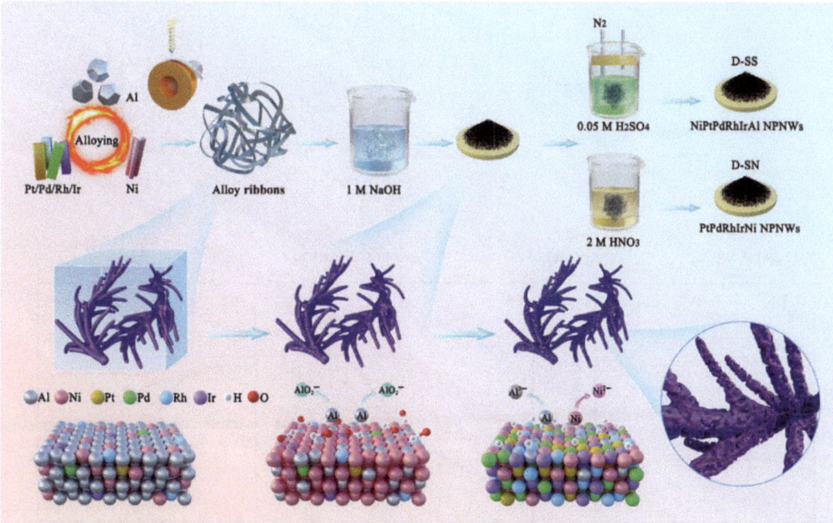

Figure 7.1 Schematic illustration showing the fabrication process of the high-entropy NiPtPdRhIrAl and PtPdRhIrNi NPNWs. Source: Reproduced with permission of Wang et al. [42], Spring nature.

role in generating abundant active sites, reducing hydrogen-binding energy on Rh, and improving charge transport. Consequently, the m-RhTe NWs serve as a versatile HER electrocatalyst throughout the entire pH range.

1D hollow geometric structure, also known as tubular nanostructure, is also a typical porous 1D nanomaterial. It is precisely because of the characteristics of hollow space with larger specific surface area, that more active sites can be exposed in HER catalysis, so it has excellent catalytic activity. As an example, Zhang et al. [44] employed a pyrolysis-displacement-reconfiguration method (shown in Figure 7.2a) to successfully synthesize electrocatalysts of PtCo bimetallic-doped porous nitrogen-doped carbon nanotubes (Pt_3Co@NCNT). In this process, a metal–organic framework (MOF) was utilized to support the confinement of ultrafine PtCo alloy on the inner and outer surfaces of the porous NCNT. The Pt_3Co@NCNT catalyst demonstrated noteworthy electrocatalytic activity and outstanding stability toward HER in acidic and alkaline media. The current density of $10\,mA\,cm^{-2}$ in 0.5 M H_2SO_4 (Figure 7.2b) and 1.0 M KOH (Figure 7.2c) requires overpotentials of only 42 and 36 mV, while the Tafel slopes are 27.2 and $34.8\,mV\,dec^{-1}$, respectively. After continuous operation at $10\,mA\,cm^{-2}$ for 100 hours, the activity attenuation of Pt_3Co@NCNT catalyst is negligible. The excellent HER performance benefited from the unique porous carbon shell structure and the synergistic effect of the bimetallic alloy components inducing the generation of a large number of active sites. Nitrogen-enriched carbon and nitrogen nanotubes can facilitate the electrocatalytic HER by enclosing a limited amount of nanoparticles. Additionally, nitrogen doping can enhance electrical conductivity, wettability, and numerous active sites of the carbon and nitrogen nanotubes.

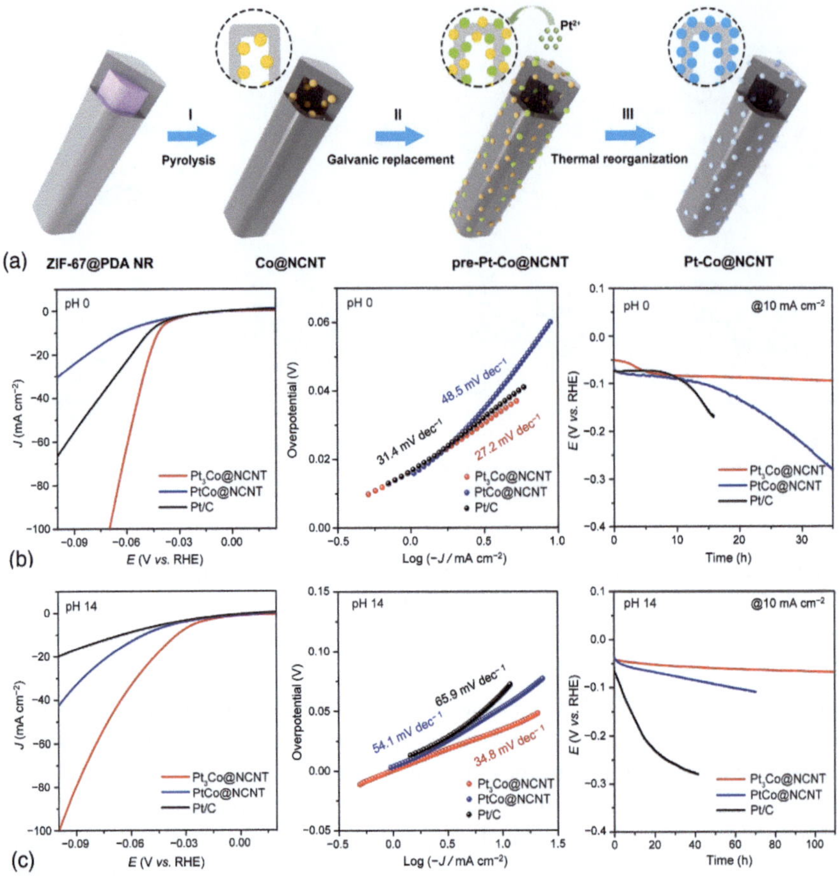

Figure 7.2 (a) Schematic illustration for the synthesis of Pt-Co@NCNT. (b) The electrocatalytic HER performance of Pt_3Co@NCNT at 0.5 M H_2SO_4. LSV curves, Tafel slopes, and chronoamperometric curves recorded at 10 mA cm^{-2}. (c) The electrocatalytic HER performance of Pt_3Co@NCNT at 1.0 M KOH. LSV curves, Tafel slopes, and chronoamperometric curves recorded at 10 mA cm^{-2}. Source: Reproduced with permission of Zhang et al. [44], Wiley-VCH.

Since nitrogen-rich doping is often accompanied by structural defects, it can easily combine with nanoparticles to produce a coordinated effect, leading to excellent HER activity without the need for noble metal nanoparticles. For example, Dai et al. [45] successfully fabricated Co nanoparticles embedded in carbonitride nanotubes with a bamboo-like structure and nitrogen-rich composition (Co@NCNT). This electrocatalyst was synthesized through a series of steps, including melamine polymerization, stripping, and thermal annealing assisted by Co^{2+} ions. The experimental findings demonstrated that Co@NCNT-800 displayed remarkable electrochemical stability (5000 cycles) and exceptional HER efficiency. A mere 200 mV overpotential was needed to attain a current density of 10 mA cm^{-2} under acidic conditions, with a Tafel slope of 82 mV dec^{-1}, suggesting that it possesses the capability to substitute platinum-based catalysts. The exceptional electrocatalytic performance may originate from the collaborative influence of inserted cobalt nanoparticles and nitrogen-rich carbon nitride nanotubes, in addition to bamboo-like carbon nitride nanotubes offering more active sites in HER. In a separate investigation, Qu et al. [46] cultivated small molybdenum diselenide ($MoSe_2$) nanosheets with (002) facet spacing extension on the exterior of porous NCNT. The samples showed exceptional HER catalytic behavior and prolonged endurance with a $10\,mA\,cm^{-2}$ current density, driven by an overpotential of only 102 mV and a Tafel slope of $53\,mV\,dec^{-1}$. The outstanding HER performance can be attributed to the small size of the $MoSe_2$ nanosheets, the expanded space of the (002) facets, the synergistic effect of the $MoSe_2$ nanosheets, and the porous NCNT substrate.

Apart from the combination of carbon and doped carbon nanotubes to form 1D porous nanotubes, another approach to obtaining such structures is through the fine-design material synthesis method. For instance, Lin et al. [47] successfully prepared a series of porous cobalt-molybdenum phosphide (CoMoP) nanotubes through in situ phosphorylation at different temperatures and employed them for the HER. Various characterization techniques confirmed the presence of CoMoP nanotubes, with the phosphorylation temperature playing a significant role in the preparation of the tubular structure. The unique porous nanotube structure of CoMoP may be attributed to the Kirkendall effect during the high-temperature phosphorylation process. Additionally, the synthesized CoMoP-600 nanotubes exhibited excellent HER catalytic performance in a 0.5 M H_2SO_4 solution. Achieving a current density of $10\,mA\,cm^{-2}$ required an overpotential of approximately 220 mV, with a Tafel slope of $136\,mV\,dec^{-1}$. The superior catalytic performance of the CoMoP-600 catalyst can be attributed to the large surface area resulting from its distinctive structure.

In a separate study, Chen et al. [48] utilized amorphous or quasi-amorphous molybdenum sulfide (MoS_x) nanosheets to fabricate hollow and porous flat boxes and nanotubes through a simple templated water and ethanol thermal process. These hollow and porous MoS_x flat boxes and nanotubes exhibited exceptional properties for HER, generating currents of $10\,mA\,cm^{-2}$ with overpotentials of approximately 206 and 210 mV, respectively. The Tafel slopes were 58 and $65\,mV\,dec^{-1}$ for the respective structures. Additionally, Ma et al. [49] synthesized novel hierarchical β-Mo_2C nanotubes with a unique tubular structure consisting of 2D ultrathin porous nanosheets using a straightforward template junction method (shown in Figure 7.3a). These hierarchical β-Mo_2C nanotubes possess several structural advantages, including ultrafine primary nanocrystals, fast charge transfer, and a large exposed active surface. These characteristics contribute to the enhancement of HER catalytic performance. Consequently, the prepared hierarchical β-Mo_2C nanotubes exhibit excellent electrocatalytic performance and remarkable stability under both acidic and alkaline conditions. In 0.5 M H_2SO_4 and 1.0 M KOH, achieving current densities of $10\,mA\,cm^{-2}$ required overpotentials of only 172 and 112 mV, with a Tafel slopes of 62 and $55\,mV\,dec^{-1}$, respectively (Figure 7.3b–d).

In addition to noble and transition metals, tin oxide (SnO_2) is a promising and cost-effective material for serving as an efficient electrocatalyst in HER. SnO_2 exhibits high stability in acidic electrolyte media, excellent charge transport

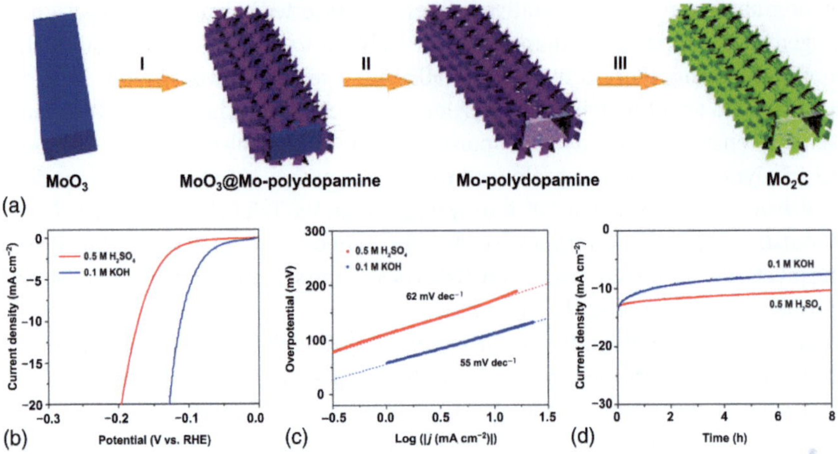

Figure 7.3 (a) Schematic diagram of the formation process of graded β-Mo_2C nanotubes. HER catalytic performance, (b) LSV curves, (c) Tafel slopes, and (d) chronoamperometric curves. (η = 170 mV, 0.5 M H_2SO_4, η = 120 mV, 0.1 M KOH). Source: Reproduced with permission of Ma et al. [49], Wiley-VCH.

properties, and abundant active centers. For example, Zhang et al. [50] successfully synthesized SnO_2 nanotubes and microporous plates using a simple solvothermal method and thermal annealing. The nanotubes, compared to the microporous plates, offer unique structural advantages, including a porous interconnected tubular structure, a large exposed active surface, and good electrical conductivity facilitated by the 1D channels. These characteristics contribute to the enhancement of the HER catalytic performance. As a result, the SnO_2 nanotubes require only 147.6 mV overpotential for a current density of 10 mA cm^{-2} in 0.5 M H_2SO_4, which is lower than the 181.3 mV for the SnO_2 microporous plate. The Tafel slope for the nanotubes was 104 mV dec^{-1}, while for the microporous plate, it was 134 mV dec^{-1}. Moreover, the SnO_2 nanotubes exhibited excellent stability, with negligible reduction in current density after a 10-hour continuous stability test in an acidic electrolyte.

In conclusion, 1D porous nanomaterials exhibit unique linear or tubular structures that can incorporate smaller grain sizes within their pore structures. Additionally, these materials possess long-distance directional channels that facilitate efficient ion diffusion. These characteristics make 1D porous nanomaterials highly promising for future research in the field of HER electrocatalysis.

7.3 Porous 2D Nanomaterials

Since the discovery of graphene, 2D nanomaterials have attracted significant attention owing to their unique intrinsic chemical structure and synergistic properties in two dimensions. These materials have been extensively studied. Unlike conventional 0D and 1D components, 2D materials exhibit strong intermolecular van der Waals forces, exceptional physical properties, and distinct structures such as folding and curling, which provide excellent performance and serve as exceptional building blocks for fabricating high-performance porous materials [80–83]. With the increasing emphasis on sustainable energy, porous 2D materials have garnered significant interest in the field of electrochemical energy conversion, owing to their distinctive structural and electronic features [84–86]. As 2D porous materials possess nanostructured pores, a high surface area with catalytic active sites, and an extensive interfacial electrolyte area, they facilitate enhanced electron and mass transfer of reactants to electrocatalysts. Therefore, it is expected that 2D porous materials will exhibit outstanding electrocatalytic performance in the HER.

An exemplary case is black phosphazene (BP), which stands as a crucial 2D material. Although it exhibits higher catalytic activity in the edge planes compared to the base planes, its performance is generally inferior to that of most existing electrocatalysts. Consequently, the incorporation of BP with more active substances is often employed to enhance its HER performance. For instance, Li et al. [51] successfully improved HER efficiency by cultivating amorphous multitransition metal oxides (CoFeO@BP) on the surface of 2D BP. The CoFeO@BP catalyst showed a remarkably low overpotential of 88 mV to achieve a current density of 10 mA cm^{-2} in 1M KOH, with a Tafel slope of only 51 mV dec^{-1}. Through a series of in situ and ex-situ investigations, the exceptional catalytic performance of CoFeO@BP is attributed to its adaptable surface structure under different redox potentials (as depicted in Figure 7.4a). Furthermore, the strong affinity between BP and transition metal ions facilitates the uniform growth of amorphous CoFe oxides on the BP

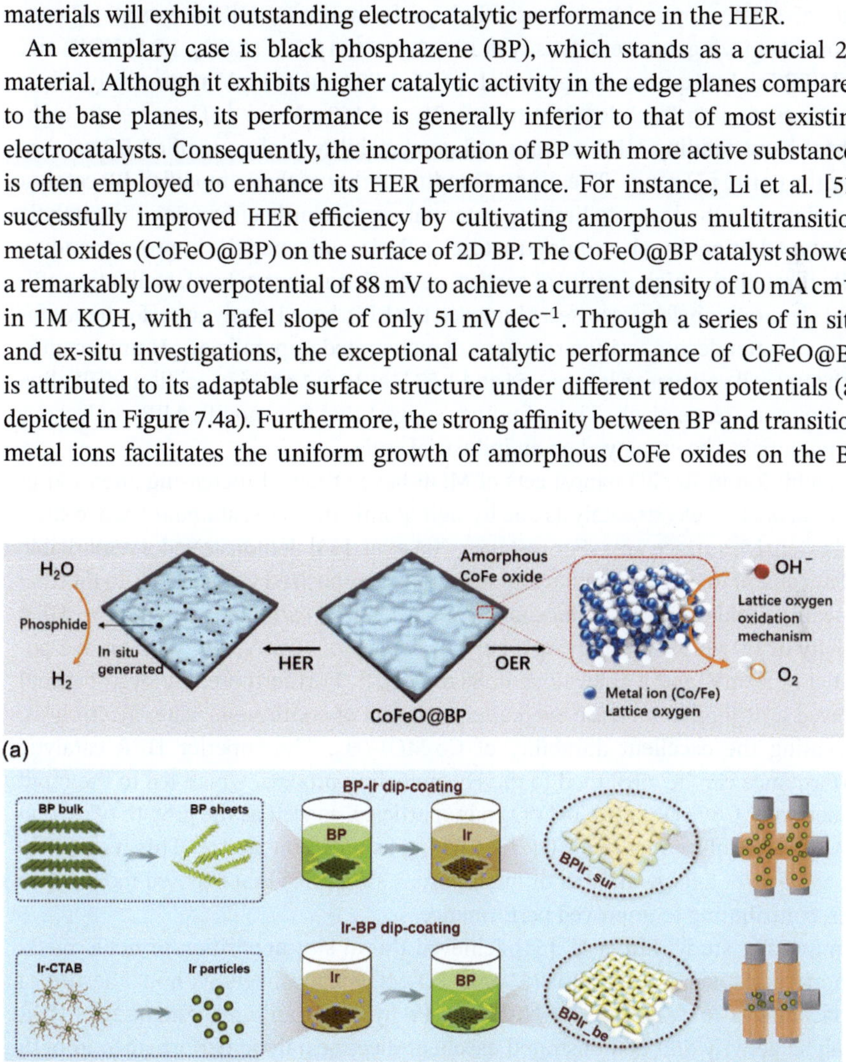

Figure 7.4 (a) Mechanism of adaptive bifunctional amorphous CoFeO@BP electrocatalyst for water-splitting. Source: Reproduced with permission of Li et al. [51], Wiley-VCH. (b) Schematic diagram of the composition of BPIr_be and BPIr_sur. Source: Reproduced with permission of Mei et al. [52], Wiley-VCH.

surface. At the reduction potential, the transient amorphous CoFe oxides react with BP to generate phosphorylated CoFe, significantly improving the HER activity.

In another study, Mei et al. [52] developed a highly efficient Ir-modified BP electrocatalyst attached to carbon cloth by strategically incorporating Ir nanomodifiers and optimizing exposure surfaces to the electrolyte (referred to as BPIr@carbon cloth (CC)_be and BPIr@CC_sur) (as shown in Figure 7.4b). The moderate binding ability of the Ir-modified BP catalysts resulted in low rate-limiting potentials, significantly enhancing the catalytic activity. As a result, BPIr@CC_be demonstrated excellent and universally applicable electrocatalytic performance for HER across all pH conditions. Achieving a current density of 10 mA cm^{-2} in 0.5 M H_2SO_4, 1.0 M KOH, and 1.0 M PBS only required overpotentials of 25, 1.98, and 121 mV, respectively, while the corresponding Tafel slopes were 30.9, 91, and 160 mV dec^{-1}. These results highlight the outstanding HER activity and good stability of the modified BP catalyst over the entire pH range. The modest binding ability of the Ir-modified BP catalyst contributed to the low rate-limiting potential, significantly enhancing the catalytic activity. Additionally, electronic structure analysis confirmed the improved conductivity through the high density of carriers at the Fermi energy level, facilitating efficient electron transfer processes. Moreover, the BPIr-based catalytic performance for the total water decomposition reactions demonstrated unparalleled pH universality, with total cell voltages as low as 1.54 and 1.57 V at 10 mA cm^{-2} in acidic and alkaline electrolyzers, respectively. These findings suggest that this modified BP catalyst has the potential to be employed on an industrial scale.

In addition to BP, 2D nanosheets of MOF have attracted increasing attention in the field of HER electrocatalysis due to their atomic thinness, abundant active sites, and extensive surface area. For instance, Xia et al. [53] demonstrated a remarkable enhancement in catalytic activity by subjecting synthesized Co-MOF/H_2 to thermal annealing under an H_2 atmosphere. Co-MOF/H_2 exhibited remarkable HER activity in 1.0 M KOH, achieving a current density of 10 mA cm^{-2} with an overpotential of 30 mV, and a Tafel slope of 88 mV dec^{-1}. Furthermore, the overpotential showed a negligible increase even after 25 hours of continuous water electrolysis, indicating the excellent durability of Co-MOF/H_2. The superior HER catalytic performance can be attributed to the H_2 annealing process, which led to the rapid reduction of Co-MOF into Co/CoO_x nanoparticles, as well as the growth of carbon nanotubes catalyzed by the MOF feed acting as a carbon source. This resulted in the creation of a large number of microporous structures that exposed more active sites, contributing to improved performance.

In another study, Rui et al. [54] achieved the in situ deposition of noble metal nanoparticles, such as Pt and Pd, onto 2D MOF nanosheets without the use of surfactants. The resulting 2D Ni-MOF@Pt hybrids exhibited remarkable HER catalytic activity and demonstrated excellent electrocatalytic performance in both acidic and alkaline conditions. 2D Ni-MOF@Pt hybrids in 0.5 M H_2SO_4 and 1.0 M KOH achieved a current density of 10 mA cm^{-2} with overpotentials of only 43 and 102 mV, respectively. The measured Tafel slopes were 30 and 88 mV dec^{-1}, respectively. The exceptional HER catalytic performance can be attributed to the successful doping of highly dispersed noble metal nanoparticles with a modulated electronic structure. This was accomplished by utilizing a large number of available oxygen sites on the defect-rich surfaces of the ultrathin MOF nanosheets, resulting from the formation of uncoordinated precious metal–oxygen bonds. Additionally, a substantial number of Pt–O bonds contributed to the enhancement of catalytic activity at the Pt centers. Furthermore, the redistribution of charge density at the Pt/Ni-MOF interface increased the electron density and mobility, favorably improving the HER catalytic activity. The uniform distribution of Pt nanoparticles maximized the exposure of active sites, while the strong interfacial interactions within the Ni-MOF@Pt heterostructure induced compressive strains. This optimization of the d-band center enhanced the HER activity. Furthermore, in a study by Guo et al. [55], 2D heterostructures of Ni-MOF and Pt nanocrystals were synthesized, and the interfacial bond-induced intermediate modulation accelerated the HER, achieving a current density of 10 mA cm^{-2} in 1.0 M KOH with an impressively low overpotential of only 25 mV. The measured Tafel slope was 42.1 mV dec^{-1}. The formation of Ni–O–Pt bonds induced changes in the charge structure, resulting in an increased electron density of Pt and the high-energy Ni 3d state in the MOF. This optimization enhanced the adsorption of H* and OH* species, respectively. The redistribution of charges within the heterostructure further optimized the adsorption of H* and OH*. Additionally, the presence of porous nanochannels increased the active channels available for HER, and the porous structure exposed a large number of active sites. As a result, the synthesized heterostructure exhibited excellent catalytic activity in the HER process.

Both MXenes and graphene are porous 2D materials with great potential for HER. MXenes, in particular, possess exceptional electronic properties due to their intrinsic metallic nature, making them highly appealing for the development of HER electrocatalysts. For example, Seh et al. [56] conducted theoretical calculations on 2D layered M_2XT_x (where M represents a metal, X represents C, N, and T_x represents a surface functional group) and identified Mo_2CT_x as a promising catalyst candidate for HER. Experimental results aligned with theoretical predictions, requiring an overpotential of 189 mV for Mo_2CT_x to achieve a current density of 10 mA cm^{-2}, with a Tafel slope of only 70 mV dec^{-1}. Additionally, theoretical analysis indicated that the basal plane of Mo_2CT_x functions as the active center for HER, distinguishing it from the widely studied active edge position of the 2H-phase MoS_2. In a separate study, Yoon et al. [57] performed high-temperature nitridation of 2D Ti_2CT_x using sodium amide ($NaNH_2$) to synthesize nitrided Ti_2CT_x (N-Ti_2CT_x). The resulting N-Ti_2CT_x exhibited remarkable HER catalytic performance, requiring an overpotential of 215 mV at a current density of 10 mA cm^{-2}, which was more than three times lower than that of pristine Ti_2CT_x (645 mV), along with a Tafel slope of just 65 mV dec^{-1}. The exceptional catalytic activity of N-Ti_2CT_x can be attributed to its unique porous nanostructure. It comprises Ti-N_x patterns on the surface or defects connected through chemical bonds, which enhance the material's stability and create an efficient pathway for electron transfer via spin polarization. Nitrogen capping or substitution simultaneously provides a significant number of active sites, while atomic embedding expands the 2D interlayers and enhances ion channels. These structural features contribute to superior electrochemical properties compared to pristine Ti_2CT_x.

Graphene, with its high conductivity and large surface area, is well-suited for HER electrocatalysis due to its porous 2D structure. It enables effective exposure of electrocatalytic active sites through doping and facilitates faster charge transport, thereby improving the kinetics of HER. As a result, graphene holds great potential as an excellent HER electrocatalyst through various doping methods. For example, Ye et al. [58] successfully prepared single platinum atoms anchored on aniline-stacked graphene (Pt SASs/AG) using a simple aniline anchoring and microwave reduction method. Pt SASs/AG exhibited exceptional HER activity, achieving an ultra-low overpotential of 12 mV at a current density of 10 mA cm^{-2}, with a Tafel slope as low as 29.3 mV dec^{-1}. Moreover, the Pt SASs/AG catalysts demonstrated higher activity and stability compared to Pt/C. The outstanding HER catalytic performance can be attributed to the atomically dispersed Pt sites, which maximize the utilization of catalytic sites, the aniline anchoring that enhances the hydrophilicity of the interface, and the excellent 2D porous structure of graphene, which enhances conductivity and provides a larger surface area.

Furthermore, Jiang et al. [59] successfully encapsulated IrCo alloy within a nitrogen-doped graphene layer (IrCo@NC) using a simple annealing method. This highly active electrocatalyst, with significantly lower Ir loading (1.56 wt%), achieved an ultra-low Tafel slope of 23 mV dec^{-1} and an overpotential of only 24 mV to reach a current density of 10 mA cm^{-2} in 0.5 M H_2SO_4. This performance surpasses that of most reported catalysts and nearly matches the performance of state-of-the-art commercial noble metal catalysts like Pt/C. The excellent catalytic performance is attributed to the strong interaction between the nitrogen-rich graphene layer and the IrCo alloy core, which facilitates faster reaction energetics, modulates the electronic structure, and enhances conductivity. DFT calculations demonstrate that electron transfer from the IrCo alloy core to the graphene shell facilitates charge accumulation on the surface's nitrogen atoms, leading to a significant reduction in ΔG_{H*} and thus improving the overall performance.

In another study, Wang et al. [60] introduced a novel hybrid electrocatalyst, $Co_{0.85}Se$ hollow nanorods uniformly anchored on nitrogen-doped graphene nanosheets ($Co_{0.85}Se$@NG), using a simple one-pot hydrothermal synthesis method. The $Co_{0.85}Se$@NG material exhibited impressive electrocatalytic performance, featuring remarkably low onset potentials of 159 and 111 mV vs. reversible hydrogen electrode (vs. RHE), along with low overpotentials of 209 and 227 mV (vs RHE) at 10 mA cm^{-2} in acidic and alkaline media, respectively. This performance was coupled with small Tafel slopes of 36.1 and 76.5 mV dec^{-1}. Furthermore, the $Co_{0.85}Se$@NG catalyst displayed remarkable durability even after undergoing 1500 cycles. The exceptional electrocatalytic performance of $Co_{0.85}Se$@NG can be attributed to its well-designed nanostructure. The synergistic effect between $Co_{0.85}Se$ nanospheres and the conductive framework of nitrogen-doped graphene nanosheets enhances the specific surface area, thereby revealing a multitude of active sites and significantly enhancing charge transfer. This results in remarkable HER catalytic activity in both acidic and alkaline electrolytes.

In conclusion, 2D porous nanomaterials possess distinctive sheet or layer structures with large surface areas, which enable them to expose a greater number of active sites. Additionally, these layers can serve as directional channels, and by incorporating dopants, the interlayer spacing can be expanded, facilitating faster diffusion rates. Consequently, the utilization of 2D porous nanomaterials holds great promise for achieving exceptional catalytic performance in HER electrocatalysis.

7.4 Porous 3D Nanomaterials

Unlike 1D and 2D porous materials, 3D porous materials exhibit more intricate channel structures, resulting in higher specific surface areas, diverse pore structures, and a wide range of pore size distributions. These characteristics provide them with distinct advantages in catalytic reaction processes [87–89]. Additionally, the various types of pore channels present in 3D porous materials can act as "micro-reactors" where uniform and stable "guest" materials at the nanoscale can be assembled, forming "host–guest materials" [62, 90, 91]. The host–guest effect, combined with potential size effects and quantum effects of the guest material, significantly enhances the catalytic HER performance.

Utilizing 3D porous carbon materials doped with elements such as nitrogen and phosphorus has emerged as a powerful technique for creating 3D porous materials with remarkable HER performance. For instance, Li et al. [61] employed an in situ double-doping technique to synthesize phosphorus and molybdenum double-doped Ru ultrasmall nanoclusters embedded in a phosphorus-doped porous carbon matrix (P, Mo-Ru@PC) (Figure 7.5). The P, Mo-Ru@PC catalyst exhibited a low overpotential of 21 mV at 10 mA cm^{-2} in an alkaline medium, along with a Tafel slope of only 21.7 mV dec^{-1}. This exceptional HER catalytic performance was attributed to the homogeneous embedding of ultrasmall nanoclusters and the hierarchical porous structure of the P, Mo-Ru@PC catalyst, which provided a large surface area and fully exposed the active centers. Furthermore, the dual doping of molybdenum and phosphorus effectively modulated the electronic structure of the ruthenium nanoclusters, inducing higher positively charged states that facilitated the water-splitting process. In another study, Xu et al. [62] employed a simple method involving freeze-drying, carbonization, and KOH activation to prepare phosphorus and nitrogen Co-doped porous carbon-encapsulated iron–copper (A-FeCoP/CPN) electrocatalysts. The A-FeCoP/CPN catalyst displayed a low overpotential of 110 mV at 10 mA cm^{-2} in a 1.0 M KOH solution, with a Tafel slope of only 101 mV dec^{-1}. This catalyst exhibited high HER catalytic activity and good stability due to its abundant pores, large surface area, and defect-rich sites that promoted electrocatalytic HER performance.

Similarly, Zhang et al. [63] successfully synthesized porous nitrogen-doped carbon-encapsulated Co-Mo_2C heterostructures (Co-Mo_2C-CN) using Co-Zn MOF. The Co and Mo_2C particles were uniformly encapsulated in highly porous leaf-shaped nitrogen-doped carbon nanosheets, creating a double Mott–Schottky heterostructure. The Co-Mo_2C-CN catalyst demonstrated excellent performance

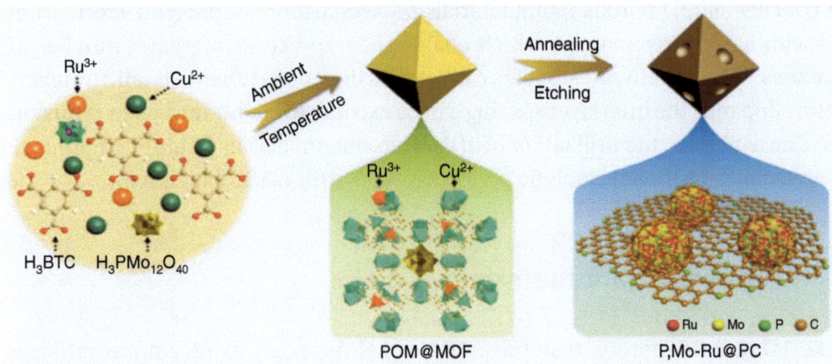

Figure 7.5 The synthesis process of P, Mo-Ru@PC. Source: Reproduced with permission of Li et al. [61], Wiley-VCH.

in both acidic and alkaline HER catalysis, requiring only 116 and 92 mV overpotentials to achieve a current density of 10 mA cm^{-2} in 0.5 M H_2SO_4 and 1.0 M KOH media, respectively. The thin leaf-like morphology and large porosity of Co-Mo_2C-CN provided abundant surface-active sites and facilitated electron transfer, ultimately enhancing water decomposition. Moreover, Lu et al. [64] successfully synthesized ultrafine carbide nanocrystals confined in porous nitrogen-doped carbon dodecahedra (MC-M_2C/PNCDs, M = Mo, W) using molybdate or tungstate annealed zeolite imidazolium salt backbone (ZIF-8). The uniform distribution of ultrafine nanocrystals exposed active sites, while the porous nitrogen-doped carbon dodecahedra acted as a carrier, promoting charge transfer and preventing nanocrystal aggregation during the HER process. The strong coupling interactions between MC and M_2C provided additional sites for water dissociation and hydrogen desorption. As a result, both MoC-Mo_2C/PNCDs and WC-W_2C/PNCDs exhibited excellent performance in alkaline HER catalysis, achieving current densities of 10 mA cm^{-2} with overpotentials of only 121 and 101 mV, and Tafel slopes of 60 and 90 mV dec^{-1}, respectively.

In another study, Liang et al. [65] synthesized robust cobalt-nitrogen/carbon catalysts (CoN_x/C) via pyrolysis of cobalt-N_4 macrocycles or cobalt/phthalimide complexes, employing silica colloids as hard templates. The CoN_x/C catalysts demonstrated excellent electrocatalytic performance and remarkable stability across a wide pH range. In 0.5 M H_2SO_4 and 1.0 M KOH, they achieved current densities of 10 mA cm^{-2} with overpotentials of only 113 and 170 mV, and Tafel slopes of 57 and 75 mV dec^{-1}, respectively. The superior performance was attributed to the well-dispersed molecular CoN_x centers generated during the water-splitting process, providing numerous active sites for HER. Additionally, the synergistic effect of the porous structure, Co–N binding properties, and larger geometrical area with more channels contributed to the excellent HER catalytic performance.

Moreover, the simultaneous synthesis of MXene, graphitic carbon nitride nanosheets, and graphene enabled the creation of 3D porous structures that served as carriers for dispersing other catalytically active materials. For example, He et al. [66] employed a bottom-up strategy to construct 3D interwoven ternary nanostructures derived from $Ti_3C_2T_x$, g-C_3N_4 nanosheets, and graphene

(MX/CN/RGO) via a co-assembly process. The resulting MX/CN/RGO nanostructures exhibited exceptional HER performance. The MX/CN/RGO electrocatalysts demonstrated HER currents with an onset potential of merely 38 mV, achieving current densities up to 100 mA cm^{-2} in 1.0 M KOH media, with overpotentials as low as 382 mV and Tafel slopes of just 76 mV dec^{-1}. The superior HER performance of the MX/CN/RGO nanostructures can be attributed to several factors. First, the 3D interwoven ternary nanoporous structure facilitates efficient electron transport and provides a multitude of catalytic active sites at different stages. Additionally, the structure offers a larger specific surface area, higher conductivity, and optimized electronic structure, all of which contribute to fast thermal reduction kinetics during the HER process.

Modulating the synthesis of 3D porous materials using the catalytic material itself, without the need for a carbon-based carrier, can also yield exceptional HER performance. For example, Li et al. [67] utilized an ultrafast one-step electrodeposition method to prepare porous nickel–cobalt–iron ternary metal phosphide nanobricks (Ni–Co–Fe–P NBs) with specific composition and morphology in less than a minute. The experimental results demonstrated that the Ni–Co–Fe–P NBs achieved a current density of 10 mA cm^{-2} in 1.0 M KOH medium with an overpotential of only 32 mV and a Tafel slope of 65 mV dec^{-1}. Moreover, even after 100 hours at 20 and 100 mA cm^{-2}, the overpotential of Ni–Co–Fe–P NBs only increased by 12 and 50 mV, respectively, showcasing excellent HER catalytic activity and stability. The porous metal foam structure not only provides unique 3D channels for rapid electrolyte diffusion and gas release but also possesses a robust integrated structure with enhanced electrical conductivity, facilitating efficient charge transfer. In another study, Shi et al. [68] employed an in situ hydroxylation method to utilize nanoporous copper-loaded intermetallic compound Co_3Mo, spontaneously isolated from a graded nanoporous Cu skeleton (Co_3Mo/Cu), as a self-supporting monolithic material for efficient electrocatalysis in alkaline HER. The exceptional intrinsic activity of the intermetallic compound Co_3Mo, combined with the fast electron and mass transfer pathways provided by the graded nanoporous Cu skeleton, resulted in an optimal hydrogen-binding energy for facilitating hydrogen adsorption/desorption onto hydrogen intermediates. Consequently, the nanoporous Cu-loaded Co_3Mo electrode exhibited impressive HER catalysis, achieving current densities of 10 mA cm^{-2} in 1 M KOH with overpotentials as low as 12 mV and a Tafel slope of only 40 mV dec^{-1}.

Furthermore, Ahn et al. [69] successfully synthesized HER catalysts with single-phase and highly porous structures using cobalt Prussian blue analog (Co-PBA, $Co_3[Co(CN)_6]_2$) as a template, converted to CoS_2 nanoparticles through sulfide annealing. These electrocatalysts exhibited high activity for total water decomposition, reaching a current density of 10 mA cm^{-2} in 1 M KOH with an overpotential as low as 196 mV and a Tafel slope of 113 mV dec^{-1}. The excellent HER catalytic activity can be attributed to the highly porous structure with nanoscale pores, which enhance the distribution and porous structure of the single-phase CoS_2 nanoparticles.

In summary, 3D porous nanomaterials offer several advantages, including a larger specific surface area and an increased number of pore channels. The presence of diverse pore sizes and shapes facilitates the effective release of H_2 bubbles from the electrodes, leading to improved electrode performance and dynamic characteristics. Additionally, the enhanced mass transfer performance and increased number of active centers resulting from more efficient channels and larger surface area further contribute to their potential as a new class of catalysts for HER.

7.5 Conclusion and Outlook

To date, significant advancements have been made in utilizing porous materials as catalysts to enhance the performance of the HER. Based on their dimensions, pore channels in porous materials can be categorized into 1D, 2D, and 3D channels. Regardless of the channel type, the abundance of porous structures and larger specific surface area provide numerous reactive active sites. Furthermore, the higher porosity and larger pore size reduce mass transfer resistance, greatly improving mass transfer performance. Additionally, the porosity and pore size of porous materials can be controlled during preparation to achieve higher selectivity in HER catalysis. Doping other components, such as heterostructures, clusters, and single atoms, into the structure of porous materials effectively enhances the activity, selectivity, and stability of HER catalysis.

Despite the great efforts, the understanding of porous materials, particularly from a mechanistic and fundamental perspective, remains incomplete. The behavior of liquid or electrolyte within nanoscale channels differs significantly from unconfined conditions, yet most published works focus solely on electrocatalysts. The impact of mass transport and other properties of the liquid or electrolyte within the channels is a crucial factor to consider. Additionally, atomic-level characterization of catalysts is vital to correlate active sites with catalytic activity. Developing in situ real-time characterization techniques, such as spectroscopy and microscopy, to monitor catalytic reactions will provide valuable insights into catalyst behavior. Therefore, future research should aim to develop comprehensive assays and gain fundamental knowledge to enable deeper investigations and fully exploit the potential of porous materials for designing efficient and stable catalysts.

To summarize, porous materials offer significant advantages, including a large specific surface area, high selectivity, and low mass transfer resistance, positioning them as promising catalysts for HER. With the ongoing progress in preparation and characterization techniques, the development of controlled synthesis strategies to obtain porous material catalysts meeting specific requirements will become increasingly mature. Consequently, the possibility of producing green hydrogen through HER, using cost-effective and high-performance porous material catalysts, is within reach for the future.

Acknowledgment

This work was supported by the National Natural Science Foundation of China (No. 52102306), and the Fundamental Research Funds for the Central Universities (Grant No. G2023KY0601).

References

1 Lewis, N.S. and D.G.N. (2006). Powering the planet: chemical challenges in solar energy utilization. *Proc. Natl. Acad. Sci. USA* 103 (43): 15729.

2 Chu, S. and Majumdar, A. (2012). Opportunities and challenges for a sustainable energy future. *Nature* 488 (7411): 294.

3 Hu, C., Zhang, L., and Gong, J. (2019). Recent progress made in the mechanism comprehension and design of electrocatalysts for alkaline water splitting. *Energy Environ. Sci.* 12 (9): 2620.

4 Morales-Guio, C.G., Stern, L.A., and Hu, X. (2014). Nanostructured hydrotreating catalysts for electrochemical hydrogen evolution. *Chem. Soc. Rev.* 43 (18): 6555.

5 Subbaraman, R., Tripkovic, D., Chang, K.C. et al. (2012). Trends in activity for the water electrolyser reactions on 3*d* M(Ni,Co,Fe,Mn) hydr(oxy)oxide catalysts. *Nat. Mater.* 11 (6): 550.

6 Chatenet, M., Pollet, B.G., Dekel, D.R. et al. (2022). Water electrolysis: from textbook knowledge to the latest scientific strategies and industrial developments. *Chem. Soc. Rev.* 51 (11): 4583.

7 Chestnut, D.H. (2009). Ten years of experience with accidental dural puncture and post-dural puncture headache in a tertiary obstetric anaesthesia department. *Energy Environ. Sci.* 2: 252.

8 Suryanto, B.H.R., Wang, Y., Hocking, R.K. et al. (2019). Overall electrochemical splitting of water at the heterogeneous interface of nickel and iron oxide. *Nat. Commun.* 10 (1): 5599.

9 Yu, M., Budiyanto, E., and Tuysuz, H. (2022). Principles of water electrolysis and recent progress in cobalt-, nickel-, and iron-based oxides for the oxygen evolution reaction. *Angew. Chem. Int. Ed.* 61 (1): e202103824.

10 Ayodele, T.R. and Munda, J.L. (2019). Potential and economic viability of green hydrogen production by water electrolysis using wind energy resources in South Africa. *Int. J. Hydrog. Energy* 44 (33): 17669.

11 Boretti, A. (2021). There are hydrogen production pathways with better than green hydrogen economic and environmental costs. *Int. J. Hydrog. Energy* 46 (46): 23988.

12 Gimbun, J. and Wongsakulphasatch, S. (2021). Recent trends in hydrogen production and utilization. *Int. J. Hydrog. Energy* 46 (48): https://doi.org/10.1016/j.ijhydene.2021.05.194.

13 Shiva Kumar, S. and Lim, H. (2022). An overview of water electrolysis technologies for green hydrogen production. *Energy Rep.* 8: 13793.

14 Megía, P.J., Vizcaíno, A.J., Calles, J.A., and Carrero, A. (2021). Hydrogen production technologies: from fossil fuels toward renewable sources. A mini review. *Energy Fuel* 35 (20): 16403.

15 Lasia, A. (2019). Mechanism and kinetics of the hydrogen evolution reaction. *Int. J. Hydrog. Energy* 44 (36): 19484.

16 Sun, F., Tang, Q., and Jiang, D.-E. (2022). Theoretical advances in understanding and designing the active sites for hydrogen evolution reaction. *ACS Catal.* 12 (14): 8404.

17 Swenson, H. and Stadie, N.P. (2019). Langmuir's theory of adsorption: a centennial review. *Langmuir* 35 (16): 5409.

18 Ooka, H., Huang, J., and Exner, K.S. (2021). The Sabatier principle in electrocatalysis: basics, limitations, and extensions. *Front. Energy Res.* 9: 654460.

19 Yang, Y., Yu, Y., Li, J. et al. (2021). Engineering ruthenium-based electrocatalysts for effective hydrogen evolution reaction. *Nano-Micro Lett.* 13 (1): 160.

20 Sarkar, S. and Peter, S.C. (2018). An overview on Pd-based electrocatalysts for the hydrogen evolution reaction. *Inorg. Chem. Front.* 5 (9): 2060.

21 Liu, Z., Nie, K., Yuan, Y. et al. (2022). Crystalline/amorphous heterophase with self-assembled hollow structure for highly efficient electrochemical hydrogen production. *CCS Chem.* 4 (10): 3391.

22 Liu, Z., Nie, K., Qu, X. et al. (2022). General bottom-up colloidal synthesis of nano-monolayer transition-metal dichalcogenides with high 1T'-phase purity. *J. Am. Chem. Soc.* 144 (11): 4863.

23 Li, B., Nie, K., Zhang, Y. et al. (2023). Engineering single-layer hollow structure of transition metal dichalcogenides with high 1t-phase purity for hydrogen evolution reaction. *Adv. Mater.* e2303285.

24 Li, C. and Baek, J.B. (2020). Recent advances in noble metal (Pt, Ru, and Ir)-based electrocatalysts for efficient hydrogen evolution reaction. *ACS Omega* 5 (1): 31.

25 Zuo, Y., Bellani, S., Ferri, M. et al. (2023). High-performance alkaline water electrolyzers based on Ru-perturbed Cu nanoplatelets cathode. *Nat. Commun.* 14 (1): 4680.

26 Yu, F., Zhou, H., Huang, Y. et al. (2018). High-performance bifunctional porous non-noble metal phosphide catalyst for overall water splitting. *Nat. Commun.* 9 (1): 2551.

27 Yin, Y., Han, J., Zhang, Y. et al. (2016). Contributions of phase, sulfur vacancies, and edges to the hydrogen evolution reaction catalytic activity of porous molybdenum disulfide nanosheets. *J. Am. Chem. Soc.* 138 (25): 7965.

28 Wu, H.B., Xia, B.Y., Yu, L. et al. (2015). Porous molybdenum carbide nano-octahedrons synthesized via confined carburization in metal-organic frameworks for efficient hydrogen production. *Nat. Commun.* 6: 6512.

29 Wang, Y., Dutta, A., Iarchuk, A. et al. (2023). Boosting nitrate to ammonia electroconversion through hydrogen gas evolution over Cu-foam@mesh catalysts. *ACS Catal.* 13 (12): 8169.

30 Wang, X., Bai, J., Zhang, M. et al. (2023). A comparison between porous to fully dense electrodeposited cuni films: insights on electrochemical performance. *Nanomaterials* 13 (3): 491.

31 Wang, H., Xiao, X., Liu, S. et al. (2019). Structural and electronic optimization of MoS_2 edges for hydrogen evolution. *J. Am. Chem. Soc.* 141 (46): 18578.

32 Singu, B.S., Chitumalla, R.K., Mandal, D. et al. (2023). Development of metal-organic framework-derived NiMo-MoO_{3-x} porous nanorod for efficient electrocatalytic hydrogen evolution reactions. *Appl. Catal. B Environ.* 328: 122421.

33 Ouyang, C., Wang, X., Wang, C. et al. (2015). Hierarchically porous Ni_3S_2 nanorod array foam as highly efficient electrocatalyst for hydrogen evolution reaction and oxygen evolution reaction. *Electrochim. Acta* 174: 297.

34 Lyu, F., Zeng, S., Jia, Z. et al. (2022). Two-dimensional mineral hydrogel-derived single atoms-anchored heterostructures for ultrastable hydrogen evolution. *Nat. Commun.* 13 (1): 6249.

35 Luo, X., Yuan, P., Luo, J. et al. (2023). The enhancing effect of stable oxygen functional groups on porous-carbon-supported Pt catalysts for alkaline hydrogen evolution. *Nanomaterials* 13 (8): 1415.

36 Li, P., Zheng, D., Gao, M. et al. (2021). Bimetallic MOF-templated fabrication of porous Zn, N Co-doped Mo_2C for an efficient hydrogen evolution reaction. *ACS Appl. Energy Mater.* 4 (9): 8875.

37 Kadam, S.R., Kawade, U.V., Bar-Ziv, R. et al. (2019). Porous MoS_2 framework and its functionality for electrochemical hydrogen evolution reaction and lithium ion batteries. *ACS Appl. Energy Mater.* 2 (8): 5900.

38 He, H., Zhang, Y., Zhang, W. et al. (2021). Dual metal-loaded porous carbon materials derived from silk fibroin as bifunctional electrocatalysts for hydrogen evolution reaction and oxygen evolution reaction. *ACS Appl. Energy Mater.* 13 (26): 30678.

39 Gao, Y., Xue, Y., Qi, L. et al. (2022). Rhodium nanocrystals on porous graphdiyne for electrocatalytic hydrogen evolution from saline water. *Nat. Commun.* 13 (1): 5227.

40 Chen, H., Liang, X., Liu, Y. et al. (2020). Active site engineering in porous electrocatalysts. *Adv. Mater.* 32 (44): e2002435.

41 Chen, B., Hu, P., Yang, F. et al. (2023). In situ porousized MoS_2 nano islands enhance HER/OER bifunctional electrocatalysis. *Small* 19 (14): e2207177.

42 Wang, Y., Yu, B., He, M. et al. (2021). Eutectic-derived high-entropy nanoporous nanowires for efficient and stable water-to-hydrogen conversion. *Nano Res.* 15 (6): 4820.

43 Zhang, H., Liu, S., Tian, P. et al. (2022). Mesoporous RhTe nanowires towards all-pH-value hydrogen evolution electrocatalysis. *Chem. Eng. J.* 435: 134798.

44 Zhang, S.L., Lu, X.F., Wu, Z.P. et al. (2021). Engineering platinum-cobalt nano-alloys in porous nitrogen-doped carbon nanotubes for highly efficient electrocatalytic hydrogen evolution. *Angew. Chem. Int. Ed.* 60 (35): 19068.

45 Dai, X., Li, Z., Ma, Y. et al. (2016). Metallic cobalt encapsulated in bamboo-like and nitrogen-rich carbonitride nanotubes for hydrogen evolution reaction. *ACS Appl. Mater. Interfaces* 8 (10): 6439.

46 Qu, B., Li, C., Zhu, C. et al. (2016). Growth of $MoSe_2$ nanosheets with small size and expanded spaces of (002) plane on the surfaces of porous N-doped carbon nanotubes for hydrogen production. *Nanoscale* 8 (38): 16886.

47 Lin, Y., Liu, M., Pan, Y., and Zhang, J. (2017). Porous Co–Mo phosphide nanotubes: an efficient electrocatalyst for hydrogen evolution. *J. Mater. Sci.* 52 (17): 10406.

48 Chen, L. and Zhu, H. (2017). Molybdenum sulfide nanosheet-based hollow porous flat boxes and nanotubes for efficient electrochemical hydrogen evolution. *ChemCatChem* 10 (2): 459.

49 Ma, F.X., Wu, H.B., Xia, B.Y. et al. (2015). Hierarchical beta-Mo_2C nanotubes organized by ultrathin nanosheets as a highly efficient electrocatalyst for hydrogen production. *Angew. Chem. Int. Ed.* 54 (51): 15395.

50 Zhang, Y., Zou, J., He, Z. et al. (2021). Facile synthesis of SnO_2 nanostructures for enhanced electrochemical hydrogen evolution reaction. *J. Alloys Compd.* 865: 158597.

51 Li, X., Xiao, L., Zhou, L. et al. (2020). Adaptive bifunctional electrocatalyst of amorphous CoFe oxide@2D black phosphorus for overall water splitting. *Angew. Chem. Int. Ed.* 59 (47): 21106.

52 Mei, J., He, T., Bai, J. et al. (2021). Surface-dependent intermediate adsorption modulation on iridium-modified black phosphorus electrocatalysts for efficient pH-universal water splitting. *Adv. Mater.* 33 (49): e2104638.

53 Xia, Q., Liu, H., Jin, M. et al. (2020). Catalysts confined inside CNTs derived from 2D metal-organic frameworks for electrolysis. *Nanoscale* 12 (16): 8969.

54 Rui, K., Zhao, G., Lao, M. et al. (2019). Direct hybridization of noble metal nanostructures on 2d metal-organic framework nanosheets to catalyze hydrogen evolution. *Nano Lett.* 19 (12): 8447.

55 Guo, C., Jiao, Y., Zheng, Y. et al. (2019). Intermediate modulation on noble metal hybridized to 2D metal-organic framework for accelerated water electrocatalysis. *Chem* 5 (9): 2429.

56 Seh, Z.W., Fredrickson, K.D., Anasori, B. et al. (2016). Two-dimensional molybdenum carbide (MXene) as an efficient electrocatalyst for hydrogen evolution. *ACS Energy Lett.* 1 (3): 589.

57 Yoon, Y., Tiwari, A.P., Lee, M. et al. (2018). Enhanced electrocatalytic activity by chemical nitridation of two-dimensional titanium carbide MXene for hydrogen evolution. *J. Mater. Chem. A* 6 (42): 20869.

58 Ye, S., Luo, F., Zhang, Q. et al. (2019). Highly stable single Pt atomic sites anchored on aniline-stacked graphene for hydrogen evolution reaction. *Environ. Sci. Technol.* 12 (3): 1000.

59 Jiang, P., Chen, J., Wang, C. et al. (2018). Tuning the activity of carbon for electrocatalytic hydrogen evolution via an iridium-cobalt alloy core encapsulated in nitrogen-doped carbon cages. *Adv. Mater.* 30 (9): 1705324.

60 Wang, H., Wang, X., Yang, D. et al. (2018). $Co_{0.85}Se$ hollow nanospheres anchored on N-doped graphene nanosheets as highly efficient, nonprecious electrocatalyst for hydrogen evolution reaction in both acid and alkaline media. *J. Power Sources* 400: 232.

61 Li, C., Jang, H., Liu, S. et al. (2022). P and Mo dual doped Ru ultrasmall nanoclusters embedded in P-doped porous carbon toward efficient hydrogen evolution reaction. *Adv. Energy Mater.* 12 (23): 2200029.

62 Xu, Y., Wang, R., Wang, J. et al. (2022). Encapsulation of Fe-CoP with P, N-co-doped porous carbon matrix as a multifunctional catalyst for wide electrochemical applications. *J. Energy Chem.* 71: 36.

63 Zhang, P., Liu, Y., Liang, T. et al. (2021). Nitrogen-doped carbon wrapped Co-Mo_2C dual Mott–Schottky nanosheets with large porosity for efficient water electrolysis. *Appl. Catal. B Environ.* 284: 119738.

64 Lu, X.F., Yu, L., Zhang, J., and Wen, X. (2019). Ultrafine dual-phased carbide nanocrystals confined in porous nitrogen-doped carbon dodecahedrons for efficient hydrogen evolution reaction. *Adv. Mater.* 31 (30): e1900699.

65 Liang, H.W., Bruller, S., Dong, R. et al. (2015). Molecular metal-N_x centres in porous carbon for electrocatalytic hydrogen evolution. *Nat. Commun.* 6: 7992.

66 He, H., Chen, Y., Yang, C. et al. (2022). Constructing 3D interweaved MXene/graphitic carbon nitride nanosheets/graphene nanoarchitectures for promoted electrocatalytic hydrogen evolution. *J. Energy Chem.* 67: 483.

67 Li, A., Zhang, L., Wang, F. et al. (2022). Rational design of porous Ni-Co-Fe ternary metal phosphides nanobricks as bifunctional electrocatalysts for efficient overall water splitting. *Appl. Catal. B Environ.* 310: 121353.

68 Shi, H., Zhou, Y.T., Yao, R.Q. et al. (2020). Spontaneously separated intermetallic Co_3Mo from nanoporous copper as versatile electrocatalysts for highly efficient water splitting. *Nat. Commun.* 11 (1): 2940.

69 Ahn, I.K., Joo, W., Lee, J.H. et al. (2019). Metal-organic framework-driven porous cobalt disulfide nanoparticles fabricated by gaseous sulfurization as bifunctional electrocatalysts for overall water splitting. *Sci. Rep.* 9 (1): 19539.

70 Zhang, B., Wang, W., Liang, L. et al. (2021). Prevailing conjugated porous polymers for electrochemical energy storage and conversion: lithium-ion batteries, supercapacitors and water-splitting. *Coord. Chem. Rev.* 436 (6): 213782.

71 Xu, X., Shao, Z., and Jiang, S.P. (2022). High-entropy materials for water electrolysis. *Energy Technol.* 10 (11): 2200573.

72 Wei, Q., Xiong, F., Tan, S. et al. (2017). Porous one-dimensional nanomaterials: design, fabrication and applications in electrochemical energy storage. *Adv. Mater.* 29 (20): 1602300.

73 Ren, J.-T., Yao, Y., and Yuan, Z.-Y. (2021). Fabrication strategies of porous precious-metal-free bifunctional electrocatalysts for overall water splitting: recent advances. *Green Energy Environ.* 6 (5): 620.

74 Qiu, T., Cheng, J., Liang, Z. et al. (2022). Unveiling the nanoalloying modulation on hydrogen evolution activity of ruthenium-based electrocatalysts encapsulated by B/N co-doped graphitic nanotubes. *Appl. Catal. B Environ.* 316: 121626.

75 Pan, J., Yu, S., Jing, Z. et al. (2021). Electrocatalytic hydrogen evolution reaction related to nanochannel materials. *Small Struct.* 2 (10): 2100076.

76 Meng, S., Li, G., Wang, P. et al. (2023). Rare earth-based MOFs for photo/electrocatalysis. *Mater. Chem. Front.* 7 (5): 806.

77 Dey, G., Shadab, and Aijaz, A. (2021). Metal-organic framework derived nanostructured bifunctional electrocatalysts for water splitting. *ChemElectroChem* 8 (20): 3782.

78 Chen, T.-W., Kalimuthu, P., Veerakumar, P. et al. (2022). Review—recent advances in the development of porous carbon-based electrocatalysts for water-splitting reaction. *J. Electrochem. Soc.* 169 (5): 054519.

79 Song, W., Li, M., Wang, C., and Lu, X. (2020). Electronic modulation and interface engineering of electrospun nanomaterials-based electrocatalysts toward water splitting. *Carbon Energy* 3 (1): 101.

80 Zubair, M., Ul Hassan, M.M., Mehran, M.T. et al. (2022). 2D MXenes and their heterostructures for HER, OER and overall water splitting: a review. *Int. J. Hydrog. Energy* 47 (5): 2794.

81 Wang, Y., Wang, S., and Ma, Z.L. (2022). Competitive coordination-oriented monodispersed ruthenium sites in conductive MOF/LDH hetero-nanotree catalysts for efficient overall water splitting in alkaline media. *Adv. Mater.* 34 (12): e2107488.

82 Nemiwal, M., Zhang, T.C., and Kumar, D. (2021). Graphene-based electrocatalysts: hydrogen evolution reactions and overall water splitting. *Int. J. Hydrog. Energy* 46 (41): 21401.

83 Liu, J., Peng, W., Li, Y. et al. (2020). 2D MXene-based materials for electrocatalysis. *Trans. Tianjin Univ.* 26 (3): 149.

84 Lu, X., Yu, L., and Xiong, L.Y. (2019). Highly crystalline Ni-doped FeP/carbon hollow nanorods as all-pH efficient and durable hydrogen evolving electrocatalysts. *Sci. Adv.* 5: eaav6009.

85 Khan, U., Nairan, A., Gao, J., and Zhang, Q. (2022). Current progress in 2D metal–organic frameworks for electrocatalysis. *Small Struct.* 4 (6): 2200109.

86 Deng, K., Zhou, T., Mao, Q. et al. (2022). Surface engineering of defective and porous ir metallene with polyallylamine for hydrogen evolution electrocatalysis. *Adv. Mater.* 34 (18): e2110680.

87 Zhang, J., Zhang, Z., Ji, Y. et al. (2021). Surface engineering induced hierarchical porous $Ni_{12}P5$-Ni_2P polymorphs catalyst for efficient wide pH hydrogen production. *Appl. Catal. B Environ.* 282: 119609.

88 Yao, H., Wang, X., Li, K. et al. (2022). Strong electronic coupling between ruthenium single atoms and ultrafine nanoclusters enables economical and effective hydrogen production. *Appl. Catal. B Environ.* 312: 119609.

89 Yang, Z., Zhao, C., Qu, Y. et al. (2019). Trifunctional self-supporting cobalt-embedded carbon nanotube films for ORR, OER, and HER triggered by solid diffusion from bulk metal. *Adv. Mater.* 31 (12): e1808043.

90 Yan, L., Cao, L., Dai, P. et al. (2017). Metal-organic frameworks derived nanotube of nickel–cobalt bimetal phosphides as highly efficient electrocatalysts for overall water splitting. *Adv. Funct. Mater.* 27 (40): 1703455.

91 Huang, X., Xu, X., Li, C. et al. (2019). Vertical CoP nanoarray wrapped by N,P-doped carbon for hydrogen evolution reaction in both acidic and alkaline conditions. *Adv. Energy Mater.* 9 (22): 1803970.

Part III

Advanced Functional Materials for Electrocatalytic Oxygen Reduction Reaction

8

Metal-Based Materials for Electrocatalytic Oxygen Reduction Reaction

Congli Qin and Aixin Fan

Langfang Normal University, School of Chemistry and Material Science, No. 100, Aimin Road, Anci District, Langfang 065000, China

8.1 Introduction

Proton exchange membrane fuel cells (PEMFCs) are of considerable interest for their high conversion efficiencies, specific energies, and low degrees of environmental pollution [1–3]. They may represent an ultimate solution for mobile power supply, such as transportation, fixed, and portable batteries [4–6]. However, their overall efficiency is primarily limited by the sluggish kinetics of the cathodic oxygen reduction reaction (ORR) [7–9], which requires a significant amount of catalyst with a substantial contribution to the overall cost [10, 11]. Therefore, developing cost-effective, high-performance electrocatalysts to improve ORR kinetics is crucial for reducing the cost of PEMFC [12, 13].

Platinum (Pt)-based catalysts are considered the state-of-the-art cathode electrocatalytic materials due to their excellent catalytic performance and represent a benchmark for ORR catalysts (Figure 8.1) [14–17]. However, there are many issues that need to be addressed with existing Pt-based catalysts, such as high platinum usage leading to high catalyst costs, easy separation and dissolution in high-temperature, and high-pressure environments [18, 19]. To solve these problems, considerable efforts have been made to improve the catalytic performance of Pt-based ORR catalysts, including the development of various Pt-based nanostructures with precisely controlled size, composition, shape, and morphology [20–23]. Besides the Pt-based catalysts, other precious metal catalysts, such as Pd [24, 25], Ru [26, 27], Ir [28], Rh [29], Au [20], and even Ag [30] catalysts, have been extensively studied over the past decades and have made remarkable progress.

Therefore, the main purpose of this chapter is provide an overview of metal-based materials for ORR. The objective is to reveal the intricate relationship between their structure and performance, while also providing essential principles for the design of ORR catalysts.

Functional Materials for Electrocatalytic Energy Conversion, First Edition.
Edited by Zhicheng Zhang, Meiting Zhao, and Yuchen Qin.

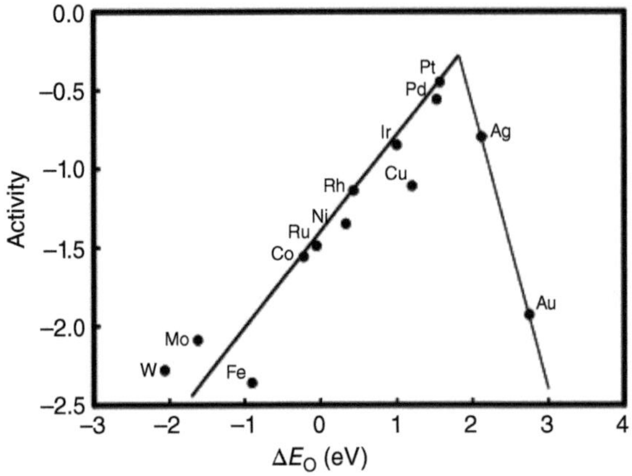

Figure 8.1 Trends in oxygen reduction activity (defined in the text) plotted as a function of the oxygen binding energy. Source: Reproduced with permission of Nørskov et al. [14], American Chemical Society.

8.2 Metal-Based Materials for ORR

8.2.1 Pt-Based Materials for ORR

At present, Pt-based materials are widely used as cathode catalysts for PEMFC due to their high oxygen reduction catalytic activity and stability in strongly acidic electrolytes [31]. Compared with other ORR catalysts, Pt-based catalysts have several outstanding advantages, such as high catalytic activity, strong environmental adaptability, simple preparation process, and good stability. Although Pt-based catalysts exhibit high catalytic activity for ORR, they suffers from some disadvantages, including low abundance, high price, and being poisoned by CO [32, 33], SO_2 [2, 34], and NO_x [35–37]. Therefore, researchers are developing advanced ORR catalysts with high catalytic activity, low cost, and anti-poisoning properties in terms of composition, morphology, crystal properties, and preparation methods.

8.2.1.1 Composition Regulation of Pt-Based ORR Catalysts

Due to the higher activity and durability of alloy catalysts compared to pure Pt, Pt-based alloys are considered the second generation of fuel cell catalysts. The alloying of transition metal (M) with Pt can not only effectively reduce the amount of Pt, but also improve the utilization rate of Pt. In addition, the original electronic structure of Pt was adjusted by the addition of transition metals, and the catalytic properties of Pt were also changed. The enhanced ORR activity of Pt-based alloy catalyst was attributed to the addition of transition metal M producing compressive strain, and the negative shift in the center of d-band, thus delaying the formation of oxides [38–40].

Among these developed Pt-based alloy catalysts, Co, Ni, and Fe alloy materials have been widely studied for their excellent electrochemical activities [41–44]. Markovich and co-workers studied the relationship in electrocatalytic trends on Pt_3M (M = Ni, Co, Fe, Ti, V) surfaces between the experimentally determined surface electronic structure (the d-band center) and activity for ORR [45]. This relationship exhibits a "volcanic" behavior, indicating that the formation of platinum

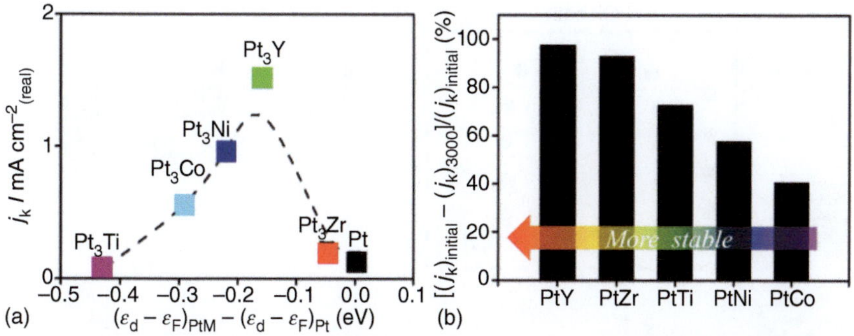

Figure 8.2 (a) Kinetic current density (j_k) at 0.9 V for ORR on Pt_3M (M = Y, Zr, Ti, Ni, and Co) alloy nanocatalysts and pure Pt nanocatalyst electrodes in 0.1 M $HClO_4$ solution, as a function of the measured d-band centers. The d-band centers were measured by high-resolution XPS. (b) Histogram showing activity change of the polarization curves for ORR on Pt_3M alloy nanocatalysts on the RDE, before and after 3000 potential cycles. Sweep rate, 10 mV s^{-1}; rotating rate, 1600 rpm. Source: Reproduced with permission of Hwang et al. [52], American Chemical Society.

alloy catalysts significantly enhances the catalytic activity of Pt. In addition, other binary and terpolymer electrocatalysts, such as Pt–Pd, Pt–Mo, Pt–Ag, and PtAuCu [20, 46–48], also have better ORR catalytic properties than Pt.

The activity and stability of Pt can also be significantly improved by alloying with rare metal catalysts such as Sc, Y, La, Ce, and Gd [49–51]. In some reports, the ORR activity of these Pt-based rare metal alloys even exceeds the catalytic activity of Pt-based transition metal alloys such as Co, Ni, and Fe. Yoo and co-workers reported a solution-based approach to the preparation of Pt_3M alloy (M = Y, Zr, Ti, Ni, and Co) (Figure 8.2) [52]. Experiment and DFT calculations revealed that the enhanced stability and activity of Pt-based alloy nanocatalysts in ORR originated from a surface Y atom surrounded by Pt atoms and the d-band vacancy of Pt on the Pt_3M alloy electrodes.

8.2.1.2 Structural Design of Pt-Based Catalysts for ORR

Numerous theoretical and experimental research have confirmed that the ORR efficiency of fuel cells is closely related to the microstructure of the catalysts. There are differences in the electronic and catalytic properties of Pt-based catalysts with different morphologies [53]. Since Stamenkovic et al. found that the ORR activity of Pt_3Ni (111) catalyst is 90 times higher than that of commercial Pt/C, controlling the synthesis of Pt-based catalysts with special morphologies has attracted extensive and in-depth research [43].

Octahedra Based on the excellent catalytic ORR kinetics of Pt_3Ni (111) crystal planes [43], many researchers have been focused on designing and synthesizing octahedral Pt_3Ni and other types of octahedral platinum alloy catalysts. Younan Xia et al. have reported the synthesis of uniform 9 nm Pt–Ni octahedra with the use of oleylamine and oleic acid as surfactants and $W(CO)_6$ as a source of CO

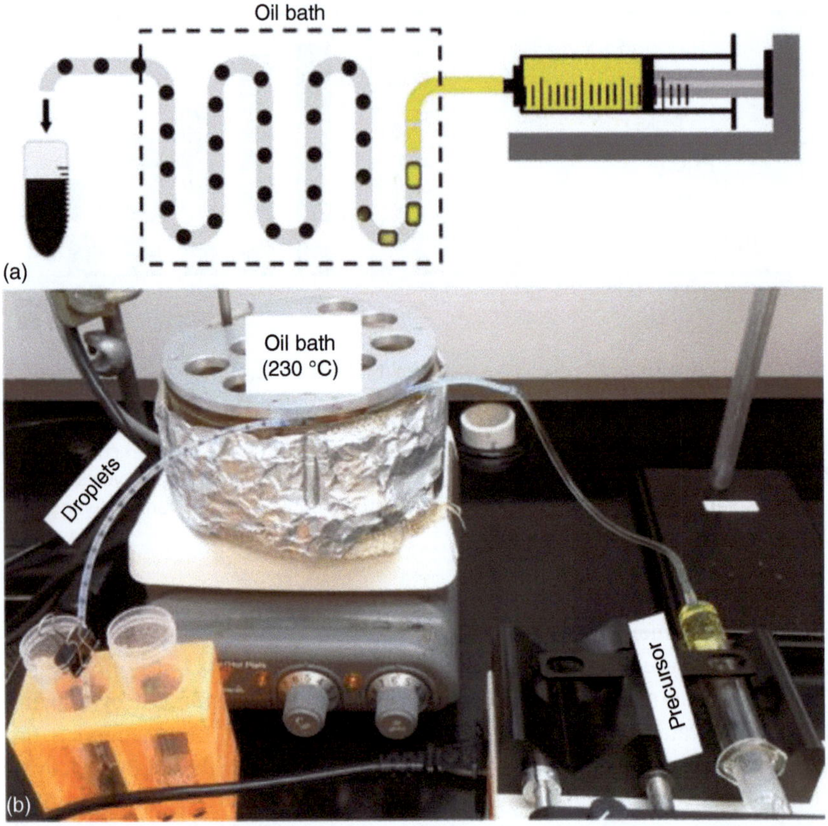

Figure 8.3 (a) Schematic illustration of the fluidic device used for generating droplets in situ, followed by the formation of Pt–Ni octahedral nanocrystals within each droplet. (b) Photograph showing the formation of droplets in a PTFE tube during the continuous pumping of a precursor solution containing $W(CO)_6$ at a concentration of 2.0 mg ml^{-1}. Source: Reproduced with permission of Niu et al. [55], American Chemical Society.

that can promote the formation of {111} facets in the presence of Ni. Through the introduction of benzyl ether as a solvent, the coverage of both surfactants on the surface of resultant Pt–Ni octahedra was significantly reduced while the octahedral shape was still attained. By further removing the surfactants through acetic acid treatment, they observed a specific activity 51-fold higher than that of the state-of-the-art Pt/C catalyst for the ORR at 0.93 V, together with a record high mass activity of 3.3 A mg_{Pt}^{-1} at 0.9 V [54]. To meet the needs of commercial evaluation, they reported the use of droplet reactors for the continuous and scalable production of Pt–Ni octahedra with high activities toward ORR. $Pt(acac)_2$, $Ni(acac)_2$, and $W(CO)_6$ were dissolved in a mixture of oleamine, oleic acid, and phenyl ether, and then they were pumped into a Teflon tube (Figure 8.3). The CO gas formed by the decomposition of $W(CO)_6$ separates the solution into individual droplets, each of which then forms a Pt–Ni octahedron. The ORR mass activity of $Pt_{2.4}Ni$ octahedral is 11 times that of commercial Pt/C catalyst at 0.9 V [55].

In addition to obtaining Pt-based catalysts with high ORR electrocatalytic activity by preparing PtNi octahedral nanoparticles with special crystal facets to change the surface electronic structure of Pt, the electronic structure on the surface of PtNi nanoparticles can also be controlled by modification with other metal atoms or nonmetal atoms, thereby improving the oxygen reduction activity and stability of PtNi catalysts.

Huang et al. revealed that the ORR activity and stability of PtNi octahedra nanocrystals were dramatically increased by introducing a trace of molybdenum to Pt_3Ni, and the mass activity was more than 70 times that of commercial Pt/C. DFT calculations indicate that the Mo atoms tend to occupy the sites on the superficial vertex/edge in an oxidizing environment [56]. It was also found by Mueller et al. that a small amount of Mo doping can reduce the equilibrium concentration of Ni on the surface and increase Pt/Ni vacancy formation energy in the surface layer, limiting the dissolution rate of Ni atoms from the particles [57]. Lim and co-workers revealed that doped with Ga on a carbon support, the Pt–Ni nanoparticles maintained their octahedral shape, even after 30,000 potential cycles. Its composition and stability are related to the oxygenophilic surface and lattice compression caused by Ga doping. This work indicates that octahedral Ga PtNi/C can be used as a highly active and durable ORR catalyst in practical applications of fuel cells [58]. Henkelman and co-workers revealed that the incorporation of In can improve the ORR activity and stability of the PtNi octahedral nanoparticle catalyst, which is based on the metal doping energy and the OH* adsorption energy [59]. Experimental and computational studies indicate that surface-doped In generates a new active site and blocks the diffusion of Ni to the surface by reducing the oxygen-philicity of Pt.

In addition, the introduction of nonmetals can also improve the durability. Yang and co-workers indicated that phosphorus treatment of the octahedral PtNi nanoparticles could significantly enhance their durability [60]. Phosphorus is adsorbed on surface Ni more strongly than on surface Pt, and the surface P is found to preserve the surface Ni and induce the ORR activity enhancement. Especially, a weaker electronegative element P can promote the electron transfer from Ni to Pt. The simple treatment significantly enhanced the long-term stability of the highly active PtNi alloy nano-octahedra. Durability testing by repeating cyclic voltammetry 10,000 cycles showed less activity loss than binary PtNi/C. Lee and co-workers prepared the octahedral PtNi nanoparticles with halide treatment, in which halides were adsorbed on surface Ni more strongly than on surface Pt [61]. The Br–PtNi NPs exhibit enhanced electrocatalytic activity and stability for ORR compared to the PtNi/C catalyst because the Br can preserve the surface Ni effectively. This study provides a paradigm for the synthesis of halide-doped Pt alloy catalysts with significantly improved ORR performance.

Nanoparticles with High-Index Crystal Facets At the nanoscale, nanomaterials with different morphologies also exhibit different catalytic activities. Pt nanoparticles with high-index crystal facets have stronger catalytic activity for oxygen reduction due to the presence of more active sites such as step atoms Pt and unsaturated Pt atoms [62].

In 2007, Sun and co-workers for the first time, prepared unusual Pt single-crystal tetrahexahedral, which is enclosed by 24 high-index facets such as {730}, {210}, and/or {520} surfaces that have a large density of atomic steps and dangling bonds [63]. Electrochemical tests have shown that the Pt nanoparticles with this morphology exhibit excellent mass and volume-specific activity. Since then, many researchers have been working on the preparation of Pt-based nanocrystals with high-index facets. Zhang and co-workers synthesized CNC and HOH Pt–Ni alloy with HIFs by a simple wet chemical method for the first time [64]. The experimental results revealed that the CNC and HOH Pt–Ni alloy NCs exhibit much higher specific activity and stability for the electrooxidation of methanol and formic acid, as well as for the ORR, than those of the Pt–Ni nanocubes and commercial Pt/C and Pt black catalysts (Figure 8.4). Xia et al. prepared concave Pt nanocubes containing {510}, {720}, and {830} high-index crystal planes using a simple aqueous solution reduction method. Electrochemical tests have shown that the area-specific activity of the cubic crystal Pt containing concave octahedra is significantly better than that of Pt cubic nanoparticles containing low-index crystal facets (100) and (111) and commercial Pt/C [65].

Core–Shell Nanostructure Usually, only a few atomic layers on the catalyst surface participate in actual electrochemical catalytic reactions, and most of the nucleating noble metal catalysts are not utilized, resulting in a low dispersion and utilization rate of Pt. For example, for a 3 nm Pt nanoparticle, only about 30% of the atoms are on the surface, and the remaining 70% of the atoms exist in the nucleus without catalytic activity. In recent years, the core–shell strategy of depositing thin layers of Pt metal on the surface of relatively inexpensive Pt alloys or non-Pt metals/alloys has received increasing attention [66]. For this structure, the Pt atoms are distributed on the outside of the nanoparticles, which greatly improves the utilization of Pt. Meanwhile, the Pt-rich shell can protect the nonplatinum core from aggregation and heterogeneous contact, thereby improving the stability of the catalyst. In addition, the catalytic activity of Pt in core–shell catalysts is mainly determined by the compression and expansion arrangement of surface atoms, which is called geometric effect [67, 68]. Therefore, the catalytic activity of the shell can be continuously regulated by the interaction between the d-band center and the surface tension [69]. Huang and co-workers reported on a class of PtPb/Pt core/shell nanoplate catalysts that exhibit large biaxial strains [70]. The stable Pt (110) facets of the nanoplates have high ORR specific and mass activities that reach 7.8 mA cm^{-2} and 4.3 A $mg_{Pt}{}^{-1}$, respectively. DFT calculations reveal that the edge-Pt and top (bottom)–Pt (110) facets undergo large tensile strains, which is helpful in optimizing the Pt-O bond strength. The intermetallic core and uniform four layers of Pt shell of the PtPb/Pt nanoplates contribute to the high durability of these catalysts.

Dealloying is an efficient technique for preparing core–shell Pt nanostructures, wherein non-Pt components are selectively dissolved from the alloy by chemical or electrochemical methods. Strasser and co-workers reported a series of PtCu core–shell catalysts by dealloying [71–73]. By electrochemical dealloying, they obtained core–shell catalysts with several layers of Pt atoms. Among them, the

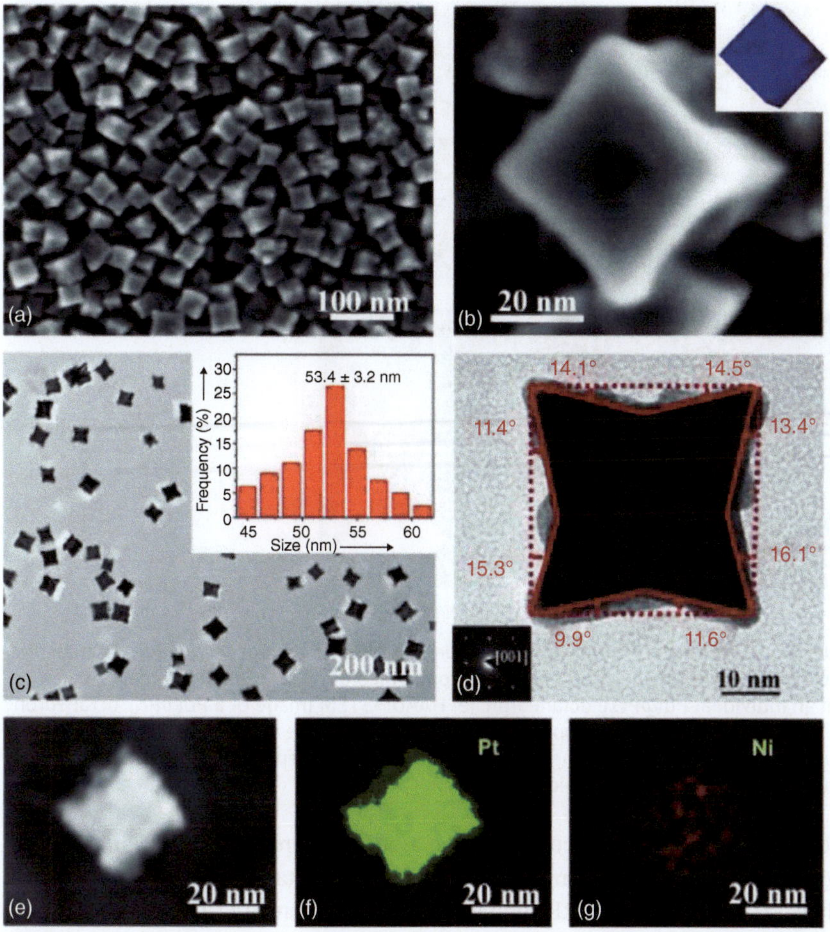

Figure 8.4 (a) SEM, (b) HRSEM (and corresponding model), (c) TEM (and size-distribution histogram), and (d) HRTEM images (and the corresponding SAED pattern) of CNC Pt–Ni alloy NCs. (e) HAADF-STEM image and (f, g) EDS-mapping images of the CNC Pt–Ni alloy NCs. Source: Reproduced with permission of Xu et al. [64], Wiley-VCH.

dealloyed $PtCu_3$ showed the best mass activity (0.56 A mg_{Pt}^{-1}) for ORR, which was attributed to the surface strain effect. It should be noted that the surface roughness of the metal particles becomes greater after dealloying, but it can be smoothed by subsequent heat treatment, thus improving the activity and stability of the catalyst [74]. The Adzic group deposited Pt monolayer on the surface of PdAu NP surfaces by the galvanic displacement method to optimize the utilization of Pt [75]. It was found that ultralow Pt content was sufficient to achieve high ORR catalytic performance. Previously, Adzic and co-workers reported a novel core-catalyzed coating strategy for the epitaxial growth of Pt atomic layer on the surface of Pd NPs, inspired by aerobic alcohol oxidation catalyzed by the Pd core [76]. In addition, a solution-phase method was successfully used to synthesize Pd@Pt_{1L} (1L: one atomic layer) core–shell octahedra (Figure 8.5) [77]. The success of this method

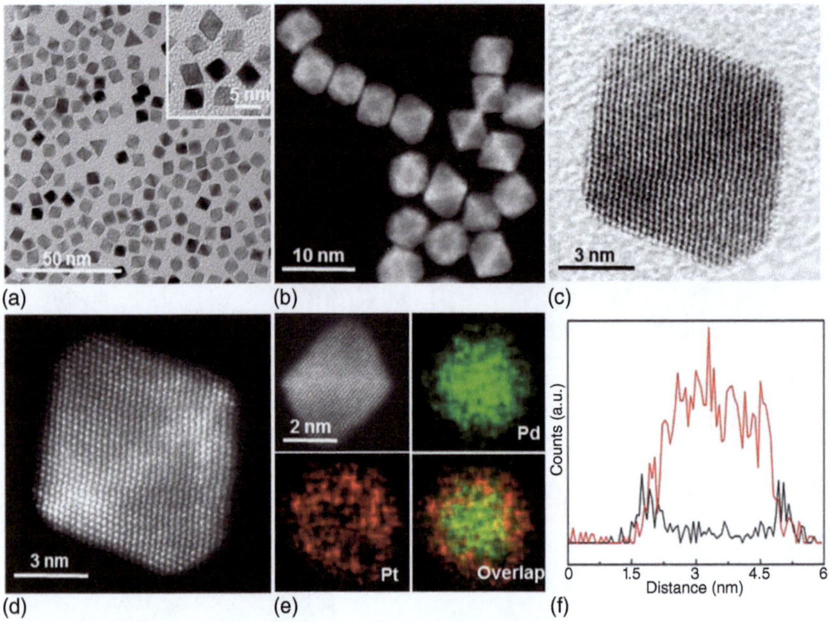

Figure 8.5 (a) TEM and (b) low-magnification HAADF-STEM images of Pd@Pt_{1L} octahedral nanocrystals. (c) Bright-field STEM and (d) atomic-resolution HAADF-STEM images taken from an individual Pd@Pt1L octahedron. (e) HAADF-STEM image of a Pd@Pt_{1L} octahedron and the corresponding EDX mapping of Pd and Pt. (f) EDX line scans of Pd (red trace) and Pt (black trace) recorded from a Pd@Pt_{1L} octahedron. Source: Reproduced with permission of Zhou et al. [77], American Chemical Society.

relies on the use of metal precursors with markedly different reduction kinetics. The Pd@Pt_{1L} core–shell octahedra showed major augmentation in terms of catalytic activity and durability for ORR compared to the commercial Pt/C catalyst.

Nanoframes The ORR electrocatalytic activity and durability of Pt-based nanoframe catalysts can be greatly improved due to the high specific surface area, open structure, and large void space. Controlling the structure at the atomic level can accurately and effectively adjust the catalytic performance of materials, thereby enhancing their activity and durability. In recent years, many research groups have focused on controlling the synthesis of Pt-based alloy nanoframe catalysts.

With a crystallized $PtNi_3$ polyhedron as the starting material, Yang and co-workers obtained Pt_3Ni nanoframes by transforming it into solution by interior erosion [78]. The edges of the Pt-rich $PtNi_3$ polyhedron remain in the final Pt_3Ni nanoframe. The internal and external catalytic surfaces of this open framework structure are composed of nanoscale Pt skin structure, which shows enhanced ORR activity. The mass activity and the specific activity of Pt_3Ni nanoframework catalyst achieved 36 times and 22 times higher than that of the state-of-the-art platinum–carbon catalysts, respectively. Xia et al. fabricated nanocagons by depositing several atomic layers of Pt on Pd nanocrystals with well-defined small planes and then etching away the Pd template [79]. The electrocatalytic performance of the octahedral nanocages

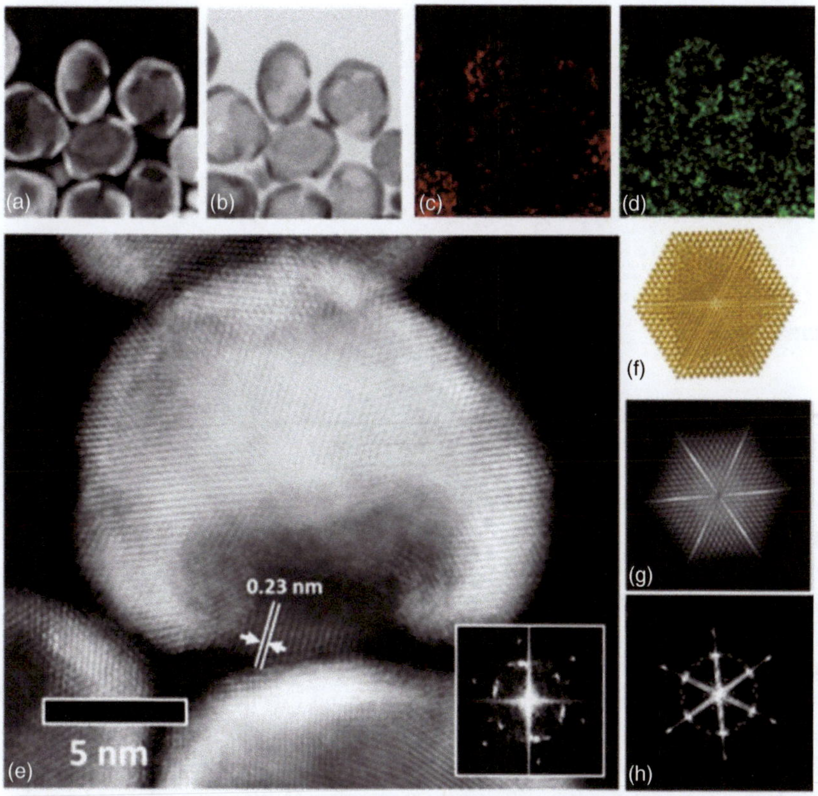

Figure 8.6 Simultaneously acquired (a) HAADF-STEM and (b) BF-STEM images of icosahedral-enriched nanocages. (c) Pd and (d) Pt signals of EDX mapping. (e) Atomically resolved HAADF-STEM image of an icosahedral Pt-enriched nanocage. The inset is the FFT of the image. (f) Atomic model of an icosahedral nanoparticle. (g) Kinematic STEM-HAADF simulation of the nanoparticle in (f) with the same direction. (h) FFT of the image in (g). Source: Reproduced with permission of He et al. [80], American Chemical Society.

had mass activity of 0.75 A mg_{Pt}^{-1} at 0.9 V, which was five times higher than that of the commercial Pt/C catalyst (0.14 A mg_{Pt}^{-1}). The results demonstrate that the transformation from Pd@Pt_{nL} NCs into Pt nanocages could greatly improve the catalytic performance toward ORR in terms of Pt mass and materials cost. The icosahedral Pt nanocage (Figure 8.6) showed extraordinary activity for ORR over Pd–Pt core–shell and commercial Pt/C with a 10 times higher specific activity and 7 times higher mass activity, which is attributed to the dense twin defects at the surface, exposure of the internal surface, and ORR-favorable Pt {111} facets [80]. This icosahedral Pt-enriched nanocage is a very promising candidate structure for Pt-based ORR catalysts, whose activity may be further enhanced by alloying with 3d metals such as Fe, Co, and Ni. Xiong and co-workers reported a unique copper-assisted selective deposition-etching approach for preparing ternary Pt-based $Pt_4PdCu_{0.4}$ nanoframes with the specific activities in ORR being 9.2 times higher than commercial Pt/C and also excellent durability [81].

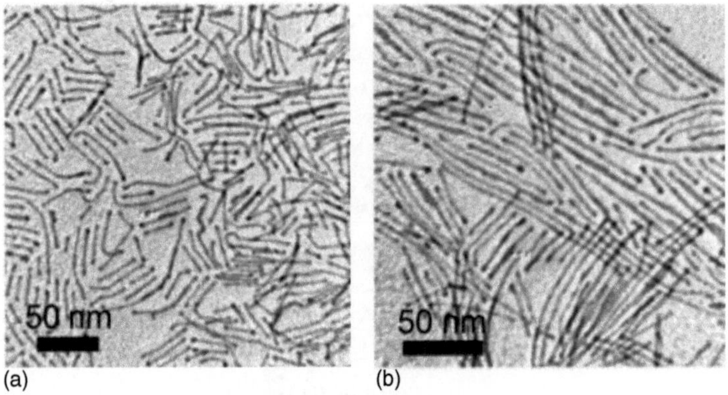

Figure 8.7 TEM images of (a) 2.5 nm wide $Fe_{56}Pt_{44}$ and (b) $Co_{63}Pt_{37}$ NWs. Source: Reproduced with permission of Guo et al. [83], Wiley-VCH.

Nanowire Benefiting from the large surface area and relatively better stability than nanoparticles, one-dimensional nanocrystal is considered an ideal structure to exhibit outstanding performance in electrocatalysis. One-dimensional nanocrystal catalysts with smooth and low-energy surfaces have higher catalytic activity than traditional Pt alloy catalysts [23, 82]. Compared with the 0-dimensional structure, the one-dimensional structure has a spatially interrelated porous aerogel structure, good electron and proton transport performance, and excellent toxicity resistance and structural stability. In addition, one-dimensional nanostructures can provide multiple anchored active sites for carbon carriers, thus avoiding agglomeration during electrochemical reactions. Recently, nanowires (NWs) have attracted much attention as unique one-dimensional structures due to their low specific surface area and volume ratio. At present, platinum-based alloy NWs with thin and serrated features have received extensive attention.

Guo and co-workers reported an advanced organic-phase synthesis of PtFe and PtCo NWs (Figure 8.7) with a diameter of 2–3 nm for enhanced ORR activity. The NWs were obtained through decomposition of metal pentacarbonyl $Fe(CO)_5$, $Co_2(CO)_8$, and $Pt(acac)_2$ and the thermal decomposition of oleamide and 1-octadecene [83]. The 2.5 nm wide PtFe NWs demonstrate exceptional specific and mass activities of $1.53\ mA\ cm^2$ and $844\ mA\ mg_{Pt}{}^{-1}$, respectively, 4.78 and 5.45 times higher than those of the benchmark Pt catalyst. The catalysts also maintained their inherent ORR catalytic activity after durability testing, due to their strong interaction with carbon supports. In order to further improve the utilization rate of Pt atoms, the research group has focused on PtNi NWs with a thickness of less than 10 nm through wet chemical reduction method in the recent research work [84]. In this work, $Mo(CO)_6$ was used as a structural guide agent and glucose served as a reducing agent. The catalytic activity of the synthesized ultra-thin PtNi NWs was 30 times higher than that of commercial Pt/C.

Previous experimental results on single-crystal electrodes have shown that the catalyst with high-index facets has better catalytic performance compared to that with lower index facets. This discovery encourages the researchers to construct a new type of one-dimensional nanostructures with high-index facets.

Huang and co-workers reported solution-synthesized Pt/NiO core/shell NWs which can be transformed into PtNi alloy NWs through a thermal annealing process and then converted into jagged Pt NWs via an electrochemical dealloying [85]. The experimental results indicate that the mass activity (13.6 A mg_{Pt}^{-1}) and specific activity (11.5 mA cm^{-2}) of the jagged Pt NWs are 50 and 32.86 times higher than that of commercial Pt/C catalysts. Theoretical calculation illustrated that the enhanced ORR activity is due to highly stressed and under-coordinated rhombohedral-rich surface. Huang et al. prepared a new class of porous Pt_3Ni NWs with Pt-rich shell and high-index facets [86]. The porous Pt_3Ni NWs exhibit an excellent mass activity of 5.60 A mg_{Pt}^{-1}, which is 37.3 higher than that of the Pt/C. They also exhibit remarkable stability with a negligible decay after 20,000 potential cycles.

8.2.2 Non-Pt-Based Metal Materials for ORR

The above-mentioned catalysts have improved the ORR performance of the existing Pt-based catalysts to varying degrees. But it is undeniable that they all contain some amount of the precious metal platinum. However, the scarcity and high price of platinum fundamentally limit the large-scale commercial application of PEMFCs. In recent years, a great deal of research has been done to find a class of electrocatalysts without platinum or other precious metals.

Some other precious metals with relatively low prices and rich reserves, such as Pd, Ru, Ir, Rh, Au, and even Ag, can be selected to prepare new ORR electrocatalysts. Due to the relatively complex reaction mechanism of ORR in acidic or alkaline media, many oxygen-containing intermediates (such as O_{ads}, OH_{ads}, and OOH_{ads}) are often produced. Nørskov et al. identified a volcanic relationship between the adsorption and desorption strength of oxygen-containing species on the metal surface and the metal type [14]. In particular, the adsorption and desorption energy of Pd to oxygen is close to that of Pt, showing high ORR activity. In addition, by forming alloys with other transition metals, the as-prepared Pd-based electrocatalyst exhibits excellent ORR activity. In these studies, the ORR catalytic activity of Pd–Co, Pd–Co–Mo, and Pd–Fe alloy catalysts was close to, or slightly higher than, that of commercial Pt/C catalysts [87–89]. Interestingly, the Pd-skin of Pd–M alloys (M = Fe, Co, Ni, etc.) is also easy to form [90, 91]. In addition, the ORR activity of chalcogenides of Ru, especially $RuSe_x$ alloys, is superior to that of Ru [92–94], but the catalytic activity of $RuSe_x$ is still lower than that of standard Pt/C catalysts so far [94]. Now, it is necessary to find a class of electrocatalysts that do not contain platinum or other precious metals. Although great progress has been made in the study of nonprecious metals as ORR catalysts, there are still many problems with their catalytic activity and stability.

8.3 Conclusion and Outlook

In a word, significant progress has been made in improving the performance of metal-based catalysts for ORR. Some recent effective strategies to rationally design metal-based catalysts, especially Pt-based catalysts with excellent ORR activity, were summarized, including structural design, surface modification, size regulation, electron configurations modulation, and lattice facets modulation. In addition, alloy, heterojunction, and core–shell structure could effectively improve the activity and stability of Pt-based catalysts toward ORR.

Despite significant progress having been made in Pt-based catalysts for ORR, there are still some challenges required to be overcome. (i) It is still challenging to further optimize the preparation conditions of low Pt-based metal catalysts and explore simple, low cost, and large-scale preparation methods to prepare highly active and stable ORR catalysts. (ii) Meanwhile, it is very important to effectively control the monodispersity of platinum layer and realize the control of platinum layer thickness in Pt-based catalysts. (iii) In addition, most of the current research on advanced catalysts relies on RDE testing technology, in which ORR activity has far exceeded the DOE target for an MA at 0.9 V in a device. However, these advanced catalysts have not been successfully applied to real fuel cell devices. (iv) It is necessary to deeply understand the active center and ORR reaction mechanism of Pt-based catalysts. Therefore, researchers need to develop more advanced in situ characterization technology to study the dynamic changes of active sites and reaction intermediates.

Together, it is urgent to develop low Pt catalysts with high activity and stability or develop efficient non-Pt oxygen reduction catalysts to replace the currently expensive Pt/C, in order to ultimately solve the problem of high fuel cell cost.

Acknowledgment

This work was supported by Hebei Natural Science Foundation (B2024408023), and the Funded by Science and Technology Project of Hebei Education Department (QN2022043, QN2021124).

References

1 Jiao, K., Xuan, J., Du, Q. et al. (2021). Designing the next generation of proton-exchange membrane fuel cells. *Nature* 595 (7867): 361.

2 Wee, J.-H. (2007). Applications of proton exchange membrane fuel cell systems. *Renew. Sust. Energ. Rev.* 11 (8): 1720.

3 Peighambardoust, S.J., Rowshanzamir, S., and Amjadi, M. (2010). Review of the proton exchange membranes for fuel cell applications. *Int. J. Hydrog. Energy* 35 (17): 9349.

4 Gencoglu, M.T. and Ural, Z. (2009). Design of a PEM fuel cell system for residential application. *Int. J. Hydrog. Energy* 34 (12): 5242.

5 Wang, Y., Diaz, D.F.R., Chen, K.S. et al. (2020). Materials, technological status, and fundamentals of PEM fuel cells-a review. *Mater. Today* 32: 178.

6 Kamarudin, S.K., Achmad, F., and Daud, W.R.W. (2009). Overview on the application of direct methanol fuel cell (DMFC) for portable electronic devices. *Int. J. Hydrog. Energy* 34 (16): 6902–6916.

7 Wang, F., Xie, W., Yang, L. et al. (2021). Revealing the importance of kinetics in N-coordinated dual-metal sites catalyzed oxygen reduction reaction. *J. Catal.* 396: 215.

8 Wan, C., Duan, X., and Huang, Y. (2020). Molecular design of single-atom catalysts for oxygen reduction reaction. *Adv. Energy Mater.* 10 (14): 1903815.

9 Tian, X., Lu, X.F., Xia, B.Y., and Lou, X.W. (2020). Advanced electrocatalysts for the oxygen reduction reaction in energy conversion technologies. *Joule* 4 (1): 45.

10 Liu, M., Zhao, Z., Duan, X., and Huang, Y. (2019). Nanoscale structure design for high-performance Pt-based ORR catalysts. *Adv. Mater.* 31 (6): 1802234.

11 Dai, L., Xue, Y., Qu, L. et al. (2015). Metal-free catalysts for oxygen reduction reaction. *Chem. Rev.* 115 (11): 4823.

12 Scofield, M.E., Liu, H., and Wong, S.S. (2015). A concise guide to sustainable PEMFCs: recent advances in improving both oxygen reduction catalysts and proton exchange membranes. *Chem. Soc. Rev.* 44 (16): 5836.

13 Hu, B., Deng, X., Zhou, L. et al. (2020). Facile synthesis of synergistic Pt/(Co-N)@C composites as alternative oxygen-reduction electrode of PEMFCs with attractive activity and durability. *Compos. Part B Eng.* 193: 108012.

14 Nørskov, J.K., Rossmeisl, J., Logadottir, A. et al. (2004). Origin of the overpotential for oxygen reduction at a fuel-cell cathode. *J. Phys. Chem. B* 108 (46): 17886.

15 Zhao, Z., Chen, C., Liu, Z. et al. (2019). Pt-based nanocrystal for electrocatalytic oxygen reduction. *Adv. Mater.* 31 (31): 1808115.

16 Kong, Z., Zhang, D., Lu, Y. et al. (2021). Advanced cathode electrocatalysts for fuel cells: understanding, construction, and application of carbon-based and platinum-based nanomaterials. *ACS Mater. Lett.* 3 (12): 1610.

17 Zhang, J., Yuan, Y., Gao, L. et al. (2021). Stabilizing Pt-based electrocatalysts for oxygen reduction reaction: fundamental understanding and design strategies. *Adv. Mater.* 33 (20): 2006494.

18 Sharma, H., Bisht, A., Sethulakshmi, N., and Sharma, S. (2024). Catalysis by substituted platinum (ionic Pt) catalysts. *Int. J. Hydrog. Energy* 768 (51): 748.

19 Wang, X.X., Tan, Z.H., Zeng, M., and Wang, J.N. (2014). Carbon nanocages: a new support material for Pt catalyst with remarkably high durability. *Sci. Rep.* 4 (1): 4437.

20 Lu, B.A., Sheng, T., Tian, N. et al. (2017). Octahedral PtCu alloy nanocrystals with high performance for oxygen reduction reaction and their enhanced stability by trace Au. *Nano Energy* 33: 65.

21 Huang, J.F. and Tseng, P.K. (2018). High performance layer-by-layer Pt_3Ni(Pt-skin)-modified Pd/C for the oxygen reduction reaction. *Chem. Sci.* 9 (28): 6134.

22 Lv, P., Jin, B., Mei, C. et al. (2018). Pt_9Ni wavelike nanowires with high activity for oxygen reduction reactions. *Chem. Eur. J.* 24 (55): 14636.

23 Li, K., Li, X., Huang, H. et al. (2018). One-nanometer-thick PtNiRh trimetallic nanowires with enhanced oxygen reduction electrocatalysis in acid media: integrating multiple advantages into one catalyst. *J. Am. Chem. Soc.* 140 (47): 16159.

24 Vo Doan, T.T., Wang, J., Poon, K.C. et al. (2016). Theoretical modelling and facile synthesis of a highly active boron-doped palladium catalyst for the oxygen reduction reaction. *Angew. Chem. Int. Ed.* 128 (24): 6956.

25 Du, C., Li, P., Yang, F. et al. (2018). Monodisperse palladium sulfide as efficient electrocatalyst for oxygen reduction reaction. *ACS Appl. Mater. Interfaces* 10 (1): 753.

26 Qin, J., Liu, H., Zou, P. et al. (2022). Altering ligand fields in single-atom sites through second-shell anion modulation boosts the oxygen reduction reaction. *J. Am. Chem. Soc.* 144 (5): 2197.

27 Guo, Y., Wu, D., Li, M. et al. (2022). Coordination engineering of ultra-uniform ruthenium nanoclusters as efficient multifunctional catalysts for zinc-air batteries. *Small Sci.* 2 (10): 2200035.

28 Neergat, M., Gunasekar, V., and Singh, R.K. (2011). Oxygen reduction reaction and peroxide generation on Ir, Rh, and their selenides-a comparison with Pt and RuSe. *J. Electrochem. Soc.* 158 (9): B1060.

29 Huang, J., Du, C., Nie, J. et al. (2019). Encapsulated Rh nanoparticles in N-doped porous carbon polyhedrons derived from ZIF-8 for efficient HER and ORR electrocatalysis. *Electrochim. Acta* 326: 134982.

30 Zhou, Y., Lu, Q., Zhuang, Z. et al. (2015). Oxygen reduction at very low overpotential on nanoporous Ag catalysts. *Adv. Energy Mater.* 5 (13): 1500149.

31 Gasteiger, H.A., Kocha, S.S., Sompalli, B. et al. (2005). Activity benchmarks and requirements for Pt, Pt-alloy, and non-Pt oxygen reduction catalysts for PEMFCs. *Appl. Catal. B Environ.* 56 (1–2): 9.

32 Farrell, C.G., Gardner, C.L., and Ternan, M. (2007). Experimental and modelling studies of CO poisoning in PEM fuel cells. *J. Power Sources* 171 (2): 282.

33 Lebedeva, N.P., Koper, M.T.M., Feliu, J.M., and van Santen, R.A. (2002). Mechanism and kinetics of the electrochemical CO adlayer oxidation on Pt (111). *J. Electroanal. Chem.* 524: 242.

34 Baturina, O.A., Gould, B.D., Korovina, A. et al. (2011). Products of SO_2 adsorption on fuel cell electrocatalysts by combination of sulfur K-edge XANES and electrochemistry. *Langmuir* 27 (24): 14930.

35 Kawase, M., Mugikura, Y., Watanabe, T. et al. (2002). Effects of NH_3 and NO_x on the performance of MCFCs. *J. Power Sources* 104 (2): 265.

36 Mohtadi, R., Lee, W.-K., and Van Zee, J.W. (2004). Assessing durability of cathodes exposed to common air impurities. *J. Power Sources* 138 (1–2): 216.

37 Yang, D., Ma, J., Xu, L. et al. (2006). The effect of nitrogen oxides in air on the performance of proton exchange membrane fuel cell. *Electrochim. Acta* 51 (19): 4039.

38 Greeley, J., Stephens, I.E.L., Bondarenko, A.S. et al. (2009). Alloys of platinum and early transition metals as oxygen reduction electrocatalysts. *Nat. Chem.* 1 (7): 552.

39 Wang, C., Markovic, N.M., and Stamenkovic, V.R. (2012). Advanced platinum alloy electrocatalysts for the oxygen reduction reaction. *ACS Catal.* 2 (5): 891.

40 Wu, Y., Zhao, Y., Liu, J., and Wang, F. (2018). Adding refractory 5d transition metal W into PtCo system: an advanced ternary alloy for efficient oxygen reduction reaction. *J. Mater. Chem. A* 6 (23): 10700.

41 Chong, L., Wen, J., Kubal, J. et al. (2018). Ultralow-loading platinum-cobalt fuel cell catalysts derived from imidazolate frameworks. *Science* 362 (6420): 1276.

42 Kaneko, S., Myochi, R., Takahashi, S. et al. (2017). Ultrahigh vacuum synthesis of strain-controlled model Pt(111)-shell layers: surface strain and oxygen reduction reaction activity. *J. Phys. Chem. Lett.* 8 (21): 5360.

43 Stamenkovic, V.R. (2007). Improved oxygen reduction activity on Pt_3Ni (111) via increased surface site availability improved oxygen reduction activity on Pt_3Ni (111) via increased surface site availability. *Science* 315 (5811): 493.

44 Wang, N., Li, Y., Guo, Z. et al. (2018). Minute quantities of hexagonal nanoplates PtFe alloy with facile operating conditions enhanced electrocatalytic activity and durability for oxygen reduction reaction. *J. Alloys Compd.* 752: 23.

45 Stamenkovic, V.R., Mun, B.S., Arenz, M. et al. (2007). Trends in electrocatalysis on extended and nanoscale Pt-bimetallic alloy surfaces. *Nat. Mater.* 6 (3): 241.

46 Wu, Z.W., Chiang, M.H., Wang, T.P., and Lee, C.-L. (2018). Dendritic Pt shells coated on concave Pd nanocrystals: synthesis and use as electrocatalysts for acidic oxygen reduction reaction. *Int. J. Hydrog. Energy* 43 (49): 22195.

47 Yoon, H., Song, H.J., Ju, B. et al. (2022). Gas-phase synthesis of PtMo alloy electrocatalysts with enhanced activity and durability for oxygen reduction reaction. *ACS Sustain. Chem. Eng.* 10 (46): 15319.

48 Zhao, Q., Wang, C., Wang, H. et al. (2020). Synthesis of a high-performance low-platinum PtAg/C alloyed oxygen reduction catalyst through the gradual reduction method. *New J. Chem.* 44 (9): 3728.

49 Johansson, T.P., Ulrikkeholm, E.T., Hernandez-Fernandez, P. et al. (2014). Pt skin versus Pt skeleton structures of Pt_3Sc as electrocatalysts for oxygen reduction. *Top. Catal.* 57 (1–4): 245.

50 Jeon, M.K. and McGinn, P.J. (2011). Carbon supported Pt-Y electrocatalysts for the oxygen reduction reaction. *J. Power Sources* 196 (3): 1127.

51 Malacrida, P., Escudero-Escribano, M., Verdaguer-Casadevall, A. et al. (2014). Enhanced activity and stability of Pt-La and Pt-Ce alloys for oxygen electroreduction: the elucidation of the active surface phase. *J. Mater. Chem. A* 2 (12): 4234.

52 Hwang, S.J., Kim, S.K., Lee, J.G. et al. (2012). Role of electronic perturbation in stability and activity of Pt-based alloy nanocatalysts for oxygen reduction. *J. Am. Chem. Soc.* 134 (48): 19508.

53 Kodama, K., Nagai, T., Kuwaki, A. et al. (2021). Challenges in applying highly active Pt-based nanostructured catalysts for oxygen reduction reactions to fuel cell vehicles. *Nat. Nanotechnol.* 16 (2): 140.

54 Choi, S., Xie, S., Shao, M. et al. (2013). Synthesis and characterization of 9 nm Pt-Ni octahedra with a record high activity of $3.3\,A/mg_{Pt}$ for the oxygen reduction reaction. *Nano Lett.* 13 (7): 3420.

55 Niu, G., Zhou, M., Yang, X. et al. (2016). Synthesis of Pt-Ni octahedra in continuous-flow droplet reactors for the scalable production of highly active catalysts toward Oxygen reduction. *Nano Lett.* 16 (6): 3850.

56 Huang, X., Cao, L., Chen, Y. et al. (2015). High-performance transition metal-doped Pt_3Ni octahedra for oxygen reduction reaction. *Science* 348 (6240): 1230.

57 Cao, L. and Mueller, T. (2016). Theoretical insights into the effects of oxidation and Mo-doping on the structure and stability of Pt-Ni nanoparticles. *Nano Lett.* 16 (12): 7748.

58 Lim, J., Shin, H., Kim, M. et al. (2018). Ga-doped Pt-Ni octahedral nanoparticles as a highly active and durable electrocatalyst for oxygen reduction reaction. *Nano Lett.* 18 (4): 2450.

59 Lim, J., Shin, K., Bak, J. et al. (2021). Outstanding oxygen reduction reaction catalytic performance of In-PtNi octahedral nanoparticles designed via computational dopant screening. *Chem. Mater.* 33 (22): 8895.

60 Wang, S., Xiong, L., Bi, J. et al. (2018). Structural and electronic stabilization of PtNi concave octahedral nanoparticles by p doping for oxygen reduction reaction in alkaline electrolytes. *ACS Appl. Mater. Interfaces* 10 (32): 27009.

61 Choi, J., Lee, Y., Kim, J., and Lee, H. (2016). Enhancing stability of octahedral PtNi nanoparticles for oxygen reduction reaction by halide treatment. *J. Power Sources* 307: 883.

62 Somorjai, G.A. and Blakely, D.W. (1975). Mechanism of catalysis of hydrocarbon reactions by platinum surfaces. *Nature* 258 (5536): 580.

63 Tian, N., Zhou, Z.Y., Sun, S.G. et al. (2007). Synthesis of tetrahexahedral platinum nanocrystals with high-index facets and high electro-oxidation activity. *Science* 316 (5825): 732.

64 Xu, X., Zhang, X., Sun, H. et al. (2014). Synthesis of Pt-Ni alloy nanocrystals with high-index facets and enhanced electrocatalytic properties. *Angew. Chem. Int. Ed.* 126 (46): 12730.

65 Yu, T., Kim, D.Y., Zhang, H., and Xia, Y. (2011). Platinum concave nanocubes with high-index facets and their enhanced activity for oxygen reduction reaction. *Angew. Chem. Int. Ed.* 50 (12): 2773.

66 Adzic, R.R. (2012). Platinum monolayer electrocatalysts: tunable activity, stability, and self-healing properties. *Electrocatalysis* 3: 163.

67 Zhou, W., Sasaki, K., Su, D. et al. (2010). Gram-scale-synthesized Pd_2Co-supported Pt monolayer electrocatalysts for oxygen reduction reaction. *J. Phys. Chem. C* 114 (19): 8950.

68 Dai, Y. and Chen, S. (2016). Au-Pt core-shell electrocatalysts for oxygen reduction reaction through combining the spontaneous Pt deposition and redox replacement of underpotential-deposited Cu. *Int. J. Hydrog. Energy* 46 (48): 22976.

69 Li, J., Alsudairi, A., Ma, Z. et al. (2017). Asymmetric volcano trend in oxygen reduction activity of Pt and non-Pt catalysts: in situ identification of the site-blocking effect. *J. Am. Chem. Soc.* 139 (4): 1384.

70 Bu, L., Zhang, N., Guo, S. et al. (2016). Biaxially strained PtPb/Pt core/shell nanoplate boosts oxygen reduction catalysis. *Science* 354 (6318): 1410.

71 Oezaslan, M. and Strasser, P. (2011). Activity of dealloyed $PtCo_3$ and $PtCu_3$ nanoparticle electrocatalyst for oxygen reduction reaction in polymer electrolyte membrane fuel cell. *J. Power Sources* 196 (12): 5240.

72 Strasser, P., Koha, S., and Greeleyb, J. (2008). Voltammetric surface dealloying of Pt bimetallic nanoparticles: an experimental and DFT computational analysis. *Phys. Chem. Chem. Phys.* 10: 3670.

73 Haschø, F., Oezaslan, M., and Strasser, P. (2011). Activity, stability, and degradation mechanisms of dealloyed $PtCu_3$ and $PtCo_3$ nanoparticle fuel cell catalysts. *ChemCatChem* 3 (11): 1805.

74 Wang, C., Chi, M., Li, D. et al. (2011). Design and synthesis of bimetallic electrocatalyst with multilayered Pt-skin surfaces. *J. Am. Chem. Soc.* 133 (36): 14396.

75 Sasaki, K., Naohara, H., Choi, Y. et al. (2012). Highly stable Pt monolayer on PdAu nanoparticle electrocatalysts for the oxygen reduction reaction. *Nat. Commun.* 3: 1115.

76 Zhang, Y., Hsieh, Y., Volkov, V. et al. (2014). High performance pt monolayer catalysts produced via core-catalyzed coating in ethanol. *ACS Catal.* 4 (3): 738.

77 Zhou, M., Wang, H., Elnabawy, A.O. et al. (2019). Facile one-pot synthesis of $Pd@Pt_{1L}$ octahedra with enhanced activity and durability toward oxygen reduction. *Chem. Mater.* 31 (4): 1370.

78 Chen, C., Kang, Y., Huo, Z. et al. (2014). Highly crystalline multimetallic nanoframes with three-dimensional electrocatalytic surfaces. *Science* 343 (6177): 1339.

79 Zhang, L., Roling, L.T., Wang, X. et al. (2015). Platinum-based nanocages with subnanometer-thick walls and well-defined, controllable facets. *Science* 349: 412.

80 He, D.S., He, D., Wang, J. et al. (2016). Ultrathin icosahedral Pt-enriched nanocage with excellent oxygen reduction reaction activity. *J. Am. Chem. Soc.* 138 (5): 1494.

81 Ye, W., Chen, S., Ye, M. et al. (2017). $Pt_4PdCu_{0.4}$ alloy nanoframes as highly efficient and robust bifunctional electrocatalysts for oxygen reduction reaction and formic acid oxidation. *Nano Energy* 39: 532.

82 Xu, G.R., Liu, F.Y., Liu, Z.H., and Chen, Y. (2015). Ethanol-tolerant polyethyleneimine functionalized palladium nanowires in alkaline media: the "molecular window gauze" induced the selectivity for the oxygen reduction reaction. *J. Mater. Chem. A* 3 (42): 21083.

83 Guo, S., Li, D., Zhu, H. et al. (2013). FePt and CoPt nanowires as efficient catalysts for the oxygen reduction reaction. *Angew. Chem. Int. Ed.* 125 (12): 3549.

84 Jiang, K., Zhao, D., Guo, S. et al. (2017). Efficient oxygen reduction catalysis by subnanometer Pt alloy nanowires. *Sci. Adv.* 3 (2): e1601705.

85 Li, M., Zhao, Z., Cheng, T. et al. (2016). Ultrafine jagged platinum nanowires enable ultrahigh mass activity for the oxygen reduction reaction. *Science* 354 (6318): 1414.

86 Jiang, K., Shao, Q., Zhao, D. et al. (2017). Phase and composition tuning of 1D platinum-nickel nanostructures for highly efficient electrocatalysis. *Adv. Funct. Mater.* 27 (28): 1700830.

87 Wang, W., Zheng, D., Du, C. et al. (2007). Carbon-supported Pd-Co bimetallic nanoparticles as electrocatalysts for the oxygen reduction reaction. *J. Power Sources* 167 (2): 243.

88 Raghuveer, V., Manthiram, A., and Bard, A.J. (2005). Pd-Co-Mo electrocatalyst for the oxygen reduction reaction in proton exchange membrane fuel cells. *J. Phys. Chem. B* 109 (48): 22909.

89 Shao, M.-H., Sasaki, K., and Adzic, R.R. (2006). Pd-Fe nanoparticles as electrocatalysts for oxygen reduction. *J. Am. Chem. Soc.* 128 (11): 3526.

90 Suo, Y., Zhuang, L., and Lu, J. (2007). First-principles considerations in the design of Pd-alloy catalysts for oxygen reduction. *Angew. Chem. Int. Ed.* 119 (16): 2920.

91 Zhou, W.P., Yang, X., Vukmirovic, M.B. et al. (2009). Improving electrocatalysts for O_2 reduction by fine-tuning the Pt-support interaction: Pt monolayer on the surfaces of a Pd_3Fe (111) single-crystal alloy. *J. Am. Chem. Soc.* 131 (35): 12755.

92 Alonso-Vante, N., Bogdanoff, P., and Tributsch, H. (2000). On the origin of the selectivity of oxygen reduction of ruthenium-containing electrocatalysts in methanol-containing electrolyte. *J. Catal.* 190 (2): 240.

93 Hilgendorff, M., Diesner, K., Schulenburg, H. et al. (2002). Preparation strategies towards selective Ru-based oxygen reduction catalysts for direct methanol fuel cells. *J. New Mater. Electrochem. Syst.* 5 (2): 71.

94 Schmidt, T.J., Paulus, U.A., Gasteiger, H.A. et al. (2000). Oxygen reduction on $Ru_{1.92}Mo_{0.08}SeO_4$, Ru/carbon, and Pt/carbon in pure and methanol-containing electrolytes. *J. Electrochem. Soc.* 147 (7): 2620.

9

Carbon-Based Materials for Electrocatalytic Oxygen Reduction Reaction

Congli Qin and Aixin Fan

Langfang Normal University, School of Chemistry & Material Science, No. 100, Aimin Road, Langfang 065000, China

9.1 Introduction

Due to the increasing global environmental problems caused by fossil fuel combustion, fuel cells that can efficiently generate electricity using clean and renewable fuels have received increasing attention in recent years [1–3]. The development of alternative energy and new energy conversion equipment has increasingly become the focus of scientific research. Hydrogen energy, a recognized low-carbon and zero-carbon energy source, along with proton exchange membrane fuel cells (PEMFCs), has attracted wide attention due to its high energy density, high performance, wide application range, and low pollutant emission [4–7].

In a PEMFC, converting the high energy of hydrogen fuel into electrical energy is achieved through two chemical processes: the hydrogen oxidation reaction (HOR) occurring at the anode and an oxygen reduction reaction (ORR) at the cathode [8, 9]. The slow kinetics of ORR reaction results in an overpotential loss of about ten times that of HOR, which requires high content and high-cost platinum-based catalysts, thus limiting the performance and large-scale commercial application of PEMFC [10, 11]. However, the durability and resource shortage of platinum-based materials are the major limiting factors preventing large-scale commercialization of fuel cells [12–14]. Therefore, it is crucial to develop low-cost, high-performance, and durable nonprecious metal catalysts.

Graphene, carbon nanotubes (CNTs), carbon nanofibers, carbon nanospheres, and other multidimensional carbon materials with excellent electrical conductivity and high specific surface area are widely used in electricity, chemistry, energy, material conversion, and other fields [15–18]. The doping of heteroatoms can regulate the surface chemical state of carbon materials, which is beneficial for improving surface activity, affecting the adsorption state of oxygen reduction intermediates, and promoting oxygen reduction [19–22].

In this chapter, we concentrate on providing a comprehensive overview of recent strategies and developments in carbon materials catalysts for ORR from both

Functional Materials for Electrocatalytic Energy Conversion, First Edition.
Edited by Zhicheng Zhang, Meiting Zhao, and Yuchen Qin.

theoretical and experimental perspectives, involving the metal-free carbon-based catalysts and metal-containing carbon-based catalysts. Specifically, the synthesis strategies, the relationship between structure and performance, and the catalytic mechanisms of the state-of-the-art carbon-based catalysts for ORR are systemically discussed. Finally, based on the above analysis, the conclusions and outlook of carbon-based catalysts for electrocatalytic ORR are provided.

9.2 Carbon-Based Materials for ORR

9.2.1 Carbon-Based Metal-Free Materials for ORR

9.2.1.1 Nitrogen-Doped Carbon Nanomaterials

Naturally, N-doped carbon materials have two advantages for ORR: (i) the atomic size of nitrogen is similar to that of carbon, but its electronic configuration is different. Doping carbon nanomaterials with nitrogen atoms can alter their electronic structure but have the least interference on the carbon lattice. (ii) N has one electron more than C, which favors reactions that need electrons like the ORR. Therefore, N is an effective element to dope carbon for ORR [23–26].

In 2009, Dai et al. reported nitrogen-containing CNT (VA-NCNTs) catalysts with better electrocatalytic activity and stability than platinum for ORR in alkaline fuel cells [22]. This article is not the first report of N-doped ORR catalysts [27, 28], but a more precise report about the activity. Since then, a series of N-doped carbon-based catalysts have been reported, which boosted drastically in this field.

Su and coworkers also prepared an N-doped CNT (CNT@NCNT) catalyst with pristine CNTs as the core and N-doped carbon layers as the shell [29]. Benefiting from integrated nitrogen atoms enriched on the surface and the undestroyed intact inner walls, the CNT@NCNT catalysts showed high electrical conductivity of $3.3\,\mathrm{S\,cm^{-1}}$ and enhanced ORR activity (≈51 mV positive shift of the onset potential and low peroxide generation).

Liu et al. synthesized nitrogen-doped graphene (N-graphene) via chemical vapor deposition (CVD) of methane in ammonia in atmosphere. As shown in Figure 9.1a. The N-graphene catalysts exhibit superb performance for ORR in alkaline fuel cells. The steady-state catalytic current of N-graphene electrode was enhanced threefold compared with that of Pt/C electrode over a large potential range. Moreover, its long-term stability, tolerance to crossover, and poison effect are also superior to Pt/C (Figure 9.1b–d) [30].

In addition to the N-doping, the morphology of the catalyst is also an important role for enhancing ORR performance. Nune et al. prepared porous nitrogen-doped carbon materials (CIRMOF-3-950) via direct carbonizing nitrogen-containing metal-organic framework-3, which exhibited excellent activity through 4-electrons reduction pathway and better stability than Pt/C in alkaline solution for ORR [31]. The experimental evidence proved that a high percentage of quaternary and pyridinic nitrogen, a high surface area as well as good conductivity improved the performance for ORR.

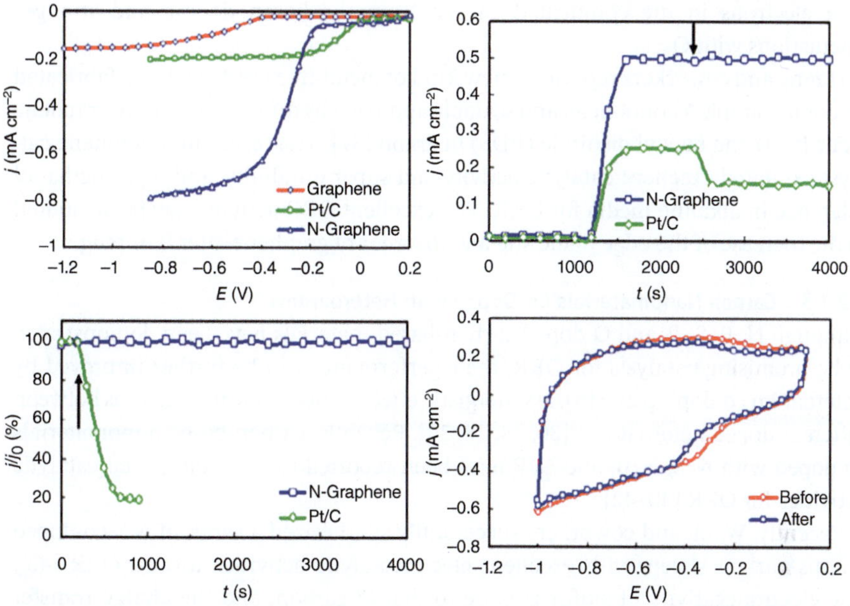

Figure 9.1 (a) RRDE voltammograms for the ORR in air-saturated 0.1 M KOH at the C-graphene electrode (red line), Pt/C electrode (green line), and N-graphene electrode (blue line). Electrode rotating rate: 1000 rpm. Scan rate: $0.01\,V\,s^{-1}$. Mass (graphene) Mass (PtC) Mass (N-graphene) 7.5 g. (b) Current density (*j*) time (*t*) chronoamperometric responses obtained at the Pt/C (circle line) and N-graphene (square line) electrodes at 0.4 V in air saturated 0.1 M KOH. The arrow indicates the addition of 2% (w/w) methanol into the air-saturated electrochemical cell. (c) Current (*j*) time (*t*) chronoamperometric response of Pt/C (circle line) and N-graphene (square line) electrodes to CO. The arrow indicates the addition of 10% (v/v) CO into air saturated 0.1 M KOH at 0.4 V; jo defines the initial current. (d) Cyclic voltammograms of N-graphene electrode in air saturated 0.1 M KOH before (circle line) and after (square line) a continuous potentiodynamic swept for 200,000 cycles at room temperature (25 °C). Scan rate: $0.1\,V\,s^{-1}$. Source: Reproduced with permission of Qu et al. [30], American Chemical Society.

9.2.1.2 Carbon Nanomaterials Doped with Non-Nitrogen Heteroatoms

Apart from nitrogen doping, carbon nanomaterials doped with other heteroatoms, such as boron, phosphorus, sulfur, and chlorine, have been synthesized via numerous methods [32–35].

With toluene as a carbon precursor and triphenylphosphine as the phosphorus source, Peng and coworkers synthesized P-doped graphite layers by a thermolysis approach [36]. The P-doped graphite/GC electrode exhibited a high electrocatalytic activity with an onset potential of −0.11 V, long-term stability, and excellent tolerance to crossover effects of methanol in the ORR in 0.1 M KOH electrolyte.

Hu and coworkers prepared boron-doped CNT (BCNT) catalysts using CVD with benzene, triphenylborane (TPB), and ferrocene as precursors and catalysts [32]. The reduction current increased and onset peak potentials positively shifted with increasing boron content. According to density functional theory (DFT) calculations, the good performance in the ORR originates from effective utilization of π electrons in the conjugated carbon from the boron doping and stronger interactions with O_2.

Huang and coworkers reported a new kind of metal-free ORR catalysts, fabricated through a simple, economical, and scalable approach by directly annealing graphene oxide (GO) and benzyl disulfide (BDS) in argon [37]. The resultant S-graphene catalyst exhibited excellent catalytic activity and super durability, and high methanol tolerance in alkaline media for ORR. The excellent ORR activity may be attributed to the increase of the edge plane site and disorder of graphene after S-doping.

9.2.1.3 Carbon Nanomaterials Co-Doped with Heteroatoms

Although N, B, S, P, and O doped carbon-based materials have been demonstrated to be promising catalysts for ORR, their performance can be further improved by heteroatom co-doping due to the synergistic effect of electronic interactions between different doped heteroatoms [30, 34, 38, 39]. Recently, carbon-based nanomaterials co-doped with N/S, N/B, and N/P have been reported to show enhanced catalytic activities for ORR [40–42].

Recently, Wang and coworkers successfully synthesized a series of N,S-co-doped porous carbon sheets with excellent electrocatalytic activities toward ORR [43]. The electronegativity of sulfur is close to that of carbon, and the charge transfer between the two is weak. However, sulfur atom doping can produce defect sites in carbon materials, while nitrogen atom doping can regulate the electronegativity and coordination environment of carbon elements, which together enhance the catalytic activity of carbon materials in ORR. The as-prepared NSPCS materials achieved excellent electrochemical performance with onset potential of 0.89 V vs. RHE and half-wave potential of 0.75 V in acidic solution. In addition, when applied as a supercapacitor electrode, the as-prepared N and S doped carbon materials also showed outstanding capacitors with high magnification capability and long-term stability.

Dai and coworkers prepared B and N co-doped CNTs (VA–BCN) by pyrolysis of melamine diborate, a single-compound source of carbon, boron, and nitrogen for BCN nanotube growth [20]. Due to a synergetic effect by the co-doping of N and B, the resultant VA–BCN nanotube electrode has higher electrocatalytic activity for ORR in alkaline medium. This is partly because the B and N dopants not only provide CNTs with excellent thermal stability but also tune the bandgap energies of the CNTs. On the other hand, the active sites of N and B atoms transfer electrons to the surrounding carbon atoms. Additionally, VA–BCN nanotube electrode displays an exceptional methanol tolerance, along with a high durability toward the ORR compared to a commercial Pt/C electrode.

Ma and coworkers synthesized the N, F, and P ternary doped macroporous carbon fibers (NFPC) by a simple electrospinning method and thermal annealing [44]. Owing to the large specific surface area with porous architecture and the synergistic effect of heteroatoms (N, F, and P), the NFPC shows extremely high ORR activity in 0.1 M KOH electrolyte, in which the half-wave potential is only 25 mV negative than that of the commercial Pt/C, demonstrating that the ternary doping method could act as an efficient strategy to enhance the electrocatalytic performance of carbon-based materials.

9.2.2 Carbon-Based Nonprecious Metal Single-Atom Catalyst

In 1964, Jasinskiet et al. first reported the cobalt phthalocyanine catalyst in the electrochemical ORR in alkaline media [45]. Since then, metal–nitrogen–carbon (M-N/C) and metal–carbide (M_3C) catalysts have attracted special attention because of their enhanced ORR activities compared to those of metal-free carbon-based catalysts. Specially, single-atom M/N/C (M = Fe, Co, Ni, Mn, and Zn) catalyst is considered an efficient alternative to Pt-based precious metal catalyst due to its high atomic utilization, low cost, and excellent ORR activity [46–49].

9.2.2.1 Fe-Based Single-Atom Catalyst

Numerous studies have been conducted on M/N/C catalysts, especially on Fe-based single-atom catalyst. Theoretical calculations show that the activity of a single FeN_x active site is comparable to that of precious metal Pt [50, 51]. By adjusting the electronic structure of the active site, the intrinsic activity of the Fe-based single-atom catalyst can be further improved. In 2009, Dodelet and coworkers obtained Fe/N/C catalyst by mixing high specific surface carbon carrier BP-2000 with phenanthroline and ferrous acetate in full ball milling, and then treated it with ammonia gas at high temperature. The amount of nitrogen-doped carbon material and number of pores was increased by the treatment of ammonia, resulting in an increase in the density of Fe/N/C catalyst active center [52]. In previous studies, the synthesis of Fe–N–C catalyst is mainly composed of a mixture of carbon substrate, nitrogen-rich material, and iron salt through pyrolysis. Later, this group prepared another Fe-based catalyst by using ZIF-8 as a microporous host for phenanthroline and ferrous acetate [53]. Compared with the previous work, this catalyst achieved an obvious enhancement of ORR activity in H_2-O_2 PEMFC with a power density of 0.75 W cm^{-2} at 0.6 V, a meaningful voltage for polymerelectrolyte-membrane fuel cells operation, comparable with that of a commercial Pt-based cathode tested under identical conditions.

Since then, as framework material of imidazolic acid molecular sieve, ZIF-8 is often used as a sacrificial template for the preparation of single-atom catalysts due to its high porosity and high specific surface area [54, 55]. Lin and coworkers prepared a high-performance Fe–N–C electrocatalyst via in situ growth of bimetallic zeolitic imidazolate frameworks on multiwalled CNTs, followed by adsorption of furfuryl alcohol and pyrolysis [56]. They found that such active Fe–N–C catalysts nest in the conducting web and utilize the electrons to reduce oxygen in the pores, leading to a current density of 880 mA cm^{-2} @ 0.6 V and a maximum power density of 820 mW cm^{-2} for the PEMFC. Jia and coworkers synthesized Fe–N–C (Figure 9.2) by CVD [57]. At 750 °C, ferric chloride vapor flows through the Zn–N–C substrate, resulting in the formation of Fe-N_4 from Zn-N_4. The characterization results showed that the density of active sites of Fe–N–C catalyst was 1.92×10^{20} g, and the utilization rate of active sites was 100%. This catalyst provides unprecedented oxygen reduction reactivity at 0.90 V, i.e. 33 mA cm^{-2}, in H_2-O_2 PEMFCs at 1.0 bar and 80 °C. Wu and coworkers prepared an Fe–N–C catalyst by adding an optimal amount of H_2 in a conventional inert atmosphere [58]. The H_2 significantly increased the overall density of FeN_4 sites and shortened the length of Fe—N bonds. In addition, the unstable pyrrolic-N-coordinated S1

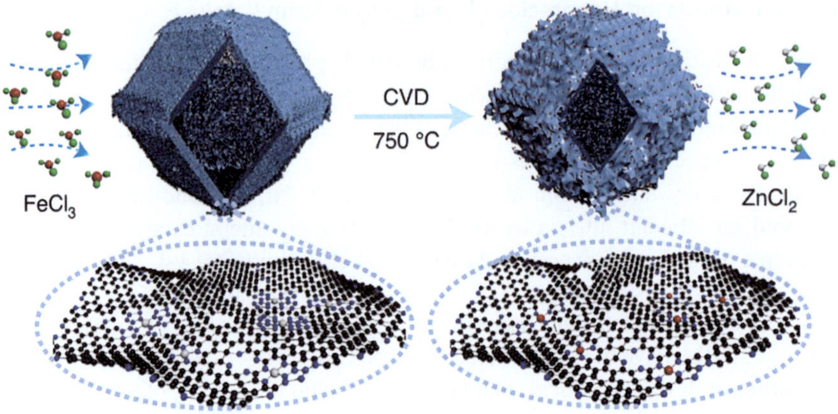

Figure 9.2 High-temperature trans-metalation. Synthesis of FeNC-CVD-750 via $FeCl_3$ CVD. Red, green, gray, purple, and black balls represent Fe, Cl, Zn, N, and C atoms, respectively. Source: Reproduced with permission of Jiao et al. [57], Springer Nature.

sites were inhibited, but the stable pyridinic-N-coordinated S2 sites were favored. The Fe–N–C catalyst demonstrated excellent stability in the membrane electrode assembly, maintaining a current density of 67 mA cm^{-2} after 30,000 voltage cycles at 0.8 V (H_2-air).

9.2.2.2 Non-Fe-Based Single-Atom Catalyst

Some non-Fe-based single-atom catalysts also have excellent ORR activities. Fu and coworkers designed and synthesized a stable Cu–N–C catalyst with an ultrahigh Cu loading of 20.9 wt%, and highly stable single-atom Cu uniformly dispersed in ultrathin carbon nitride nanoplates [59]. The researchers confirmed that the introduced nitrogen could trap and protect such high levels of copper atoms. The mass activity of ORR for the single-atom Cu catalyst is 54 times that of the Cu nanoparticles at 0.85 V (vs. RHE), and the Tafel slope is only 37 mV dec^{-1}, which is even better than the commercial Pt/C catalyst.

Wang et al. obtained Co-N-C@F127 catalyst by calcining the metal-organic framework coated with surfactant F123 [60]. The confining effect of the outer surfactant inhibits the agglomeration of Co atoms and inhibits the collapse of the internal microporous structure of the ZIF. The results of DFT calculation show that the atomically dispersed Co-N-C@F127 catalyst contains a large number of CoN_{2+2} sites, which have high activity and thermodynamic stability, which are conducive to improving the ORR activity of the catalyst under acidic conditions.

Wu et al. reported an atomically dispersed MnN_4 catalyst through a two-step synthesis step involving preparing Mn-doped ZIF-8 precursors, followed by a subsequent thermal activation [61]. The atomically dispersed MnN_4 group is embedded in the exposed base surface of the carbon surface. Electrochemical tests showed that the half-wave potential of the Mn–N–C catalyst is only 0.80 V vs. RHE, which is close to that of Fe–N–C catalysts, and the stability of Mn–N–C catalyst in acidic media is significantly improved.

Wei and coworkers reported an atomically dispersed Zn–N–C catalyst with Zn content up to 9.33 wt% [62]. The Zn–N–C catalyst thus synthesized exhibited exceptional activity approaching that of Fe-N–C catalysts along with significantly enhanced stability in acidic and alkaline environments. The excellent stability of the catalyst is attributed to the fact that the energy required to convert the Zn sites (*Zn) to *Zn(OH) is higher than the free energy required to form *Fe(OH), which is certificated by DFT calculations.

9.2.2.3 Bimetallic Single-Atom Catalyst

In addition to single-atom catalysts, bimetallic-atom electrocatalysts received extensive attention recently [63–65]. The electronic synergism between the adjacent metals is generated by introducing the second metal into the single transition metal M–N–C, and the ORR activity is better than that of the single-atom catalyst by regulating the coordination environment of the active site. M–N–C catalysts with Fe and Co as centered metals have been widely studied for their optimal ORR activity.

Wang and coworkers constructed a novel Fe–Co dual-site catalyst supported by nitrogen-doped CNTs, by precisely controlling the bonding between Fe^{3+} precursors and Co sites in the zinc/cobalt bimetallic organic skeleton [66]. The bimetallic organic skeleton with Fe–Co bimetallic sites is transformed into bimetallic site catalyst supported by nitrogen-doped CNTs after high-temperature carbonization. The half-wave potential of the (Fe, Co)/CNT catalyst is 0.954 V vs. RHE, which is higher than that of the Pt/C electrode (0.842 V vs. RHE). In addition, the research of Ni and Fe bimetallic single-atom catalysts for ORR has made some progress.

9.2.3 Carbon-Based Non-Noble Metals for ORR

In the process of preparing ORR catalyst, the transition metal precursor is easy to agglomerate into metal nanoparticles at high temperatures, and the defective carbon is formed due to the incorporation of heteroatoms into the carbon material [67, 68]. The structure and composition of carbon materials have a great influence on ORR activity. For example, the porous structure of carbon material affects the electron transport and the exposure of active sites. In addition, the ORR stability of the catalyst significantly improved because the nanoparticles were coated with carbon shells, which could avoid the aggregation and dissolution of the active site.

Qian and coworkers prepared Fe_7C_3-doped in-situ grown CNTs and N-doped hollow carbon (Fe_x-CNT@NHC) by using a simple and robust preparation method, which used cubic ZIF-8-derived zinc oxide cubes as a template for secondary MOFs regrowth followed by the final carbonization [69]. In the 0.1 M KOH, the as-pyrolyzed $Fe_{0.1}$-CNT@NHC electrocatalyst value of half-wave potential is 0.92 V and the value of diffusion-limited current density is 6.08 mA cm^{-2}, respectively, which are close to the corresponding electrochemical values of the standard commercial Pt/C (0.89 V and 5.89 mA cm^{-2}).

Many transition metal oxides, such as Fe-, Co-, and Mn-based oxides, have also been reported with excellent ORR activity in alkaline media [70–72]. However, due to their poor electrical conductivity, transition metal oxides are usually combined with functionalized carbon materials for further application [73, 74]. Dai and coworkers used a simple two-step method to synthesize cobalt oxide/CNT strongly coupled hybrid as efficient ORR catalyst by directly growing nanocrystals on oxidized multiwalled CNTs [75]. They proposed that the ORR activity of CoO/NCNT hybrid originates from the strong coupling of the inorganic/nanocarbon hybrid.

Transition metal phosphide/carbon composites, as promising electrocatalysts for ORR, have attracted much attention, in which Fe-based phosphide catalysts generally show better ORR catalytic activity than other transition metal-based phosphide catalysts. Xu and coworkers proposed a facile method for fabricating myricetus-like N, P-rich C NSs embedded in Fe_2P nanocrystals (Fe_2P/NPCs) by synthesizing Myrica-like PANI-NSs through microemulsion templating, incorporating Fe into the PANI-NS during hydrothermal treatment, and obtaining FexP nanocrystals and N, P-rich carbonaceous support through pyrolysis of hydrothermally treated PANI-NS [76]. The Fe_2P/NPCs exhibit outstanding ORR activity with a half-wave potential of 0.820 V (vs. RHE) and a diffusion-limited current density of 5.58 mA cm^{-2}, comparable with commercial Pt/C. Moreover, Fe_2P/NPCs possess a good duration stability and higher methanol tolerance than commercial Pt/C.

The study of the ORR property of transition metal nitrides and oxynitrides can be traced back to 1963. Mazza and Trassatti used TiN compound obtained by high-temperature cementation to exhibit ORR catalytic activity in alkaline electrolytes. The formation of nitrides advantageously changes the electronic structure of the catalyst, resulting in the d-band contraction in TMNs and increasing the electron density near the Fermi level. So, transition metal nitrides have the electronic properties of noble metals, which facilitate the transfer of electrons to oxygen. The formation of nitrides or oxynitrides on the transition metal surface may accelerate the kinetics of the ORR [77, 78]. Recently, studies have shown that TaO_xN_y, N-Co-W/C, and TiO_xN_y exhibit moderate ORR activity and excellent chemical stability [79–81].

9.3 Conclusion and Outlook

After decades of research, significant progress has been reported in improving the performance of carbon-based catalysts for ORR. General strategies for the design of carbon-based catalysts with excellent catalytic activity have been summarized, including elemental doping, structural design, and surface modification. Despite significant progress having been made in carbon-based catalysts for ORR, several significant scientific problems remain unresolved: (i) It is still challenging to accurately control the active site of carbon-based catalysts. For example, the electronic configuration and coordination environment of the active center in Fe-based single-atom catalysts are diverse and have different catalytic effects on ORR. In addition, the low metal loading capacity of the catalyst and the low yield of carbon-based catalyst are difficult to meet the requirements of large-scale preparation at the industrial level. (ii) The onset and half-wave potentials of carbon-based catalysts under alkaline conditions are higher than that of the commercial Pt/C catalyst, however, the activity and stability in acidic media of carbon-based catalysts are insufficient. The preparation of carbon-based catalysts with better activity and stability than platinum-carbon catalysts under acidic conditions remains challenging. (iii) The understanding of the active center and reaction mechanism of carbon-based composite catalysts is still controversial, and theoretical research and more advanced characterization techniques are not perfect. (iv) In-depth understanding of the active center and ORR reaction mechanism of carbon-based composite catalysts is required. Therefore, researchers need to develop more advanced characterization techniques, such as in situ X-ray absorption spectroscopy, X-ray photoelectron spectroscopy, and infrared spectroscopy, to study the dynamic changes of active sites and reaction intermediates. (v) In addition, the poor stability of carbon-based catalysts in practical applications of PEMFCs has seriously hindered the commercialization of the materials. The poor stability may be due to the corrosion of carbon substrate, the removal of metal active center, the oxidation reaction of intermediate H_2O_2, and micropore flooding.. These problems may be solved by multiple carbonizations to form a highly graphitized carbon structure, constructing a graphite core–shell structure, and using membrane materials with better hydrophobicity.

At present, most carbon-based metal catalysts still have a long way to go before practical application. Therefore, it is necessary to comprehensively consider factors such as catalyst activity and stability to promote the large-scale commercial application of carbon-based nonprecious metal catalysts in the field of fuel cells.

Acknowledgment

This work was supported by Hebei Natural Science Foundation (B2024408023), and the Funded by Science Research Project of Hebei Education Department (QN2022043, QN2021124).

References

1 Stambouli, A.B. and Traversa, E. (2002). Solid oxide fuel cells (SOFCs): a review of an environmentally clean and efficient source of energy. *Renew. Sust. Energ. Rev.* 6 (5): 433.

2 Hu, C. and Dai, L. (2016). Carbon-based metal-free catalysts for electrocatalysis beyond the ORR. *Angew. Chem. Int. Ed.* 55 (39): 11736.

3 Liu, X. and Dai, L. (2016). Carbon-based metal-free catalysts. *Nat. Rev. Mater.* 1 (11): 1.

4 Thompson, S.T. and Papageorgopoulos, D. (2019). Platinum group metal-free catalysts boost cost competitiveness of fuel cell vehicles. *Nat. Catal.* 2 (7): 558.

5 Cao, S., Sun, T., Li, J.-R. et al. (2023). The cathode catalysts of hydrogen fuel cell: From laboratory toward practical application. *Nano Res.* 16 (4): 4365.

6 Zhao, J., Liu, H., and Li, X. (2023). Structure, property, and performance of catalyst layers in proton exchange membrane fuel cells. *Electrochem. Energy Rev.* 6 (1): 13.

7 Yang, L., Shui, J., Du, L. et al. (2019). Carbon-based metal-free ORR electrocatalysts for fuel cells: past, present, and future. *Adv. Mater.* 31 (13): 1804799.

8 Zhou, X., Qiao, J., Yang, L., and Zhang, J. (2014). A review of graphene-based nanostructural materials for both catalyst supports and metal-free catalysts in PEM fuel cell oxygen reduction reactions. *Adv. Energy Mater.* 4 (8): 1301523.

9 Othman, R., Dicks, A.L., and Zhu, Z. (2012). Non precious metal catalysts for the PEM fuel cell cathode. *Int. J. Hydrog. Energy* 37 (1): 357.

10 Nie, Y., Li, L., and Wei, Z. (2015). Recent advancements in Pt and Pt-free catalysts for oxygen reduction reaction. *Chem. Soc. Rev.* 44 (8): 2168.

11 Yang, X., Priest, C., Hou, Y., and Wu, G. (2022). Atomically dispersed dual-metal-site PGM-free electrocatalysts for oxygen reduction reaction: opportunities and challenges. *SusMat* 2 (5): 569.

12 Ren, X., Wang, Y., Liu, A. et al. (2020). Current progress and performance improvement of Pt/C catalysts for fuel cells. *J. Mater. Chem. A* 8 (46): 24284.

13 de Sá, M.H., Moreira, C.S., Pinto, A.M.F.R., and Oliveira, V.B. (2022). Recent advances in the development of nanocatalysts for direct methanol fuel cells. *Energies* 15 (17): 6335.

14 Lang, P., Yuan, N., Jiang, Q. et al. (2020). Recent advances and prospects of metal-based catalysts for oxygen reduction reaction. *Energy Technol.* 8 (3): 1900984.

15 Nieto-Márquez, A., Romero, R., Romero, A., and Valverde, J.L. (2011). Carbon nanospheres: synthesis, physicochemical properties and applications. *J. Mater. Chem.* 21 (6): 1664.

16 Shi, Q., Wang, Y., Wang, Z. et al. (2016). Three-dimensional (3D) interconnected networks fabricated via in-situ growth of N-doped graphene/carbon nanotubes on Co-containing carbon nanofibers for enhanced oxygen reduction. *Nano Res.* 9: 317.

17 Li, J., Hou, P., and Liu, C. (2017). Heteroatom-doped carbon nanotube and graphene-based electrocatalysts for oxygen reduction reaction. *Small* 13 (45): 1702002.

18 Zhu, Q., Cai, D., Lan, X. et al. (2018). Design of multidimensional nanocomposite material to realize the application both in energy storage and electrocatalysis. *Sci. Bull.* 63: 152.

19 Xing, T., Zheng, Y., Li, L.H. et al. (2014). Observation of active sites for oxygen reduction reaction on nitrogen-doped multilayer graphene. *ACS Nano* 8 (7): 6856.

20 Wang, S., Iyyamperumal, E., Roy, A. et al. (2011). Vertically aligned BCN nanotubes as efficient metal-free electrocatalysts for the oxygen reduction reaction: a synergetic effect by co-doping with boron and nitrogen. *Angew. Chem. Int. Ed.* 50 (49): 11756.

21 Zhang, J., Zhao, Z., Xia, Z., and Dai, L. (2015). A metal-free bifunctional electrocatalyst for oxygen reduction and oxygen evolution reactions. *Nat. Nanotechnol.* 10 (5): 444.

22 Gong, K., Du, F., Xia, Z. et al. (2009). Nitrogen-doped carbon nanotube arrays with high electrocatalytic activity for oxygen reduction. *Science* 323 (5915): 760.

23 Zhou, X., Yang, Z., Nie, H. et al. (2011). Catalyst-free growth of large scale nitrogen-doped carbon spheres as efficient electrocatalysts for oxygen reduction in alkaline medium. *J. Power Sources* 196 (23): 9970.

24 Singh, S.K., Takeyasu, K., and Nakamura, J. (2019). Active sites and mechanism of oxygen reduction reaction electrocatalysis on nitrogen-doped carbon materials. *Adv. Mater.* 31 (13): 1804297.

25 Daems, N., Sheng, X., Vankelecom, I.F.J., and Pescarmona, P.P. (2014). Metal-free doped carbon materials as electrocatalysts for the oxygen reduction reaction. *J. Mater. Chem. A* 2 (12): 4085.

26 Wei, Q., Tong, X., Zhang, G. et al. (2015). Nitrogen-doped carbon nanotube and graphene materials for oxygen reduction reactions. *Catalysts* 5 (3): 1574.

27 Erable, B., Duteanu, N., Kumar, S.M.S. et al. (2009). Nitric acid activation of graphite granules to increase the performance of the non-catalyzed oxygen reduction reaction (ORR) for MFC applications. *Electrochem. Commun.* 11 (7): 1547.

28 Matter, P.H., Wang, E., Arias, M. et al. (2007). Oxygen reduction reaction activity and surface properties of nanostructured nitrogen-containing carbon. *J. Mol. Catal. A Chem.* 264 (1–2): 73.

29 Tian, G., Zhang, Q., Zhang, B. et al. (2014). Toward full exposure of "active sites": nanocarbon electrocatalyst with surface enriched nitrogen for superior oxygen reduction and evolution reactivity. *Adv. Funct. Mater.* 24 (38): 5956.

30 Qu, L., Liu, Y., Baek, J.-B., and Dai, L. (2010). Nitrogen-doped graphene as efficient metal-free electrocatalyst for oxygen reduction in fuel cells. *ACS Nano* 4 (3): 1321.

31 Fu, S., Zhu, C., Zhou, Y. et al. (2015). Metal-organic framework derived hierarchically porous nitrogen-doped carbon nanostructures as novel electrocatalyst for oxygen reduction reaction. *Electrochim. Acta* 178: 287.

32 Yang, L., Jiang, S., Zhao, Y. et al. (2011). Boron-doped carbon nanotubes as metal-free electrocatalysts for the oxygen reduction reaction. *Angew. Chem. Int. Ed.* 50 (31): 7132.

33 Zhang, J., Qu, L., Shi, G. et al. (2016). N, P-codoped carbon networks as efficient metal-free bifunctional catalysts for oxygen reduction and hydrogen evolution reactions. *Angew. Chem. Int. Ed.* 55 (6): 2230.

34 Ma, Z., Dou, S., Shen, A. et al. (2015). Sulfur-doped graphene derived from cycled lithium–sulfur batteries as a metal-free electrocatalyst for the oxygen reduction reaction. *Angew. Chem. Int. Ed.* 127 (6): 1908.

35 Ding, S., Barr, J.A., Shi, Q. et al. (2022). Engineering atomic single metal–FeN_4Cl sites with enhanced oxygen-reduction activity for high-performance proton exchange membrane fuel cells. *ACS Nano* 16 (9): 15165.

36 Liu, Z., Peng, F., Wang, H. et al. (2011). Phosphorus-doped graphite layers with high electrocatalytic activity for the O_2 reduction in an alkaline medium. *Angew. Chem. Int. Ed.* 50 (14): 3257.

37 Yang, Z., Yao, Z., Li, G. et al. (2012). Sulfur-doped graphene as an efficient metal-free cathode catalyst for oxygen reduction. *ACS Nano* 6 (1): 205.

38 Wu, M., Li, J., Liu, Z. et al. (2023). Boron doped partially wrapped hierarchical porous carbon materials towards oxygen reduction reaction. *J. Electroanal. Chem.* 938: 117456.

39 Zheng, Y., Chen, K., Jiang, K. et al. (2022). Progress of synthetic strategies and properties of heteroatoms-doped (N, P, S, O) carbon materials for supercapacitors. *J. Energy Storage* 56 (PB): 105995.

40 Zhang, Y., Zhuang, X., Su, Y. et al. (2014). Polyaniline nanosheet derived B/N co-doped carbon nanosheets as efficient metal-free catalysts for oxygen reduction reaction. *J. Mater. Chem. A* 2 (21): 7742.

41 Li, J.-C., Qin, X., Hou, P.-X. et al. (2019). Identification of active sites in nitrogen and sulfur co-doped carbon-based oxygen reduction catalysts. *Carbon.* 147: 303.

42 Xie, X., Shi, J., Pu, Y. et al. (2020). Cellulose derived nitrogen and phosphorus co-doped carbon-based catalysts for catalytic reduction of p-nitrophenol. *J. Colloid Interface Sci.* 571: 100.

43 Yang, C., Jin, H., Cui, C. et al. (2018). Nitrogen and sulfur co-doped porous carbon sheets for energy storage and pH-universal oxygen reduction reaction. *Nano Energy* 54: 192.

44 Wu, M., Wang, Y., Wei, Z. et al. (2018). Ternary doped porous carbon nanofibers with excellent ORR and OER performance for zinc–air batteries. *J. Mater. Chem. A* 6 (23): 10918.

45 Jasinski, R. (1964). A new fuel cell cathode catalyst. *Nature* 201 (4925): 1212.

46 Xu, H., Wang, D., Yang, P. et al. (2020). Atomically dispersed M–N–C catalysts for the oxygen reduction reaction. *J. Mater. Chem. A* 8 (44): 23187.

47 Tong, M., Wang, L., and Fu, H. (2021). Designed synthesis and catalytic mechanisms of non-precious metal single-atom catalysts for oxygen reduction reaction. *Small Methods* 5 (10): 2100865.

48 Wang, Y., Zhang, X., Xi, S. et al. (2020). Rational design and synthesis of hierarchical porous Mn–N–C nanoparticles with atomically dispersed mnn_x moieties for highly efficient oxygen reduction reaction. *ACS Sustain. Chem. Eng.* 8 (25): 9367.

49 Du, J., Cheng, F., Wang, S. et al. (2014). M (Salen)-derived nitrogen-doped M/C (M = Fe, Co, Ni) porous nanocomposites for electrocatalytic oxygen reduction. *Sci. Rep.* 4 (1): 4386.

50 Liang, W., Chen, J., Liu, Y., and Chen, S. (2014). Density-functional-theory calculation analysis of active sites for four-electron reduction of O_2 on Fe/N-doped graphene. *ACS Catal.* 4 (11): 4170.

51 Orellana, W. (2013). Catalytic properties of transition metal-N_4 moieties in graphene for the oxygen reduction reaction: evidence of spin-dependent mechanisms. *J. Phys. Chem. C* 117 (19): 9812.

52 Lefèvre, M., Proietti, E., Jaouen, F., and Dodelet, J.P. (2009). Iron-based catalysts with improved oxygen reduction activity in polymer electrolyte fuel cells. *Science* 324 (5923): 71.

53 Proietti, E., Jaouen, F., Lefèvre, M. et al. (2011). Iron-based cathode catalyst with enhanced power density in polymer electrolyte membrane fuel cells. *Nat. Commun.* 2 (1): 416.

54 Zhong, H., Wang, J., Zhang, Y. et al. (2014). ZIF-8 derived graphene-based nitrogen-doped porous carbon sheets as highly efficient and durable oxygen reduction electrocatalysts. *Angew. Chem. Int. Ed.* 53 (51): 14235.

55 Zhang, L., Su, Z., Jiang, F. et al. (2014). Highly graphitized nitrogen-doped porous carbon nanopolyhedra derived from ZIF-8 nanocrystals as efficient electrocatalysts for oxygen reduction reactions. *Nanoscale* 6 (12): 6590.

56 Zhang, C., Wang, Y., An, B. et al. (2017). Networking pyrolyzed zeolitic imidazolate frameworks by carbon nanotubes improves conductivity and enhances oxygen-reduction performance in polymer-electrolyte-membrane fuel cells. *Adv. Mater.* 29 (4): 1604556.

57 Jiao, L., Li, J., Richard, L.L.R. et al. (2021). Chemical vapour deposition of Fe–N–C oxygen reduction catalysts with full utilization of dense Fe–N_4 sites. *Nat. Mater.* 20 (10): 1385.

58 Zeng, Y., Li, C., Li, B. et al. (2023). Tuning the thermal activation atmosphere breaks the activity–stability trade-off of Fe–N–C oxygen reduction fuel cell catalysts. *Nat. Catal.* 6: 1215.

59 Li, F., Han, G.-F., Noh, H.-J. et al. (2018). Boosting oxygen reduction catalysis with abundant copper single atom active sites. *Energy Environ. Sci.* 11 (8): 2263.

60 Wang, X.X., Cullen, D.A., Pan, Y. et al. (2018). Nitrogen-coordinated single cobalt atom catalysts for oxygen reduction in proton exchange membrane fuel cells. *Adv. Mater.* 30 (11): 1706758.

61 Li, J., Chen, M., Cullen, D.A. et al. (2018). Atomically dispersed manganese catalysts for oxygen reduction in proton-exchange membrane fuel cells. *Nat. Catal.* 1 (12): 935.

62 Li, J., Chen, S., Yang, N. et al. (2019). Ultrahigh-loading zinc single-atom catalyst for highly efficient oxygen reduction in both acidic and alkaline media. *Angew. Chem. Int. Ed.* 131 (21): 7109.

63 Li, H., Di, S., Niu, P. et al. (2022). A durable half-metallic diatomic catalyst for efficient oxygen reduction. *Energy Environ. Sci.* 15 (4): 1601.

64 Du, J., Han, G., Zhang, W. et al. (2023). CoIn dual-atom catalyst for hydrogen peroxide production via oxygen reduction reaction in acid. *Nat. Commun.* 14 (1): 4766.

65 Chen, Z., Su, X., Ding, J. et al. (2022). Boosting oxygen reduction reaction with Fe and Se dual-atom sites supported by nitrogen-doped porous carbon. *Appl. Catal. B Environ.* 308: 121206.

66 Wang, J., Liu, W., Luo, G. et al. (2018). Synergistic effect of well-defined dual sites boosting the oxygen reduction reaction. *Energy Environ. Sci.* 11 (12): 3375.

67 He, L., Weniger, F., Neumann, H., and Beller, M. (2016). Synthesis, characterization, and application of metal nanoparticles supported on nitrogen-doped carbon: catalysis beyond electrochemistry. *Angew. Chem. Int. Ed.* 55 (41): 12582.

68 Lian, Y., Shi, K., Yang, H. et al. (2020). Elucidation of active sites on S, N codoped carbon cubes embedding Co–Fe carbides toward reversible oxygen conversion in high-performance zinc–air batteries. *Small* 16 (23): 1907368.

69 Chai, L., Hu, Z., Wang, X. et al. (2021). Fe_7C_3 nanoparticles with in situ grown CNT on nitrogen doped hollow carbon cube with greatly enhanced conductivity and ORR performance for alkaline fuel cell. *Carbon N. Y.* 174: 531.

70 Wu, G., Li, N., Zhou, D.R. et al. (2004). Anodically electrodeposited Co+Ni mixed oxide electrode: preparation and electrocatalytic activity for oxygen evolution in alkaline media. *J. Solid State Chem.* 177 (10): 3682.

71 Liang, Y., Li, Y., Wang, H. et al. (2011). Co_3O_4 nanocrystals on graphene as a synergistic catalyst for oxygen reduction reaction. *Nat. Mater.* 10 (10): 780.

72 Liang, Y., Wang, H., Zhou, J. et al. (2012). Covalent hybrid of spinel manganese-cobalt oxide and graphene as advanced oxygen reduction electrocatalysts. *J. Am. Chem. Soc.* 134 (7): 3517.

73 Zhang, Y., Hu, Y., Li, S. et al. (2011). Manganese dioxide-coated carbon nanotubes as an improved cathodic catalyst for oxygen reduction in a microbial fuel cell. *J. Power Sources* 196 (22): 9284.

74 Gorlin, Y., Chung, C.J., Nordlund, D. et al. (2012). Mn_3O_4 supported on glassy carbon: an active non-precious metal catalyst for the oxygen reduction reaction. *ACS Catal.* 2 (12): 2687.

75 Liang, Y., Wang, H., Diao, P. et al. (2012). Oxygen reduction electrocatalyst based on strongly coupled cobalt oxide nanocrystals and carbon nanotubes. *J. Am. Chem. Soc.* 134 (38): 15849.

76 Wang, R., Yuan, Y., Zhang, J. et al. (2021). Embedding Fe_2P nanocrystals in bayberry-like N, P-enriched carbon nanospheres as excellent oxygen reduction electrocatalyst for zinc-air battery. *J. Power Sources* 501: 230006.

77 Mazza, F. and Trassatti, S. (1963). Tungsten, titanium, and tantalum carbides and titanium nitrides as electrodes in redox systems. *J. Electrochem. Soc.* 110 (7): 847.

78 Ham, D.J. and Lee, J.S. (2009). Transition metal carbides and nitrides as electrode materials for low temperature fuel cells. *Energies* 2 (4): 873.

79 Ishihara, A., Doi, S., Mitsushima, S., and Ota, K.I. (2008). Tantalum (oxy)nitrides prepared using reactive sputtering for new nonplatinum cathodes of polymer electrolyte fuel cell. *Electrochim. Acta* 53 (16): 5442.

80 Ando, T., Izhar, S., Tominaga, H., and Nagai, M. (2010). Ammonia-treated carbon-supported cobalt tungsten as fuel cell cathode catalyst. *Electrochim. Acta* 55 (8): 2614.

81 Chisaka, M. (2018). Creation of oxygen reduction reaction active sites on titanium oxynitride without increasing the nitrogen doping level. *Phys. Chem. Chem. Phys.* 20 (23): 15613.

10

Porous Materials for Electrocatalytic Oxygen Reduction Reaction

Jing Li[1]*, Chang Long*[2]*, and Jun Guo*[1,3]

[1] *Tiangong University, State Key Laboratory of Separation Membranes and Membrane Processes, School of Materials Science and Engineering, Jingwu Street, Xiqing Zone, Tianjin 300387, China*
[2] *University of Electronic Science and Technology of China, Institute of Fundamental and Frontier Sciences, Xiyuan Avenue, West Hi-Tech Zone, Chengdu 610054, China*
[3] *Tiangong University, School of Chemistry, Jingwu Street, Xiqing Zone, Tianjin 300387, China*

10.1 Introduction

Porous materials refer to solids that contain a certain number of pores/holes. According to the definition of the International Union of Pure and Applied Chemistry (IUPAC), porous materials can be mainly divided into three categories: micropore (pore diameter < 2 nm), mesopore (2 nm < pore diameter < 50 nm), and macropore (pore diameter > 50 nm) materials according to their pore sizes. Among recently reported oxygen reaction reduction (ORR) catalysts, porous ones have attracted extensive attention thanks to, but not limited to, the following two aspects. Microporous and mesoporous catalysts whose pore sizes are below 10 nm can provide large specific surface areas for loading exterior catalytically active moieties. Conversely, mesopores and macropores with larger sizes belonging to 10–50 nm can be beneficial for the mass transfer of reactants and products.

10.2 Noble Metal-Based Porous ORR Catalysts

Noble metal-based porous catalysts are extensively studied in ORR due to their high catalytic activity, rich active sites, and fast electron and mass transport [1]. Commonly used noble metal catalysts include Pt-based, Pd-based, and other noble metal-based ones [1, 2]. During the past decade, there have been hard template, dealloying, soft template, and self-assembly methods developed to synthesize unique porous noble metal-based ORR catalysts [3–5].

10.2.1 Pt-Based Porous ORR Catalysts

The hard template method has been widely used to construct Pt-based porous ORR catalysts. As its name suggests, the judicious selection of a kind of material

Functional Materials for Electrocatalytic Energy Conversion, First Edition.
Edited by Zhicheng Zhang, Meiting Zhao, and Yuchen Qin.

with a certain shape or/and structure is the prerequisite of the template. Following this, a precursor solution is injected into the template pores and the template is removed finally to obtain the porous product. To date, the commonly used hard templates include SiO_2, polystyrene spheres, metal oxides, biomaterials, and so on. For instance, Wu and co-workers [6] successfully constructed the nitrogen-doped carbon (NC) coated Pt frameworks (FM) using porous ZrO_2 FM as a hard template. Cubic ZrO_2 FW with a size of 200–300 nm prepared by a solvothermal method was employed as the template. After impregnation and adsorption in a chloroplatinic acid (CPA)–polydopamine (PDA) buffer solution, a PDA layer with adsorption of Pt^{4+} precursor was successfully polymerized and coated on the ZrO_2 FW surface (denoted as Pt^{4+}-ZrO_2@PDA). Subsequently, the as-prepared brown powder was subjected to reduction and carbonization, resulting in the generation of NC layers and Pt nanoparticles (NPs) on the surface of ZrO_2 FW (denoted as Pt-ZrO_2@NC). After etching in HF solution for eight hours, the template was removed and Pt-Zr FM@NC featuring abundant pore structures was finally constructed. The mass activity and specific activity of this Pt-Zr FM@NC were evaluated to be 1.26 mg Pt^{-1} and 1.44 mA cm^{-2}, respectively, which were 4.5 and 5.8 times higher than that of commercial Pt/C (20 wt%). In addition, negligible activity attenuation was observed for the constructed Pt-Zr FM@NC after 30k catalytic cycles.

Alternatively, the soft template method is considered to be a more flexible and efficient method for preparing porous materials. Soft template refers to molecules or molecular aggregates with a flexible structure, such as surfactants, micelles, microemulsions, etc. Choi et al. [7] used block PS-b-PDMS copolymer microspheres as the raw template. Then, the PS blocks in PS-B-PDMS were cross-linked triggered by UV irradiation to form original BCP particles (oBCP), and were further cross-linked through the Friedel–Crafts reaction. Furthermore, the PDMS chain segments were selectively removed to obtain mesoporous hyper-cross-linked BCP particles (hBCP), and the resultant polymer was carbonized at 700–1000 °C to form carbonized BCP particles (cBCP) with uniform pore distribution. The PtFe nanoparticles were further loaded in the pores to obtain cBCP–PtFe catalysts with a very low Pt loading (1 wt%). The resultant spherical cBCP–PtFe showed a specific surface area of 500 m^2 g^{-1}. In single-cell tests, cBCP–PtFe also achieved the same output power as commercial Pt/C (5 wt%) but with only a 5 wt% Pt dosage of the latter.

Dealloying is another method for the fabrication of porous noble metal nanomaterials. Namely, a certain type of metal is dissolved from the alloy by chemical or electrochemical methods [8–10] with formations of extensive defects and vacancies. Fang and co-workers [11] prepared a stratified Pt-based hierarchical alloy by the combinational electrochemical deposition and dealloying method (Figure 10.1), which showed excellent ORR electrocatalytic activity in an alkaline medium. The ORR activity was enhanced by the synergistic effect of doped Au and Pt and the introduced interpenetrating voids formed by Cu dealloying. Apart from better performance durability, the resultant $Pt_{37}Cu_{56}Au_7$ exhibited excellent mass and specific activities (c. 0.871 A mg_{Pt}^{-1}, 1.85 mA cm^{-2}), which were 7.20 and 8.15 higher than those of commercial Pt/C (40 wt% Pt), respectively.

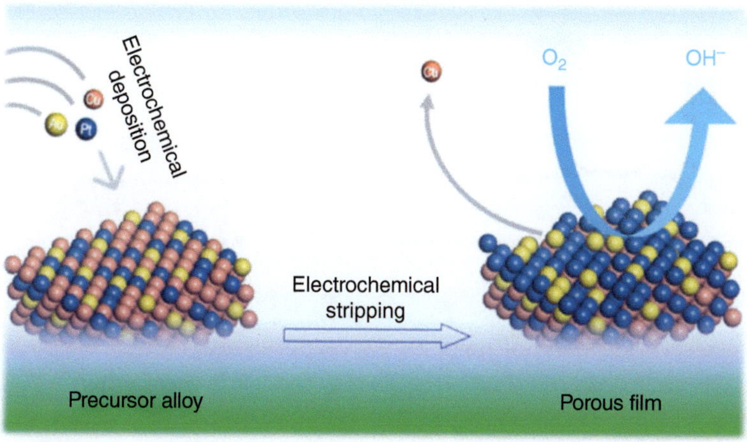

Figure 10.1 Schematic of the synthesis of Pt–Cu–Au ternary alloy porous thin film catalysts. Source: Reproduced with permission of Fang et al. [11], American Chemical Society.

Limited by the surface atoms exposed to the electrocatalytic reaction only, conventional Pt electrocatalysts have to suffer from very low atom utilization [12–14]. For example, for a 3 nm Pt nanoparticle, only about 30 wt% of the atoms are present on the surface, with the rest 70 wt% wasted inside the particle. Pt nanocatalysts with core–shell structures are expected to improve atom utilization by fully exposing the active Pt outside another cheap metal. Liu and co-workers [15] prepared a mesoporous Pt@Pt-skin Pt_3Ni core–shell framework (CSFWs) nanowires (NWs) with one-dimensional (1D) ultrafine Pt NWs (~3 nm) as the core and mesoporous Pt–Ni framework as the shell. They synthesized CSFWs by the solvothermal reduction of platinum(II) acetylacetonate [$Pt(acac)_2$] and nickel(II) acetylacetonate [$Ni(acac)_2$]. Then, the Pt precursor gradually reduced to yield ultrathin pure jagged Pt NWs with abundant surface atomic steps. Such intrinsic surface atomic steps allowed further site-selective nucleation of the Pt–Ni alloy phase on Pt NWs and induced the formation of uniform nano gourd-string-like Pt@Pt-Ni alloy core–shell NWs (CSNWs). After treatment under acidic conditions, a Ni-rich phase within the nano gourd-string-like Pt–Ni alloy shell was selectively etched, forming well-defined anisotropic mesoporous Pt@Pt_3Ni CSFWs. After annealing to simultaneously form a Pt-skin surface and remove organic surfactants, well-defined mesoporous Pt@Pt-skin Pt_3Ni CSFWs were finally obtained. In the ORR catalytic reaction, the mass and specific activity of catalysts were as high as 6.69 A mg_{pt}^{-1} and 8.42 mA cm^{-2}, which were 29 and 26 times higher than those of commercial Pt nanocatalysts, respectively. The catalyst also showed excellent stability with activity attenuation of less than 3 wt% after 50,000 cycles.

Wang and co-workers [16] prepared Pt@C/C via a low-temperature decomposition process as shown in Figure 10.2, and used malic acid ($C_4H_6O_5$) as a carbon precursor and mixed it with H_2PtCl_6 and Vulcan carbon. The Pt carboxylate complex was first formed during the impregnation stage due to the abundant carboxylic

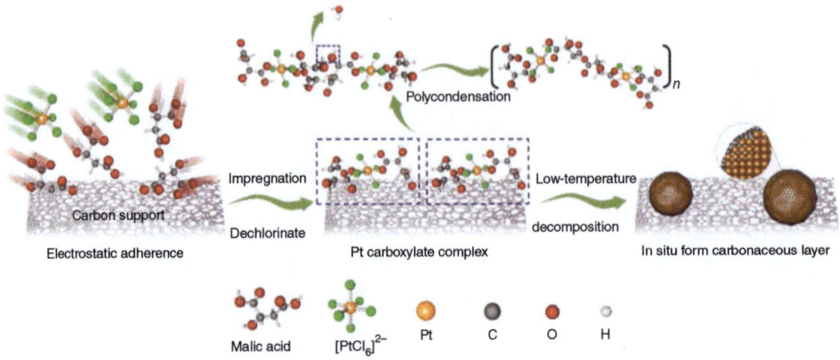

Figure 10.2 Schematic synthesis of Pt@C/C obtained by one-step low-temperature heat treatment. Wiley-VCH. Source: Reproduced with permission of Wang et al. [16], Wiley-VCH.

acid group (–COOH) from malic acid. Then, the formed malic acid-Pt/Vulcan was treated under low temperature to further reduce the Pt^{4+} ions into Pt nanoparticles, and $C_4H_6O_5$ was decomposed to form an amorphous carbon layer adjacent to Pt nanoparticles via a subsequent polymerization and dehydration process. The electrochemical results showed that the carbon layer not only effectively improved the ORR stability of the catalyst, but also weakened the adsorption strength between the catalyst and intoxicated small molecules, thereby resulting in superior antiseptic properties. XPS tests before and after the cycle showed that the oxidation state of Pt did not change significantly, indicating that the carbon shell can also effectively inhibit the reduction of the active site. Further, a 40 wt% high-load Pt@C/C catalyst was synthesized and applied in a high-temperature proton exchange membrane fuel cell, which exhibited a peak power density of $662\ mW\ cm^{-2}$ at 160 °C.

10.2.2 Pd-Based Porous ORR Catalysts

In addition to the study of noble metal Pt-based catalysts, many researchers have explored Pd-based catalysts. In alkali conditions, Pd-based catalysts have even greater ORR activity than Pt due to the upward shift of the d-band center of Pd with respect to Pt beneficial for stronger adsorption of oxygen species. The addition of Cu metal to Pd changes the value of the d-band center, making the adsorption energy of oxygen more favorable to ORR. Yamauchi and co-workers [17] used PSb–PEO block copolymer as a soft template to prepare three-dimensional (3D) spherical palladium copper (PdCu) alloy nanoparticles with medium porosity (~10 nm). The porous nature enabled the distribution of more active sites over the entire inner and outer surface of the PdCu nanoparticles. As expected, the mesoporous PdCu exhibited higher electrocatalytic activity and stability in ORR than single Pd nanoparticle catalysts.

In addition to forming alloys with metals, doping with nonmetallic elements is also a viable alternative. Liu and co-workers [18] synthesized ternary palladium–boron–phosphorus alloy mesoporous nanospheres (PdBP MSs) with a 3D dendritic backbone by using the surfactant dioctadecyl dimethylammonium chloride (DODAC) as the soft template, and dimethylamino-borane and sodium hypophosphite as the B and P sources, respectively. By controlling the amount of $NH_3{\cdot}H_2O$ and the ratio of DODAC/$PdCl_4^{2-}$, the final diameter of PdBP MSs was controlled at 125 nm. The catalysts showed high mass activity (1.45 A mg_{Pd}^{-1}) at 0.9 V, which was 1.77, 2.93, and 4.97 times higher than that of the PdB, commercial Pt, and commercial Pd nanoparticles, respectively. The introduction of B and P weakened the adsorption of *OOH on the Pd surface and reduced the overpotential of electrocatalytic ORR consequently.

10.2.3 Other Precious Metal-Based Porous ORR Catalysts

To enhance the activity of other noble metal-based catalysts such as Ag, Ir, and Au, researchers have used various methods including modulation of catalyst morphology and dimensions, alloying with other metals, and supporting with transition metal oxides. Fang and co-workers [19] prepared porous $Co(OH)_2$ nanosheets with H-vacancies by pulsed laser ablation process. The element cobalt was heated to a vapor state by laser beam, and the highly reactive cobalt atoms combined with the hydroxide ions in solution to form $Co(OH)_2$ nanosheet. Due to the quenching effect of the laser, the cobalt was simultaneously hydroxylated and oxidized to form porous $Co(OH)_2$ nanosheets. Next, the P–Ag–$Co(OH)_2$ catalyst was obtained by locating Ag atoms around the porous $Co(OH)_2$ nanosheets in the presence of the H-vacancies. The abundant nanopores provided sufficient diffusion channels for gas molecules, and the special Ag–O–Co–OH catalytic groups around the nanopores showed high intrinsic catalytic activity. In the alkaline electrolyte, P–Ag–$Co(OH)_2$ showed high redox half-wave potential (0.902 V vs. RHE) and low overpotential (235 mV at 10 mA cm^{-2}), which was superior to that of the Pt/C (Ir/C) catalysts in previous studies.

Wang and co-workers [20] prepared bimetallic IrAu nanovesicles (mIrAu-NVs) composed of coiled mesoporous nanosheets (~12 nm) by using poly(styrene-ethylene oxide) as a soft template. The obtained mIrAu-NVs exhibited Pt-like catalytic activity for ORRas well as enhanced durability under alkaline conditions, due to the synergistic effect of the bimetallic composition of IrAu.

10.3 Transition Metal-Based Porous ORR Catalysts

Over the past few decades, the usage of precious metal-based catalysts has been substantially studied in ORR thanks to their high catalytic activity. However, precious metal catalysts suffer from high cost, scarcity, and low durability [21]. As a result, research seeking replacements for precious metal catalysts has become a worldwide topic. Currently, a variety of nonprecious metal-based catalysts have been developed, such as transition metal/carbon composites, transition metal oxides/carbon composites, transition metal/carbon nitrogen composites, and so on. In particular, porous transition metal ORR catalyst materials have become one of the research hotspots nowadays because of their low cost, high catalytic activity, and long-term stability.

10.3.1 Transition Metal/Carbon Composite Porous ORR Catalysts

Since Jasinski [21] discovered that phthalocyanine-metal complexes can be used as ORR electrocatalysts in 1964, transition metals have been extensively studied for their good electrocatalytic activity and low cost. Li et al. [22] used glucose as the carbon source, cobalt nitrate as the metal source, and thiourea as the sulfur source to prepare porous carbon nanorods (Co_9S_8/CS) containing Co_9S_8 by hydrothermal method (Figure 10.3). The high ORR activity of Co_9S_8/CS arose from the strong interaction between the S- and N-Co-doped carbon nanorods and Co_9S_8/CS, which altered the electronic structure of the cobalt site.

Metal–organic frameworks (MOFs) are another new type of porous material formed by metal sites and organic ligands through periodic coordination. The MOFs have a wide range of applications in catalysis due to their high specific surface area, regular pore structure, and controllable composition. MOF-derived transition metal/carbon composites thus have high ORR activity by taking advantage of the following: (i) The nanoparticles formed by metal precursors in the MOF structure have better dispersibility and thus have higher electrochemical activity and structural stability. (ii) Fully doped heteroatoms in the ligand precursor can be used as ORR active sites. (iii) Graphite–carbon after the thermal decomposition of the organic ligand improves the electrical conductivity of the material and accelerates the electron transfer.

Xu and co-workers [23] introduced transition metal ions into MOFs by the impregnation method (Figure 10.4). Specifically, Zn-based triazole-rich CoFe@MET-6 doped with Co and/or Fe ions were used as precursors. Under the inert atmosphere

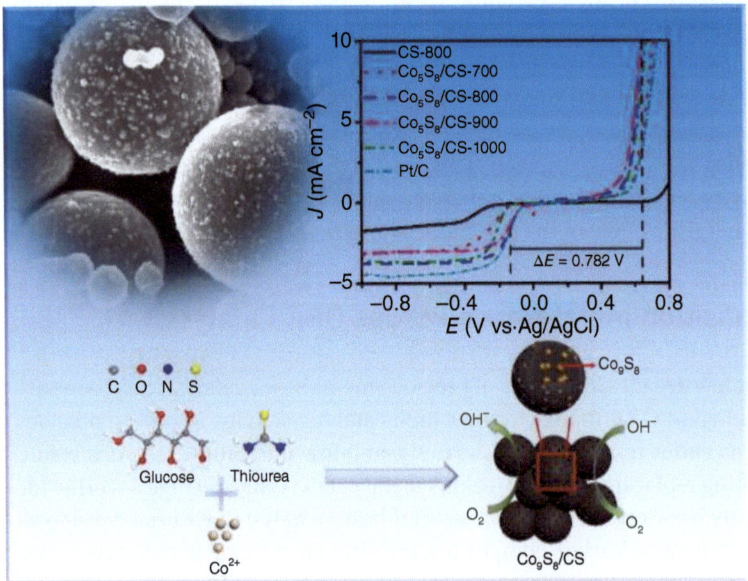

Figure 10.3 Synthesis process and electrocatalytic activity of Co9S8/CS. Source: Reproduced with permission of Li et al. [22], Elsevier.

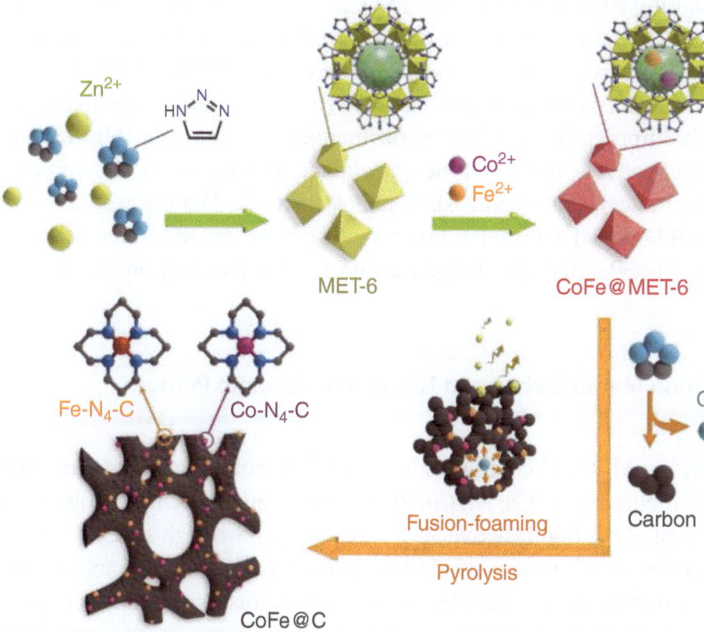

Figure 10.4 Schematic of the synthesis of CoFe@C. Source: Reproduced with permission of Xu et al. [23], Wiley-VCH.

of high-temperature heat treatment (1000 °C), the organic linkers in MOF decomposed to release a large amount of gas and generate a carbon substrate, and the Zn component subsequently evaporated during heating. While Co(II) and Fe(II) doped in MOFs were transformed into atomically dispersed metals on the carbon substrate to obtain the CoFe@C catalyst. Thanks to extensive hierarchical pore structures, the material showed better catalytic activity and stability than Pt/C in ORR.

10.3.2 Transition Metal Oxide/Carbon Composite Porous ORR Catalysts

By taking advantage of abundant reserves, simple synthesis, stable structure, etc., transition metal oxides are expected to replace precious metal catalysts in alkaline systems. However, most transition metal oxides are semiconducting materials with large band gaps, causing poor conductivity and inferior catalytic activity. To solve this problem, researchers increased the ORR activity of metal oxide catalysts by combining them with porous carbon substrates. Commonly used conductive carbon substrates are carbon black, carbon nanotubes, graphene, and so on. Xu and co-workers [24] used Al_2O_3 as an anode, surface-modified Fe–Co oxide as a hard template, and nitrogen-doped carbon NWs (NCWs) as a supporter to construct CoO/NCWs composite catalyst. As there were rich pores, defect sites, and highly active Co and N sites on the surface of the material, CoO/NCWs exhibited higher ORR catalytic activity and stability than the counterpart without surface pores.

The dynamic limiting current density (30.3 mA cm^{-2}, 0.7 V) was almost three times that of the latter. The authors believed that the high ORR catalytic performance resulted from the increased active sites during the preparation of CoO/NCWs.

The synergistic effect between metal nanocatalysts with oxide supporters can also significantly improve the catalytic performances. For example, Muhler and co-workers [25] effectively improved electrocatalytic activity by in situ forming carbon polyhedral encapsulated core–shell nanostructures. The encapsulation of Co_3O_4 shells with N-doped carbon nanotube shells formed an additional Schottky barrier and hence accelerated the charge transport and separation on the catalyst surface.

10.3.3 Transition Metal/Carbon and Nitrogen Composite Porous ORR Catalysts

M–N–C catalysts are M–Nx–C complexes formed by the coordination of transition metal atoms (M, including Fe, Co, Ni, Cu, Zn, etc.) with surrounding nitrogen and carbon atoms. Most of the highly active M–N–C catalysts have high specific surface areas, reasonable hierarchical pore structures, and a dense number of active sites. Taking Fe–N–C catalysts as an example, the nitrogen atoms located at the edge of graphene are more likely to coordinate with Fe atoms to form Fe–Nx active sites [26, 27]. However, these active sites are usually located in the microporous pores in which molecular oxygen cannot directly enter. Li et al. [28] synthesized Fe–N–CNT electrocatalyst containing multistage mesoporous/microporous for the first time. First, they filled $Fe(NO)_3$ into the nanochannels of the AAO template. Fe/FeO_x NPs were formed on the surface of the AAO nanochannels, and then NC was deposited on the exposed surface of AAO nanochannels and Fe/FeO_x NPs by the chemical vapor deposition (CVD). Next, they removed the AAO template and Fe/FeO_x NPs by immersing them in a mixture of HF and HCl to obtain the meso-Fe-N-CNTs. Subsequent NH_3 activation was performed to obtain meso/micro-Fe-N-CNTs with a large number of micropores and active sites. This hybrid material had a highly conductive carbon nanoskeleton, an ultra-high surface area of 2137 m^2 g^{-1}, a high density of Fe–N–C catalytic active sites, and abundant mesopores for efficient mass transfer. The experimental results showed that the catalyst had excellent oxygen reduction performance in acidic conditions and comparable performances to Pt/C in alkaline conditions.

In addition, using polymers as soft templates has also received research attention. Since polymers can be used as carbon or nitrogen sources, catalysts obtained by pyrolysis of polymer templates tend to have a higher graphitization range, which is beneficial to the conductivity of the catalysts. Li and co-workers [29] developed N-doped and Fe/N double-doped porous carbon as catalysts for ORR by simple high-temperature pyrolysis of porphyrin-based conjugated microporous polymers (CMPs) (Figure 10.5). Through the combination of heteroatom doping, porous structure, and tubular structure, the prepared carbon samples exhibited high electrocatalytic activity with a 4-electron transfer mechanism, which was close to that of commercial Pt/C catalysts. Compared with the Pt/C catalyst, Fe/N-CMP-1000

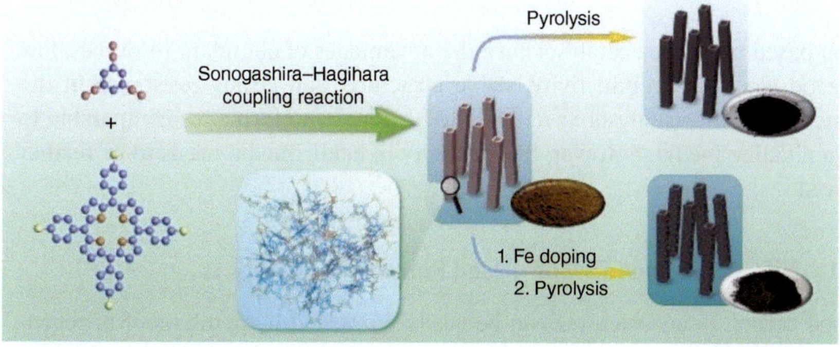

Figure 10.5 Schematic of the synthesis of CMP-based porous carbon materials. Source: Reproduced with permission of Li et al. [29], Wiley-VCH.

even had higher onset potential (0.95 eV), positive half-wave potential (0.85 eV), and limiting current density value (5.1 mA cm^{-2}), as well as better durability and methanol resistance. It has been found that many MOFs are available as desirable precursors for M–N–C catalyst preparation. Ahn et al. [30] prepared Fe,N-doped carbon materials with a bi-porous distribution. First, 1D mesoporous Te nanotubes were used as cores for the growth of ZIF-8 shells to obtain Te NT@ZIF-8. Afterward, dopamine was in situ polymerized at room temperature in the presence of $FeCl_3$ to form Te NT@ZIF-8@Fe-PDA. Finally, Fe,N-embedded porous carbon nanotubes were constructed after the pyrolysis process at 950 °C for three hours under argon flow. Among them, the 1D nanotubes contained abundant mesoporous pores while the mosaic particles contained a large number of micropores, resulting in a specific surface area of 1380 cm^2 g^{-1} (Figure 10.6). Meanwhile, the ultrathin graphite layer coated on the surface of the material prevented the active centers from degradation and agglomeration during long-term operation.

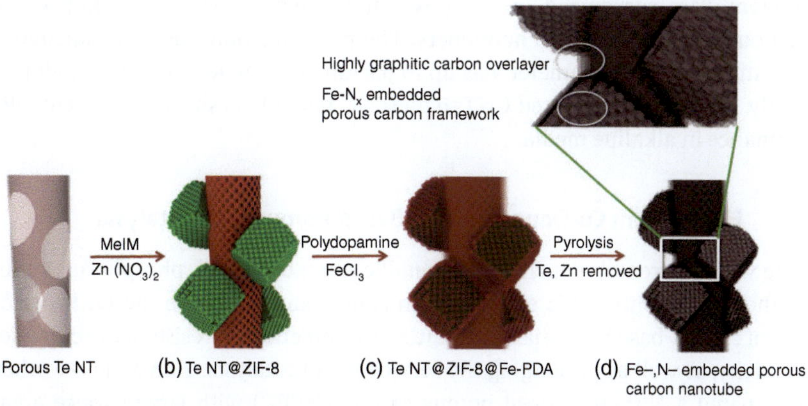

Figure 10.6 Schematic diagram of the synthesis of Fe- and N-doped carbon nanotubes. Source: Reproduced with permission of Ahn et al. [30], Wiley-VCH.

10.4 Carbon-Based Metal-Free Porous ORR Catalyst

Carbon-based metal-free catalysts have the advantages of abundant resources, low cost, good electronic conductivity, stable structure, and strong resistance in the alkaline media. NC catalysts have exhibited catalytic performance comparable to Pt/C in alkaline media, however, their activity in acidic media needs to be further improved.

10.4.1 Nitrogen-Doped Carbon-Based Porous ORR Catalysts

N-doped carbon nanostructures can be easily fabricated by pyrolysis of nitrogen-containing precursors, annealing in ammonia, or nitrogen-containing gas plasmas. Peng and co-workers [31] prepared NC nanospheres with large mesoporous pores by the nano-emulsion assembly method. F127/TMB/dopamine was assembled into nanoemulsions in an ethanol/water system, and NC nanospheres with smooth, multi-cavity, dendrite, and other new structures were obtained after polymerization and carbonization. The dendritic mesoporous carbon nanorods have the characteristics of large mesoporous (about 37 nm), small particle size (about 128 nm), high specific surface area (about $635\,m^2\,g^{-1}$), rich N content (about 6.8 wt%, mass fraction), and high oxygen reduction current density in the alkaline solutions.

The pore length has a strong influence on mass transfer. Namely, short-range pores are more favorable for mass transfer as well as interfacial reaction dynamics. Zhao and co-workers [32] synthesized carbon nanotubes@mesoporous NC (CNTs@mesoNC) core–shell structured nanofibers by molecularly mediated interfacial co-assembly strategy. Firstly, the dopamine molecules were polymerized under the catalysis of ammonium hydroxide, then coassembled with Pluronic F127 into composited micelles via the mediation of the TMB. Next, the micelles could self-assemble into mesostructured layers on the surface of the CNTs to form CNTs@Close packing micelles. Finally, the as-prepared carbon nanotubes@mesostructured polydopamine (CNTs@mesoPDA) core–shell structured nanofibers were carbonized in a hypoxic atmosphere at 800 °C, leading to the formation of CNTs@mesoNC nanofibers. The pore direction was perpendicular to the substrate, the pore diameter was up to 6.9 nm, and the length of the shell pore was only 28 nm. The fabricated CNTs@mesoNC nanofibers showed excellent ORR performance in alkaline media.

10.4.2 Heteroatom Co-Doped Carbon-Based Porous ORR Catalysts

Doping with heteroatoms such as nonmetallic nitrogen, boron, phosphorus, sulfur, and other light nonmetallic elements can significantly improve the ORR performance of carbon-based materials, in addition to their effective resistance to corrosion and toxicity, as well as prolonging the service life of catalysts. Han and co-workers [33] prepared a N,P dual-doped porous carbon (NPPC) with large surface areas. N,P-doped hydrogel (NPH) was obtained by a one-step hydrothermal reaction of a homogenous dispersion that contained P123, phytic acid, and pyrrole. N,P-doped aerogel (NPA) was obtained after hydrothermally reacting and freeze-drying. Then N and P double-doped porous carbon NPPC was obtained by CO_2-activated roasting. The CO_2-assisted roasting process not only created abundant micropores and mesopores in NPPC but also facilitated the conversion of N species from pyrrole N to pyrrole N and graphite N and promoted the enhancement of ORR activity. The results showed that NPPC had a larger specific surface area (2850 $m^2\ g^{-1}$), higher ORR activity, and better stability than the material without CO_2-assisted carbonization.

10.4.3 Undoped Carbon-Based Porous ORR Catalysts

Introducing defects into the graphite–carbon skeleton has also been proven to alter the electron distribution of graphite–carbon nanomaterials, thereby optimizing their catalytic activity for ORR electrocatalysis [34]. Chen and co-workers [35] prepared defect-rich mesoporous carbon nanoribbons NR-900 by self-assembly and pyrolysis strategies. Zn^{2+} salt was first treated with 2,5-dihydroxyterephthalic acid (DHTA) ligand in methanol at room temperature to produce MOF-nanoparticles (Zn-MOF-NP), which were then hydrothermally heated to produce uniform MOF-nanorods (Zn-MOF-NR) (Figure 10.7a). Finally, NR-900 was obtained by pyrolysis strategies. Due to the effective modulation of the nanostructures, a large number of defective catalytic sites (edge defects and cavities) were exposed, which

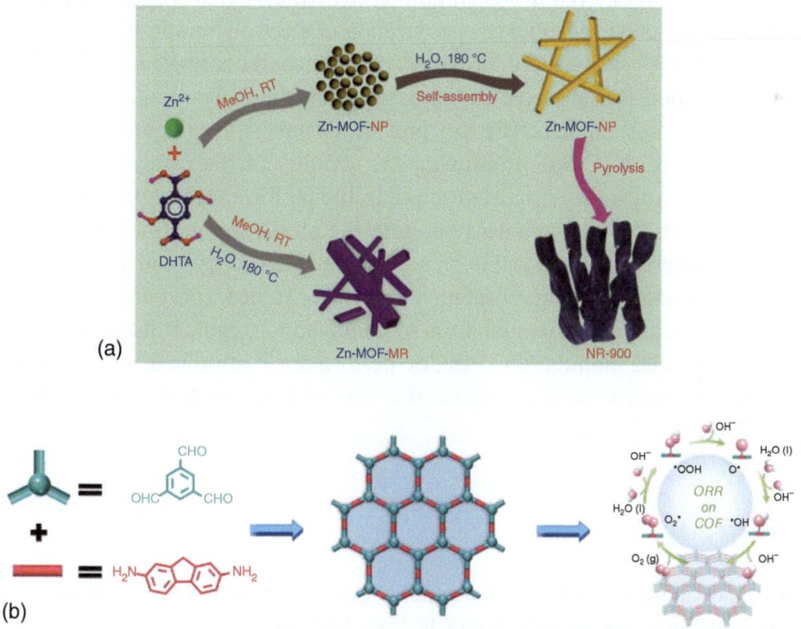

Figure 10.7 (a) Schematic of NR-900 synthesis. Source: Reproduced with permission of Chen et al. [35], Wiley-VCH. (b) Schematic diagram of ORR process on DAF–COF. Source: Reproduced with permission of Xia et al. [36], Wiley-VCH.

improved the electron transfer kinetics and catalytic activity. The catalytic reduction activity of pure mesoporous nanocarbon (NR-900, $E_{1/2} = 0.837\,V$) was superior to that of most undoped metal-free electrocatalysts and comparable to that of commercial Pt/C.

Covalent organic frameworks (COFs) are polycrystalline polymers connected by covalent bonds, with a modular structure allowing well-defined functions inserted into their backbones. Xia and co-workers [36] introduced methyl groups into metal-free carbon-based COFs, which effectively modulated the band gaps, enhanced the adsorption of intermediates, and improved the kinetics for ORR (Figure 10.7b). This proved that controlling the structure of active sites in carbon-based catalysts to further improve their ORR catalytic activity is also a proven strategy. In general, good electrical conductivity and porosity are the basis for the ORR performance of carbon-based catalysts, and structural defects are the effective catalytic sites for ORR reactions.

10.5 Summary

During the catalytic process, the effective transfer of reaction intermediates, products, and electrons at the catalyst surface are key steps in determining the overall reaction rate. This chapter briefly analyzes the constitutive relationship between three types of pore structures, namely microporous, mesoporous, and macroporous, with ORR performances. Well-designed porous structures, such as interconnected layered micropores, can accelerate the penetration of electrolytes and the diffusion of reactant O_2. Mesoporous structures can provide transport channels for reactants/intermediates, which is conducive to the catalytic reaction. For transition metal composites, the core–shell structure can protect the metal from aggregation and nonhomogeneous contact, thus improving the stability of the catalyst. In brief, pore structures occupy a very important place in the performance improvement of ORR catalysts. Over the past two decades, with the proliferation of new synthesis methods, researchers have been able to adjust the microscopic and mesoscopic structure of materials with unprecedented precision. Hence, the pore structures of ORR catalysts can be generated from random and irregular distribution to a much more precise population. Besides performance improvement, the fast development of porous ORR catalysts also enabled an in-depth mechanism understanding.

References

1 Su, Z.P. and Chen, T.H. (2021). Porous noble metal electrocatalysts: synthesis, performance, and development. *Small* 17 (22): 2005354.

2 Matin, M.A., Lee, J., Kim, G.W. et al. (2020). Morphing Mn_{core} @Pt_{shell} nanoparticles: effects of core structure on the ORR performance of Pt shell. *Appl. Catal. B-Environ.* 267: 118727.

3 Hu, Y.Z., Guo, X.Y., Shen, T. et al. (2022). Hollow porous carbon-confined atomically ordered $PtCo_3$ intermetallics for an efficient oxygen reduction reaction. *ACS Catal.* 12 (9): 5380.

4 Yi, S., Jiang, H., Bao, X. et al. (2019). Recent progress of Pt-based catalysts for oxygen reduction reaction in preparation strategies and catalytic mechanism. *J. Electroanal Chem.* 848: 113279.

5 Lai, J., Huang, B., Tang, Y. et al. (2018). Barrier-free interface electron transfer on PtFe-Fe_2C janus-like nanoparticles boosts oxygen catalysis. *Chem* 4 (5): 1153.

6 Guo, W.X., Cheng, L.M., Gao, X.P. et al. (2023). Hierarchical Porous Pt/ZrO_2 nanoframework for efficient oxygen reduction reaction. *ACS Catal.* 13 (8): 5397.

7 Choi, J., Lee, Y.J., Park, D.M. et al. (2020). Highly durable fuel cell catalysts using crosslinkable block copolymer-based carbon supports with ultralow Pt loadings. *Energy Environ. Sci.* 13 (12): 4921.

8 Huang, J., Sementa, L., Liu, Z. et al. (2022). Experimental Sabatier plot for predictive design of active and stable Pt-alloy oxygen reduction reaction catalysts. *Nat. Catal.* 5: 513.

9 Guo, W.X., Gao, X.P., Zhu, M.Z. et al. (2023). A closely packed $Pt_{1.5}Ni_{1-x}$/Ni–N–C hybrid for relay catalysis towards oxygen reduction. *Energy Environ. Sci.* 16 (1): 148.

10 Xie, M.H., Lyu, Z.H., Chen, R.H. et al. (2021). Pt–Co@Pt octahedral nanocrystals: enhancing their activity and durability toward oxygen reduction with an intermetallic core and an ultrathin shell. *J. Am. Chem. Soc.* 143 (22): 8509.

11 Xie, Y.X., Yang, Y., Muller, D.A. et al. (2020). Enhanced ORR kinetics on Au-doped Pt-Cu porous films in alkaline media. *ACS Catal.* 10 (17): 9967.

12 Yildirim, M.H., Curòs, A.R., Motuzas, J. et al. (2009). Nafion®/H-ZSM-5 composite membranes with superior performance for direct methanol fuel cell. *J. Membr. Sci.* 338 (1): 75.

13 Da, H.C., Lee, J., Kwon, O. et al. (2008). Sulfonated poly (fluorinated arylene ether)s/poly (N-vinylimidazole) blend polymer and PTFE layered membrane for operating PEMFC at high temperature. *J. Power Sources* 178 (2): 677.

14 Rusanov, A.L., Likhatchev, D., Kostoglodov, P.V. et al. (2005). Proton-exchanging electrolyte membranes based on aromatic condensation polymers. *Adv. Polym. Sci.* 179: 83.

15 Jin, H., Xu, Z.W., Hu, Z.Y. et al. (2023). Mesoporous Pt@Pt-skin Pt_3Ni core–shell framework nanowire electrocatalyst for efficient oxygen reduction. *Nat. Commun.* 14: 1518.

16 Hu, Y.Z., Zhang, J.J., Shen, T. et al. (2021). A low-temperature carbon encapsulation strategy for stable and poisoning-tolerant electrocatalysts. *Small Methods* 5 (11): 2100937.

17 Iqbal, M., Kim, Y., Li, C.L. et al. (2019). Tailored design of mesoporous PdCu nanospheres with different compositions using polymeric micelles. *ACS Appl. Mater. Interfaces.* 11 (40): 36544.

18 Lv, H., Xu, D.D., Sun, L.Z. et al. (2019). Ternary palladium-boron-phosphorus alloy mesoporous nanospheres for highly efficient electrocatalysis. *ACS Nano.* 13 (10): 12052.

19 Li, Z., Kang, W.J., Liu, H. et al. (2023). Ag anchored atomically around nanopores of porous $Co(OH)_2$ for efficient bifunctional oxygen catalysis. *Adv. Funct. Mater.* 2301947.

20 Li, C.J., Xu, Y., Dai, Z.C. et al. (2020). Bimetallic IrAu mesoporous nanovesicles. *Chem. Eng. J.* 395: 125135.

21 Zhao, C.X., Liu, J.N., Wang, J. et al. (2021). Recent advances of noble-metal-free bifunctional oxygen reduction and evolution electrocatalysts. *Chem. Soc. Rev.* 50 (13): 7745.

22 Li, W.Q., Li, Y.H., Wang, H.G. et al. (2018). Co_9S_8-porous carbon spheres as bifunctional electrocatalysts with high activity and stability for oxygen reduction and evolution reactions. *Electrochimica Acta.* 265: 32.

23 Zhao, R., Liang, Z.B., Gao, S. et al. (2019). Puffing up energetic metal-organic frameworks to large carbon networks with hierarchical porosity and atomically dispersed metal sites. *Angew. Chem. Int. Ed.* 131 (7): 1997.

24 Zhang, H.M., Zhao, Y., Zhang, Y.J. et al. (2019). Fe_3O_4 encapsulated in porous carbon nanobowls as efficient oxygen reduction reaction catalyst for Zn-air batteries. *Chem. Eng. J.* 375: 122058.

25 Aijaz, A., Masa, J., Rösler, C. et al. (2016). Co@Co_3O_4 encapsulated in carbon nanotube-grafted nitrogen-doped carbon polyhedra as an advanced bifunctional oxygen electrode. *Angew. Chem. Int. Ed.* 55 (12): 4087.

26 Li, J.K., Ghoshal, S., Liang, W.T. et al. (2016). Structural and mechanistic basis for the high activity of Fe–N–C catalysts toward oxygen reduction. *Energy Environ. Sci.* 9 (7): 2418.

27 Patniboon, T. and Hansen, H.A. (2021). Acid-stable and active M–N–C catalysts for the oxygen reduction reaction: the role of local structure. *ACS Catal.* 11 (21): 13102.

28 Li, J.C., Hou, P.X., Shi, C. et al. (2016). Hierarchically porous Fe-N-doped carbon nanotubes as efficient electrocatalyst for oxygen reduction. *Carbon* 109: 632.

29 Sun, H.X., Zhou, P.L., Tian, Z.Y. et al. (2022). Non-precious metal-doped carbon materials derived from porphyrin-based porous organic polymers for oxygen reduction electrocatalysis. *ChemPlusChem* 87 (7): 202200168.

30 Ahn, S.H., Yu, X.W., and Manthiram, A. (2017). "Wiring" Fe-N_x-embedded porous carbon framework onto 1D nanotubes for efficient oxygen reduction reaction in alkaline and acidic media. *Adv. Mater.* 29 (26): 1606534.

31 Zhao, L., Sui, X.L., Li, J.Z. et al. (2018). Supramolecular assembly promoted synthesis of three-dimensional nitrogen doped graphene frameworks as efficient electrocatalyst for oxygen reduction reaction and methanol electrooxidation. *Appl. Catal. B-Environ.* 231 (5): 224.

32 Peng, L., Hung, C.T., Wang, S.W. et al. (2019). A versatile nanoemulsion assembly approach to synthesize functional mesoporous carbon nanospheres with tunable pore sizes and architectures. *J. Am. Chem. Soc.* 141 (17): 7073.

33 Zhu, X.H., Xia, Y., Zhang, X.M. et al. (2019). Synthesis of carbon nanotubes@-mesoporous carbon core–shell structured electrocatalysts via a molecule mediated interfacial co-assembly strategy. *J. Mater. Chem. A* 7 (15): 8975.

34 Xie, C., Yan, D.F., Chen, W. et al. (2019). Insight into the design of defect electrocatalysts: from electronic structure to adsorption energy. *Mater. Today* 31: 47.

35 Srinivas, K., Liu, D.W., Ma, F. et al. (2023). Defect-engineered mesoporous undoped carbon nanoribbons for benchmark oxygen reduction reaction. *Small* 19 (34): 2301589.

36 You, Z.H., Wang, B.B., Zhao, Z.J. et al. (2023). Metal-free carbon-based covalent organic frameworks with heteroatom-free units boost efficient oxygen reduction. *Adv. Mater.* 35 (7): 2209129.

Part IV

Advanced Functional Materials for Electrocatalytic Oxygen Evolution Reaction

11

Metal-Based Materials for Electrocatalytic Oxygen Evolution Reaction

Mengyi Xu[1], Xu Li[1], Yurong Su[2], Meirong Song[1], Xianfu Zheng[1], Yuchen Qin[1], Guangxin Ru[3], Xiuhong Zhu[3], Shun Wang[1], and Xia Sheng[1]

[1]*Henan Agricultural University, College of Science, Department of applied chemistry, Nongye Road 63, Zhengzhou 450002, China*
[2]*Zhengzhou University, Green Catalysis Center, and College of Chemistry, 100 Science Avenue, Zhengzhou 450001, China*
[3]*Henan Agricultural University, College of Forestry, Department of environment, Nongye Road 63, Zhengzhou 450002, China*

11.1 Introduction

In order to alleviate the energy crisis and climate problems, clean energy hydrogen has become a promising alternative to fossil fuels. Compared to the traditional complex process of hydrogen production, electrochemical water decomposition is a clean and efficient method to obtain hydrogen through the hydrogen evolution reaction (HER) on the cathode side and to produce oxygen simultaneously through the oxygen evolution reaction (OER) on the anode side (Figure 11.1a) [1]. HER is a double electron transfer process, which is the simplest electrochemical reaction and is relatively easy to occur. In contrast, OER reactions involving four-electron transfer are complex processes that generate O_2 molecules through a series of proton–electron coupling steps. Depending on the pH value of the electrolyte, the reaction pathway to produce O_2 is also completely different. In an alkaline electrolyte, the hydroxyl group (OH^-) is oxidized and converted to H_2O and O_2. In an acidic medium, two water molecules (H_2O) are oxidized to produce four protons (H^+) and O_2 (Figure 11.1b). The specific mechanism of OER in acidic and alkaline electrolytes is shown in Figure 11.1c and d. Due to the complexity of the mechanism and the slow reaction kinetics, the overpotential is high, which is the key factor limiting the efficiency of electrolytic water. Therefore, the development of high-performance OER electrocatalysts is the key to achieving fast energy conversion, helping to lower the energy barrier and promote the dynamics of electrochemical processes. To date, various advanced electrocatalysts have been designed and synthesized to enhance acidic OER properties. This chapter will focus on the research and progress of metal-based heterogeneous OER electrocatalysts, which are mainly divided into metal monoatom and alloy materials.

Functional Materials for Electrocatalytic Energy Conversion, First Edition.
Edited by Zhicheng Zhang, Meiting Zhao, and Yuchen Qin.

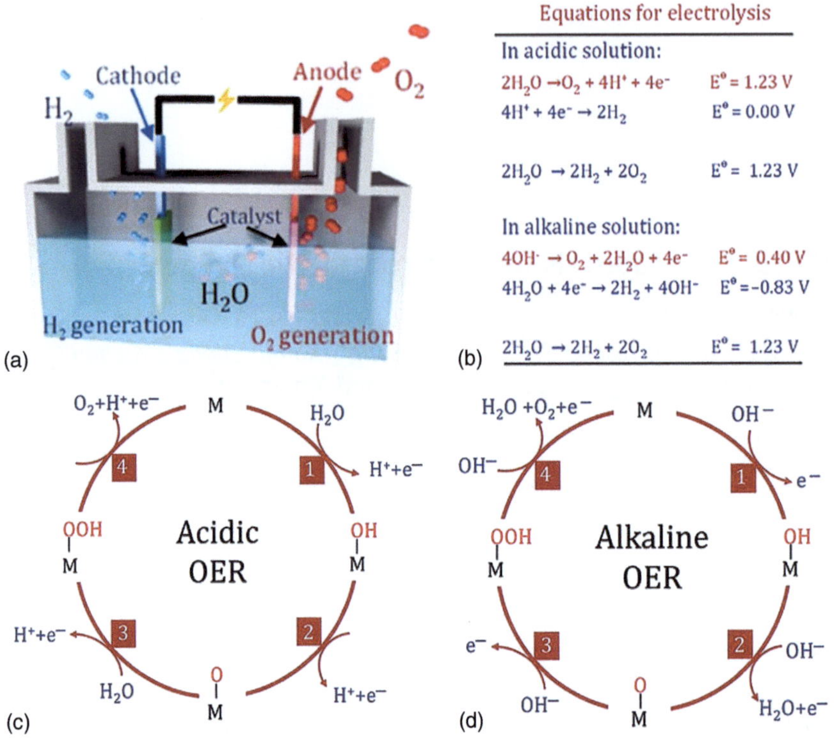

Figure 11.1 (a) Schematic illustration of the water splitting cell. (b) The reactions at the anode and cathode under the acidic and alkaline conditions of water-splitting. (c) Conventional OER mechanism for acidic and (d) alkaline conditions. Source: Yan et al. [1]/Royal Society of Chemistry/CC BY 3.0 DEED.

11.2 Metal Single-Atom Materials

Transition metal-based electrocatalysts in a pure metallic state have recently gained attention as alternative options for OER. The catalytic behavior of metal species, varying in size from single atoms to nanoclusters and nanoparticles (NPs), plays a crucial role in heterogeneous catalytic reactions [2]. Due to its strong corrosion effect, it often cannot directly survive in acidic and alkaline electrolytes. Therefore, it is usually embedded or decorated in a relatively stable host (such as a carbon material). For example, Xu et al. encapsulated nickel NPs in N-doped graphene with an efficient OER electrocatalyst through an easily processable metal–organic framework (MOF) material [3]. Chen et al. obtained 2D N-doped carbon nanosheets and ultrafine Co NPs in three-dimensional N-doped hollow carbon polyhedra (Co@NHCP@CC) by pyrolysis of 2D/3D ZIF-67@carbon cloth [4]. After C is coupled with Co, the energy barrier of *OH adsorption is effectively reduced, and the transformation from *O to *OOH becomes RDS. The charge transfer of Co to the carbon layer in the heterogeneous interface improves the binding strength of C–OOH. The doping of N in the composite further improves the binding strength of C–OOH and decreases the energy barrier, showing the synergistic effect between Co, N, and C. The same phenomenon can be observed in Co–N-doped graphitic carbon (Co–NGC).

However, the size effect of transition metal NPs on electrocatalytic performance remains unclear, especially when reducing the size to the atomic level. Han et al. successfully achieved spatial isolation of cobalt species at the atomic level by fine-tuning the zinc dopant content in bimetallic ZnCo–ZIFs precursors [5]. This manipulation resulted in the generation of Co NPs, atomic Co clusters, and Co single atoms on N-doped carbon, denoted as Co-NPs@NC, Co-ACs@NC, and Co-SAs@NC, respectively (Figure 11.2). This innovative approach allowed for an in-depth investigation into the size influence on the electrochemical activity of transition metal catalysts, extending down to the single-atom scale. Preliminary results indicate that Co-SAs@NC exhibited remarkable OER activity. The exceptional electrocatalytic performance of Co-SAs@NC can be attributed to several factors. First, the isolated single Co atoms offer maximized atomic utilization, providing a greater number of catalytic sites with superior reactivity. Simultaneously, the

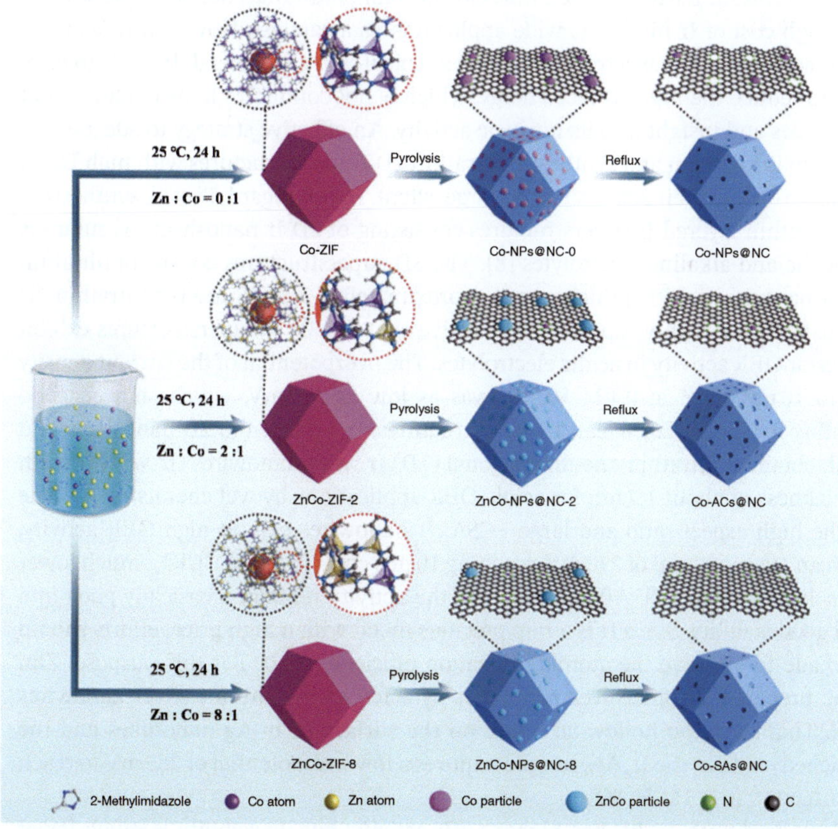

Figure 11.2 Preparation of the Co-NPs@NC, Co-ACs@NC, and Co-SAs@NC catalysts. Source: Adapted from Han et al. [5].

coordination with nitrogen species in the N-doped carbon support stabilizes the Co–SAs. Secondly, the large surface area and mesoporous structure of the carbon matrix expose accessible atomic sites, facilitating the fast diffusion of O_2 and electrolytes. Finally, the high conductivity of the carbon framework enhances the catalytic process by enabling efficient electron transport.

Metal single-atom catalysts have garnered significant attention in the field of electrocatalysis. Reducing metal NPs to nanoclusters or even individual atoms can greatly enhance the catalytic activity and selectivity of electrochemical reactions. However, single atoms possess high surface free energy, making them prone to aggregation. Therefore, it is necessary to anchor these isolated atoms onto various substrates to form stable structures. The high dispersion of metal atoms in metal single-atom catalysts not only facilitates high activity and selectivity for various electrocatalytic reactions but also improves atom utilization efficiency and reduces the cost of large-scale applications. Reducing the scale of noble metals to the atomic level for the synthesis of metal single-atom catalysts represents an alluring strategy. This approach allows for the utmost utilization of metal atoms and cost reduction, all while preserving catalytic activity [6, 7]. Ir is a rare metal that occurs in very low amounts on Earth, even ten times lower than platinum. Therefore, the scarcity and high cost of Ir hinder its wide application as a catalyst. However, Ir is one of the few highly active electrocatalysts. As a typical transition metal, Ir tends to form NPs to reduce the high surface energy, which is not conducive to its enhancement properties and insight into its intrinsic activity. An effective strategy to address this problem is to design and synthesize ultrathin 2D Ir nanostructures with high ECSA values to expose their active sites to a large extent. In this regard, Pi et al. synthesized 3D ultrathin layered Ir superstructures consisting of 2D Ir nanosheets as subunits in acidic and alkaline electrolytes [8]. The 3D superstructures consist of ultrathin 2D Ir nanosheets with a thickness of approximately 1.3 nm. Since the ultrathin 2D nanosheets are highly exposed to the active site, the 3D Ir superstructures exhibit superior OER activity in acidic electrolytes. The overpotential of the current density up to $10\,\mathrm{mA\,cm^{-2}}$ in 0.1 M $HClO_4$ was as low as 270 mV, which showed good stability for the surface-clean Ir superstructures. In addition to 2D nanosheets, Fu et al. obtained ultrathin one-dimensional (1D) Ir wavy nanowires (Ir WNWs) with a thickness of about 1.7 nm for acidic OER applications by wet chemistry [9]. Due to the high aspect ratio and large ECSA, Ir nanowires exhibit high OER activity, with an overpotential of 270 mV, reaching $10\,\mathrm{mA\,cm^{-2}}$ in 0.5 M $HClO_4$, much lower than Ir NPs (298 mV). After 500 cycles, the activity remained basically good and had good stability. Since Ir is a rare precious metal with a high price, efforts should be made to improve the atomic utilization efficiency of Ir. For this purpose, Zhu et al. prepared chrome-silver nanotubes by acid-etching chrome-silver nanowires [10]. Thanks to the hollow structure on the surface of Ir–Ag nanotubes and the enriched oxide Ir, the Ir_6Ag_9 sample requires a low overpotential of 285 mV to reach $10\,\mathrm{mA\,cm^{-2}}$ in 0.5 M H_2SO_4.

Compared with Ir, Ru has higher OER activity, but its stability is much lower than that of Ir. Therefore, Ru and its related compounds have been extensively studied to obtain excellent OER properties. For example, Kong et al. synthesized independent 2D Ru nanosheets by solvothermal method [11]. 2D oxide derivatives of ruthenium nanosheets showed excellent OER properties in 0.5 M H_2SO_4 and reached 10 mA cm^{-2} at an overpotential of 260 mV, showing reasonable stability. Compared with the polycrystalline Ru NPs, the multifaceted branched Ru NPs with controllable crystallinity synthesized by Poerwoprajitno et al. showed significantly improved activity and stability [12]. Branched ruthenium is a single crystal with a hexagonal close-packed (hcp) crystal structure extending along the *c*-axis. An overpotential as low as 180 mV can reach 10 mA cm^{-2} and maintain it for several hours, which represents the high activity and stability of Ru nanocrystals. As proposed, Ru atoms on the low refractive index plane are more stable to dissolution because there are fewer high-energy defect sites on the surface [13]. For branched-chain ruthenium nanocrystals, only edge atoms with high energy are likely to dissolve into solution. For polycrystalline Ru NPs, Ru atoms are in low coordination, in an unstable state, and are easily oxidized and dissolved. This work suggests that designing and synthesizing Ru nanocrystals with low exponential surfaces is a potential strategy for obtaining high-performance acidic OER electrocatalysts.

In addition, replacing the transition metal sites with precious metal atoms is an effective method to increase the metal load and protect the active sites. For example, Cai et al. designed an ultrathin CoO_x amorphous nanosheet ($Ir_1Co_{13.3}O_{20.1}$ ANSs) with high-density Ir single-atom doping, with a layered microstructure composed of coiled nanosheets [14]. The selected electron diffraction (SAED) reveals the amorphous properties. CoO_x ANSs are only 1 nm thick, and thanks to their ultrathin characteristics, $Ir_1Co_{13.3}O_{20.1}$ has a large surface area of 213.3 m^2 g^{-1}. The high-magnification STEM image shows uniformly dispersed Ir monatoms on CoO_x ANSs. $Ir_1Co_{13.3}O_{20.1}$ exhibits better OER activity than ANSs and IrO_2 of CoO_x over the whole potential range. $Ir_1Co_{13.3}O_{20.1}$ requires a low overpotential of 152 ± 5.2 mV to reach 10 mA cm^{-2}, which is much lower than that of pure CoO_x ANSs (330.1 ± 12.5 mV) and commercial IrO_2 (320.0 ± 10.1 mV). At an RHE of 1.485 V, the specific current density of $Ir_1Co_{13.3}O_{20.1}$ is 160 mA cm^{-2} (mass activity 1412 A g^{-1}), which is 160 times that of commercial IrO_2 (1 mA cm^{-2}). Qiao et al. replaced the Co^{4+} site by integrating Ir atoms into the Co_3O_4 lattice through an ion exchange pyrolysis process [15]. As shown in Figure 11.3a, the brighter Ir single-atom can be clearly identified from the lattice of Co_3O_4, and the Co-edge Fourier transform (FT) extended-ray absorption fine structure (EXAFS) reveals three different interatomic distance features of the octahedron and tetrahedron in the $Ir_{0.06}Co_{2.94}O_4$ phase (Figure 11.3b). The prepared electrocatalyst $Ir_{0.06}Co_{2.94}O_4$ has higher mass activity and better stability than the commodity IrO_2. The OER current density of $Ir_{0.06}Co_{2.94}O_4$ catalyst is 5.99 mA $cm_{ox}{}^{-2}$ at an overpotential of 300 mV (1.53 V vs. RHE). The mass activity is 2511 A $g_{Ir}{}^{-1}$ (Figure 11.3c). Similarly, Lin et al. substituted the Mn atoms on the surface for Ru atoms and prepared a crystal α-MnO_2 nanofibril-supported electrocatalyst with an array or chain of Ru atoms with a lower overpotential of 161 mV at 10 mA cm^{-2} (Figure 11.3e). The in situ reconstruction process of Ru/MnO_2 is also revealed (Figure 11.3d).

To date, many atomically dispersed nonprecious metal-based single-atom catalysts have been developed to improve OER performance. Pei and colleagues

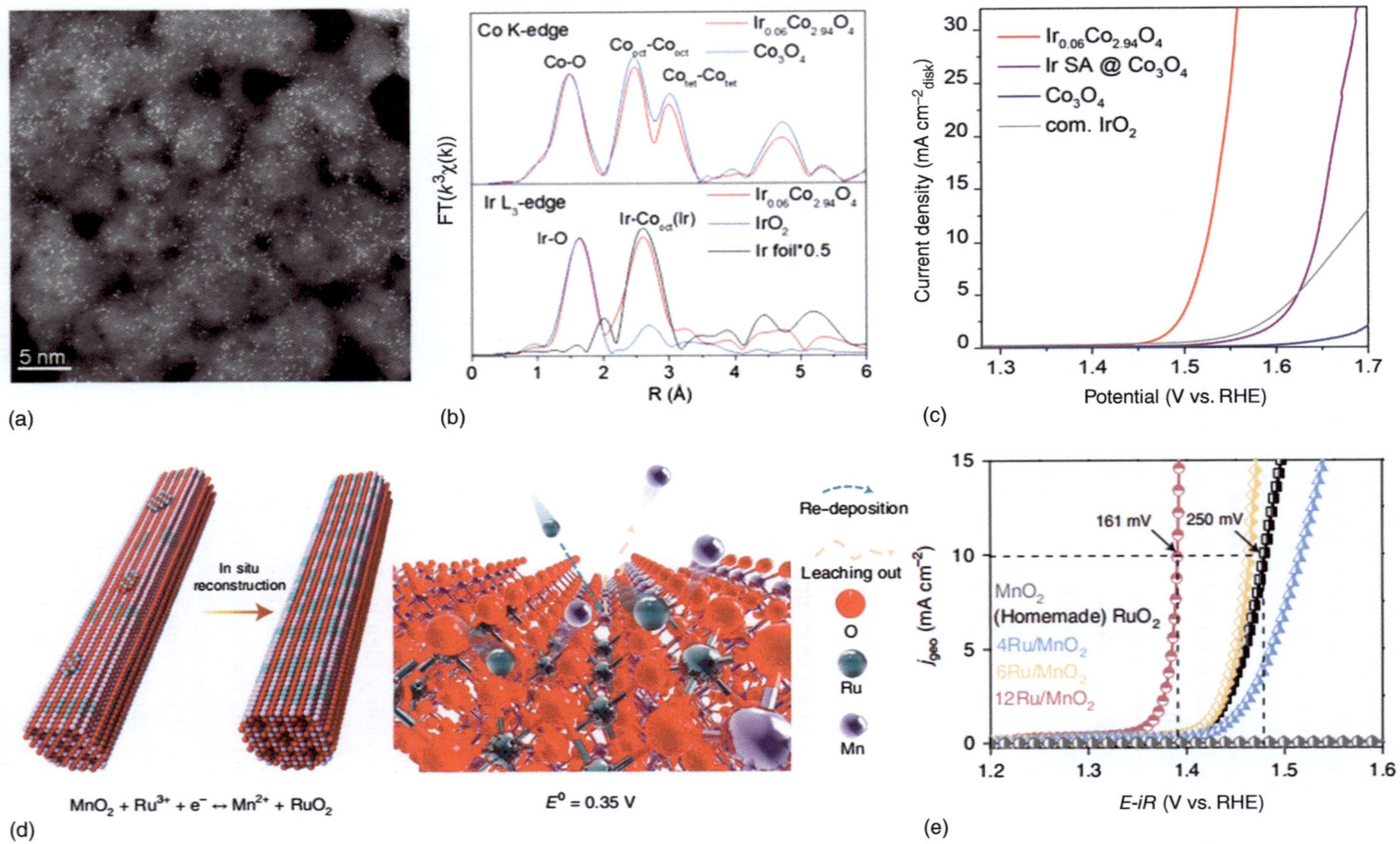

Figure 11.3 (a) HAADF–HRSTEM image of $Ir_{0.06}Co_{2.94}O_4$. (b) FT–EXAFS spectra at the Co K-edge and Ir L_3-edge of $Ir_{0.06}Co_{2.94}O_4$, and reference samples. (c) Normalized polarization curves of surface area of a disk electrode made of $Ir_{0.06}Co_{2.94}O_4$, Ir SA@Co_3O_4, Co_3O_4, and commercial IrO_2 catalysts in 0.1 M $HClO_4$ saturated with O_2. (d) Schematic illustration for the in situ reconstruction process of Ru/MnO_2. (e) Electrocatalytic OER performance of MnO_2, Ru/MnO_2, and homemade RuO_2 in 0.1 M $HClO_4$. Source: Reproduced with permission of Shan et al. [15], American Chemical Society.

applied single-atom ferrothermal loading to graphite porous N-doped carbon (FeN-embedded PNC) [16], showing superior OER (η = 395 mV, 10 mA cm^{-2}) properties in alkaline media. The single-atom Fe catalyst was assembled into a compressed Zn–air battery, which showed excellent charge–discharge performance. At 5 mA cm^{-2} and a maximum power density of 118 mW cm^{-2}, a low charge and discharge voltage difference of 0.78 V compared to RHE can be achieved, and the power density does not decrease significantly under different compression conditions. In addition, the Zn–air battery assembled with the single-atom iron cathode has stable charge–discharge behavior and high durability. In addition to atomically dispersed Fe–Nx catalysts, doping S atoms in Fe–N–C catalysts can improve the catalytic activity of OER. Zhu et al. synthesized an S, N Co-doped atomically dispersed Fe catalyst (Fe-NSDC) by pyrolysis technology, which can effectively promote OER [17]. The superior performance in OER (η = 410 mV at 10 mA cm^{-2}) is attributed to its abundant active sites and high surface structure, which can induce rapid electron transfer and is a superior energy source. Fe-NSDC can be used in Zn–air batteries to achieve a low charge–discharge voltage gap of 1.0 V and a high power density (2211 mW cm^{-2}).

Pan et al. reported that an atomically dispersed Fe-N_4 catalyst (Fe-N_4SAs/NPC) had better OER performance [18]. After pyrolysis of bimetallic Zn/Fe polyphthalocyanine, C, N, and Fe atoms are evenly distributed on the carbon-based carrier. In 1 M KOH, Fe-N_4SAs /NPC has an overpotential of 430 mV at 10 mA cm^{-2} and a Tafel slope of 95 mV dec^{-1}. Fe-N_4SAs/NPC can be assembled into water electrolyzers and Zn–air battery cells. In the water electrolyzer, a low voltage of ≈1.67 V and high stability at 10 mA cm^{-2} were achieved. In the Zn–air battery, a smaller charge–discharge voltage gap and a larger power density (232 mW cm^{-2}) are obtained. After 108 cycles, there is no significant voltage drop, indicating that Fe-N_4SAs/NPC has good stability.

11.3 Metal Alloys Materials

An alloy is a material with metallic properties, which is composed of two or more metals and metal or nonmetal by a certain method. There are many kinds of metal alloys, which can be divided into metal–metal and metal–nonmetal according to the element category. According to the number of elements, it can be divided into binary alloy, ternary alloy, multicomponent alloy, and so on. In recent years, the research of nanoalloy electrode materials has attracted extensive attention. The selectivity and reaction rate of metal alloy catalyst can be changed by changing the d-band hole of metal catalyst by alloying method. After forming a nanostructure, its size decreases, and its special surface area increases, showing quantum size effects and macroscopic quantum tunneling effects [19], and therefore it has unique electron transfer and electrocatalytic properties [20].

In binary metal alloys, the study of precious metals is essential. The design of single-metal Ir nanostructures has improved the performance of OER to some extent, but its activity and stability are still far from satisfactory, and people have been working on improving the performance of OER through various strategies. A simple and commonly used approach is to design and synthesize polymetallic Ir-based nanostructures by alloying them with nonprecious transition metals such as nickel, cobalt, and iron [21–23]. Through alloying and composition optimization, the activity and stability of Ir-based nanostructures can be significantly improved compared with pure IrO_2 or commercial IrO_2 [24]. The reason for this is that alloying can greatly alter the electronic structure of Ir, thereby optimizing the adsorption and desorption of adsorbents on the catalyst surface, thereby improving OER performance. At the same time, the introduction of foreign atoms can also reduce the use of rare and precious metals.

A wide range of Ir group nanostructures, such as nanoclusters, nanodendrites, nanoflowers, and dodecahedrons, have been designed and synthesized by chemical synthesis methods developed. For example, monodisperse bimetallic IrM (M = Ni, Co, Fe) nanoclusters are grown directly on the toner by solvothermal method to ensure surface cleanliness, with an average size of 1.5 nm in the OER acidic electrolyte [23]. The introduction of Ni, Co, and Fe into the $HClO_4$ or H_2SO_4 electrolyte makes it have good OER activity and long-lasting stability, of which Ni is the most effective element for performance enhancement. The promoting effect of the alloy can be explained by the d-band center theory, which describes the relationship between the d-band electrons and the adsorption energy. There is a lot of scope to enhance the performance of Ir through alloying, which can optimize the interaction between catalysts and oxygen intermediates. In addition to increasing the intrinsic activity of each active site, increasing the density of the active site can also improve the catalytic activity through morphological and structural control. Feng et al. synthesized a series of IrM (M = Co, Ni, and CoNi) porous hollow nanocrystals (PHNCs) by etching Ir-based solid nanocrystals (SNCs) with Fe^{3+} ions [25]. IrM SNCs were prepared by a two-step process, first by using oil amine (OAm) at high temperature, and then by etching in a hexane/ethanol mixture with Fe^{3+} ions at room temperature. The prepared IrCoNi PHNCs are monodisperse and highly porous. According to density functional theory (DFT) simulations, the ligand effect of alloying results in a shift in the density of states (DOS) from Fermi levels. Therefore, the center of the d-band is also far away from the Fermi level, which leads to the weakening of the adsorption of oxygen intermediates and the enhancement of catalytic performance. The OER activity of IrM PHNCs in acidic electrolytes is greatly improved due to the reduction of activation energy by alloying. IrCoNi PHNCs reached an overpotential of 303 mV to 10 mA cm^{-2} in 0.1 M $HClO_4$, indicating excellent OER activity in acidic media. Lv et al. achieved a balance between adsorption energy regulation and corrosion resistance by designing porous Ir–W nanodendrites with high activity and durable stability in acids [26]. At 1.53 V, the current density of Ir–W nanodendrites is three times that of Ir/C, and it maintains good stability after 3000 cycles in 0.1 M $HClO_4$. The DFT calculation results show that the adsorption energy of the oxygen intermediate is lower than that of IrO_2 due to the incorporation of W, and W can also stabilize the active site formed during OER. Therefore, the dissolution of Ir results in accelerated OER kinetics and enhanced stability. In addition, IrNi and

IrCo nanoflowers have also been reported to effectively improve OER activity and stability in 0.1 M $HClO_4$.

Since monometallic Ru is unstable as an OER electrocatalyst, the properties of Ru have also been improved by alloying to improve the stability of OER and further increase the activity. Alloying with transition metals can significantly change the adsorption of key oxygen intermediates of OER, thereby improving OER activity and long-lasting stability [27]. Nonprecious metal transition metals, such as Ni, Co, Fe, Cu, Mn, and Mo, are often allotted with Ru to form bimetallic catalysts to improve acidic OER properties. 2D materials, as highly efficient electrocatalysts, have the characteristics of high specific surface area, high aspect ratio, high conductivity, and high density of unsaturated atoms on the surface, and have received extensive attention and effort. Ru and its related metal materials tend to form 2D nanostructures under certain growth conditions, which is conducive to performance enhancement. Yao et al. recently synthesized 2D RuCu nanosheets with abundant channels by wet chemistry using OAm as solvent and surfactant for water cracking (Figure 11.4a,b) [28]. The structural characterization shows that the 2D RuCu nanosheets are composed of crystalline Ru and amorphous Cu. Near the channel region, the lattice is distorted. OER in 0.5 M H_2SO_4 achieves a current density of 10 mA cm^{-2}, requiring only 236 mV overpotential, which is much lower than Ir/C and RuCu NPs (Figure 11.4c), indicating the effectiveness of Cu alloying and the advantages of 2D nanosheets over NPs. Both copper alloying and abundant channels with lattice distortion contribute to the superior acidic OER activity of RuCu nanosheets. The results show that amorphous copper may optimize the adsorption energy of oxygen intermediates, thereby improving the OER performance. The high ratio of Ru^{4+}/Ru^0 to $Cu^{2+}/Cu^0/^{1+}$ and the synergistic effect between Ru and Cu may be the key factors for the enhanced performance of OER. DFT calculations show that the Fortis channel properties induce surface electron activation (Figure 11.4d) and that short-range disorder minimizes the intermediate energy expenditure for lattice relaxation bond dissociation. The downward shift of the 2p band of water indicates that the tight adsorption of water promotes the acidic OER process (Figure 11.4e). In addition, near the channel region, the strong splitting effect of the Ru site is weakened, and active electron transfer is ensured by overcoming the Coulomb barrier. In summary, all these factors lead to the good acidic OER activity of RuCu nanosheets. According to the proposed acidic OER reaction pathway, the final O_2 formation is the rate-determining step that spontaneously forms at a small application potential (Figure 11.4f), resulting in superior acidic OER activity for RuCu nanosheets. In addition to 2D RuCu nanosheets, other Ru-based bimetallic nanostructures have been constructed, all of which exhibit highly enhanced acidic OER properties, such as Ru–Ni nanosheet assemblies [29], Ru–Pb wavy nanowires [30], and Ru–Ni sandwiched-in nanoplates [31]. The OER stability of these Ru-based polymetallic nanostructured electrocatalysts is much longer than that of commercial catalysts. However, no mechanistic studies have been provided in these works to elucidate the underlying causes of enhanced stability. Therefore, more experimental and theoretical efforts are needed to help understand the OER stability of Ru-based polymetallic nanoelectrocatalysts.

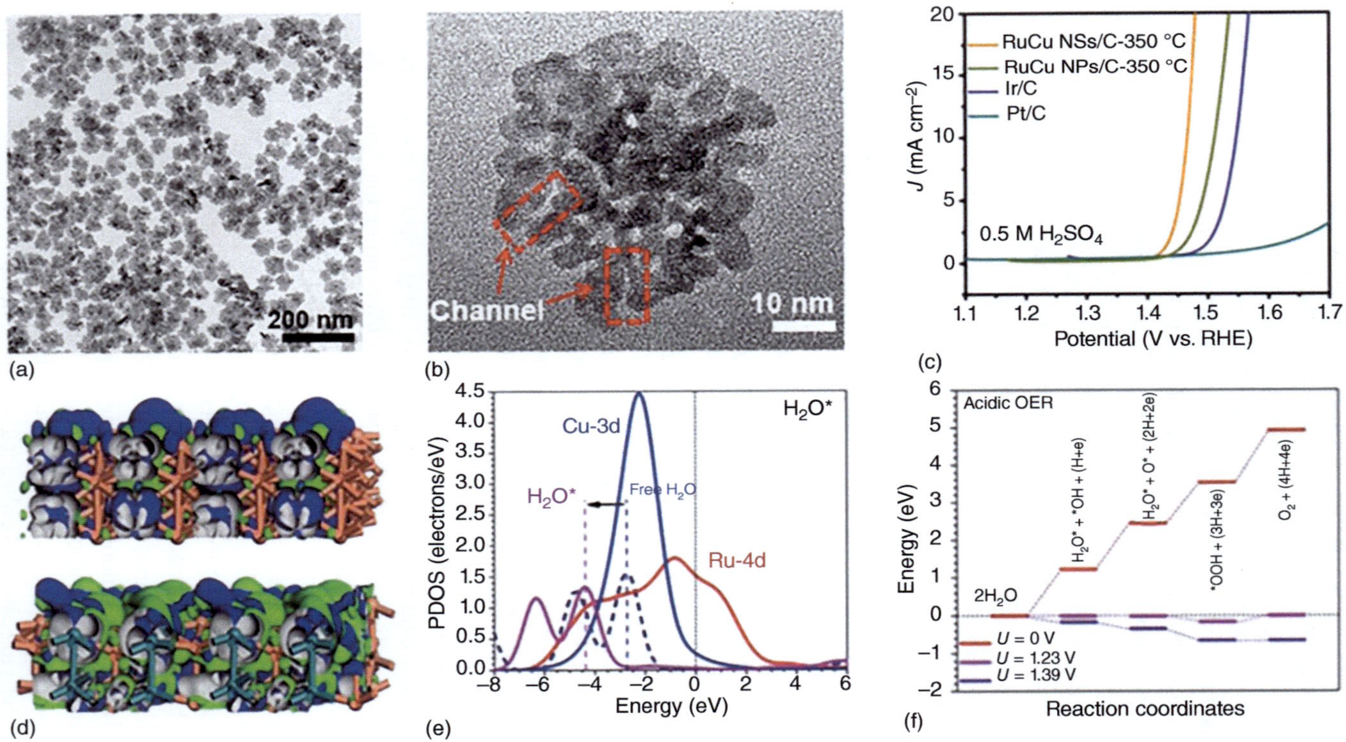

Figure 11.4 (a,b) TEM images of channel-rich RuCu nanosheets. (c) OER polarization curves of the catalysts in 0.5 M H_2SO_4. (d) Real-spatial contour plots of the bonding and anti-bonding orbitals near the Fermi level for (upper) pristine RuCu nanoparticles and (lower) channel-rich RuCu nanosheets from the side view. (e) The partial density of states of H_2O adsorption on channel-rich RuCu nanosheets. (f) Reaction pathways of acidic OER on channel-rich RuCu nanosheets. Source: Reproduced with permission of Yao et al. [28], Wiley-VCH.

In addition to precious metal Ir-based catalysts and Ru-based catalysts, the researchers also studied the catalytic effects of nonprecious metal catalysts in OER. The Rainey effect shows that the dissolution of nonprecious metals added to the alloy catalyst roughens the surface of the catalyst [32], increasing the effective active surface area of the catalyst, thereby improving the performance of the catalyst. Yan et al. used a simple in situ surface modification strategy to develop NiS/Ni NPs with excellent catalytic properties for electrolytic water [33]. For OER, a small overpotential of 301 mV is required to drive a current density of $10\,mA\,cm^{-2}$, and the corresponding Tafel slope is $46\,mV\,dec^{-1}$. The strong coupling between the electron enrichment S at the interface of NiS and Ni promotes important electron transfer. At the same time, its good electrochemical stability was also proved. Good electrical conductivity and regulation of electronic structure help accelerate the dissociation of water and improve its intrinsic activity. Li et al. prepared amorphous nickel monoxide nanosheets (Figure 11.5a) combining ultrafine Ni and molybdenum trioxide NPs (MoO_3/Ni-NiO) with two heterostructures (Ni-NiO and MoO_3-NiO) using a new sequential electrodeposition method [34]. MoO_3/Ni-NiO composites have abundant heterogeneous interfaces, fully exposed active centers, promote charge transfer and synergistic effects, and have excellent electrocatalytic properties. The synthesized MoO_3/Ni-NiO composite has a current density of

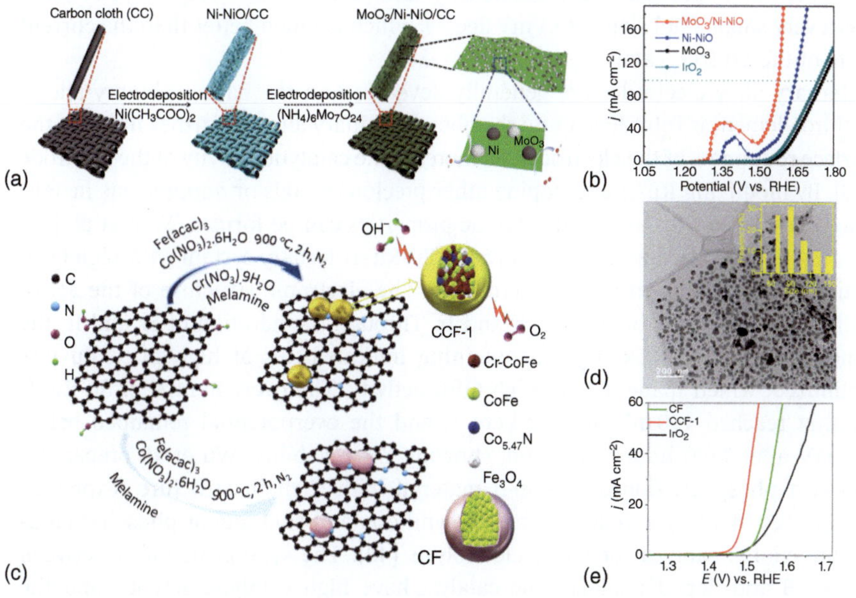

Figure 11.5 (a) Schematic illustration for the fabrication of MoO_3/Ni-NiO on carbon cloth. (b) OER polarization curves at $1\,mV\,s^{-1}$ in 1 M KOH. Source: Reproduced with permission of Li et al. [34], Wiley-VCH. (c) Schematic preparation process of the CCF-1 and CF for OER electrocatalysis. The active sites in the catalyst are represented by light green, dark red, dark blue, and white balls. (d) TEM image (inset: NP-size distribution). (e) Linear sweep voltammogram (LSV) curves of catalysts on glassy carbon electrode in O_2 saturated 1 M KOH at a scan rate of $2\,mV\,s^{-1}$. Source: Reproduced with permission of Dang et al. [35], Elsevier.

347 mV with a current density of 100 mA cm^{-2} in catalytic OER (Figure 11.5b). In addition, the prepared catalyst has excellent catalytic durability, significantly superior to precious metal catalysts and many previously reported materials. Bao et al. designed a series of spinel nanosheets with hypoxia and ultrathin thickness to improve the catalyst's reactivity and the number of active centers and used it as a good platform to promote the water oxidation process [36]. The results show that the oxygen vacancy on the ultrathin nanosheets can reduce the adsorption energy of water and improve the efficiency of OER. $NiCo_2O_4$ ultrathin nanosheets rich in oxygen vacancies have a high current density of 285 mA cm^{-2} and a small overpotential of 0.32 V at 0.8 V. Hou et al. synthesized a catalyst with high OER activity by controlling $Co(AC)_2$ and adding the core–shell heterostructure of $AgNO_3$ solution Ag@CoxP [37].

Due to the strong electron interaction between the silver core and the CoxP shell, the synthesized catalyst has high catalytic activity. It can provide a current density of 10 mA cm^{-2} at a small overpotential of 310 mV. In addition, the catalyst has high catalytic stability. Dang et al. reported an OER catalyst consisting of Cr-CoFe, CoFe, $Co_{5.47}N$, and Fe_3O_4 sites embedded in nitrogen-doped graphene (labeled CCF-1) (Figure 11.5c) [35]. Figure 11.5d shows that NPs in CCF-1 are uniformly distributed on the surface of N-doped graphene oxide. The diameter is 0.05–0.15 μm. The catalyst has a significant effect in alkaline media, requiring only 210/290 mV overpotential to produce a current density of 10/500 mA cm^{-2} (Figure 11.5e), and has a very small Tafel slope (39.1 mV dec^{-1}), which is much better than the current state-of-the-art IrO_2 catalyst.

Ternary alloy electrodes are generally developed on the basis of binary alloys. A third element is introduced into the obtained binary alloy to further improve the surface roughness of the electrode and increase the catalytic activity of the electrode [38]. By modifying Ru/Ir and doping other precious metals or nonprecious metals, alloy complexes with excellent catalytic properties can be formed. Wen et al. [39] studied and prepared an electrocatalyst of Sr–Ru–Ir ternary system with high OER activity and stability in acidic electrolytes. The electronic structure of the active ruthenium site is modulated by Sr and Ir. Through the addition of Sr and Ir, the binding energy of OER oxygen-containing intermediates at high price Ru was optimized, which made it have high OER activity. The overpotential of Sr–Ru–Ir catalyst reached 190 mV at 10 mA cm^{-2}, and the overpotential remained below 225 mV after 1500 hours operation, showing good stability. Wu et al. prepared a low-cost phosphate-doped Co–Fe–B material with a chain structure (expressed as Co_1–Fe_1–B–P) by simple water bath synthesis and subsequent phosphating as a bifunctional catalyst for water electrolysis [40]. The synergistic effect between Co–Fe–B and doped p makes the catalyst have high catalytic activity, and the 1D chain structure provides the catalytic active site rich in Co_1–Fe_1–B–P, which improves the atomic utilization efficiency. For OER, the prepared Co_1–Fe_1–B–P nanochains require a very low overpotential of about 225 mV at a current density of 10 mA cm^{-2} and have a small Tafel slope of 40 mV dec^{-1} in alkaline media. Its performance is better than that of non-phosphorus Co–Fe–B nanochains.

High-entropy alloys (HEAs) have emerged as a promising class of materials for electrocatalysis due to their rich compositional diversity, tunable electronic structure, and exceptional structural stability [41, 42]. These attributes stem from the synergistic effects of multiple metallic sites within the HEAs, resulting in significantly enhanced catalytic activity and stability. However, the presence of multiple competing adsorption sites in HEAs can lead to suboptimal reaction selectivity. Rational control over the metallic sites and reaction intermediates in HEA catalysts to favor competitive preferential adsorption remains a formidable challenge, constraining further enhancement of their catalytic performance [43–46].

Cui et al. embark on the challenge of enhancing OER efficiency by synthesizing high-entropy metal sulfide (HEMS) NPs, specifically (CrMnFeCoNi)S_x [47]. These NPs are synthesized through a pulse thermal decomposition method, aiming to overcome the immiscibility challenges of multiple metallic constituents. Remarkably, these HEMS NPs exhibit one of the best reported activities for OER, with an impressively low overpotential of 295 mV at a current density of 100 mA cm^{-2} in 1 M KOH solution and notable durability, maintaining performance with only slight polarization after 10 hours (Figure 11.6a). The study posits that the

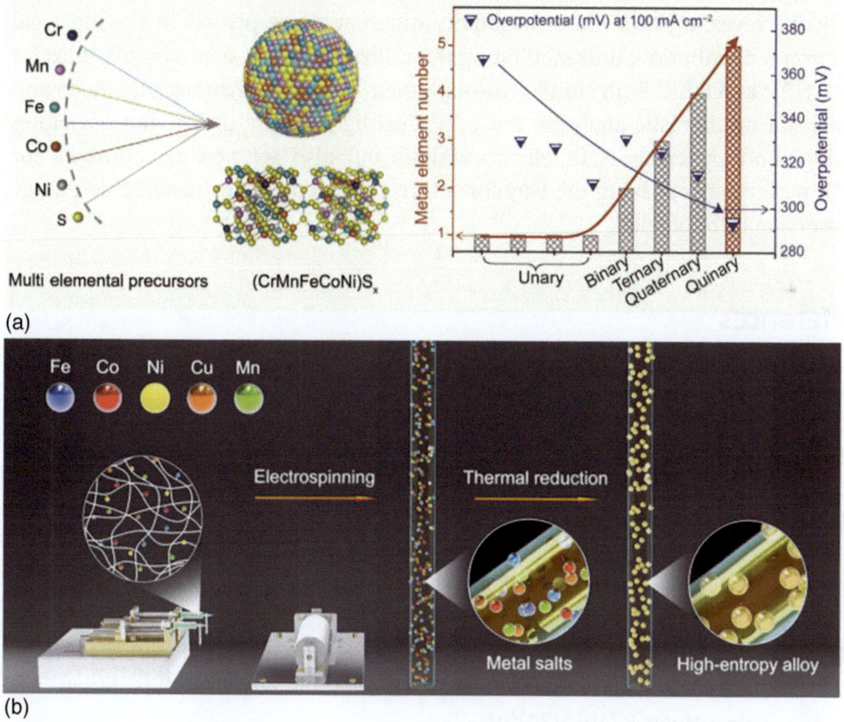

Figure 11.6 (a) Schematic demonstrating the structure of HEMS (CrMnFeCoNi)Sx nanoparticles and their application as the OER catalyst. Mixing immiscible metallic elements (i.e. Cr, Mn, Fe, Co, and Ni) into a homogeneous and high-entropy sulfide nanoparticle. Comparison of overpotentials and metal element numbers among unary, binary, ternary, quaternary materials, and quinary HEMS. Source: Reproduced with permission of Cui et al. [47], Wiley-VCH. (b) Synthesis procedure of HEA/CNF hybrid nanomaterials through a polymer fiber nanoreactor. Source: Reproduced with permission of Zhu et al. [48], Royal Society of Chemistry.

outstanding performance is attributed to a synergistic effect among the metal atoms in the HEMS, leading to desirable electronic states that enhance OER activity. Computational efforts and X-ray photoelectron spectroscopy analysis support this notion, suggesting that the multielemental synergy in HEMS fine-tunes the charge state in metal sulfides, significantly bolstering the catalytic activity. Zhu et al. delve into the synthesis of a novel FeCoNiCuMn HEA NP through a sophisticated polymer fiber nanoreactor approach. This approach ensures the homogeneous mixing of metal ions, embedded in polymer nanofibers, followed by controlled thermal treatments, resulting in a unique FeCoNiCuMn HEA/CNF structure. This HEA exhibits exceptional electrocatalytic activity for both HER and OER. In contrast, the FeCoNiCuMn HEA/CNF catalyst notably requires an overpotential of only 386 mV at 200 mA cm^{-2} for OER, significantly outperforming benchmark catalysts(Figure 11.6b) [48]. The study attributes this remarkable performance to the high-entropy atomic environment within the HEAs, which converts traditionally inactive Cu sites into electron-enriched active sites, effectively lowering the adsorption energies of reactants, intermediates, and products. This phenomenon is thoroughly elucidated through DFT calculations and in situ Raman spectroscopy analyses, revealing that electronegativity differences are pivotal in driving local electron redistribution, thus facilitating the utilization of Cu sites as active sites for both HER and OER. Both studies, through their innovative synthesis methods and insightful mechanistic analyses, not only shed light on the underlying principles of high-entropy catalysts in electrocatalysis but also set new benchmarks for OER performance, paving the way for future advancements in sustainable energy conversion technologies.

References

1 Yan, Z., Liu, H., Hao, Z. et al. (2020). Electrodeposition of (hydro)oxides for an oxygen evolution electrode. *Chem. Sci.* 11 (39): 10614.

2 Liu, L. and Corma, A. (2018). Metal catalysts for heterogeneous catalysis: from single atoms to nanoclusters and nanoparticles. *Chem. Rev.* 118 (10): 4981.

3 Xu, Y., Tu, W., Zhang, B. et al. (2017). Nickel nanoparticles encapsulated in few-layer nitrogen-doped graphene derived from metal-organic frameworks as efficient bifunctional electrocatalysts for overall water splitting. *Adv. Mater.* 29 (11): 1605957.

4 Chen, Z., Ha, Y., Jia, H. et al. (2019). Oriented transformation of Co-LDH into 2D/3D ZIF-67 to achieve Co–N–C hybrids for efficient overall water splitting. *Adv. Energy Mater.* 9 (19): 1803918.

5 Han, X., Ling, X., Wang, Y. et al. (2019). Generation of nanoparticle, atomic-cluster, and single-atom cobalt catalysts from zeolitic imidazole frameworks by spatial isolation and their use in zinc-air batteries. *Angew. Chem. Int. Ed.* 58 (16): 5359.

6 Zhao, D., Zhuang, Z., Cao, X. et al. (2020). Atomic site electrocatalysts for water splitting, oxygen reduction and selective oxidation. *Chem. Soc. Rev.* 49 (7): 2215.

7 Xu, Q., Zhang, J., Wang, D., and Li, Y. (2021). Single-atom site catalysts supported on two-dimensional materials for energy applications. *Chin. Chem. Lett.* 32 (12): 3771.

8 Pi, Y., Zhang, N., Guo, S. et al. (2016). Ultrathin laminar Ir superstructure as highly efficient oxygen evolution electrocatalyst in broad pH range. *Nano Lett.* 16 (7): 4424.

9 Fu, L., Yang, F., Cheng, G., and Luo, W. (2018). Ultrathin Ir nanowires as high-performance electrocatalysts for efficient water splitting in acidic media. *Nanoscale* 10 (4): 1892.

10 Zhu, M., Shao, Q., Qian, Y., and Huang, X. (2019). Superior overall water splitting electrocatalysis in acidic conditions enabled by bimetallic Ir-Ag nanotubes. *Nano Energy* 56: 330.

11 Kong, X., Xu, K., Zhang, C. et al. (2016). Free-standing two-dimensional Ru nanosheets with high activity toward water splitting. *ACS Catal.* 6 (3): 1487.

12 Poerwoprajitno, A.R., Gloag, L., Benedetti, T.M. et al. (2019). Formation of branched ruthenium nanoparticles for improved electrocatalysis of oxygen evolution reaction. *Small* 15 (17): e1804577.

13 Chang, S.H., Danilovic, N., Chang, K.-C. et al. (2014). Functional links between stability and reactivity of strontium ruthenate single crystals during oxygen evolution. *Nat. Commun.* 5 (1): 4191.

14 Cai, C., Wang, M., Han, S. et al. (2020). Ultrahigh oxygen evolution reaction activity achieved using Ir single atoms on amorphous CoO_x nanosheets. *ACS Catal.* 11 (1): 123.

15 Shan, J., Ye, C., Chen, S. et al. (2021). Short-range ordered iridium single atoms integrated into cobalt oxide spinel structure for highly efficient electrocatalytic water oxidation. *J. Am. Chem. Soc.* 143 (13): 5201.

16 Ma, L., Chen, S., Pei, Z. et al. (2018). Single-site active iron-based bifunctional oxygen catalyst for a compressible and rechargeable zinc-air battery. *ACS Nano* 12 (2): 1949.

17 Zhang, J., Zhang, M., Zeng, Y. et al. (2019). Single Fe atom on hierarchically porous S, N-Co doped nanocarbon derived from porphyra enable boosted oxygen catalysis for rechargeable Zn-air batteries. *Small* 15 (24): e1900307.

18 Pan, Y., Liu, S., Sun, K. et al. (2018). A bimetallic Zn/Fe polyphthalocyanine-derived single-atom Fe-N_4 catalytic site: a superior trifunctional catalyst for overall water splitting and Zn-Air batteries. *Angew. Chem. Int. Ed.* 57 (28): 8614.

19 Fang, M., Dong, G., Wei, R., and Ho, J.C. (2017). Hierarchical nanostructures: design for sustainable water splitting. *Adv. Energy Mater.* 7 (23): 1700559.

20 Ghosh, S. and Basu, R.N. (2018). Multifunctional nanostructured electrocatalysts for energy conversion and storage: current status and perspectives. *Nanoscale* 10 (24): 11241.

21 Park, J., Sa, Y.J., Baik, H. et al. (2017). Iridium-based multimetallic nanoframe@nanoframe structure: an efficient and robust electrocatalyst toward oxygen evolution reaction. *ACS Nano* 11 (6): 5500.

22 Alia, S.M., Shulda, S., Ngo, C. et al. (2018). Iridium-based nanowires as highly active, oxygen evolution reaction electrocatalysts. *ACS Catal.* 8 (3): 2111.

23 Pi, Y., Shao, Q., Wang, P. et al. General formation of monodisperse IrM (M = Ni, Co, Fe) bimetallic nanoclusters as bifunctional electrocatalysts for acidic overall water splitting. *Adv. Funct. Mater.* 27 (27): 1700886.

24 Lv, F., Zhang, W., Yang, W. et al. (2019). Ir-based alloy nanoflowers with optimized hydrogen binding energy as bifunctional electrocatalysts for overall water splitting. *Small Methods* 4 (6): 1900129.

25 Feng, J., Lv, F., Zhang, W. et al. (2017). Iridium-based multimetallic porous hollow nanocrystals for efficient overall-water-splitting catalysis. *Adv. Mater.* 29 (47): 1703798.

26 Lv, F., Feng, J., Wang, K. et al. (2018). Iridium-tungsten alloy nanodendrites as pH-universal water-splitting electrocatalysts. *ACS Cent. Sci.* 4 (9): 1244.

27 Forgie, R., Bugosh, G., Neyerlin, K.C. et al. (2010). Bimetallic Ru electrocatalysts for the OER and electrolytic water splitting in acidic media. *Electrochem. Solid-State Lett.* 13 (4): B36.

28 Yao, Q., Huang, B., Zhang, N. et al. (2019). Channel-rich RuCu nanosheets for pH-universal overall water splitting electrocatalysis. *Angew. Chem. Int. Ed.* 58 (39): 13983.

29 Yang, J., Shao, Q., Huang, B. et al. (2019). pH-universal water splitting catalyst: Ru-Ni nanosheet assemblies. *iScience* 11: 492.

30 Yang, J., Ji, Y., Shao, Q. et al. (2018). A universal strategy to metal wavy nanowires for efficient electrochemical water splitting at pH-universal conditions. *Adv. Funct. Mater.* 28 (41): 1803722.

31 Ding, J., Shao, Q., Feng, Y., and Huang, X. (2018). Ruthenium-nickel sandwiched nanoplates for efficient water splitting electrocatalysis. *Nano Energy* 47: 1.

32 Chi, J.Q., Yan, K.L., Xiao, Z. et al. (2017). Trimetallic Ni Fe Co selenides nanoparticles supported on carbon fiber cloth as efficient electrocatalyst for oxygen evolution reaction. *Int. J. Hydrog. Energy* 42 (32): 20599.

33 Yan, C., Huang, J., Wu, C. et al. (2020). In-situ formed NiS/Ni coupled interface for efficient oxygen evolution and hydrogen evolution. *J. Mater. Sci. Technol.* 42: 10.

34 Li, X., Wang, Y., Wang, J. et al. (2020). Sequential electrodeposition of bifunctional catalytically active structures in MoO_3/Ni-NiO composite electrocatalysts for selective hydrogen and oxygen evolution. *Adv. Mater.* 32 (39): e2003414.

35 Dang, N.K., Tiwari, J.N., Sultan, S. et al. (2021). Multi-site catalyst derived from Cr atoms-substituted CoFe nanoparticles for high-performance oxygen evolution activity. *Chem. Eng. J.* 404: *126513.*

36 Bao, J., Zhang, X., Fan, B. et al. (2015). Ultrathin spinel-structured nanosheets rich in oxygen deficiencies for enhanced electrocatalytic water oxidation. *Angew. Chem. Int. Ed.* 54 (25): 7399.

37 Hou, Y., Liu, Y., Gao, R. et al. (2017). Ag@Co_xP core–shell heterogeneous nanoparticles as efficient oxygen evolution reaction catalysts. *ACS Catal.* 7 (10): 7038.

38 Rauscher, T., Müller, C.I., Schmidt, A. et al. (2016). Ni–Mo–B alloys as cathode material for alkaline water electrolysis. *Int. J. Hydrog. Energy* 41 (4): 2165.

39 Wen, Y., Chen, P., Wang, L. et al. (2021). Stabilizing highly active Ru sites by suppressing lattice oxygen participation in acidic water oxidation. *J. Am. Chem. Soc.* 143 (17): 6482.

40 Wu, Z., Nie, D., Song, M. et al. (2019). Facile synthesis of Co-Fe-B-P nanochains as an efficient bifunctional electrocatalyst for overall water-splitting. *Nanoscale* 11 (15): 7506.

41 George, E.P., Raabe, D., and Ritchie, R.O. (2019). High-entropy alloys. *Nat. Rev. Mater.* 4 (8): 515–534.

42 Yao, Y., Dong, Q., Brozena, A. et al. (2022). High-entropy nanoparticles: synthesis-structure-property relationships and data-driven discovery. *Science* 376 (6589): eabn3103.

43 Batchelor, T.A.A., Pedersen, J.K., Winther, S.H. et al. (2019). High-entropy alloys as a discovery platform for electrocatalysis. *Joule* 3 (3): 834–845.

44 Li, H., Lai, J., Li, Z., and Wang, L. (2021). Multi-sites electrocatalysis in high-entropy alloys. *Adv. Funct. Mater.* 31 (47): 2106715.

45 Ma, Y., Ma, Y., Wang, Q. et al. (2021). High-entropy energy materials: challenges and new opportunities. *Energy Environ. Sci.* 14 (5): 2883–2905.

46 Han, X., Wu, G., Zhao, S. et al. (2023). Nanoscale high-entropy alloy for electrocatalysis. *Matter* 6 (6): 1717–1751.

47 Cui, M., Yang, C., Li, B. et al. (2020). High-entropy metal sulfide nanoparticles promise high-performance oxygen evolution reaction. *Adv. Energy Mater.* 11 (3): 2002887.

48 Zhu, H., Sun, S., Hao, J. et al. (2023). A high-entropy atomic environment converts inactive to active sites for electrocatalysis. *Energy Environ. Sci.* 16: 619–628. DOI: 10.1039/d2ee03185j.

12

Metallic Compounds for Electrocatalytic Oxygen Evolution Reaction

Xiaobo Zheng[1] *and Yao Wang*[2]

[1] *University of Wollongong, Institute for Superconducting and Electronic Materials, Australia Institute for Innovation Material, Squires Way, Wollongong, 2522, NSW, Australia*
[2] *Jiangnan University, Ministry of Education, School of Chemical and Material Engineering, International Joint Research Center for Photoresponsive Molecules and Materials, Key Laboratory of Synthetic and Biological Colloids, Lihu Street 1800, Wuxi, 214122, China*

12.1 Introduction

Hydrogen energy stands as an ideal, environmentally friendly energy source, completely free from carbon emissions [1]. Harnessing hydrogen technology has emerged as a pivotal pathway toward achieving carbon neutrality [2, 3]. Electrochemical hydrogen production represents a promising and compelling technique for generating hydrogen gas. It possesses the advantages of being pollution-free, highly efficient, and straightforward to operate [4–6]. Nevertheless, the widespread adoption of this technology encounters limitations, primarily attributable to the significant electricity costs resulting from the elevated overpotential required for the oxygen evolution reaction (OER) [7]. This challenge stems from the high theoretical equilibrium potential (1.23 V) and the sluggish kinetics inherent in the four-electron process, contributing to over 80% of the electricity consumption cost originating from OER [8, 9]. In light of this, the development of efficient OER electrocatalysts to mitigate this overpotential has been of paramount importance. Noble metal-based catalysts, particularly IrO_2 and RuO_2, owing to their excellent conductivity and distinct electronic structure, serve as benchmark catalysts in OER reactions [10]. Specifically, Ir-based catalysts demonstrate excellent activity and stability under acidic OER conditions, making them well-suited for use in commercial proton exchange membrane water electrolyzers (PEMWE) [11]. However, the considerable cost and limited availability of noble metal-based catalysts pose significant barriers to their broader large-scale commercial application [12]. Consequently, the pursuit of efficient, cost-effective, and durable catalysts has gained heightened significance.

Boosting intrinsic catalytic activity and increasing the count of active sites are considered two of the most potent strategies for enhancing catalyst performance [13]. Consequently, a range of corresponding methodologies, including vacancy engineering, heterostructure engineering, nanostructuring, and spatial confinement,

Functional Materials for Electrocatalytic Energy Conversion, First Edition.
Edited by Zhicheng Zhang, Meiting Zhao, and Yuchen Qin.

have been developed [14, 15]. Building upon these principles, researchers have ingeniously designed a variety of advanced catalysts, involving both carbon-based and non-carbon-based materials. Carbon-based catalysts, particularly under OER conditions with elevated potentials, often exhibit structural instability and susceptibility to dissolution, thereby compromising the integrity of the catalyst architecture. This limitation considerably restricts their practical application in OER-related devices. As a result, non-carbon-based catalysts, including metals, alloys, and metal compounds, have undergone thorough investigation, leading to noteworthy advancements [16]. Diverse metal-based compounds, including metal chalcogenides [17], carbides [18], phosphides [19], and metal hydroxides [20], have been synthesized and employed in OER to enhance the energy utilization efficiency of devices. Notably, metal oxides and metal hydroxides have emerged as some of the most prominent catalysts, due to their exceptional structural stability, high thermal endurance, and cost-effectiveness. These attributes are a direct outcome of their advantages in terms of controllable structure and composition. Therefore, in this chapter, our primary emphasis will be on the application of metal oxides and metal hydroxides, along with their supported single-atom/nanoparticle derivatives, in the context of OER reactions and associated devices. The objective is to unravel the intricate relationship between their structure and performance, while also providing essential principles for catalyst design.

12.2 Metal Oxides and Their Supported Single-Atom/Nanoparticle Materials

Metal oxides, encompassing both noble metal oxides and non-noble metal oxides, represent one of the most prevalent categories among OER catalysts, playing a pivotal role in the realm of oxygen electrocatalysis. Among them, IrO_2, one of the most emblematic metal oxides, is regarded as the benchmark catalyst for OER and has been widely applied in commercial PEMWE [21]. Furthermore, owing to the strong electronic interactions between metal oxide supports and supported metal atoms, metal oxides single-atom/nanoparticle materials exhibit remarkable activity and cyclic stability in OER [22, 23]. Hence, in the subsequent discussion, we will extensively delve into the application of metal oxides and their supported single-atom/nanoparticle counterparts in the OER. The primary objective is to unveil the inherent structure–performance relationships and catalytic mechanisms, providing the design concept for other electrocatalysts.

12.2.1 Metal Oxides

Noble metal-based oxides, such as IrO_2 and RuO_2, play a crucial role in the OER due to their excellent electrical conductivity and unique electronic structure. Theoretical studies have shown that noble metal-based oxides like RuO_2 possess outstanding oxygen adsorption capabilities with neither too weak nor too strong binding energy for oxygen species (O*) (Figure 12.1a and b) [24]. This specific characteristic gives them a distinct advantage in achieving the low overpotentials required to drive the OER. The balanced oxygen adsorption energy on noble metal-based oxides facilitates the efficient splitting of water molecules into oxygen gas during the OER, making them highly desirable catalysts in various electrochemical applications, especially in proton exchange membrane water electrolysis. Therefore, a variety of noble metal-based oxides are designed by manipulating different strategies [26]. Li and co-workers developed defect-rich RuO_2 nanosheets (RuO_2 NSs) for the OER via a facile molten salt method [27]. The obtained RuO_2 NSs delivered a current density of 10 mA cm^{-2} with only 199 mV overpotential and exhibited a high mass activity of 0.52 A $mg_{Ru}{}^{-1}$ at an overpotential of 230 mV in 0.5 M H_2SO_4 solution, much higher than that of commercial RuO_2 NPs. In addition, the RuO_2 NSs also showed decent stability with only 32 mV decay after six hours operation at the current density of 10 mA cm^{-2}. Theoretical calculations revealed that the Ru vacancy on the surface of RuO_2 NS could optimize the binding energy between Ru active sites and O* intermediate, thereby resulting in the reduction of the energy barrier and overpotential. Except for defect engineering, the doping strategy has been regarded as an effective way to tune the electronic structure and coordination environment to achieve enhanced OER activity. Recent reports have shown that doping transitional metal elements (e.g. Ni, Sn, Nb, and Rh) into RuO_2 can significantly enhance its OER activity and cyclic stability in the acidic solution [28, 29]. Wang et al. prepared a nickel-doped ruthenium oxide (Ni-RuO_2) catalyst by a three-step process strategy (Figure 12.1c) [25]. Benefiting from the incorporation of Ni into the RuO_2 lattice, the stability of lattice Ru and oxygen on the surface was enhanced substantially, which endowed Ni-RuO_2 with significantly improved durability in the acidic media. The PEMWE assembled with Ni-RuO_2 delivered higher activity (1.5 A cm^{-2} @2.1 V) and excellent stability without obvious degradation after more than 1000 hours operation at the current density of 200 mA cm^{-2}, illustrating its great potential in practical application (Figure 12.1d and e). Similarly, Wu and co-workers found that incorporating the Rh atoms into the RuO_2 (Rh-RuO_2) lattice resulted in considerably enhanced acidic OER performance [30]. The Rh-RuO_2 catalyst exhibited a current density of 10 mA cm^{-2} with only 161 mV and outstanding durability with 700 hours operation. Therefore, the incorporation of transition metal atoms into the Ru-based oxides holds great promise in enhancing the OER activity and stability.

Iridium-based oxide (e.g. IrO_2) holds greater potential in practical PEMWE due to its superior structural stability in a highly acidic and oxidative environment. However, its stability falls far short of the demanding requirements of practical production. Furthermore, its lower activity compared to RuO_2, coupled with the unaffordable cost due to the low Ir abundance, further hinders its large-scale application. Therefore, researchers have devised a series of methods to enhance the activity and stability of Ir-based oxide, aiming to meet the demands of industrial applications [31, 32]. Nanostructure engineering has been considered an effective strategy to enhance the activity and durability of iridium-based oxide [31]. For instance, Jin et al. developed a torsion-strained GB-$Ta_{0.1}Tm_{0.1}Ir_{0.8}O_{2-\delta}$ catalyst with rich grain boundaries via a fast pyrolysis strategy for the PEMWE [33]. Benefiting from the synergistic effects between grain boundaries and ligand effect derived

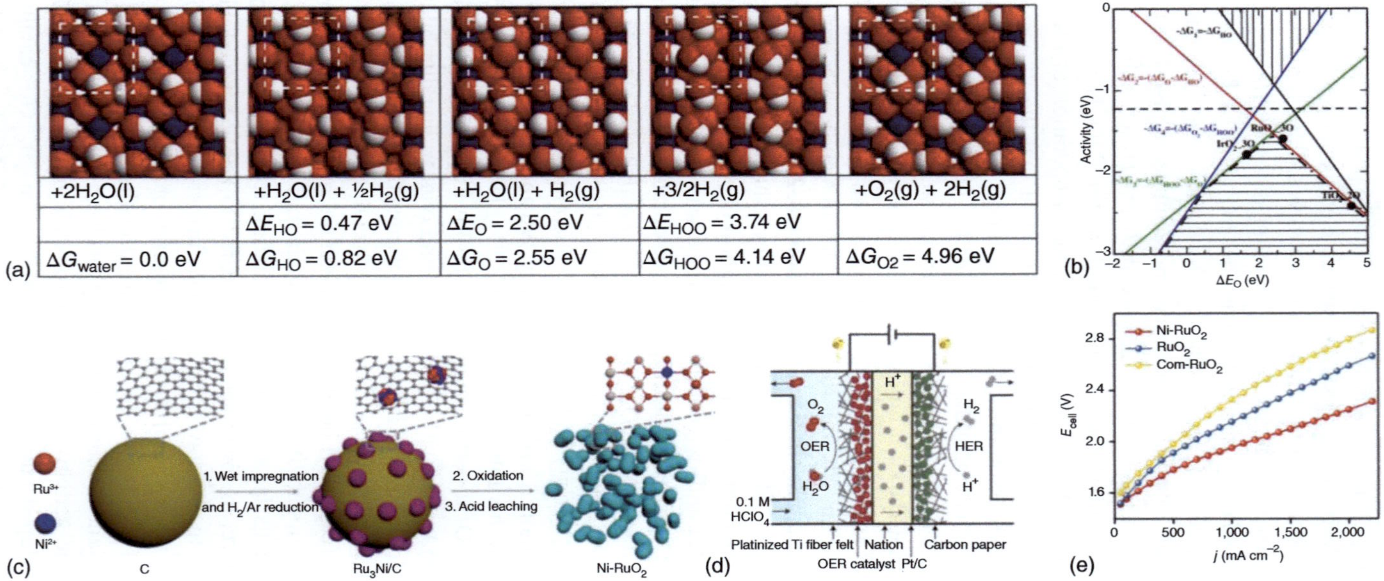

Figure 12.1 (a) The structural illustration of RuO_2 (110) with oxygen-related intermediates. Blue spheres stand for Ru atoms, red oxygen, and white hydrogen. (b) The theoretical activity of OER for different catalysts depicted as a function of the oxygen binding energy. Source: Reproduced with permission of Rossmeisl et al. [24], Elsevier. (c) Schematic diagram of the synthesis process of Ni–RuO_2. (d) Schematic illustration of PEMWE device. (e) *I*–*V* curves of PEMWE assembly with Ni–RuO_2, RuO_2, or Com-RuO_2. Source: Reproduced with permission of Wu et al. [25], Springer Nature.

from Ta/Tm dual metal dopant, the GB-$Ta_{0.1}Tm_{0.1}Ir_{0.8}O_{2-\delta}$ exhibited superior OER activity (10 mA cm^{-2} @1.428 V) and outstanding durability in acid media. Importantly, the PEMWE assembling with GB-$Ta_{0.1}Tm_{0.1}Ir_{0.8}O_{2-\delta}$ can reliably sustain an operational current density of 1.5 A cm^{-2} for 500 hours.

This achievement comes at an estimated cost of US$1 per kilogram of H_2, significantly surpassing the target of US$2 per kg of H_2 set by the US Department. In addition, phase engineering exhibits great potential in improving the performance of catalysts. Shao and co-workers have undertaken numerous pioneering efforts, synthesizing several new phases of IrO_2 by mechano-thermal method [21, 34]. For example, they successfully synthesized a 1T phase IrO_2 by combining the mechanochemistry and thermal treatment [21]. The obtained 1T–IrO_2 featured 2D nanosheet morphology with untrain thin thickness. Furthermore, the 1T–IrO_2 delivered excellent acidic OER activity with only 197 mV overpotential reaching 10 mA cm^{-2}, superior to that of commercial Ir/C and IrO_2.

Despite its advantages of noble metal-based oxides, considering the cost concern and the abundance in the Earth's crust, the development of inexpensive non-noble metal-based oxide catalysts has become crucial and pressing. In recent years, considerable non-noble metal-based oxides such as spinel oxides [35], and perovskites [36] are therefore developed and designed. To meet the practical production requirements, their performance can be further enhanced by increasing the number of active sites and intrinsic activity. The strategy of two-dimensionalization of materials and coupling with doping or defect engineering is considered highly effective for improving material performance. Sun and his colleagues have developed a facile and universal mechanical shear-assisted approach for 2D material synthesis [37]. By combining Mg doping and mechanical shear-assisted strategies, they successfully prepared 2D $LiCo_{0.95}Mg_{0.05}O_2$ nanosheets (Figure 12.2a) [38]. The synergistic effect between them not only significantly increases the exposure of Co active sites but also optimizes the electronic structure of active sites, thereby enhancing the adsorption energy of active sites for intermediate products, greatly improving the alkaline OER performance of the material (Figure 12.2b).

Additionally, although non-noble metal-based oxides traditionally exhibit significantly lower durability in acidic media compared to their precious metal counterparts, researchers have discovered that their durability can be significantly boosted by incorporating suitable transition metal atoms into their metal oxide hosts [40]. For instance, Liu and co-workers reported tensile-strained iridium atoms modified manganese oxide (TS–Ir/MnO_2) for acidic OER (Figure 12.2c). [39] This unique TS–Ir/MnO_2 catalyst exhibited superior OER activity with a lower overpotential of 198 mV to reach the current density of 10 mA cm^{-2} (Figure 12.2d). More importantly, the PEMWE device assembled with TS–Ir/MnO_2 delivered excellent durability with small decay after 200 hours operation at the high current density of 500 mA cm^{-2} (Figure 12.2e). It is expected that much more advanced non-noble metal-based oxides can be developed for the practical PEMWE under continuous efforts.

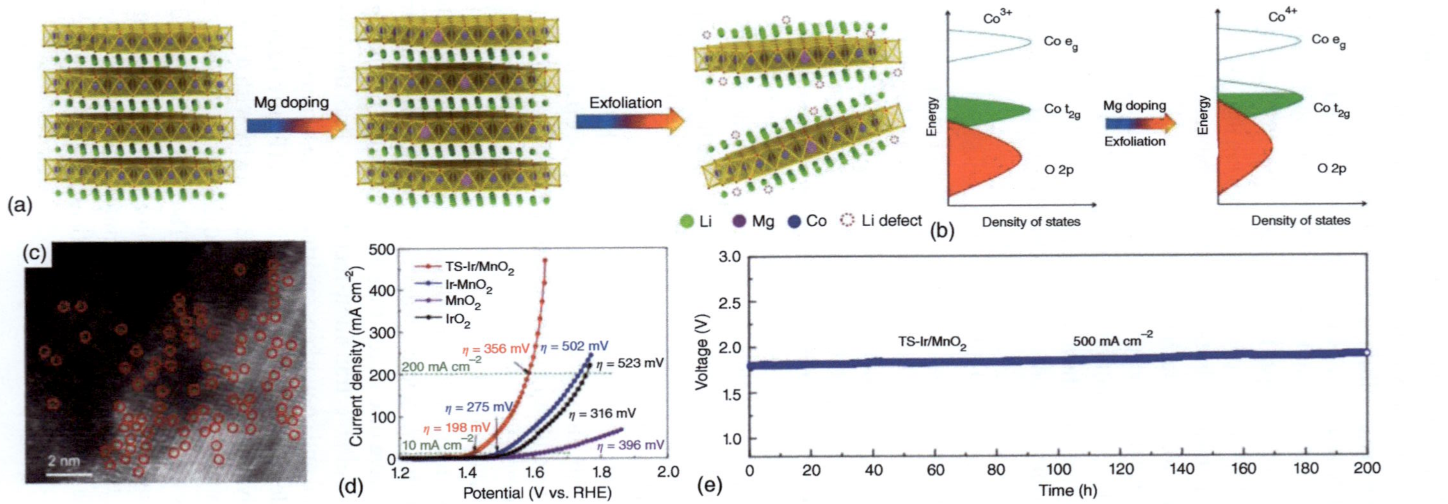

Figure 12.2 (a) Schematic illustration of the preparation process of 2D $LiCo_{0.95}Mg_{0.05}O_2$ nanosheets. (b) The qualitative one-electron density of states diagram. Source: Reproduced with permission of Zheng et al. [38], Wiley-VCH. (c) HAADF–STEM image of the TS–Ir/MnO_2. (d) LSV curves of TS–Ir/MnO_2, Ir-MnO_2, MnO_2, and IrO_2. (e) The cell potential of the PEMWE assembled with TS–Ir/MnO_2 at 500 mA cm^{-2}. Source: Su al. [39]/Springer Nature/CC BY 4.0.

12.2.2 Metal Oxide-Supported Nanoparticle Materials

Supported catalysts have garnered significant attention from researchers and have found successful applications in various fields, including batteries, electrocatalysis, thermocatalysis, and photocatalysis, among others [41, 42]. Within supported catalysts, the strong metal–support interactions between the support and the metal play a pivotal role in tuning the electronic structure of active sites, thereby optimizing the binding energy with intermediate products and reducing reaction barriers [43]. Furthermore, the well-defined support serves to disperse and stabilize the metal species, preventing nanoparticle aggregation and migration, ultimately reducing metal usage and enhancing catalyst stability. This holds immense significance, especially for precious metals, as it has the potential to significantly reduce costs. Sun and co-workers have conducted extensive research on the synthesis of supported heterostructures and their applications in electrocatalysis [44, 45]. They developed a multifunctional heterostructured Pt/$LiCoO_2$ for overall water-splitting [46]. The Pt/$LiCoO_2$ was prepared by a two-step method by mechanical shear exfoliation treatment followed by a wet-chemical process (Figure 12.3a). Microscopic characterizations confirmed the successful synthesis of Pt/$LiCoO_2$ heterostructure consisting of ~2.2 nm Pt NPs anchored on the $LiCoO_2$ nanosheets (Figure 12.3b–e). The optimized Pt/$LiCoO_2$ heterostructure not only exhibited excellent HER activity but also the OER activity with only 285 mV to achieve 10 mA cm^{-2} in the alkaline media, enabling a promising bifunctional electrocatalysts for the overall water-splitting (Figure 12.3f). Also, the full cells assembled with Pt/$LiCoO_2$ exhibited superior activity and stability than the IrO_2||Pt/C cell. Interestingly, in this heterostructure system, the active center alternatively transferred between Pt and $LiCoO_2$ for HER and OER (Figure 12.3g). In particular, Pt species function as the active centers in the HER, with $LiCoO_2$ serving as a co-catalyst to promote the Volmer step and optimize Pt–H_{ad} affinity. Conversely, in the OER, the active center shifts to $LiCoO_2$, and Pt transitions into the role of a co-catalyst to promote the generation of oxygen vacancies and under-coordinated Co centers, ultimately enhancing the adsorption kinetics and reducing the energy barrier of the rate-determining step of OER.

In addition to their extensive application in alkaline OER, metal oxide-supported nanoparticles also have widespread utility in acidic OER. Apart from dispersing and anchoring metal particles, the interaction between the metal and the support effectively fine-tunes the metal's electronic structure, thereby enhancing its resistance to dissolution under highly acidic and oxidative conditions [47]. Therefore, the choice of support material plays a crucial role in enhancing the structural durability of catalysts. In general, support materials should possess the following properties: (i) good resistance to oxidation and acid corrosion; (ii) excellent electronic conductivity to facilitate electron transfer during reactions; and (iii) a substantial surface area to accommodate a higher loading of metal species. Previous studies have identified certain oxides such as Sn-based and Mn-based materials as promising support candidates due to their excellent corrosion resistance under acidic OER conditions [48]. Additionally, lowering the metal's oxidation state has been found to contribute to enhanced stability. Based on this knowledge, Li et al. developed

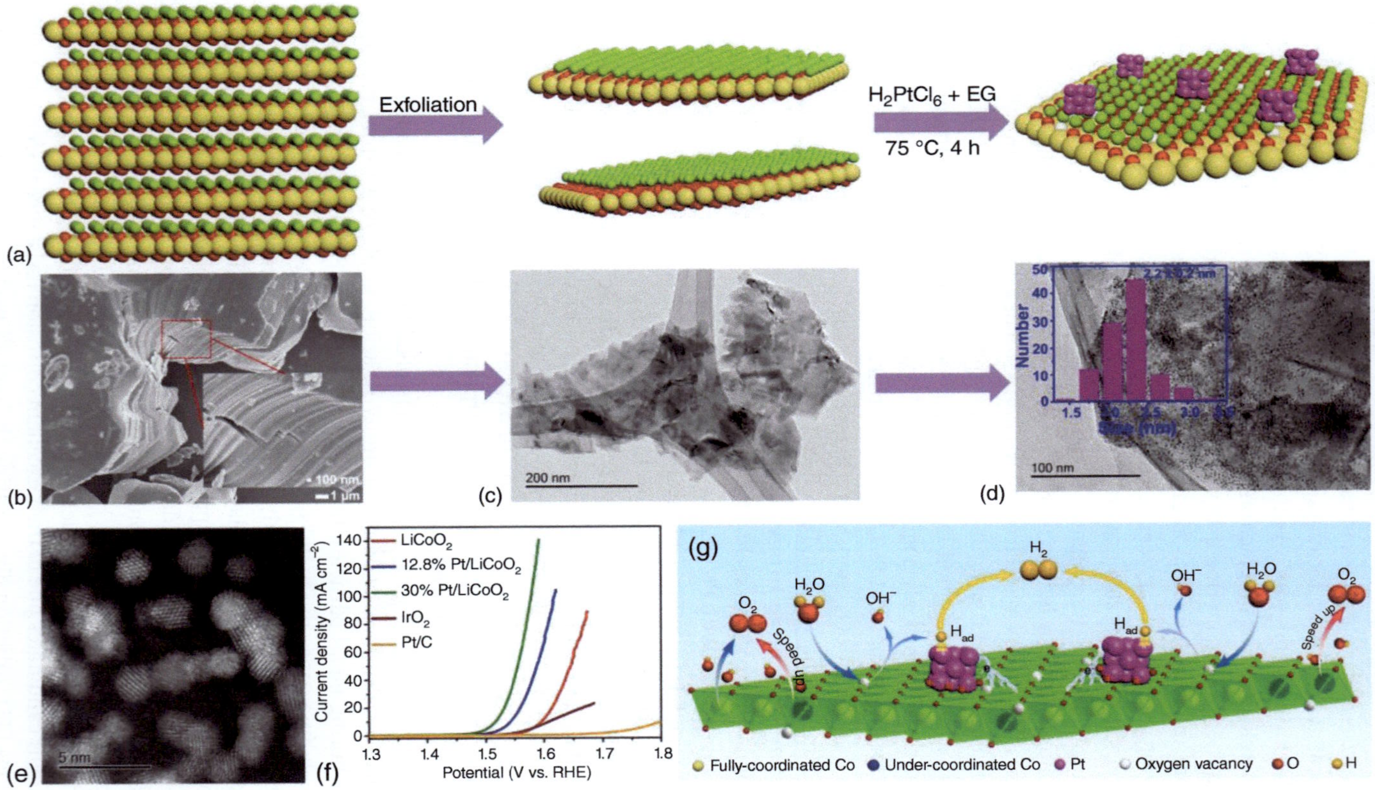

Figure 12.3 Schematic diagram of the synthesis process of Pt/$LiCoO_2$ heterostructure. (b) SEM image of bulk $LiCoO_2$. (c) SEM image of 2D $LiCoO_2$ nanosheets. (d) HRTEM, and (e) HAADF–STEM image of Pt/$LiCoO_2$. (f) LSV curves of $LiCoO_2$, 12.8% Pt/$LiCoO_2$, 30% Pt/$LiCoO_2$, IrO_2, and Pt/C. (g) Schematic illustration of the proposed reaction mechanisms. Source: Reproduced with permission of Zheng et al. [46], Wiley-VCH.

a heterostructured Ir–Sn pair-site catalyst (Ir–Sn PSC) comprising metallic Ir NPs anchored on antimony doped tin oxide (SSO) nanocrystalline for acidic OER [49]. Microscopic characterizations confirmed the formation of heterostructure. The generation of heterointerface could trigger the strong electronic interactions between Ir–Sn/Sb, which endow the Ir species with the decreased d-band holes during the fabrication and OER process. The optimized Ir–Sn PSC demonstrated outstanding performance and durability, owing to its distinctive heterostructured configuration. This configuration provides it with a favorable electronic structure and coordination environment, allowing for nearly optimal oxygen adsorption energetics. Furthermore, strong electronic interactions induced the generation of oxygen radicals, effectively restraining Ir overoxidation. Consequently, the PWMWE using this catalyst delivered excellent activity with only 1.711 V to reach 2 A cm^{-2} and minimal degradation during the long-term durability test. In addition, other Co/Mo oxide-based heterostructures have shown great potential in the acidic OER with enhanced cyclic durability [50, 51]. For instance, Lee et al. reported an Ir/MoO_3 heterostructure by electrospinning approach [52]. The strong metal–support interactions between Ir and MoO_3 substrate induced the electron-deficient metal surface, which resulted in enhanced OER activity and durability in an acidic solution. Therefore, it is expected that many novel metal oxide-supported nanoparticle materials will be designed and developed intelligently for acidic water electrooxidation.

12.2.3 Metal Oxide-Supported Single-Atom Materials

Single-atom materials have received substantial attention in recent years due to their high atomic utilization, mass activity, and selectivity [53–58]. They have achieved great success in many fields such as rechargeable batteries [59], electrocatalysis [60, 61], thermocatalysis [62], environmental remedy [63], and photocatalysis [64]. Metal oxide-supported single-atom materials, as a category of non-carbon-supported single-atom catalysts (SACs), have garnered exceptional attention since the inception of the single-atom catalyst concept. Owing to the high chemical and thermal stability of their oxide supports, metal oxide-supported SACs have played a pivotal role in numerous catalytic reactions, particularly in thermocatalysis. Furthermore, they have proven their significance in catalyzing the OER, particularly under challenging reaction conditions, where their chemical stability ensures the structural integrity and long-term performance of the catalyst throughout the OER process. Therefore, the development of novel metal oxide-supported SACs and the exploration of the underlying structure–activity relationships and reaction mechanisms are both vital and urgent pursuits.

Single-atom doping and surface modification to the metal oxides have been regarded as effective ways to construct metal oxide-supported SACs. Wang and co-workers constructed a Ru–Co/$LiCoO_2$ SAC by single-site Ru cation coordination engineering (Figure 12.4a) [65]. Due to the robust electronic coupling between Ru and Co sites, the catalyst exhibits improved electrical conductivity and attains nearly optimal oxygen adsorption energies. Moreover, the Ru–Co/$LiCoO_2$ delivered high mass activity (1000 A g_{oxide}^{-1}) and outstanding long-term durability. Also, it showed great potential in a zinc–air battery with high energy density.

Enhancing the mass loading of single metal atoms is of paramount significance for improving catalyst performance and practical applications. Gu and co-workers have developed a straightforward and efficient method for fabricating Ir–NiO SAC [66]. Significantly, they achieved an impressive Ni single-atom mass loading of up to 18% by weight. As a result, Ir–NiO demonstrated remarkable OER activity, showcasing a low overpotential of 215 mV at 10 mA cm^{-2}. This performance surpassed that of both NiO and commercial IrO_2. Theoretical calculations unveiled that the introduction of the Ir single-atom could modify the OER reaction mechanism and energy barrier (Figure 12.4d). The rate-determining step transitioned from *OH on the Ni(001) surface to *O to *OOH on the surface of Ir–NiO(001). The authors attributed the enhanced OER performance to the high density of active sites and the synergistic effects between Ir centers and the NiO support.

As mentioned above, some metal oxide substrates could exhibit excellent acid and oxidation resistance. Consequently, some metal oxide-supported SACs display commendable OER activity and stability in acidic media. What is even more intriguing is that when these metal single-atoms are introduced into oxide supports, it has the potential to trigger novel OER reaction pathways, leading to new reaction mechanisms. This, in turn, enhances the acidic OER activity and stability of the catalyst. Lee and co-workers developed an α-MnO_2-supported Ru-atom-array catalyst for acidic OER [68]. They found that the existence of Ru-atom-array could trigger the new reaction mechanism involving only *O and *OH intermediates, which is different from the adsorbate evolution mechanism (AEM). The new mechanism enables direct O–O coupling, and therefore endows the catalyst with enhanced OER activity and durability. Similarly, Ge and co-workers reported that confined Ir single-atoms in γ-MnO_2 could trigger the occurrence of a lattice-oxygen-mediated mechanism (Figure 12.4c) [67]. Notably, the locally triggered LOM would not result in the structural degradation of catalysts. As a result, the Ir–MnO_2 showed extremely high mass activity and outstanding stability with negligible decay after 650 hours operation (Figure 12.4d). As discussed above, metal oxide-supported SACs hold significant potential for practical applications in acidic OER and PEMWE. Therefore, further continued research efforts are urgently needed to realize their industrial applications.

12.3 Metal Hydroxides and Their Supported Single-Atom/Nanoparticle Materials

Metal hydroxides have attracted significant attention in the field of electrocatalysis recently, due to their distinctive structural and electronic properties [69]. Numerous metal hydroxides comprise readily available elements and are simple to synthesize, thus offering significant potential for large-scale applications. Moreover, previous studies have illustrated that metal hydroxides can act as a supporting framework for immobilizing nanoparticles and individual metal atoms. These

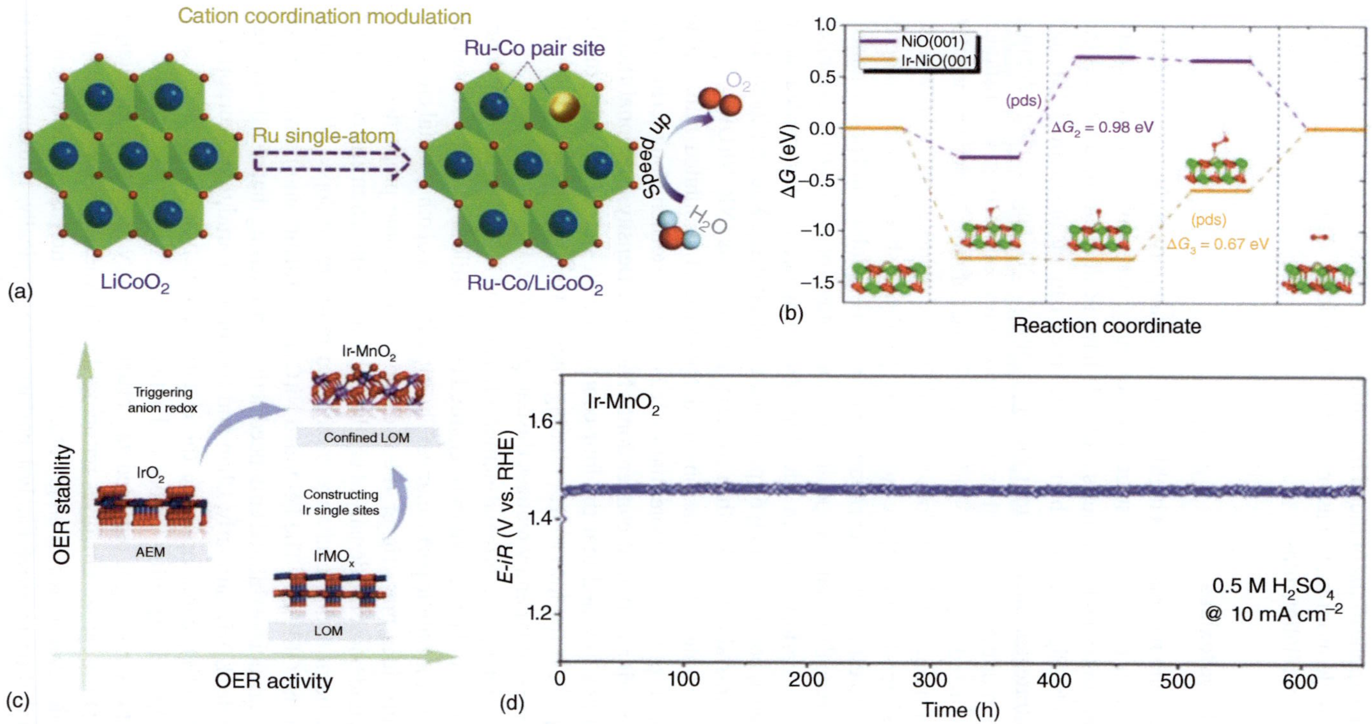

Figure 12.4 (a) Schematic of Ru–Co/LiCoO$_2$ SAC. Source: Reproduced with permission of Zheng et al. [65], Wiley-VCH. (b) Gibbs free energy diagram of OER for Ni(001) and Ir–NiO(001). (c) OER reaction mechanism on the Ir–MnO$_2$. Source: Reproduced with permission of Wang et al. [66], American Chemical Society. (d). Chronopotentiometric test of Ir–MnO$_2$ for OER at 10 mA cm^{-2} in 0.5 M H$_2$SO$_4$ solution. Source: Reproduced with permission of Shi et al. [67], Elsevier.

supported materials, featuring metal hydroxides as a host, exhibit enhanced performance in water electrooxidation. Therefore, in the following sections, we will illustrate the applications of metal hydroxides and metal hydroxides-supported nanoparticles/single-atom materials in the OER, shedding light on their intrinsic structure–function relationships.

12.3.1 Metal Hydroxides

Metal hydroxides are regarded as a highly promising category of electrocatalysts for alkaline OER. Their composition, structural flexibility, accessibility of raw materials, and simple, controllable synthesis methods have positioned them as a central focus of research. Significantly, their performance can be augmented through modification strategies such as morphology control, 2D engineering, doping, heterostructure modulation, as well as defect and strain engineering. For instance, Luan et al. conducted a systematic study on the impact of morphology on the OER performance of monometallic α-$Co(OH)_2$ [70]. They employed a lamellar reverse micelle synthesis method to prepare bud, flower, petal, and sheet-like morphologies of α-$Co(OH)_2$, and established a correlation between morphology and performance. Their findings revealed that the petal-like morphology of α-$Co(OH)_2$ exhibited the most superior performance. In addition, strain engineering is regarded as an effective strategy to boost the catalytic performance of catalysts. Xue and co-workers fabricated a strain-stabilized ultrathin $Ni(OH)_2$ nanoribbons (NR–$Ni(OH)_2$) by an electrooxidation strategy [71]. The obtained NR–$Ni(OH)_2$ exhibited remarkable OER activity with only an overpotential of 162 mV and excellent long-term durability, much superior to that of conventional $Ni(OH)_2$. Density functional theory (DFT) calculations unveiled that the four-coordinated Ni sites at the edge in the NR–$Ni(OH)_2$ possess the lower OH^- binding energy, which facilitates the OOH* adsorption process and thereby reduces the energy barrier.

The incorporation of various metal elements (e.g. Fe and Mn) with distinct valence states into monometallic hydroxides is regarded as a versatile approach for boosting the inherent electrocatalytic performance. The introduction of other metal elements into metal hydroxides may trigger the evolution of electronic properties, create defective sites, and boost the density of catalytically active sites. Sun and co-workers developed a general cation exchange method to fabricate Fe-doped Ni/$Co(OH)_2$ nanosheets for the OER [69]. The Fe-doped $Ni(OH)_2$ featuring nanoporous surface with enriched vacancies exhibited enhanced activity, surpassing that of NiFe layered double hydroxide (LDH). Similarly, Chen and co-workers found that incorporating Cu/Mn into $Co(OH)_2$ could optimize the electronic structure of Co catalytically active sites, ultimately resulting the boosted activity [72].

Dimensional engineering is regarded as a crucial strategy for enhancing catalyst performance. This approach involves reducing the thickness of the catalyst to the nanometer scale, even down to single atomic layers, to increase the exposure of active sites and thus improve catalyst performance. Typically, two-dimensional nanosheets are synthesized using both bottom-up and top-down methods. Hu and co-workers successfully synthesized ultrathin CoMn-LDH nanoplates through a one-pot coprecipitation method [73]. These CoMn-LDH nanoplates exhibited outstanding OER activity and durability in alkaline environments, comparable to the best Ni- or Fe-based catalysts. They also synthesized single-layer NiFe and NiCo LDH by liquid phase exfoliation strategy from their bulk materials [74]. Due to the much more exposed active sites and increased electronic conductivity, the single-layer nanosheets showed considerably enhanced OER activity and durability, outperforming that of their bulk materials and commercial IrO_2.

12.3.2 Metal Hydroxide-Supported Nanoparticle Materials

Many metal hydroxides feature a high ability to cleave HO—H bonds, and they could act as the co-catalyst to the metal nanoparticles for boosted water-splitting. Markovic et al. reported that $Ni(OH)_2$ could result in an eight times HER activity increase of Pt due to the significant role of $Ni(OH)_2$ in facilitating the water dissociation step [75]. In addition, the strong electronic interaction between metal hydroxide support and metal nanoparticles would alter the electronic and geometric effects, leading to better performance than the single components. Sun and co-workers constructed a novel $Ir/Ni(OH)_2$ heterostructure compromising of uniform Ir NPs embedded in the $Ni(OH)_2$ nanosheets for accelerated water oxidation (Figure 12.5a,b) [76]. They discovered the presence of electronic interactions between Ir and $Ni(OH)_2$, which favors the stabilization of the electrophilic Ir^{4+}. This, in turn, effectively promotes the formation of O—O bonds. What is even more intriguing is that this electronic interaction can break the scaling relationship, significantly reducing the reaction energy barrier and accelerating reaction kinetics. Consequently, the optimized $Ir/Ni(OH)_2$ exhibits remarkably high alkaline OER activity, requiring only a 270 mV overpotential to achieve a current density of 100 mA cm^{-2} in the 1 M KOH solution. Simultaneously, $Ir/Ni(OH)_2$ demonstrated outstanding cycling stability.

In addition to loading precious metals, various heterogeneous structures featuring nonprecious metals supported by metal hydroxides have been developed to reduce the cost of catalysts. For example, Zhang et al. employed a partial reduction method to create a 2D Ni(0)-embedded $Ni(OH)_2$ heterostructure for water-splitting [77]. It is worth noting that in this particular heterogeneous structure, distinct active sites are employed for the HER and OER. The presence of Ni(0)–Ni(II) active interfaces facilitates the occurrence of HER, while Ni(II) defects serve as active sites for OER. As a result, the $Ni/Ni(OH)_2$ catalyst exhibits outstanding bi-functionality for HER and OER, requiring only 77 and 270 mV of overpotential, to achieve a current density of 10 mA cm^{-2}, respectively. Moreover, the full cell composed of this catalyst demonstrates excellent performance, requiring just 1.59 V to achieve a current density of 10 mA cm^{-2}.

12.3.3 Metal Hydroxide-Supported Single-Atom Materials

Metal hydroxides possess a high specific surface area, a significant number of cation/anion defect sites, and Brønsted OH^- coordination sites, which endow them with a distinct advantage in anchoring single-metal-atom species. Furthermore,

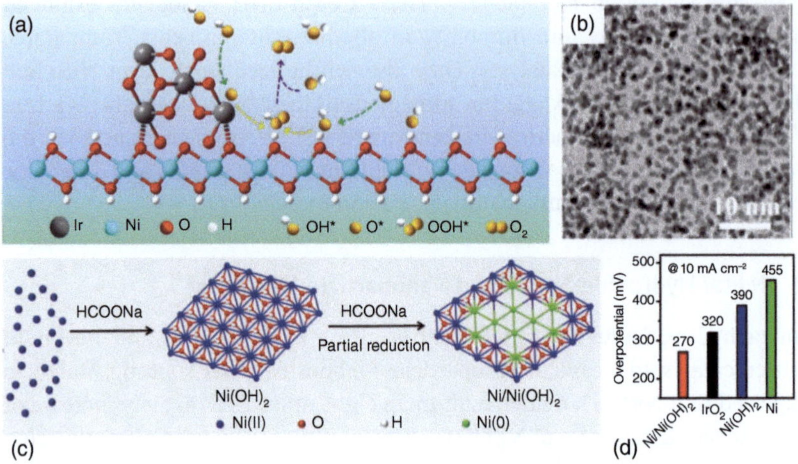

Figure 12.5 (a) Schematic illustration of the proposed reaction mechanism on the $Ir/Ni(OH)_2$. (b) TEM image of $Ir/Ni(OH)_2$. Source: Reproduced with permission of Zhao et al. [76], Wiley-VCH. (c) Schematic of the synthesis process of $Ni/Ni(OH)_2$. (d) The overpotential required for 10 mA cm^{-2} for $Ni/Ni(OH)_2$, $Ni(OH)_2$, Ni, and IrO_2. Source: Reproduced with permission of Dai et al. [77], Wiley-VCH.

the synergy achieved by combining metal hydroxides with single-atom materials in metal hydroxide-supported single-atom materials capitalizes on the strengths of both, resulting in a substantial improvement in material performance while simultaneously reducing catalyst costs. Consequently, metal hydroxide-supported single-atom materials are widely applied in catalytic reactions, particularly in alkaline OER. Hou and co-workers developed a defect-rich Ni–Fe layered double hydroxide nanosheets supported Ru single-atom site catalyst (Ru_1/D–NiFe LDH) for overall water-splitting by a facile electroposition and subsequent etching process [78]. Benefiting from the synergistic effects of defect and coordination environment modulation, Ru_1/D–NiFe LDH not only exhibited excellent HER activity (−18 mV @10 mA cm^{-2}) but also boosted OER activity with a small overpotential of 189 mV to reach 10 mA cm^{-2}. Inspired by the remarkable HER and OER activity, a full cell using the Ru_1/D–NiFe LDH as anode and cathode delivered a low potential to reach the current density of 100 mA cm^{-2} with only 1.54 V. Notably, it also showed good long-term stability without obvious decay after 100 hours operation at 100 mA cm^{-2}. Theoretical insights revealed that the unique configuration of Ru_1/D–NiFe LDH could not only optimize the H adsorption energy but also accelerate the O–O coupling to boost the oxygen gas evolution.

The industrial water electrolysis for hydrogen production takes place under high current conditions, posing significant challenges to the activity and stability of catalysts. For SACs, increasing the loading of individual metal atoms is a crucial approach to achieving high current density hydrogen evolution. Therefore, the surface area of the support and the density of metal anchoring sites on it becomes particularly important. Two-dimensional metal hydroxides with defects are well-suited for meeting this requirement, and therefore many metal hydroxides-supported single-atom materials have been designed. Additionally, Zeng and co-workers have developed a one-step and universally applicable electroposition strategy for creating high-density metal hydroxides-supported SACs [79]. The mass loading of Ir in the fabricated $Ir_1(CoOH)_2$ can reach 32 atom nm^{-2} based on the microscopic characterizations. Due to its high density of Ir sing-atom sites, the $Ir_1(CoOH)_2$ exhibited an outstanding OER activity with only 178 mV at 10 mA cm^{-2}. Surprisingly, it can realize the large current density of 2000 mA cm^{-2} under 1.65 V. Remarkably, an electrolysis cell assembled with $Pt_1(CoOH)_2$ and $Ir_1(CoOH)_2$ exhibited a high mass activity of 4.9 $mg_{noble\ metal}{}^{-1}$ at the voltage of 2 V, which was more than two orders of magnitudes of $Pt/C//IrO_2$. More importantly, the method can be largely applicable to synthesizing other supported SACs. Therefore, constructing high-density metal hydroxides supportedhydroxides-supported SACs can be an effective approach to achieving the practical application of water electrolyzers.

12.4 Conclusion and Perspective

Hydrogen energy is an ideal source of energy and plays a crucial role in achieving carbon neutrality. Water electrolysis has been considered an essential technique for producing green hydrogen. However, its relatively high hydrogen production cost significantly hinders its large-scale application. Among these costs, the OER accounts for over 80% of electricity consumption. Therefore, there is a pressing need to develop high-performance, cost-effective, and durable OER catalysts. Various metallic compounds have been developed and applied in OER, leading to significant advancements.

In this chapter, we mainly focus on the applications of metal oxides and metal hydroxides, as well as their supported single-atom/nanoparticle materials, in the OER. Through illustrative examples, we aim to elucidate their structure–performance relationships, providing researchers with fresh perspectives for designing novel and advanced OER catalysts.

Despite significant advancements in the field of OER with metallic compounds, there still exist numerous challenges and opportunities. First, developing scalable and controllable synthesis approaches for the commercialization of metallic compound catalysts remains a formidable scientific challenge. Many synthesis methods exhibit significant limitations, with lengthy processes and low yields, resulting in high production costs and constraints on large-scale applications. Therefore, the development of one-step and universal approaches such as electrodeposition methods is of paramount importance. Additionally, integrating machine learning and high-throughput calculations to accelerate the catalyst screening process holds great promise [80]. This approach can help circumvent extensive trial-and-error synthesis methods, significantly reducing synthesis costs and shortening catalyst development cycles.

Second, the majority of metallic compounds are primarily applied in alkaline OER, with relatively limited research in acidic OER, especially in nonprecious metal-based compounds. This significantly hinders the large-scale practical utilization of PEMWE. Hence, there is a need to develop many more cost-effective compounds that exhibit high activity and stability in highly acidic and oxidative environments. In addition, for Ru-based oxides, doping with other transition metal elements is an effective means of enhancing catalyst stability. In the case of nonprecious metal oxides and similar materials, selecting appropriate metal oxides to construct heterostructured architecture might contribute to boosting the material's corrosion resistance under acidic OER conditions.

Last, to identify the underlying OER mechanisms and the dynamic evolution of electronic and coordination structures, it is imperative to develop more advanced *Operando* characterization techniques such as Operando Raman spectroscopy and X-ray absorption spectroscopy. These advanced characterization methods can identify the real active sites and track their dynamic evolution during synthesis and reaction processes. This will provide essential assurance for revealing catalytic mechanisms and understanding structure–performance relationships, ultimately offering fundamental guidance for the rational design of advanced OER catalysts. Despite these challenges, it is anticipated that significant breakthroughs will be realized in the rational design, characterization, and mechanistic understanding of OER catalysts through persistent and unwavering efforts.

Acknowledgments

This work was supported by the National Natural Science Foundation of China (22305101), the Natural Science Foundation of Jiangsu Province (BK20231032), and Fundamental Research Funds for the Central Universities (JUSRP123020).

References

1 Yang, Y., Li, P., Zheng, X. et al. (2022). Anion-exchange membrane water electrolyzers and fuel cells. *Chem. Soc. Rev.* 51 (23): 9620.

2 Li, L., Wang, P., Shao, Q. et al. (2020). Metallic nanostructures with low dimensionality for electrochemical water splitting. *Chem. Soc. Rev.* 49 (10): 3072.

3 Liu, S. and Shui, J. (2022). Mechanism and properties of emerging nanostructured hydrogen storage materials. *Battery Energy* 1 (4): 20220033.

4 Jiao, S., Fu, X., Wang, S., and Zhao, Y. (2021). Perfecting electrocatalysts via imperfections: towards the large-scale deployment of water electrolysis technology. *Energy Environ. Sci.* 14 (4): 1722.

5 Lagadec, M.F. and Grimaud, A. (2020). Water electrolysers with closed and open electrochemical systems. *Nat. Mater.* 19: 1140.

6 Wang, T., Cao, X., and Jiao, L. (2021). Ni_2P/NiMoP heterostructure as a bifunctional electrocatalyst for energy-saving hydrogen production. *eScience* 1 (1): 69.

7 Wu, X., Wang, Y., and Wu, Z. (2022). Design principle of electrocatalysts for the electrooxidation of organics. *Chem* 8 (10): 2594.

8 Wang, Y., Zheng, X., and Wang, D. (2022). Design concept for electrocatalysts. *Nano Res.* 15: 1730.

9 Cai, S., An, Y., Feng, Y. et al. (2023). Recent progress in porous carbon-supported materials as efficient oxygen electrocatalysts for zinc-air batteries. *Sci. China Mater.* 66: 3381.

10 Jiang, S., Song, H., Zheng, X. et al. (2022). Lightest metal leads to big change: lithium-mediated metal oxides for oxygen evolution reaction. *Adv. Energy Mater.* 12 (33): 2201934.

11 Zu, L., Qian, X., Zhao, S. et al. (2022). Self-assembly of Ir-based nanosheets with ordered interlayer space for enhanced electrocatalytic water oxidation. *J. Am. Chem. Soc.* 144 (5): 2208.

12 Huang, L., He, Z., Guo, J. et al. (2020). Photodeposition fabrication of hierarchical layered Co-doped Ni oxyhydroxide ($Ni_xCo_{1-x}OOH$) catalysts with enhanced electrocatalytic performance for oxygen evolution reaction. *Nano Res.* 13: 246.

13 She, Z.W., Kibsgaard, J., Dickens, C.F. et al. (2017). Combining theory and experiment in electrocatalysis: insights into materials design. *Science* 355 (6321): 146.

14 Jiang, Z., Song, S., Zheng, X. et al. (2022). Lattice strain and schottky junction dual regulation boosts ultrafine ruthenium nanoparticles anchored on a N-modified carbon catalyst for H_2 production. *J. Am. Chem. Soc.* 144 (42): 19619.

15 Zhang, E., Tao, L., An, J. et al. (2022). Engineering the local atomic environments of indium single-atom catalysts for efficient electrochemical production of hydrogen peroxide. *Angew. Chem. Int. Ed.* 61 (12): e202117347.

16 Wang, B., Wang, C., Yu, X. et al. (2022). General synthesis of high-entropy alloy and ceramic nanoparticles in nanoseconds. *Nature Synthesis* 1: 138.

17 Jaramillo, T.F., Jorgensen, K.P., Bonde, J. et al. (2007). Identification of active edge sites for electrochemical H_2 evolution from MoS_2 nanocatalysts. *Science* 317: 100.

18 Zhang, J., Zhao, Y., Guo, X. et al. (2018). Single platinum atoms immobilized on an MXene as an efficient catalyst for the hydrogen evolution reaction. *Nat. Catal.* 1: 985.

19 Song, H., Wu, M., Tang, Z. et al. (2021). Single atom ruthenium-doped CoP/CDs nanosheets via splicing of carbon-dots for robust hydrogen production. *Angew. Chem. Int Ed.* 60 (13): 7234.

20 Zhang, J., Liu, J., Xi, X. et al. (2021). Single-atom Au/NiFe Layered double hydroxide electrocatalyst: probing the origin of activity for oxygen evolution reaction. *J. Am. Chem. Soc.* 140 (11): 3876.

21 Dang, Q., Lin, H., Fan, J. et al. (2021). Iridium metallene oxide for acidic oxygen evolution catalysis. *Nat. Commun.* 12: 6007.

22 Yang, J., Li, W., Wang, D., and Li, Y. (2020). Electronic metal-support interaction of single-atom catalysts and applications in electrocatalysis. *Adv. Mater.* 32 (49): e2003300.

23 Li, W., Yang, J., and Wang, D. (2022). Long-range interactions in diatomic catalysts boosting electrocatalysis. *Angew. Chem. Int. Ed.* 61 (52): e202213318.

24 Rossmeisl, J., Qu, Z., Zhu, H. et al. (2007). Electrolysis of water on oxide surfaces. *J. Electroanal. Chem.* 607: 83.

25 Wu, Z., Chen, F., Li, B. et al. (2023). Non-iridium-based electrocatalyst for durable acidic oxygen evolution reaction in proton exchange membrane water electrolysis. *Nat. Mater.* 22: 100.

26 Jin, H., Choi, S., Bang, G. et al. (2022). Safeguarding the RuO_2 phase against lattice oxygen oxidation during acidic water electrooxidation. *Energy Environ. Sci.* 15 (3): 1119.

27 Zhao, Z., Wang, Q., Huang, X. et al. (2020). Boosting the oxygen evolution reaction using defect-rich ultra-thin ruthenium oxide nanosheets in acidic media. *Energy Environ. Sci.* 13 (12): 5143.

28 Liu, H., Zhang, Z., Fang, Z. et al. (2023). Eliminating over-oxidation of ruthenium oxides by niobium for highly stable electrocatalytic oxygen evolution in acidic media. *Joule* 7 (3): 558.

29 Shi, Z., Wang, Y., Liu, S. et al. (2023). Customized reaction route for ruthenium oxide towards stabilized water oxidation in high-performance PEM electrolyzers. *Nat. Commun.* 14: 843.

30 Wang, Y., Yang, R., Liu, Y. et al. (2023). Unraveling oxygen vacancy site mechanism of Rh-doped RuO_2 catalyst for long-lasting acidic water oxidation. *Nat. Commun.* 14: 1412.

31 Zheng, Y., Vernieres, J., Wang, Z. et al. (2021). Monitoring oxygen production on mass-selected iridium–tantalum oxide electrocatalysts. *Nat. Energy* 7: 55.

32 Zhao, F., Wen, B., Liu, W. et al. (2021). Increasing iridium oxide activity for the oxygen evolution reaction with hafnium modification. *J. Am. Chem. Soc.* 143 (38): 15616.

33 Hao, S., Sheng, H., Liu, M. et al. (2021). Torsion strained iridium oxide for efficient acidic water oxidation in proton exchange membrane electrolyzers. *Nat. Nanotechnol.* 16: 1371.

34 Fan, Z., Li, Y., Shao, Q. et al. (2021). Extraordinary acidic oxygen evolution on new phase 3R-iridium oxide. *Joule* 5 (12): 3221.

35 Zhao, Q., Yan, Z., and Chen, C. (2017). Spinels: controlled preparation, oxygen reduction/evolution reaction application, and beyond. *Chem. Rev.* 117 (15): 10121.

36 Nam-In, K., Young, J., Tae, S. et al. (2018). Oxygen-deficient triple perovskites as highly active and durable bifunctional electrocatalysts for oxygen electrode reactions. *Sci. Adv.* 4 (6): eaap9360.

37 Zheng, X., Chen, Y., Lai, W. et al. (2022). Enriched d-band holes enabling fast oxygen evolution kinetics on atomic-layered defect-rich lithium cobalt oxide nanosheets. *Adv. Funct. Mater.* 32 (28): 2200663.

38 Zheng, X., Chen, Y., Zheng, X. et al. (2019). Electronic structure engineering of $LiCoO_2$ toward enhanced oxygen electrocatalysis. *Adv. Energy Mater.* 9 (12): 1803482.

39 Su, H., Yang, C., Liu, M. et al. (2024). Tensile straining of iridium sites in manganese oxides for proton-exchange membrane water electrolysers. *Nat. Commun.* 15: 95.

40 Li, A., Kong, S., Guo, C. et al. (2022). Enhancing the stability of cobalt spinel oxide towards sustainable oxygen evolution in acid. *Nat. Catal.* 5: 109.

41 Zhang, Q., Zhang, X., Wang, J., and Wang, C. (2021). Graphene-supported single-atom catalysts and applications in electrocatalysis. *Nanotechnology* 32: 032001.

42 Wang, W., Wang, Z., Hu, Y. et al. A potential-driven switch of activity promotion mode for the oxygen evolution reaction at Co_3O_4/NiO_xH_y interface. *eScience* 2: 438.

43 Chen, H., Yang, Z., Wang, X. et al. (2021). Photoinduced strong metal-support interaction for enhanced catalysis. *J. Am. Chem. Soc.* 143 (23): 8521.

44 Chen, J., Cui, P., Zhao, G. et al. (2019). Low-coordinate iridium oxide confined on graphitic carbon nitride for highly efficient oxygen evolution. *Angew. Chem. Int. Ed.* 58 (36): 12540.

45 Chen, Y., Zhou, Q., Zhao, G. et al. (2018). Electrochemically inert g-C_3N_4 promotes water oxidation catalysis. *Adv. Funct. Mater.* 28 (5): 1705583.

46 Zheng, X., Cui, P., Qian, Y. et al. (2020). Multifunctional active-center-transferable platinum/lithium cobalt oxide heterostructured electrocatalysts towards superior water splitting. *Angew. Chem. Int. Ed.* 59 (34): 14533.

47 Oh, H.S., Nong, H., Reier, T. et al. (2016). Electrochemical catalyst-support effects and their stabilizing role for IrO_x nanoparticle catalysts during the oxygen evolution reaction. *J. Am. Chem. Soc.* 138 (38): 12552.

48 Oh, H.-S., Nong, H.N., and Strasser, P. (2015). Preparation of mesoporous Sb-, F-, and In-doped SnO_2 bulk powder with high surface area for use as catalyst supports in electrolytic cells. *Adv. Funct. Mater.* 25 (7): 1074.

49 Zheng, X., Yang, J., Li, P. et al. (2023). Ir-Sn pair-site triggers key oxygen radical intermediate for efficient acidic water oxidation. *Sci. Adv.* 9 (42): eadi8025.

50 Huang, J., Sheng, H., Ross, R. et al. (2021). Modifying redox properties and local bonding of Co_3O_4 by CeO_2 enhances oxygen evolution catalysis in acid. *Nat. Commun.* 12: 3036.

51 Du, K., Zhang, L., Shan, J. et al. (2022). Interface engineering breaks both stability and activity limits of RuO2 for sustainable water oxidation. *Nat. Commun.* 13: 5448.

52 Liu, X., Xi, S., Kim, H. et al. (2021). Restructuring highly electron-deficient metal-metal oxides for boosting stability in acidic oxygen evolution reaction. *Nat. Commun.* 12: 5676.

53 Zheng, X., Li, B., Wang, Q. et al. (2022). Emerging low-nuclearity supported metal catalysts with atomic level precision for efficient heterogeneous catalysis. *Nano Res.* 15: 7806.

54 Zheng, X., Yang, J., Li, P. et al. (2023). Dual-atom support boosts nickel-catalyzed urea electrooxidation. *Angew. Chem. Int. Ed.* 62 (22): e202217449.

55 Zheng, X., Yang, J., and Wang, D. (2023). Advanced dual-atom catalysts for efficient oxygen evolution reaction. *EES Catal.* 1: 665.

56 Ji, J., Hou, Y., Zhou, S. et al. (2022). Oxygen-coordinated low-nucleus cluster catalysts for enhanced electrocatalytic water oxidation. *Carbon Energy* 5 (2).

57 Shu, C., Tan, Q., Deng, C. et al. (2021). Hierarchically mesoporous carbon spheres coated with a single atomic Fe–N–C layer for balancing activity and mass transfer in fuel cells. *Carbon Energy* 4 (1): 1.

58 Wang, L., Wang, D., and Li, Y. (2022). Single-atom catalysis for carbon neutrality. *Carbon Energy* 4: 1021.

59 Zhang, E., Hu, X., Meng, L. et al. (2022). Single-atom yttrium engineering Janus electrode for rechargeable Na-S batteries. *J. Am. Chem. Soc.* 144 (41): 18995.

60 Li, X., Liu, L., Ren, X. et al. (2020). Microenvironment modulation of single-atom catalysts and their roles in electrochemical energy conversion. *Sci. Adv.* 6: eabb6833.

61 Su, H., Soldatov, M.A., Roldugin, V., and Liu, Q. (2022). Platinum single-atom catalyst with self-adjustable valence state for large-current-density acidic water oxidation. *eScience* 2: 102.

62 Wang, Q., Zheng, X., Wu, J. et al. (2022). Recent progress in thermal conversion of CO_2 via single-atom site catalysis. *Small Structures* 3 (9): 2200059.

63 Shang, Y., Xu, X., Gao, B. et al. (2021). Single-atom catalysis in advanced oxidation processes for environmental remediation. *Chem. Soc. Rev.* 50 (8): 5281.

64 Gao, C., Low, J., Long, R. et al. (2020). Heterogeneous single-atom photocatalysts: fundamentals and applications. *Chem. Rev.* 120 (21): 12175.

65 Zheng, X., Yang, J., Xu, Z. et al. (2022). Ru-Co pair sites catalyst boosts the energetics for the oxygen evolution reaction. *Angew. Chem. Int. Ed.* 134 (32): e202205946.

66 Wang, Q., Huang, X., Zhao, J. et al. (2020). Ultrahigh-loading of Ir single atoms on NiO matrix to dramatically enhance oxygen evolution reaction. *J. Am. Chem. Soc.* 142 (12): 7425.

67 Shi, Z., Wang, Y., Li, J. et al. (2021). Confined Ir single sites with triggered lattice oxygen redox: toward boosted and sustained water oxidation catalysis. *Joule* 5: 2164.

68 Lin, C., Li, J., Li, X. et al. (2021). In-situ reconstructed Ru atom array on α-MnO_2 with enhanced performance for acidic water oxidation. *Nat. Catal.* 4: 1012.

69 Zhou, Q., Chen, Y., Zhao, G. et al. (2018). Active-site-enriched iron-doped nickel/cobalt hydroxide nanosheets for enhanced oxygen evolution reaction. *ACS Catal.* 8 (6): 5382.

70 Luan, C., Liu, G., Liu, Y. et al. (2018). Structure effects of 2D materials on alpha-nickel hydroxide for oxygen evolution reaction. *Acs Nano* 12 (4): 3875.

71 Wang, X., Wu, H., Xi, S. et al. (2020). Strain stabilized nickel hydroxide nanoribbons for efficient water splitting. *Energy Environ. Sci.* 13 (1): 229.

72 Chen, L., Zhang, H., Chen, L. et al. (2017). Facile synthesis of Cu doped cobalt hydroxide (Cu–$Co(OH)_2$) nano-sheets for efficient electrocatalytic oxygen evolution. *J. Mater. Chem. A* 5 (43): 22568.

73 Song, F. and Hu, X. (2014). Ultrathin cobalt-manganese layered double hydroxide is an efficient oxygen evolution catalyst. *J. Am. Chem. Soc.* 136 (47): 16481.

74 Song, F. and Hu, X. (2014). Exfoliation of layered double hydroxides for enhanced oxygen evolution catalysis. *Nat. Commun.* 5: 4477.

75 Subbaraman, R., Dusan, T., Strmcnik, D. et al. (2014). Enhancing hydrogen evolution activity in water splitting by tailoring Li^{+}-$Ni(OH)_2$-Pt interfaces. *Science* 334: 1256.

76 Zhao, G., Li, P., Cheng, N. et al. (2020). An Ir/$Ni(OH)_2$ heterostructured electrocatalyst for the oxygen evolution reaction: breaking the scaling relation, stabilizing iridium(V), and beyond. *Adv. Mater.* 32 (24): e2000872.

77 Dai, L., Chen, Z., Li, L. et al. (2020). Ultrathin Ni(0)-embedded $Ni(OH)_2$ heterostructured nanosheets with enhanced electrochemical overall water splitting. *Adv. Mater.* 32 (8): e1906915.

78 Zhai, P., Xia, M., Wu, Y. et al. (2021). Engineering single-atomic ruthenium catalytic sites on defective nickel-iron layered double hydroxide for overall water splitting. *Nat. Commun.* 12: 4587.

79 Cao, D., Zhang, Z., Cui, Y. et al. (2022). One-step approach for constructing high density single-atom catalysts toward overall water splitting at industrial current densities. *Angew. Chem. Int .Ed.* 62 (9): e202214259.

80 Liu, T., Sang, J., Li, H. et al. (2022). Towards understanding of CO_2 electroreduction to C_{2+} products on copper-based catalysts. *Battery Energy* 1: 20220012.

13

Porous Materials for Electrocatalytic Oxygen Evolution Reaction

Zhongjie Yang and Danfei Fu

Guizhou Normal University, School of Chemistry and Materials, Universities Town, Huaxi District, Guiyang, 550025, China

13.1 Introduction

To address a range of environmental concerns and the worldwide energy crisis resulting from the burning and depletion of fossil fuels, there is growing interest in metal–air batteries, fuel cells, and water-splitting [1–3]. These technologies are gaining prominence in emerging fields of energy conversion and storage due to their exceptional efficiency, ability to enhance energy security, and positive impact on the environment. It is imperative to highlight that devices used for energy conversion and storage involve several vital electrochemical processes that enable the conversion of chemical energy into electrical energy and vice versa. A prominent example of such processes is the oxygen evolution reaction (OER), involving the oxidation of water or hydroxyl ions on the surfaces of electrocatalysts, resulting in the concurrent production of gaseous oxygen [4, 5]. However, OER follows a four-electron transfer mechanism while also breaking forming the O—O bond. This kinetic process is characterized by its slow nature and requires a significant overpotential to overcome the energy barrier [6]. Therefore, it has become imperative to explore novel electrocatalysts for OER to surmount technological obstacles and meet the demand for extensive device manufacturing.

Porous materials, mainly metal–organic frameworks (MOFs), covalent–organic frameworks (COFs), and derivatives, have roles in established catalytic processes and emerging energy storage and production technologies [7]. Particularly, these porous materials have garnered considerable attention as competitive electrocatalysts for OER due to their expansive specific surface area and optimization of active sites encompassing not only density and accessibility but also intrinsic activity [8].

The OER electrochemical activity of MOFs is enhanced by the presence of redox-active metal centers, typically transition metals such as Fe, Co, and Ni [9]. The periodic structures of MOFs ensure a uniform dispersion of abundant active sites. Additionally, the well-defined porous structures of MOFs facilitate easier access to these active sites, thereby enhancing their electrocatalytic performances

Functional Materials for Electrocatalytic Energy Conversion, First Edition.
Edited by Zhicheng Zhang, Meiting Zhao, and Yuchen Qin.

[10–12]. Moreover, the surface characteristics of pristine MOFs can be altered by integrating specific functional groups (such as $-NH_2$, $-OH$, etc.) into organic ligands or employing modified organic ligands with coordinative bonds involving nitrogen or oxygen atoms within metal nodes [13–15]. This modification enhances the exposure of densely packed active sites in the frameworks, resulting in enhanced OER properties. Further, the electrocatalytic properties can be enhanced by modifying the structure of as-prepared MOFs, offering an alternative approach. Instead of utilizing 3D bulk MOFs, transforming them into 2D ultrathin nanosheets is considered a viable strategy [16–18]. This transformation facilitates efficient mass transport and electron transfer across the exposed catalytic active surface. Lastly, an encouraging method to augment the intrinsic electrocatalytic activity of pure MOFs involves their combination with other functional materials, yielding active composites [19]. These composites harness the synergistic effects of both the MOFs and functional species, while also exhibiting unique structures that promote efficient electron transfer [20].

Similarly, COFs are a novel class of framework materials interconnected by robust covalent bonds [21]. These materials exhibit exceptional crystallinity and possess expansive surface areas, along with adjustable pore sizes, thereby facilitating the development of highly organized network structures with outstanding mass transport capabilities. Moreover, COFs demonstrate tremendous potential in terms of precise control over metal or nonmetal heteroatom doping and the strategic positioning of electrochemically active sites [22, 23]. By judiciously selecting appropriate building blocks, linkage motifs, and synthesis routes, it becomes feasible to synthesize a programmable COF electrocatalyst [24]. Based on the distinctive physical and chemical properties of COFs mentioned above, to enhance the OER performance of COFs, the construction strategies for COF-based catalysts primarily involve: (i) Introducing heteroatoms (such as B, N, O, S atoms) into the COFs structure to regulate the electronic structure of carbon active sites within the framework, thereby augmenting OER performance [21]. (ii) Presynthetic modification is commonly employed as a bottom-up approach for the synthesis of hybrid materials based on metal-doped pristine COFs. Precoordinating metal ions to specific sites in the building units or linkers, which do not hinder the formation of COFs during the reaction, enables one-step synthesis [25–27]. (iii) Efforts have been dedicated to the exploration of alternative hybrids based on COF materials, which can be effectively employed for electrocatalysis or potential electrocatalytic applications. A promising strategy involves the utilization of porous carbons derived from organic polymers as a supportive matrix, enabling their combination with electrochemically active components such as metal oxide (MO) or hydroxide species [28, 29].

The primary purpose of this chapter is to present a comprehensive overview of recent advancements and strategies in the field of catalysts for OER based on MOFs and COFs. This includes an examination of the performance and structure aspects, encompassing pristine MOF/COF catalysts, composite catalysts incorporating MOF/COF materials, as well as derivative catalysts derived from MOFs/COFs. The synthesis methods, the correlation between structure and performance, and the underlying catalytic mechanisms governing OER in these MOF/COF-based catalysts are thoroughly discussed. Finally, we address both the opportunities and challenges associated with utilizing MOF/COF-based catalysts for electrocatalytic OER.

13.2 Metal–Organic Frameworks (MOFs) for OER

13.2.1 Pristine MOFs for OER

Pristine MOFs possess a porous structure, redox-active metal centers, and functionalized organic ligands, which facilitate mass and electron transfer. Enhancing the OER activities of pristine MOFs can be achieved through strategies such as mixed-metal node manipulation, modification of ligands, and engineering of the structure.

13.2.1.1 Mixed-Metal Node Engineering

The incorporation of mixed-metal nodes within pristine MOFs as electrocatalysts for OER holds promise in enhancing intrinsic properties by leveraging synergies between different metals and modifying the electronic framework at active sites [30]. Notably, 3D transition metal ions (such as Fe^{3+}, Ni^{2+}, Co^{2+}, Mn^{2+}, and Cu^{2+}) find common use in crafting bi-/tri-/ or high-entropy metallic MOFs, resulting in heightened electroactivity [31, 32]. Recently, a group of researchers led by Zhou successfully synthesized a range of Co/Ni–MOFs with molecules using the solvothermal method [33]. They optimized the hierarchical nanobelts structure of Co/Ni–MOFs (CTGU-10) and observed remarkable performance in the OER. Specifically, they achieved a low overpotential at a current density of $10\,mA\,cm^{-2}$, which can be attributed to the unsaturated metal sites and the synergistic effect between Ni and Co. Following that, Li et al. conducted a study where they synthesized a range of tri-metallic MIL-53 structures with precisely controlled Fe/Ni/Co molar ratios [34]. The introduction of a third metal ion (Co^{2+}) in varying proportions did not compromise the exceptional stability exhibited by the FeNi–MIL-53 structure. Among these tri-metallic variants, the optimized $Fe/Ni_{2.4}/Co_{0.4}$-MIL-53 demonstrated superior OER activity with an $\eta_j = 10\,mA\,cm^{-2}$ value of 219 mV, surpassing even the bimetallic FeNi–MIL-53 counterparts. Lastly, Zhao and colleagues reported on the synthesis of high nanosheets using a mixed solution containing five different metal ions (Fe^{3+}, Co^{2+}, Ni^{2+}, Mn^{2+}, and Cu^{2+}) along with 1,4-bezendicarboxylic acid [35]. These nanosheets displayed remarkable electrocatalytic OER activity characterized by a $\eta_j = 10\,mA\,cm^{-2}$ value of 245 mV and an impressively low Tafel slope of $54\,mV\,dec^{-1}$, outperforming the aforementioned MOFs. While numerous MOFs with pristine mixed-metal nodes have been synthesized with success, the progress in creating effective electrocatalysts for OER is significantly hindered by challenges such as imprecise control over mixing ratios and uneven distribution of metals.

13.2.1.2 Ligand-Based Modification Engineering

The surface properties of pristine MOFs can be modified by incorporating specific functional groups (such as $-NH_2$ and –OH) into organic ligands or utilizing modified organic ligands with nitrogen or oxygen atoms coordinative bonds within metal

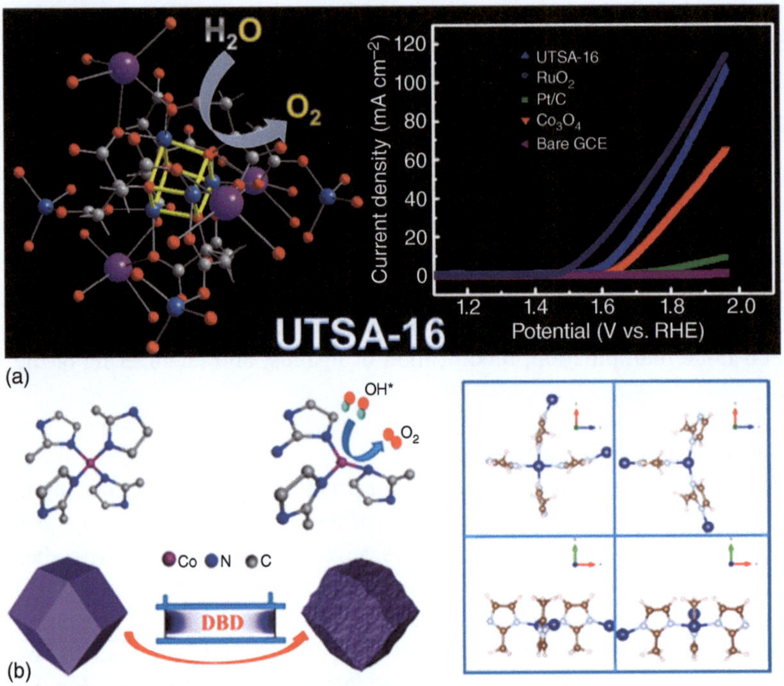

Figure 13.1 (a) The representative structures of Co_4O_4 cubane in the UTSA-16, and OER catalytic performances. Source: Reproduced with permission of Jiang et al. [36], American Chemical Society. (b) The preparation and OER catalytic performances of CUMSs-ZIF-67. Source: Reproduced with permission of Tao et al. [37], Elsevier.

nodes [13–15]. This modification enhances the exposure of high-density active sites in the frameworks, leading to improved OER properties. For instance, Jiang et al. reported a novel cobalt–citrate MOF where citrate ligands chelate octahedral cobalt atoms, resulting in a unique structure that exhibits exceptional catalytic activity for OER in an alkaline environment (Figure 13.1a) [36]. Another effective approach to activate pristine MOFs through ligand modification involves the selective extraction of specific organic ligands from the frameworks while preserving their original structures and compositions. This extraction process generates vacancies, leading to increased pore volumes that enhance mass transfer [38]. Furthermore, by eliminating organic ligands that coordinate with metal ions, a greater number of coordinatively unsaturated metal sites are exposed [39]. These exposed sites can then function as active sites for reactant adsorption and facilitate the OER process [40]. Wang et al. initially produced active ZIF-67 by subjecting it to dielectric barrier discharge plasma treatment, which preserved its structure and increased the number of coordinately unsaturated metal sites [37]. The plasma etching process resulted in the partial removal of imidazole ligands, leading to a higher abundance of CUMSs. Compared to the pristine ZIF-67, the active CUMS-67 demonstrated superior performance in the OER due to factors such as its larger surface area, presence of mesopores, and even distribution of CUMSs. Additionally, the positively charged unsaturated metal sites facilitated strong adsorption of O_2, O, OH, and OOH species, thereby enhancing the efficiency of the OER process (Figure 13.1b).

13.2.1.3 Structure Engineering

The enhancement of electrocatalytic properties can be achieved by modifying the structure of as-prepared MOFs, which provides an alternative approach [16–18]. Instead of using 3D bulk MOFs, transforming them into 2D ultrathin nanosheets is considered a viable strategy. This transformation facilitates rapid mass transport and electron transfer across the exposed catalytic active surface. To prepare 2D MOFs, various methods such as solvothermal and exfoliation techniques are extensively employed [16]. The authors, Li et al., utilized a solvothermal technique to synthesize ultrathin 2D heterometallic Ni–M–MOF (M = Fe, Al, Co, Mn, Zn, and Cd) nanosheets with extremely reduced thicknesses [41]. The resulting ultrathin Ni–Fe–MOF NSs exhibited exceptional electrocatalytic alkaline solution: a mere overpotential of 221 mV at a current density of $10\,mA\,cm^{-2}$ and a minimal Tafel slope of $56\,mV\,dec^{-1}$. Furthermore, these nanosheets displayed outstanding stability without any noticeable activity decay for at least 20 hours. Tang and colleagues employed ultrasonic synthesis to create a range of bimetallic NiCo–MOF nanosheets (NiCo–UMOFN) with metal sites that coordinated [42]. The optimized NiCo–UMOFN exhibited impressive activity in the OER, surpassing even RuO_2, with a low overpotential of 189 mV at a current density of $10\,mA\,cm^{-2}$. Furthermore, Huang et al. demonstrated an electrochemical exfoliation method for producing two-dimensional MOFs by utilizing 3D MOFs with redox-active backbones for oxidation and selective removal during OER tests at pH 13, they successfully prepared ultrathin MOF nanosheets (2D–Co–NS) [17]. Remarkably, these resulting 2D–Co–NS displayed exceptional performance as an electrocatalyst for the OER, exhibiting an impressively low overpotential of only 211 mV at a current density of $10\,mA\,cm^{-2}$.

13.2.2 MOF Composites for OER

An effective approach to heighten the intrinsic electrocatalytic activity of pure MOFs involves their fusion with other functional materials to produce active MOF composites. These composites capitalize on the synergies between the MOFs and functional elements, boasting distinct structures that facilitate efficient electron transfer [43, 44]. One strategy integrates pristine MOFs with supportive materials, like conductive substrates and porous substances, utilizing MOFs as supports or templates for their growth. Furthermore, incorporating diverse functional MOF materials into composite structures holds promise for further enhancing OER performance. Another impactful method to augment electrocatalytic activity is combining pure MOFs with OER-active species, such as metal complexes or quantum dots, thereby generating advanced materials [20].

13.2.2.1 MOF/Support Composites

MOF composites can utilize two types of support materials: metallic substrates with high conductivity (e.g. Ni foam (NF), NiFe alloy foam, and Fe foils) and carbon-based conductive materials such as graphene and carbon nanotubes (CNTs). Integrating MOFs with these materials enhances electrocatalytic activity for several reasons [45]. First, the presence of conductive substrates enhances electrical conductivity, thereby facilitating synergistic effects, electron transportation, and maintaining structural stability. Furthermore, the utilization of substrates such as NF and carbon cloth in the growth of MOFs eliminates the necessity for binders that could potentially obstruct active sites or lead to catalyst detachment during OER. Additionally, the utilization of supporting material MOFs facilitates the exposure of a larger quantity of active sites, thereby enhancing the overall activity. Lastly, incorporating complex hierarchical structures within MOF/support composites enhances both mass transfer and charge transport capabilities, rendering them promising candidates for achieving efficient and stable OER electrodes.

Typically, the growth of MOFs is facilitated by utilizing metallic substrates, such as NF or alloy foam (e.g. NiFe and NiCo alloy foam), due to their exceptional strength in mechanics, flexibility, and superior electrical properties. A solvothermal process can be employed in a single step to synthesize MOF/NF composites with nanosheet array structures by means of the surface deposition of metal ions and organic ligands, NF undergoes self-assembly. [46–48]. Duan et al. employed a solvothermal method to synthesize NiFe–MOF with an array structure of ultrathin nanosheets on various substrates, utilizing a dissolution–crystallization mechanism [48]. Huang et al. utilized an in situ self-dissociation-assembly technique to fabricate ultrathin MOF nanosheet arrays (NA) on CoNi alloy foam, facilitating the release of metal ions for coordination with benzene dicarboxylic acid [49]. The MOF/support composites exhibited exceptional electrocatalytic characteristics due to their remarkable attributes, including compact and easily accessible active sites, metal unsaturated coordination, and the synergistic effects of electronic coupling between two metals. Moreover, they also demonstrated enhanced charge transportation capabilities. Similarly, the integration of MOFs with carbon-based substrates, which exhibit exceptional electrical conductivity, high surface area, and stability, represents a promising approach for enhancing OER performance. Various MOF/carbon composites have been successfully synthesized by combining MOFs with 1D, 2D, and 3D carbon materials such as CNTs, MWCNTs, and graphene. For instance, a team of researchers led by Srinivas successfully synthesized composites comprising MOFs and CNTs [50]. In this study, they effectively immobilized $FeNi_3$–Fe_3O_4 nanoparticles (NPs) onto the MOF nanosheets and CNTs, resulting in $FeNi_3$–Fe_3O_4 NPs/MOF–CNT composites that exhibited enhanced performance in the OER, characterized by a low overpotential value of $\eta_j = 10\ \mathrm{mA\ cm^{-2}}$ at 234 mV and a small Tafel slope of $37\ \mathrm{mV\ dec^{-1}}$, even without CNT support. Zheng et al. successfully developed a composite material consisting of CoNi–MOF/rGO, where rGO effectively encapsulated the uniformly distributed MOF nanosheets [51]. The resulting catalyst exhibited superior performance compared to commercial Ir/C due to the synergistic effect of CoNi–MOF nanosheets and rGO, enhanced exposure of active sites, and improved electron conductivity. Similarly, the incorporation of 3D graphene with a higher specific surface area also significantly contributed to enhancing the OER performance of Ni–MOF. The unique structure, impressive conductive properties, and area of alternative carbon-based materials such as carbon nanohorns and carbon spheres are acknowledged for their ability to enhance the performance [52].

13.2.2.2 MOF/Active Species Composites

The second approach to enhance the performance of OER involves integrating pristine MOFs with active species for OER, such as metal NPs, metal quantum dots, and conductive molecules. MOFs are an excellent option for these substances due to their remarkable porous structures, significant specific surface areas, customizable pore sizes, and adaptable structural properties [53]. Moreover, the surface of MOFs can be decorated with active species, potentially enhancing OER activity by fostering interfacial and synergistic effects between these active species and MOFs. The utilization of composites consisting of MOF incorporated with metallic NPs is widespread due to the presence of highly desirable electrocatalytic metals in this application. With the exception of metal NPs, metal complexes encompassing MOs, metal sulfides (MSs), and their composites are commonly employed in the fabrication of MOF-based composites [54].

13.2.3 MOF Derivatives for OER

Metal ions, organic ligands and unique structures in MOFs are the precursors or templates for the synthesis of functional materials with different components and desired morphology [55, 56]. Lately, there has been a worldwide focus on exploring the potential of MOF derivatives as efficient electrocatalysts. Controlled pyrolysis of MOF composites under precise conditions, including specific temperatures and atmospheres, enables their direct transformation into functional derivatives of MOFs. Through this process, the organic ligands within MOFs can metamorphose into conductive carbon porous materials like graphene and CNTs, while the metal nodes undergo conversion into uniformly embedded metal NPs or compounds within the carbon porous matrix [57, 58]. Furthermore, the inherent periodic structures of MOFs prevent metal NPs/compounds from aggregating during pyrolysis by effectively hindering their coalescence through the presence of organic ligands. The incorporation of metals into N-doped carbon porous matrix (M–N–PC) materials, as well as MOs, and other composites for OER electrocatalysts derived from MOF are discussed in this section [59, 60].

13.2.3.1 M–N–PC for OER

In recent years, there has been a swift emergence of electrocatalysts consisting of single metal atoms [61–63]. These materials are gaining more and more attention due to the distribution of active sites, high selectivity, almost complete utilization efficiency of atoms, and strong interaction between atomic active sites and supports that can enhance catalytic performance. The general strategies for fabricating electrocatalysts at the atomic level involve improving the precursor materials and supports while effectively controlling the synthetic techniques. Some recent advancements have led to the development of M–N–PC which exhibits exceptional OER performance due to the presence of well-isolated active metal atoms on support [64]. These materials are achieved by utilizing suitable MOFs as precursors or implementing precise post-treatment techniques. For instance, Liu et al. successfully produced hybrid nanocages with dual shells by employing a solvent reaction method to grow Co-based ZIF-67 on Zn-based ZIF-8. The resulting product consisted of inner shells made of N-functionalized microporous carbon and outer shells composed of Co-decorated N-functionalized mesoporous graphitic carbon. These hybrid nanocages exhibited a low overpotential (0.387 V) for the OER. Furthermore, although atomically dispersed transition metals confined within nitrogen atoms on a carbon support have shown remarkable electrochemical performance, their mass activity is limited due to the tendency of metal atoms to aggregate during the sintering process (Figure 13.2a). In order to address this concern, Wu et al. employed potassium chloride particles as a template for the fabrication of densely populated cobalt atoms on an N-functionalized carbon support resembling graphene [65]. In their study, they synthesized a thin layer of cobalt-based organic framework on potassium chloride particles through annealing treatment with cobalt(II) salt and 2-methylimidazole under an argon atmosphere, followed by washing with hydrochloric acid and water. The resulting catalyst exhibited a mass activity of 0.278 A mg_{Co}^{-1} for OER (Figure 13.2b).

The MOFs with diverse compositions and structures serve as highly efficient platforms or precursors for conveniently obtaining oxygen electrocatalysts with targeted electrocatalytic performances. Generally, these derived electrocatalysts possess intricate structures and are adjustable, which offer advantages in increasing the quantity of active sites and enhancing intrinsic activity. The achievement of comparable structures, compositions, and electrocatalytic activities using certain traditional synthetic techniques remains a challenging task. Therefore, the utilization of MOFs as platforms or precursors holds great promise in the development of highly efficient oxygen electrocatalysts for energy storage and conversion methodologies.

13.2.3.2 MOs for OER

MOs and their composites, which possess enhanced electron transportation capabilities, have been extensively investigated as electrocatalysts for the OER in recent years [66]. This is particularly true for transition metals such as iron (Fe), cobalt (Co), and nickel (Ni). Most electrocatalysts based on transition MOs (TMOs) derived from 3D MOFs are typically synthesized through high-temperature pyrolysis. Recent studies have extensively explored the utilization of Co–MOFs and their composites as electrocatalysts for OER [67]. Straightforward techniques involve subjecting Co–MOFs to pyrolysis either in an air environment or in the presence of ligands containing oxygen under a nonreactive atmosphere. This conversion process alters the Co nodes within MOFs into Co_3O_4 [68]. Moreover, mixed-MOs, incorporating elements with diverse valence states, often exhibit enhanced electrocatalytic activity in comparison to single MOs. $NiCo_2O_4$ and $CoFe_2O_4$ serve as exemplary oxides, successfully synthesized by controlled pyrolysis of bimetallic CoNi–MOF and FeCo–MOF, respectively [69]. These oxides demonstrate exceptional performance OER. Shen et al. have reported a multidimensional Ni_xCoO_4/NF material with rapid charge transfer achieved through air calc, exhibiting superior OER

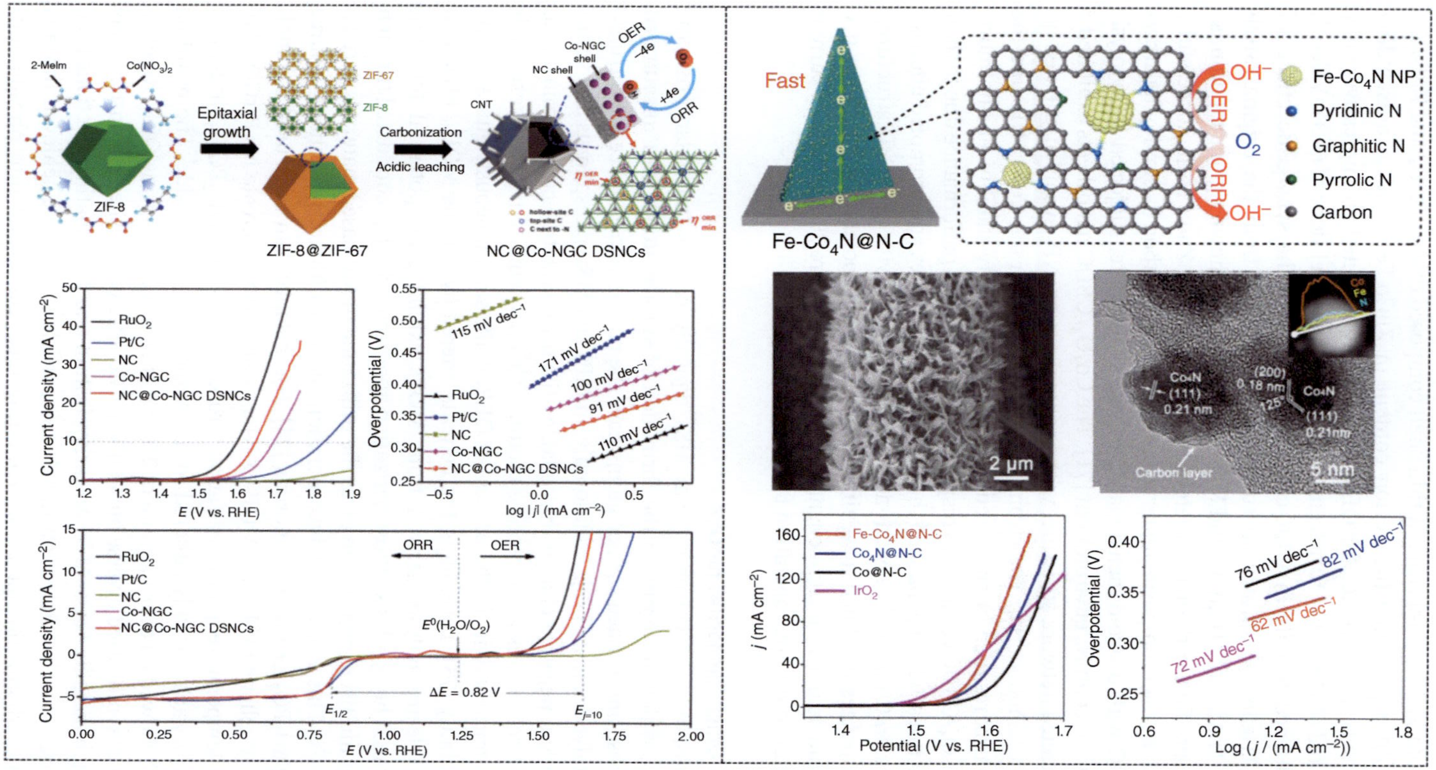

Figure 13.2 (a) Schematic illustration of the synthesis of NC@Co-NGC DSNCs, and *iR*-corrected OER polarization curves and Tafel plots of NC, Co-NGC, NC@Co-NGC DSNC, Pt/C, and RuO_2 catalysts in O_2-saturated 0.1 M KOH solution at a scan rate of $10\,mV\,s^{-1}$ at 1600 rpm. Source: Reproduced with permission of Liu et al. [63], WILEY-VCH. (b) Schematic illustration, SEM image, high-resolution TEM images of the Fe-Co_4N@N-C nanosheet grown on carbon cloth for oxygen evolution reaction. Source: Reproduced with permission of Xu et al. [65], Elsevier.

activity compared to similar materials. Fu et al. synthesized core–shell nanocages derived from MOFs, comprising a thin shell composed of a $CoFe_2O_4$ and $ZnFe_2O_4$ mixture [70]. Optimizing the Fe ion content in ZnCo-Fe-20 led to outstanding electrocatalytic performance for OER. This variant exhibited an impressively low overpotential of 176 mV at 10 mA cm^{-2} and a minimal Tafel slope of 69.3 mV dec^{-1}. The heightened activity in OER can be attributed to the synergistic effects of mixed-MOs and the substantial surface area inherited from the original MOF structure. The use of 2D MOFs and MOF composites has shown significant promise as precursors for synthesizing materials based on MOs. Zhou et al. successfully engineered a novel composite material comprising MO/carbon by employing arrays of Co–BDC with a two-dimensional structure as template, subjected to pyrolysis treatment under an N_2 environment [71]. The Co_3O_4/CBDC nanosheets exhibited a well-connected network structure, with the small Co_3O_4 NPs evenly distributed on the carbon nanomaterial. The strong interaction between the Co_3O_4 NPs and the carbon matrix resulted in enhanced electrical conductivity stability. Furthermore, the 2D Co_3O_4/CBDC composites displayed improved OER performance due to their array structures, which enhanced the number of active sites and promoted effective mass transfer. Compared to MOF-derived counterparts with 1D or 3D structures, these composites exhibited a significantly reduced overpotential of only 208 mV for OER and demonstrated excellent durability with a mere 3% decrease in current density after continuous operation for 36 hours.

13.2.3.3 Other Composites for OER

Metal phosphides (MPs) and metal MSs typically exhibit triangular prismatic structures, which are prone to the formation of surface atoms with unsaturated bonds. These unsaturated surface atoms have the potential to enhance the intrinsic electrocatalytic activity [72]. Additionally, in MPs and MSs, phosphorus and sulfur atoms carry a negative charge that enables them to attract electrons from metal atoms and act as sites for accepting protons. This particular attribute boosts their electrochemical activity in the process of OER, while simultaneously guaranteeing their durability [73]. Leveraging the benefits offered by various MOFs, MPs, and MSs, the electrocatalysts derived from MOFs exhibit great potential for OER applications. The pyrolysis–phosphorization approach, whether conducted through a direct or multistep process, involves the utilization of MP-based materials derived from MOFs. This entails the incorporation of an additional phosphorus source such as NaH_2PO_2 or red phosphorous [74]. For instance, Yan et al. used a combination of graphene oxide and MOF-74-Ni as a precursor to create Ni_2P/rGO [75]. In a low-temperature phosphorization process involving NaH_2PO_2, ultra-small Ni_2P nanocrystals strongly adhered to the reduced graphene oxide. The resulting Ni_2P/rGO showed an overpotential of 260 mV at 10 mA cm^{-2} during OER. Xu et al. successfully synthesized diverse bimetallic Fe_xCo_y-P/C composites by consecutively subjecting bimetallic FeCo–MOF to carbonization and phosphating reactions [76]. The precise control of transition metal nanoparticle size can be achieved through the initial carbonization process, facilitating their conversion into phosphorous compounds without encountering the common problem of agglomeration observed in one-step phosphorization precursors. Similarly, by manipulating the calcination and sulfuration processes, it is possible to achieve a robust attachment between the MSs and the conductive carbon substrates derived from MOFs. This effectively addresses the issue of insufficient electronic conductivity observed in unmodified MSs enhancements in OER activity [77]. The synthesis of Co–NSOMOF (Co–metal–organic framework) was conducted by Du et al., where Co ions were coordinated with ligands containing sulfur (H_2tda, 2,5-thiophenedicarboxylic acid) and nitrogen (bpp, 1,3-bis(4-pyridyl)propane) [78]. Through a one-step pyrolysis process applied to the Co–NSOMOF, a distinct hybrid material named Co/Co_9S_8@NSOC was obtained. This material comprises NPs of Co/Co_9S_8 embedded within a porous carbon matrix that is doped with nitrogen, sulfur, and oxygen. Subsequently, this resulting material was employed for the process of OER. A comparable approach involving multiple metals can be considered for MSs [79, 80]. Liu et al. employed bimetallic MOF (CoFe-ZIF@GO) encapsulated in graphene oxide as precursor materials to synthesize $Co_{9-x}Fe_xS_8$/Co, Fe–N–C wrapped by reduced graphene oxide through a semi-sulfuration and calcination process [81]. By incorporating abundant OER-active CoFeS, the resulting S-$Co_{9-x}Fe_xS_8$@rGO-10 catalyst exhibited an exceptionally low overpotential (η_j = 10 mA cm^{-2} of 290 mV).

13.3 Covalent–Organic Frameworks (COFs) for OER

13.3.1 Pristine COFs for OER

The distinguishing feature of COFs lies in their ability to form robust covalent bonds using a diverse range of organic molecules as building blocks. These frameworks exhibit a uniform pore structure, offer high accessibility to their extensive surface area, and demonstrate exceptional chemical stability [82]. A notable advantage is the ease with which they can be tailored and precisely modified through the selection of appropriate functionalized organic units. Although pristine COFs typically exhibit poor conductivity and low electrocatalytic activity due to the absence of metal sites, the incorporation of heteroatoms (such as N, O, and S) into the carbon framework of pristine COFs as direct active centers also promotes OER. Additionally, introducing a catalysis-active metal center into the porphyrin/phthalocyanine/bipyridine unit within pristine COF enhances its performance in OER [83, 84].

13.3.1.1 Metal-Free COFs for OER

The catalytic active sites of metal-free COFs are predominantly derived from the carbon or nitrogen atoms within the framework. Pristine metal-free COFs exhibit relatively modest electrocatalytic performance in the OER due to their restricted availability of active sites and inadequate electrical conductivity. The number of available reports on electrocatalytic OER using metal-free COF-based materials is limited [21]. For instance, a recently published study utilized theoretical calculations to suggest a unique planar phenazine-linked COF (CO-FC_4N) with appropriate CN ratio, N placement, and band characteristics to enhance the effectiveness of electrocatalytic OER [85]. The combined contribution of theoretical and experimental findings suggests that the exceptional performance in OER can be attributed to factors such as enhanced crystallinity, remarkable durability, well-optimized band structure, and the presence of highly reactive carbon active sites. Another study conducted by Mondal et al. presented a novel thiadiazole-based COF (C_4-SHz COF) that was synthesized via Schiff base condensation polymerization involving 1,3,5-tris(4-formylphenyl)benzene (C_4-CHO) and 2,5-dihydrazinyl-1,3,4-thiadiazole (SHz) [86]. The OER performance of the C_4-SHz COF in alkaline environments was remarkable due to its exceptional structural features, including a large specific surface area, abundant porosity, and extended π-conjugation. These characteristics facilitated rapid charge and mass transfer. The C atom in the cof skeleton serves as the active site, while the nitrogen atom in the structure also exhibits remarkable catalytic activity. For instance, Kathiresan et al. recently synthesized a filamentous network structure of CTF (PD–CTF) using cyanuric chloride and 1,4-phenylenediamine in a basic environment [87]. The exceptional performance in OER can be attributed to the presence of enriched pyridinic-N sites within the resulting material. However, it is important to note that metal-free COFs exhibit significantly lower catalytic activity compared to metal-based catalysts. Therefore, metallization of COFs becomes crucial for enhancing their OER efficiency.

13.3.1.2 Metal Sites Containing COFs for OER

Presynthetic modification is commonly employed as a bottom-up approach to synthesize hybrid materials based on metal-doped pristine COFs. Precoordinating metal ions to specific sites in the building units or linkers, which do not interfere with the formation of COFs during the reaction, enables one-step synthesis. Various prevalent metal complexation units, such as metalloporphyrins [25, 26], metalphthalocyanines [27], and metallobipyridine [88] have been investigated for direct fabrication of metal-COFs due to their unique structures that offer M–N–C centers as active sites. Particularly, the field of OER catalyst design has witnessed significant attention toward redox-active Co due to its capacity to exist in multiple stable redox states. In 2020, a Schiff base reaction was utilized to synthesize a bifunctional Co(II) Por-based covalent–organic polymer (CoCOP), which exhibited remarkable electrocatalytic properties in OER [89]. The resulting CoCOP demonstrated significantly reduced overpotentials for OER (350 mV@10 mA cm^{-2}) in an alkaline solution of 1 M KOH. The exceptional performance of CoCOP in OER can be attributed to its extensive surface area and the synergistic effect between the C=N bonds and Co(II) Por moieties. For instance, a metallophthalocyanine-based COF (CoCMP) was synthesized utilizing a Schiff base condensation reaction [90]. When utilized as an electrocatalyst for the OER, the Co-containing phthalocyanine-based COF, coordinated with N_4, displayed an onset potential at 1.57 V and a Tafel slope of 87 mV per decade, surpassing CMP. Interestingly, COFs, a type of crystalline porous polymers, are recognized for their strong chemical stability, large accessible surface area, and customizable building units [91, 92]. By carefully selecting appropriate structural components with desired functional groups and topological architectures, COFs can be intricately designed. It has been proven that COFs have the potential to serve as a platform for creating non-noble metal NPs that are smaller than 2 nm in size and heterogeneous single atoms. For example, the immersion of a cobalt-containing precursor solution into TpBpy can modify the porous framework of TpBpy with numerous N atoms possessing lone electron pairs to incorporate Co single atoms [88]. The COF (TpBpy) incorporating bipyridine demonstrates effective electrocatalytic activity toward OER, achieving an overpotential of 400 mV at a current density of 1 mA cm^{-2} in a neutral buffer solution. Furthermore, it exhibits remarkable stability with approximately 94% retention in the OER current even after undergoing 1000 cycles.

13.3.2 COF Composites for OER

To ensure effective electrocatalytic activity alongside strategically engineered active sites, the comprehensive electron transfer capacity of the complete electrocatalyst plays a crucial role in diverse electrochemical OER processes. As COFs typically demonstrate limited electrical conductivity, it is crucial to incorporate intrinsic π-conjugated frameworks or enhance π-conjugation through carbonization to achieve satisfactory electrochemical performance [82]. However, the process of pyrolysis can lead to undesired alterations in the structure and potentially even disrupt more intricate framework structures, thereby introducing uncertainty into the mechanism of the electrocatalytic process. In this context, it is crucial to eliminate the involvement of pyrolysis in the procedure of acquiring COF-based electrocatalytic materials that possess precisely defined framework architectures and manageable active components. To address this obstacle while preserving the ability to precisely manipulate the structure at an atomic level, it is possible to combine well-defined COFs with conductive support materials (e.g. CPs, graphene, and CNTs) [93]. This composite approach offers sufficient conduits for conductivity and enhances the surface area, thereby facilitating the formation of hybrid systems based on COFs. An alternative strategy to improve the performance of OER involves incorporating active components, such as metal NPs, metal quantum dots, and conductive molecules, into pristine COFs. The exceptional porous structures, large specific surface areas, adjustable pore sizes, and adaptable structures make COFs an excellent candidate for hosting these species.

13.3.2.1 Carbon Materials Supporting COF for OER

In order to improve the electrical conductivity of catalysts, hybrid systems based on COFs have utilized various carbon matrices with a large surface area, abundant functional sites, and excellent structural stability. For instance, COF-based carbon composites have been constructed using GO [94], CNT [95], and graphitic carbon nitride ($g\text{-}C_3N_4$) [96]. As an example, researchers synthesized a group of COF-coated MWCNTs (COF-366–Co@CNT_x) through imine condensation. Subsequently, Co ions were introduced to coordinate with the porphyrin units as active centers for electrocatalytic OER in an alkaline environment [97]. The COF-366–Co@CNT-0.4 sample exhibited the lowest overpotential of 358 mV at a current density of 10 mA cm^{-2}. Moreover, the Co^{2+}-saturated sample with a ratio of 2/3 demonstrated superior catalytic activity for OER, resulting in an overpotential of 378 mV. These findings indicate that optimizing the thickness of the COF layer to ensure efficient mass/charge transfer and arranging active sites appropriately to prevent competitive masking behavior are crucial factors in enhancing electrocatalytic performance. CTFs exhibit remarkable stability in aqueous solutions and electrochemical environments, making them a promising candidate for the development of metal-doped CTFs as exceptionally durable electrocatalysts [98]. These CTFs consist of triazine units that possess an abundance of nitrogen atoms, enabling effective coordination with metals. Furthermore, the electronic state of these coordinated metals within the CTF structure can be tailored by carefully selecting the building blocks incorporated into the framework. Yamaguchi et al. successfully developed a robust catalyst by modifying CTFs with Ru atoms, which exhibited excellent selectivity for the electrooxidation of benzyl alcohol over OER in aqueous solutions [99]. It was proposed that the absence of repetitive Ru–O–Ru sites in Ru–CTFs significantly hindered the OER process, while still allowing the Ru_2O sites to function as active centers for C–H oxidation at high-potential regions. The presence of an extended π conjugated structure in CTF ensured that Ru maintained a high-valency state during electrocatalysis, thereby resulting in enhanced selectivity toward carboxylic acid products.

13.3.2.2 Metal Hybrids Containing COF for OER

Efforts have been made to explore alternative COF-based hybrids that can be utilized for electrocatalysis or potential electrocatalysis applications. One promising approach involves utilizing porous carbons derived from organic polymers as a support material, which can then be combined with electrochemically active components like MO or hydroxide [28]. Chakraborty et al. utilized the IISERP–COF as a carrier to encapsulate Ru^{3+} within nanopores [29]. Subsequently, through a gentle pyrolysis process, they successfully synthesized RuO_2 NPs with uniform distribution primarily ranging from 2.5 to 4 nm on the COF-derived micro/mesoporous carbons without employing any capping agent, thanks to the restraining effects of the COF carrier. The catalytic activity of the IISERP–COF1_RuO_2 hybrid toward OER was significantly enhanced compared to RuO_2, achieving an overpotential of only 210 mV at 10 mA cm^{-2}. However, a slight decline in activity was observed after 12 hours during chronoamperometry measurement, possibly due to either the release of RuO_2 from the COF framework or reduced adherence between the catalyst and electrode. Additionally, the exceptional adaptability of COFs allows for the integration of numerous active sites. Mullangi and colleagues conducted a study where they successfully incorporated very tiny $Co_xNi_y(OH)_2$ active NPs (approximately 2 nm in size) into the mesoporous structure of a tetrahedral N-rich IISERP–COF support [100]. The resulting nanocomposite, based on pseudo-3D COF, exhibited remarkable performance as an OER catalyst in alkaline conditions. Specifically, at a current density of 10 mA cm^{-2}, the $Co_xNi_y(OH)_2$–COF composite with 4 wt% Co and 12 wt% Ni demonstrated an impressively low overpotential of only 0.258 V. The specific reasons for the relatively higher catalytic activity of this COF-based $Co_xNi_y(OH)_2$ hybrid compared to certain other Ni/Co-based electrocatalysts were not explicitly specified, potentially due to localized electronic interactions occurring between the NPs and the flexible sp^3-nitrogen-rich COF framework.

13.3.3 COF Derivatives for OER

Although carbons doped with heteroatoms and free of metals, nanohybrids of nonprecious transition metal–nitrogen–carbon (M–N–C) obtained through the pyrolysis of precursors containing nonprecious transition metals, nitrogen, and carbon have demonstrated exceptionally high electrocatalytic performance and hold promising prospects in alkaline environments, the occurrence of OER often takes place at elevated oxidation potentials, resulting in inevitable carbon corrosion and subsequent catalyst agglomeration as well as carbonate formation that obstructs the active surface [101]. A metal-free electrocatalyst for OER was synthesized by carbonizing a porous COF derived from ethylene diamine. When applied to carbon paper, the resulting carbon material exhibited significantly improved catalytic activity for OER in alkaline conditions. It demonstrated an onset potential of 1.527 V (vs. RHE) and an overpotential of 297 mV at a current density of 10 mA cm^{-2} [102]. Moreover, even at a high current density of 300 mA cm^{-2}, the COF-derived carbon material showed a low overpotential of 580 mV and maintained stability for more than 10 hours, surpassing certain metal-based OER electrocatalysts. To enhance conductivity, hybrid materials were obtained by pyrolyzing metal-infused COFs, leading to uniform dispersion of metal NPs on the derived carbon frameworks from COFs. Jia et al. conducted a research study on cobalt NPs incorporated into nitrogen-enriched porous carbons, which were derived from cobalt-porphyrin-based COF [103]. CoP-*n*ph-CMP-800, where $n = 2$, 3, or 4, was synthesized and evaluated. Among them, CoP-2ph-CMP-800 exhibited exceptional catalytic performance for the OER. It demonstrated a low overpotential of 370 mV at 10 mA cm^{-2} and displayed remarkable stability.

13.4 Summary and Perspective

Significant advancements have been made in recent times to enhance the efficiency of catalysts based on porous materials for OER. The improved performance of electrocatalysts derived from MOF and COF primarily stems from employing strategies such as adjusting coordination environment, engineering nodes/ligands, regulating geometry, incorporating composite active species, utilizing isolated active sites, and employing carbon derivatives.

Although great progress has been made, several foreseeable challenges should also be significantly emphasized and addressed in the future.

As for MOF-based electrocatalysts, the persisting challenges are as follows:

1. Improving the stability of MOF-based materials is crucial for enhancing their catalytic performance during the OER process. The collapse of these materials' structure could result in agglomeration of metal NPs, loss of active sites, and hindered mass transfer efficiency, all of which have a detrimental impact on their catalytic activity.
2. During the process of OER, a phase transformation occurs in most MOF-based materials leading to the formation of active catalytic sites. Hence, it is imperative

to employ advanced in situ detection techniques for accurate identification of these active sites, which greatly aids in comprehending the actual reaction mechanism on MOF-based materials.

3. The precise manipulation of the coordination environment of metal atoms is required. However, there has been a lack of thorough investigation into the correlation between synthesis parameters and the resulting coordination environment of metal atoms in materials based on MOFs. Hence, it is crucial to acquire a thorough comprehension of the coordination surroundings and chemical conversion mechanisms occurring in unmodified MOFs as well as their derived forms, necessitating further investigation.

As for COF-based electrocatalysts, the persisting challenges are as follows:

1. Greater emphasis should be placed on understanding the electrocatalytic mechanisms. Presently, there is ongoing debate surrounding the electrocatalytic mechanisms of carbon-based materials primarily due to the uncertain and poorly defined active sites responsible for catalysis. By leveraging advancements in in situ characterization techniques, it would be valuable to investigate COF-based materials as model catalysts, as they can offer deeper insights into the relationship between structure and activity. Therefore, it is recommended to dedicate continuous efforts toward studying the in situ characterization of COF-based materials during OER electrocatalysis. Armed with this newfound knowledge, more effective electrocatalysts can be designed and fabricated.
2. The enhancement of intrinsic activity in active sites is crucial, particularly when operating under acidic conditions. Currently, COF-based electrocatalysts exhibit significant electrocatalytic performance exclusively in alkaline electrolytes, while they still fall short compared to noble metal-based benchmarks in acidic environments. Hence, the investigation of high-efficiency active sites for OER electrocatalysis and the understanding of reaction mechanisms in acidic electrolytes hold the utmost importance.
3. Urgent efforts are required to explore the potential of 3D COF-based electrocatalysts. Currently, the majority of reported COF-based electrocatalysts consist of 2D stacked structures. To further enhance the exposure of active sites and establish more rational channels and pore microstructures, it is highly recommended to synthesize 3D COFs.

In conclusion, despite encountering these obstacles, the future prospects for the advancement of electrocatalysts based on MOF and COF materials are highly promising due to their distinctive structures and performance benefits. Moving forward, more efficient approaches for material design and preparation will be proposed in order to create innovative, high-performing, and stable electrocatalysts using pristine MOFs and COFs that do not require carbonization at high temperatures. Additionally, efforts will be made to develop single-atom TM–NC materials derived from MOFs and COFs with exceptional electroactivity. These strategies aim to overcome the aforementioned challenges while maximizing their immense potential in practical applications of electrocatalysis.

Acknowledgment

The authors acknowledge financial support from the National Natural Science Foundation of China ([2023] No. 22309033 Z.J.Y.), the Guizhou Provincial Science and Technology Foundation ([2024] No. ZK General 425, Z.J.Y.), the Academic Young Talent Foundation of Guizhou Normal University ([2022] No. B06, Z.J.Y.), and the Natural Science Research Project of Guizhou Provincial Department of Education ([2022] No. 125 Z.J.Y.).

References

1 Li, Y. and Dai, H. (2014). Recent advances in zinc-air batteries. *Chem. Soc. Rev.* 43 (15): 5257–5275.

2 Wang, J., Zhang, H., and Wang, X. (2017). Recent methods for the synthesis of noble-metal-free hydrogen-evolution electrocatalysts: from nanoscale to sub-nanoscale. *Small Methods* 1 (6): 1700118.

3 Shrivastav, V., Sundriyal, S., Goel, P. et al. (2019). Metal-organic frameworks (MOFs) and their composites as electrodes for lithium battery applications: novel means for alternative energy storage. *Coord. Chem. Rev.* 393: 48–78.

4 Han, L., Dong, S., and Wang, E. (2016). Transition-metal (Co, Ni, and Fe)-based electrocatalysts for the water oxidation reaction. *Adv. Mater.* 28 (42): 9266–9291.

5 Grimaud, A., Diaz-Morales, O., Han, B. et al. (2017). Activating lattice oxygen redox reactions in metal oxides to catalyse oxygen evolution. *Nat. Chem.* 9 (5): 457–465.

6 Lu, F., Zhou, M., Zhou, Y., and Zeng, X. (2017). First-row transition metal-based catalysts for the oxygen evolution reaction under alkaline conditions: basic principles and recent advances. *Small* 13 (45): 1701931.

7 Slater, A. and Cooper, A. (2015). Function-led design of new porous materials. *Science* 348 (6238): aaa8075.

8 He, X., Yin, F., Wang, H. et al. (2018). Metal-organic frameworks for highly efficient oxygen electrocatalysis. *Chin. J. Catal.* 39 (2): 207–227.

9 Wu, H. and Lou, X. (2017). Metal-organic frameworks and their derived materials for electrochemical energy storage and conversion: promises and challenges. *Sci. Adv.* 3 (12): eaap9252.

10 Liu, J., Chen, L., Cui, H. et al. (2014). Applications of metal-organic frameworks in heterogeneous supramolecular catalysis. *Chem. Soc. Rev.* 43 (16): 6011–6061.

11 Shen, K., Zhang, L., Chen, X. et al. (2018). Ordered macro-microporous metal-organic framework single crystals. *Science* 359 (6372): 206–210.

12 Lu, X., Xia, B., Zang, S. et al. (2020). Metal-organic frameworks based electrocatalysts for the oxygen reduction reaction. *Angew. Chem. Int. Ed.* 132 (12): 4662–4678.

13 Li, W., Xue, S., Watzele, S. et al. (2020). Advanced bifunctional oxygen reduction and evolution electrocatalyst derived from surface-mounted metal-organic frameworks. *Angew. Chem. Int. Ed.* 132 (14): 5886–5892.
14 Wen, T., Zheng, Y., Xu, C. et al. (2018). A boron imidazolate framework with mechanochromic and electrocatalytic properties. *Mater. Horiz.* 5 (6): 1151–1155.
15 Liang, Z., Qu, C., Xia, D. et al. (2018). Atomically dispersed metal sites in MOF-based materials for electrocatalytic and photocatalytic energy conversion. *Angew. Chem. Int. Ed.* 57 (31): 9604–9633.
16 Dhakshinamoorthy, A., Asiri, A., and Garcia, H. (2019). 2D metal–organic frameworks as multifunctional materials in heterogeneous catalysis and electro/photocatalysis. *Adv. Mater.* 31 (41): 1900617.
17 Huang, J., Li, Y., Huang, R. et al. (2018). Electrochemical exfoliation of pillared-layer metal-organic framework to boost the oxygen evolution reaction. *Angew. Chem. Int. Ed.* 130 (17): 4722–4726.
18 Li, Y., Fu, Z., and Xu, G. (2019). Metal-organic framework nanosheets: preparation and applications. *Coord. Chem. Rev.* 388: 79–106.
19 Sohrabi, S., Dehghanpour, S., and Ghalkhani, M. (2018). A cobalt porphyrin-based metal organic framework/multi-walled carbon nanotube composite electrocatalyst for oxygen reduction and evolution reactions. *J. Mater. Sci.* 53 (5): 3624–3639.
20 Cai, G., Zhang, W., Jiao, L. et al. (2017). Template-directed growth of well-aligned MOF arrays and derived self-supporting electrodes for water splitting. *Chem* 2 (6): 791–802.
21 Lin, C., Zhang, D., Zhao, Z. et al. (2018). Covalent organic framework electrocatalysts for clean energy conversion. *Adv. Mater.* 30 (5): 1703646.
22 Nagai, A., Guo, Z., Feng, X. et al. (2011). Pore surface engineering in covalent organic frameworks. *Nat. Commun.* 2 (1): 536.
23 Xiang, Z., Xue, Y., Cao, D. et al. (2014). Highly efficient electrocatalysts for oxygen reduction based on 2D covalent organic polymers complexed with non-precious metals. *Angew. Chem. Int. Ed.* 53 (9): 2433–2437.
24 Spitler, E., Colson, J., Uribe-Romo, F. et al. (2011). A 2D covalent organic framework with 4.7-nm pores and insight into its interlayer stacking. *Angew. Chem. Int. Ed.* 133 (48): 19416–19421.
25 Lin, S., Diercks, C., Zhang, Y. et al. (2015). Covalent organic frameworks comprising cobalt porphyrins for catalytic CO_2 reduction in water. *Science* 349 (6253): 1208–1213.
26 Lin, Q., Bu, X., Kong, A. et al. (2015). Heterometal-embedded organic conjugate frameworks from alternating monomeric iron and cobalt metalloporphyrins and their application in design of porous carbon catalysts. *Adv. Mater.* 27 (22): 3431–3436.
27 Yao, C., Li, J., Gao, W., and Jiang, Q. (2018). An integrated design with new metal-functionalized covalentorganic frameworks for the effective electroreduction of CO_2. *Chem. Eur. J.* 24 (43): 11051–11058.
28 Gu, W. and Yushin, G. (2014). Review of nanostructured carbon materials for electrochemical capacitor applications: advantages and limitations of activated

carbon, carbide-derived carbon, zeolite-templated carbon, carbon aerogels, carbon nanotubes, onion-like carbon, and graphene. *Wiley Interdiscip. Rev.: Energy Environ.* 3 (5): 424–473.

29 Chakraborty, D., Nandi, S., Illathvalappil, R. et al. (2019). Carbon derived from soft pyrolysis of a covalent organic framework as a support for small-sized RuO_2 showing exceptionally low overpotential for oxygen evolution reaction. *ACS Omega* 4 (8): 13465–13473.

30 Zheng, F., Zhang, Z., and Zhang, C. (2019). Advanced electrocatalysts based on metal-organic frameworks. *ACS Omega* 5 (6): 2495–2502.

31 Xue, J., Li, C., Li, F. et al. (2020). Recent advances in pristine tri-metallic metal-organic frameworks toward the oxygen evolution reaction. *Nanoscale* 12 (8): 4816–4825.

32 Syzgantseva, M., Ireland, C., Ebrahim, F. et al. (2019). Metal substitution as the method of modifying electronic structure of metal-organic frameworks. *J. Am. Chem. Soc.* 141 (15): 6271–6278.

33 Zhou, W., Huang, D., Wu, Y. et al. (2019). Stable hierarchical bimetal–organic nanostructures as high performance electrocatalysts for the oxygen evolution reaction. *Angew. Chem. Int. Ed.* 58 (13): 4227–4231.

34 Li, F., Shao, Q., Huang, X. et al. (2018). Nanoscale trimetallic metal-organic frameworks enable efficient oxygen evolution electrocatalysis. *Angew. Chem. Int. Ed.* 130 (7): 1906–1910.

35 Zhao, X., Xue, Z., Chen, W. et al. (2019). Ambient fast, large-scale synthesis of entropy-stabilized metal-organic framework nanosheets for electrocatalytic oxygen evolution. *J. Mater. Chem. A* 7 (46): 26238–26242.

36 Jiang, J., Huang, L., Liu, X. et al. (2017). Bioinspired cobalt–citrate metal–organic framework as an efficient electrocatalyst for water oxidation. *ACS Appl. Mater. Interfaces* 9 (8): 7193–7201.

37 Tao, L., Lin, C., Dou, S. et al. (2017). Creating coordinatively unsaturated metal sites in metal-organic-frameworks as efficient electrocatalysts for the oxygen evolution reaction: insights into the active centers. *Nano Energy* 41: 417–425.

38 Momeni, M. and Cramer, C. (2019). Computational screening of roles of defects and metal substitution on reactivity of different single-vs double-node metal–organic frameworks for sarin decomposition. *J. Phys. Chem. C* 123 (24): 15157–15165.

39 Morris, R. and Brammer, L. (2017). Coordination change, lability and hemilability in metal–organic frameworks. *Chem. Soc. Rev.* 46 (17): 5444–5462.

40 Vos, A., Hendrickx, K., Van Der Voort, P. et al. (2017). Missing linkers: an alternative pathway to UiO-66 electronic structure engineering. *Chem. Mater.* 29 (7): 3006–3019.

41 Li, F., Wang, P., Huang, X. et al. (2019). Large-scale, bottom-up synthesis of binary metal-organic framework nanosheets for efficient water oxidation. *Angew. Chem. Int. Ed.* 131 (21): 7125–7130.

42 Zhao, S., Wang, Y., Dong, J. et al. (2016). Ultrathin metal-organic framework nanosheets for electrocatalytic oxygen evolution. *Nat. Energy* 1 (12): 1–10.

43 Li, Z., Ge, X., Li, C. et al. (2020). Rational microstructure design on metal–organic framework composites for better electrochemical performances: design principle, synthetic strategy, and promotion mechanism. *Small Methods* 4 (3): 1900756.

44 Yi, F., Zhang, R., Wang, H. et al. (2017). Metal-organic frameworks and their composites: synthesis and electrochemical applications. *Small Methods* 1 (11): 1700187.

45 Wang, C., Liu, H., Bian, G. et al. (2019). Metal-layer assisted growth of ultralong quasi-2D MOF nanoarrays on arbitrary substrates for accelerated oxygen evolution. *Small* 15 (51): 1906086.

46 Sun, F., Wang, G., Ding, Y. et al. (2018). NiFe-based metal-organic framework nanosheets directly supported on nickel foam acting as robust electrodes for electrochemical oxygen evolution reaction. *Adv. Energy Mater.* 8 (21): 1800584.

47 Senthil Raja, D., Chuah, X., and Lu, S. (2018). In situ grown bimetallic MOF-based composite as highly efficient bifunctional electrocatalyst for overall water splitting with ultrastability at high current densities. *Adv. Energy Mater.* 8 (23): 1801065.

48 Duan, J., Chen, S., and Zhao, C. (2017). Ultrathin metal-organic framework array for efficient electrocatalytic water splitting. *Nat. Commun.* 8 (1): 15341.

49 Huang, L., Gao, G., Zhang, H. et al. (2020). Self-dissociation-assembly of ultrathin metal-organic framework nanosheet arrays for efficient oxygen evolution. *Nano Energy* 68: 104296.

50 Srinivas, K., Lu, Y., Chen, Y. et al. (2020). $FeNi_3$-Fe_3O_4 heterogeneous nanoparticles anchored on 2D MOF nanosheets/1D CNT matrix as highly efficient bifunctional electrocatalysts for water splitting. *ACS Sustain. Chem. Eng.* 8 (9): 3820–3831.

51 Zheng, X., Cao, Y., Liu, D. et al. (2019). Bimetallic metal-organic-framework/reduced graphene oxide composites as bifunctional electrocatalysts for rechargeable Zn–air batteries. *ACS Appl. Mater. Interfaces* 11 (17): 15662–15669.

52 Guo, Y., Zhou, Y., Nan, Y. et al. (2020). Ni-based nanoparticle-embedded N-doped carbon nanohorns derived from double core-shell CNH@PDA@NiMOFs for oxygen electrocatalysis. *ACS Appl. Mater. Interfaces* 12 (11): 12743–12754.

53 Wang, M., Wang, P., Li, C. et al. (2018). Boosting electrocatalytic oxygen evolution performance of ultrathin Co/Ni-MOF nanosheets via plasmon-induced hot carriers. *ACS Appl. Mater. Interfaces* 10 (43): 37095–37102.

54 Xia, Z., Fang, J., Zhang, X. et al. (2019). Pt nanoparticles embedded metal-organic framework nanosheets: a synergistic strategy towards bifunctional oxygen electrocatalysis. *Appl. Catal. B Environ.* 245: 389–398.

55 Shen, K., Chen, X., Chen, J., and Li, Y. (2016). Development of MOF-derived carbon-based nanomaterials for efficient catalysis. *ACS Catal.* 6 (9): 5887–5903.

56 Chen, Y., Wang, C., Wu, Z. et al. (2015). From bimetallic metal-organic framework to porous carbon: high surface area and multicomponent active dopants for excellent electrocatalysis. *Adv. Mater.* 27 (34): 5010–5016.

57 Du, L., Xing, L., Zhang et al. (2020). Metal-organic framework derived carbon materials for electrocatalytic oxygen reactions: recent progress and future perspectives. *Carbon* 156: 77–92.

58 Sun, T., Xu, L., Wang, D., and Li, Y. (2019). Metal organic frameworks derived single atom catalysts for electrocatalytic energy conversion. *Nano Res.* 12: 2067–2080.

59 Li, D., Gu, Z., Zhang, W. et al. (2017). Epitaxial encapsulation of homodispersed CeO_2 in a cobalt–porphyrin network derived thin film for the highly efficient oxygen evolution reaction. *J. Mater. Chem. A* 5 (38): 20126–20130.

60 Li, Y., Zhao, T., Lu, M. et al. (2019). Enhancing oxygen evolution reaction through modulating electronic structure of trimetallic electrocatalysts derived from metal-organic frameworks. *Small* 15: 1901940.

61 Zhang, W., Yao, X., Zhou, S. et al. (2018). ZIF-8/ZIF-67-derived Co-N_x-embedded 1D porous carbon nanofibers with graphitic carbon-encased Co nanoparticles as an efficient bifunctional electrocatalyst. *Small* 14 (24): 1800423.

62 Zhao, J., Hu, H., and Wu, M. (2020). N-doped-carbon/cobalt-nanoparticle/N-doped-carbon multi-layer sandwich nanohybrids derived from cobalt MOFs having 3D molecular structures as bifunctional electrocatalysts for on-chip solid-state Zn-air batteries. *Nanoscale* 12 (6): 3750–3762.

63 Liu, S., Wang, Z., Zhou, S. et al. (2017). Metal-organic-framework-derived hybrid carbon nanocages as a bifunctional electrocatalyst for oxygen reduction and evolution. *Adv. Mater.* 29 (31): 1700874.

64 Wu, J., Zhou, H., Li, Q. et al. (2019). Densely populated isolated single Co-N site for efficient oxygen electrocatalysis. *Adv. Energy Mater.* 9 (22): 1900149.

65 Xu, Q., Jiang, H., Li, Y. et al. (2019). In-situ enriching active sites on co-doped Fe-Co_4N@N-C nanosheet array as air cathode for flexible rechargeable Zn-air batteries. *Appl. Catal. B Environ.* 256: 117893.

66 Li, Y., Xu, Y., Yang, W. et al. (2018). MOF-derived metal oxide composites for advanced electrochemical energy storage. *Small* 14 (25): 1704435.

67 Lyu, F., Wang, Q., Choi, S., and Yin, Y. (2019). Noble-metal-free electrocatalysts for oxygen evolution. *Small* 15 (1): 1804201.

68 Kuang, X., Luo, Y., Kuang, R. et al. (2018). Metal organic framework nanofibers derived Co_3O_4-doped carbon-nitrogen nanosheet arrays for high efficiency electrocatalytic oxygen evolution. *Carbon* 137: 433–441.

69 Shen, Y., Guo, S., Du, F. et al. (2019). Prussian blue analogue-derived Ni and Co bimetallic oxide nanoplate arrays block-built from porous and hollow nanocubes for the efficient oxygen evolution reaction. *Nanoscale* 11 (24): 11765–11773.

70 Fu, Y., Wang, W., Wang, J. et al. (2019). MOFs-derived ZnCo-Fe core–shell nanocages with remarkable oxygen evolution reaction performance. *J. Mater. Chem. A* 7 (29): 17299–17305.

71 Zhou, J., Dou, Y., Zhou, A. et al. (2018). Layered metal-organic framework-derived metal oxide/carbon nanosheet arrays for catalyzing the oxygen evolution reaction. *ACS Energy Lett.* 3 (7): 1655–1661.

72 Wang, Q., Liu, Z., Zhao, H. et al. (2018). MOF-derived porous Ni_2P nanosheets as novel bifunctional electrocatalysts for the hydrogen and oxygen evolution reactions. *J. Mater. Chem. A* 6 (38): 18720–18727.

73 Li, X., Wang, X., Zhou, J. et al. (2018). Ternary hybrids as efficient bifunctional electrocatalysts derived from bimetallic metal-organic-frameworks for overall water splitting. *J. Mater. Chem. A* 6 (14): 5789–5796.

74 Bu, F., Chen, W., Aly Aboud, M. et al. (2019). Microwave-assisted ultrafast synthesis of adjustable bimetal phosphide/graphene heterostructures from MOFs for efficient electrochemical water splitting. *J. Mater. Chem. A* 7 (24): 14526–14535.

75 Yan, L., Jiang, H., Xing, Y. et al. (2018). Nickel metal–organic framework implanted on graphene and incubated to be ultrasmall nickel phosphide nanocrystals acts as a highly efficient water splitting electrocatalyst. *J. Mater. Chem. A* 6 (4): 1682–1691.

76 Hong, W., Kitta, M., and Xu, Q. (2018). Bimetallic MOF-derived FeCo-P/C nanocomposites as efficient catalysts for oxygen evolution reaction. *Small Methods* 2 (12): 1800214.

77 He, P., Xie, Y., Dou, Y. et al. (2019). Partial sulfurization of a 2D MOF array for highly efficient oxygen evolution reaction. *ACS Appl. Mater. Interfaces* 11 (44): 41595–41601.

78 Du, J., Wang, R., Lv, Y. et al. (2019). One-step MOF-derived Co/Co_9S_8 nanoparticles embedded in nitrogen, sulfur and oxygen ternary-doped porous carbon: an efficient electrocatalyst for overall water splitting. *Chem. Commun.* 55 (22): 3203–3206.

79 Wang, D., Tian, L., Huang, J. et al. (2020). "One for two" strategy to prepare MOF-derived $NiCo_2S_4$ nanorods grown on carbon cloth for high-performance asymmetric supercapacitors and efficient oxygen evolution reaction. *Electrochim. Acta* 334: 135636.

80 Zou, H., He, B., Kuang, P. et al. (2018). Metal-organic framework-derived nickel-cobalt sulfide on ultrathin mxene nanosheets for electrocatalytic oxygen evolution. *ACS Appl. Mater. Interfaces* 10 (26): 22311–22319.

81 Liu, T., Yang, F., Cheng, G., and Luo, W. (2018). Reduced graphene oxide-wrapped $Co_{9-x}Fe_xS_8/Co$, Fe–N–C composite as bifunctional electrocatalyst for oxygen reduction and evolution. *Small* 14 (10): 1703748.

82 Guo, J., Lin, C., Xia, Z., and Xiang, Z. (2018). A pyrolysis-free covalent organic polymer for oxygen reduction. *Angew. Chem. Int. Ed.* 57 (38): 12567–12572.

83 Zhu, C., Li, H., Fu, S. et al. (2016). Highly efficient nonprecious metal catalysts towards oxygen reduction reaction based on three-dimensional porous carbon nanostructures. *Chem. Soc. Rev.* 45 (3): 517–531.

84 Aiyappa, H., Thote, J., Shinde, D. et al. (2016). Cobalt-modified covalent organic framework as a robust water oxidation electrocatalyst. *Chem. Mater.* 28 (12): 4375–4379.

85 Yang, C., Yang, Z., Dong, H. et al. (2019). Theory-driven design and targeting synthesis of a highly-conjugated basal-plane 2D covalent organic framework for metal-free electrocatalytic OER. *ACS Energy Lett.* 4 (9): 2251–2258.

86 Mondal, S., Mohanty, B., Nurhuda, M. et al. (2020). A thiadiazole-based covalent organic framework: a metal-free electrocatalyst toward oxygen evolution reaction. *ACS Catal.* 10 (10): 5623–5630.

87 Wu, Z., Sun, S., Tan, Y. et al. (2012). Three-dimensional graphene-based macro-and mesoporous frameworks for high-performance electrochemical capacitive energy storage. *J. Am. Chem. Soc.* 134 (48): 19532–19535.

88 Zhao, X., Pachfule, P., Li, S. et al. (2019). Macro/microporous covalent organic frameworks for efficient electrocatalysis. *J. Am. Chem. Soc.* 141 (16): 6623–6630.

89 Wang, A., Cheng, L., Zhao, W. et al. (2020). Electrochemical hydrogen and oxygen evolution reactions from a cobalt-porphyrin-based covalent organic polymer. *J. Colloid Interface Sci.* 579: 598–606.

90 Singh, A., Roy, S., Das, C. et al. (2018). Metallophthalocyanine-based redox active metal-organic conjugated microporous polymers for OER catalysis. *Chem. Commun.* 54 (35): 4465–4468.

91 Cote, A., Benin, A., Ockwig, N. et al. (2005). Porus, crystalline, covalent organic frameworks. *Science* 310 (5751): 1166–1170.

92 Feng, X., Ding, X., and Jiang, D. (2012). Covalent organic frameworks. *Chem. Soc. Rev.* 41 (18): 6010–6022.

93 Zhang, H., Zhu, M., Schmidt, O. et al. (2021). Covalent organic frameworks for efficient energy electrocatalysis: rational design and progress. *Adv. Energy Sustain. Res.* 2 (4): 2000090.

94 Ma, B., Guo, H., Wang, M. et al. (2020). Electrocatalysis and simultaneous determination of hydroquinone and acetaminophen using PNCOF/graphene oxide modified electrode. *Microchem. J.* 155: 104776.

95 Jia, H., Sun, Z., Jiang, D., and Du, P. (2015). Covalent cobalt porphyrin framework on multiwalled carbon nanotubes for efficient water oxidation at low overpotential. *Chem. Mater.* 27 (13): 4586–4593.

96 Hou, Y., Cui, C., Zhang, E. et al. (2019). A hybrid of gC_3N_4 and porphyrin-based covalent organic frameworks via liquid-assisted grinding for enhanced visible-light-driven photoactivity. *Dalton Trans.* 48 (40): 14989–14995.

97 Gan, Z., Lu, S., Qiu, L. et al. (2020). Fine tuning of supported covalent organic framework with molecular active sites loaded as efficient electrocatalyst for water oxidation. *Chem. Eng. J.* 415: 127850.

98 Huang, D., Wang, Q., Dong, W. et al. (2022). Atomic Co embedded in a covalent triazine framework for efficient oxygen evolution catalysis. *Energy Fuel* 36 (19): 11601–11608.

99 Yamaguchi, S., Kamiya, K., Hashimoto, K., and Nakanishi, S. (2017). Ru atom-modified covalent triazine framework as a robust electrocatalyst for selective alcohol oxidation in aqueous electrolytes. *Chem. Commun.* 53 (75): 10437–10440.

100 Mullangi, D., Dhavale, V., Shalini, S. et al. (2016). Low-overpotential electrocatalytic water splitting with noble-metal-free nanoparticles supported in a sp^3 N-rich flexible COF. *Adv. Energy Mater.* 6 (13): 1600110.

101 Wang, W., Luo, J., and Chen, S. (2017). Carbon oxidation reactions could misguide the evaluation of carbon black-based oxygen-evolution electrocatalysts. *Chem. Commun.* 53 (84): 11556–11559.

102 Gopi, S., Giribabu, K., and Kathiresan, M. (2018). Porous organic polymer-derived carbon composite as a bimodal catalyst for oxygen evolution reaction and nitrophenol reduction. *ACS Omega* 3 (6): 6251–6258.

103 Jia, H., Yao, Y., Gao, Y. et al. (2016). Pyrolyzed cobalt porphyrin-based conjugated mesoporous polymers as bifunctional catalysts for hydrogen production and oxygen evolution in water. *Chem. Commun.* 52 (92): 13483–13486.

Part V

Advanced Functional Materials for Electrocatalytic CO_2 Reduction Reaction

14

Cu-Based Metal Materials for Electrocatalytic CO_2 Reduction Reaction

Mingliang Hu[1,2,], Junjun Li[1,2,*], Yongxia Shi[1,2,*], Man Hou[1,2,*], Xinyi Wang[1,2,*], Zhenwei Zhao[1,2,*], and Zhicheng Zhang[1,2]*

[1] *Tianjin University, School of Science, Tianjin Key Laboratory of Molecular Optoelectronic Sciences, Department of Chemistry, No.92 Weijin Road, Tianjin, 300072, China*
[2] *Tianjin University, School of Science, Collaborative Innovation Center of Chemical Science and Engineering (Tianjin), No.92 Weijin Road, Tianjin, 300072, China*

14.1 Introduction

The development of modern society has consumed large amounts of fossil fuels, and excessive CO_2 emissions have caused serious environmental issues [1, 2]. The high level of CO_2 in the atmosphere has disturbed the balance of the carbon cycle in nature, leading to a variety of environmental problems such as climate warming and ocean acidification [3, 4]. Both human living space and balanced natural ecosystems can be seriously affected by these problems. Researchers are working on converting CO_2 into valuable chemicals and fuels through chemical processes, which has become a research hotspot for reducing CO_2 emissions and tackling the energy crisis. Typical conversion methods, such as thermocatalysis, photocatalysis, electrocatalysis, biomass conversion, and mineralization, have been successfully applied to achieve the goal of chemical conversion of CO_2 [5–8]. Among them, the electrochemical CO_2 reduction reaction (CO_2RR) has become a popular method due to its easy operation, mild operation conditions, and ability to produce a diverse range of reduction products. The CO_2RR offers a solution to convert CO_2 into basic chemicals and fuels more efficiently [9, 10]. In addition, as the cost of converting renewable energy sources (wind energy, solar energy, tidal energy, etc.) into electricity gradually decreases, CO_2RR shows a high industrial application potential and favorable economics. In conclusion, CO_2RR can convert CO_2 into key chemicals and fuels, realizing efficient storage of green energy and providing a promising way to close the carbon cycle [11].

The highly linear and symmetric molecular structure makes CO_2 remain strongly chemically inert, with a C=O bond energy of 750 kJ mol^{-1}. It is very necessary to

* These authors contributed equally to this work.

Functional Materials for Electrocatalytic Energy Conversion, First Edition.
Edited by Zhicheng Zhang, Meiting Zhao, and Yuchen Qin.

activate the CO_2 molecules for highly efficient CO_2RR [12]. Typically, high overpotential is used to overcome the high energy barrier of activated CO_2 molecules. However, over-negative potentials are inevitable to result in a competitive hydrogen evolution reaction (HER) in an aqueous electrolyte environment, which leads to a significant decrease in Faraday efficiency (FE) for CO_2RR [13]. In addition, the wide range of products of electrocatalytic CO_2RR, including CO, CH_4, CH_3OH, C_2H_4, C_2H_5OH, C_3H_6, etc., is formed through multiple proton-coupled and electron transfer steps [14–17]. However, the complex pathways lead to slow reaction kinetics and complicated reaction mechanisms. Therefore, it is crucial to design electrocatalysts with high catalytic activity, selectivity, and stability, for inhibiting HER and accelerating electron–proton transfer. Since Hori et al. first discovered the unique electrocatalytic CO_2 reduction properties of copper (Cu) catalysts, the research and design of Cu-based catalysts have progressed [18]. The outstanding electrocatalytic activity of Cu-based catalysts, owing to suitable binding energy for *CO intermediates, favorable C–C coupling, and the ability to produce various multicarbon products [19, 20]. Sixteen different products have been reported for Cu-based catalysts applied to the CO_2RR process [21]. Despite obtaining many different products on Cu-based catalysts, the presence of shared intermediates reduces the selectivity of CO_2RR. For example, when *CO is the intermediate, a variety of products, such as CO, CH_4, CH_3OH, C_2H_4, CH_3CH_2OH, etc. can be generated [22, 23]. In recent years, the performance of Cu-based catalysts has been continuously improved, and the FE of C_1 products (CO, HCOOH) can reach nearly 100% under industrial-grade current densities. However, for other products, especially multicarbon (C_{2+}) products, the selectivity, yield, and energy efficiency are still too low and far from reaching the industrial application level [24].

In recent years, researchers have devoted themselves to improving the activity and selectivity of Cu-based catalysts for CO_2RR. Various strategies such as size tuning [25, 26], crystal surface engineering [27], structural design [28, 29], and surface modification [30, 31] have been used to enhance the efficiency of Cu catalysts. Size effects can modulate the chemisorption of reactant molecules on the surface of Cu-based catalysts, thus affecting the final reactivity and product selectivity. The different Cu crystalline surfaces can influence the formation of the energy barrier of intermediates, and different types of intermediates play a crucial role in the formation of final products. Designing Cu-based catalysts with porous structure significantly increases the specific surface area, which allows more active sites to be fully exposed, leading to enhanced activity and yield of CO_2RR by enabling more reactants to participate in the reaction process. As for the surface modification strategy, modification of functional groups, construction of defects, or grain boundaries on the surface of Cu-based catalysts can directly affect reactant activation and the formation of intermediates and products. Introducing other metals to obtain Cu-based multimetallic catalysts is a strategy to enhance the performance of CO_2RR. The interaction of different metals with Cu adjusts the electronic structure of Cu-based catalysts, optimizes the binding strength between intermediates and active sites, and adjusts the final selectivity of products. Besides Cu, other metals can also participate in the reaction as active sites, providing more possibilities for CO_2RR [32–34].

The purpose of this chapter is to offer an overview of the latest strategies and developments in Cu-based metal catalysts for CO_2RR. The application of Cu-based catalysts for CO_2RR is discussed through both theoretical and experimental approaches, including Cu-based bimetallic catalysts and Cu-based trimetallic catalysts. This discussion covers the synthesis strategies, the relationship between structure and performance, and the advanced catalytic mechanisms of Cu-based metal catalysts for electrocatalytic CO_2RR. Finally, the challenges and opportunities of Cu-based metal catalysts for electrocatalytic CO_2RR are proposed.

14.2 Cu-Based Metal Materials for Electrocatalytic CO_2 Reduction

14.2.1 Cu-Based Bimetal Materials for Electrocatalytic CO_2 Reduction

14.2.1.1 Cu–Co

Group VIII B metals, such as Co, are attractive candidates for CO_2 electroreduction owing to their various electronic structures, adjustable composition, and low cost. Therefore, constructing Cu–Co bimetallic catalysts is expected to improve the selectivity of specific products over Cu-based catalysts. For example, Bernal et al. precisely controlled the size and composition of Cu–Co NPs via inverse micellar encapsulation and systematically studied the effect of size and composition of Cu–Co bimetallic NPs on CO_2RR performance [35]. Electrochemical results showed that an increase in particle size is favorable for CO_2RR over HER, while small amounts of Co could improve the activity of Cu–Co NPs. Operando XAS and quasi in situ XPS were carried out to investigate the evolution of the Cu–Co NPs surface and the local atomic structure during CO_2RR. A drastic difference between the as-prepared state of the bimetallic NPs and the working catalyst was observed. Specifically, both initially oxidized metallic species in Cu–Co NPs were reduced, and the Cu underwent surface segregation. DFT results showed that the expansion and compression of the Cu–Cu interatomic distance could weaken the binding energy between the surface and key intermediates, such as COOH*, CO*, and H*, thus affecting the CO_2RR performance. Recently, well-dispersed Cu–Co bimetallic NPs on porous carbon were synthesized via a simple ZIF-derived method, which showed high selectivity of CO [36]. The maximum FE of CO reached 97.4% with a current density of 62.1 mA cm^{-2} at −2.1 V vs. Ag/Ag^+ in MeCN containing 0.5 M $[Bmim]PF_6$. The high activity and selectivity toward CO could be attributed to the synergistic effect between Cu and Co, i.e. higher electrochemical surface area, more space for mass diffusion, strong CO_2 adsorption, and appropriate $CO_2^{\bullet-}$ adsorption.

14.2.1.2 Cu–Ni

Due to strong CO affinity, Ni surface tends to be poisoned during electrochemical CO_2RR. Alloying Ni with other metals can effectively destabilize the Ni–CO interaction, thus alleviating poisoning. However, it has been reported that combining Cu with hydrogen generation elements such as Ni could increase the activity for hydrogen production compared with pure Cu [13, 37]. In terms of higher H^+ adsorption energy, Ni bounding on the surface or adjacent to the Cu sites is in favor of the formation of C_2 products on Cu. Thanks to the stronger CO affinity to Ni rather than Cu, the dropped Ni–Cu surface would possess increased CO coverage at low overpotential, thus facilitating the kinetics of the crucial C–C coupling step [38].

Nonuniform distribution of Ni on alloy surface from surface segregation tendency of Ni limits further application of Cu–Ni bimetal catalysts [39]. To address this issue, Zhang et al. proposed that highly dispersed Ni on Cu surface could be more efficient for C–C coupling and the formation of C_2 products. [38]. Atomic nickel cluster decorated defect-rich copper was successfully prepared via coordination-enabled galvanic replacement method. Compared with nondecorated Cu, Ni-decorated Cu nanowires (NWs) a showed sevenfold increase in the selectivity of C_2 products. DFT results showed that the RDS for *CO formation is the formation of *COOH on Cu with Ni–Ni bridge as binding sites for $*CO_2$ and *COOH. For Ni-decorated surfaces, the chemisorption of CO_2 is RDS with Ni–Cu bridge as binding sites for *CO, which is favorable for the key C–C coupling step.

Besides surface segregation of Ni, severe self-aggregation of small-sized nanoparticles during CO_2RR, resulting from high surface energy, could also lead to degradation or deactivation of the active sites of the catalyst. To balance the surface energy of ultrasmall nanoparticles, Fu et al. proposed a graphdiyne (GDY)-stabilization strategy to improve dispersity of the catalyst, stabilize nanoparticles, and promote mass transfer [40]. As a new type of two-dimensional (2D) carbon material, GDY possesses intrinsic charge carrier mobility. Cu/Ni@GDY, in which GDY acts as a substrate to stabilize the Cu–Ni alloy NPs, shows high selectivity toward CO (a FE of 95.2%) and exhibits ultralong stability of 26 hours. In addition, Cu–Ni alloy nanoparticles anchored on nitrogen-doped carbon nanotubes could also achieve highly efficient CO_2 electroreduction to CO (a FE of 94.8%) [41].

14.2.1.3 Cu–Ga

Recently, Bagchi et al. developed two Cu–Ga intermetallic compounds, $CuGa_2$ and Cu_9Ga_4 [42]. $CuGa_2$ possesses a unique 2D-flake-type structure that could maintain surface and subsurface oxide species even in the reduction atmosphere. While for Cu_9Ga_4, three-dimensional (3D) spherical morphology hinders the surface oxide formation (Figure 14.1a,b). Compared with Cu_9Ga_4, $CuGa_2$ showed a higher selectivity of methanol (77.26%) at ultralow potential (−0.3 V vs. RHE), contributing to the formation of a higher amount of surface Ga_2O_3 (Figure 14.1c,d). Note that $CuGa_2$ exhibited lower catalytic activity for methanol at higher negative potential. In situ XAFS, quasi in situ powder XRD, and ex situ XPS measurements demonstrated that the Ga_2O_3 layer was reduced to metallic Ga accompanying the lattice expansion, which is responsible for the decreased FE for methanol at higher potential (Figure 14.1e,f).

Besides C_1 products, strong adsorption of *COOH and *CO by Ga atoms makes it promising to enhance the adsorption of *CO and *COOH and subsequent C–C coupling to C_2 products over Cu–Ga bimetallic catalysts. For example, Sun et al. reported Cu-5Ga alloy plates with large-scale manufacturing capability, and after further heat treatment and electroreduction, the as-obtained catalysts achieved an FE of 36% for

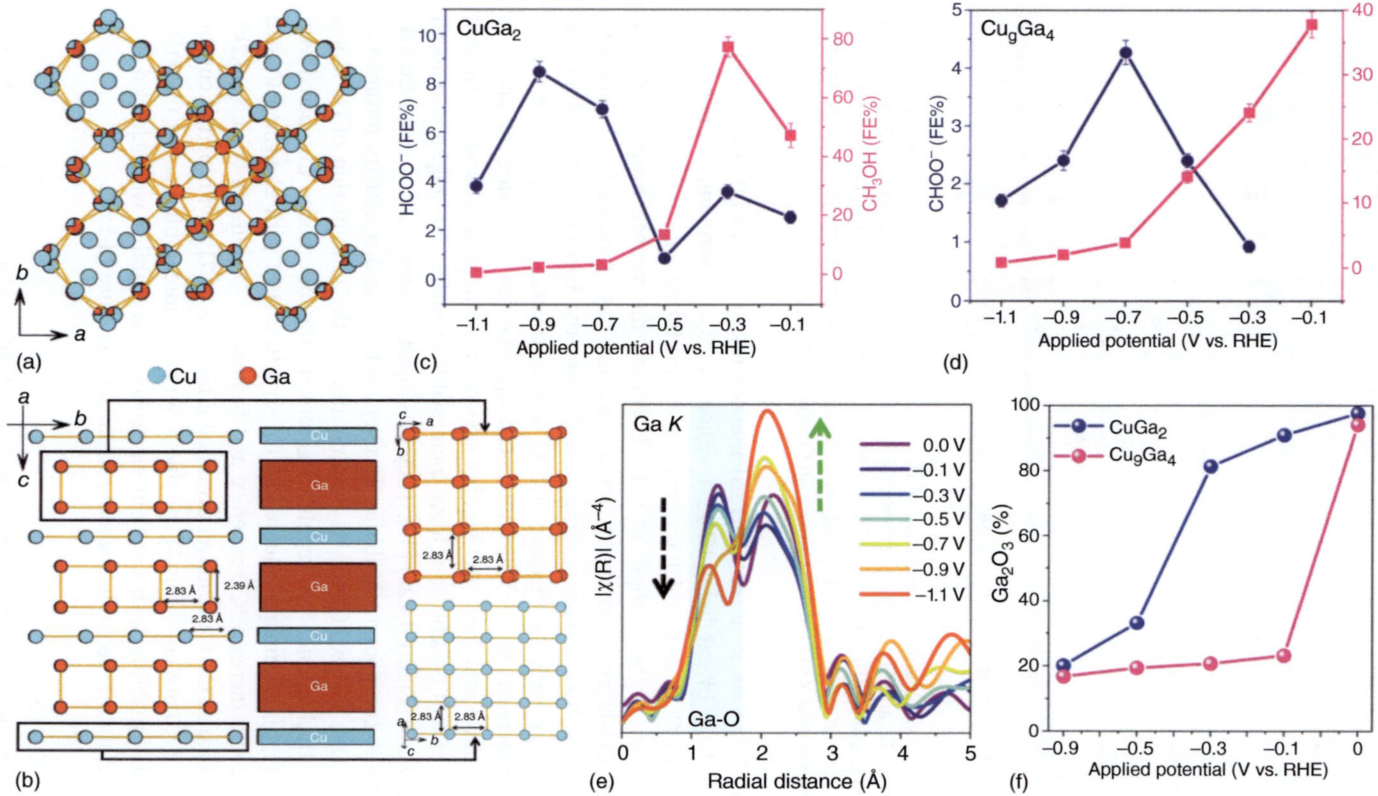

Figure 14.1 Crystal structure of (a) Cu_9Ga_4 and (b) $CuGa_2$. (c) Fes of methanol and formate over (c) $CuGa_2$ and (d) Cu_9Ga_4. (e) FT R-space of Ga K-edge for $CuGa_2$ during in situ XAS. (f) The change of surface Ga_2O_3 on $CuGa_2$ and Cu_9Ga_4 at applied potentials. Source: Reproduced with permission of Bagchi et al. [42], Wiley-VCH.

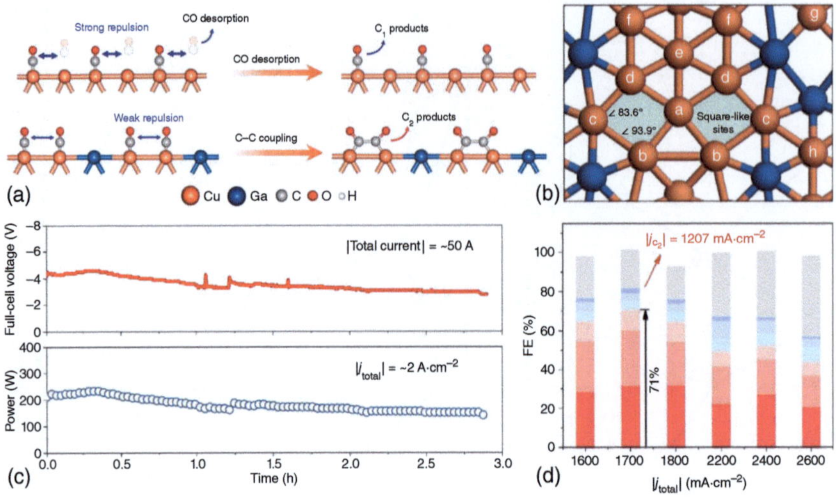

Figure 14.2 (a) Schematic illustration of the CO_2RR pathway with strong or weak *CO repulsion on Cu and Cu_9Ga_4, respectively. (b) Schematic of different Cu and Ga atoms on the optimized $Cu_9Ga_4(110)$ surface. (c) FEs of different products over Cu_9Ga_4 catalyst. (d) High-power (~200 W) CO_2RR of Cu_9Ga_4 in a 25 cm^{-2} MEA electrolyzer. Source: Reproduced with permission of Yan et al. [45], American Chemical Society.

C_2H_6 at −0.45 V vs. RHE [43]. DFT calculations reveal that the introduction of Ga atom is in favor of stabilizing *COOH, enhancing *CO adsorption, and promoting *CO*CO-*COCOH step, which is crucial for ethane formation.

In addition to the strength of CO adsorption, the surface coverage of local adsorbed CO is an important factor dominating C–C kinetics [44]. Generally, the increased *CO surface coverage could lower the barrier for *CO dimerization. However, with the increase of *CO surface coverage, *CO repulsion would inevitably increase, which will limit C–C coupling efficiency. Yan et al. proposed that both active and inert *CO binding sites on a highly ordered and interspaced structure make it possible to achieve increased *CO surface coverage and reduced *CO repulsion (Figure 14.2a) [45]. They developed an atomically ordered Cu_9Ga_4 intermetallic compound, exhibiting excellent activity and selectivity toward C_{2+} products. Due to low *CO or *H binding affinity, the introduction of Ga atoms could effectively reduce *CO repulsion as catalytically inert nodes. Compared with pure Cu (100), the unique square-like Cu structure sites with elongated Cu–Cu distance could increase *CO coverage and facilitate C–C coupling (Figure 14.2b). The as-prepared Cu_9Ga_4 catalyst exhibited a high FE of 71% with partial current density of 1207 mA cm^{-2} for C_{2+} products. What is more, a high-power (~200 W) electrolysis capability with excellent electrochemical stability was achieved in a 25 cm^2 membrane electrode assembly electrolyzer (Figure 14.2c,d).

14.2.1.4 Cu–Ce

Due to the distinctive electronic structure of $4f^{0-14}$ electrons, rare earth (RE) elements such as Ce are extensively used for applications in electrochemical utilizations. It is expected that the introduction of Ce into Cu will form oxygen vacancies and modify the electronic structure of Cu, thus improving the selectivity and activity toward C_{2+} products. Shan et al. synthesized Ce-doped Cu nanoparticles (Ce–Cu NPs) via simple one-step coreduction method, which showed a maximum FE of 53% for C_2H_4 with a current density of 150 mA cm^{-2} [46]. The enhanced performance could be attributed to shrinking particle sizes and more catalytic sites by introducing Ce into Cu. More importantly, Ce could change the electronic state of Cu, boost local electronic redistribution of Cu, and form oxygen defects into the neighboring positions, thus improving the selectivity of C_2H_4. In situ spectroscopies confirmed that the stabilized CuO_x species and enhanced adsorption of *CO are beneficial to C–C coupling to C_2H_4.

14.2.1.5 Cu–Bi

Among all the products of electrochemical CO_2RR, formate (HCOOH) is commonly used in hydrogen storage carriers and fuel cells due to its high energy density and high hydrogen density [47]. In recent years, people have been committed to developing advanced and efficient electrocatalysts. However, due to the inertness of CO_2 and the complexity of its reduction pathways, it has become difficult to prepare highly active and selective electrochemical CO_2 reduction catalysts for formate. Metal-based materials have been widely reported as effective electrocatalysts for CO_2 reduction to formic acid, including Pb, Sn, In, Cd, and Bi-based materials [48–51]. Among them, Bi-based materials are considered the most suitable electrocatalysts for the conversion of CO_2 to formate due to their low toxicity, low cost, good stability, environmental friendliness, and high overpotential in HER [33, 52]. However, there are still some challenges to be solved for industrial applications, and there are still very few electrocatalysts with long-term stability at high current densities, which seriously hinders the industrial application of catalysts. Therefore, the development of Bi-based electrocatalysts with efficient comprehensive catalytic performance is urgently needed.

Introducing guest metals to form bimetallic active sites is a recognized method for improving catalyst performance. In our previous introduction, Cu has a unique electronic structure and good conductivity, which is often chosen as a guest metal. Introducing Cu into Bi-based catalysts can not only adjust the electronic structure of the catalyst but also optimize the binding energy of reaction intermediates at active sites, thereby promoting the formation of formate. Li and coworkers [53] prepared a Cu_1Bi_1 bimetallic ginger-shaped composite catalyst ($CuO/CuBi_2O_4$) using a coprecipitation calcination method. The FE of the catalyst can maintain over 85% over a wide potential range (500 mV), and the current density of formate can reach −56.12 mA cm^{-2} at −1.28 V vs. RHE. The catalyst exhibits stability for up to 60 hours at −0.98 V vs. RHE. Research has shown that the main components in Cu_1Bi_1 are CuO and $CuBi_2O_4$, and their electronic interactions are close. Electrons in Cu_1Bi_1 can transfer from Cu with lower electronegativity to Bi with higher electronegativity, indicating a clear interface structure between CuO and $CuBi_2O_4$. When using the classic three-electrode system for electrochemical carbon dioxide reduction performance testing, Cu_1Bi_1 exhibits high formate selectivity, with only carbon monoxide and hydrogen gas in its meteorological products, and both have low Faraday efficiencies, while only formate is present in the liquid phase products. This also indicates that the composite structure of CuO and $CuBi_2O_4$ can promote the formation of formate while inhibiting the occurrence of HER. Further mechanistic studies have shown that the abundant $CuO/CuBi_2O_4$ interface in Cu_1Bi_1 is conducive to charge transfer and provides a large number of active sites. It also facilitates the stable adsorption of CO_2^{*-} intermediates on the active sites of Cu_1Bi_1, promoting the adsorption and activation of CO_2, and ultimately converting it into formate.

In recent years, single-atom alloys (SAA) have become an emerging strategy for constructing catalysts. Single-atom catalysts (SACs) loaded on solid substrates with monodisperse single atoms have attracted increasing attention from scholars due to their maximum atomic utilization efficiency. Recently, Zhang and coworkers [54] prepared a single-atom bismuth modified copper alloy (BiCu-SAA) through in situ electrochemical reconstruction method. They reported and demonstrated for the first time that the SAA catalyst can promote C–C coupling during CO_2 electroreduction, selectively converting CO_2 to C_{2+} products, and achieving FE_{C2+} over 70% in the CO_2RR at a current density of 400 mA cm^{-2}. Research has shown that dispersed Bi single atoms are formed by replacing the position of Cu atoms. When evaluating the CO_2RR performance of the catalyst in a three-electrode H-type battery with 0.1 M $KHCO_3$ as the electrolyte, it was found that the total FE of C_{2+} product could reach 73.4%. DFT calculations show that the free energy of CO_2 adsorption to generate $*CO_2$ on Cu(111)-nanocatalyst modified with single-atom Bi is greatly reduced. This also indicates that CO_2 is more easily activated on BiCu(111)-SAA, thereby accelerating the C–C coupling step and generating C_{2+} products.

In addition, Li and coworkers [55] developed a multistep *in situ* construction strategy, designing a 3D porous core–shell nanostructure consisting of a lattice defect Bi nanosheet shell and a Cu NW (Cu NWs) core (Cu@Bi NWs/Cu). The synergistic effect of lattice defects Bi Ns and 3D porous core–shell nanostructures result in a $FE_{formate}$ of up to 90.1%. Theoretical studies have shown that due to the presence of atomic defects on the Bi (110) surface, four low coordination atoms are generated around the Bi atom, resulting in more active sites, which is beneficial for the reduction of CO_2.

14.2.1.6 Cu–In

Among the various products of CO_2RR, CO provides a multifunctional compound to produce liquid fuels and plastics through well-established industrial processes. Previously, we summarized that Cu–Bi-based materials are commonly used in the production of formates or C_{2+} products in CO_2RR, while the Cu–In materials, we are referring to now are mainly used in the production of liquid fuel CO [56–58]. Cu–In (or oxide) materials are widely used in the conversion of CO_2 to CO due to their low cost advantages. For example, Takanabe and coworkers [58, 59] developed Cu–In alloy catalysts to suppress HER and achieve high CO selectivity in the CO_2RR process, achieving FE_{CO} of up to 90% at −0.6 V vs. RHE.

More importantly, the interface between In and Cu has attracted widespread attention from scholars, who believe that this may be one of the key factors for the high efficiency of electrochemical CO_2 to CO. Luo et al. [60] covered a thin In layer on Cu NWs, achieving ~93% FE_{CO} at −0.6 to −0.8 V vs. RHE and exhibiting stability up to 60 hours. Afterward, Dai and coworkers [56] synthesized a composite Cu–In_2O_3 catalyst through an oleylamine method. Theoretical calculations and some in situ characterizations have shown that strong electronic interactions and coupling occur at the Cu/In_2O_3 interface. In_2O_3 undergoes oxidation–reduction reactions during the CO_2RR process, preventing Cu from being oxidized, protecting the metallic state of Cu, and promoting the formation of COOH* while reducing the adsorption of CO*, resulting in up to 95% FE_{CO} at −0.7 V vs. RHE. In addition, the novel perforated nitrogen-rich graphene-like carbon nanolayers (PNGC) serve as a good carrier. Du and coworkers [61] prepared a porous nitrogen-rich graphene layered carbon using biomass derivatives as raw materials to promote the directional deposition of In-doped $Cu_2(OH)_3(NO_3)$ nanosheets and prepared a Cu–In/PNGC catalyst using in situ electrochemical reduction method. Among them, PNGC can serve as a powerful CO_2 enrichment carrier combined with Cu–In nanosheets as an active component, exhibiting high CO_2 gas adsorption capacity and effectively reducing the free energy generated by key intermediates (COOH*), thereby promoting the activation and conversion of CO_2. DFT calculations indicate that there is a significant amount of charge transfer and accumulation around Cu–In. The redistribution of this charge between PNGC and Cu–In can improve the electron mobility in the composite catalyst, increase the charge transfer rate at the binding site, and achieve high FE_{CO} over a wide potential range.

In addition, in recent years, there have been studies on the production of ethanol through In–Cu bimetallic sites. Wang and coworkers [62] synthesized $CuInS_2$ with sulfur vacancies using an electrodriven intercalation method. Research has shown that Cu^{2+} can be inserted into the crystal structure of In_2S_3, which is conducive to the formation of Cu–In double active sites, leading to delocalization of electron density, enhanced orbital hybridization between Cu—C and In—H bonds, which is conducive to the adsorption of CO and H intermediates, promotes the formation of *CHO intermediates, reduces the energy barrier generated by the key intermediate *OCHCHO*, and further promotes the C–C coupling process to generate ethanol during CO_2RR, achieving 71 ± 3% $FE_{C_2H_5OH}$, simultaneously, it has a stability of 120 hours.

14.2.1.7 Cu–Zn

Among all the copper- based bimetallic catalysts, zinc–copper (Zn–Cu) bimetallic electrocatalyst stands out due to its eco-friendliness and cost-friendliness, as well as its significant ability to selectively regulate CO_2RR products, which have been considered as alternatives to precious Cu–Au and Cu–Ag catalysts [13, 63]. Due to the higher binding energy of Cu to *COOH and *CO compared to Zn, the combination of Cu and Zn is expected to optimize the adsorption of key intermediates on the bimetallic surface through synergistic effect, thereby improving the CO_2RR performance. More importantly, due to the poor activity of Zn to HER, Cu–Zn bimetallic electrode is expected to inhibit the formation of H_2 [16].

For example, Wang and coworkers investigated an active Zn–Cu bimetallic electrocatalyst prepared by a facile galvanic procedure for CO_2 reduction to CO and which shows higher intrinsic activity than a planar Ag catalyst within the operating potential window from −0.8 to −1.0 V vs. RHE. Physical and electrochemical characterization and first principles theory reveal that the enhanced intrinsic activity for CO formation is an electronic effect from the Cu adatoms on the Zn surface. The incorporation of Cu onto the Zn surface lowers the reaction energy (from 0.44 to 0.15 eV) of the PDS to form COOH*, stabilizes the carboxyl intermediate during CO_2 reduction to CO, and improves CO production rates. Meanwhile, due to the increased coverage of CO, it hinders the adsorption of H and suppresses HER [64]. A small change in Cu doping amount can significantly affect the morphology of Zn–Cu alloy, thereby affecting product selectivity and hydrogen evolution kinetics. Recently, a 3D porous Zn–Cu alloy electrode was synthesized with different Cu content using a simple hydrogen bubble template method. The regular pore structure of $Cu_{0.3}Zn_{9.7}$ can not only increase the active sites on the catalyst surface and enhance the interaction between intermediates and the catalyst but also significantly reduces the formation energy barrier of *CO intermediates, resulting in a high CO_2 reduction to CO [65]. Zn–Cu bimetallic electrocatalyst also exhibits unique advantages in the electrocatalytic generation of C_{2+} products from CO_2RR. Zhang and coworkers used CO* adsorption energy as a descriptor to select a series of Cu-based binary active sites and then calculated the reaction energy in the limiting steps. As a result, Zn was found as an appropriate candidate for alloying with Cu, displaying a weaker CO* binding on Cu–Zn_{Zn} site (the subscript indicates the adsorption site) than Cu to create asymmetric surface CO* binding energies with the lowest limiting energy for C–C coupling. Then they developed an alloying–dealloying strategy to fabricate experimentally a series of Zn-incorporated Cu catalysts. As a consequence, Zn with Cu enables asymmetric surface CO* bindings on surface binary sites to reduce the reaction energy for C–C coupling and the consequent hydrogenation, realizing a combined performance of 91 ± 2% C_{2+} FE with notable 73 ± 2% ethylene FE [66].

From that research, Zn–Cu catalysts show great potential for application in CO_2RR. However, the conclusions of previous studies on its catalytic selectivity are controversial. It is crucial to figure out the structure–activity relationship of catalysts and get high product selectivity for practical applications of CO_2RR. Cuenya and coworkers have employed size-controlled (∼5 nm) $Cu_{100-x}Zn_x$ nanoparticles supported on carbon to investigate the correlation between their structure and composition and catalytic performance. It was found that Cu–Zn NPs with Zn content of 10–50 can produce CH_4 with high selectivity (about 70% FE). With the increase of Zn concentration, the formation of Cu–Zn alloy exhibits selective conversion of reduction products from CH_4 to CO. Specifically, time-dependent XAFS data shows that the metal Cu in close proximity to ZnO leads to the production of CH_4 in the initial stage of the reaction. Under CO_2RR conditions, the gradual reduction of ZnO species and the concomitant enhanced Cu–Zn interaction, as well as the enhancement of brass alloy formation, results in the selective transition to only producing CO and H_2 [67]. Another study found that the metal distribution plays a crucial role, together with the composition in adjusting the selectivity of Cu–Zn catalysts.

Seyedeh et al. tailored Cu/ZnO precatalysts to investigate the compositional and structural sensitivities of the CO_2RR in bimetallic Cu–Zn catalysts to the promotion of ethanol vs. methane. The results indicate that ZnO@Cu-derived catalysts with a higher degree of surface alloying and a higher concentration of metallic Zn contribute to producing ethanol while Cu@ZnO-derived catalysts exhibit selectivity to methane. The DFT results indicate that the dilution amount of Zn impurities can affect the electronic structure of the catalyst and improve the selectivity of CH_4, and an increase in Zn content will increase the local CO concentration and achieve a tandem conversion mechanism, determining the selectivity from CH_4 to EtOH [68]. Yeo and coworkers investigated the selectivity of CO_2RR toward C_2H_5OH on a series of Cu_xZn catalysts. By adjusting the amount of Zn in Cu–Zn catalysts, the ratio of $FE_{C_2H_5OH}/FE_{C_2H_4}$ could be greatly tuned by a factor of up to ~12.5. A two-site mechanism is used to explain the formation process of C_2H_5OH as follows. The CO_2 molecules could first bind to Cu or Zn sites and be reduced to CO (1→2). At the Cu site, CO can be further reduced to CHO or CH_x (x = 1–3) intermediates. The adsorption of CO on Zn sites is weak and is likely to desorb. The desorbed CO can diffuse and overflow to Cu sites (2→3). The spilled CO may then be inserted into the binding between the Cu surface and $*CH_2$ to form $*COCH_2$ (3→4). Further reduction of $*COCH_2$ will produce acetaldehyde (4→5), followed by ethanol (5→6). In addition, Operando Raman spectroscopy has recorded the signals of CO adsorbed on Cu sites, further proving the tandem two-site mechanism [69]. In short, several factors, such as the striking structural differences, size regime, and composition of the catalysts certainly play an important role in directing selectivity.

Despite the former encouraging experimental results, the understanding of accurate atomistic structure–performance relations over Cu–Zn NPs is still missing. By combining neural network potential, DFT calculations and experimental measurements, Zhen et al. reveal the atomic level active sites of CO_2RR on a set of Cu–Zn NPs catalysts with different Zn doping concentrations (from 5% to 30%) and prove that Cu–Zn NPs driven by moderate Zn doping can promote CO_2RR to C_{2+} products because they have the highest abundance of adsorption sites of optimal ΔE_{*CO}. In addition, the Cu–Zn atomic arrangement can influence the interaction between the key intermediate *CO and the Cu–Zn surface, which affects the electron transfer between the *CO and the active sites. This work provides new insight into the identification of active sites and atom-level structure–performance relationship in Cu–Zn NP catalysts [70].

14.2.1.8 Cu–Al

Due to the wide variety of Cu alloys, the composition and proportion of elements can be adjusted; quickly seeking the optimal alloy composition is an important scientific challenge. Zhong et al. constructed a screening framework that combines machine learning with traditional DFT calculations to directly predict reactions and achieve the goal of predicting catalytic performance. Researchers first established 2D activity and selectivity volcano plots for CO_2 reduction based on the adsorption energy of CO and determined that the highest catalytic activity can be achieved when the optimal adsorption energy of CO is around −0.67 eV. Among multiple copper alloy candidates, Cu–Al alloys exhibit the highest adsorption sites and site types, with ΔE_{CO} values approaching the optimal range, indicating that Cu–Al alloys may have activity across a relatively wide range of surface compositions and site types. On the basis of a simple model, nanoporous Cu–Al alloy catalysts were prepared on a large-area gas diffusion electrode, and the content of the surface components of the alloy was adjusted. Notably, Cu–Al alloy achieved a FE of 80% for C_2H_4 at an ultrahigh current density of 600 mA cm^{-2}. The DFT results indicate that the addition of Al in Cu–Al alloy reduces the reaction energy of C–C bond formation, which is lower than the corresponding pure Cu surface. Meanwhile, the reaction energy for the formation of ethanol intermediate HO(CH)CH is higher than that for the formation of C_2H_4 intermediate CCH, thereby inhibiting the production of ethanol and promoting the production of C_2H_4. In addition, in situ X-ray absorption measurements indicate that Cu and Al can provide a favorable Cu coordination environment, thereby enhancing C–C dimerization [71]. At present, there is little research on the effect of Al atoms on the selective reduction of CO_2 to ethanol. Inspired by the work of Sargent et al., Wang et al. introduced the oxophilic Al atom into the Cu (111) surface to form $Al^{\delta+}$-$(Cu_6)^{\delta-}$ configuration on Al/Cu (111) bimetallic catalysts for electrochemical CO_2RR. The results showed that compared with the original Cu (111) surface, introducing Al atoms on the Cu (111) surface is thermodynamically and dynamically beneficial for the CO dimerization reaction. The Al center serves as an electron transfer medium for coupling two CO molecules on the Al/Cu (111) surface. At the same time, the presence of hydrophilic Al atoms on the Cu (111) surface effectively improves the selectivity of ethanol products by forming Al—O bond on the hydrogenation path [72].

14.2.1.9 Cu–Au

Au nanocrystals have attracted great interest in the fields of optics [73], photocatalysis [74, 75], and biomedical technology [76, 77] due to their controllable morphology and plasmonic properties. In addition, Au nanocrystals have shown great potential as effective catalysts with good selectivity for CO_2 reduction to CO. Therefore, Au–Cu bimetallic systems exhibit interesting electrocatalytic properties for CO_2 reduction [78, 79].

It is widely believed that *CO is a key intermediate in the electrochemical CO_2RR [13, 80, 81]. In CO_2RR, protonation of *CO to *CHO is a potential determining step during CH_4 generation while inhibiting HER and C–C coupling can improve the selectivity of CH_4 [82]. Sargent and coworkers [83] experimentally constructed a set of Au–Cu catalysts and controlled the availability of *CO by adjusting the concentration of CO_2 and reaction rate. The results showed that under low *CO coverage, introducing Au into Cu was beneficial for *CO protonation and C–C coupling, weakening the adsorption energy of *CO on the surface of *H, resulting in a decrease in hydrogen evolution.

There are synergistic interactions in Au–Cu bimetallic catalysts. Dai et al. [84] developed an elaborate Au–Cu catalyst (Cu@AuCu) to achieve efficient electrochemical conversion of CO_2 to CO. The synthesized Au_2–Cu_8 catalyst exhibits higher

FE_{CO} and CO_2RR mass activity than pure Au catalyst, but the Au content is as low as 17%. The results indicate that the role of Cu is to locate the d-band of surface Au atoms, thereby reducing the formation energy of COOH*. On the contrary, the role of Au is to reduce the desorption energy of CO* and recover the active sites of CO_2RR. In addition, Kuang et al. [85] prepared Cu/Au heterojunctions through systematically designed electrostatic self-assembly of positively charged Cu nanoparticles and negatively charged Au nanoparticles. Based on the asymmetric adsorption of OCCO* on Cu/Au surface, the selectivity of C_2H_5OH has been improved. In the intermediate product, one carbon atom binds to Au and the other carbon atom binds to Cu, enhancing the adsorption energy of Cu. The joint interaction between sites leads to the preferential hydrogenation of OCCO* at the carbon atom after binding with Cu, resulting in the intermediate OCCOH*. The FE of C_2H_5OH is 60% at a current density exceeding 500 mA cm^{-2}. In the modified catalyst, the ratio of ethanol to ethylene increased by 200 times compared to Cu catalyst.

So far, research on Cu-based bimetallic catalysts has mostly understood the active sites and reaction mechanisms from a theoretical perspective, lacking research on the real active sites in the CO_2RR process, especially under the premise of unstable metal structure dynamics. In view of this, Zhu et al. [86] studied the reconstruction/phase transition of bimetallic electrocatalysts during CO_2RR process using epitaxial Au–Cu heterojunctions as model structures. Electrochemical tests showed that compared to the elemental Cu, the epitaxial Au–Cu effectively increased the yield and overpotential of C_{2+} alcohol and inhibited the generation of hydrocarbons. Atomic scale characterization indicates that after CO_2RR reaction, the initial phase-separated Au–Cu interface transforms into unique AuCu alloy-supported epitaxial Au@Cu core–shell nanoclusters. Combining with in situ spectroscopy, finite-element simulation, and DFT calculation, the authors demonstrated that the in situ formed AuCu alloy and Cu shell serve as active phases for CO accumulation and CO–C_{2+} alcohol transition, respectively. Enriching CO is a key step to maintaining the sustained production of C_{2+} alcohols. This work fills the research gap in the dynamic reconstruction of CO_2RR bimetallic interfaces and establishes an example of cascade catalysis from CO_2 to C_{2+} alcohols at the atomic scale.

14.2.1.10 Cu–Ag

Constructing Cu-based multimetal catalysts is an important strategy to improve the poor selectivity of individual Cu catalysts toward multicarbon products. Ag is considered one of the promising candidate metals for constructing Cu-based bimetallic catalysts due to its ability to reduce CO_2 to CO with high selectivity. According to the different interactions between Cu and Ag, Cu–Ag bimetallic catalysts can be roughly divided into three categories [87]. The first type of CuAg catalyst is formed by physical mixing of Ag and Cu, where Ag creates a local environment for CO enrichment, and Cu can use the enriched CO to improve selectivity for multicarbon products [88]. The second type of CuAg catalyst is rich in Cu and Ag interfaces, which can further enhance the spillover of CO from Ag sites to the Cu surface [89–91]. In the third type of CuAg catalyst, the mixed energy changes caused by the large surface area of Cu and Ag nanostructures may result in slight alloying of the two at the nanoscale.

In recent years, researchers have made significant progress in the application of CuAg bimetallic catalysts for electrocatalytic formation of multicarbon products from CO_2RR [44, 92, 93]. For example, Ma et al. prepared three CuAg Janus nanostructures with {100} faces for electrocatalytic CO_2 reduction and found that CuAg heterogeneous nanocatalysts exhibited high selectivity for C_{2+} products [94]. Density functional theory calculations show that the CuAg Janus nanostructure with the best CO_2RR performance has the optimal electronic structure, promoting CO spillover effect, which is the reason for its high selectivity and low energy barrier to generate C_{2+} products.

In addition to highly selective preparation of C_2 products, CuAg bimetallic alloys have also made progress in electrocatalytic reduction of CO_2 to C_3 products. Due to the direct synthesis of C_3 products involving the coupling steps of C_1–C_1 and C_1–C_2, the energy barrier for direct conversion of CO_2 to C_3 molecules is high, and the FE of C_3 products is very low. Qi et al. developed an electrolyte supersaturation strategy to achieve electrocatalytic reduction of CO_2 to C_3 products [95]. They found that in the state of CO_2 supersaturation, the electro-substitution reaction of Cu and Ag separation was suppressed, achieving controlled co-deposition of CuAg alloy with priority exposure of CO_2 active surface (100). The experimental results and DFT theoretical calculations indicate that *CO plays an important role in C–C coupling, and the $*OCH_2CH_3$ intermediate is a key intermediate for selective generation of 2-propanol. The supersaturation strategy in this work can increase the local CO_2 concentration, improve the generation rate of *CO intermediates, and promote the electrocatalytic CO_2 reduction of CuAg alloy to prepare 2-propanol products. Compared to the absence of C_3 products detected in CO_2-saturated electrolytes, CuAg alloy exhibits excellent selectivity for 2-propanol products in highly carbonated electrolytes, with FE reaching a maximum value of 39.6% at −0.73 V vs. RHE.

Although many CuAg bimetallic catalysts have been developed for electrocatalytic CO_2RR, there is relatively little research on their evolution process in CO_2RR. Exploring the dynamic catalytic active centers in electrocatalytic CO_2RR is crucial for the rational design of CuAg bimetallic catalysts. Yang and coworkers prepared intermixed and phase-separated CuAg catalysts and revealed their chemical and structural evolution during the electrochemical CO_2RR process [87]. Due to the high mobility of Cu under CO_2RR conditions and the thermodynamic immiscibility of Ag and Cu, they usually tend to phase separate during the reaction, and the two CuAg catalysts evolve into similar structures. Specifically, Cu in the intermixed CuAg catalyst will leach out of the catalyst and then accumulate on the surface of the particles or recrystallize elsewhere during the CO_2RR process. The Cu in the phase-separated CuAg catalyst will be corroded and migrate to the Ag surface, resulting in a similar structure to that produced during the evolution process of the intermixed CuAg catalyst. Finally, the two metals in the CuAg catalyst undergo phase separation and exhibit a slightly mixed state, forming Ag rich and Cu rich domains ($Ag_{0.88}Cu_{0.12}$ and $Ag_{0.05}Cu_{0.95}$) within the thermodynamic solubility limit.

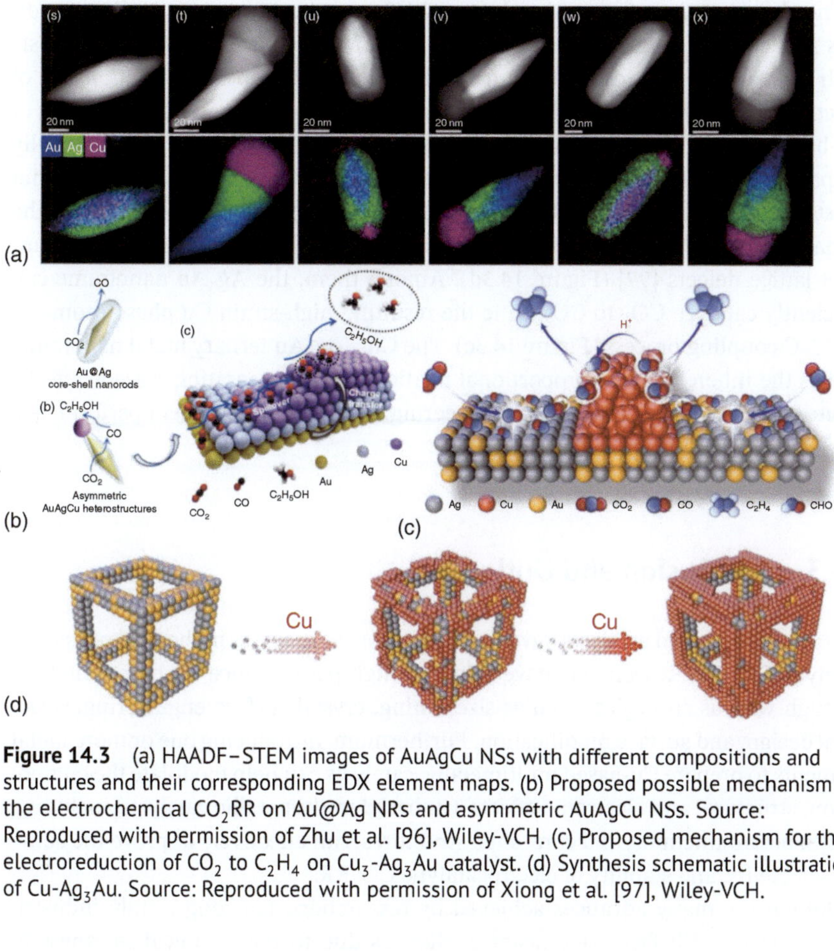

Figure 14.3 (a) HAADF–STEM images of AuAgCu NSs with different compositions and structures and their corresponding EDX element maps. (b) Proposed possible mechanism for the electrochemical CO_2RR on Au@Ag NRs and asymmetric AuAgCu NSs. Source: Reproduced with permission of Zhu et al. [96], Wiley-VCH. (c) Proposed mechanism for the electroreduction of CO_2 to C_2H_4 on Cu_3-Ag_3Au catalyst. (d) Synthesis schematic illustration of Cu-Ag_3Au. Source: Reproduced with permission of Xiong et al. [97], Wiley-VCH.

14.2.2 Cu-Based Trimetallic Materials for Electrocatalytic CO_2 Reduction

In addition to constructing Cu-based bimetallic catalysts, Cu-based trimetallic catalysts have also been proven to have advantages in improving the catalytic performance of CO_2RR. Zhu and coworkers prepared ternary asymmetric metal AuAgCu nanostructures with adjustable morphology and composition through a multistep seed mediated method for electrocatalytic CO_2RR [96]. Research has shown that changing the molar ratio of Au/Ag/Cu components can achieve structural regulation of asymmetric AuAgCu, thereby effectively regulating the selectivity of C_2 products. Specifically, Cu can selectively deposit on Au@Ag nanorods (Figure 14.3a). At −0.8 V vs. RHE, the $FE_{C_2H_5OH}$ of asymmetric $Au_1Ag_1Cu_5$ with three metal separation domains can reach 37.5%. The author believes that the increase in ethanol selectivity of ternary asymmetric metal AuAgCu nanostructures can be attributed to electronic effects and tandem catalytic mechanisms (Figure 14.3b).

The electrocatalytic CO_2RR involves multiple electrons from reaction intermediates participating in electrode reactions. Therefore, the key to designing catalysts with high CO_2RR performance is to break the linear proportion relationship of reaction intermediates [98, 99]. Another study also demonstrated the advantages of Cu-based trimetallic materials in electrocatalytic CO_2RR. Xiong et al. successfully prepared Cu_3–Ag_3Au ternary metal nanoframe materials with high strain and twisted Cu phases by carefully designing the composition and structure of the ternary metal and epitaxial growth of Cu on an alloyed Ag/Au substrate with abundant lattice defects [97] (Figure 14.3d). Among them, the Ag_3Au nanoframe can efficiently catalyze CO_2 to CO, while the resulting high-strain Cu phase promotes the C–C coupling process (Figure 14.3c). The Cu_3–Ag_3Au ternary metal nanoframe breaks the inherent linear proportional relationship by integrating cascade effects, electronic modulation, and defect engineering, and achieves excellent performance in electrocatalytic CO_2 reduction to C_2H_4.

14.3 Conclusion and Outlook

Up to now, Cu-based catalysts have shown significant progress in the field of electrocatalytic CO_2RR. Researchers have achieved high-performance Cu-based catalysts through various strategies, such as size tuning, crystal surface engineering, structural design, and surface modification. Furthermore, introducing one or more metal elements to prepare Cu-based multimetallic catalysts can help modulate the composition, structure, and electronic effects. Cu-based multimetallic catalysts, including alloys, heterojunctions, and core–shell structures, can effectively improve the activity, selectivity, and stability of electrocatalytic CO_2RR.

Despite the many advances achieved by researchers, realizing a fully industrialized CO_2RR still faces significant challenges due to the chemical inertness of CO_2 and the complex multistep proton–electron transfer pathway. It is necessary to investigate the coupling of key intermediates and clarify the electron transfer process in the reaction. Tandem catalytic system is a promising method to produce targeted products. The system allows for intermediate coupling and continuous electron–proton transfer. The microstructure of the catalyst constructed and regulated at an atomic level provided explicit reaction sites for the CO_2RR process, which has great significance for revealing the structure–performance relationship. Advanced in situ characterization methods have successfully observed the dynamic evolution of the fine structure at the atomic scale. Isotope tracing, spectroscopy analysis, and electronic characterization support understanding individual components under operando conditions. The combination of advanced characterization and theoretical calculations can provide a more comprehensive understanding of the intrinsic CO_2RR mechanism. Moreover, machine learning, high-throughput synthesis, and testing techniques can effectively guide experiments, reducing time costs and human resources, thus accelerating the development of advanced Cu-based metal catalysts for CO_2RR.

Cu-based metal catalysts have not yet reached a stage of wide-scale application. However, in-depth exploration in specific directions can help to industrialize the technology. In the future, more research needs to focus on the structural reconfiguration of Cu-based catalysts and avoiding the generation of carbonates. Furthermore, it is crucial to optimize Cu-based catalysts to reduce overpotentials, enhance energy utilization efficiency, and achieve high one-way conversion and FE under industrial-scale current densities.

Acknowledgment

This work was supported by the National Natural Science Foundation of China (22071172 and 22375142), 2022 Subsidized Project of Tianjin University Graduate Arts and Science Excellence Innovation Award Program (B2-2022-002).

References

1 Gattuso, J.-P., Magnan, A., Billé, R. et al. (2015). Contrasting futures for ocean and society from different anthropogenic CO_2 emissions scenarios. *Science* 349 (6243): 4722.

2 Wang, L., Chen, W., Zhang, D. et al. (2019). Surface strategies for catalytic CO_2 reduction: from two-dimensional materials to nanoclusters to single atoms. *Chem. Soc. Rev.* 48 (21): 5310–5349.

3 Sanz-Pérez, E.S., Murdock, C.R., Didas, S.A., and Jones, C.W. (2016). Direct capture of CO_2 from ambient air. *Chem. Rev.* 116 (19): 11840.

4 Weiss, L., Ludwig, W., Heussner, S. et al. (2021). The missing ocean plastic sink: Gone with the rivers. *Science* 373 (6550): 107.

5 Wang, W., Song, S., Wang, P. et al. (2023). Chemical bonding of g-C_3N_4/UiO-66(Zr/Ce) from Zr and Ce single atoms for efficient photocatalytic reduction of CO_2 under visible light. *ACS Catal.* 13 (7): 4597–4610.

6 Su, J., Liu, Y., Song, Y. et al. (2022). Recent development of nanomaterials for carbon dioxide electroreduction. *SmartMat* 3 (1): 35–53.

7 Wu, W., Wang, Y., Luo, L. et al. (2022). CO_2 hydrogenation over copper/ZnO single-atom catalysts: water-promoted transient synthesis of methanol. *Angew. Chem. Int. Ed.* 61 (48): e202213024.

8 Studt, F., Sharafutdinov, I., Abild-Pedersen, F. et al. (2014). Discovery of a Ni-Ga catalyst for carbon dioxide reduction to methanol. *Nat. Chem.* 6 (4): 320.

9 Zhang, L., Zhao, Z.-J., Wang, T., and Gong, J. (2018). Nano-designed semiconductors for electro- and photoelectro-catalytic conversion of carbon dioxide. *Chem. Soc. Rev.* 47 (14): 5423.

10 Lees, E.W., Mowbray, B.A.W., Parlane, F.G.L., and Berlinguette, C.P. (2022). Gas diffusion electrodes and membranes for CO_2 reduction electrolysers. *Nat. Rev. Mater.* 7 (1): 55.

11 Xie, Y., Ou, P., Wang, X. et al. (2022). High carbon utilization in CO_2 reduction to multi-carbon products in acidic media. *Nat. Catal.* 5: 564–570.

12 Xiao, C. and Zhang, J. (2021). Architectural design for enhanced C_2 product selectivity in electrochemical CO_2 reduction using Cu-based catalysts: a review. *ACS Nano* 15 (5): 7975–8000.

13 Nitopi, S., Bertheussen, E., Scott, S.B. et al. (2019). Progress and perspectives of electrochemical CO_2 reduction on copper in aqueous electrolyte. *Chem. Rev.* 119 (12): 7610.

14 Wang, Y., Liu, J., and Zheng, G. (2021). Designing copper-based catalysts for efficient carbon dioxide electroreduction. *Adv. Mater.* 33 (46): 2005798.

15 Sun, B., Dai, M., Cai, S. et al. (2023). Challenges and strategies towards copper-based catalysts for enhanced electrochemical CO_2 reduction to multi-carbon products. *Fuel* 332: 126114.

16 Jia, Y., Li, F., Fan, K., and Sun, L. (2022). Cu-based bimetallic electrocatalysts for CO_2 reduction. *Adv. Powder Mater.* 1 (1): 100012.

17 Morales-Guio, C.G., Cave, E.R., Nitopi, S.A. et al. (2018). Improved CO_2 reduction activity towards C_{2+} alcohols on a tandem gold on copper electrocatalyst. *Nat. Catal.* 1 (10): 764.

18 Cao, B., Li, F.-Z., and Gu, J. (2022). Designing Cu-based tandem catalysts for CO_2 electroreduction based on mass transport of CO intermediate. *ACS Catal.* 12 (15): 9735–9752.

19 Duan, G.-Y., Li, X.-Q., Ding, G.-R. et al. (2022). Highly efficient electrocatalytic CO_2 reduction to C_{2+} products on a poly(ionic liquid)-based Cu^0-Cu^1 tandem catalyst. *Angew. Chem. Int. Ed.* 61 (9): e202110657.

20 Su, X., Jiang, Z., Zhou, J. et al. (2022). Complementary operando spectroscopy identification of in-situ generated metastable charge-asymmetry Cu_2-CuN_3 clusters for CO_2 reduction to ethanol. *Nat. Commun.* 13 (1): 1322.

21 Kuhl, K.P., Cave, E.R., Abram, D.N., and Jaramillo, T.F. (2012). New insights into the electrochemical reduction of carbon dioxide on metallic copper surfaces. *Energy Environ. Sci.* 5 (5): 7050.

22 Ahmad, T., Liu, S., Sajid, M. et al. (2022). Electrochemical CO_2 reduction to C_{2+} products using Cu-based electrocatalysts: a review. *Nano Res. Energy* 1: 9120021.

23 Wang, H., Zhan, G., Tang, C. et al. (2023). Scalable edge-oriented metallic two-dimensional layered Cu_2Te arrays for electrocatalytic CO_2 methanation. *ACS Nano* 17 (5): 4790–4799.

24 Zhang, J., Wang, Y., Li, Z. et al. (2022). Grain boundary-derived Cu^+/Cu^0 interfaces in CuO nanosheets for low overpotential carbon dioxide electroreduction to ethylene. *Adv. Sci.* 9 (21): 2200454.

25 Mistry, H., Behafarid, F., Reske, R. et al. (2016). Tuning catalytic selectivity at the mesoscale via interparticle interactions. *ACS Catal.* 6 (2): 1075.

26 Reske, R., Mistry, H., Behafarid, F. et al. (2014). Particle size effects in the catalytic electroreduction of CO_2 on Cu nanoparticles. *J. Am. Chem. Soc.* 136 (19): 6978.

27 Jiang, K., Sandberg, R.B., Akey, A.J. et al. (2018). Metal ion cycling of Cu foil for selective C-C coupling in electrochemical CO2 reduction. *Nat. Catal.* 1 (2): 111.

28 Wang, H., Bi, X., Yan, Y. et al. (2023). Efficient electrocatalytic reduction of CO_2 to ethanol enhanced by spacing effect of Cu-Cu in $Cu_{2-x}Se$ nanosheets. *Adv. Funct. Mater.* 33 (25): 2214946.

29 Song, H., Im, M., Song, J.T. et al. (2018). Effect of mass transfer and kinetics in ordered Cu-mesostructures for electrochemical CO_2 reduction. *Appl. Catal. B Environ.* 232: 391.

30 Zhang, B., Zhang, J., Hua, M. et al. (2020). Highly electrocatalytic ethylene production from CO_2 on nanodefective Cu nanosheets. *J. Am. Chem. Soc.* 142 (31): 13606.

31 Mariano, R.G., McKelvey, K., White, H.S., and Kanan, M.W. (2017). Selective increase in CO_2 electroreduction activity at grain-boundary surface terminations. *Science* 358 (6367): 1187.

32 Hoang, T.T.H., Verma, S., Ma, S. et al. (2018). Nanoporous copper-silver alloys by additive-controlled electrodeposition for the selective electroreduction of CO_2 to ethylene and ethanol. *J. Am. Chem. Soc.* 140 (17): 5791.

33 Zhang, X., Sun, X., Guo, S.-X. et al. (2019). Formation of lattice-dislocated bismuth nanowires on copper foam for enhanced electrocatalytic CO_2 reduction at low overpotential. *Energy Environ. Sci.* 12 (4): 1334.

34 Zhu, W., Zhang, L., Yang, P. et al. (2018). Formation of enriched vacancies for enhanced CO_2 electrocatalytic reduction over AuCu alloys. *ACS Energy Lett.* 3 (9): 2144.

35 Bernal, M., Bagger, A., Scholten, F. et al. (2018). CO_2 electroreduction on copper-cobalt nanoparticles: size and composition effect. *Nano Energy* 53: 27.

36 Guo, W., Bi, J., Zhu, Q. et al. (2020). Highly selective CO_2 electroreduction to CO on Cu-Co bimetallic catalysts. *ACS Sustain. Chem. Eng.* 8 (33): 12561.

37 Suzuki, T.M., Ishizaki, T., Kosaka, S. et al. (2020). Electrochemical CO_2 reduction over nanoparticles derived from an oxidized Cu-Ni intermetallic alloy. *Chem. Commun.* 56 (95): 15008.

38 Zhang, X., Liu, C., Zhao, Y. et al. (2021). Atomic nickel cluster decorated defect-rich copper for enhanced C_2 product selectivity in electrocatalytic CO_2 reduction. *Appl. Catal. B Environ.* 291: 120030.

39 Sachtler, W.M.H. and Dorgelo, G.J.H. (1965). The surface of copper-nickel alloy films: I. Work function and phase composition. *J. Catal.* 4 (6): 654.

40 Fu, X., Zhu, A., Chen, X. et al. (2021). Stabilization of Cu/Ni alloy nanoparticles with graphdiyne enabling efficient CO_2 reduction. *Chem. Res. Chinese Univ.* 37 (6): 1328.

41 Ren, Y., Bai, Y., Wang, G. et al. (2023). Cu-Ni alloy nanoparticles anchored on nitrogen-doped carbon nanotubes for efficient CO_2 electroreduction to CO. *Energy Fuel* 37 (13): 9289–9296.

42 Bagchi, D., Raj, J., Singh, A.K. et al. (2022). Structure-tailored surface oxide on Cu-Ga intermetallics enhances CO_2 reduction selectivity to methanol at ultralow potential. *Adv. Mater.* 34 (19): e2109426.

43 Sun, L., Hao, P., Wang, H. et al. (2023). Selective ethane production from carbon dioxide enabled by Cu-Ga plates with large-scale manufacturing capability. *Appl. Surf. Sci.* 638: 158025.

44 Du, C., Mills, J.P., Yohannes, A.G. et al. (2023). Cascade electrocatalysis via AgCu single-atom alloy and Ag nanoparticles in CO_2 electroreduction toward multicarbon products. *Nat. Commun.* 14 (1): 6142.

45 Yan, S., Chen, Z., Chen, Y. et al. (2023). High-power CO_2-to-C_2 electroreduction on Ga-spaced, square-like Cu sites. *J. Am. Chem. Soc.* 145 (48): 26374.

46 Shan, J., Shi, Y., Li, H. et al. (2022). Effective CO_2 electroreduction toward C_2H_4 boosted by Ce-doped Cu nanoparticles. *Chem. Eng. J.* 433: 133769.

47 Tsujiguchi, T., Kawabe, Y., Jeong, S. et al. (2021). Acceleration of electrochemical CO_2 reduction to formate at the Sn/reduced graphene oxide interface. *ACS Catal.* 11 (6): 3310.

48 Kumar, B., Atla, V., Brian, J.P. et al. (2017). Reduced SnO_2 porous nanowires with a high density of grain boundaries as catalysts for efficient electrochemical CO_2-into-HCOOH conversion. *Angew. Chem. Int. Ed.* 56 (13): 3645.

49 Zhang, B., Sun, L., Wang, Y. et al. (2020). Well-dispersed SnO_2 nanocrystals on N-doped carbon nanowires as efficient electrocatalysts for carbon dioxide reduction. *J. Energy Chem.* 41: 7.

50 Zhang, J., Yin, R., Shao, Q. et al. (2019). Oxygen vacancies in amorphous InO_x nanoribbons enhance CO_2 adsorption and activation for CO_2 electroreduction. *Angew. Chem. Int. Ed.* 58 (17): 5609.

51 Chang, S., Xuan, Y., Duan, J., and Zhang, K. (2022). High-performance electroreduction CO_2 to formate at Bi/Nafion interface. *Appl. Catal. B Environ.* 306: 121135.

52 Wu, D., Wang, X., Fu, X.Z., and Luo, J.-L. (2021). Ultrasmall Bi nanoparticles confined in carbon nanosheets as highly active and durable catalysts for CO_2 electroreduction. *Appl. Catal. B Environ.* 284: 119723.

53 Ren, H., Wang, X., Zhou, X. et al. (2023). In-situ constructing Cu_1Bi_1 bimetallic catalyst to promote the electroreduction of CO_2 to formate by synergistic electronic and geometric effects. *J. Energy Chem.* 79: 263.

54 Cao, Y., Chen, S., Bo, S. et al. (2023). Single atom Bi decorated copper alloy enables C-C coupling for electrocatalytic reduction of CO_2 into C_{2+} products. *Angew. Chem. Int. Ed.* 62 (30): e202303048.

55 Hu, Y., Lu, D., Zhou, W. et al. (2023). In situ construction of 3D low-coordinated bismuth nanosheets@Cu nanowire core-shell nanoarchitectures for superior CO_2 electroreduction activity. *J. Mater. Chem. A* 11 (4): 1937.

56 Jia, Y., Hsu, H.S., Huang, W.C. et al. (2023). Probing the roles of indium oxides on copper catalysts for enhanced selectivity during CO_2-to-CO electrochemical reduction. *Nano Lett.* 23 (6): 2262.

57 Jedidi, A., Rasul, S., Masih, D. et al. (2015). Generation of Cu-In alloy surfaces from $CuInO_2$ as selective catalytic sites for CO_2 electroreduction. *J. Mater. Chem. A* 3 (37): 19085.

58 Larrazábal, G.O., Martín, A.J., Mitchell, S. et al. (2016). Enhanced reduction of CO_2 to CO over Cu–In electrocatalysts: catalyst evolution is the key. *ACS Catal.* 6 (9): 6265.

59 Rasul, S., Anjum, D.H., Jedidi, A. et al. (2015). A highly selective copper–indium bimetallic electrocatalyst for the electrochemical reduction of aqueous CO_2 to CO. *Angew. Chem. Int. Ed.* 54 (7): 2146.

60 Luo, W., Xie, W., Mutschler, R. et al. (2018). Selective and stable electroreduction of CO_2 to CO at the copper/indium interface. *ACS Catal.* 8 (7): 6571.

61 Zhang, X., Zhu, Y., Liu, Z. et al. (2022). Perforated nitrogen-rich graphene-like carbon nanolayers supported Cu-In catalyst for boosting CO_2 electroreduction to CO. *J. Energy Chem.* 75: 383.

62 Wen, G., Ren, B., Zhang, X. et al. (2023). Cu-In dual sites with sulfur defects toward superior ethanol electrosynthesis from CO_2 electrolysis. *Adv. Mater.* 2023: e2310822.

63 Kuhl, K.P., Hatsukade, T., Cave, E.R. et al. (2014). Electrocatalytic conversion of carbon dioxide to methane and methanol on transition metal surfaces. *J. Am. Chem. Soc.* 136 (40): 14107.

64 Wang, L., Peng, H., Lamaison, S. et al. (2021). Bimetallic effects on Zn-Cu electrocatalysts enhance activity and selectivity for the conversion of CO_2 to CO. *Chem Catal.* 1 (3): 663.

65 Zhang, Z., Li, S., Rao, Y. et al. (2023). Three-dimension porous Zn-Cu alloy: an inexpensive electrocatalyst for highly selective CO_2 reduction to CO in non-aqueous electrolyte. *Chem. Eng. J.* 479: 147376.

66 Zhang, J., Guo, C., Fang, S. et al. (2023). Accelerating electrochemical CO_2 reduction to multi-carbon products via asymmetric intermediate binding at confined nanointerfaces. *Nat. Commun.* 14: 1298.

67 Jeon, H.S., Timoshenko, J., Scholten, F. et al. (2019). Operando insight into the correlation between the structure and composition of CuZn nanoparticles and their selectivity for the electrochemical CO_2 reduction. *J. Am. Chem. Soc.* 141 (50): 19879.

68 Seyedeh, B.V., Dragos, S., Jan, V. et al. (2021). Elucidating the structure-dependent selectivity of CuZn towards methane and ethanol in CO_2 electroreduction using tailored Cu/ZnO precatalysts. *Chem. Sci.* 12: 14484.

69 Ren, D., Ang, B.S.-H., and Yeo, B.S. (2016). Tuning the selectivity of carbon dioxide electroreduction toward ethanol on oxide-derived CuxZn catalysts. *ACS Catal.* 6 (12): 8239.

70 Zhen, S., Zhang, G., Cheng, D. et al. (2022). Nature of the active sites of copper zinc catalysts for carbon dioxide electroreduction. *Angew. Chem. Int. Ed.* 61 (22): e202201913.

71 Zhong, M., Tran, K., Min, Y. et al. (2020). Accelerated discovery of CO_2 electrocatalysts using active machine learning. *Nature* 581 (7807): 178.

72 Wang, D., Qin, R., Qiu, Y., and Liu, C.-G. (2023). The first-principles calculation to predict electroreduction of CO_2 to ethanol over Al/Cu(111) bimetallic catalyst. *Mol. Catal.* 547: 113340.

73 Knight, M.W., Sobhani, H., Nordlander, P., and Halas, N.J. (2011). Photodetection with active optical antennas. *Science* 332 (6030): 702.

74 Jia, H., Du, A., Zhang, H. et al. (2019). Site-selective growth of crystalline ceria with oxygen vacancies on gold nanocrystals for near-infrared nitrogen photofixation. *J. Am. Chem. Soc.* 141 (13): 5083.

75 Mubeen, S., Lee, J., Singh, N. et al. (2013). An autonomous photosynthetic device in which all charge carriers derive from surface plasmons. *Nat. Nanotechnol.* 8: 247.

76 Dreaden, E.C., Alkilany, A.M., Huang, X. et al. (2012). The golden age: gold nanoparticles for biomedicine. *Chem. Soc. Rev.* 41 (7): 2740.

77 Zhu, R., Li, J., Lin, L. et al. (2020). Emerging plasmonic assemblies triggered by DNA for biomedical applications. *Adv. Funct. Mater.* 31 (10): 2005709.

78 Kim, D., Resasco, J., Yu, Y. et al. (2014). Synergistic geometric and electronic effects for electrochemical reduction of carbon dioxide using gold–copper bimetallic nanoparticles. *Nat. Commun.* 5: 4948.

79 Xu, Z., Lai, E., Shao-Horn, Y., and Hamad-Schifferli, K. (2012). Compositional dependence of the stability of AuCu alloy nanoparticles. *Chem. Commun.* 48 (111): 5626.

80 Ross, M.B., De Luna, P., Li, Y.F. et al. (2019). Designing materials for electrochemical carbon dioxide recycling. *Nat. Catal.* 2: 648.

81 Li, F., Li, Y., Wang, Z. et al. (2020). Cooperative CO_2-to-ethanol conversion via enriched intermediates at molecule–metal catalyst interfaces. *Nat. Catal.* 3: 75.

82 Peterson, A.A., Abild-Pedersen, F., Studt, F. et al. (2010). How copper catalyzes the electroreduction of carbon dioxide into hydrocarbon fuels. *Energy Environ. Sci.* 3: 1311.

83 Wang, X., Ou, P., Wicks, J. et al. (2021). Gold-in-copper at low *CO coverage enables efficient electromethanation of CO_2. *Nat. Commun.* 12: 3387.

84 Dai, S., Huang, T.-H., Liu, W.-I. et al. (2021). Enhanced CO_2 electrochemical reduction performance over Cu@AuCu catalysts at high noble metal utilization efficiency. *Nano Lett.* 21 (21): 9293.

85 Kuang, S., Su, Y., Li, M. et al. (2023). Asymmetrical electrohydrogenation of CO_2 to ethanol with copper-gold heterojunctions. *Proc. Natl. Acad. Sci. USA* 120: e2214175120.

86 Zhu, C., Zhou, L., Zhang, Z. et al. (2022). Dynamic restructuring of epitaxial Au–Cu biphasic interface for tandem CO_2-to-C_{2+} alcohol conversion. *Chem* 8: 3288.

87 Chen, P.-C., Chen, C., Yang, Y. et al. (2023). Chemical and structural evolution of AgCu catalysts in electrochemical CO_2 reduction. *J. Am. Chem. Soc.* 145 (18): 10116.

88 Chen, C., Li, Y., Yu, S. et al. (2020). Cu-Ag tandem catalysts for high-rate CO_2 electrolysis toward multicarbons. *Joule* 4 (8): 1688.

89 Huang, J., Mensi, M., Oveisi, E. et al. (2019). Structural sensitivities in bimetallic catalysts for electrochemical CO_2 reduction revealed by Ag-Cu nanodimers. *J. Am. Chem. Soc.* 141 (6): 2490.

90 Ting, L.R.L., Piqué, O., Lim, S.Y. et al. (2020). Enhancing CO_2 electroreduction to ethanol on copper–silver composites by opening an alternative catalytic pathway. *ACS Catal.* 10 (7): 4059.

91 Chen, P.-C., Gao, M., Yu, S. et al. (2021). Revealing the phase separation behavior of thermodynamically immiscible elements in a nanoparticle. *Nano Lett.* 21 (15): 6684.

92 Frisch, M.L., Wu, L., Atlan, C. et al. (2023). Unraveling the synergistic effects of Cu-Ag tandem catalysts during electrochemical CO_2 reduction using nanofocused X-ray probes. *Nat. Commun.* 14: 7833.

93 Choukroun, D., Pacquets, L., Li, C. et al. (2021). Mapping composition-selectivity relationships of supported sub-10 nm Cu-Ag nanocrystals for high-rate CO_2 electroreduction. *ACS Nano* 15 (9): 14858.

94 Ma, Y., Yu, J., Sun, M. et al. (2022). Confined growth of silver-copper janus nanostructures with {100} facets for highly selective tandem electrocatalytic carbon dioxide reduction. *Adv. Mater.* 34 (19): 2110607.

95 Qi, K., Zhang, Y., Onofrio, N. et al. (2023). Unlocking direct CO_2 electrolysis to C_3 products via electrolyte supersaturation. *Nat. Catal.* 6: 319.

96 Xiong, L., Zhang, X., Yuan, H. et al. (2021). Breaking the linear scaling relationship by compositional and structural crafting of ternary Cu-Au/Ag nanoframes for electrocatalytic ethylene production. *Angew. Chem. Int. Ed.* 60 (5): 2508.

97 Zhu, Y., Gao, Z., Zhang, Z. et al. (2022). Selectivity regulation of CO_2 electroreduction on asymmetric AuAgCu tandem heterostructures. *Nano Res.* 15 (9): 7861.

98 Pérez-Ramírez, J. and López, N. (2019). Strategies to break linear scaling relationships. *Nat. Catal.* 2 (11): 971.

99 Seh, Z.W., Kibsgaard, J., Dickens, C.F. et al. (2017). Combining theory and experiment in electrocatalysis: Insights into materials design. *Science* 355 (6321): eaad4998.

15

Non-Cu Metal-Based Materials for Electrocatalytic CO_2 Reduction Reaction

Liu-Liu Shen[1] *and Gui-Rong Zhang*[2]

[1] *Tiangong University, School of Chemistry, Binshuixi Road 399, Tianjin, 300387, China*
[2] *Tiangong University, School of Chemical Engineering and Technology, Binshuixi Road 399, Tianjin, 300387, China*

15.1 Introduction

The electrocatalytic reduction of carbon dioxide (CO_2RR) is a process that uses electricity to convert carbon dioxide into a wide range of value-added chemicals and fuels, including carbon monoxide (CO), formic acid/formate, ethylene, ethanol, acetic acid, and propanol. This process can not only utilize carbon dioxide as a resource, reducing greenhouse gas emissions, but also store surplus renewable energy, alleviating the energy crisis [1]. However, carbon dioxide reduction reaction (CO_2RR) faces many challenges, such as low reaction activity, poor selectivity, poor stability, and low energy efficiency, which greatly compromise the economic viability of CO_2RR [2]. The key to overcome these challenges lies in the development of high-performance electrocatalysts for the CO_2RR. The electrochemical CO_2RR has a long history that can date back to the 1900s, and until the 1980s, several metal electrodes, such as Zn, Pb, and Hg, were shown to convert CO_2 to formate [3]. Thereafter, some metals (e.g. Mo, Ru) and semiconductors (e.g. p-GaP, n-Si) were reported to convert CO_2 to methanol [4, 5]. The most exciting breakthrough was made in 1985 when Cu was found to be capable of reducing CO_2 to hydrocarbons, which sparked intensive research activities in this field [6].

The electrochemical CO_2RR involves several sequential proton and electron transfer steps. Diverse reaction pathways may become operative based on the catalyst's nature and the applied potential, leading to a broad product spectrum. The ultimate goal of developing high-performing CO_2RR electrocatalysts is to rationally steer the reaction toward specific desired products via a particular pathway [7], which turns out to be a rather challenging task. Although the thermodynamic stability of CO_2 is akin to that of water, initiating its reductive activation via the first electron transfer to form $CO_2^{\cdot-}$ radicals without using a catalyst necessitates an extremely negative potential (i.e. −1.9 V vs. standard hyrogen electrode (SHE)). Nevertheless, numerous electrocatalysts are capable of activating CO_2 and stabilizing $CO_2^{\cdot-}$ by forming chemical bonds, thereby facilitating the first electron transfer at a significantly

Functional Materials for Electrocatalytic Energy Conversion, First Edition.
Edited by Zhicheng Zhang, Meiting Zhao, and Yuchen Qin.

reduced overpotential. The reactivity of the formed $CO_2^{\cdot-}$ on distinct catalytic surfaces governs the ultimate product distribution. A benchmark work in this field was conducted by Hori et al. in the 1980s and 1990s, who comprehensively studied the CO_2RR over a wide range of metal electrodes, including Pb, Hg, Tl, In, Sn, Cd, Bi, Au, Ag, Zn, Pd, Ga, Cu, Ni, Fe, Pt, and Ti [8]. These metals are categorized into four groups based on their predominant product distribution: (i) Cu, the only metal proficient in producing hydrocarbons/alcohols with appreciable Faradaic efficiency (FE); (ii) Au, Ag, Zn, Pd, and Ga, capable of selectively producing CO; (iii) Pb, Hg, In, Sn, Cd, Tl, and Bi, predominantly yielding formate; and (iv) Ni, Fe, Pt, and Ti, favoring competitive hydrogen evolution reaction (HER). Notably, Cu distinguishes itself from the list due to its exceptional ability to produce more valuable products, thereby garnering considerable research attention in this field.

Despite the exciting progress in innovating the electrocatalytic systems, the classification of the electrode materials and their associated product distribution largely holds its validity. Through comprehensive spectroscopic characterization and theoretical analyses, it has been uncovered that not only the interaction between $CO_2^{\cdot-}$ and catalysts but also the binding strength of diverse intermediates, including *CO, *COOH, and *OCOH, may decisively influence the resultant product distribution [9]. Notably, certain metal electrodes like Zn, Ag, Au, and Pd exhibit thermodynamic favorability for adsorption/stabilization of *COOH, leading to the predominant formation of CO. Conversely, Ga, In, Sn, and Bi are adept at stabilizing the critical intermediate *OCOH, consequently favoring formate formation. Cu uniquely stabilizes both *COOH and *CO, positioning itself as the sole metal capable of producing C_1 and C_{2+} products, as outlined in Figure 15.1. Given the pivotal role of binding strengths of key intermediates on metal electrode surfaces, there is a growing number of works devoted to rationally tuning the geometric/electronic structures of various metal-based electrocatalysts. This involves constructing innovative nanostructures featuring single-sites, core–shell configurations, heterojunctions, and altering surface compositions through heteroatom doping or alloy formation, as summarized in Figure 15.1.

Although Cu-based CO_2RR catalysts have attracted most of the research attention in this field, they typically exhibit a broad product spectrum. Despite the capability of producing value-added C_{2+} products over Cu-based catalysts, recent techno-economic assessments imply a high production cost for typical C_{2+} products like ethylene and ethanol, which compromises the short-term economic viability of this appealing CO_2 valorization route [10]. In contrast, many non-Cu metal-based electrode materials exhibit notably high selectivity toward CO and formate formation, suggesting promising economic feasibility for CO_2RR on these materials. Therefore, this chapter aims to provide a comprehensive overview of the latest advancements in non-Cu metal-based catalysts for electrochemical CO_2RR by highlighting the performance of cutting-edge catalytic systems. We also summarize the efforts to establish the structure–performance relationship and the mechanistic understanding in this field. Additionally, the updated perspective on technological challenges concerning the practical application of non-Cu metal-based catalysts in electrocatalytic CO_2RR will be discussed.

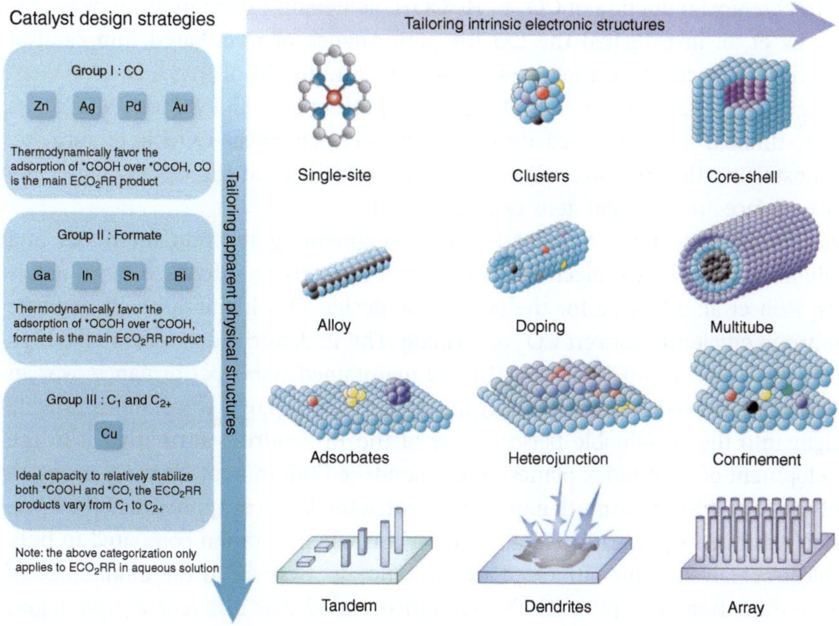

Figure 15.1 Typical catalyst categories toward the electrochemical CO_2RR and the emerging strategies for catalyst engineering. Source: Reproduced with permission of Li et al. [9], the Royal Society of Chemistry.

15.2 Non-Cu Metal-Based Catalyst for Electrocatalytic CO_2 Reduction

15.2.1 Monometallic catalysts

The product selectivity of non-copper (Cu) single metal catalysts in the CO_2RR can be broadly categorized into two groups based on their major catalytic products: formate (e.g. Sn, Bi, In, Pb, Hg, Cd, and Pd) and CO (e.g. Zn, Ag, Pd, and Au) [8, 9, 11–16]. Sn and Bi are the most extensively studied monometallic materials for CO_2RR to formate due to their low toxicity, environmentally friendly features, and cost-effectiveness. In the initial stages, bulk electrodes, such as metal plates/foils and metal single crystals, were predominantly employed for CO_2RR catalysis, demonstrating certain selectivity but with relatively large overpotentials and low current densities [6, 17–21]. Over the past two decades, it has been observed that reducing the sizes and/or dimensions of the metal catalysts to the nanoscale significantly enhances their performance in CO_2RR in terms of selectivity, overpotential, and current density [9, 16]. For example, Won et al. synthesized hierarchical Sn dendrites with a multibranched conifer-like structure, featuring an enlarged surface area. In comparison to Sn foil, the Sn dendrites exhibited a considerably higher formic acid (HCOOH) FE of 83.4% at −1.36 V vs. reversible hydrogen electrode (RHE) and displayed good stability over 18 hours of operation [22]. Sinha et al. utilized graphite rods deposited with Sn nanocatalysts as cathodes for the electrochemical reduction of CO_2 to HCOOH, achieving a maximum FE of 84.25% [23]. Lv et al. investigated the CO_2RR performance of a Sn-based film catalyst prepared by electrodeposition, revealing a maximum FE of 91.5% at −1.8 V vs. Ag/AgCl on a ~325 nm-thick Sn-based film [24]. Qiu et al. synthesized a series of size-tunable nano-Bi-based catalysts for electrochemical CO_2RR to formate, demonstrating that the size effect significantly influenced the electrochemically active surface area, current density, and FE [25].

Nanotechnology plays a crucial role in augmenting reaction interfaces and modulating the surface electronic structure of metals to introduce more active sites. Koh et al. advocate for the systematic design of a hierarchical Bi dendrite catalyst to efficiently convert CO_2 to formate. The Bi dendrite demonstrated a high selectivity (~89% at −0.74 V vs. RHE) and maintained stable performance over an extended period (~12 hours). Density functional theory (DFT) calculations provide insight into the sustainable performance of the Bi dendrite, attributing it to the development of high-index planes on the dendrite [26]. In another study, Su et al. synthesized ultrathin bismuth nanosheets with a thickness of approximately 10 nm, which exhibited significantly higher FE for formate generation compared to bulk Bi electrocatalysts. This enhanced performance is ascribed to the abundance of low-coordination sites [27]. Kim's research revealed that different morphologies of Bi-based catalysts can be obtained using various electrodeposition methods. The Bi nanoflakes electrode, produced by a novel pulse electrodeposition method, demonstrated numerous edge and corner sites, leading to stronger local electric fields conducive to electrochemical CO_2RR [28]. Beyond particle size and shape, the creation of defects introduces active sites. For example, Gong et al. reported a defective bismuth nanotube prepared through the electroreduction of bismuth oxide double-walled nanotubes with a fragmented surface. This nanotube exhibited excellent activity, selectivity, and stability, achieving a remarkable current density of ~288 mA cm^{-2} at −0.61 V vs. RHE in a flow cell reactor [29].

Metal catalysts commonly employed for the CO_2RR to CO include Zn, Ag, Pd, and Au. To attain high CO selectivity at low overpotential, various-sized metallic materials with nanostructures have been developed. The CO_2RR performance over these nanostructures has been thoroughly investigated, focusing on the correlation between catalyst shape/size and activity. For instance, Kauffman et al. examined CO_2 electroreduction on Au_{25} clusters, promoting CO formation within 90 mV of the formal potential and achieving an approximately 100% CO FE at −1 V vs. the RHE. Compared to Au nanoparticles (NPs) and bulk Au electrodes, Au25 clusters exhibit a 200–300 mV improvement in onset overpotential [30]. Zhu et al. synthesized a series of monodisperse polyhedral Au NPs via burst nucleation, with particle sizes of 4, 6, 8, and 10 nm. The 8-nm NPs (Figure 15.2a) demonstrated the highest selectivity for CO, reaching an FE of 90% at −0.67 V vs. RHE in 0.5 M $KHCO_3$ (Figure 15.2b). The ratio of edges, corners, and plane sites varied with the catalyst size, with the 8-nm NPs possessing a higher ratio of edge sites (Figure 15.2c). DFT calculations revealed that the presence of dominant edge sites facilitates the stabilization of the first reduction intermediate (COOH*) and its subsequent conversion to CO [31]. Subsequent studies on Au nanowires (NWs) were informed

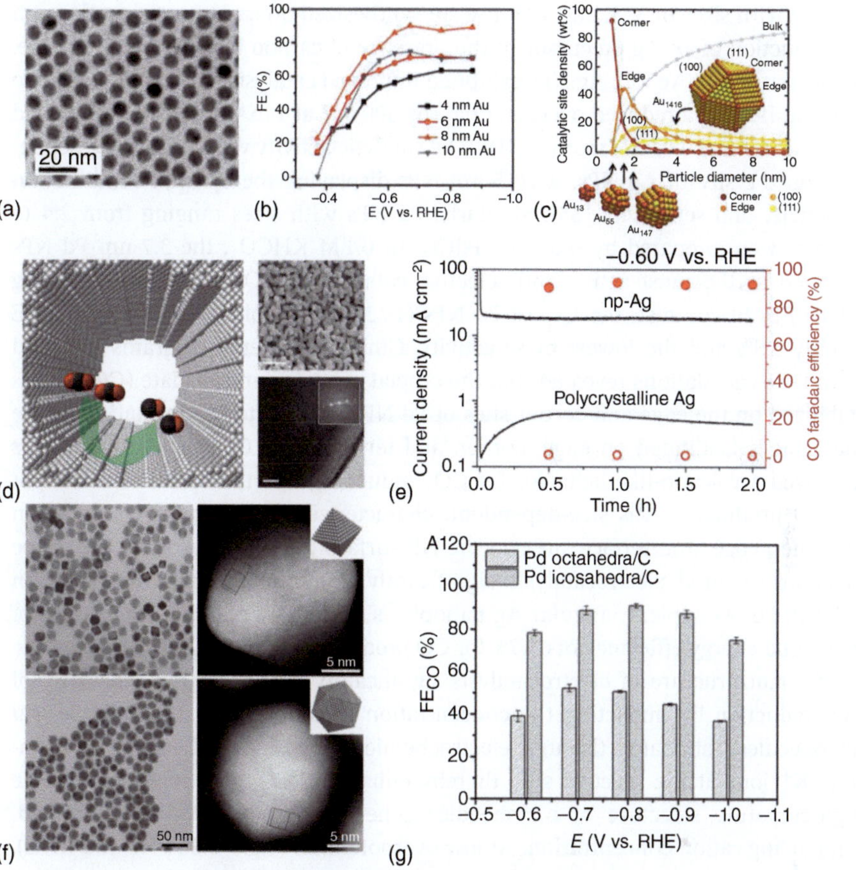

Figure 15.2 (a) TEM images of the 8 nm Au NPs. (b) Potential-dependent FE's of the C–Au on electrocatalytic reduction of CO_2 to CO. (c) Density of adsorption sites for (111), (001), edge, or corner on-top sites on closed-shell cuboctahedra Au clusters vs. the cluster diameter. Source: Reproduced with permission of Zhu et al. [31], American Chemical Society. (d) Structure and morphology of nanoporous Ag. (e) CO_2 reduction activity of nanoporous Ag and polycrystalline silver at −0.60 V. Source: Reproduced with permission of Lu et al. [32], Springer Nature. (f) TEM and atomic-resolution HAADF-STEM images of the Pd octahedral and Pd icosahedra. (g) Reduction potential dependence on FEs for CO production over Pd octahedra/C and Pd icosahedra/C catalysts. Source: Reproduced with permission of Huang et al. [33], Wiley-VCH.

by the results from Au NPs. Given the same total number of Au atoms, a 2-nm-wide Au NW has significantly more edge sites than an equivalently sized polyhedral NP. Consequently, this Au NW is expected to be more active than the Au NP for the conversion of CO_2 to CO. The NW-length-dependent activity was also explored, confirming that longer NWs were much more active, exhibiting an onset potential for CO_2 reduction to CO at −0.2 V (with 37% FE) and a maximum FE of 94%, along with a mass activity of 1.84 Ag_{Au}^{-1} at −0.35 V [34].

In addition to Au NPs, Ag and Pd NPs also demonstrate size- and geometry-dependent catalysis for the electrochemical reduction of CO_2 to CO. To illustrate,

Ag NPs with sizes of 3, 5, and 10 nm were synthesized on carbon support through the reduction of an Ag precursor in the presence of carbon black and cysteamine. In comparison to Ag foil, Ag NPs exhibited improved catalysis for CO_2 reduction to CO, manifesting a reduced overpotential (by 300 mV at 1 mA cm^{-2}) and a fourfold increase in FE at −0.75 V vs. the RHE. The catalytic activity was found to be dependent on the size of Ag NPs, with 5-nm NPs displaying the optimal combination of activity and selectivity [35]. Similarly, Pd NPs with sizes ranging from 2.4 to 10.3 nm were prepared by reducing $PdCl_2$. In 0.1 M $KHCO_3$, the 3.7-nm Pd NPs emerged as the most active and selective catalyst for CO formation, achieving a 90% FE. In contrast, the largest Pd NPs (10.3 nm) exhibited the lowest CO FE at only 5.8% and the lowest mass activity. Gibbs free energy diagrams obtained from DFT calculations revealed that the crucial reaction intermediate (COOH*) is stabilized on the edge and corner sites of Pd NPs, leading to CO formation, while the HER is facilitated on edge, corner, and terrace sites. Consequently, the NPs displayed a volcano-like behavior for CO_2 reduction, and the optimized hot spot for CO production was size-dependent, characterized by the right combination of corner, edge, and terrace sites on the NP surface [36]. Furthermore, the shape of nanostructured electrocatalysts significantly influences their performance in CO_2RR. For example, triangular Ag nanoplates exhibited notably enhanced FE of 96.8% and energy efficiency of 61.7% for CO production from CO_2RR [37].

The nanostructure of electrocatalysts significantly influences the selectivity of CO_2 reduction by impacting the concentration of cations near the cathode. Au nanoneedles, fabricated through electrochemical deposition, effectively concentrate K^+ ions at the reactive site, thereby enhancing CO_2 electroreduction. The high-curvature structure of nanoneedles generates a substantial electric field, influencing cation concentration. At low overpotentials (e.g. −0.35 V vs. the RHE), Au nanoneedles create a local high electric field that concentrates K^+ ions, resulting in an elevated CO_2 concentration near the Au surface. Consequently, this configuration yields a high current density of 22 mA cm^{-2}, a Tafel slope of 42 mV dec^{-1}, and an outstanding FE (>90%) for CO formation [38].

Mesoscale phenomena arising from porous architectures or controlled interparticle spacing play a pivotal role in regulating interparticle reactant diffusion, generating diffusional gradients, and adsorbing intermediates, thus significantly influencing the selectivity of CO_2 electroreduction. For instance, Surendranath et al. reported a 99% selectivity for CO evolution with a 10-fold reduction in hydrogen evolution on mesoporous Au inverse opal thin films at an exceptionally low overpotential of 0.4 V. They observed that while the selectivity for CO formation remained relatively constant, the HER activity decreased with increasing film thickness [39]. The same research group investigated CO_2 electroreduction on mesoporous Ag inverse opal electrodes. By adjusting the deposition potential, they could modulate the roughness factor of the mesoporous Ag film from 43 to 110, leading to enhanced CO formation FE from less than 5% to over 80%. Similar to mesoporous Au inverse opal thin films, this heightened CO selectivity and suppressed H_2 evolution can be ascribed to the elevated local pH, depleting the local concentration of proton donors at the electrocatalyst's surface [40].

The selectivity of CO_2RR on metallic catalysts exhibits a strong dependency on crystal facet orientation. For example, among the low-index crystal facets, Ag(110) exhibits the highest selectivity for CO evolution, followed by Ag(111) and Ag(100) [19]. This can be attributed to the rougher and active-site-abundant nature of Ag(110) compared to these other facets, providing the highest catalytic activity in the CO_2RR. Hence, increasing the density of low-coordination surface sites on Ag holds promise for promoting C–C coupling in multicarbon product formation during the CO_2RR. Recently, Dutta et al. reported highly porous Ag-nanofoam electrodes rich in low-coordination Ag sites capable of reducing CO_2 to C_2H_4 with an FE of 8.6% [41]. Additionally, Jiao and co-workers developed nanoporous Ag electrocatalysts (Figure 15.2d) displaying a selectivity of ≈92% for CO production in CO_2RR, with a current over 3000-fold higher than polycrystalline Ag under moderate overpotentials of <0.50 V (Figure 15.2e) [32]. Park et al. reported that an Ag dendrite catalyst predominantly featuring Ag (220) facets achieved a CO FE close to 65% at −0.8 V vs. NHE compared to Ag (111) facets [42]. Meanwhile, Won et al. synthesized hierarchical hexagonal-shaped Zn catalysts over Zn foils, rich in (101) facets, exhibiting high FE_{CO} of 85.4% in CO_2-saturated $KHCO_3$ and 95.4% in KCl electrolytes at −0.95 V vs. RHE [43].

Beyond facet type, surface strain also significantly influences CO_2RR. Huang et al. designed a platform using Pd octahedra and icosahedra to investigate their mechanisms (Figure 15.3f). The interplanar spacing of (111) planes in the Pd icosahedron measured 2.32 Å, slightly larger than the octahedron's 2.24 Å, indicating tensile strain in the Pd icosahedron. Electronic band structures with varying strains showed an upward shift in the d-band center for tensile strain. This led to reduced occupation and stronger adsorbate bonding due to the antibonding states being pushed above the Fermi level, thereby enhancing CO_2 and COOH* intermediate adsorption. Experimentally, the Pd icosahedra displayed a maximum FE_{CO} of 91.1% at −0.8 V vs. RHE in 0.1 M $KHCO_3$, 1.7-fold higher than the maximum FE of Pd octahedra at −0.7 V vs. RHE (Figure 15.3g) [33].

15.2.2 Multimetallic Catalysts

Creating alloys or nanocomposites stands as a powerful technique to fine-tune the activity and selectivity of catalysts in CO_2RR [13, 45]. For instance, enhancing Sn-catalysis in CO_2RR involves depositing Bi NPs onto Sn nanosheets. This modification at the Sn–Bi interface notably enhanced the catalyst's selectivity for formate (96% FE, 0.74 mmol h^{-1} cm^{-2} at −1.1 V vs. RHE). The interaction between Bi and Sn is believed to elevate the Fermi level of Sn, consequently boosting the formation of the HCOO* intermediate and subsequent formate production [46]. Park and coworkers reported a range of Sn–Pb compositions for formate selectivity. X-ray photoelectron spectroscopy (XPS) and cyclic voltammetric (CV) analyses of these Sn–Pb alloys showed that Sn facilitates the formation of oxidized tin (SnO_x) and metallic lead (Pb^0) on the surface by inhibiting the formation of a low-conductivity lead oxide (PbO). The alloys are characterized by higher electrical conductivities and electron-donating capabilities compared to each of the constituent metals. The

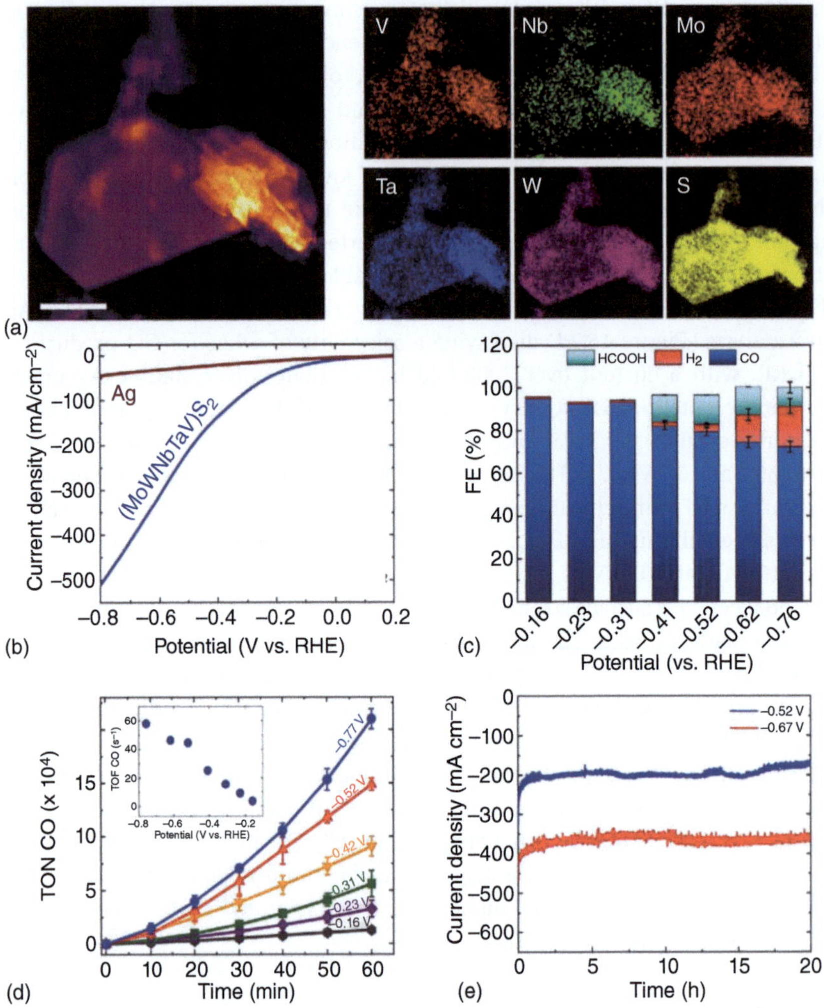

Figure 15.3 (a) STEM-HAADF image and EDS elemental maps of the high-entropy TMDC flake. (b) Polarization curves of the high-entropy TMDC and Ag nanoparticles. (c) FE of the high-entropy TMDC at different potentials. (d) Turn over numbers of CO formation over the course of an hour at various voltages with the turnover frequency (TOF) of the CA inset. (e) Long-term chronoamperometry at 2 different potentials. Source: Reproduced with permission of Cavin et al. [44], Wiley-VCH.

alloy $Sn_{56.3}Pb_{43.7}$ exhibited the highest formate FE of 80% and the highest formate partial current density of 45.7 mA cm^{-2} at −2.0 V (vs. Ag/AgCl). These values are about 20% higher compared to those measured at exclusively Sn or Pb electrodes [47]. Sun et al. reported a Sn–Pd alloy electrocatalyst for the exclusive conversion of CO_2 into formic acid in an aqueous solution. Activated carbon (AC)-supported Pd–Sn alloy NPs were synthesized via a modified wetting chemistry reduction method with sodium citrate as a stabilizing agent and sodium borohydride as a reductive agent. The composition of Pd–Sn NPs was controlled by varying the molar ratios of the precursors, $PdCl_2$ and $SnCl_2$. The catalysts showed nearly quantitative formation of formate at a very low overpotential of −0.26 V. Both CO formation and hydrogen evolution were completely suppressed. Both metallic Pd and oxidized Pd were present in the alloy, and systematically varying the Pd/Sn composition of the NPs resulted in a valley-shaped curve of Pd(0)/Pd(II) correlating to CO_2 electrocatalytic activity and selectivity. First-principles DFT calculations suggested that the surface Pd–Sn–O configuration with the highest oxygen occupancy facilitated the formation of the key reaction intermediate HCOO* and the generation of formic acid as the only product of the second proton–electron transfer [48].

Lewis et al. investigated Ni–Ga alloys with various compositions prepared through drop-casting followed by a temperature-programmed reduction method. The current densities for hydrocarbon products were relatively low, with the partial cathodic current density for CH_4 production surpassing 140 μA cm^{-2} at −1.18 V vs. RHE, achieving an FE_{CH4} of 2%. Additionally, partial cathodic current densities for C_2H_6 production reached 100 μA cm^{-2} at the same potential. The sum of the partial cathodic current densities for all three hydrocarbons surpassed >200 μA cm^{-2} at −1.18 V. A notable discovery was made in aqueous bicarbonate electrolytes at neutral pH, revealing an onset potential for methane, ethylene, and ethane production on all three phases at −0.48 V vs. the RHE. These values represent the lowest reported onset potentials to date for C_2 product production from CO_2. Notably, this value exceeds polycrystalline copper by >250 mV and is approximately equal to that of single crystals of copper, renowned for having some of the lowest overpotentials for CO_2 reduction to C_2 and methane products. Cyclic voltammetry analysis using CO-saturated solutions for Ni catalysts showed an unchanged CO desorption wave at 0.55 V for every cycle, whereas for the Ni–Ga sample, this peak vanished in the second scan. This suggests that the introduction of Ga in Ni films destabilizes the Ni–CO interaction, retarding surface poisoning and significantly enhancing yields of highly reduced products [49].

Pd and Pt have shown effectiveness as formic acid oxidation catalysts, making Pd/Pt mixed-metal systems particularly attractive for catalyzing both the forward and backward reactions of this process. In an experimental study, a series of Pd_xPt_{100-x}/C samples were meticulously prepared and analyzed with precise stoichiometric control of intermetallic Pd and Pt fcc structures. The cyclic voltammograms of the electro-oxidation of formic acid on the synthesized $Pd_xPt_{(100-x)}$/C NPs exhibited an enhancement in the oxidation wave compared to the respective monometals on carbon. However, the best-performing formic acid oxidation catalysts did not demonstrate the best CO_2 electrochemical reduction performance. The $Pd_{70}Pt_{30}$/C catalyst exhibited the most optimal performance with an FE_{HCOOH} of 88% after one hour of electrolysis at −0.4 V vs. RHE and an average current density of ~5 mA cm^{-2}. Remarkably, these Pd*x*Pt(100−*x*)/C structures displayed a very low onset potential for the reduction of CO_2 to formic acid, approaching the theoretical equilibrium potential of 0.02 V vs. RHE for this reaction [50].

Recently, Jaramillo and colleagues demonstrated that CO_2 could be converted into hydrocarbons without using a copper catalyst. They discovered that polycrystalline Ag–Zn foils exhibited enhanced methane and methanol evolution activities, despite

Ag and Zn not being competent CO_2 reduction catalysts for conversions requiring $>2e^-$. The alloy foil was slightly enriched in zinc, both at the surface and in the bulk, with a surface alloy composition of 61.3% zinc and a predominant Ag_5Zn_8 bulk phase. The CH_3OH FE and FE_{CH_4} values at 1.43 V vs. RHE using these alloys were fivefold and threefold higher, respectively, than those produced at pure Ag and Zn electrodes. Although the yields remain low, these results underscore an unequivocal synergistic effect between the metals of the alloy [51]. Zhang et al. detailed a cold plasma-activated AgCo surface alloy capable of selectively reducing CO_2 to C_2H_5OH. Notably, C_2H_5OH was the major product with an FE of 72.3% and a partial current density of 5.4 mA cm^{-2} at −0.80 V vs. RHE [53].

15.3 Non-Cu Metal Compounds

15.3.1 Metal Oxides

Oxide-derived materials obtained via the electroreduction of oxide precursors are increasingly recognized as prospective catalysts for CO_2RR. In contrast to their metallic bulk counterparts, these materials exhibit advantageous characteristics including metastable electronic structures, enhanced surface roughness, elevated local pH, and augmented grain boundary formation [54–57]. For instance, Zhang et al. synthesized oxidized Sn nanocrystals distributed evenly on graphene support with a high surface area using a straightforward hydrothermal method. Controlled potential electrolysis revealed that the formate FE increased with the potential, peaking at 86.2% at −1.8 V (vs. SCE) before declining sharply while the current density continued to rise. The size-dependent electrochemical CO_2RR to formate on SnO_2 NPs exhibited a maximum FE of 93% at −1.8 V vs. SCE on 5 nm SnO_2 NPs, as shown in [58]. In another study, Liang et al. produced SnO_2 NPs with numerous exposed grain boundaries and corner/step sites by oxidizing SnS_2 sheet precursors, yielding an exceptional total CO_2RR FE of 97% at −0.95 V vs. RHE [59]. Similarly, reduced SnO_2 NWs possessing a highly porous structure with densely packed grain boundaries further improved formate activity and selectivity [60], reaching a maximum FE of 87% for formate and 92% for overall CO_2RR at −0.8 V vs. RHE. Furthermore, oxide-derived Pb nanocrystalline films exhibited over a 700-fold lower HER activity in 0.25 m Na_2CO_3 electrolyte, resulting in an FE of ≈100% for formate between −1.0 and −0.75 V vs. RHE in CO_2-saturated 0.5 m $NaHCO_3$ solution [61]. Additionally, Au NPs derived from the reduction of thick Au oxide films demonstrated highly selective CO_2 to CO conversion at an overpotential as low as 140 mV. This remarkable catalytic performance correlated strongly with the high density of grain boundaries in oxide-derived Au nanocrystalline structures, facilitating the stabilization of $CO_2^{\cdot-}$ intermediates on their surfaces [57]. Ma et al. prepared nanostructured Ag catalysts by electrochemically reducing Ag_2O, revealing significantly enhanced CO_2RR activity and selectivity compared to polycrystalline Ag. They proposed that this enhancement primarily arose from an increase in local pH, suppressing the HER at the electrolyte-nanostructured Ag interface due to oxide derivation [56]. More recently, a metallic oxide-derived Zn catalyst, prepared by Li electrochemical tuning with freshly exposed grain boundaries, displayed a CO partial current density of ≈23 mA cm^{-2} at an overpotential of 948 mV. This represented a fivefold increase compared to Zn fabricated by conventional electroreduction, with the highest FE_{CO} recorded at 91.1% at −1.17 V vs. RHE [62].

Bimetallic materials, derived from metal oxide composites, have been explored as potential catalysts for CO_2RR due to their ability to offer high activity and stability. For instance, the BiIn hybrid electrocatalyst, originating from the Bi_2O_3/In_2O_3 heterojunction, displays exceptional catalytic performance in CO_2RR. It achieves a formate FE exceeding 90% over a wide potential range from −0.4 to −1.4 V vs. RHE while generating a formate partial current density of approximately 136.7 mA cm^{-2} at −1.4 V in a flow cell without IR compensation. Moreover, this hybrid electrocatalyst exhibits robust stability with high formate selectivity at 100 mA cm^{-2}. Theoretical calculations validate that the hybrid structure, housing dual Bi/In metal sites, contributes to optimal free energy of *H and *OCHO intermediates, expediting the formation and desorption steps of *HCOOH toward formate production. Such catalysts offer promise for effective CO_2RR [63].

Apart from oxide-derived materials, metal oxides find utility as CO_2RR catalysts, often enhanced by carbon-based materials for increased electroconductivity. For instance, a 3D hierarchical structure comprising mesoporous SnO_2 nanosheets on carbon cloth (SnO_2/CC) efficiently and selectively converts CO_2 to formate in aqueous media. It exhibits an unprecedented partial current density of about 45 mA cm^{-2} at a moderate overpotential (0.88 V) with high FE (87%). The superior performance attributes to its robust and highly porous hierarchical structure, providing ample surface area and aiding charge and mass transfer [64]. The influence of the carbon support on the activity of metal oxides is significant. Yuan et al. developed undoped, N-doped, S-doped, and N, S-co-doped porous carbon supports (C, NC, SC, NSC) hosting high-density SnO_2 NPs (over 60 wt%) of small size (~3.5 nm) and even distribution. The strong metal oxide–support interaction (SMOSI) was observed between SnO_2 NPs and carbon supports in the order of SnO_2/NSC > SnO_2/NC > SnO_2/SC > SnO_2/C. This SMOSI facilitates effective electron transfer from the carbon support to SnO_2 NPs, bolstering the adsorption of $CO_2^{\cdot-}$, thus promoting CO_2RR. Among these, SnO_2/NSC demonstrates significantly enhanced selectivity and activity for CO_2 reduction to HCOOH, showcasing a high FE of 94.4% and an extraordinary partial current density of 56.0 mA cm^{-2} in an H-cell, outperforming the majority of Sn-based catalysts [65].

Transition metal oxides, notably CoO_x, FeO_x, and NiO_x, are widely recognized for their low cost, low toxicity, and abundant nature across various electrochemical reactions. However, their efficacy in CO_2RR is largely limited due to their inherent inertness. Nitrogen-doped carbon nanotubes (N-CNTs) surprisingly exhibit an activation effect on the activity and selectivity of transition metal-oxide (MO_x, where M = Fe, Ni, and Co) nanoclusters in CO_2RR. MO_x supported on N-CNT displays a remarkable CO yield of 2.6–2.8 mmol cm^{-2} min^{-1} at an overpotential of −0.55 V, marking a two-orders-of-magnitude increase compared to MO_x supported on acid-treated CNTs (MO_x/O-CNT) and a fourfold enhancement over pristine

N-CNT. The faraday efficiency for electrochemical CO_2-to-CO conversion reaches an impressive 90.3% at an overpotential of 0.44 V. Insights from in situ XAS measurements and DFT calculations unveil that MO_x nanoclusters undergo hydration in CO_2-saturated $KHCO_3$, while the N defects of N-CNT effectively stabilize these metal hydroxyl species during CO_2RR conditions. This stabilization facilitates the splitting of water molecules, providing local protons to prevent active site poisoning during the CO_2RR process [66].

The synthesis of metal and metal oxide composites has emerged as an effective strategy to enhance the CO_2RR activity of metal oxides. For instance, Ni NP-decorated MnO_2 nanodendrites supported on carbon fibers have been identified as efficient electrocatalysts for CO_2 reduction to formate, demonstrating remarkable formate selectivity and a high FE of over 85%. Further investigations highlight that the strong electronic interactions between Ni and MnO_2 contribute to achieving high hydrogen storage capacity and reducing energy barriers for intermediates and products, thereby significantly promoting the selective electroreduction of CO_2 to formate [67]. The Bi NPs/Bi_2O_3 nanosheets with abundant grain boundaries synthesized by Li et al. also showed a FE_{HCOOH} of ~100% [68]. Carbon-supported BiO_x NPs (BiO_x/C) were investigated as a potential candidate for CO_2RR. In bicarbonate solutions, the BiO_x/C catalysts exhibited a high FE of 93.4% for formate from −1.37 to −1.70 V vs. Ag/AgCl with a negligible amount of CO and H_2 [69].

An effective strategy for constructing high-performance metal oxide-based catalysts for CO_2RR involves the introduction of heteroatoms or defects. For instance, the addition of bismuth (Bi) stabilizes the *OCOH intermediate on the SnO (001) surface, reducing the energy barrier for HCOOH formation and enhancing selectivity [70]. Besides metal element doping, nonmetal-doped electrocatalysts like N-doped SnO_2 nanostructured catalysts and F-doped SnO_2 thin films have also been developed for electrochemical CO_2RR [71, 72]. In a study by Daiyan et al., defective ZnO nanomaterials were synthesized using flame spray, exhibiting concurrent H_2 and CO generation during CO_2RR. These engineered ZnO materials achieved an H_2/CO ratio of ≈1 at a high current density (j) of 40 mA cm^{-2} during long-term continuous reactions at a cell voltage of 2.6 V. Investigating the stability of these ZnO structures through in situ atomic pair distribution function analysis addressed the knowledge gap in understanding oxide catalyst dynamics during CO_2RR. Synthesis condition optimization led to modulated ZnO facets, influencing reaction selectivity in accordance with theoretical calculations. These findings and insights into manipulating active sites in defective metal oxides can serve as guiding principles for developing active catalysts for syngas production in renewable power-to-X applications, enabling the generation of various fuels and chemicals [73].

15.3.2 Metal Chalcogenides

Metal chalcogenides, particularly 2D transition metal dichalcogenides (TMDCs) and transition metal carbides/nitrides (MXenes) serve as promising catalysts for CO_2RR. Theoretical simulations have indicated that the under-coordinated active sites on 2D

TMDCs decrease the activation energy barriers for CO_2 molecule chemisorption and hydrogenation, effectively addressing the challenges of sluggish kinetics and poor selectivity in these reactions [74]. For instance, in the case of MoS_2, the presence of Mo-terminated edges with reduced coordination numbers on the surface of MoS_2 bulk exhibits remarkable electroactivity in converting CO_2 into CO, especially in an ionic liquid-based electrolyte, achieving a current density of 65 mA cm^{-2} at −0.764 V vs. RHE with an FE of ≈98 % [75].

TMDCs offer substantial tunability owing to their capability to combine various transition metal atoms with non-transition metal atoms. The surface structures and electronic states of TMDCs can be readily adjusted to target specific reactions and minimize undesired side reactions, such as the HER [76–78]. As illustrated by Asadi et al., they observed an increase in the current density for CO_2RR in the order of $MoS_2 < WS_2 < MoSe_2 < WSe_2$, aligning with their electron transfer properties. Notably, WSe_2, with the lowest work function, demonstrated the most effective CO_2 activation among these TMDCs [79].

TMDCs offer extensive tunability through alloying strategies. For instance, in a study by Xu et al., MoS_2 and $MoSe_2$ were alloyed to produce the MoSSe monolayer via a liquid–liquid mediated synthesis method. This synthesis led to lengthened Mo—Se bonds and shortened Mo—S bonds, resulting in an improved density of states near the conduction band edge and a partially delocalized charge around Mo atoms. The augmented density of states at the conduction band edge facilitated charge transfer, as evidenced by the reduced work functions of MoSSe. The partially delocalized charge around Mo disrupted the linear scaling relations by stabilizing *COOH and weakening the binding with *CO. Consequently, the MoSSe monolayer exhibited remarkable CO_2RR performance, achieving an FE of 45.2% in CO evolution at −1.15 V vs. RHE, outperforming both MoS_2 (16.6%) and $MoSe_2$ (30.5%) monolayers. A synergy between the metals and the chalcogens also contributed to the CO_2RR pathways [80]. Similarly, ternary TMDC alloys were explored to enhance CO_2 reduction for CO generation. Xie and coworkers reported on a MoSeS alloy demonstrating a higher CO FE (45.2%) compared to MoS_2 (16.6%) and $MoSe_2$ (30.5%) monolayers at 1.15 V [80].

Furthermore, high-entropy alloys amalgamating multiple principal elements in near-equal fractions enable a broad compositional space, leading to exceptional functionalities absent in alloys with fewer principal elements. Here, a study reports the prediction, synthesis, and multiscale characterization of 2D high-entropy TMDC alloys incorporating four/five transition metals. Among these, a five-component alloy, $(MoWVNbTa)S_2$, with the highest configurational entropy, was examined for CO_2 conversion to CO, revealing an excellent current density of 0.51 A cm^{-2} and a turnover frequency (TOF) of 58.3 s^{-1} at around −0.8 V vs. RHE (as depicted in Figure 15.3). First-principles calculations underscored the superior CO_2 electroreduction attributed to a multisite catalysis where atomic-scale disorder optimized the rate-limiting step of CO desorption by facilitating isolated transition metal edge sites with weak CO binding. These 2D high-entropy TMDC alloys offer a versatile material platform to design superior catalysts across various electrochemical systems [44].

Manipulating electronic properties through elemental doping stands as a direct and effective strategy in the design of efficient catalysts for CO_2 electroreduction. Theoretical studies on S-edge MoS_2, doped with various metals like Fe, Co, and Ni, have indicated that the presence of dual binding sites involving metallic and S sites leads to significant enhancements in CO_2RR efficiency [81]. In another study, Zhang et al. observed substantial improvements in the current density and FE for carbonaceous products upon introducing Ni into SnS_2 nanosheets compared to pristine SnS_2 nanosheets. At a 5 atm% Ni content, the Ni-doped SnS_2 nanosheets demonstrated an impressive FE of 93 % for carbonaceous products, coupled with a current density of 19.6 mA cm^{-2} at −0.9 V vs. RHE. Mechanistic insights unveiled that Ni doping induced a defect level, consequently reducing the work function of SnS_2 nanosheets, thereby enhancing CO_2 activation and improving CO_2 electroreduction performance [82]. Similarly, MoS_2 doped with 5% niobium (Nb) exhibited an order of magnitude higher TOF of CO compared to pristine MoS_2 at the studied overpotentials (50–150 mV) [83].

15.3.3 Metal Carbides and Nitrides

Transition metal carbides and nitrides, particularly 2D transition metal carbides/nitrides known as MXenes, represent compelling nonprecious materials capable of replacing or reducing the reliance on precious metals in catalyzing CO_2RR [84]. MXenes boast ample surface metal atoms with vacant d-orbitals, facilitating the activation of CO_2 molecules, alongside a metallic electronic structure that ensures efficient electron transport [85, 86]. In investigations involving Mo_2C and Ti_3C_2 MXenes as earth-abundant electrocatalysts for CO_2RR, notable FEs were observed. Mo_2C exhibited a 90% efficiency at 250 mV overpotential, while Ti_3C_2 showed 65% at 650 mV overpotential for CO_2 to CO reduction in acetonitrile using an ionic liquid electrolyte. The employment of ionic liquid 1-ethyl-2-methylimidazolium tetrafluoroborate as an electrolyte in the organic solvent effectively suppressed the competing HER. Computational analyses through DFT propose that oxygen vacancy sites on both MXene surfaces serve as catalytically active sites. Moreover, spontaneous dissociation of adsorbed COOH species to a water molecule and adsorbed CO on Mo_2C promoted CO_2RR [87]. Handoko et al. reported formic acid as the primary CO_2 reduction product on Ti_2CT_x and Mo_2CT_x MXenes, with Ti_2CT_x achieving a maximum FE exceeding 56% at 1.73 V vs. RHE. Their DFT findings suggested that CO_2RR activity correlated with the presence of Ti_2CT_x -F and -O surface termination groups, where fewer -F and more -O groups correlated with lower potential energy steps for formic acid generation [88]. Esmaeilirad et al. synthesized four members of 2D transition metal carbides with a formula of M_2C, i.e. W_2C, Mo_2C, Nb_2C, and V_2C nanoflakes using the carburization method followed by the liquid exfoliation technique. The electrocatalytic performance of CO_2RR shows these materials are mainly selective for CH_4 formation with W_2C nanoflakes having the best CO_2RR activity compared to the studied catalysts. A maximum CH_4 current density of −421.63 mA cm^{-2} at the potential of −1.05 V vs. RHE was observed for W_2C nanoflakes. The electronic property analyses reveal that the superior performance of W_2C NFs is a result of the low work function, small charge transfer resistance in the electrochemical double layer region, and heavily reduced tungsten atoms at the surface [89].

Transition metal carbides, nitrides, and carbon nitride serve as versatile platforms capable of modulating the product selectivity of CO_2RR by adjusting the binding energy of reaction intermediates [84, 90]. Recent studies have showcased the utility of transition metal carbides and nitrides as support materials, not only reducing the reliance on precious metals but also enhancing activity and selectivity. For example, investigating Pd, Au, and Ag catalysts supported by cost-effective transition metal carbides has aimed to reduce precious metal consumption while fine-tuning CO_2RR activity and selectivity toward synthesizing gas [91–93]. Particularly, Pd/TaC exhibits superior CO_2RR activity, stability, and CO FE compared to commercial Pd/C, with 10 wt% Pd/TaC showcasing higher CO_2RR activity than the commercial 40% Pd/C. In situ measurements validate the transformation of Pd into hydride (PdH) in the CO_2RR environment. DFT calculations reveal that the TaC substrates modulate the binding energies of crucial intermediates on supported PdH [91]. Zhang et al. found that C_3N_4-supported Au NPs (Au/C_3N_4) outperform carbon black-supported Au NPs (Au/C) in CO_2RR performance. The selectivity for CO_2RR reaches 90% over Au/C_3N_4 at a low potential of −0.45 V vs. RHE (overpotential of −340 mV). Experimental and theoretical insights highlight that the interaction between Au and C_3N_4 enhances the electron richness of the Au surface, strengthening the adsorption of the key reaction intermediate *COOH. Similar performance enhancements in CO_2RR are observed for C_3N_4-supported Ag NPs (Ag/C_3N_4), which also exhibit an electron-enriched Ag surface. Both Au/C_3N_4 and Ag/C_3N_4 signify successful engineering of metal catalysts to enhance CO_2RR activity [94].

Moreover, composite materials involving metal carbides/nitrides have been developed for CO_2RR. Zhao et al. reported Ni single atoms (SANi) supported on N-doped carbon nanotubes (NCNT), partially nitrided Ni nanoclusters (NiN_x), and Ni NPs with SANi (SANi-Ni) on NCNT as catalysts for CO_2RR. Among these, NiNx/NCNT with 10.2 wt% Ni exhibits a CO FE surpassing 99.0% at a potential as low as −0.272 V (vs. RHE) and a CO generation rate of 43.0 mol g^{-1} h^{-1} at a potential of −0.572 V (vs. RHE), showcasing superior performance compared to SANi/NCNT and SANi-Ni/NCNT catalysts. DFT calculations suggest that NiNx nanoclusters favor *COOH formation with a lower Gibbs free energy than isolated SANi [95]. Additionally, Chen et al. developed Fe–N–C nanofibers featuring iron nitride (Fe_xN) NPs encapsulated by Fe and N codoped carbon layers (Fe_xN@Fe–N–C) as an electrocatalyst for aqueous CO_2RR. This catalyst demonstrates efficient CO_2 reduction to CO with a high FE of 95% and a significant current density of 4.71 mA cm^{-2} at −0.53 V vs. RHE. Theoretical calculations indicate that the Fe–N_4–C model could be easily deactivated by the generated CO. However, encapsulating Fe_xN NPs dramatically reduces CO bonding strength, rendering the Fe atom (on Fe–N_4–C) an efficient active center for CO_2RR [96].

15.4 Non-Cu Metal-Based Molecular Catalysts

15.4.1 Molecular Catalysts

Molecular catalysts, distinguished by their refined electronic structures and precise atomic coordination, constitute an important category alongside conventional metal-based catalysts for CO_2RR. Many molecular catalysts can exhibit superior performance toward the CO_2RR, mainly benefiting from atomically dispersed catalytic sites and versatile coordination structures, which can be tailored at a molecular level. Similar to other non-Cu metal-based electrocatalysts, most molecular catalysts primarily facilitate the 2-electron reduction of CO_2, yielding CO or formate [97]. However, their practical application is often hindered by low current densities. The metal-based molecular catalysts toward the CO_2RR mainly include (i) homogeneous molecular catalysts, typically organometallic complexes dissolved in organic solvents capable of electrochemical CO_2RR; (ii) heterogenized molecular catalysts, which are usually immobilized on the electrode for improving electrical contact between the catalyst and electrode, and also for facilitating the catalyst recycling and separation; (iii) reticular 3D materials like metal–organic frameworks (MOFs), encapsulating active molecular catalysts within porous (metal)organic networks; and (iv) single-site catalysts, where metal atoms disperse atomically on supports, notably carbon materials, stabilized via coordination structures with heteroatoms. Further discussion on this topic will follow in the subsequent section.

Among various molecular catalysts for electrochemical CO_2RR, metal phthalocyanines (Pc) and porphyrins stand as some of the earliest and most extensively researched systems. For instance, Meshitsuka et al. conducted pioneering research evaluating the electrochemical CO_2RR performance of various M–Pc (M = Fe, Ni, Co, Mn, Cu, and Pd) in nonaqueous electrolytes, and found that Co–Pc and Ni–Pc were active toward the CO_2RR [98]. They also proposed that the occupied d_{z2} orbital of the metal center in the M–Pc complex plays a key role in determining the CO_2RR activity. Subsequent studies have delved into polypyridyl-coordinated precious metals like Re and Ru [99, 100]. For instance, Lehn et al. highlighted Re(bipy)$(CO)_3$Cl (bipy = 2,2′-bipyridine) as a highly active homogeneous molecular catalyst toward the electrochemical CO_2RR, with CO as the major product with high FEs (>90%) [101]. Similarly, Kubiak et al. reported on a Ru-based molecular catalyst with a bipyridine ligand, exhibiting notably high intrinsic activity toward electrochemical CO_2RR and also favoring CO as the dominant product (FE > 95%) [102]. Recent attention has shifted back to cost-effective 3d transition metals such as Mn, Ni, and Co [100]. The CO_2RR over Mn-based complexes was first reported by Deronzier et al., who demonstrated that Mn complexes with bipyridine-based ligands were active toward the CO_2RR with fairly high stability and selectivity toward the CO formation (FE = 85%) in a mixture of acetonitrile and water [103]. Daasbjerg and co-workers made another interesting attempt by studying CO_2RR over a series of Mn-based carbonyl complexes modified with bipyridine and phenanthroline. They found that amines properly positioned in the secondary coordination sphere of the complexes could change the product formation from CO to HCOOH [104].

Beyond metal complexes, MOF -based materials and covalent organic framework (COF)-based materials, known for their expansive structural and chemical diversity [105–107] and having already been widely employed as active catalysts for photocatalytic CO_2 reduction [108, 109], emerge as another class of promising catalysts for electrochemical CO_2RR. For example, Zhuang et al. reported a nickel (Ni) porphyrin-based covalent triazine framework (NiPor-CTF) with atomically dispersed NiN_4 centers as an efficient electrocatalyst for CO_2RR. [110] The NiPor-CTF showed high selectivity for CO_2RR with a high FE (>90%) at potentials from −0.6 to −0.9 V and achieved a maximum FE of 97% at −0.9 V with a high current density of 52.9 mA cm^{-2}. DFT calculations revealed that the superior CO_2RR performance resulted from the decreased energy barrier for $^*CO_2$ to $^*COOH$ over Ni–N_4 sites. Some iron (Fe) porphyrin-based MOFs were also reported to reduce CO_2 to CO in both nonaqueous and aqueous electrolytes. For instance, Hupp et al. reported that Fe–MOF-525 reduced CO_2 to CO in acetonitrile electrolyte with a CO FE of 54% [111]. Dong et al. constructed an Fe-based MOF (PCN-222) and demonstrated that the Fe–MOF mixed with carbon black exhibits high catalytic performance for electrochemical reduction of CO_2 to CO with 494 mV overpotential and a maximum FE_{CO} of 91% in a CO_2-saturated 0.5 M $KHCO_3$ aqueous solution, achieving a TOF of 0.336 $site^{-1}$ s^{-1} [112]. However, most of the molecular catalysts still suffer from low current densities during the CO_2RR. Berlinguette and co-workers addressed this issue by studying the CO_2RR performance of a molecular Co–Pc catalyst using a membrane electrode assembly (MEA) setup, where the cathode was fed with CO_2 gas and the anode was fed with KOH electrolyte [113]. The resultant MEA delivered an impressively high current density of 150 mA cm^{-2} with an FE_{CO} >95% and maintained long-term stability for over 100 hours. This result greatly demonstrates that these widely available and abundant molecular catalysts are capable of delivering high performance under practical operating conditions.

15.4.2 Single-Site M–N–C Catalysts

In recent years, nitrogen-doped carbon-supported atomically dispersed single metal sites (M–N–C) have emerged as a promising category of high-performance CO_2RR catalysts, owing to their efficient utilization of metal sites and the unique electronic structure conferred by singular metal active sites [114]. The inclusion of nitrogen dopants within the carbon matrix serves a dual role: providing additional defect sites to anchor metal sites and offering the potential to modulate the electronic structure of both surface carbon and metal sites [52]. This has prompted a growing number of studies focusing on M–N–C catalysts for CO_2RR, showcasing high FE_{CO}, enhanced stability, and relatively low production costs. The synthesis of M–N–C samples typically involves high-temperature pyrolysis of transition metal compounds/mixtures and C/N-containing precursors. During pyrolysis, metal atoms tend to form coordination structures with carbon and nitrogen, resulting in the formation of M–N_x moieties. Tailoring the geometric and electronic structure of single metal sites, and consequently, the ensuing CO_2RR performance can be achieved by deliberately varying metal identities and their coordination environments, as depicted in Figure 15.4. While altering metal centers can be readily accomplished by changing metal salts,

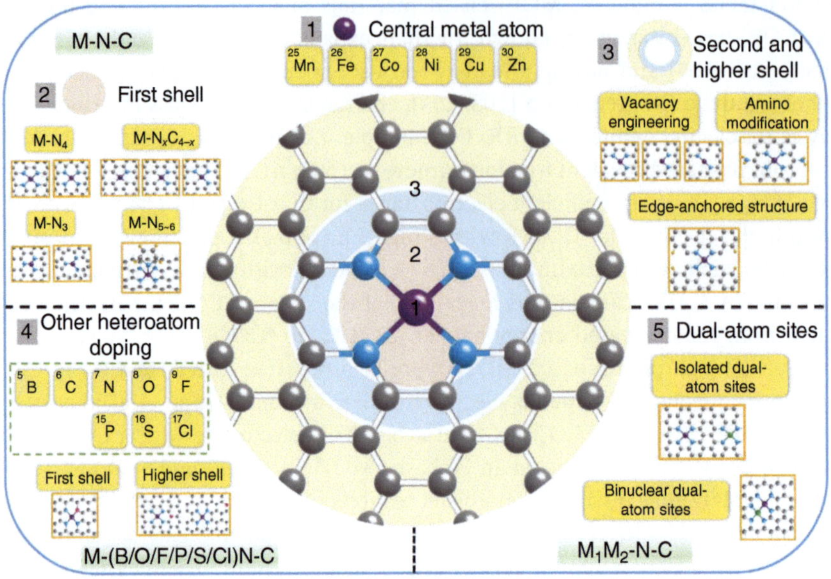

Figure 15.4 Schematic illustration of the coordination structure engineering strategies of M–N–C catalysts toward the CO_2RR. Source: Liang, S et al., [52]/Wiley-VCH/CC BY 4.0.

regulating the coordination environment poses greater challenges, primarily due to the difficulty in controlling pyrolysis conditions, leading to a wide distribution of surface nitrogen functionalities. Recent efforts have also aimed at introducing additional heteroatoms alongside nitrogen or constructing dual-atom sites during M–N–C catalyst synthesis. These strategies have shown promise in reducing energy barriers for the formation of crucial intermediates during CO_2RR.

A benchmark work conducted by Strasser et al. systematically investigated the electrochemical CO_2RR performance across a broad spectrum of M–N–C catalysts (M = Mn, Fe, Co, Ni, and Cu). Their findings emphasized the pivotal role of *COOH binding strength over metal centers in dictating the catalytic conversion of CO_2 to CO [115]. Specifically, M–N–C catalysts (e.g. M = Fe, Mn, and Co), exhibiting strong *COOH binding, favored CO formation at relatively low overpotentials. However, due to their robust binding to *H, they promoted competitive HER at high overpotentials, consequently diminishing FEs toward CO. In contrast, the Ni–N–C catalyst displayed weak *H binding, leading to the highest FE_{CO} at elevated overpotentials among all examined catalysts. Similarly, Jiang and co-workers synthesized a series of MOF-derived M–N–C catalysts (M = Ni, Fe, Co, and Cu) with comparable metal coordination structures (pyridinic-type $M\text{-}N_4$), revealing Ni–N–C achieving a maximum CO FE of up to 96.8%, followed by Fe–N–C and Co–N–C. Furthermore, they demonstrated that even with diluted CO_2 gas (down to 15 vol%), a high FE_{CO} could be maintained, signifying practical industrial applicability of CO_2RR [116]. Zhang and co-workers reported the construction of a series of M–N–C catalysts (M = Cr, Mn, Fe, Co, Ni, Cu, Zn, Ru, and Pt) containing exclusively $M\text{-}N_4$ structural moieties [117]. They disclosed an intriguing volcano relation between

FE_{CO} and metal atomic number, which might be related to the d-band center of metal in the M–N–C catalysts. Among these catalysts, Ni–N–C emerged as the most efficient system for electrochemical CO_2RR to form CO, consistently achieving CO FEs exceeding 90% in the potential range from −0.7 to −0.9 V. Besides metal center identities, the specific coordination structure significantly influences the electrochemical CO_2RR performance of M–N–C catalysts. Sun et al. showcased the distinct activity between pyridinic Ni–N_3 and pyrrolic Ni–N_3 structural moieties supported on carbon nanotubes during electrochemical CO_2RR to form CO [118]. The pyrrolic Ni–N_3 exhibited superior catalytic activity due to its lower binding energy of the key reaction intermediate (*COOH) [119]. Wu and co-workers synthesized Co–N–C catalysts with atomically dispersed Co of different coordination structures, i.e. Co–N_4, Co–N_3–C, and Co–N_2C_2 [120]. It is disclosed that the maximum FE of CO (94%) was obtained on Co–N_2C_2, which is significantly higher than that on Co–N_4 and Co–N_3–C, probably due to the lower coordination number, which favors the activation of CO_2. However, Lin et al. reported that Co–N_4 exhibited both superior activity and stability compared to its Co–N_x–C_{4-x} counterparts [121].

Despite these exciting progresses, the current density reported in these works is usually too small to be employed for practical applications. Recently, several reports have focused on M–N–C catalysts for large-scale electrochemical CO_2RR at industrial current densities utilizing flow cells, achieving substantial CO partial current densities exceeding 100 mA cm^{-2}. These findings underscore the high promise of M–N–C catalysts for efficient CO_2RR to CO at an industrial scale. Moreover, beyond selective CO formation, the direct production of syngas (CO + H_2), combining CO_2RR and HER, is considered a prospective avenue for future industrial application. Nonetheless, ensuring electrode stability for CO or syngas production in a flow cell system remains an urgent challenge. Encouragingly, Liu and co-workers demonstrated that Ni(I)–N–C catalysts exhibited superior FE_{CO} up to 98% [114]. More importantly, the catalyst maintained 98% of its initial activity after 100 hours of continuous reaction, showcasing significant promise for practical application. The catalytic properties and the identification of the active sites in M–N–C catalysts are still highly debated. While some reports consider M–N_4 moieties as active sites, others suggest higher activity from unsaturated M–N_x moieties. Moreover, debates exist around whether Ni coordinated with carbon or intrinsic defects, instead of Fe–N_4 sites, serves as the active site for CO_2RR. Overall, substantial scope remains for exploring the potential of central metal atoms and coordination atoms to enhance CO_2RR efficiency.

15.5 Concluding Remarks and Outlook

In this chapter, we focus on non-Cu-based metal catalysts for electrochemical CO_2RR, particularly emphasizing selective catalysis toward specific products such as CO and formate, illuminates essential facets of this evolving field. The overview aims to facilitate a deeper grasp of fundamental science and the latest advancements in CO_2RR over non-Cu metal-based catalysts. Additionally, we identify some of the key challenges and opportunities for further developing high-performing CO_2RR catalysts, as detailed in the following context.

Numerous metals (e.g. Au, Ag, Pd, and Zn) and metal–carbon-based molecular catalysts can exhibit high FEs for CO production, while many of them can only acquire a small electrolysis current density, which limits their practical application. The integration of high-performance electrocatalysts with fast kinetics and high selectivity into gas diffusion electrodes or membrane electrodes is crucial to overcome these challenges. The durability of electrocatalysts remains a pressing concern. While only a few reported catalysts have demonstrated stability beyond 100 hours, atomically distributed catalysts, particularly single-atom catalysts, exhibit not just high catalytic activity and selectivity but also longevity, sustaining continuous electrolysis for over 120 hours without notable decay in efficiency. This highlights a promising avenue for exploring robust electrocatalysts. One prominent avenue involves exploring emerging catalyst classes capable of disrupting scaling relations between reaction intermediates. These encompass stable alloys, single-atom catalysts, heteroatom-doped carbon materials, MOFs/COFs, materials featuring 3D active sites for confinement effects, and functionalized surfaces. Ongoing material discovery endeavors must continue along these trajectories to identify novel catalysts with enhanced properties and efficiency. Enhancing spectroscopic characterization techniques to augment spatial resolution and sensitivity during monitoring of catalyst surface changes in electrochemical testing remains pivotal. Moreover, beyond designing and preparing high-performance catalysts, understanding the impact of reaction conditions – such as reactor design, temperature, electrolyte pH, and pressure – on CO_2RR electrocatalysis is indispensable for optimizing catalytic efficiency.

In summary, addressing these multifaceted challenges through sustained research endeavors and interdisciplinary collaborations holds the key to advancing the field of electrochemical CO_2 conversion, steering it closer toward practical applications and sustainable energy solutions.

Acknowledgment

This work was supported by the National Natural Science Foundation of China (22272117).

References

1 Birdja, Y.Y., Pérez-Gallent, E., Figueiredo, M.C. et al. (2019). Advances and challenges in understanding the electrocatalytic conversion of carbon dioxide to fuels. *Nat. Energy* 4 (9): 732.

2 Zhang, G.-R., Straub, S.-D., Shen, L.-L. et al. (2020). Probing CO_2 reduction pathways in copper catalysts using ionic liquid as a chemical trapping agent. *Angew. Chem. Int. Ed.* 59 (41): 18095.

3 Wang, G., Chen, J., Ding, Y. et al. (2021). Electrocatalysis for CO_2 conversion: from fundamentals to value-added products. *Chem. Soc. Rev.* 50 (8): 4993.

4 Frese, K.W. and Leach, S. (1985). Electrochemical reduction of carbon dioxide to methane, methanol, and co on ru electrodes. *J. Electrochem. Soc.* 132 (1): 259.

5 Summers, D.P., Leach, S., and Frese, K.W. (1986). The electrochemical reduction of aqueous carbon dioxide to methanol at molybdenum electrodes with low overpotentials. *J. Electroanal. Chem. Interfacial Electrochem.* 205 (1): 219.

6 Yoshio, H., Katsuhei, K., and Shin, S. (1985). Production of Co and CH_4 in electrochemical reduction of CO_2 at metal electrodes in aqueous hydrogencarbonate solution. *Chem. Lett.* 14 (11): 1695–1698.

7 Yin, Z., Palmore, G.T.R., and Sun, S. (2019). Electrochemical reduction of CO_2 catalyzed by metal nanocatalysts. *Trends Chem.* 1 (8): 739.

8 Hori, Y. (2008). Electrochemical CO_2 reduction on metal electrodes. In: *Modern Aspects of Electrochemistry* (ed. C.G. Vayenas, R.E. White, and M.E. Gamboa-Aldeco), 89–189. New York, NY: Springer New York.

9 Li, L., Li, X., Sun, Y., and Xie, Y. (2022). Rational design of electrocatalytic carbon dioxide reduction for a zero-carbon network. *Chem. Soc. Rev.* 51 (4): 1234.

10 Shin, H., Hansen, K.U., and Jiao, F. (2021). Techno-economic assessment of low-temperature carbon dioxide electrolysis. *Nat. Sustainability* 4 (10): 911.

11 Li, M., Garg, S., Chang, X. et al. (2020). Toward excellence of transition metal-based catalysts for CO_2 electrochemical reduction: an overview of strategies and rationales. *Small Methods* 4 (7): 2000033.

12 Ross, M.B., De Luna, P., Li, Y. et al. (2019). Designing materials for electrochemical carbon dioxide recycling. *Nat. Catal.* 2 (8): 648.

13 Wang, Y., Liu, J., Wang, Y. et al. (2017). Tuning of CO_2 reduction selectivity on metal electrocatalysts. *Small* 13 (43): 1701809.

14 Jin, S., Hao, Z., Zhang, K. et al. (2021). Advances and challenges for the electrochemical reduction of CO_2 to co: from fundamentals to industrialization. *Angew. Chem.* 133 (38): 20795.

15 Ding, P., Zhao, H., Li, T. et al. (2020). Metal-based electrocatalytic conversion of CO_2 to formic acid/formate. *J. Mater. Chem. A* 8 (42): 21947.

16 An, X., Li, S., Hao, X. et al. (2021). Common strategies for improving the performances of tin and bismuth-based catalysts in the electrocatalytic reduction of CO_2 to formic acid/formate. *Renew. Sust. Energ. Rev.* 143: 110952.

17 Ito, K., Murata, T., and Ikeda, S. (1976). Electrochemical reduction of carbon dioxide to organic compounds. *Bull. Nagoya Inst. Technol.* 27: 209.

18 Hori, Y., Wakebe, H., Tsukamoto, T., and Koga, O. (1994). Electrocatalytic process of co selectivity in electrochemical reduction of CO_2 at metal electrodes in aqueous media. *Electrochim. Acta* 39 (11): 1833.

19 Hoshi, N., Kato, M., and Hori, Y. (1997). Electrochemical reduction of CO_2 on single crystal electrodes of Ag (111), Ag (100) and Ag (110). *J. Electroanal. Chem.* 440 (1–2): 283.

20 Hori, Y., Murata, A., Kikuchi, K., and Suzuki, S. (1987). Electrochemical reduction of carbon dioxides to carbon monoxide at a gold electrode in aqueous potassium hydrogen carbonate. *J. Chem. Soc. Chem. Commun.* (10): 728.

21 Komatsu, S., Yanagihara, T., Hiraga, Y. et al. (1995). Electrochemical reduction of CO_2 at Sb and Bi electrodes in $KHCO_3$ solution. *Denki Kagaku oyobi Kogyo Butsuri Kagaku* 63 (3): 217.

22 Won, D.H., Choi, C.H., Chung, J. et al. (2015). Rational design of a hierarchical tin dendrite electrode for efficient electrochemical reduction of CO_2. *ChemSusChem* 8 (18): 3092.

23 Sinha, R., Bisht, A., Rarotra, S., and Mandal, T.K. (2020). Continuous semi-micro reactor prototype for the electrochemical reduction of CO_2 into formic acid. *Ind. Eng. Chem. Res.* 59 (5): 1737.

24 Lv, W., Zhou, J., Kong, F., and Wang, W. (2016). Porous tin-based film deposited on copper foil for electrochemical reduction of carbon dioxide to formate. *Int. J. Hydrog. Energy* 41 (3): 1585.

25 Qiu, Y., Du, J., Dong, W. et al. (2017). Selective conversion of CO_2 to formate on a size tunable nano-Bi electrocatalyst. J. CO_2 Util. 20: 328.

26 Koh, J.H., Won, D.H., Eom, T. et al. (2017). Facile CO_2 electro-reduction to formate via oxygen bidentate intermediate stabilized by high-index planes of bi dendrite catalyst. *ACS Catal.* 7 (8): 5071.

27 Su, P., Xu, W., Qiu, Y. et al. (2018). Ultrathin bismuth nanosheets as a highly efficient CO_2 reduction electrocatalyst. *ChemSusChem* 11 (5): 848.

28 Kim, S., Dong, W.J., Gim, S. et al. (2017). Shape-controlled bismuth nanoflakes as highly selective catalysts for electrochemical carbon dioxide reduction to formate. *Nano Energy* 39: 44.

29 Gong, Q., Ding, P., Xu, M. et al. (2019). Structural defects on converted bismuth oxide nanotubes enable highly active electrocatalysis of carbon dioxide reduction. *Nat. Commun.* 10 (1): 2807.

30 Kauffman, D.R., Alfonso, D., Matranga, C. et al. (2012). Experimental and computational investigation of Au_{25} clusters and CO_2: a unique interaction and enhanced electrocatalytic activity. *J. Am. Chem. Soc.* 134 (24): 10237.

31 Zhu, W., Michalsky, R., Metin, O. et al. (2013). Monodisperse au nanoparticles for selective electrocatalytic reduction of CO_2 to CO. *J. Am. Chem. Soc.* 135 (45): 16833.

32 Lu, Q., Rosen, J., Zhou, Y. et al. (2014). A selective and efficient electrocatalyst for carbon dioxide reduction. *Nat. Commun.* 5 (1): 3242.

33 Huang, H., Jia, H., Liu, Z. et al. (2017). Understanding of strain effects in the electrochemical reduction of CO_2: using Pd nanostructures as an ideal platform. *Angew. Chem.* 129 (13): 3648.

34 Zhu, W., Zhang, Y.-J., Zhang, H. et al. (2014). Active and selective conversion of CO_2 to CO on ultrathin au nanowires. *J. Am. Chem. Soc.* 136 (46): 16132.

35 Kim, C., Jeon, H.S., Eom, T. et al. (2015). Achieving selective and efficient electrocatalytic activity for CO_2 reduction using immobilized silver nanoparticles. *J. Am. Chem. Soc.* 137 (43): 13844.

36 Gao, D., Zhou, H., Wang, J. et al. (2015). Size-dependent electrocatalytic reduction of CO_2 over Pd nanoparticles. *J. Am. Chem. Soc.* 137 (13): 4288.

37 Liu, S., Tao, H., Zeng, L. et al. (2017). Shape-dependent electrocatalytic reduction of CO_2 to CO on triangular silver nanoplates. *J. Am. Chem. Soc.* 139 (6): 2160.

38 Liu, M., Pang, Y., Zhang, B. et al. (2016). Enhanced electrocatalytic CO_2 reduction via field-induced reagent concentration. *Nature* 537 (7620): 382.

39 Hall, A.S., Yoon, Y., Wuttig, A., and Surendranath, Y. (2015). Mesostructure-induced selectivity in CO_2 reduction catalysis. *J. Am. Chem. Soc.* 137 (47): 14834.

40 Yoon, Y., Hall, A.S., and Surendranath, Y. (2016). Tuning of silver catalyst mesostructure promotes selective carbon dioxide conversion into fuels. *Angew. Chem. Int. Ed.* 55 (49): 15282.

41 Dutta, A., Morstein, C.E., Rahaman, M. et al. (2018). Beyond copper in CO_2 electrolysis: effective hydrocarbon production on silver-nanofoam catalysts. *ACS Catal.* 8 (9): 8357.

42 Park, H., Choi, J., Kim, H. et al. (2017). AgIn dendrite catalysts for electrochemical reduction of CO_2 to CO. *Appl. Catal.* B 219: 123.

43 Won, D.H., Shin, H., Koh, J. et al. (2016). Highly efficient, selective, and stable CO_2 electroreduction on a hexagonal Zn catalyst. *Angew. Chem.* 128 (32): 944.

44 Cavin, J., Ahmadiparidari, A., Majidi, L. et al. (2021). 2D high-entropy transition metal dichalcogenides for carbon dioxide electrocatalysis. *Adv. Mater.* 33 (31): 2100347.

45 He, J., Johnson, N.J., Huang, A., and Berlinguette, C.P. (2018). Electrocatalytic alloys for CO_2 reduction. *ChemSusChem* 11 (1): 48.

46 Wen, G., Lee, D.U., Ren, B. et al. (2018). Orbital interactions in Bi-Sn bimetallic electrocatalysts for highly selective electrochemical CO_2 reduction toward formate production. *Adv. Energy Mater.* 8 (31): 1802427.

47 Choi, S.Y., Jeong, S.K., Kim, H.J. et al. (2016). Electrochemical reduction of carbon dioxide to formate on tin–lead alloys. *ACS Sustain. Chem. Eng.* 4 (3): 1311.

48 Bai, X., Chen, W., Zhao, C. et al. (2017). Exclusive formation of formic acid from CO_2 electroreduction by a tunable Pd-Sn alloy. *Angew. Chem.* 129 (40): 12387.

49 Torelli, D.A., Francis, S.A., Crompton, J.C. et al. (2016). Nickel–gallium-catalyzed electrochemical reduction of CO_2 to highly reduced products at low overpotentials. *ACS Catal.* 6 (3): 2100.

50 Kortlever, R., Peters, I., Koper, S., and Koper, M.T.M. (2015). Electrochemical CO_2 reduction to formic acid at low overpotential and with high faradaic efficiency on carbon-supported bimetallic Pd–Pt nanoparticles. *ACS Catal.* 5 (7): 3916.

51 Hatsukade, T., Kuhl, K.P., Cave, E.R. et al. (2017). Carbon dioxide electroreduction using a silver–zinc alloy. *Energy Technol.* 5 (6): 955.

52 Liang, S., Huang, L., Gao, Y. et al. (2021). Electrochemical reduction of CO_2 to CO over transition metal/n-doped carbon catalysts: the active sites and reaction mechanism. *Adv. Sci.* 8 (24): 2102886.

53 Zhang, Q., Tao, S., Du, J. et al. (2020). A cold plasma-activated in situ agco surface alloy for enhancing the electroreduction of CO_2 to ethanol. *J. Mater. Chem. A* 8 (17): 8410.

54 Dutta, A., Kuzume, A., Rahaman, M. et al. (2015). Monitoring the chemical state of catalysts for CO_2 electroreduction: an in operando study. *ACS Catal.* 5 (12): 7498.

55 Chen, Y., Li, C.W., and Kanan, M.W. (2012). Aqueous CO_2 reduction at very low overpotential on oxide-derived au nanoparticles. *J. Am. Chem. Soc.* 134 (49): 19969.

56 Ma, M., Trześniewski, B.J., Xie, J. et al. (2016). Selective and efficient reduction of carbon dioxide to carbon monoxide on oxide-derived nanostructured silver electrocatalysts. *Angew. Chem.* 128 (33): 9900.

57 Feng, X., Jiang, K., Fan, S., and Kanan, M.W. (2015). Grain-boundary-dependent CO_2 electroreduction activity. *J. Am. Chem. Soc.* 137 (14): 4606.

58 Zhang, S., Kang, P., and Meyer, T.J. (2014). Nanostructured tin catalysts for selective electrochemical reduction of carbon dioxide to formate. *J. Am. Chem. Soc.* 136 (5): 1734.

59 Liang, C., Kim, B., Yang, S. et al. (2018). High efficiency electrochemical reduction of CO_2 beyond the two-electron transfer pathway on grain boundary rich ultra-small SnO_2 nanoparticles. *J. Mater. Chem. A* 6 (22): 10313.

60 Kumar, B., Atla, V., Brian, J.P. et al. (2017). Reduced SnO_2 porous nanowires with a high density of grain boundaries as catalysts for efficient electrochemical CO_2-into-HCOOH conversion. *Angew. Chem. Int. Ed.* 56 (13): 3645.

61 Lee, C.H. and Kanan, M.W. (2015). Controlling H^+ vs CO_2 reduction selectivity on Pb electrodes. *ACS Catal.* 5 (1): 465.

62 Fang, G., Wang, Q., Zhou, J. et al. (2019). Metal organic framework-templated synthesis of bimetallic selenides with rich phase boundaries for sodium-ion storage and oxygen evolution reaction. *ACS Nano* 13 (5): 5635.

63 Ye, R., Zhu, J., Tong, Y. et al. (2023). Metal oxides heterojunction derived Bi-In hybrid electrocatalyst for robust electroreduction of CO_2 to formate. *J. Energy Chem.* 83: 180.

64 Li, F., Chen, L., Knowles, G.P. et al. (2017). Hierarchical mesoporous SnO_2 nanosheets on carbon cloth: a robust and flexible electrocatalyst for CO_2 reduction with high efficiency and selectivity. *Angew. Chem. Int. Ed.* 56 (2): 505.

65 Yuan, L.-P., Jiang, W.-J., Liu, X.-L. et al. (2020). Molecularly engineered strong metal oxide–support interaction enables highly efficient and stable CO_2 electroreduction. *ACS Catal.* 10 (22): 13227.

66 Cheng, Y., Chen, J., Yang, C. et al. (2023). Activation of transition metal (Fe, co and Ni)-oxide nanoclusters by nitrogen defects in carbon nanotube for selective CO_2 reduction reaction. *Energy Environ. Mater.* 6 (1): e12278.

67 He, X.-J., Feng, J.-X., Ren, Q., and Li, G.-R. (2018). Ni nanoparticle-decorated-MnO_2 nanodendrites as highly selective and efficient catalysts for CO_2 electroreduction. *J. Mater. Chem. A* 6 (40): 19438.

68 Li, L., Ma, D.-K., Qi, F. et al. (2019). Bi nanoparticles/Bi_2O_3 nanosheets with abundant grain boundaries for efficient electrocatalytic CO_2 reduction. *Electrochim. Acta* 298: 580.

69 Lee, C.W., Hong, J.S., Yang, K.D. et al. (2018). Selective electrochemical production of formate from carbon dioxide with bismuth-based catalysts in an aqueous electrolyte. *ACS Catal.* 8 (2): 931.

70 Li, F. (2020). Efficient electrochemical reduction of CO_2 to formate using Sn-polyaniline film on Ni foam. *Electrochim. Acta* 332: 135457.

71 Li, Q., Wang, Z., Zhang, M. et al. (2017). Nitrogen doped tin oxide nanostructured catalysts for selective electrochemical reduction of carbon dioxide to formate. *J. Energy Chem.* 26 (5): 825.

72 Shiratsuchi, R., Hongo, K., Nogami, G., and Ishimaru, S. (1992). Reduction of CO_2 on fluorine-doped SnO_2 thin-film electrodes. *J. Electrochem. Soc.* 139 (9): 2544.

73 Daiyan, R., Lovell, E.C., Huang, B. et al. (2020). Uncovering atomic-scale stability and reactivity in engineered zinc oxide electrocatalysts for controllable syngas production. *Adv. Energy Mater.* 10 (28): 2001381.

74 Zhao, S., Xue, J., and Kang, W. (2014). Gas adsorption on MoS_2 monolayer from first-principles calculations. *Chem. Phys. Lett.* 595: 35.

75 Asadi, M., Kumar, B., Behranginia, A. et al. (2014). Robust carbon dioxide reduction on molybdenum disulphide edges. *Nat. Commun.* 5 (1): 4470.

76 Fu, Z., Wang, N., Legut, D. et al. (2019). Rational design of flexible two-dimensional mxenes with multiple functionalities. *Chem. Rev.* 119 (23): 11980.

77 Choi, W., Choudhary, N., Han, G.H. et al. (2017). Recent development of two-dimensional transition metal dichalcogenides and their applications. *Mater. Today* 20 (3): 116.

78 Li, Z., Attanayake, N.H., Blackburn, J.L., and Miller, E.M. (2021). Carbon dioxide and nitrogen reduction reactions using 2D transition metal dichalcogenide (TMDC) and carbide/nitride (MXene) catalysts. *Energy Environ. Sci.* 14 (12): 6242.

79 Asadi, M., Kim, K., Liu, C. et al. (2016). Nanostructured transition metal dichalcogenide electrocatalysts for CO_2 reduction in ionic liquid. *Science* 353 (6298): 467.

80 Xu, J., Li, X., Liu, W. et al. (2017). Carbon dioxide electroreduction into syngas boosted by a partially delocalized charge in molybdenum sulfide selenide alloy monolayers. *Angew. Chem. Int. Ed.* 56 (31): 9121.

81 Hong, X., Chan, K., Tsai, C., and Nørskov, J.K. (2016). How doped MoS_2 breaks transition-metal scaling relations for CO_2 electrochemical reduction. *ACS Catal.* 6 (7): 4428.

82 Zhang, A., He, R., Li, H. et al. (2018). Nickel doping in atomically thin tin disulfide nanosheets enables highly efficient CO_2 reduction. *Angew. Chem. Int. Ed.* 57 (34): 10954.

83 Abbasi, P., Asadi, M., Liu, C. et al. (2017). Tailoring the edge structure of molybdenum disulfide toward electrocatalytic reduction of carbon dioxide. *ACS Nano* 11 (1): 453.

84 Tian, D., Denny, S.R., Li, K. et al. (2021). Density functional theory studies of transition metal carbides and nitrides as electrocatalysts. *Chem. Soc. Rev.* 50 (22): 12338.

85 Zhang, X., Zhang, Z., Li, J. et al. (2017). Ti_2CO_2 mxene: a highly active and selective photocatalyst for CO_2 reduction. *J. Mater. Chem. A* 5 (25): 12899.

86 Chen, H., Handoko, A.D., Xiao, J. et al. (2019). Catalytic effect on CO_2 electroreduction by hydroxyl-terminated two-dimensional MXenes. *ACS Appl. Mater. Interfaces* 11 (40): 36571.

87 Attanayake, N.H., Banjade, H.R., Thenuwara, A.C. et al. (2021). Electrocatalytic CO_2 reduction on earth abundant 2D Mo_2C and Ti_3C_2 MXenes. *Chem. Commun.* 57 (13): 1675.

88 Handoko, A.D., Chen, H., Lum, Y. et al. (2020). Two-dimensional titanium and molybdenum carbide mxenes as electrocatalysts for CO_2 reduction. *iScience* 23 (6): 101181.

89 Esmaeilirad, M., Baskin, A., Kondori, A. et al. (2021). Gold-like activity copper-like selectivity of heteroatomic transition metal carbides for electrocatalytic carbon dioxide reduction reaction. *Nat. Commun.* 12 (1): 5067.

90 Back, S. and Jung, Y. (2017). Tic-and tin-supported single-atom catalysts for dramatic improvements in CO_2 electrochemical reduction to CH_4. *ACS Energy Lett.* 2 (5): 969.

91 Wang, J., Kattel, S., Hawxhurst, C.J. et al. (2019). Enhancing activity and reducing cost for electrochemical reduction of CO_2 by supporting palladium on metal carbides. *Angew. Chem. Int. Ed.* 58 (19): 6271.

92 Liu, Y., Tian, D., Biswas, A.N. et al. (2020). Transition metal nitrides as promising catalyst supports for tuning CO/H_2 syngas production from electrochemical CO_2 reduction. *Angew. Chem. Int. Ed.* 59 (28): 11345.

93 Nian, Y., Wang, Y., Biswas, A.N. et al. (2021). Trends and descriptors for tuning CO_2 electroreduction to synthesis gas over Ag and Au supported on transition metal carbides and nitrides. *Chem. Eng. J.* 426: 130781.

94 Zhang, L., Mao, F., Zheng, L.R. et al. (2018). Tuning metal catalyst with metal–C_3N_4 interaction for efficient CO_2 electroreduction. *ACS Catal.* 8 (12): 11035.

95 Zhao, R., Wang, Y., Ji, G. et al. (2023). Partially nitrided Ni nanoclusters achieve energy-efficient electrocatalytic CO_2 reduction to CO at ultralow overpotential. *Adv. Mater.* 35 (5): 2205262.

96 Cheng, Q., Mao, K., Ma, L. et al. (2018). Encapsulation of iron nitride by Fe–N–C shell enabling highly efficient electroreduction of CO_2 to CO. *ACS Energy Lett.* 3 (5): 1205.

97 Franco, F., Rettenmaier, C., Jeon, H.S., and Cuenya, B.R. (2020). Transition metal-based catalysts for the electrochemical CO_2 reduction: from atoms and molecules to nanostructured materials. *Chem. Soc. Rev.* 49 (19): 6884.

98 Meshitsuka, S., Ichikawa, M., and Tamaru, K. (1974). Electrocatalysis by metal phthalocyanines in the reduction of carbon dioxide. *J. Chem. Soc. Chem. Commun.* (5): 158.

99 Zhang, H., Xu, C., Zhan, X. et al. (2022). Mechanistic insights into CO_2 conversion chemistry of copper bis-(terpyridine) molecular electrocatalyst using accessible operando spectrochemistry. *Nat. Commun.* 13 (1): 6029.

100 Elgrishi, N., Chambers, M.B., Wang, X., and Fontecave, M. (2017). Molecular polypyridine-based metal complexes as catalysts for the reduction of CO_2. *Chem. Soc. Rev.* 46 (3): 761.

101 Hawecker, J., Lehn, J.-M., and Ziessel, R. (1984). Electrocatalytic reduction of carbon dioxide mediated by $Re(bipy)(CO)_3Cl$ (bipy = 2,2′-bipyridine). *J. Chem. Soc. Chem. Commun.* (6): 328.

102 Machan, C.W., Sampson, M.D., and Kubiak, C.P. (2015). A molecular ruthenium electrocatalyst for the reduction of carbon dioxide to CO and formate. *J. Am. Chem. Soc.* 137 (26): 8564.

103 Bourrez, M., Molton, F., Chardon-Noblat, S., and Deronzier, A. (2011). $[Mn(bipyridyl)(CO)_3Br]$: an abundant metal carbonyl complex as efficient electrocatalyst for CO_2 reduction. *Angew. Chem. Int. Ed.* 50 (42): 9903.

104 Rønne, M.H., Cho, D., Madsen, M.R. et al. (2020). Ligand-controlled product selectivity in electrochemical carbon dioxide reduction using manganese bipyridine catalysts. *J. Am. Chem. Soc.* 142 (9): 4265.

105 Yong, C., Xu, Y., Yu, H. et al. (2023). Toward quantification of active site density and size-dependent ORR activity of ZIF-derived carbons in alkaline electrolyte. *J. Catal.* 428: 115148.

106 Zhang, G.-R., Yong, C., Shen, L.-L. et al. (2023). Increasing accessible active site density of non-precious metal oxygen reduction reaction catalysts through ionic liquid modification. *ACS Appl. Mater. Interfaces* 15: 18781.

107 Wu, H.B. and Lou, X.W. (2017). Metal-organic frameworks and their derived materials for electrochemical energy storage and conversion: promises and challenges. *Sci. Adv.* 3 (12): 9252.

108 Xu, W., Zhang, G.-R., Wang, J. et al. (2023). Enhanced intermolecular electron transfer in fluorinated metal–organic framework photocatalysts for efficient CO_2 reduction. *Adv. Funct. Mater.* 34: 2312691.

109 Wang, J.-L., Wang, C., and Lin, W. (2012). Metal–organic frameworks for light harvesting and photocatalysis. *ACS Catal.* 2 (12): 2630.

110 Lu, C., Yang, J., Wei, S. et al. (2019). Atomic Ni anchored covalent triazine framework as high efficient electrocatalyst for carbon dioxide conversion. *Adv. Funct. Mater.* 29 (10): 1806884.

111 Hod, I., Sampson, M.D., Deria, P. et al. (2015). Fe-porphyrin-based metal–organic framework films as high-surface concentration, heterogeneous catalysts for electrochemical reduction of CO_2. *ACS Catal.* 5 (11): 6302.

112 Dong, B.-X., Qian, S.-L., Bu, F.-Y. et al. (2018). Electrochemical reduction of CO_2 to CO by a heterogeneous catalyst of Fe–porphyrin-based metal–organic framework. *ACS Appl. Energy Mater.* 1 (9): 4662.

113 Ren, S., Joulié, D., Salvatore, D. et al. (2019). Molecular electrocatalysts can mediate fast, selective CO reduction in a flow cell. *Science* 365 (6451): 367.

114 Yang, H.B., Hung, S.-F., Liu, S. et al. (2018). Atomically dispersed Ni(i) as the active site for electrochemical CO_2 reduction. *Nat. Energy* 3 (2): 140.

115 Ju, W., Bagger, A., Hao, G.-P. et al. (2017). Understanding activity and selectivity of metal-nitrogen-doped carbon catalysts for electrochemical reduction of CO_2. *Nat. Commun.* 8 (1): 944.

116 Jiao, L., Yang, W., Wan, G. et al. (2020). Single-atom electrocatalysts from multivariate metal–organic frameworks for highly selective reduction of CO_2 at low pressures. *Angew. Chem. Int. Ed.* 59 (46): 20589.

117 Yang, H., Shang, L., Zhang, Q. et al. (2019). A universal ligand mediated method for large scale synthesis of transition metal single atom catalysts. *Nat. Commun.* 10 (1): 4585.

118 Fan, Q., Hou, P., Choi, C. et al. (2020). Activation of Ni particles into single Ni–N atoms for efficient electrochemical reduction of CO_2. *Adv. Energy Mater.* 10 (5): 1903068.

119 Zhang, C., Yang, S., Wu, J. et al. (2018). Electrochemical CO_2 reduction with atomic iron-dispersed on nitrogen-doped graphene. *Adv. Energy Mater.* 8 (19): 1703487.

120 Wang, X., Chen, Z., Zhao, X. et al. (2018). Regulation of coordination number over single Co sites: triggering the efficient electroreduction of CO_2. *Angew. Chem. Int. Ed.* 57 (7): 1944.

121 Geng, Z., Cao, Y., Chen, W. et al. (2019). Regulating the coordination environment of Co single atoms for achieving efficient electrocatalytic activity in CO_2 reduction. *Appl. Catal. B* 240: 234.

16

Carbon-Based Materials for Electrocatalytic CO_2 Reduction Reaction

Jingjing Wang[1], Haodong Zheng[1], Kai Ge[1], Boxiong Shen[1], Hui Liu[2], Xiwen Du[2], and Pengfei Yin[2]

[1] *Hebei University of Technology, School of Chemical Engineering, Department of Applied Chemistry, No. 8 Guangrong Road, Hongqiao District, Tianjin 300130, China.*
[2] *Tianjin University, Institute of New Energy Materials, School of Materials Science and Engineering, No. 92 Weijin Road, Nankai District, Tianjin 300072, China.*

16.1 Introduction

In the last century, conventional fossil fuels have become the primary feedstock for global energy production due to their unmatched energy and power density. However, excessive emission of CO_2 and other greenhouse gases due to fossil energy consumption, have caused serious global environmental issues, such as ocean acidification and sea-level rise [1]. To address the potential ecological risks and existential crises arising from climate change, the Paris Climate Agreement advocates limiting the increase in global temperature (compared to preindustrial periods) to 2 °C. However, the most recent Emission Gap Report from the United Nations Environment Programme, released in 2020, underscores a significant shortfall in meeting this climate objective. Instead, current trends indicate a trajectory toward a temperature rise exceeding 3 °C by the end of the century. Therefore, the development of CO_2 conversion and utilization technology is of great importance.

CO_2, with a valence state of +4, is the most oxidized form of carbon. To convert it into more valuable products, the key step is transferring electrons to reduce its oxidation state, a process known as CO_2 reduction. Various methods for CO_2 reduction exist, such as thermal catalytic hydrogenation, natural photocatalysis or biomimetic catalysis, and electrocatalytic CO_2 reduction (CO_2RR). Among them, electrocatalytic CO_2RR stands out as the most feasible and eco-friendly strategy.

Operating at room temperature, electrocatalytic CO_2RR uses renewable electricity to efficiently convert carbon dioxide into valuable chemicals and fuels [2]. Its highly controllable steps and potential for high conversion efficiencies make it a key focus of current research, offering intrinsic advantages in harnessing renewable energy and contributing significantly to sustainable carbon dioxide utilization [3]. However, the CO_2RR process encounters notable challenges. On the one hand, CO_2 is an extraordinarily stable molecule, with a high dissociation energy of 803 kJ mol^{-1}

Functional Materials for Electrocatalytic Energy Conversion, First Edition.
Edited by Zhicheng Zhang, Meiting Zhao, and Yuchen Qin.

for the C=O bond, resulting in a substantial kinetic barrier for electron transfer to the adsorbed CO_2 [4]. On the other hand, electrochemical reduction of CO_2 often involves multi-proton-coupled electron transfer processes, generating a notable number of intermediates and leading to low product selectivity. Additionally, CO_2RR in aqueous conditions also faces competition from the hydrogen evolution reaction (HER) [5]. Therefore, highly effective catalysts are imperative to drive the conversion process with high selectivity and energy efficiency.

So far, various CO_2RR electrocatalysts, including metals [6–8], metal alloys [9, 10], oxides [11–14], metal phosphides [15], metal phosphates [16], single-atom [17–19], and molecular catalysts [20], have demonstrated outstanding performances. Notably, carbon-based materials have attracted attention for their distinctive properties, including an ultrahigh surface area, facilitating precise tuning of composition and structure, excellent electrical conductivity, stability, and cost-effectiveness [21–23]. However, it is essential to acknowledge that the intrinsic catalytic activity of pure carbon is relatively modest. Hence, strategic approaches, including heteroatom doping, group functionalization, defect engineering, and the construction of composite materials through anchoring or loading metals, are imperative to enhance the catalytic activity of pure carbon.

This chapter introduces the classification and electrochemical properties of carbon-based CO_2RR catalysts and summarizes the latest research results in this field. It clarifies the reaction mechanism and active sites of CO_2RR in different types of carbon-based materials, summarizes the mainstream strategies for enhancing the catalytic activity of carbon materials in CO_2RR, and finally, outlines the challenges and prospects that carbon-based materials face in future basic research and practical applications of CO_2RR.

16.2 Fundamentals of Electrochemical CO_2 Reduction

CO_2RR can progress through multistep pathways involving different numbers of electrons and protons, leading to various end products. The equations and standard potentials for the conversion of CO_2 into main C_1–C_3 products are shown in Table 16.1 [24]. Specifically, species such as CO and formate require fewer electrons to be transferred and have exhibited high rates of formation and Faradaic efficiency (FE) values above 90% [25, 26]. Compared to those C_1 species, the electrochemical conversion of CO_2 into multicarbon products is more desirable due to their higher energy density and market price [27]. However, these processes pose thermodynamic and kinetic challenges [28]. Generally, the highest FEs for C_2 products (e.g. ethylene and ethanol) are around 60% [29–31], while the FEs for C_3 (e.g. n-propanol) production are limited to less than 30% according to reported works so far [32–34]. Moreover, high overpotentials are typically required. In addition, the standard potential for H_2 evolution also falls within the same range, posing competition with CO_2RR when an aqueous electrolyte is used, leading to a decrease in the current efficiency of the overall reaction [35]. Therefore, understanding the principles of CO_2RR is crucial to paving a favorable kinetic pathway that overcomes the

Table 16.1 Main products and standard potential of CO_2 reaction in aqueous solutions (V vs. SHE) at 1.0 atm and 25 °C Adapted from [24].

Product	Half reaction	E_0 (V)
H_2	$2H^+ + 2e^- \rightarrow H_2$	−0.42
CO	$CO_2 + 2H^+ + 2e^- \rightarrow CO + H_2O$	−0.53
HCOOH	$CO_2 + 2H^+ + 2e^- \rightarrow HCOOH$	−0.61
HCHO	$CO_2 + 4H^+ + 4e^- \rightarrow HCHO + H_2O$	−0.48
CH_3OH	$CO_2 + 6H^+ + 6e^- \rightarrow CH_3OH + H_2O$	−0.38
CH_4	$CO_2 + 8H^+ + 8e^- \rightarrow CH_4 + 2H_2O$	−0.24
C_2H_6	$2CO_2 + 14H^+ + 14e^- \rightarrow C_2H_6 + 4H_2O$	−0.27
C_2H_4	$2CO_2 + 12H^+ + 12e^- \rightarrow C_2H_4 + 4H_2O$	−0.34
C_2H_5OH	$2CO_2 + 12H^+ + 12e^- \rightarrow C_2H_5OH + 3H_2O$	−0.33
CH_3CHO	$2CO_2 + 10H^+ + 10e^- \rightarrow CH_3CHO + 3H_2O$	−0.35
CH_3COOH	$2CO_2 + 8H^+ + 8e^- \rightarrow CH_3COOH + 2H_2O$	−0.30
C_3H_7OH	$3CO_2 + 18H^+ + 18e^- \rightarrow C_3H7OH + 5H_2O$	−0.32
C_2H_5CHO	$3CO_2 + 16H^+ + 16e^- \rightarrow C_2H_5CHO + 5H_2O$	−0.32

challenges of high overpotential, suppresses the process of the HER, and enhances selectivity toward C_{2+} products.

While the exact reaction mechanism remains unclear to date and is likely to vary under different reaction conditions, there is a widely acknowledged consensus on four fundamental steps when modeling this heterogeneous catalytic process [28]:

Step 1: Chemisorption of CO_2 onto the surface of an electrocatalyst leads to the formation of a partially charged $*CO_2^{\delta-}$ species, which can readily be protonated to generate a bidentate intermediate (*OCHO) or an adsorption intermediate (*COOH) [36, 37]. This step is recognized as the rate-determining step (RDS) for the CO_2RR process. The involvement of the bending of the linear and highly stable CO_2 molecule contributes to a substantial kinetic barrier in this step [38, 39].

Step 2: Electron transfer and/or proton coupling to cleave the C—O bond, forming further reduced intermediates, such as the most important *CO.

Step 3: The formation of C—C bonds is crucial for producing a variety of C_{2+} species, including hydrocarbons and oxygenates. This step is particularly challenging and requires a high coverage of C_1 intermediates generated in step 2.

Step 4: Rearrangement of the carbon products followed by desorption from active sites and diffusion into the bulk electrolyte solution.

According to the most possible pathways during CO_2RR (Figure 16.1), *CO serves as the common intermediate for the formation of almost all products (except formate), and its subsequent reactions determine the type of final products [40, 41]. For C_1 products, when the protons are accepted by O in *CO first, the *COH is

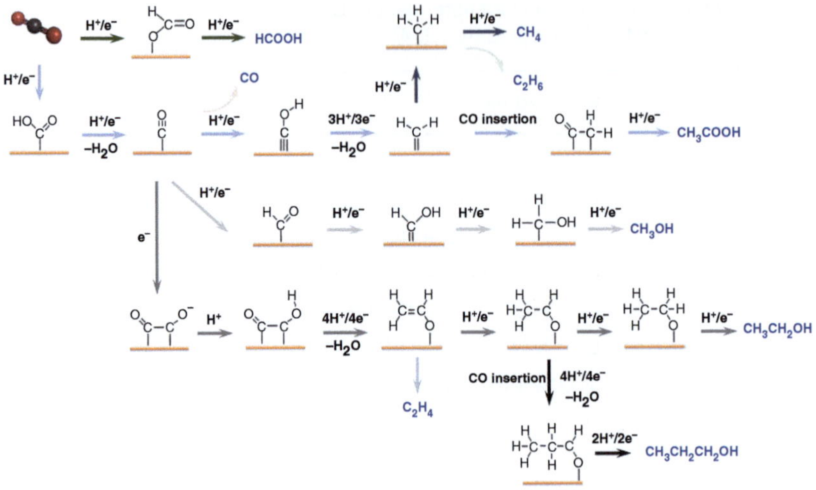

Figure 16.1 Possible pathways for CO_2RR. Source: Reproduced with permission of Zheng et al. [40] American Chemical Society.

generated and then it combines with several protons to produce CH_4 by losing one water molecule. When C in *CO accepts protons, the protonated *CO generates *CHO and finally produces methanol through several subsequent hydrogenation processes [42].

For the formation of C_2 products, two main pathways exist. On the one hand, $*CH_2$ intermediate can be generated from the proton-coupled electron transfer steps of *CO. Subsequent protonation of $*CH_2$ results in the $*CH_3$ intermediate, leading to the formation of C_2H_6 through $*CH_3$ dimerization. Correspondingly, CH_3COO^- can be formed by CO insertion into $*CH_2$. On the other hand, the *CO dimerization is considered the RDS for C_2H_4, CH_3CH_2OH, and *n*-propanol (*n*-C_3H_7OH) formation. Following a series of electron transfer and protonation steps, the *CO–CO transforms into the $*CH_2CHO$ intermediate, which serves as the selectivity-determining step for C_2H_4 and C_2H_5OH [41]. It is noteworthy that the reduction of $*CH_2CHO$ to C_2H_4 has a lower energy barrier than transforming $*CH_3CHO$ to C_2H_5OH, potentially explaining the higher FE for C_2H_4 over C_2H_5OH observed on most copper catalysts. Additionally, stabilized C_2 intermediates ($*CH_3CHO$) can contribute to the formation of *n*-C_3H_7OH through CO insertion [43]. Considering the intricate and uncontrollable reaction pathways during CO_2RR, the design of highly selective and active electrocatalysts is imperative for achieving specific product formation at high yields.

16.3 Categories of Carbon-Based Electrocatalysts

Carbon-based materials play a pivotal role in a wide range of electrochemical reactions due to their high electrical conductivity, good stability, and cost-effectiveness [21, 22]. In aqueous reaction conditions, carbonaceous materials display chemical inertness at negative potentials, and their onset potential for the HER is higher than that of metal materials. This inherent characteristic aids in suppressing H_2 evolution, showcasing their potential as high-performance CO_2RR catalysts [44, 45]. Additionally, carbon-based materials with different hybridization states of carbon atoms or nanostructures demonstrate diverse CO_2 conversion selectivity and activity. These properties can be finely tuned by incorporating heteroatoms (e.g. N, S, P, and B) and constructing metal/carbon composites [46]. Based on these features, carbon-based materials have been developed and extensively utilized for CO_2RR in recent years. These electrocatalysts can be categorized into metal-free carbon, metal–N–C single-atom catalysts (SACs), and metal/carbon composites.

16.3.1 Metal-Free Carbon

16.3.1.1 Graphene Materials

Graphene consists of a thin layer of sp^2 hybridized carbon atoms arranged in a honeycomb crystal structure with unique highly crystalline and electronic qualities. It can be synthesized through high-temperature pyrolysis of graphene oxide (GO) or chemical vapor deposition (CVD) methods [47]. Since its first laboratory synthesis in 2004 by Geim and co-workers, graphene has become a focal point in materials science and nanotechnology [48]. Pristine graphene, however, is typically inactive toward the CO_2 reduction reaction due to its well-delocalized π-π bonding framework, making it ineffective for CO_2 activation and adsorption of related intermediates (e.g. *COOH and *CO). Fortunately, with the application of appropriate strategies or post-treatments, the electronic structure of graphene can be modulated, leading to excellent CO_2 electroreduction activity [44].

Adding heteroatoms into sp^2-bonded carbon frameworks is the most extensively studied method, which can not only increase the number/activity of active sites but also stabilize the intermediate products of CO_2 reduction better [45]. For example, Wu and coworkers synthesized nanometer-size N-doped graphene quantum dots (NGQDs) and used them for electrocatalytic conversion of CO_2. The NGQDs showed high total FE of CO_2 reduction (up to 90%) and good selectivity for C_{2+} products, including hydrocarbons (CH_4 and C_2H_4) and oxygenates (CH_3COO^-, C_2H_5OH, and n-C_3H_7OH) [49]. Zou et al. studied the underlying mechanisms governing these processes. They found the introduction of N atoms into edges of graphene quantum dots enhanced their bonding with *COOH, effectively promoting the reduction of CO_2 to CO. Further investigations elucidate that the selective generation of CH_4, as opposed to CH_3OH, is influenced by kinetics, specifically through the formation of the key intermediate $*CH_2$. In the subsequent step, the effective coupling between $*CH_2$ and CO gives rise to various C_2 products, including C_2H_4 and C_2H_5OH, with a higher yield of C_2H_4 [50].

Besides N, boron (B) atom doping is also reported to alter carbon electronic work bands and promote the chemical activities of carbon materials. Sreekanth et al. prepared B-doped graphene (B-GO) by mixing GO with boric acid followed by annealing the resulting uniform powder at 900 °C at Ar atmosphere [51]. The synthesized

B-graphene showed good performance for the reduction of CO_2 to formate with an FE of 66% at −1.4 V vs. SCE in 0.1 M $KHCO_3$ aqueous solutions.

Moreover, compared with single heteroatom doping, co-doping with two different heteroatoms shows greater potential for CO_2RR due to the synergistic effect between these introduced atoms that can provide more active sites and more flexible reaction routes. For instance, Yang and coworkers synthesized N and S co-doped hierarchically porous carbon nanofibers (NSHCF) via electrospinning of a mixture solution of ZIF-8 nanoparticles, trithiocyanuric acid (TA), and polyacrylonitrile (PAN) and subsequent annealing at 900 °C under Ar atmosphere. The NSHCF catalyst with abundant heteroatom doping was highly efficient for CO_2 reduction with a maximum CO FE of 94% at −0.7 V vs. RHE and had superior durability for 36 hours. Experiments and DFT calculations supported the fact that N and S co-doping lowered the Gibbs free energy for the first step of CO_2 to CO_2^{*-} intermediate [52].

16.3.1.2 Carbon Nanotubes

Carbon nanotubes (CNTs) are helical microtubules of graphitic carbon with a one-dimensional (1D) structure and diameters ranging from a few nanometers to hundred nanometers [53, 54]. CNTs possess a unique architecture with inner and outer layered walls. These two layers comprise different structures, the inner and outer layers show hollow fishbone-type and disordered grown structures, respectively [55]. Recently, CNTs have emerged as promising electrocatalytic materials for CO_2 reduction, owing to their advantages of facile doping and the provision of more functionalization sites [47]. For example, Zhou group prepared N-doped CNTs with various surface structures and nitrogen contents by controlling the precursors (triethylamine, acetonitrile, and dimethylformamide) and the heating temperatures (750, 850, and 950 °C) [56]. Finally, the NCNTs-ACN-850 (acetonitrile, 850 °C) exhibited the best CO_2RR performance with the smallest overpotential of −0.18 V and highest maximal FE of 80% toward CO, owing to its highest pyridinic and graphitic N content, which were both active sites for CO_2RR. A similar result was reported by Xu et al.'s work, in which a maximum FE of 90% for CO production and excellent stability were achieved by N-doped CNTs [57].

In another work, Zhang et al. reported polyethylenimine (PEI) group-modified N-doped CNTs achieving a high activity for CO_2 conversion into HCOOH with a maximum FE of 87 % at −1.8 V vs. SCE. First, ammonia plasma treatment on N-doped CNTs was applied to enlarge surface area and N doping. Then a PEI overlayer was grown on NCNT/GC electrodes by dip-coating, which can increase CO_2 adsorption at the surface due to the introduction of amine groups. The combination of N-doping and a PEI overlayer played a synergistic effect, creating a local environment in which the reduction of CO_2 to $CO_2^{\bullet-}$ occurs at a greatly reduced overpotential [58].

Computational studies have shown the effect of curvature causes partial sp^3 hybridization in the graphitic lattice and as a result, can bind *CO and *HCOOH intermediates more strongly [59–61]. Marepally et al. prepared oxide carbon nanotubes (O-CNTs) by preliminary treatment of commercial CNTs with concentrated nitric acid, which created various oxygen-containing groups on O-CNT electrocatalysts. Finally, the O-CNTs as electrocatalyst converted CO_2 into formic acid and acetic acid with production rates of 0.51 and 0.32 μmol h^{-1}, respectively. Moreover, CNTs showed a strong tendency to self-assemble into well-configured 3D CNTs with built-in interconnections and structural robustness [62]. In addition, a variety of catalyst configurations supported by 3D CNTs were designed and their electrochemical behavior was further improved by heterogeneous doping and catalyst substrate interactions [63].

16.3.1.3 Nanodiamond

Diamond nanocrystals possess tetrahedral bonding units with sp^3-hybridized carbon atoms and show great stability in harsh environments. The unique sp^3 bonding of the diamond is beneficial to achieve C–C coupling and shows high selectivity toward C_{2+} oxygenates production from CO_2 reduction [61]. Moreover, their electrical conductivity and CO_2 electrochemical activity can substantially increase after doping with heteroatoms, such as N and B species [47]. Nakata et al. investigated the electrochemical reduction of CO_2 in seawater using a boron-doped diamond (BDD) electrode under ambient conditions to produce formaldehyde. In comparison with other electrode materials, BDD electrodes exhibited very high FE (74%) for the production of formaldehyde as well as a wide potential window and high electrochemical stability, using either methanol, aqueous NaCl, or seawater as the electrolyte. They ascribed the high FE to the sp^3-bonded carbon of the BDD [64]. In another work, Zhao et al. synthesized N-doped nanodiamond (NDD/Si RA) by depositing NDD on a Si rod array substrate through microwave plasma-enhanced CVD. NDD/Si RA can be used to reduce CO_2 to acetate with the onset potential of −0.36 V vs. RHE. The highest FE of acetate was around 77.3–77.6% at −0.8 V. Electrochemical kinetic analysis and in situ Fourier transform infrared (FTIR) spectroscopy revealed the main pathway might be C–C coupling via dimerization of two $CO_2^{\bullet-}$ radicals and the formed *OOC–COO was further reduced to acetate by hydrogenation process [65].

Subsequently, the same group synthesized a B and N co-doped nanodiamond (BND) on a Si substrate by using a hot filament CVD. Serving as CO_2RR electrocatalyst, boron BND showed an ultrahigh selectivity to C_2H_5OH with the highest FE reaching 93.2% (−1.0 V vs. RHE) in CO_2 saturated 0.1 M $NaHCO_3$ electrolyte, which seems to be the highest selectivity for ethanol so far [66]. Furthermore, Tomisaki et al. reported that the product selectivity of CO_2RR can be switched by controlling the electrolytes and boron concentration of BDD catalysts. They found that with the use of 1% boron BDD electrode as the cathode and $KClO_4$ for the catholyte, CO production was high, whereas, in contrast by using 0.1% boron BDD electrode as the cathode and KCl for the catholyte, HCOOH production was dominant [67]. In in situ attenuated total reflectance-infrared measurements, it was observed that the $CO_2{\bullet}^-$ intermediates were adsorbed on the surface of BDD in $KClO_4$ aqueous solution. In contrast, these intermediates existed as free species in the case of KCl aqueous solution.

16.3.1.4 Graphitic Carbon Nitride (g-C_3N_4)

Graphitic carbon nitride (g-C_3N_4) is formed through C—N atom bonds, featuring a 2D structure consisting of a sp^2 single layer. It exhibits a visible bandgap of

2.7 eV and is held together by van der Waals forces. It is usually prepared by a polycondensation process of cyanamide, which contains a small amount of hydrogen at its end edges in the form of primary and secondary amines [68]. As one kind of metal-free catalyst, g-C_3N_4 attracts intense research attention as it has a large number of pyridinic-like nitrogen active sites. In addition, g-C_3N_4 has shown the selective adsorption ability of large amounts of CO_2 by effectively polarizing CO_2 molecules, leading to enhanced adsorption which could facilitate the CO_2RR process [61, 69]. The CO_2RR performances on the pure g-C_3N_4 catalyst was studied by Zhang et al [70]. Polarized few-layer, ultrathin 2D graphitic carbon nitrides (2D-pg-g-C_3N_4) were synthesized by using the hydrothermal thermal exfoliation method with a thickness of about 1 nm. Serving as CO_2RR electrocatalyst, the prepared 2D-pg-g-C_3N_4 catalyst was selective toward CO with a high FE of 80% at an applied potential of −1.1 $V_{Ag/AgCl}$, whereas in contrast, mesoporous g-C_3N_4 at the same applied conditions showed a negligible reduction product.

Besides, the combination of g-C_3N_4 and metal-free carbon composites has emerged as an efficient catalyst for CO_2RR. Lu et al. first reported the potential use of covalently attached graphitic carbon nitride onto multiwall carbon nanotubes (g-C_3N_4/MWCNTs) composite as a selective electrocatalyst for CO_2RR in 0.1 M $KHCO_3$. A maximum of 60% FE was achieved for this composite with selectivity toward the target products CO at an applied potential of −0.75 V RHE [71]. Jhong synthesized a similar pyrolyzed carbon nitride and multiwall carbon nanotube composite, enhancing the selectivity for CO to approximately 98% and pushing the current density up to 90 mA cm^{-2} using an electrochemical flow cell [63]. These works revealed that the electronic structures of g-C_3N_4 can be effectively modified by metal-free carbon materials [63].

Subsequently, Zhi et al. reported the interfacial electron transfer effect of g-C_3N_4/doped graphene materials that could significantly boost CO_2 reduction for producing methane. With the assistance of computational methods, the impact of heteroatom doping (including B, N, O, and P) was comprehensively explored, and variable adsorption energetics and electronic structures among different doping cases were observed, demonstrating that the interfacial electron transfer modulation can promote the catalytic performance [72].

16.3.2 Metal–N–C SACs

Metal–N–C materials, a typical kind of SAC consisting of metal atoms anchored with N atoms on a carbon support, have become attractive for electrochemical CO_2 conversion. One key advantage of metal–N–C materials lies in their significant kinetic barrier to *H adsorption, as revealed by DFT studies [73]. This property reduces the competitiveness of the HER on metal–N–C materials, enabling CO_2RR to proceed with higher FE. Moreover, the selectivity and activity of metal–N–C materials can be finely tuned by selecting the metal center, optimizing coordination numbers, and configuring the structure of the carbon material [18, 74, 75]. Recently, various metal–N–C structures centered around transition metals have been developed for CO_2RR, leveraging the rich *d* orbital electrons and variable

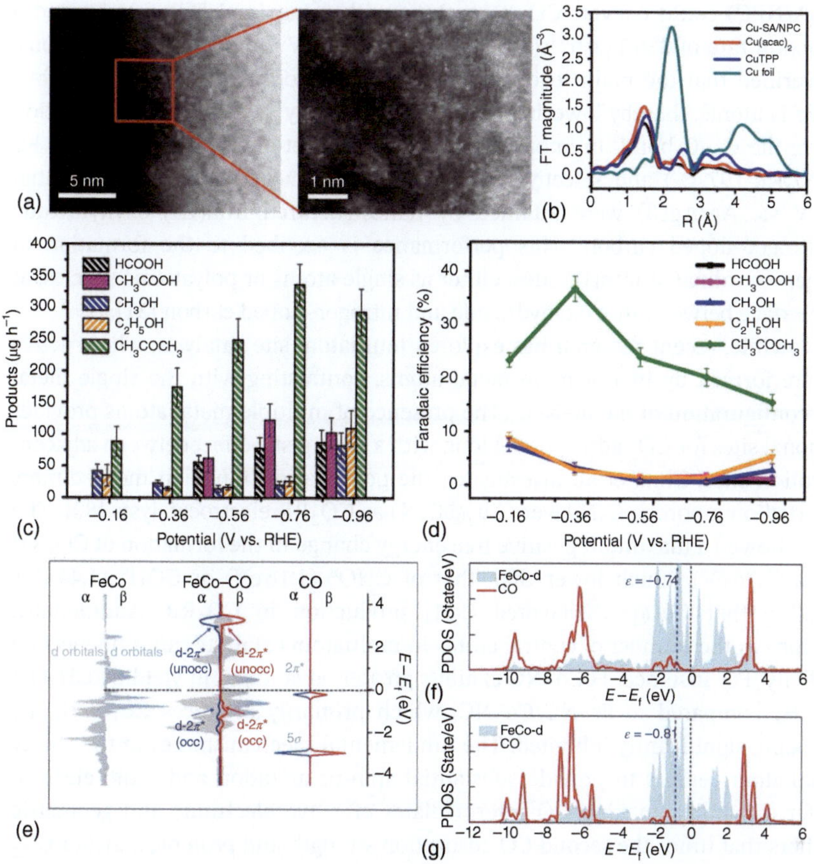

Figure 16.2 (a) HAADF-STEM images and enlarged images of Cu-SA/NPC. (b) Fourier transforms magnitude of the experimental EXAFS spectra of Cu-SA/NPC, Cu $(acac)_2$, CuTPP, and Cu foil. (c) Production rate of CO_2 reduction products and (d) Faradaic efficiency of CO_2 reduction products on Cu-SA/NPC. Source: Zhao et al., [33]/Springer Nature/CC BY 4.0. (e) The DOS of CO adsorption on FeCo–NC. (f) D-band center analysis of FeCo–NC (above) and CO assisted FeCo–NC (below). Source: Reproduced with permission of Cao et al. [81], Wiley-VCH.

valence states of transition metals. For instance, Ju et al. investigated five different transition metals (Mn, Fe, Co, Ni, and Cu) centered M–N–C structures as the CO_2RR electrocatalysts. Among these, the Ni–N–C catalyst exhibited the highest activity toward CO at high overpotentials, while the Fe–N–C catalyst demonstrated good activity at low overpotentials compared to other counterparts [76]. These findings have been corroborated by several studies, establishing that Ni–N–C species are the most active for converting CO_2 to CO [45, 77]. So far, the reported maximum FE_{CO} can reach 99 % (or almost 100%) and the highest partial current density reaches 308.4 mA cm^{-2} when flow cells are applied [78–80].

Notably, metal–N–C catalysts have also been reported to be capable of producing C_{2+} products. Zhao et al. demonstrated that Cu-anchored N-doped porous carbon

(Cu-SA/NPC) could convert CO_2 to acetone with a maximal FE of 36.7% and a production rate of 336.1 μg h^{-1} (Figure 16.2a–d) [33]. Via theoretical calculations, they verified that the main reaction sites were the coordination of Cu with four pyrrole-N atoms, thereby lowering the activation energy needed for CO_2 reduction to form the C—C bond. In another study, Karapinal et al. reported that high FE for CO_2RR (97.4%) and selectivity to acetic acid (61%) at a very low potential (−0.5 V vs. Ag/AgCl) were attained by nanostructured iron(III) oxyhydroxide on nitrogen-doped carbon. This performance is ascribed to the formation of nitrogen-coordinated iron(II) sites, either as single atoms or polyatomic species, at the interface between iron oxyhydroxide and nitrogen-doped carbon [31].

Meanwhile, recent research has explored multiatom site catalysts, where active sites are formed by two or more metal atoms, contrasting with the single metal atom configuration of metal–N–C. The presence of multiple metal atoms provides additional sites for CO_2 adsorption, along with a synergistic effect between adjacent adsorption sites. Zhao et al. investigated the potential of transition metal dimers supported on a porous C_2N layer (Cu_2@C_2N) as CO_2RR electrocatalysts [82]. The results showed a maximum positive free energy change in the formation of COCO* species (0.76 eV), much lower than that of CHO* (1.16 eV) or COH* (1.44 eV), indicating that Cu_2@C_2N favored C_2H_4 production in CO_2RR. Additionally, variations in the number of central atoms in multiatom catalysts can influence the selectivity. For instance, FeCo–NC exhibits greater advantages in yielding CH_3OH and CH_4 compared to Fe–NC/Co–NC, which primarily produces CO, with the HER being significantly inhibited. The fundamental mechanism lies in the ability of dual-atom centers to provide substantial spin-polarization and multi-electron transfer capability, making CO intermediates effective electronic and geometric modifiers that tunes the second CO adsorption strength and promotes further CO_2 reductions in CO_2RR (Figure 16.2e,f) [81]. In another study, He et al. investigated the activity and selectivity of a series of multi-Fe atoms (Fe_n, n = 1–4) doped graphdiyne catalysts for CO_2 electrochemical reduction using *ab initio* studies. Theoretical calculations revealed that Fe_2 and Fe_3 exhibited a better performance in CO_2-to-CH_4 and CO_2-to-HCOOH, respectively [83].

16.3.3 Metal/Carbon Composites

16.3.3.1 Carbon-Supported Metal/Alloy

The formation of metal/carbon composites is an ideal strategy to improve catalytic activity and selectivity toward specific products benefiting from the synergistic effect between metals and carbon materials, where the metal elements (Au, Pd, Ag, Fe, Co, Ni, Cu, Sn, and In) can efficiently adsorb and activate CO_2 to lower the energy barrier, and carbon materials with the high surface areas and tunable porous structure supply a large electrochemical active area to decrease the diffusion resistance [47, 84, 85]. It is reported that the structures of the metal will significantly affect the CO_2RR performance [86–88]. For example, Gao et al. studied the effect of Pd particle size in Pd/carbon black composites on the selectivity and activity toward CO_2RR [89]. The CO FE increased from 5.8 to a peak of 91.2% with an 18.4-fold increase in CO partial current density as the Pd size decreased from 10.3 to 3.7 nm. DFT calculation revealed that the Gibbs free energies for both CO_2 adsorption and *COOH radical formation on small Pd NPs (2.4–4.5 nm) were found to be lower than those on large Pd NPs (4.5–10.3 nm). In another work, Hwang immobilized Ag nanoparticles supported on carbon using cysteamine as the anchoring agent, DFT calculations enlightened that the specific interaction between Ag nanoparticle and the anchoring agents (Ag-S) modified the catalyst surface to have a selectively higher affinity to the intermediate *COOH over CO, which effectively lowers the overpotential [90].

Moreover, the interaction between metal and carbon substrate also plays an important role in affecting the CO_2RR process. The research results indicate that electron flow occurs at the interface before the alignment of the Fermi energy levels between the metal and carbon carriers, allowing atoms at interface sites to have unique electronic structures and forming electron-rich/electron-deficient sites. These sites can significantly regulate the adsorption of reactants/reaction intermediates at interface sites, thereby affecting reaction performance [91, 92]. Gong group constructed a molybdenum carbide/N-CNT catalyst that promotes the formation and adsorption of oxygen-bound intermediates benefiting from the hybridization between the metal (Mo) d states and the carbon s states. Finally, a high FE of 80.4% for methanol was achieved at −1.1 V vs. SHE [93].

In addition, by converting metal to alloys, metal oxides, or tuning the type of carbon materials, the catalytic activity of each active site can be further improved. For example, Zhang's group prepared ultrafine tin oxide (SnO_2) nanocrystals supported on carbon materials with controllable particle size, which were very promising for electrochemical CO_2 reduction to formate [94]. Surface analysis suggested that the electrogenerated metallic Sn was the active component. Moreover, they found that the supporting carbon materials used in the study affected catalysis performance. Graphene-supported nano tin catalysts gave higher FEs for formate production compared to carbon black in 0.1 M $NaHCO_3$, which can be ascribed to the higher electron donating ability of graphene-facilitated adsorption of CO_2 and the stabilization of $CO_2^{\bullet-}$ intermediates as they formed on the catalyst surface. Finally, the FE for formate reached 93% at current densities of 10 mA cm^{-2} with the electrodes exhibiting long-term stabilities for 18 hours.

16.3.3.2 Graphitic-Layer-Encapsulated Metal/Alloy

As a special type of metal/carbon composite, carbon-encapsulated metal (M@C) can facilitate electrocatalytic performance due to its unique structure. It has been widely reported that carbon shells can enhance the stability of electrocatalysts in acidic and alkaline solutions by avoiding direct contact between nanoparticles and solutions. For example, in Zhuo group's work, Co@C, Co@NC, Ni@C, and Ni@NC catalysts were synthesized and employed in CO_2RR. The Ni@NC catalyst exhibited an industry-level current density of 220 mA cm^{-2} and a high FE of 98% for CO production at −0.87 V vs. RHE for 100 hours. The N species, especially the pyrrolic N in the shell of Ni@NC material provided active sites for adsorbing and activating CO_2 molecules, and metal nanoparticles improved the electron density of the N–C skeleton, thereby improving the interaction between pyrrolic N active site and *COOH, *CO, and *H intermediates [95].

The significance of the proper carbon layer in the balance between activity and stability is crucial, as thicker carbon layers can significantly impact the electrocatalyst's performance. It is an essential consideration in the preparation of M@C electrocatalysts. While there is some controversy regarding whether the shell serves as the sole catalytic site or if both the core and shell contribute to catalysis, it is undeniable that the nanoconfinement effect achieved through carbon encapsulation plays a unique role. This effect modulates the electronic structure of the outer carbon layer, providing a combination of protection and stable dispersion for metal cores, accelerating chemical reactions, enhancing or altering selectivity, and ultimately promoting electrochemical performance [96–98].

16.4 Strategies for Modulation of Carbon-Based Electrocatalysts

16.4.1 Group Functionalization

The presence of functional groups plays a crucial role in optimizing the electronic structure of catalyst surfaces and the active sites [86]. In a study by the Xu group, metal-free single-layer graphene nanodisks (GNDs) with varying surface contents of oxygen-containing groups were investigated for the CO_2RR to produce formate ($HCOO^-$). The research revealed that the carbon atoms adjacent to oxygen-containing groups are the real active sites for CO_2RR, while the catalytic performance of CO_2RR was positively correlated only with the surface content of carboxyl groups. DFT calculations further indicated that the high observed CO_2RR catalytic activity did not originate solely from carboxyl groups but rather from the synergistic effect between carboxyl groups and other adjacent types of groups, namely, hydroxyl, epoxide, and carbonyl, on GNDs [99].

Moreover, some functional groups can also act as promoters by meriting the catalysts with several functions. Take the amino group as an example, it has the capability to capture CO_2 to the local surface and further stabilize the intermediates, such as CO* and COOH* [100–102]. Chen and co-workers reported a generalized strategy of amino-functionalized carbon supports to regulate electronic structure of M–N–C catalysts (M = Ni, Fe, and Zn) to significantly increase current density of CO production. The aminated Ni single-atom catalyst achieved a remarkable CO partial current density of 447.6 mA cm^{-2} with a nearly 90% CO FE at a moderate overpotential of 0.89 V [103].

16.4.2 Heteroatom Doping

The doping of heteroatoms, such as nitrogen (N), boron (B), and sulfur (S), show different electronegativities with carbon atoms, which can break the electroneutrality of pure carbon structures, induce charge redistribution, and create reactive sites for adsorbing and activating CO_2 [47]. Furthermore, by exact regulation of the structure of doped atoms, the activity can be controlled. Taking the extensively studied N-doped carbon as an example, pyridinic N possesses an electron pair and provides a p electron to the π system, pyrrolic N provides two p electrons to the π system, which render pyridinic N and pyrrolic N active for CO_2 reduction. While graphitic N coordinates with three C atoms, losing long electron pair near N atoms, resulting in low activity to CO_2 reduction [104, 105].

More specifically, the N species influence free energies for CO_2 activation and adsorption energy of intermediates. Specifically, Song et al. investigated the reaction mechanisms on pure graphene, pyridinic N, graphitic N, and pyrrolic N using DFT calculations, which demonstrated the key initial intermediate *COOH formation on pyridinic N and pyrrolic N were more favorable than that on pure carbon and graphite N. Sequentially, the reaction energy for the further conversion of *CO to *CHO was 0.82 eV on pyridinic N, while it was a slight exothermicity of −0.24 eV on pyrrolic N, revealing that *CHO formation was thermodynamically favorable on pyrrolic N [106]. Wu et al. calculated free energies for CO_2 conversion on above four N species. The results showed that graphene requires the highest free energy for adsorption of *COOH. Furthermore, the triple-pyridinic N led to a decrease in the *COOH adsorption barrier, followed by single-pyridinic N and graphitic N. These results confirmed pyridinic N as the most active site for CO_2 reduction, consistent with experimental results and other reported works [107].

16.4.3 Coordination Environment Control

For metal–N–C materials, the main active sites were identified as isolated metal atoms coordinated by pyridine/pyrrole nitrogen atoms, carbon atoms, or hybrid of C and N [75]. The electronic structure of active sites can be optimized by tuning the type of metal elements and their coordination environments, such as species of coordinated atoms and the coordination numbers. Yang and coworkers investigated N-doped graphene with atomically dispersed Ni for CO_2 reduction. Based on operando X-ray absorption and photoelectron spectroscopy measurements, they found the delocalization of an unpaired electron in Ni $3d_{x2-y2}$ orbital can transfer charge from Ni(I) to the carbon $2p$ orbital in CO_2 to form a $CO_2^{\delta-}$, thus reducing the energy barrier for CO_2 reduction [108]. To further investigate the role of coordinated N species, the reaction energy of CO_2 reduction on Cu–pyrrolic–N_4 and Cu–pyridinic–N_4 structures were calculated using DFT calculations. The results showed that the free energy required for the basic elementary steps of CO_2 conversion on Cu–pyrrolic–N_4 and Cu–pyridinic–N_4 were different. Specifically, the reaction energies of CO_2 activation to form *COOH and *CO dimerization on Cu–pyrrolic–N_4 were 1.06 and −1.23 eV, respectively, which were lower than those on Cu–pyridinic–N_4 (1.30 and 1.67 eV, respectively), indicating that Cu–pyrrolic–N_4 could reduce the required energy for activating CO_2 and forming C—C bond [33]. These phenomena demonstrated that tuning coordination environment can optimize the adsorption energy of CO_2-related intermediates, thus changing the selectivity of CO_2 conversion.

Cheng et al.'s work reveals the impact of the N-coordination number. Various structures with Ni–N coordination numbers ranging from 2 to 4 were constructed, including four-coordinated NiN_4 models (edge-anchored 4-coordinate NiN_4 and in-plane NiN_4), three-coordinated edge-anchored models ($NiN_2(NH_2)$ and NiN_3), and a two-coordinated model (NiN_2). The binding energy of CO intermediates was calculated based on these structures. The results indicated that NiN_4 structures were unfavorable for CO_2 adsorption, whereas the NiN_2 structure was unfavorable for CO desorption. The structures with 3-coordinate edge-anchored configurations, specifically edge NiN_3 and $NiN_2(NH_2)$, demonstrate the highest reactivity. This is attributed to the favorable energy conditions for both CO_2 adsorption and CO desorption. In contrast, the in-plane NiN_3 configuration is not conducive to CO desorption. These findings highlight the crucial role of the number of coordination N atoms in adsorbing CO_2-related intermediates [109].

16.4.4 Defects Control

Defective carbon-based nanomaterials have attracted extensive attention due to the unbalanced electron distribution and electronic structural distortion caused by the defects in the carbon skeleton. Commonly, defects have various types including edges, vacancies, holes, or topological defects which can be constructed by mechanical ball milling, chemical oxidation, or etching method.

Overall, almost all defect types exhibit better CO_2RR activity compared to perfect carbon materials [110]. However, recent studies still reveal that specific defect types may have better promoting effects. Zhang group's work demonstrated that defective porous carbon catalysts containing no active heteroatom dopants can show excellent catalytic performance for CO_2RR. Notably, the exceptional activity was found to be significantly influenced by the presence of sp^2 defects (octagonal and pentagonal), rather than edge defects [111]. Kang and co-workers reported a hierarchical porous carbon catalyst rich in intrinsic defects with an FE_{CO} reaching 99.5% at −0.5 V vs. RHE in 0.5 M $KHCO_3$ electrolyte. Experimental results and DFT calculation showed that the appropriate CO_2 and intermediate adsorption energy at specific carbon defect sites (e.g. C5 + H) decreased the reaction energy barriers thus facilitating the CO_2RR process remarkably [112]. Similarly, Chen et al. prepared biomass-derived carbon-based electrocatalysts with intrinsic defects that could catalyze CO_2 reduction to CO with an FE of ~89% and maintained good selectivity for about 10 days. Their study underscored the pivotal role of intrinsic defects in the carbon matrix, especially those containing pentagons, as the primary active sites expediting direct electron transfer for CO_2 reduction [113].

16.5 Challenges for Carbon-Based CO_2RR Electrocatalysts

16.5.1 *In Situ* Characterization

While significant progress has been made in carbon-based electrocatalysts for CO_2 reduction, the precise mechanisms underlying these processes remain elusive.

Understanding these mechanisms is crucial for the design of novel catalysts. Fortunately, a combination of theoretical calculations and the advancement of *in situ*/operando characterization techniques holds promise in providing deeper insights. The establishment of theoretical models, encompassing reaction and process conditions, electrode morphology, and (sub)surface structure, is vital for a comprehensive understanding of the actual reaction pathways. *In situ* spectroscopic characterization, including *in situ* X-ray diffraction (XRD), surface-enhanced infrared adsorption spectroscopy, surface-enhanced Raman spectroscopy, X-ray photoelectron spectroscopy (XPS), X-ray absorption spectroscopy (XAS), as well as *in situ* transmission electron microscopy (TEM), emerges as a valuable tool, offering detailed information on catalyst structure, surface states, chemical properties, and binding configurations of surface-adsorbed intermediates under realistic working conditions [5].

To date, several operando studies have been reported on CO_2RR electrodes, but there are still many challenges and opportunities in this field. Since catalysts often exhibit topographic surface inhomogeneities, which could cause spatially heterogeneous reactivity, it is highly desirable to identify local reacting species with atomic precision during the CO_2 reduction process [114]. In addition, considering the various reaction environments as well as evolutions of catalysts and intermediates during the CO_2RR, more precise and surface-sensitive in situ techniques must be developed to achieve the direct observation of their states and evolution processes at atomic resolution [115].

16.5.2 Commercialization

In the pursuit of commercializing CO_2 reduction technology, the current stage of CO_2 electrolyzers is still in its infancy. Adding to the complexity, carbon-based catalysts often exhibit lower current densities compared to their metal-based counterparts, posing challenges in terms of scalability. Therefore, the engineering of electrocatalysts requires a heightened focus on achieving practical targets for industrial current density and actual CO_2 conversion efficiency, with the help of the aforementioned strategies.

Furthermore, the membrane as the key module in membrane electrode assemblies (MEA) and alkaline polymer electrolyte membrane (APEM) still faces grand challenges for stability, due to the detrimental process for products crossing and ions shuttle [116, 117]. Thus, a stable membrane that can inhibit product crossing and be very selective for passing ions, is highly required.

Electrolytes are also important factors in the CO_2RR process. It can impact the reaction and process conditions in the vicinity of the electrode by blocking the active site of catalysts, changing the local pH value, and buffering the equilibrium of CO_2/HCO_3^-, H_2CO_3/HCO_3^-, and HCO_3^-/CO_3^{2-} [5]. Additionally, electrolytes can induce catalyst restructuring and generate active species for CO_2RR by interacting with intermediates and products on the electrode surface [116]. While alkaline electrolytes, such as KOH and NaOH, are commonly used for CO_2RR, they inevitably lead to the conversion of CO_2 into carbonate, causing a significant portion of CO_2

to be absorbed by the solution rather than undergoing electroreduction. Hence, the replacement of alkaline electrolytes with neutral or acidic counterparts, facilitated by the utilization of acid-resistant carbon-based catalysts, stands as a promising approach to enhance CO_2 utilization efficiency. However, this transition poses a substantial challenge that is yet to be fully addressed.

16.6 Summary and Outlook

The unique advantages of CO_2RR alongside its potential for scalability have drawn significant attention to this technology as a viable strategy to close the global carbon cycle. As catalyst is at the heart of a catalytic reaction, a tremendous amount of work has been devoted to exploring robust electrocatalyst materials for the selective reduction of CO_2 in the past few years. In this review, we focus on carbon-based electrocatalysts and systematically present their categories, electrochemical properties and current achievements on CO_2RR. Importantly, the modulating strategies are discussed in depth on the basis of identification of their true active sites. Given that the current densities are typically low with carbon-based electrocatalysts, the direction shifts to improving reaction kinetics through group functionalization, heteroatom doping, defect introduction, synergistic effect, SACs and coordination environment control.

In addition, the importance and capability of in situ and operando characterization techniques are identified to provide insight into the reaction pathways and dynamic nature of structure-activity relationships. Moreover, the necessity for engineering of electrolyzer configuration, electrolyte, and membrane is emphasized to achieve the practical target of industrial current density and actual CO_2 conversion efficiency. Prospectively, there is no doubt that continued research and development in this area should be effective in improving fuel economy and contributing to an early carbon neutrality blueprint. We expect the pathway from academic studies to industrial production and commercialization to be realized in the future.

References

1 Panwar, N.L., Kaushik, S.C., and Kothari, S. (2011). Role of renewable energy sources in environmental protection: a review. *Renew. Sust. Energ. Rev.* 15 (3): 1513.

2 Kibria, M.G., Edwards, J.P., Gabardo, C.M. et al. (2019). Electrochemical CO_2 reduction into chemical feedstocks: from mechanistic electrocatalysis models to system design. *Adv. Mater.* 31 (31): e1807166.

3 Kondratenko, E.V., Mul, G., Baltrusaitis, J. et al. (2013). Status and perspectives of CO_2 conversion into fuels and chemicals by catalytic, photocatalytic and electrocatalytic processes. *Energy Environ. Sci.* 6 (11): 3112.

4 Zhang, S., Fan, Q., Xia, R., and Meyer, T.J. (2020). CO_2 reduction: from homogeneous to heterogeneous electrocatalysis. *Acc. Chem. Res.* 53 (1): 255.

5 Jin, S., Hao, Z., Zhang, K. et al. (2021). Advances and challenges for the electrochemical reduction of CO_2 to CO: from fundamentals to industrialization. *Angew. Chem. Int. Ed.* 60 (38): 20627.

6 Luc, W., Fu, X., Shi, J. et al. (2019). Two-dimensional copper nanosheets for electrochemical reduction of carbon monoxide to acetate. *Nat. Catal.* 2 (5): 423.

7 Luo, W., Zhang, J., Li, M., and Züttel, A. (2019). Boosting CO production in electrocatalytic CO_2 reduction on highly porous Zn catalysts. *ACS Catal.* 9 (5): 3783.

8 Abeyweera, S.C., Yu, J., Perdew, J.P. et al. (2020). Hierarchically 3D porous Ag nanostructures derived from silver benzenethiolate nanoboxes: enabling CO_2 reduction with a near-unity selectivity and mass-specific current density over 500 A/g. *Nano Lett.* 20 (4): 2806.

9 Yu, F., Liu, G., Zhan, J. et al. (2023). Self-driven electron enrichment of ultrafine PdAu nanoparticles for electrochemical CO_2 reduction: high applicability of work function as an activity descriptor. *Appl. Catal. B: Environ.* 338: 122931.

10 Xu, Y., Li, C., Xiao, Y. et al. (2022). Tuning the selectivity of liquid products of CO_2RR by Cu-Ag alloying. *ACS Appl. Mater. Interfaces* 14 (9): 11567.

11 Wang, P., Yang, H., Tang, C. et al. (2022). Boosting electrocatalytic CO_2-to-ethanol production via asymmetric C-C coupling. *Nat. Commun.* 13 (1): 3754.

12 Li, C.W. and Kanan, M.W. (2012). CO_2 reduction at low overpotential on cu electrodes resulting from the reduction of thick Cu_2O films. *J. Am. Chem. Soc.* 134 (17): 7231.

13 Li, Q., Fu, J., Zhu, W. et al. (2017). Tuning Sn-catalysis for electrochemical reduction of CO_2 to CO via the core/shell Cu/SnO_2 structure. *J. Am. Chem. Soc.* 139 (12): 4290.

14 Liu, S., Xiao, J., Lu, X.F. et al. (2019). Efficient electrochemical reduction of CO_2 to HCOOH over sub-2 nm SnO_2 quantum wires with exposed grain boundaries. *Angew. Chem. Int. Ed.* 58 (25): 8499.

15 Calvinho, K.U.D., Laursen, A.B., Yap, K.M.K. et al. (2018). Selective CO_2 reduction to C_3 and C_4 oxyhydrocarbons on nickel phosphides at overpotentials as low as 10 mV. *Energy Environ. Sci.* 11 (9): 2550.

16 Zhou, Y., Martín, A.J., Dattila, F. et al. (2022). Long-chain hydrocarbons by CO_2 electroreduction using polarized nickel catalysts. *Nat. Catal.* 5 (6): 545.

17 Liu, S., Yang, H.B., Hung, S.F. et al. (2020). Elucidating the electrocatalytic CO_2 reduction reaction over a model single-atom nickel catalyst. *Angew. Chem. Int. Ed.* 59 (2): 798.

18 Gao, D., Liu, T., Wang, G., and Bao, X. (2021). Structure sensitivity in single-atom catalysis toward CO_2 electroreduction. *ACS Energy Lett.* 6 (2): 713.

19 Li, M., Wang, H., Luo, W. et al. (2020). Heterogeneous single-atom catalysts for electrochemical CO_2 reduction reaction. *Adv. Mater.* 32 (34): 1848.

20 Morlanés, N., Takanabe, K., and Rodionov, V. (2016). Simultaneous reduction of CO_2 and splitting of H_2O by a single immobilized cobalt phthalocyanine electrocatalyst. *ACS Catal.* 6 (5): 3092.

21 Zhang, L., Xiao, J., Wang, H., and Shao, M. (2017). Carbon-based electrocatalysts for hydrogen and oxygen evolution reactions. *ACS Catal.* 7 (11): 7855.

22 Zheng, Y., Jiao, Y., and Qiao, S.Z. (2015). Engineering of carbon-based electrocatalysts for emerging energy conversion: from fundamentality to functionality. *Adv. Mater.* 27 (36): 5372.

23 Paul, R., Dai, Q., Hu, C., and Dai, L. (2019). Ten years of carbon-based metal-free electrocatalysts. *Carbon Energy* 1 (1): 19.

24 Masood Ul Hasan, I., Peng, L., Mao, J. et al. (2020). Carbon-based metal-free catalysts for electrochemical CO_2 reduction: activity, selectivity, and stability. *Carbon Energy* 3 (1): 24.

25 Ren, S., Joulié, D., Salvatore, D. et al. (2019). Molecular electrocatalysts can mediate fast, selective CO_2 reduction in a flow cell. *Science* 365 (6451): 367.

26 Gong, Q., Ding, P., Xu, M. et al. (2019). Structural defects on converted bismuth oxide nanotubes enable highly active electrocatalysis of carbon dioxide reduction. *Nat. Commun.* 10 (1): 2807.

27 Bushuyev, O.S., De Luna, P., Dinh, C.T. et al. (2018). What should we make with CO_2 and how can we make it? *Joule* 2 (5): 825.

28 Xiao, C. and Zhang, J. (2021). Architectural design for enhanced c(2) product selectivity in electrochemical CO_2 reduction using cu-based catalysts: a review. *ACS Nano* 15 (5): 7975.

29 Wang, Y., Shen, H., Livi, K.J.T. et al. (2019). Copper nanocubes for CO_2 reduction in gas diffusion electrodes. *Nano Lett.* 19 (12): 8461.

30 Wang, X., Wang, Z., García de Arquer, F.P. et al. (2020). Efficient electrically powered CO_2-to-ethanol via suppression of deoxygenation. *Nat. Energy* 5 (6): 478.

31 Genovese, C., Schuster, M.E., Gibson, E.K. et al. (2018). Operando spectroscopy study of the carbon dioxide electro-reduction by iron species on nitrogen-doped carbon. *Nat. Commun.* 9 (1): 935.

32 Wang, X., Ou, P., Ozden, A. et al. (2022). Efficient electrosynthesis of n-propanol from carbon monoxide using a Ag–Ru–Cu catalyst. *Nat. Energy* 7 (2): 170.

33 Zhao, K., Nie, X., Wang, H. et al. (2020). Selective electroreduction of CO_2 to acetone by single copper atoms anchored on n-doped porous carbon. *Nat. Commun.* 11 (1): 2455.

34 Quan, Y., Zhu, J., and Zheng, G. (2021). Electrocatalytic reactions for converting CO_2 to value-added products. *Small Sci.* 1 (10): 2100043.

35 Goyal, A., Marcandalli, G., Mints, V.A., and Koper, M.T.M. (2020). Competition between CO_2 reduction and hydrogen evolution on a gold electrode under well-defined mass transport conditions. *J. Am. Chem. Soc.* 142 (9): 4154.

36 Favaro, M., Xiao, H., Cheng, T. et al. (2017). Subsurface oxide plays a critical role in CO_2 activation by Cu(111) surfaces to form chemisorbed CO_2, the first step in reduction of CO_2. *Proc. Natl. Acad. Sci. USA* 114 (26): 6706.

37 Feaster, J.T., Shi, C., Cave, E.R. et al. (2017). Understanding selectivity for the electrochemical reduction of carbon dioxide to formic acid and carbon monoxide on metal electrodes. *ACS Catal.* 7 (7): 4822.

38 Appel, A.M., Bercaw, J.E., Bocarsly, A.B. et al. (2013). Frontiers, opportunities, and challenges in biochemical and chemical catalysis of CO_2 fixation. *Chem. Rev.* 113 (8): 6621.

39 Kortlever, R., Shen, J., Schouten, K.J. et al. (2015). Catalysts and reaction pathways for the electrochemical reduction of carbon dioxide. *J. Phys. Chem. Lett.* 6 (20): 4073.

40 Zheng, Y., Vasileff, A., Zhou, X. et al. (2019). Understanding the roadmap for electrochemical reduction of CO_2 to multi-carbon oxygenates and hydrocarbons on copper-based catalysts. *J. Am. Chem. Soc.* 141 (19): 7646.

41 Fan, L., Xia, C., Yang, F. et al. (2020). Strategies in catalysts and electrolyzer design for electrochemical CO_2 reduction toward C_{2+} products. *Sci. Adv.* 6 (8): 3111.

42 Fu, L., Wang, R., Zhao, C. et al. (2021). Construction of Cr-embedded graphyne electrocatalyst for highly selective reduction of CO_2 to CH_4: a DFT study. *Chem. Eng. J.* 414: 128857.

43 Pan, F. and Yang, Y. (2020). Designing CO_2 reduction electrode materials by morphology and interface engineering. *Energy Environ. Sci.* 13 (8): 2275.

44 Liu, S., Yang, H., Su, X. et al. (2019). Rational design of carbon-based metal-free catalysts for electrochemical carbon dioxide reduction: a review. *J. Energy Chem.* 36: 95.

45 Jia, C., Dastafkan, K., Ren, W. et al. (2019). Carbon-based catalysts for electrochemical CO_2 reduction. *Sustain. Energy Fuels* 3 (11): 2890.

46 Xiao, W., Cheng, M., Liu, Y. et al. (2023). Functional metal/carbon composites derived from metal–organic frameworks: insight into structures, properties, performances, and mechanisms. *ACS Catal.* 13 (3): 1759.

47 Zhao, K. and Quan, X. (2021). Carbon-based materials for electrochemical reduction of CO_2 to C_{2+} oxygenates: recent progress and remaining challenges. *ACS Catal.* 11 (4): 2076.

48 Novoselov, K.S., Geim, A.K., Morozov, S.V. et al. (2004). Electric field effect in atomically thin carbon films. *Science* 306 (5696): 666.

49 Wu, J., Ma, S., Sun, J. et al. (2016). A metal-free electrocatalyst for carbon dioxide reduction to multi-carbon hydrocarbons and oxygenates. *Nat. Commun.* 7: 13869.

50 Zou, X., Liu, M., Wu, J. et al. (2017). How nitrogen-doped graphene quantum dots catalyze electroreduction of CO_2 to hydrocarbons and oxygenates. *ACS Catal.* 7 (9): 6245.

51 Sreekanth, N., Nazrulla, M.A., Vineesh, T.V. et al. (2015). Metal-free boron-doped graphene for selective electroreduction of carbon dioxide to formic acid/formate. *Chem. Commun.* 51 (89): 16061.

52 Yang, H., Wu, Y., Lin, Q. et al. (2018). Composition tailoring via N and S co-doping and structure tuning by constructing hierarchical pores: metal-free catalysts for high-performance electrochemical reduction of CO_2. *Angew. Chem. Int. Ed.* 57 (47): 15476.

53 Iijima, S. (1991). Helical microtubules of graphitic carbon. *Nature* 354: 56.

54 Wu, Y., Zhao, X., Shang, Y. et al. (2021). Application-driven carbon nanotube functional materials. *ACS Nano* 15 (5): 7946.

55 Genovese, C., Ampelli, C., Perathoner, S., and Centi, G. (2013). Electrocatalytic conversion of CO_2 to liquid fuels using nanocarbon-based electrodes. *J. Energy Chem.* 22 (2): 202.

56 Sharma, P.P., Wu, J., Yadav, R.M. et al. (2015). Nitrogen-doped carbon nanotube arrays for high-efficiency electrochemical reduction of CO_2: on the understanding of defects, defect density, and selectivity. *Angew. Chem. Int. Ed.* 54 (46): 13701.

57 Xu, J., Kan, Y., Huang, R. et al. (2016). Revealing the origin of activity in nitrogen-doped nanocarbons towards electrocatalytic reduction of carbon dioxide. *ChemSusChem* 9 (10): 1085.

58 Zhang, S., Kang, P., Ubnoske, S. et al. (2014). Polyethylenimine-enhanced electrocatalytic reduction of CO_2 to formate at nitrogen-doped carbon nanomaterials. *J. Am. Chem. Soc.* 136 (22): 7845.

59 Duan, X., Tian, W., Zhang, H. et al. (2019). sp^2/sp^3 framework from diamond nanocrystals: a key bridge of carbonaceous structure to carbocatalysis. *ACS Catal.* 9 (8): 7494.

60 Deng, D., Novoselov, K.S., Fu, Q. et al. (2016). Catalysis with two-dimensional materials and their heterostructures. *Nat. Nanotechnol.* 11 (3): 218.

61 Vasileff, A., Zheng, Y., and Qiao, S.Z. (2017). Carbon solving carbon's problems: recent progress of nanostructured carbon-based catalysts for the electrochemical reduction of CO_2. *Adv. Energy Mater.* 7 (21): 759.

62 Marepally, B.C., Ampelli, C., Genovese, C. et al. (2017). Role of small cu nanoparticles in the behaviour of nanocarbon-based electrodes for the electrocatalytic reduction of CO_2. *J. CO_2 Util.* 21: 534.

63 Jhong, H.M., Tornow, C.E., Smid, B. et al. (2017). A nitrogen-doped carbon catalyst for electrochemical CO_2 conversion to CO with high selectivity and current density. *ChemSusChem* 10 (6): 1094.

64 Nakata, K., Ozaki, T., Terashima, C. et al. (2014). High-yield electrochemical production of formaldehyde from CO_2 and seawater. *Angew. Chem. Int. Ed.* 53 (3): 871.

65 Liu, Y., Chen, S., Quan, X., and Yu, H. (2015). Efficient electrochemical reduction of carbon dioxide to acetate on nitrogen-doped nanodiamond. *J. Am. Chem. Soc.* 137 (36): 11631.

66 Liu, Y., Zhang, Y., Cheng, K. et al. (2017). Selective electrochemical reduction of carbon dioxide to ethanol on a boron- and nitrogen-Co-doped nanodiamond. *Angew. Chem. Int. Ed.* 56 (49): 15607.

67 Tomisaki, M., Kasahara, S., Natsui, K. et al. (2019). Switchable product selectivity in the electrochemical reduction of carbon dioxide using boron-doped diamond electrodes. *J. Am. Chem. Soc.* 141 (18): 7414.

68 Cheng, Y., Wang, H., Qian, T., and Yan, C. (2022). Interfacial engineering of carbon-based materials for efficient electrocatalysis: recent advances and future. *EnergyChem* 4 (3): 74.

69 Tan, X., Kou, L., Tahini, H.A., and Smith, S.C. (2015). Conductive graphitic carbon nitride as an ideal material for electrocatalytically switchable CO_2 capture. *Sci. Rep.* 5: 17636.

70 Zhang, B., Zhao, T.J., Feng, W.J. et al. (2018). Polarized few-layer g-C_3N_4 as metal-free electrocatalyst for highly efficient reduction of CO_2. *Nano Res.* 11 (5): 2450.

71 Lu, X., Tan, T.H., Ng, Y.H. et al. (2016). Highly selective and stable reduction of CO_2 to CO by a graphitic carbon nitride/carbon nanotube composite electrocatalyst. *Chemistry* 22 (34): 11991.

72 Zhi, X., Jiao, Y., Zheng, Y., and Qiao, S.-Z. (2019). Impact of interfacial electron transfer on electrochemical CO_2 reduction on graphitic carbon nitride/doped graphene. *Small* 15 (10): e1804224.

73 Bagger, A., Ju, W., Varela, A.S. et al. (2017). Single site porphyrine-like structures advantages over metals for selective electrochemical CO_2 reduction. *Catal. Today* 288: 74.

74 Qu, Q., Ji, S., Chen, Y. et al. (2021). The atomic-level regulation of single-atom site catalysts for the electrochemical CO_2 reduction reaction. *Chem. Sci.* 12 (12): 4201.

75 Xu, H., Cheng, D., Cao, D. et al. (2018). A universal principle for a rational design of single-atom electrocatalysts. *Nat. Catal.* 1 (5): 339.

76 Ju, W., Bagger, A., Hao, G.P. et al. (2017). Understanding activity and selectivity of metal-nitrogen-doped carbon catalysts for electrochemical reduction of CO_2. *Nat. Commun.* 8 (1): 944.

77 Pan, F., Zhang, H., Liu, K. et al. (2018). Unveiling active sites of CO_2 reduction on nitrogen-coordinated and atomically dispersed iron and cobalt catalysts. *ACS Catal.* 8 (4): 3116.

78 Yang, H., Lin, Q., Zhang, C. et al. (2020). Carbon dioxide electroreduction on single-atom nickel decorated carbon membranes with industry compatible current densities. *Nat. Commun.* 11 (1): 593.

79 Li, X., Bi, W., Chen, M. et al. (2017). Exclusive Ni-N_4 sites realize near-unity CO selectivity for electrochemical CO_2 reduction. *J. Am. Chem. Soc.* 139 (42): 14889.

80 Jiang, K., Siahrostami, S., Zheng, T. et al. (2018). Isolated Ni single atoms in graphene nanosheets for high-performance CO_2 reduction. *Energy Environ. Sci.* 11 (4): 893.

81 Cao, S., Zhou, S., Chen, H. et al. (2022). How can the dual-atom catalyst FeCo–NC surpass single-atom catalysts Fe–NC/Co–NC in CO_2RR? – CO intermediate assisted promotion via a synergistic effect. *Energy Environ. Mater.* 6 (1): e12287.

82 Zhao, J., Zhao, J., Li, F., and Chen, Z. (2018). Copper dimer supported on a C_2N layer as an efficient electrocatalyst for CO_2 reduction reaction: a computational study. *J. Phys. Chem. C* 122 (34): 19712.

83 He, T., Zhang, L., Kour, G., and Du, A. (2020). Electrochemical reduction of carbon dioxide on precise number of fe atoms anchored graphdiyne. *J. CO_2 Util.* 37: 272.

84 Baturina, O.A., Lu, Q., Padilla, M.A. et al. (2014). CO_2 electroreduction to hydrocarbons on carbon-supported cu nanoparticles. *ACS Catal.* 4 (10): 3682.

85 Daiyan, R., Saputera, W.H., Masood, H. et al. (2020). A disquisition on the active sites of heterogeneous catalysts for electrochemical reduction of CO_2 to value-added chemicals and fuel. *Adv. Energy Mater.* 10 (11): 106.

86 Li, Z., Wu, R., Zhao, L. et al. (2021). Metal-support interactions in designing noble metal-based catalysts for electrochemical CO_2 reduction: recent advances and future perspectives. *Nano Res.* 14 (11): 3795.

87 Hoang, V.C., Gomes, V.G., and Kornienko, N. (2020). Metal-based nanomaterials for efficient CO_2 electroreduction: recent advances in mechanism, material design and selectivity. *Nano Energy* 78: 105311.

88 Zhang, B., Jiang, Y., Gao, M. et al. (2021). Recent progress on hybrid electrocatalysts for efficient electrochemical CO_2 reduction. *Nano Energy* 80: 105504.

89 Gao, D., Zhou, H., Wang, J. et al. (2015). Size-dependent electrocatalytic reduction of CO_2 over Pd nanoparticles. *J. Am. Chem. Soc.* 137 (13): 4288.

90 Kim, C., Jeon, H.S., Eom, T. et al. (2015). Achieving selective and efficient electrocatalytic activity for CO_2 reduction using immobilized silver nanoparticles. *J. Am. Chem. Soc.* 137 (43): 13844.

91 Rao, R.G., Blume, R., Greiner, M.T. et al. (2022). Oxygen-doped carbon supports modulate the hydrogenation activity of palladium nanoparticles through electronic metal–support interactions. *ACS Catal.* 12 (12): 7344.

92 Huo, Y., Peng, X., Liu, X. et al. (2018). High selectivity toward C_2H_4 production over cu particles supported by butterfly-wing-derived carbon frameworks. *ACS Appl. Mater. Interfaces* 10 (15): 12618.

93 Zhang, G., Wang, T., Zhang, M. et al. (2022). Selective CO_2 electroreduction to methanol via enhanced oxygen bonding. *Nat. Commun.* 13 (1): 7768.

94 Zhang, S., Kang, P., and Meyer, T.J. (2014). Nanostructured tin catalysts for selective electrochemical reduction of carbon dioxide to formate. *J. Am. Chem. Soc.* 136 (5): 1734.

95 Liang, M., Liu, Y., Zhang, J. et al. (2022). Understanding the role of metal and N species in M@NC catalysts for electrochemical CO_2 reduction reaction. *Appl. Catal. B* 306: 121115.

96 Shen, L., Ying, J., Ozoemena, K.I. et al. (2021). Confinement effects in individual carbon encapsulated nonprecious metal-based electrocatalysts. *Adv. Funct. Mater.* 32 (15): 851.

97 Lu, J., Yin, S., and Shen, P.K. (2018). Carbon-encapsulated electrocatalysts for the hydrogen evolution reaction. *Electrochem. Energy Rev.* 2 (1): 105.

98 Varnell, J.A., Tse, E.C., Schulz, C.E. et al. (2016). Identification of carbon-encapsulated iron nanoparticles as active species in non-precious metal oxygen reduction catalysts. *Nat. Commun.* 7: 12582.

99 Yang, F., Ma, X., Cai, W.B. et al. (2019). Nature of oxygen-containing groups on carbon for high-efficiency electrocatalytic CO_2 reduction reaction. *J. Am. Chem. Soc.* 141 (51): 20451.

100 Jakobsen, J.B., Ronne, M.H., Daasbjerg, K., and Skrydstrup, T. (2021). Are amines the holy grail for facilitating CO_2 reduction? *Angew. Chem. Int. Ed.* 60 (17): 9174.

101 Ahn, S., Klyukin, K., Wakeham, R.J. et al. (2018). Poly-amide modified copper foam electrodes for enhanced electrochemical reduction of carbon dioxide. *ACS Catal.* 8 (5): 4132.

102 Thevenon, A., Rosas-Hernandez, A., Peters, J.C., and Agapie, T. (2019). In-situ nanostructuring and stabilization of polycrystalline copper by an organic salt additive promotes electrocatalytic CO_2 reduction to ethylene. *Angew. Chem. Int. Ed.* 58 (47): 16952.

103 Jia, C., Li, S., Zhao, Y. et al. (2021). Nitrogen vacancy induced coordinative reconstruction of single-atom Ni catalyst for efficient electrochemical CO_2 reduction. *Adv. Funct. Mater.* 31 (51): 2107072.

104 Liu, S., Yang, H., Huang, X. et al. (2018). Identifying active sites of nitrogen-doped carbon materials for the CO_2 reduction reaction. *Adv. Funct. Mater.* 28 (21): 499.

105 Shang, Y., Ding, Y., Zhang, P. et al. (2022). Pyrrolic N or pyridinic N: the active center of N-doped carbon for CO_2 reduction. *Chin. J. Catal.* 43 (9): 2405.

106 Song, Y., Wang, S., Chen, W. et al. (2020). Enhanced ethanol production from CO_2 electroreduction at micropores in nitrogen-doped mesoporous carbon. *ChemSusChem* 13 (2): 293.

107 Wu, J., Liu, M., Sharma, P.P. et al. (2016). Incorporation of nitrogen defects for efficient reduction of CO_2 via two-electron pathway on three-dimensional graphene foam. *Nano Lett.* 16 (1): 466.

108 Yang, H.B., Hung, S.F., Liu, S. et al. (2018). Atomically dispersed Ni(i) as the active site for electrochemical CO_2 reduction. *Nat. Energy* 3 (2): 140.

109 Cheng, Y., Zhao, S., Li, H. et al. (2019). Unsaturated edge-anchored Ni single atoms on porous microwave exfoliated graphene oxide for electrochemical CO_2. *Appl. Catal. B: Environ.* 243: 294.

110 Xue, D., Xia, H., Yan, W. et al. (2020). Defect engineering on carbon-based catalysts for electrocatalytic CO_2 reduction. *Nano-Micro Lett.* 13 (1): 5.

111 Wang, W., Shang, L., Chang, G. et al. (2019). Intrinsic carbon-defect-driven electrocatalytic reduction of carbon dioxide. *Adv. Mater.* 31 (19): e1808276.

112 Wu, Q., Gao, J., Feng, J. et al. (2020). A CO_2 adsorption dominated carbon defect-based electrocatalyst for efficient carbon dioxide reduction. *J. Mater. Chem. A* 8 (3): 1205.

113 Chen, M., Wang, S., Zhang, H. et al. (2020). Intrinsic defects in biomass-derived carbons facilitate electroreduction of CO_2. *Nano Res.* 13 (3): 729.

114 Li, X., Wang, S., Li, L. et al. (2020). Progress and perspective for in situ studies of CO_2 reduction. *J. Am. Chem. Soc.* 142 (21): 9567.

115 Zou, Y. and Wang, S. (2021). An investigation of active sites for electrochemical CO_2 reduction reactions: from in situ characterization to rational design. *Adv. Sci.* 8 (9): 2003579.

116 Tan, X., Yu, C., Ren, Y. et al. (2021). Recent advances in innovative strategies for the CO_2 electroreduction reaction. *Energy Environ. Sci.* 14 (2): 765.

117 Sánchez, O.G., Birdja, Y.Y., Bulut, M. et al. (2019). Recent advances in industrial CO_2 electroreduction. *Curr. Opin. Green Sustainable Chem.* 16: 47.

17

Porous Materials for CO_2RR

Zibo Zhao, Shun Zhang, and Meiting Zhao

Tianjin University, Tianjin Key Laboratory of Molecular Optoelectronic Sciences, Institute of Molecular Aggregation Science, Department of Chemistry, No. 92 Weijin road, Nankai district, Tianjin, 300072, China

17.1 Introduction

With the rapid development of economy and society, human productivity has greatly improved, and the dependence on fossil fuels is increasing day by day. This makes the concentration of carbon dioxide in the atmosphere increase exponentially and thus leads to a series of environmental crises, including greenhouse effect, ocean acidification, species extinction, and extreme weather. These environmental problems will seriously affect human survival and development [1–5]. Research shows that about 50% of CO_2 has been emitted into the air in the past 45 years, the atmospheric CO_2 level has reached 420 ppm nowadays, and the upward trend has hardly slowed down [6–9]. Therefore, the work of CO_2 emission reduction is extremely urgent, and it is an important task for all countries in the world to achieve the goal of carbon neutralization. At present, the simplest way to save energy is to avoid the abuse of fossil fuels as well as use alternative energy sources, such as solar energy, wind energy, hydro energy, and geothermal energy, to gradually replace fossil fuels [10, 11]. At present, there are three main ways to reduce CO_2 emissions: (i) carbon capture and storage; (ii) reducing the consumption of fossil fuels; and (iii) CO_2 conversion and recycling [12]. Among the strategies, carbon capture and sequestration technology is currently a high-potential solution to deal with CO_2 emissions, which can effectively separate CO_2 gas from its mixture, but the high implementation cost and leakage during the storage process are the main obstacles to the realization of large-scale application of this technology [13–15].

Recently, the transformation of CO_2 has attracted great attention from researchers [16–20], and the approach of electrocatalytic carbon dioxide reduction reaction (CO_2RR) shows great application potential [21]. The main reasons are as follows: (i) the reaction conditions are mild and can be carried out at room temperature and pressure; (ii) renewable electrons from wind, solar, and tidal energy can

Functional Materials for Electrocatalytic Energy Conversion, First Edition.
Edited by Zhicheng Zhang, Meiting Zhao, and Yuchen Qin.

be stored in new energy materials or chemical products; and (iii) the reduction products are abundant and can be tuned by the catalysts, external parameters such as electrolyte and applied voltage. CO_2RR is a complex process; through multiple proton–electron-coupled transfer processes, many different gas and liquid phase products can be obtained, such as C_1 products (carbon monoxide (CO), methane (CH_4), formic acid (HCOOH), and methanol (CH_3OH)) and C_2 products (ethylene (C_2H_4), ethane (C_2H_6), ethanol (C_2H_5OH), and acetic acid (CH_3COOH)) [22–24]. However, there are still a series of challenges in CO_2RR, such as poor selectivity, low current density, and high overpotential, mainly due to the diversity of reaction paths, complex reaction mechanisms, and slow reaction kinetics of CO_2RR [25]. In addition, hydrogen evolution reaction (HER) is the main competitive reaction in CO_2RR reaction, which greatly limits its further application. In order to improve the activity of CO_2RR, the catalyst with good stability, activity, and selectivity should be reasonably designed to restrain the direct effect of the above limitations on the CO_2RR reaction [26].

Previously, homogeneous catalysts were widely used in CO_2RR. However, these catalysts are easily affected by protons and water solvents, which necessitate the selection of right organic solvent [27]. Subsequently, many experiments were carried out and contributed to the understanding of catalytic mechanism of CO_2RR, laying the foundation for future scientific research work. Hereafter, researchers first investigated the heterogeneous catalysts like metal single crystals and polycrystalline materials because of their high catalytic activity and stability in catalytic reactions. In particular, precious metals such as Pd, Au, and Ag have been widely reported to show high selectivity and activity in the transformation from CO_2 to CO with a clear structure–activity relationship [28, 29]. In recent years, various types of heterogeneous catalysts, especially the crystalline porous materials led by metal–organic framework (MOF) and covalent organic framework (COF), and their derived composites have been widely used in the catalytic reaction of CO_2. As a new type of functional porous material, they have many advantages, such as modified framework, adjustable charge transfer, designable porosity or surface area, flexible control of synthetic active centers [30, 31], and the easy recovery of catalysts [32, 33]. In order to improve the catalytic activity of functional porous materials for CO_2RR, researchers have proposed three strategies, such as: (i) combining them with conductive substrates to improve their stability and conductivity; (ii) pyrolyzing into carbon-based materials with regular internal metal coordination environment to enhance electron transfer ability and intrinsic activity; and (iii) constructing 2D/3D coordination mode-based materials [34]. Up to now, the research on functional porous framework materials as CO_2RR catalysts has been mainly focused on the structural design of metal single crystal or polycrystalline materials, but the mechanism has not been explored enough. Therefore, it is necessary to introduce the role of functional crystalline porous materials in CO_2RR in all aspects logically. This chapter comprehensively introduces the research progress of functional porous framework materials in CO_2RR according to the order of MOFs, COFs, and MOF/COF-derived porous materials to provide new ideas for the design of efficient catalysts in this field.

17.2 Metal–Organic Frameworks (MOFs)

MOF is a kind of crystalline functional porous material that can cleverly combine inorganic metal ions and organic ligands through coordination bonds. It can be precisely designed as 1D, 2D, and 3D network structure in space [35]. MOF materials have many advantages, such as (i) the dense distribution of metal nodes in MOF materials and the unique spatial coordination mode; (ii) the metal nodes in MOF materials can be arranged periodically to avoid agglomeration and improve the utilization of metal nodes; and (iii) the unique highly ordered and periodic arrangement of pores results in larger specific surface area and higher porosity compared to other homogeneous catalysts [36, 37]. These advantages have attracted researchers' attention in recent years in the fields of catalysis [38, 39], magnetism and molecular recognition [40], gas storage and separation [41], and so on. Especially in the field of electrocatalysis, MOF materials with heterostructures based on porphyrin and phthalocyanine catalytic units are promising electrocatalysts [42].

17.2.1 Porphyrin-Based Metal–Organic Framework

Yang and co-workers early reported the application of MOF materials in electrocatalytic CO_2RR. They chose MOF films as nanomaterial catalysts. First, porous MOF materials with catalytic bonding units were designed and synthesized, then coated on conductive substrates to form thin films, and then studied their electrochemical properties [43]. It is found that Co-porphyrin MOF catalyst ($Al_2[OH]_2$TCPP-Co) has good catalytic activity and selectivity in aqueous electrolytes. CO_2 can be efficiently converted to CO in the potential range of −0.5 to −0.9 V (vs. RHE), and the maximum FE_{CO} can reach 76%. The stability test results show that the current density and FE_{CO} have no obvious attenuation trend within seven hours. It has been proven that the catalyst has good stability. In addition, they also studied the oxidation state of cobalt in CO_2RR process by in situ spectroelectrochemical measurement. When the applied potential changes continuously from 0.2 to −0.7 V (vs. RHE), the Soret band of Co(II) begins to decrease at 422 nm, while the Co(I) Soret band at 408 nm increases significantly. This change is quantified to clarify that most of the active sites in the MOF catalyst are redox Co(II) sites that can be easily reduced to Co(I) during the reaction.

Yin and co-workers adopted a new synthesis strategy to further deposit 2D conjugated Co-based MOF (Co-[Co + Cu]PMOF) obtained by mixed preparation on g-C_3N_4 nanosheets and synthesized a series of g-C_3N_4@Co-(Co + Cu)PMOF composite catalysts [44]. Among them, compared with the single material, the g-C_3N_4Co-(Co + Cu)PMOF-50% composite has larger electrochemical active surface area, faster charge transfer rate in the reaction process, and more excellent catalytic activity than the original MOF material and other porphyrin-based MOF derivatives. Through the FE test to detect the selectivity of the catalyst, it can be seen that the g-C_3N_4Co-(Co + Cu)PMOF-50% material has excellent selectivity for converting CO_2 to CO compared with other catalysts, and the maximum FE_{CO} is as high as 97.8%, which is better than most of the reported MOF catalysts. At the same time, the g-C_3N_4Co-(Co + Cu)PMOF-50% catalyst has the highest double-layer capacitance (C_{dl} = 35.5 mF cm^{-2}) and the lowest slope of the Tafel curve, indicating that the material has the highest electrochemical active surface area and the fastest reaction kinetics, which further shows excellent catalytic activity. Long-term electrochemical stability tests show that the g-C_3N_4Co-(Co + Cu)PMOF-50% catalyst has no obvious attenuation of current density in the electrolysis experiment for 12 hours, and the FE_{CO} can be maintained at more than 85%, indicating that it has excellent stability for CO_2 reduction. This work provides a new idea for the reasonable design of new porphyrin-based conjugated MOF materials for CO_2RR.

Lu and co-workers adopted a novel and simple synthesis method to further deliver Cd metal sites to zirconium-based porphyrin MOF hollow nanotubes for electrocatalytic CO_2 reaction (Cd-PCN-222HT_S) [45]. Due to the unique structure of porphyrin units, Cd sites can coordinate with N atoms and be anchored and dispersed on zirconium-based MOF, effectively avoiding agglomeration and maximizing the utilization of metal sites (Figure 17.1a). It is found that Cd-PCN-222HTs catalyst has excellent selectivity for the conversion of CO_2 to CO in ionic liquid–water-acetonitrile electrolyte. In the potential range of −1.9 to 2.4 V (vs. Ag/Ag^+), the FE_{CO} of Cd-PCN-222HTs is significantly better than that of PCN-222HTs and Au-PCN-222HTs and reached a maximum of 97.0% (Figure 17.1b).

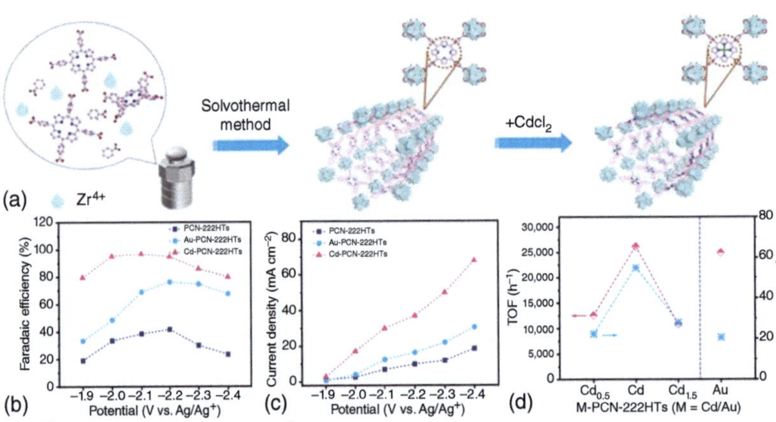

Figure 17.1 (a) The schematic diagram of the synthesis of Cd-PCN-222HTs catalyst. (b) FE_{CO} and (c) current density curves of different catalysts at different applied potentials. (d) TOF value and j_{CO} of different types of catalysts at −2.4 V (vs. Ag/Ag+) applied potential. Different catalysts (e) current density vs. scanning rate curves and (f) Nyquist curves. (g) Long-term stability test of Cd-PCN-222HTs catalyst. Reproduced with permission of Lu et al. [45], Wiley-VCH.

In the whole potential window, the current density of Cd-PCN-222HTs catalyst is also significantly higher than that of the other two materials. At −2.1 V (vs. RHE), the current density reaches 30.0 mA cm^{-2}, which is 2.5 and 1.6 times higher than that of PCN-222HTs and Au-PCN-222HTs at this potential, respectively (Figure 17.1c). Accordingly, the maximum TOF value (up to 26,220 h^{-1}) and CO partial current density of Cd-PCN-222HTs catalyst are significantly higher than those of other types of catalysts, indicating that Cd-PCN-222HTs material has the best Cd site content and better electron transfer ability (Figure 17.1d). According to the C_{dl} curves, it can be seen that Cd-PCN-222HTs has a higher C_{dl} value than the other two catalysts, indicating that it has the largest electrochemical active surface area (Figure 17.1e). At the same time, the electrochemical impedance spectrum shows that the Cd-PCN-222HTs catalyst has the smallest semicircle diameter, indicating that it has lower interfacial transfer resistance and faster electron transfer rate in the reaction process, which further improves its catalytic activity (Figure 17.1f). The long-term stability test result shows that the current density and FE_{CO} of Cd-PCN-222HTs catalyst have no obvious attenuation trend in six hours electrolysis experiment, indicating that the catalyst has good stability (Figure 17.1g). Density functional theory calculations show that the dispersed Cd active sites in PCN-222HTs materials can not only reduce the formation barrier of *COOH intermediates but also effectively inhibit the occurrence of hydrogen evolution side reactions in the reaction process, thus improving the activity of electrocatalytic conversion of CO_2 to CO.

17.2.2 Zinc Imidazolate Frameworks (ZIFs)

ZIF materials are a kind of building unit containing zinc clusters and imidazolium ligands. The homologs of ZIF-8 species have strong CO_2 adsorption properties, which can give full play to the advantages of MOF-based materials in CO_2RR reaction and show excellent catalytic activity. Kim and co-workers adopted a general synthesis method for doping various transition metal ions into the ZIF-8 framework at different concentrations. According to the doping ratio, the Zn metal nodes can be properly replaced by tetrahedral transition metal nodes and used in the CO_2RR process. The synthesis strategy is easy to operate and does not need further processing (Figure 17.2a) [46]. LSV curves indicated the three catalysts can effectively inhibit the occurrence of competitive hydrogen evolution side reaction in CO_2RR and have good electrochemical reduction activity for CO_2. It is worth noting that the $Cu_{0.5}Zn_{0.5}$/ZIF-8 material also has a lower overpotential, indicating that it has the best catalytic activity (Figure 17.2b). According to the selectivity of the generated products, the activity of various catalysts is further evaluated, and it is concluded that $Cu_{0.5}Zn_{0.5}$/ZIF-8 material has the best catalytic activity. At the applied potential of −1.0 V (vs. RHE), FE_{CO} reaches the maximum value (88.5%) with the highest local current density (11.57 mA cm^{-2}) of CO, which is higher than those of other catalysts at the same potential (Figure 17.2c,d). The TOF of $Cu_{0.5}Zn_{0.5}$/ZIF-8 material at −1.0 V (vs. RHE) is 1860 h^{-1}, which is higher than

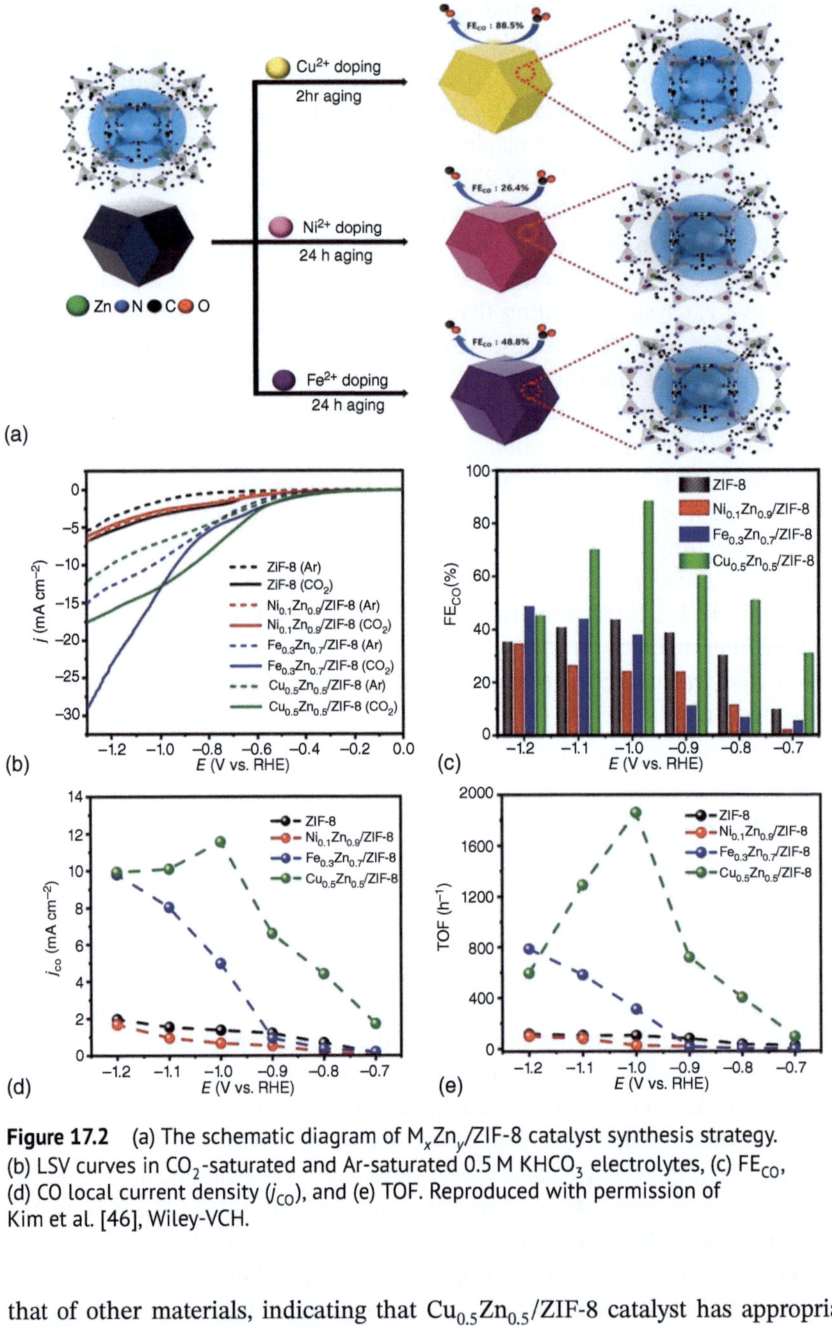

Figure 17.2 (a) The schematic diagram of M_xZn_y/ZIF-8 catalyst synthesis strategy. (b) LSV curves in CO_2-saturated and Ar-saturated 0.5 M $KHCO_3$ electrolytes, (c) FE_{CO}, (d) CO local current density (j_{CO}), and (e) TOF. Reproduced with permission of Kim et al. [46], Wiley-VCH.

that of other materials, indicating that $Cu_{0.5}Zn_{0.5}$/ZIF-8 catalyst has appropriate intrinsic catalytic activity (Figure 17.2e).

Qiu and co-workers synthesized porous Fe, N-co-doped carbon nanoparticles (Fe-CNPs) electrocatalyst using ZIF-8 as raw material and modified its surface with amorphous SiO_2 coating for CO_2RR [47]. The activity on Fe-CNPs shows a dependence of porosity that porous Fe-CNPs are better than that of low-porous

Fe-CNP material for the conversion of CO_2 to CO. From the LSV curves, it can be seen that compared with Fe-CNPs-w/o, Fe-CNPs material shows higher current density in both CO_2- and Ar-saturated 1 M $KHCO_3$ electrolytes, which stems from the fact that Fe-CNPs material has larger porosity and more active sites. At the same time, compared with Ar-saturated electrolyte, Fe-CNPs catalyst has higher current density and lower initial potential in CO_2-saturated 1 M $KHCO_3$ electrolyte, which reveals that Fe-CNPs catalyst has good electrocatalytic activity for CO_2 reduction. In the high concentration CO_2-saturated 1 M $KHCO_3$ electrolyte, when the applied potential changes from −0.38 to −0.88 V (vs. RHE), the FE_{CO} of Fe-CNPs material shows a volcanic trend and reaches its maximum (FE_{CO} is close to 100%) at −0.58 V (vs. RHE), showing excellent CO selectivity. Moreover, in the total current density curves of the reaction process, it can be seen that both Fe-CNPs-w/o and Fe-CNPs catalysts show an increasing trend of current density with the increase of applied potential, but at −0.88 V (vs. RHE) potential. The total current density of Fe-CNPs catalyst reaches the maximum value of 15.1 mA cm^{-2}, which is about three times higher than that of Fe-CNPs-w/o material, further proving Fe-CNPs catalyst has better selectivity and catalytic activity. In addition, the long-term electrolysis experiment was carried out at the applied potential of −0.68 V (vs. RHE). It can be seen that the FE_{CO} of the catalyst can be kept above 95% in the eight hours electrolysis experiment, and there is no obvious attenuation trend. However, the current density decreases by about 27%, which can be attributed to the shedding of the catalyst from the substrate during the electrolysis process. Therefore, it is proven that the Fe-CNPs catalyst has not only high selectivity but also good stability.

Ashraf and co-workers synthesized copper-doped ZIF-8 catalysts with different proportions by a simple solvothermal method for CO_2RR. When the copper content was 30%, the catalyst showed the best activation activity and selectivity [48]. From the LSV curves, it can be seen that the current density of $Cu_{30\%}$ZIF-8 catalyst is higher than that of the other two catalysts in the wide potential range from −0.8 to −2.1 V (vs. Ag/AgCl), and the highest current density is 40 mA cm^{-2} at the applied potential of −2.1 V (vs. Ag/AgCl). This can be attributed to the fact that $Cu_{30\%}$ZIF-8 material has more copper active sites. As a result, the catalytic performance of the CO_2RR process is effectively improved. According to the calculation of Faraday efficiency, the HER is dominant at lower potential, and when the applied potential is −1.2 V (vs. Ag/AgCl), the FE_{H2} of the three catalysts reaches its maximum. With the increase IN potential, CO_2 reduction reaction dominates, and FE_{H2} decreases gradually. CO is the first C_1 product produced in the CO_2RR process. In the whole potential range, it can be seen that the FE_{CO} of the three catalysts reaches its maximum at −1.5 V (vs. Ag/AgCl). It is worth noting that the $Cu_{10\%}$ZIF-8 catalyst shows the largest FE_{CO} (62.26%), which is significantly higher than that of the other two materials. This may be due to the highest N content in the $Cu_{10\%}$ZIF-8 catalyst facilitating the transfer of electrons during the reaction, thus reducing the formation energy barrier of the intermediate *COOH, and the abundant Zn sites also contribute to the selectivity of CO products. However, the $Cu_{30\%}$ZIF-8 catalyst has the highest selectivity for CH_4 products, and its methane Faraday efficiency reaches 35.21% when the applied potential is −1.6 V (vs. Ag/AgCl). The results show that with the decrease of zinc sites in the molecular sieve imidazole skeleton and the increase of copper content, the selectivity of the catalyst for methane products is significantly enhanced in the reaction process, overcoming the energy barrier for the production of hydrocarbons.

17.3 Covalent Organic Frameworks (COFs)

Reticular materials can combine discrete molecular building blocks together to make them into periodic and scalable material scaffolds with considerable porosity and specific surface area. COF is a kind of important network structure material, which is often prepared by the cross-linking of aromatic molecules and reversible covalent bonds. It has many advantages, such as: (i) the composition is easy to regulate; (ii) the porosity and specific surface area are large; and (iii) the structure range of p–π conjugation is wide. In view of the above advantages, COF materials can skillfully make up for the differences between homogeneous and heterogeneous catalysts and have been widely used in optoelectronics, gas storage and separation, catalysis, and other fields. More importantly, as a well-defined M-N_4 coordination mode, metal-porphyrin/phthalocyanine has been widely explored to create solid materials with adjustable reactivity and high catalytic activity for CO_2RR.

17.3.1 Porphyrin Covalent Organic Framework

Zhang and co-workers designed and synthesized a series of covalently linked MWCNTPor-COF-M catalysts by combining porphyrin COF with carbon nanotubes (CNT) for CO_2RR. In view of the fact that carbon nanotubes can effectively promote the transfer of electrons from porphyrin planes to metal active sites, CNT is often used as an ideal carrier in the electrocatalytic CO_2 reduction process. In order to better realize the rapid electron transfer process, they chose MWCNT modified by $-NH_2$ group to bond porphyrin COF uniformly anchored on the surface of MWCNT (Figure 17.3a) [49]. Compared with the original Por-COF-M materials, the MWCNT-Por-COF-M series has higher selectivity and catalytic activity. The LSV curves indicate MWCNT-Por-COF-Co catalyst has lower initial potential and higher current density than the other three catalysts, indicating higher catalytic activity in the CO_2RR process (Figure 17.3b). Faraday efficiency results for different types of materials show that MWCNT-Por-COF-Co catalysts have the optimal activity for the conversion of CO_2 to CO. In the whole potential range, FE_{CO} is higher than that of MWCNT-Por-COF-Ni, MWCNT-Por-COF-Fe, and Por-COF-Co materials. And FE_{CO} can be maintained above 90% in the potential range of −0.6 to −0.9 V (vs. RHE) with a maximum of 99.3%. At the same time, the total current density of the MWCNT-Por-COF-Co catalyst is also higher than that of the other three catalysts, with the maximum current density of $-18.77\ \mathrm{mA\ cm^{-2}}$ occurring at −1.0 V (vs. RHE) (Figure 17.3c,d). According to the calculation results of the Tafel slope, the reaction kinetics of different catalysts in CO_2RR process were clarified. It can be seen that MWCNT-Por-COF-Co catalyst shows the smallest Tafel slope of

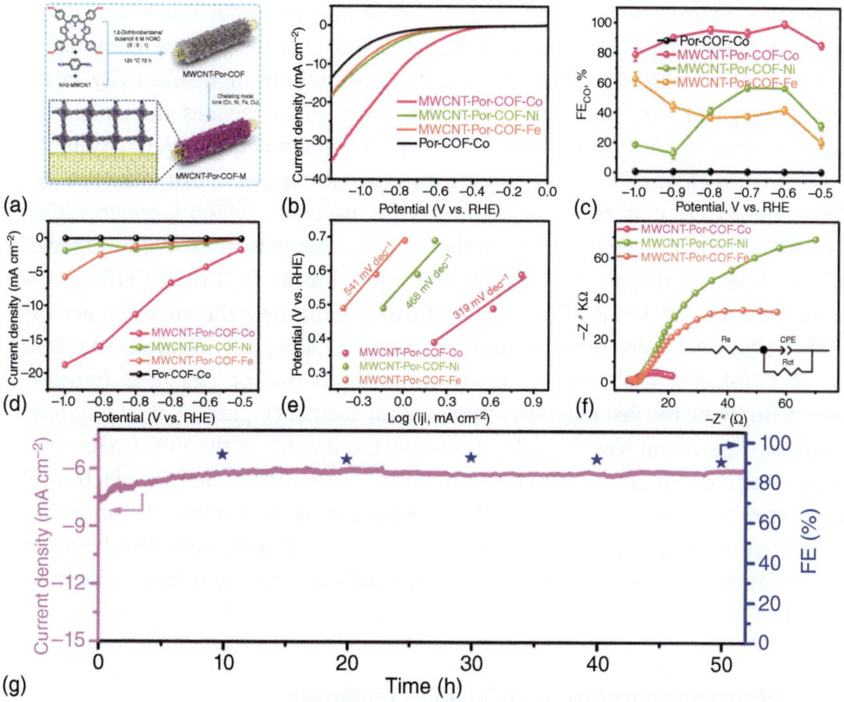

Figure 17.3 (a) The schematic diagram of synthesis of MWCNT-Por-COF-M catalyst. (b) LSV curves in CO_2-saturated 0.5 M $KHCO_3$ electrolyte, (c) FE_{CO}, and (d) total current density of the four catalysts. (e) Tafel slopes and (f) Nyquist curves of MWCNT-Por-COF-Co, MWCNT-Por-COF-Ni and MWCNT-Por-COF-Fe catalysts. (g) Long-term stability test of MWCNT-Por-COF-Co catalyst at −0.7 V (vs. RHE) applied potential. Reproduced with permission of Zhang et al. [49], Elsevier.

319 mV dec^{-1}, indicating that it has the fastest reaction kinetics in the process of electrochemical reduction of CO_2 (Figure 17.3e). In addition, the electrochemical impedance spectra show that the MWCNT-Por-COF-Co material has the smallest semicircle diameter, indicating that it has the lowest interfacial transfer resistance and the most efficient charge transfer rate and thus has a faster reaction kinetics, which is consistent with the results of the Tafel slope (Figure 17.3f). Additionally, long-term durability experiment of MWCNT-Por-COF-Co catalyst shows that the current density and FE_{CO} of the catalyst kept stable for at least 50 hours without notable attenuation, proving its excellent stability (Figure 17.3g).

Jiang and co-workers adopted a novel strategy to synthesize two 2D COFs based on tetraphenylethylene and cobalt porphyrin blocks for the first time, in which tetraphenylethylene can be used as the electron donor and cobalt porphyrin as the electron acceptor. This can enhance the electron transfer rate in the internal structure of the catalyst and further enhance the catalytic activity and selectivity in the CO_2RR process (Figure 17.4a) [50]. LSV measurements, as illustrated in Figure 17.4b, indicate the hydrogen evolution side reaction during CO_2RR could be well suppressed. According to the chronopotentiometry tests, the FE_{CO}

of the TPE-CoPor-COF catalyst exceeds 90% in the potential range of −0.6 to −1.0 V (vs. RHE), and the maximum value of FE_{CO} is 95% at the potential of −0.7 V (vs. RHE). TPTPE-CoPor-COF materials show ultrahigh selectivity in the potential range of −0.7 to −1.0 V (vs. RHE), and FE_{CO} reaches the maximum value of 96% at −0.8 V (vs. RHE). Both materials show excellent selectivity in the CO_2RR process (Figure 17.4c). However, the partial current density diagram j_{CO} of TPE-CoPor-COF material is as high as $-30.4\,mA\,cm^{-2}$, which is much higher than that of TPTPE-CoPor-COF material and also exceeds most of the reported COF-based electrocatalysts. Accordingly, the TOF value of TPE-CoPor-COF catalyst also exceeds that of TPTPE-CoPor-COF, further indicating the superior activity of TPE-CoPor-COF over other materials (Figure 17.4d,e). In addition, the Tafel slope of TPE-CoPor-COF catalyst is also lower than that of TPTPE-CoPor-COF material, implying the fast reaction kinetics of the former (Figure 17.4f). Long-term electrolysis experiment was carried out to verify the stability of the TPE-CoPor-COF catalyst at a potential of −0.7 V (vs. RHE). It can be seen that when the electrolysis time reaches 10 hours, the FE_{CO} of the catalyst can be maintained at more than 75% without obvious attenuation trend, but its current density decreases by about 20%, indicating that the TPE-CoPor-COF material has a medium level of stability (Figure 17.4g).

17.3.2 Phthalocyanine Covalent Organic Framework

COF materials are constructed by covalent bonds and hence exhibit more attractive stability than MOF materials. In particular, for phthalocyanine COF, its extensive planar π conjugation in the structure endows it with high electrical conductivity, which is suitable for the field of electrocatalysis. Additionally, the fully exposed active sites in the COF signify excellent catalytic activity. Therefore, in recent years, new metal-Pc-based COF materials have been widely designed and synthesized for CO_2RR applications.

Han and co-workers designed and prepared a new type of phthalocyanine derivative catalyst by a two-step synthesis strategy. After pyrolysis of the well-dispersed polymeric nickel phthalocyanine species, both $Ni\text{–}N_4$ active sites and highly dispersed Ni nanoclusters were formed, which endowed it with high conductivity during CO_2RR process and hence the enhanced catalytic activity [51]. AC impedance tests under the open circuit voltage showed that the NiPPc-CB-p800 catalyst has the smallest semicircle diameter, indicating that it has the smallest interfacial transfer resistance in the reaction process, namely a faster charge transfer rate. Obviously, it can also be found that the charge transfer rate in CO_2RR varied with the increase in pyrolysis temperature. Under the applied potential of −0.6 V (vs. RHE), the NiPPc-CB-p800 catalyst exhibits the maximum current density, which is much higher than that of the other three materials. In addition, at this potential, all four materials show ultrahigh selectivity for CO_2 reduction to CO (FE_{CO} exceeding 97%) with NiPPc-CB-p800 material delivering an excellent CO selectivity (FE_{CO} close to 100%) under a wider potential window (−0.55 to −0.7 V (vs. RHE)). Subsequently, the intrinsic catalytic activity of various catalysts was

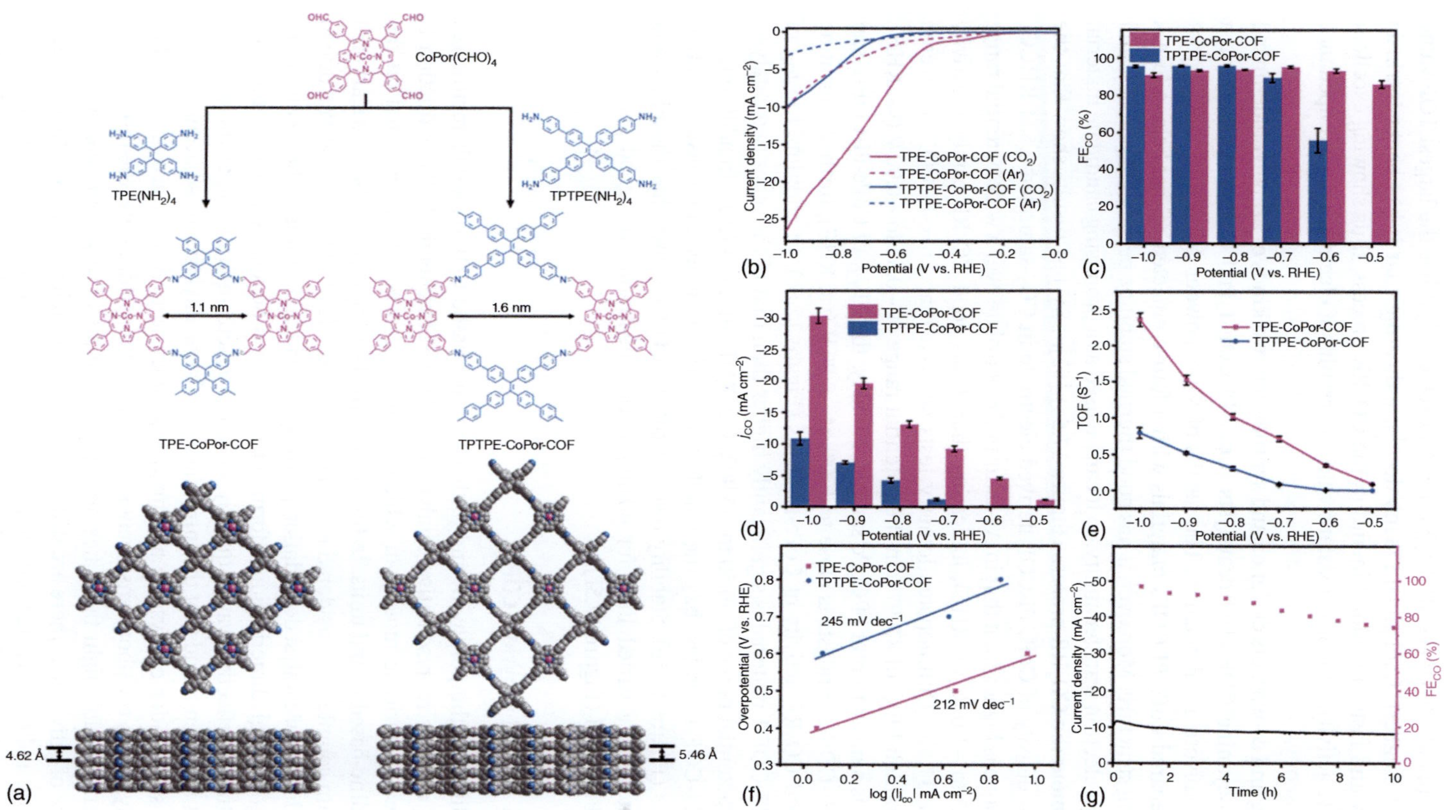

Figure 17.4 (a) The synthesis diagram, (b) LSV curves in 0.5 M $KHCO_3$ electrolyte under CO_2 saturation and Ar saturation conditions, (c) FE_{CO}, (d) j_{CO}, and (e) Tafel slope of TPE-CoPor-COF and TPTPE-CoPor-COF catalysts. (f) The long-term stability of TPE-CoPor-COF catalyst was tested at −0.7 V (vs. RHE) overpotential. Reproduced with permission of Jiang et al. [50], The Royal Society of Chemistry.

evaluated by calculating the turnover frequency of CO conversion. It is worth noting that at −0.6 V (vs. RHE), NiPPc-CB-p800 catalyst has the largest TOF value and the smallest Tafel slope (73 mV dec^{-1}), indicating that it has better intrinsic activity and faster reaction kinetics in the CO_2RR process, thus showing excellent catalytic activity. This is consistent with the results of electrochemical impedance spectroscopy.

Jiang and co-workers designed and synthesized two kinds of 2D polyimide-linked phthalocyanine COF electrocatalysts (CoPc-PI-COF-1 and CoPc-PI-COF-2) via a simple solvothermal reaction. The results of X-ray powder diffraction showed that the prepared CoPc-PI-COFs materials with a four-connected sql net belong to AA stacking structure. Moreover, it has good thermal stability, permanent porosity, and efficient electron transfer ability, so it can be used as a promising candidate material to promote the CO_2RR process (Figure 17.5a) [52]. LSV test results show that the current density of CoPc-PI-COF-derived electrode in CO_2-saturated 0.5 M $KHCO_3$ electrolyte is higher than that in Ar-saturated electrolyte over a wide potential range of −0.2 to −1.0 V (vs. RHE), indicating that it has good CO_2RR catalytic activity (Figure 17.5b). Chronopotentiometry tests showed that CoPc-PI-COFs delivered an ultrahigh FE_{CO} of 87–97% in the potential range of −0.60 to −0.90 V (vs. RHE), which is similar to that of CoPc sample. Notably, the FE_{CO} of CoPc-PI-COF-1 and CoPc-PI-COF-2 catalysts is close to 100% at −0.80 V (vs. RHE), further indicating the good CO_2RR activity of CoPc-PI-COFs (Figure 17.5c). In addition, the j_{CO} of CoPc-PI-COF-1 material is significantly higher than that of the other two catalysts in the potential range of −0.50 to −0.90 V (vs. RHE, Figure 17.5d). In addition, the CoPc-PI-COF-1 catalyst has the smallest Tafel slope, indicating a faster reaction kinetics (Figure 17.5e). Stability test of CoPc-PI-COF-1 shows that the FE_{CO} and current density varied little for 40 hours test, illustrating the good stability of CoPc-PI-COF-1 (Figure 17.5f).

17.3.3 Metal Bipyridyl COFs

In addition to the porphyrin and phthalocyanine-based COF materials mentioned above, another representative catalytic unit for electrochemical CO_2 reduction is the metal-bipyridine moiety, in which the metal can be firmly anchored on the bipyridine-based ligand units. In this regard, the [Re(bpy)$(CO)_3$Cl]-type tricarbonyl rhenium complex is considered to be a typical example of CO_2RR. However, due to the influence of catalyst degradation pathway, individual Re units in these complexes often go through dimerization to form inactive dimers. So researchers are committed to finding an effective way to quickly solve the deactivation in homogeneous solution. It has been found that the introduction of large alkyl groups adjacent to metal active sites under catalysis is an effective strategy to prevent deactivation. Additionally, due to the photosensitiveness of the metal-bipyridine site, the introduction of external visible light fields has proven efficient for CO_2RR. But given the short lifetime of active sites, there are still some difficulties in implementing this strategy.

Huang and co-workers used the post-synthesis method to integrate $Ru(bpy)_3Cl_2$, which possesses long lifetime and good excitation donor characteristics, with cobalt

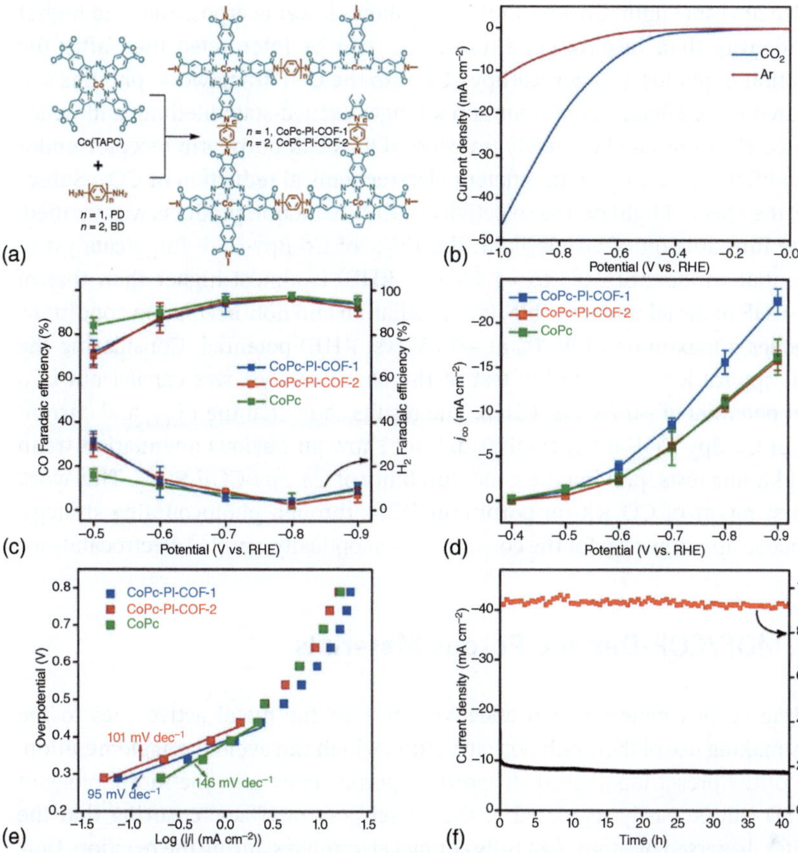

Figure 17.5 (a) The schematic diagram of synthesis of CoPc-PI-COF-1 and CoPc-PI-COF-2 catalysts. (b) LSV curves in 0.5 M $KHCO_3$ electrolyte under CO_2 saturation and Ar saturation conditions, respectively, (c) FE_{CO}, (d) j_{CO}, and (e) Tafel slope of different types of electrocatalysts. (f) The long-term stability of CoPc-PI-COF-1 catalyst was tested at −0.7 V (vs. RHE) potential. Reproduced with permission of Jing et al. [52], American Chemical Society.

porphyrin-based COF functionalized by 2,2′-bipyridine. The CO_2RR experiments of Co-Bpy-COF-Ru_x (x represents the molar ratio of Ru to Co species, $x = 1/2$ or $1/3$) electrocatalysts were carried out under optoelectronic conditions. Compared with the unmodified Co-Bpy-COF materials, the catalysts designed by the post-synthesis strategy have higher crystallinity and can form a huge internal electric field. This could accelerate the transfer of photoinduced excited electrons from the donor to the cobalt porphyrin receptor and greatly increase the excited-state lifetime of the catalytic active sites [53]. A series of electrochemical tests revealed a significant improvement in Co-Bpy-COF-Ru_x catalyst for CO_2RR. The LSV results demonstrate that the current density of Co-Bpy-COF-$Ru_{1/2}$ and Co-Bpy-COF catalysts in CO_2-saturated 0.5 M $KHCO_3$ electrolyte is higher than that in N_2-saturated electrolyte, indicating the good suppression of HER. More importantly, with the assistance of visible light, Co-Bpy-COF-$Ru_{1/2}$ shows lower overpotential and higher current density than in dark conditions. This can be interpreted that, after the introduction of photosensitizer $Ru(bpy)_3Cl_2$ into the COF framework, photons can be captured more effectively to maintain a longer excited-state lifetime, and hence the excited electrons can be quickly transferred to cobalt porphyrin receptor under light conditions, resulting in an efficient electrochemical reduction of CO_2. Subsequently, the effect of light on the selectivity of CO_2 reduction products was verified. After the introduction of visible light, the FE_{CO} of Co-Bpy-COF-$Ru_{1/2}$ catalyst in the potential window of −0.5 to −1.1 V (vs. RHE) is almost higher than that of Co-Bpy-COF material under visible light irradiation and non-irradiation conditions and reaches a maximum of 96.7% at −0.7 V (vs. RHE) potential. Considering the practical application, the stability test of the above catalyst was carried out at a constant potential of −0.7 V (vs. RHE). The results show that the FE_{CO} and current density of Co-Bpy-COF-$Ru_{1/2}$ catalyst did not show an obvious attenuation trend during 13 hours tests, proving the good durability of Co-Bpy-COF-$Ru_{1/2}$. This work is the first report of CO_2RR on porphyrin COFs through photocoupling strategy, which opens up a new way for the construction of optically coupled electrocatalysts.

17.4 MOF/COF-Derived Porous Materials

Crystalline porous materials can effectively restrict the metal active sites to the pores by making use of their rich pore structure, which can avoid the agglomeration. In view of the preset topological structure of porous materials, the morphology of 1D/2D/3D can be easily regulated at the preset pore interface, ensuring that the active sites dispersed by atoms can fully contact electrolytes during the reaction, thus effectively promoting the activity and selectivity of electrocatalytic CO_2 reduction. However, the original crystalline porous materials still face the problems of poor thermal stability/chemical stability, and low electrical conductivity, which limits their further application. MOF/COF-derived materials can effectively solve the above problems via their greatly improved thermal/electrochemical stability while maintaining the advantages of the original crystalline porous structure. Besides, the derived porous materials have a wider choice of metal elements as active centers in the CO_2RR process, which could broaden the scope of the metal species with good activity. Moreover, the mass transfer can be optimized to a certain extent by effectively utilizing the pores and cavity microstructure of crystalline porous-derived materials, and therefore improves the catalytic activity of CO_2 reduction.

17.4.1 Copper-Based MOF/COF-Derived Porous Materials

In order to make up for the shortcomings of the original MOF/COF materials, the composite strategy is a simple and effective method. It can be realized by direct mixing with conductive carriers, in situ synthesis of MOF/COF catalysts on self-supporting electrodes, and electrochemical direct reduction of porous materials. In the process of electrocatalytic reduction of CO_2 to C_1 products, because of the variety of C_1 products and the similar reduction potentials of C_1 products, it is difficult to maintain the ultrahigh selectivity of a single product. However, if a single type of active site was introduced on the basis of porous conductive materials to enhance the stability of key reaction intermediates, it would be applicable to achieve a high selectivity of single reduction products of CO_2RR. According to the above consideration, Cao and co-workers chose porous and conductive copper-based MOF (CuHHTP) as the substrate. Which part of the copper ions in the framework were reduced to a single-type and well-dispersed Cu_2O (average particle size 3.5 nm) by a simple electrochemical reduction technique? The reason for choosing CuHHTP as the growth substrate is that it possesses large number of periodic Cu—O_4 sites in its interior, which is an ideal platform for the reduction of Cu^{2+} to Cu^{1+} in the electrochemical reduction process (Figure 17.6a) [54]. According to the LSV curves, a reduction peak at −0.93 V (vs. RHE) potential is observed in the LSV curves of CuHHTP catalyst, indicating the transition process of Cu^{2+} to Cu^{1+}. In addition, CuHHTP and Cu_2O@CuHHTP materials show lower initial potential and higher current density in CO_2-saturated electrolyte than in Ar-saturated electrolyte, implying both catalysts have CO_2RR activity. Besides, the current density of Cu_2O@CuHHTP sample is higher than CuHHTP, indicating Cu_2O quantum dots are the real active sites during CO_2RR process (Figure 17.6b). Subsequently, the selectivity of CO_2RR reduction products was further determined. The reduction products are all gas phase (the sum of FE efficiency is almost 100%). No liquid phase products are detected. It is obvious that Cu_2O@CuHHTP catalyst has high selectivity for CH_4 reduction products in the wide potential range of −1.2 to −1.5 V (vs. RHE), and the FE is as high as 73% at −1.4 V (vs. RHE) applied potential. In addition, at this potential, the local current density of CH_4 on CuHHTP catalyst reaches $-10.8\ mA\ cm^{-2}$, which is better than that of most copper-based electrocatalysts reported at present. The higher current density can be attributed to the easy transfer of electrons from the conductive substrate to the Cu_2O active sites during the reaction, thus accelerating the CO_2 reduction reaction (Figure 17.6c,d). In order to fully demonstrate the excellent selectivity of Cu_2O@CuHHTP catalyst for CH_4 products, Cu_2O@CCB (conductive carbon black) and commercial Cu_2O were tested under the same conditions. The average particle size of Cu_2O quantum dots in both samples is 5 nm. It can be seen that the CH_4 Faraday efficiency of Cu_2O@CuHHTP catalyst is much higher than that of the other two materials at −1.4 V (vs. RHE). The results show that, the smaller particle size of Cu_2O has more active sites, and the highly exposed crystal plane plays a key role in improving the selectivity of CH_4 products (Figure 17.6e). It could be concluded that, after the copper ion active centers are reduced to Cu_2O species, many hydroxyl groups near the active sites are released. These free groups could form hydrogen bonds with the reaction intermediates, which could effectively reduce the energy barrier of forming CH_4 product intermediates and finally improve the selectivity of CO_2 electrochemical reduction to a single product.

The porous MOF/COF derivative can not only maintain the porosity of the original crystalline porous materials but also possess some new properties, such as excellent electrical conductivity and good thermochemical stability. According to the existing forms of active sites of Cu element, the catalysts synthesized by this strategy are

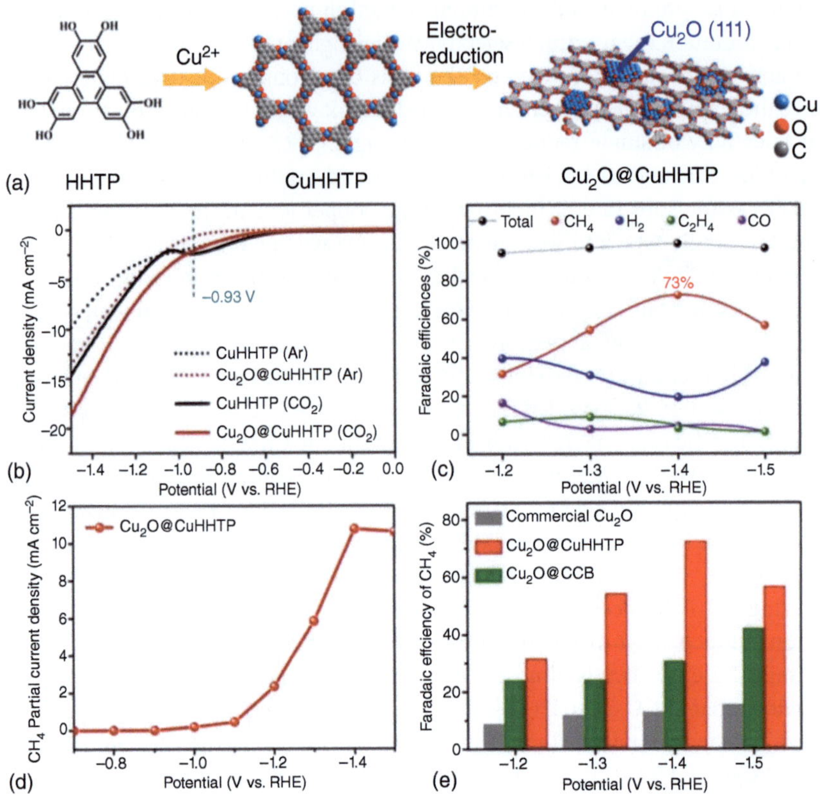

Figure 17.6 (a) The diagram of synthesis of Cu_2O@CuHHTP catalyst. (b) LSV curves of CuHHTP and Cu_2O@CuHHTP catalysts in 0.1 M KCl/0.1 M $KHCO_3$ mixed electrolyte saturated with CO_2 and Ar, respectively. (c) CH_4 Faraday efficiency and (d) CH_4 partial current density of Cu_2O@CuHHTP material in the process of CO_2RR. Reproduced with permission of Cao et al. [54], Wiley-VCH.

mainly divided into Cu monatomic catalysts and Cu-based compound catalysts. It is worth noting that in view of the unique electronic structure, adjustable local coordination environment, and quantum size effect of SACs, these kinds of catalysts have become the most rapidly developing candidate materials in recent years.

He and co-workers reported a large-scale preparation of porous carbon nanofiber electrocatalysts (CuSAs/TCNFs) with rich and uniformly distributed monatomic active sites of Cu by a simple method for the first time and applied it to the efficient electrochemical reduction of CO_2 [55]. In view of the fact that copper and its compounds have been widely proven to be efficient materials for the reduction of CO_2 to methanol, CuSAs/TCNFs catalyst show excellent selectivity for methanol reduction products in the CO_2RR process. The performance of CuSAs/TCNFs catalyst for CO_2RR was first tested by LSV measurement. In CO_2-saturated 0.1 M $KHCO_3$ electrolyte, CuSAs/TCNFs show lower initial potential and higher current density than the two contrast samples of CuSAs/CNFs and TCNFs, indicating that CuSAs/TCNFs catalyst has good catalytic activity in CO_2RR reaction. Because of the good mechanical properties of CuSAs/TCNFs catalyst, it can be directly used as a working electrode for electrochemical tests without any other treatment. Through the detection of the reduction products, it can be observed that there are only CO and H_2 in the reduction products of TCNF catalyst, and there are no liquid phase products. However, the other two kinds of Cu monatomic materials not only have the same two kinds of gas phase products but also detect the formation of liquid phase product methanol, indicating that Cu single atoms are the effective active sites for the formation of methanol. When the applied potential varies from −0.4 to −1.2 V (vs. RHE), the CH_4 Faraday efficiency of both catalysts increases at first and then decreases. At the potential of −0.9 V (vs. RHE), CH_4 Faraday efficiency reaches the maximum, but the selectivity of CuSAs/TCNFs (44%) is slightly higher than that of CuSAs/CNFs (41%). More interestingly, the CuSAs/TCNFs catalyst almost obtained relatively pure reduction products CH_4 (FE = 44%) and CO (FE = 56%) at this potential. In addition, the local current density of C_1 product of CuSAs/TCNFs catalyst is much higher than that of CuSAs/CNFs material, proving that the unique through-pore structure of CuSAs/TCNFs material plays a key role in improving the performance of CO_2RR. It can be clearly seen that the current density and the FE of C_1 reduction product are relatively stable in the electrolysis test for as long as 50 hours, and there is no obvious attenuation trend, further implying good stability and long-term application potential. This work provides a new idea for large-scale design and synthesis of highly efficient and stable monatomic catalysts for the efficient electrochemical reduction of CO_2.

17.4.2 Nickel-Based MOF/COF-Derived Porous Material

As a kind of electrocatalyst without noble metal, nickel-based MOF/COF-derived SAC materials are featured with a single Ni atom uniformly anchored in the pores of porous materials by coordinating with N atoms on the carbon carrier, which effectively avoids the formation of large-size particles. Considering the advantages of the original crystalline porous materials, researchers have made an in-depth study on the derived monatomic catalyst materials and adopted a new synthesis strategy to prepare electrocatalysts with a large number of exposed Ni active sites by adjusting the coordination environment and electronic structure of Ni monomers.

Jiang and co-workers proposed a new synthesis method, that is, using the strategy of post-synthetic metal substitution. The controllable synthesis (Ni-N_x-C) of Ni monoatomic catalysts with different N coordination numbers was realized on the nitrogen-doped carbon skeleton derived from MOF materials (Figure 17.7a) [56]. Electrochemical experiments show that the Ni-N_3-C catalyst delivered the largest current response compared with Ni-N_4-C and N-C, indicating that the lower N coordination number is more beneficial to improving the reaction activity of CO_2RR (Figure 17.7b). Then, the selectivity of reduction products of various catalysts was analyzed. The results show that the FE_{CO} of the three catalysts shows a volcanic trend in the whole potential range, but the FE_{CO} of Ni-N_3-C material is better than that of other two samples in the potential range of −0.5 to −1.0 V (vs. RHE), and the FE_{CO} of Ni-N_3-C catalyst reaches 95.6% at −0.65 V (vs. RHE), which is much

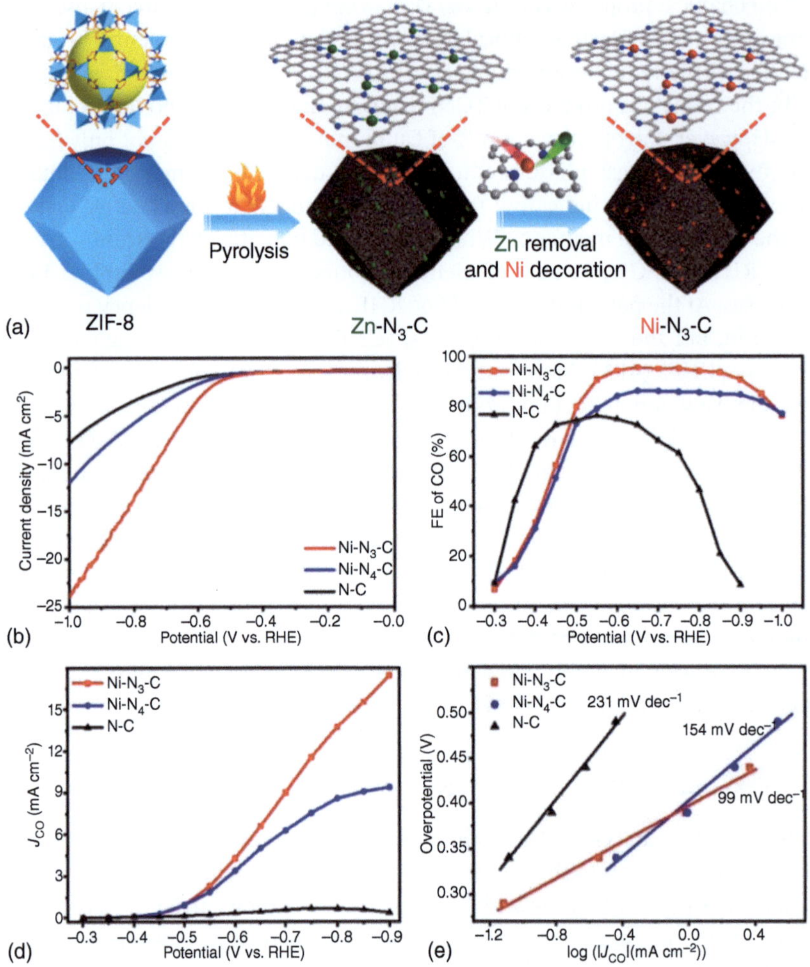

Figure 17.7 (a) The synthesis diagram of Ni-N_x-C catalyst. (b) LSV curves in CO_2-saturated 0.5 M $KHCO_3$ electrolyte, (c) FE_{CO}, (d) j_{CO}, and (e) Tafel slope of Ni-N_3-C, Ni-N_4-C, and N-C catalysts. Reproduced with permission of Jiang et al. [56], Wiley-VCH.

higher than that of the other two materials. Besides, the j_{CO} of Ni-N_3-C delivered the largest current density in the whole potential window, further confirming the better catalytic performance of Ni-N_3-C catalyst (Figure 17.7c,d). In addition, the Ni-N_3-C catalyst has the lowest Tafel slope (99 mV dec^{-1}), which highlights its fastest reaction kinetics in the catalytic reaction, thus showing excellent CO_2 reduction performance and highlighting the importance of the regulation of nickel monoatomic coordination microenvironment in electrocatalysis (Figure 17.7e).

Because of the periodic arrangement and easy adjustment of porous structure, COF materials are considered to be ideal materials for accurately anchoring metal active sites. Strasser and co-workers prepared a series of Ni monatomic electrocatalysts (C-TpDt-Ni) derived from COF materials with different Ni-N_x contents using nitrogen-rich triazolyl COF (TpDt-COF and TpDt-COF-Ni) as precursors. The Ni-N_x content can be adjusted according to the pyrolysis temperature (Figure 17.8a) [57]. Electrochemical tests of a series of prepared catalysts were carried out in a typical H-type electrolytic cell filled with CO_2-saturated 0.5 M $KHCO_3$ electrolyte. Experiments showed that the activity of unpyrolyzed TpDt-Ni material in the CO_2RR process is negligible. However, after pyrolysis treatment, the transformation process from CO_2 to CO occurred. The test results show that CO is dominant C_1 reduction product and H_2 is the main side reaction product, and no liquid phase product is detected. The FE_{CO} of C-TpDt-Ni-900 catalyst is higher than that of other contrast samples in the potential range of −0.4 to −0.9 V (vs. RHE). When the potential reaches −0.7 V (vs. RHE), the FE_{CO} is more than 90%, demonstrating good CO selectivity. In addition, C-TpDt-Ni-900 material exhibits j_{CO} of about 22 mA cm^{-2} at this potential, which is also better than other kinds of catalysts. Interestingly, the sample treated at 800 °C shows inferior catalytic performance than C-TpDt-Ni-1000 material, but the catalytic performance of these materials is better than that of the catalysts synthesized with precursors with lower N content (C-TpPa-Ni-900) (Figure 17.8b–e).

17.5 Summary and Prospect

Electrocatalytic CO_2 reduction has always been considered as one of the most effective ways to achieve the strategic goal of "carbon peak, carbon neutralization." However, there are still some inevitable problems in practical application, such as poor selectivity, low current density, and high overpotential due to the complexity of CO_2 reduction path and the diversity of reduction products. Here, we introduce in detail the latest research on functional porous framework materials as a new type of catalyst for CO_2RR, including crystalline porous materials (MOFs/COFs) and their derived porous composites (Cu-based and Ni-based porous composites). As we all know, this kind of material has become the focus of researchers in recent years because of the following advantages: (i) the original MOF/COF materials with adjustable framework structure can provide abundant active trapping sites of the metal active center with unique pore structure. The formation of uniform atomic dispersion active sites can effectively avoid the excessive agglomeration of metal atoms and can not only improve the atomic utilization but also significantly enhance the activity of CO_2RR; (ii) the composite strategy can effectively enhance the ability of chemical interaction and electron transfer and make up for the deficiency of the original crystalline porous materials. Typical composite strategies include direct in situ synthesis of MOF/COF electrocatalysts on conductive substrates and in situ reduction of MOF/COF materials by electrochemical methods. These could significantly enhance the performance of the catalysts during CO_2RR; (iii) the graphitization of the material by high-temperature pyrolysis can effectively improve the conductivity and stability of the original material.

With the rapid progress of modern science and technology, people have adopted a variety of innovative technical means to comprehensively consider the inherent

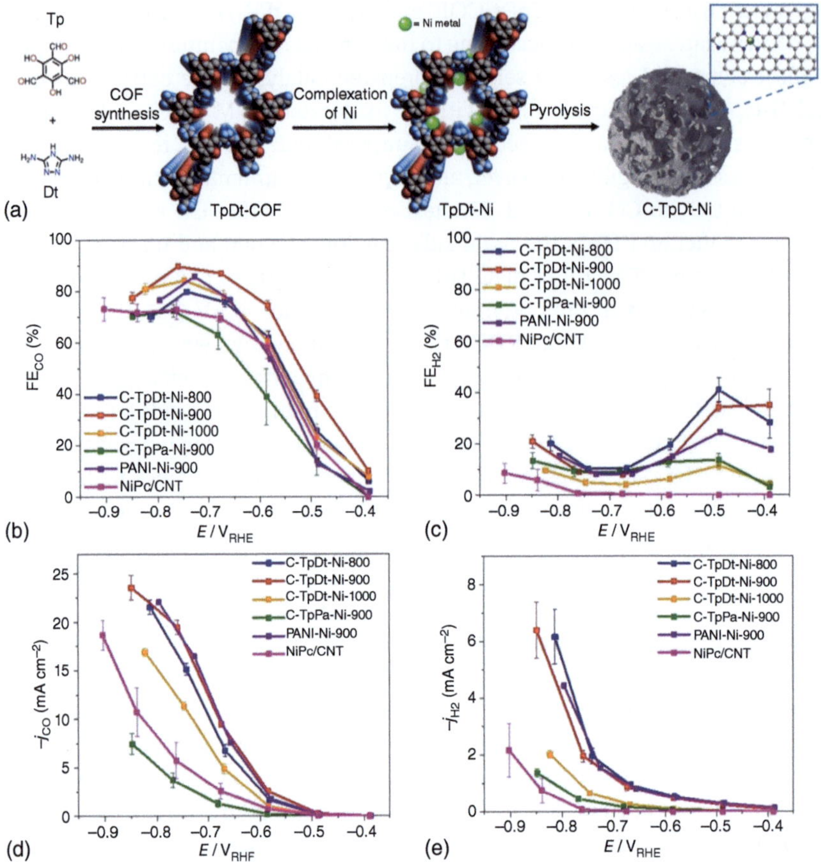

Figure 17.8 (a) The synthesis diagram of C-TpDt-Ni catalyst. (b) CO Faraday efficiency, (c) H_2 Faraday efficiency, (d) CO partial current density, and (e) H_2 partial current density in the wide potential range from −0.4 to −0.9 V (vs. RHE) of different catalysts in 0.5 M $KHCO_3$ electrolyte saturated with CO_2. Strasser et al. [57]/Wiley-VCH/CC BY 4.0.

structure–activity relationship of the excellent performance of the catalyst in the CO_2RR process to clarify the catalyst mechanism for the electrochemical reduction of CO_2 in more detail. By selecting appropriate metal elements and adjusting the size, coordination environment, and microstructure of metal active centers, serious agglomeration can be effectively avoided, thus maximizing the utilization of metal atoms. These are of great significance to improve the activity, selectivity, and stability of porous material electrocatalysts. In a word, although the research on electrocatalytic CO_2 reduction is still in its infancy, we still need to make more efforts to solve some problems. We believe that crystalline porous materials and their derived composites, as typical candidate materials for electrocatalyst synthesis, have broad application prospects because of their unique advantages. It will help us achieve the long-term goal of decarbonization.

Acknowledgments

This work was supported by the National Natural Science Foundation of China (Nos. 22473083, 21905195), the Natural Science Foundation of Tianjin City (No. 20JCYBJC00800), and the PEIYANG Young Scholars Program of Tianjin University (No. 2020XRX-0023).

References

1 Yang, H., Han, X., Douka, A.I. et al. (2021). Advanced oxygen electrocatalysis in energy conversion and storage. *Adv. Funct. Mater.* 31 (12): 2007602.

2 She, X., Wang, Y., Xu, H. et al. (2022). Challenges and opportunities in electrocatalytic CO_2 reduction to chemicals and fuels. *Angew. Chem. Int. Ed.* 61 (49): e202211396.

3 De Luna, P., Hahn, C., Higgins, D. et al. (2019). What would it take for renewably powered electrosynthesis to displace petrochemical processes? *Science* 364 (6438): eaav3506.

4 Lai, W., Qiao, Y., Zhang, J. et al. (2022). Design strategies for markedly enhancing energy efficiency in the electrocatalytic CO_2 reduction reaction. *Energy Environ. Sci.* 15 (9): 3603–3629.

5 Wang, Y., Zhou, J., Lin, C. et al. (2022). Decreasing the overpotential of aprotic Li-CO_2 batteries with the in-plane alloy structure in ultrathin 2D Ru-based nanosheets. *Adv. Funct. Mater.* 32 (30): 2202737.

6 Jeong, K., Hong, T., and Kim, J. (2018). Development of a CO_2 emission benchmark for achieving the national CO_2 emission reduction target by 2030. *Energy Build.* 158: 86–94.

7 Ito, K. (2017). CO_2 emissions, renewable and non-renewable energy consumption, and economic growth: evidence from panel data for developing countries. *International Economics.* 151: 1–6.

8 Liu, S., Yang, H.B., Hung, S.-F. et al. (2020). Elucidating the electrocatalytic CO_2 reduction reaction over a model single-atom nickel catalyst. *Angew. Chem. Int. Ed.* 59 (2): 798–803.

9 Gabrielli, P., Gazzani, M., and Mazzotti, M. (2020). The role of carbon capture and utilization, carbon capture and storage, and biomass to enable a net-zero-CO_2 emissions chemical industry. *Ind. Eng. Chem. Res.* 59 (15): 7033–7045.

10 Garba, M.D., Usman, M., Khan, S. et al. (2021). CO_2 towards fuels: a review of catalytic conversion of carbon dioxide to hydrocarbons. *J. Environ. Chem. Eng.* 9 (2): 104756.

11 Zhang, Z., Liu, P., Song, Y. et al. (2022). Heterostructure engineering of 2D superlattice materials for electrocatalysis. *Adv. Sci.* 9 (35): 2204297.

12 Burkart, M.D., Hazari, N., Tway, C.L., and Zeitler, E.L. (2019). Opportunities and challenges for catalysis in carbon dioxide utilization. *ACS Catal.* 9 (9): 7937–7956.

13 Minx, J.C., Callaghan, M., Lamb, W.F. et al. (2017). Learning about climate change solutions in the IPCC and beyond. *Environ. Sci. Pol.* 77: 252–259.

14 Devès, M.H., Lang, M., Bourrelier, P.-H., and Valérian, F. (2017). Why the IPCC should evolve in response to the UNFCCC bottom-up strategy adopted in Paris? An opinion from the French Association for Disaster Risk Reduction. *Environ. Sci. Pol.* 78: 142–148.

15 Hammond, G.P., Akwe, S.O., and Williams, S. (2011). Techno-economic appraisal of fossil-fuelled power generation systems with carbon dioxide capture and storage. *Energy* 36 (2): 975–984.

16 Nair, M.M. and Abanades, S. (2018). Experimental screening of perovskite oxides as efficient redox materials for solar thermochemical CO_2 conversion. *Sustainable Energy Fuels* 2 (4): 843–854.

17 Zhao, T.-T., Feng, G.-H., Chen, W. et al. (2019). Artificial bioconversion of carbon dioxide. *Chin. J. Catal.* 40 (10): 1421–1437.

18 Habisreutinger, S.N., Schmidt-Mende, L., and Stolarczyk, J.K. (2013). Photocatalytic reduction of CO_2 on TiO_2 and other semiconductors. *Angew. Chem. Int. Ed.* 52 (29): 7372–7408.

19 Jhong, H.-R.M., Ma, S., and Kenis, P.J.A. (2013). Electrochemical conversion of CO_2 to useful chemicals: current status, remaining challenges, and future opportunities. *Curr. Opin. Chem. Eng.* 2 (2): 191–199.

20 Zhao, J., Wang, X., Xu, Z., and Loo, J.S. (2014). Hybrid catalysts for photoelectrochemical reduction of carbon dioxide: a prospective review on semiconductor/metal complex co-catalyst systems. *J. Mater. Chem. A* 2 (37): 15228–15233.

21 Luo, Y.-H., Dong, L.-Z., Liu, J. et al. (2019). From molecular metal complex to metal-organic framework: the CO_2 reduction photocatalysts with clear and tunable structure. *Coord. Chem. Rev.* 390: 86–126.

22 Han, N., Ding, P., He, L. et al. (2020). Promises of main group metal-based nanostructured materials for electrochemical CO_2 reduction to formate. *Adv. Energy Mater.* 10 (11): 1902338.

23 Tan, X., Yu, C., Ren, Y. et al. (2021). Recent advances in innovative strategies for the CO_2 electroreduction reaction. *Energy Environ. Sci.* 14 (2): 765–780.

24 Lei, Q., Zhu, H., Song, K. et al. (2020). Investigating the origin of enhanced C_{2+} selectivity in oxide–/hydroxide-derived copper electrodes during CO_2 electroreduction. *J. Am. Chem. Soc.* 142 (9): 4213–4222.

25 Wagner, A., Sahm, C.D., and Reisner, E. (2020). Towards molecular understanding of local chemical environment effects in electro- and photocatalytic CO_2 reduction. *Nat. Catal.* 3 (10): 775–786.

26 Zou, L., Wei, Y.-S., Hou, C.-C. et al. (2021). Single-atom catalysts derived from metal-organic frameworks for electrochemical applications. *Small* 17 (16): 2004809.

27 Rakowski Dubois, M. and Dubois, D.L. (2009). Development of molecular electrocatalysts for CO_2 reduction and H_2 production/oxidation. *Acc. Chem. Res.* 42 (12): 1974–1982.

28 Ma, W., He, X., Wang, W. et al. (2021). Electrocatalytic reduction of CO_2 and CO to multi-carbon compounds over cu-based catalysts. *Chem. Soc. Rev.* 50 (23): 12897–12914.

29 Zhang, Z., Bian, L., Tian, H. et al. (2022). Tailoring the surface and interface structures of copper-based catalysts for electrochemical reduction of CO_2 to ethylene and ethanol. *Small* 18 (18): 2107450.

30 Liang, Z., Wang, H.-Y., Zheng, H. et al. (2021). Porphyrin-based frameworks for oxygen electrocatalysis and catalytic reduction of carbon dioxide. *Chem. Soc. Rev.* 50 (4): 2540–2581.

31 Freund, R., Canossa, S., Cohen, S.M. et al. (2021). 25 years of reticular chemistry. *Angew. Chem. Int. Ed.* 60 (45): 23946–23974.

32 Huang, X. and Zhang, Y.-B. (2021). Reticular materials for electrochemical reduction of CO_2. *Coord. Chem. Rev.* 427: 213564.

33 Zhang, Y., Xia, B., Ran, J. et al. (2020). Atomic-level reactive sites for semiconductor-based photocatalytic CO_2 reduction. *Adv. Energy Mater.* 10 (9): 1903879.

34 Wang, C., Lv, Z., Yang, W. et al. (2023). A rational design of functional porous frameworks for electrocatalytic CO_2 reduction reaction. *Chem. Soc. Rev.* 52 (4): 1382–1427.

35 He, Y.-C., Xiao, L.-Y., Yuan, Z.-H. et al. (2019). Two coordination polymers constructed by 5-[(4-carboxyphenoxy)methyl]benzene-1, 3-dicarboxylic acid and 2,2′-bipyridine: syntheses, structures and luminescence properties. *Acta Crystallogr., Sect. C Struct. Chem.* 75 (12): 1562–1568.

36 Wen, J., Li, Y., and Gao, J. (2020). Two-dimensional metal-organic frameworks and derivatives for electrocatalysis. *Chem. Res. Chin. Univ.* 36: 662–679.

37 Huang, J., Wu, J.-Q., Shao, B. et al. (2020). Ion-induced delamination of layered bulk metal-organic frameworks into ultrathin nanosheets for boosting the oxygen evolution reaction. *ACS Sustain. Chem. Eng.* 8 (28): 10554–10563.

38 Zengqiang, G., Congyong, W., Junjun, L. et al. (2021). Conductive metal-organic frameworks for electrocatalysis: achievements, challenges, and opportunities. *Acta Phys. -Chim. Sin.* 37 (7): 2010025.

39 Jin, H., Xiong, L., Zhang, X. et al. (2021). Cu-based catalyst derived from nitrogen-containing metal organic frameworks for electroreduction of CO_2. *Acta Phys. -Chim. Sin.* 37 (11): 2006017.

40 Yang, Y., Shi, Z., Chang, Y. et al. (2021). Surface molecularly imprinted magnetic MOFs: a novel platform coupled with magneto electrode for high throughput electrochemical sensing analysis of oxytetracycline in foods. *Food Chem.* 363: 130337.

41 Fu, J., Ben, T., and Qiu, S. (2020). Fabrication of polymer-supported metal organic framework membrane and its gas separation performance. *Acta Phys. -Chim. Sin.* 36: 1901079.

42 Yang, D., Zuo, S., Yang, H. et al. (2022). Tailoring layer number of 2D porphyrin-based MOFs towards photocoupled electroreduction of CO_2. *Adv. Mater.* 34 (7): 2107293.

43 Kornienko, N., Zhao, Y., Kley, C.S. et al. (2015). Metal-organic frameworks for electrocatalytic reduction of carbon dioxide. *J. Am. Chem. Soc.* 137 (44): 14129–14135.

44 Li, J., Shen, H., Ma, C. et al. (2023). Assembly and electrocatalytic CO_2 reduction of two-dimensional bimetallic porphyrin-based conjugated cobalt metal-organic framework. *Electrochim. Acta* 443: 141896.

45 Yang, J., Yu, J., Dong, W. et al. (2023). Enhanced the efficiency of electrocatalytic CO_2-to-CO conversion by Cd species anchored into metal-organic framework. *Small* 19 (35): 2301319.

46 Cho, J.H., Lee, C., Hong, S.H. et al. (2022). Transition metal ion doping on ZIF-8 enhances the electrochemical CO_2 reduction reaction. *Adv. Mater.* 35 (43): 2208224.

47 Hu, C., Bai, S., Gao, L. et al. (2019). Porosity-induced high selectivity for CO_2 electroreduction to CO on Fe-doped ZIF-derived carbon catalysts. *ACS Catal.* 9 (12): 11579–11588.

48 Ahmad, A., Iqbal, N., Noor, T. et al. (2021). Cu-doped zeolite imidazole framework (ZIF-8) for effective electrocatalytic CO_2 reduction. *J. CO_2 Util.* 48: 101523.

49 Dong, H., Lu, M., Wang, Y. et al. (2022). Covalently anchoring covalent organic framework on carbon nanotubes for highly efficient electrocatalytic CO_2 reduction. *Appl. Catal. B* 303: 120897.

50 Gong, L., Gao, Y., Wang, Y. et al. (2022). Efficient electrocatalytic carbon dioxide reduction with tetraphenylethylene- and porphyrin-based covalent organic frameworks. *Catal. Sci. Technol.* 12 (21): 6566–6571.

51 Chen, J., Li, J., Xu, J. et al. (2023). Phthalocyanine-derived catalysts decorated by metallic nanoclusters for enhanced CO_2 electroreduction. *Green Energy Environ.* 8 (2): 444–451.

52 Han, B., Ding, X., Yu, B. et al. (2021). Two-dimensional covalent organic frameworks with cobalt (II)-phthalocyanine sites for efficient electrocatalytic carbon dioxide reduction. *J. Am. Chem. Soc.* 143 (18): 7104–7113.

53 Wu, Q.-J., Si, D.-H., Ye, S. et al. (2023). Photocoupled electroreduction of CO_2 over photosensitizer-decorated covalent organic frameworks. *J. Am. Chem. Soc.* 145 (36): 19856–19865.

54 Yi, J.-D., Xie, R., Xie, Z.-L. et al. (2020). Highly selective CO_2 electroreduction to CH_4 by in situ generated Cu_2O single-type sites on a conductive MOF: stabilizing key intermediates with hydrogen bonding. *Angew. Chem. Int. Ed.* 59 (52): 23641–23648.

55 Yang, H., Wu, Y., Li, G. et al. (2019). Scalable production of efficient single-atom copper decorated carbon membranes for CO_2 electroreduction to methanol. *J. Am. Chem. Soc.* 141 (32): 12717–12723.

56 Zhang, Y., Jiao, L., Yang, W. et al. (2021). Rational fabrication of low-coordinate single-atom Ni electrocatalysts by MOFs for highly selective CO_2 reduction. *Angew. Chem. Int. Ed.* 60 (14): 7607–7611.

57 Li, C., Ju, W., Vijay, S. et al. (2022). Covalent organic framework (COF) derived Ni-N-C catalysts for electrochemical CO_2 reduction: unraveling fundamental kinetic and structural parameters of the active sites. *Angew. Chem. Int. Ed.* 61 (15): e202114707.

18

Cu-Based Compounds for Electrocatalytic CO_2 Reduction Reaction

Teng Wang and Yong Yang

Northwestern Polytechnical University, Center of Advanced Lubrication and Seal Materials, State Key Laboratory of Solidification Processing, School Of Materials Science and Engineering. 127 West Youyi Road, Beilin District, Xi'an, Shaanxi, 710072, P.R. China

18.1 Cu-Based Compounds for Electrocatalytic CO_2 Reduction

The electroreduction of carbon dioxide involves complex multiple electron transfer processes and is a typical multiphase catalytic reaction. The proton/ion transport in the catalytic reaction, as well as the adsorption and desorption characteristics of the catalyst toward the reaction intermediates, have a significant impact on the catalytic performance. Specifically, there are many factors closely related to catalyst performance, such as catalyst structure (morphology, pore structure, and interface structure), electrolyte type and composition, and reaction gas source [1–4]. In addition, the competitive hydrogen evolution reaction (HER), due to its comparable thermodynamic potential, also inhibits the activity of the CO_2RR. Therefore, developing CO_2RR electrocatalysts with high activity, selectivity, and stability is crucial for promoting the development of CO_2 electroreduction. Therefore, developing novel CO_2RR electrocatalysts with high activity, selectivity, and stability is crucial for promoting the development of CO_2 electroreduction.

Copper-based compound catalysts are considered one of the most promising material systems for CO_2 electroreduction to formate due to their low cost, simple synthesis, and good catalytic performance. According to existing research, copper-based compounds prepared by CO_2 electroreduction of formates are mainly concentrated in copper-based chalcogenides (Cu_xS) and copper-based oxides (CuO, Cu_2O) [5, 6]. Compared to those of metals, the surfaces of metal oxides usually exhibit more complex crystal structures and surface morphologies. The crystal faces exposed to Cu_2O not only determine its microstructure but also strongly affect its catalytic activity. Therefore, optimizing the performance of Cu_2O-based catalysts through surface morphology engineering, element doping, heterojunctions construction, catalyst loading, and coating protective layers is the key to achieving efficient conversion of CO_2 to high-value products.

Functional Materials for Electrocatalytic Energy Conversion, First Edition.
Edited by Zhicheng Zhang, Meiting Zhao, and Yuchen Qin.

This chapter mainly summarizes the recent research progress on copper-based compound catalysts in the field of electrocatalytic CO_2RR. By adjusting and controlling the preparation process, the geometric shape of copper-based compounds can be reasonably designed to expose different crystal planes, expose more active sites, improve CO_2 activation, and enhance the adsorption of key intermediates to improve their CO_2 reduction performance.

18.1.1 Copper(II) Oxide (CuO)

Among a wide variety of Cu-based catalysts for electrochemical CO_2 reduction reaction (CO_2RR), copper oxides have demonstrated superior activity for catalyzing the carbon dioxide to hydrocarbons and oxygenates with favorable selectivity. Considering the extremely high industrial value and limited sources of C_2H_4, the CO_2RR to ethylene (C_2H_4) with high current density and FE is intensively studied, and oxide-derived Cu catalyst presents the highest product ratio of C_2H_4 to CH_4 [7]. However, the self-evolution of surface species and morphologies resulting in the disappearance of Cu/$Cu^{\delta+}$ heterointerfaces can lead to the instability of C_2H_4 production. To address the issue, a facile and scalable anodic corrosion method was employed to synthesize oxygen-rich ultrathin CuO nanoplate arrays (CuO-NPs), and a dense vertical lamellate Cu nanostructure (DVL-Cu) was further generated via in situ electrochemical reduction of CuO-NPs [8]. In the electrochemical CO_2RR tests, the catalyst exhibits a high C_2H_4 FE of 84.5% and stable electrolysis for ~55 hours in a flow cell using a neutral KCl electrolyte. During the CO_2RR, the high oxidation degree of Cu foil with vertically arranged Cu nanoplates prevents the agglomeration of nanostructures and preserves stable Cu/Cu_2O interfaces. Density functional theory (DFT) calculations suggest that the formation of Cu/Cu_2O interfaces in DVL-Cu facilitates C_2H_4 formation due to their reduced C–C dimerization energy in the C_2H_4 formation pathway.

The initial morphology of the catalysts has significant effects on the eventual electrocatalytic performance. To readily observe morphology-dependent properties, Song and co-workers synthesized two different morphologies of copper oxide nanoparticles, branched CuO and cubic Cu_2O, to investigate the influence of surface morphology on product selectivity [9]. After activation, Cu^+ became the major copper species in these two nanoparticles. In comparison with cubic Cu_2O, branched CuO achieved a higher Faradaic efficiency (FE) of ethylene reaching more than 70% and a hydrogen FE of 30% without any byproducts. The superior performance was attributed mainly to the fact that the branched morphology could result in the production of highly active domains along with a high density of interfaces and junctions between them, but the cubic particles did not generate a fully disintegrated morphology. In addition to shape-controlled Cu oxide cubes, new freestanding two-dimensional (2D) Cu(II) oxide nanosheet catalysts featuring a highly preferred (001) facet orientation were reported for the CO_2RR under neutral pH conditions.

For the neutral or basic electrolyte used in CO_2RR, H_2O serves as the hydrogen source, and the catalytic activity should be related to the activation of H_2O in CO_2RR. The Sabatier principle indicates that the energy barrier for the activation

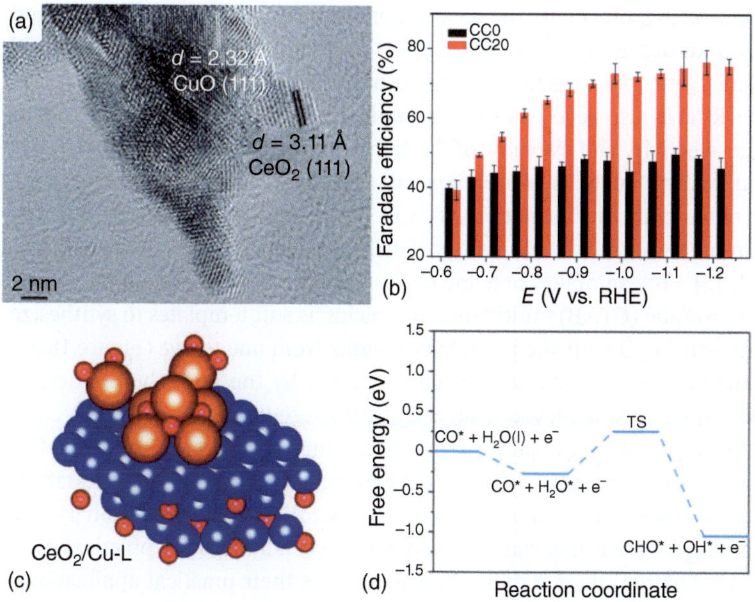

Figure 18.1 (a) The HR-TEM image of the CC20. (b) The average FEs of C_{2+} products at various potentials in 1 M KOH over CC0 and CC20, respectively. (c) The side views of CeO_2/Cu-L model, in which the blue balls, red balls and orange balls stand for Cu, oxygen, and carbon and hydrogen, respectively. (d) The reaction energy diagram for *CO hydrogenation to *COH on CeO_2/Cu-L. Source: Han, et al., [13]/Royal Society of Chemistry/CC BY 3.0 DEED.

of water should be controlled to provide enough hydrogen for the hydrogenation of intermediates and reduce the excessive production of H_2. CeO_2 has a high activity for water activation in CO_2 hydrogenation and shows poor HER activity [10–12]. Based on the earlier discussion, Han and co-workers combined CeO_2 with CuO to obtain CeO_2/CuO catalysts, which achieved an FE of 75.2% for the C_{2+} products with a total current density of 1.21 A cm^{-2} in a flow-cell system (Figure 18.1b) [13]. As shown in the high-resolution transmission electron microscopy (HRTEM) image (Figure 18.1a), the coexistence of CeO_2 and CuO in the catalysts was confirmed, and the CeO_2/CuO catalysts were named CCX (X = the molar ratio of Ce and Cu times 100). In addition, the DFT calculation results (Figure 18.1c,d) revealed the energy of generation of *CHO was thermodynamically reduced by the interfacial effect compared to CeO_2-modified CuO catalysts, and the rapid activation of water around CeO_2 accelerated the formation of *CHO kinetically; thus, the C–C coupling step was facilitated via the *CHO route.

18.1.2 Cuprous(I) Oxide (Cu_2O)

Cuprous(I) oxide (Cu_2O), which has controllable crystal planes and relatively high stability, has been recognized as a good platform for obtaining efficient CO_2RR electrocatalysts and understanding the catalytic mechanism of the CO_2RR. Surface and interface engineering, especially the creation of abundant Cu^0/Cu^+ interfaces and nanograin boundaries, is known to facilitate C^{2+} production during electrochemical CO_2 reduction over copper-based catalysts. At present, Cu_2O-based electrocatalysts for the CO2RR have also been extensively studied, but there are problems such as poor product selectivity and overflow of the cathodic competitive HER.

Significant efforts, such as morphological and facet engineering, have been devoted to enhancing the formation of *CO intermediates and promoting the CO–CO coupling step, with the ultimate goal of improving the selectivity toward C^{2+} products [14, 15]. For example, Chen and co-workers employed hexadecyltrimethylammonium bromide (CTAB) multilamellar vesicles as soft templates to synthesize hollow multishell Cu_2O with shell numbers ranging from one to five (Figure 18.2a) [16]. With the help of the confinement effect caused by the multishell structure, the 4.4-shell Cu catalyst achieved a high C_{2+} FE exceeding 80% at 900 mA cm^{-2} (Figure 18.2b), suppressing the performance of commercial Cu catalyst (<60% C_{2+} FE). During CO_2R experiments, increasing the number of shells in the Cu catalysts considerably enhanced the C_{2+}/C_1 ratio (Figure 18.2c). Although the activity and selectivity of Cu_2O-based materials have shown great potential for practical electrochemical production, their stability largely hinders their practical applications. To investigate the origin of instability and structural evolution of Cu_2O nanocube electrocatalysts, Luo and co-workers utilized Cu_2O nanocubes as a model catalyst to study the origin of instability and structural evolution of Cu_2O nanocubes as CO_2RR electrocatalysts; these properties were tracked by combining detailed electron microscopic analysis with theoretical calculations [17]. They found that the structural instability was attributed to the O loss induced by H attacking under the electrochemical environment, which was followed by a series of structural and morphological transitions and, finally, transformation to aggregates of smaller nanoparticles with mixed Cu and Cu oxide phases (Figure 18.2d). This study provides insights into rational design of long-term Cu_2O-based electrocatalysts for the CO_2RR.

In addition to morphological engineering, crystal facet regulation, especially the construction of high-index facets with significant amounts of atomic steps, edges, and unsaturated coordinated sites, provides additional opportunities for obtaining desired catalysts with high activity and selectivity. Li's group systematically studied the correlations between the high-index facets of Cu-based catalysts and their CO_2RR performance [18]. Several Cu_2O nanocrystals (NCs) enclosed by different crystal facets were fabricated, including cubic Cu_2O (denoted as c-Cu_2O) with {100} facets, rhombic dodecahedral Cu_2O (denoted as d-Cu_2O) with {110} facets, and concave octahedral Cu_2O (denoted as o-Cu_2O) with {511} high-index facets. Benefiting from the newly formed crystal defects/grain boundaries, o-Cu_2O, as a CO_2RR catalyst, exhibited an outstanding partial current density of 17.7 mA cm^{-2} and a maximum FE of 48.3% for C^{2+} products at −1.1 V vs. reversible hydrogen electrode (RHE) (Figure 18.2e), which are greater than those of c-Cu_2O (32.8%) and d-Cu_2O (31.6%).

The oxidation state of Cu is also an important factor affecting the adsorption and conversion of key reaction intermediates (*CO) during the CO_2RR [19, 20]. To study the valence state-activity relationship, Li and co-workers designed and synthesized three Cu-based electrocatalysts with different oxidation states by a

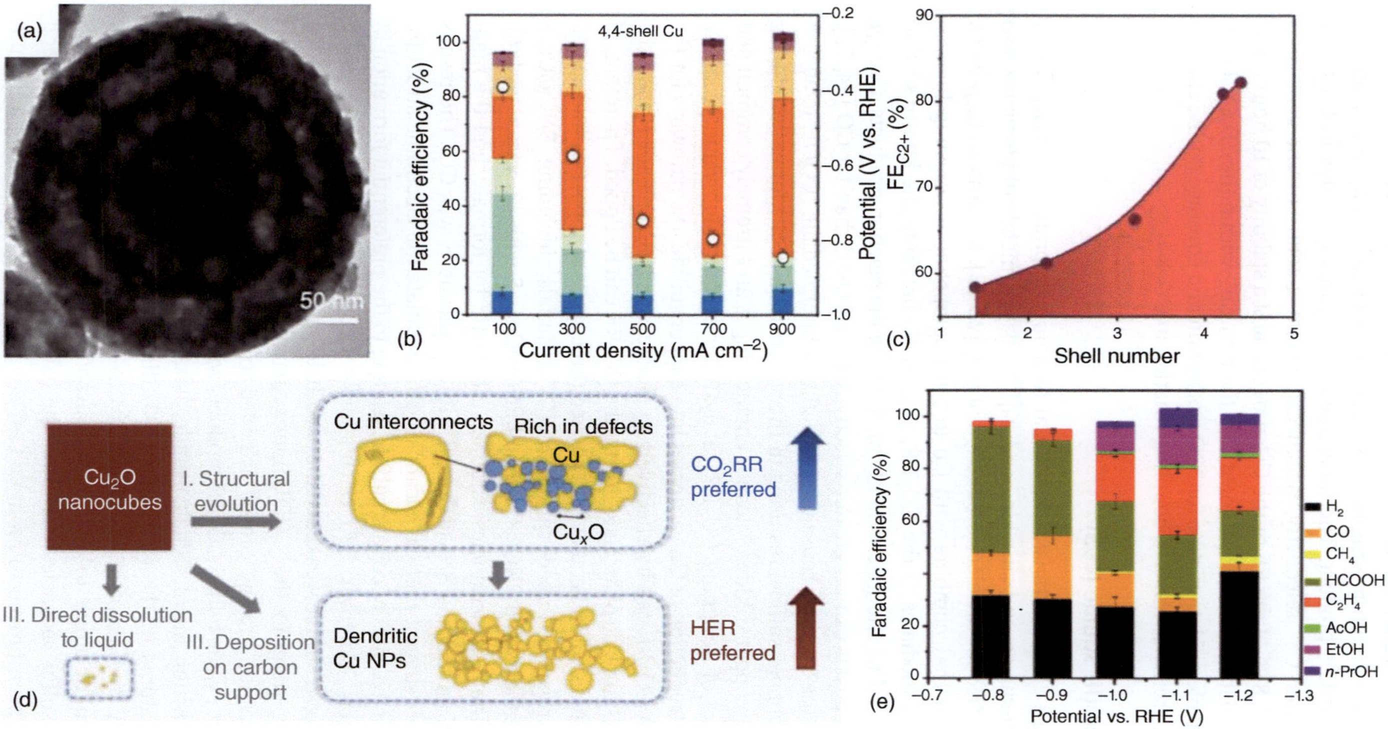

Figure 18.2 (a) TEM images of the quadruple-shell Cu_2O in 4.2-shell Cu_2O. (b) CO_2R products distribution at different current densities on 4.4-shell Cu. (c) Correlations between Faradic efficiency of C_{2+} products and the shell number of the Cu catalysts. Source: Reproduced with permission of Chen et al. [16], Wiley-VCH. (d) The schematic shows the different pathways of degradation lead to two typical final structures evolved from Cu_2O nanocubes. Reproduced with permission of Luo et al. [17], Elsevier. (e) Reduction potential-dependent FEs of various products for CO_2 electroreduction measured on o-Cu_2O. Source: Reproduced with permission of Luo et al. [18], American Chemical Society.

simple coprecipitation method [21]. The ultrathin Sub1 nm Cu_2O nanosheets (Cu_2O-NS), with coordination defects and oxygen vacancies, stabilize the oxidation state of the surface Cu, promoting the adsorption of *CO and finally improving the selectivity for the C^{2+} product. The moderate Cu-CO binding energy of Cu_2O-NS results in superior C^{2+} selectivity, which can obtain a total current density of $-187.6\,mA\,cm^{-2}$, 81% of the FE of the C^{2+} product, and a stability of 10 hours. The theoretical calculations further revealed that the rich Cu^+ within the 2D Cu_2O-NS is the most suitable oxidation state for *CO adsorption and coverage on the catalyst surface, which promotes the C–C coupling reaction in CO_2RR.

18.1.3 Copper(II) Sulfide (CuS)

Due to their fast electron transfer, structural flexibility, appropriate adsorption of reaction intermediates, and suitable band gap, transition metal sulfides have also been widely studied for the electroreduction of CO_2 [22, 23]. By virtue of its p-block properties, sulfur may combine with Cu to form competing or assisting binding sites for the intermediates of the CO_2RR, potentially breaking the knotty scaling relationships and thus energy-efficiently modulating the reaction products. Copper sulfide-based catalysts exhibit greater selectivity for HCOOH than for CO [24, 25]. This is because the S anions cause the CO pathway intermediate *COOH to adsorb too weakly on copper sulfide, resulting in the suppression of CO formation during the CO_2RR. For copper sulfides, vacancy engineering is an extremely common and effective method of designing CO_2RR catalysts for specific high-value products. Through the introduction of Cu vacancies into copper sulfide, the pathways for the electrochemical CO_2RR can be effectively modulated, and the balance toward the formation of CO in place of formate during the CO_2RR can be tipped. Therefore, Li et al. designed and simultaneously synthesized three kinds of copper sulfide-coated sulfur and nitrogen-co-doped carbon (SNC) materials, including SNC@Cu_2S without Cu vacancies and SNC@Cu_xS and SNC@$Cu_{1.96}S$ with Cu vacancies [26]. SNC@Cu_2S without Cu vacancies exhibited a high FE for formate, and the main product of SNC@Cu_xS and SNC@$Cu_{1.96}S$ with Cu vacancies was CO. The results presented earlier suggest that Cu vacancies can modulate the electronic structure of the surface-active sites of copper sulfide, leading to the preferential formation of different products. Moreover, DFT calculations revealed that the formation of Cu vacancies in copper sulfide weakened the adsorbate–metal interaction and provided suitable binding energy for the *COOH intermediate toward the efficient production of CO. This vacancy strategy can be extended to other copper-based catalysts to produce different value-added CO_2RR products with high activity and selectivity.

As a 2D transition metal dichalcogenide, hexagonal-phase CuS has abundant surface sulfur atoms in the (001) and (100) directions in its layered structure and a relatively low dissociation energy of Cu—S bonds [27, 28]. By removing these surface sulfur atoms to form sulfur vacancies, designated $CuS_x(0 < x < 1)$, the controlled sulfur vacancies can lower the oxidation of copper between 0 and +2, which has been proven to promote C–C coupling. Therefore, using a lithium electrochemical tuning approach, Zheng and co-workers developed a CuS_x catalyst with a double-sulfur vacancy (DSV) structure, which exhibited a much greater

$FE_{n\text{-}PrOH}$ value of 15.4 ± 1% for *n*-propanol production at −1.05 V vs. RHE in 0.1 M $KHCO_3$ [29]. The presence of two sulfur vacancies enabled the stabilization of the CO* and OCCO* dimers and further coupled CO-OCCO to form the key *C_3 intermediate of n-propanol.

18.1.4 Copper(I) Sulfide (Cu_2S)

$Cu^{\delta+}$ ($0 < \delta < 1$) species, which are dynamically formed on Cu-based catalysts during the CO_2RR process, have been revealed as the veritable active site for ethanol formation owing to their moderate electron-donating ability to neutralize positively charged C atoms [30, 31]. However, the selectivity and energy efficiency of ethanol over Cu-based electrocatalysts are still low. Besides, the in situ-formed $Cu^{\delta+}$ is insufficient and easily reduced to Cu^0. Therefore, the rational design of Cu-based materials with stable $Cu^{\delta+}$ active sites still remains challenging. Engineering anion vacancies is regarded as a tool to adjust the electronic structures of metal cations, achieving the efficient production of ethanol. For example, Zhang and co-workers reported Cu_2S hollow nanocubes with abundant sulfur vacancies (Cu_2S_{1-x} HN) synthesized by the electrochemical reduction of CuS hollow nanocubes (Figure 18.3a) [32]. It exhibited an ultralow overpotential of 0.19 V for ethanol production. The catalyst not only delivers an outstanding ethanol selectivity of 86.9% and an FE of 73.3% but also provides long-term stability to $Cu^{\delta+}$ (Figure 18.3b,c). The combined results of in situ ATR-FTIR spectra and DFT calculations (Figure 18.3d–f) demonstrated that the strong electron-donating ability of $Cu^{\delta+}$ sites can decrease the ΔG_{PDS} of the C–C coupling step toward ethanol generation and increase the ΔG_{PDS} toward formate formation.

Notably, the rapid reduction of Cu–S compounds and rapid structural evolution during the initial reduction of CO_2 led to an unclear reaction mechanism and natural reaction active sites on the S-derived Cu catalysts. Lei and co-workers reported a Cu_2S nanorod/copper mesh (Cu_2S/CM) for the CO_2RR synthesized through chemical oxidation of CM to $Cu(OH)_2$/CM and further gas-phase sulfurization [33]. The Cu_2S/CM catalyst exhibited high performance in the electroreduction of CO_2 to ethanol. At −0.8 V vs. RHE, an FE of 13.5%, a partial current density of 13.3 mA cm^{-2}, and a yield of 986.4 μmol l^{-1} h^{-1} were achieved for ethanol. Furthermore, DFT calculations demonstrated that the existence of Cu^0/Cu^+ sites in Cu/Cu_2S enhances the adsorption of the key intermediate *CH_2CHO to C_2 products, and the Cu^0/Cu^+ sites favor the C_2H_5OH pathway. In this work, the study of the structural evolution and reaction mechanism provides a new understanding of CO_2 electroreduction with Cu_2S.

18.1.5 Copper Nitrides (Cu_3N)

As mentioned earlier, copper(I) oxide (Cu_2O) has been mostly used as a precursor to Cu-based CO_2 reduction catalysts and is a candidate for Cu^+ support. Nevertheless, the instability of Cu^+ from Cu_2O under CO_2 reduction conditions is inevitable. Instead of Cu_2O, copper(I) nitrides can be employed not only as stable catalytically active species but also as supports. To investigate whether copper(I) nitride (Cu_3N)

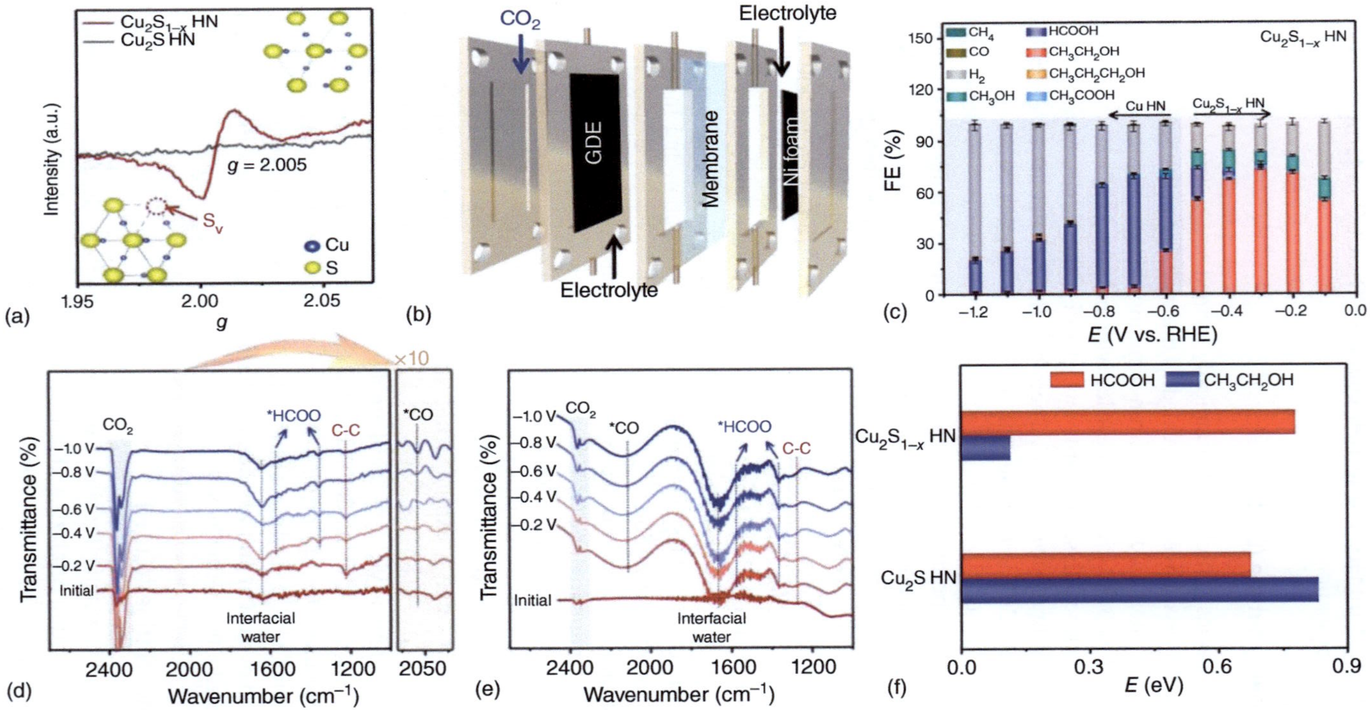

Figure 18.3 (a) EPR spectra of Cu_2S_{1-x} HN and Cu_2S HN. (b) Schematic illustration for CO_2RR in a flow-type cell. Potential-dependent FE of different products over (c) Cu_2S_{1-x} HN and (d) Cu_2S HN. Potential-dependent ATR-FTIR spectra of (d) Cu_2S_{1-x} HN and (e) Cu_2S HN with CO_2 bubbling. (f) Energy barriers of potential-determining step in ethanol and formate generation. Source: Reproduced with permission of Zhang et al. [32], Wiley-VCH.

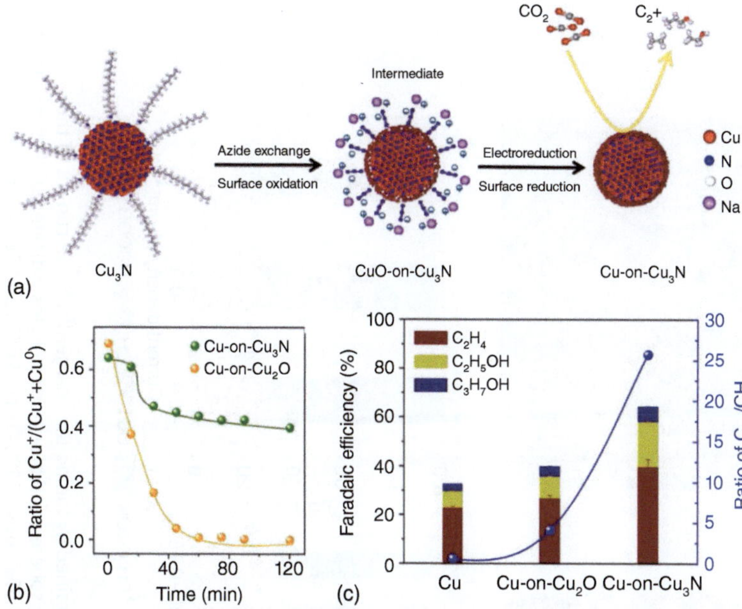

Figure 18.4 (a) Schematic of preparing the Cu-on-Cu_3N catalyst. (b) Ratio of Cu^+ relative to the reaction time at −0.95 V vs. RHE, (c) comparison of FE for C_{2+} and the ratio of C_{2+}/CH_4 at −0.95 V vs. RHE on Cu, Cu-on-Cu_2O, and Cu-on-Cu_3N. Source: Liang, et al. [34]/Springer Nature/CC BY 4.0.

could be used as a Cu^+ support during CO_2 reduction, Liang et al. synthesized a copper-on-nitride catalyst (Cu-on-Cu_3N), and Cu-on-Cu_2O and pure Cu catalysts were used for comparison (Figure 18.4a) [34]. Among the three catalysts tested, Cu-on-Cu_3N produced the most C_{2+} and achieved a 6.3-fold increase in the ratio of C_{2+} to CH_4 compared to that of Cu-on-Cu_2O and a 40-fold increase over that of pure Cu (Figure 18.4b,c). Cu_3N was chosen as the inner support to modify the electronic structure of the surface metal, thereby affecting the adsorption and dimerization of intermediate CO properly during CO_2 reduction. Mechanistic studies demonstrated that the use of copper nitride contributes to reducing the CO dimerization energy barrier.

Introducing ion vacancies into the lattice of copper-based catalysts has been demonstrated to be an effective approach for enhancing C_2 products [15, 35]. Compared with oxygen vacancies in copper oxides, nitrogen vacancies on transition metal nitrides have been suggested to present greater stability because of the high metal-N bond energies [36, 37]. It has been reported that transition metal nitrides can be prepared as lithium-ion battery anodes in which N^3 anions can be controllably removed to form Li_3N during the partially reversible lithiation process while retaining the nanostructures of the electrodes [38]. Inspired by this, Zheng and co-workers fabricated a Cu_3N_x catalyst with different densities of nitrogen vacancies by an electrochemical lithium tuning (Figure 18.5a,b) [39]. The Cu_3N_x catalyst with abundant nitrogen vacancies presents one of the highest FEs toward C_2 products of 81.7 ± 2.3% at −1.15 V vs. RHE (without ohmic correction), corresponding to

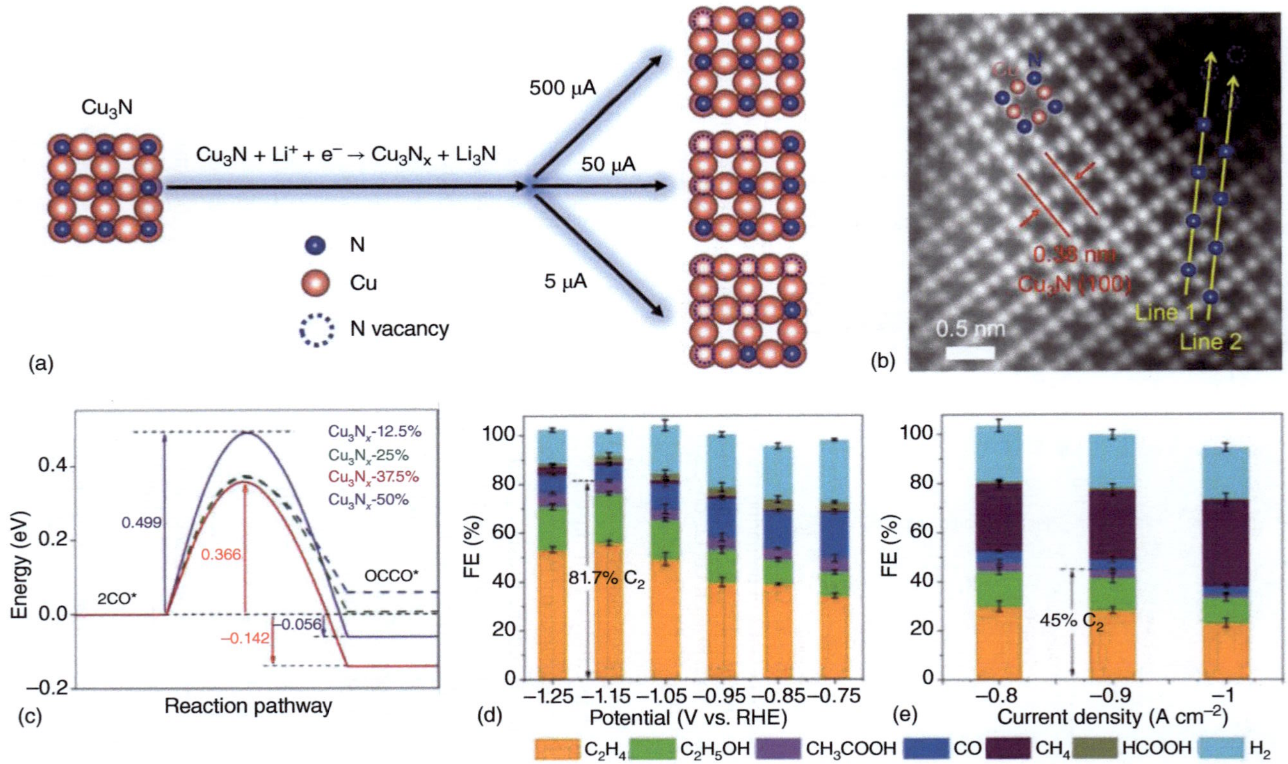

Figure 18.5 (a) Schematic diagram of nitrogen vacancies with different concentrations obtained by tuning the charge/discharge current of the lithium-ion battery. (b) Spherical aberration-corrected STEM of Cu_3N_x-50-μA. The pink, blue spheres, and blue dotted circles indicate Cu, N atoms, and N vacancies, respectively. (c) Corresponding energy diagrams of CO–CO coupling on Cu_3N_x-12.5%, Cu_3N_x-25%, Cu_3N_x-37.5%, and Cu_3N_x-50% at 0 V versus RHE, respectively. (d,e) CO_2RR product distributions using Cu_3N_x-50-μA catalysts at various applied constant potentials (without ohmic correction) (d) and various current densities (e). Reproduced with permission of Zheng et al. [39], Wiley-VCH.

a partial current density for C_2 production as $-307 \pm 9\,mA\,cm^{-2}$ (Figure 18.5d,e). Besides, DFT calculations indicated that the adsorption energies of CO* and the energy barriers for forming key C_2 intermediates are strongly correlated with nitrogen vacancy density. In addition, adjacent multiple nitrogen vacancies on the Cu_3N_x (100) planes increased the probability of CO–CO coupling (Figure 18.5c).

18.1.6 Conclusion and Outlook

In this review, we systematically discussed various kinds of Cu-based compounds, including CuO, Cu_2O, CuS, Cu_2S, and Cu_3N, for CO_2 electroreduction, and several typical strategies, such as vacancy engineering and morphological regulation, are highlighted. In addition, several important suggestions and expectations for further researching Cu-based compounds as CO_2RR electrocatalysts are proposed as follows:

(i) Through extensive investigations, it was found that the structural design of copper compounds plays a crucial role in their application in the CO_2RR, and the instability of catalysts during the CO_2RR usually originates from the structural evolution and decomposition of compounds. Therefore, a more rational design of the nanostructure of copper compounds combined with theoretical calculations is necessary to develop novel Cu-based CO_2RR catalysts.
(ii) Currently, the selectivity of Cu-based compounds for specific CO_2RR products is still relatively low. Thus, further optimization of these materials is needed, especially for newly developed Cu-based amorphous and high-entropy compounds. The unique electronic structure and unsaturated coordination bonds of these materials may provide abundant active sites, which is conducive to enhancing the selectivity of specific C_2 products.
(iii) With regard to copper oxides, sulfides, and nitrides, vacancy engineering is obviously an important tool for modulating the selectivity and number of active sites on Cu-based catalysts. Due to the different degrees of vacancy, Cu-based compounds with an adjustable number of vacancies can achieve the specific goal for the desired products. Thus, innovative methods for creating vacancies in Cu-based compounds are crucial topics.
(iv) A thorough understanding of the transformation of catalysts and reaction intermediates during the CO_2RR is propitious for designing more effective Cu-based electrocatalysts. To improve the understanding of the CO_2RR mechanism, a combination of in situ/operando characterization techniques such as in situ spectroscopy and transmission electron microscopy (TEM) with theoretical calculations is necessary to evaluate the potential value of Cu-based catalysts.

References

1 Liu, S., Zhang, B., Zhang, L., and Sun, J. (2022). Rational design strategies of Cu-based electrocatalysts for CO_2 electroreduction to C_2 products. *J. Energy Chem.* 71: 63–82.

2 Woldu, A.R., Huang, Z., Zhao, P. et al. (2022). Electrochemical CO_2 reduction (CO_2RR) to multi-carbon products over copper-based catalysts. *Coord. Chem. Rev.* 454: 214340.

3 Grosse, P., Yoon, A., Rettenmaier, C. et al. (2021). Dynamic transformation of cubic copper catalysts during CO_2 electroreduction and its impact on catalytic selectivity. *Nat. Commun.* 12 (1): 6736.

4 Kunze, S., Tănase, L.C., Prieto, M.J. et al. (2021). Plasma-assisted oxidation of Cu(100) and Cu(111). *Chem. Sci.* 12 (42): 14241–14253.

5 Mistry, H., Varela, A.S., Bonifacio, C.S. et al. (2016). Highly selective plasma-activated copper catalysts for carbon dioxide reduction to ethylene. *Nat. Commun.* 7 (1): 12123.

6 Wang, Y., Wang, Z., Dinh, C.-T. et al. (2020). Catalyst synthesis under CO_2 electroreduction favours faceting and promotes renewable fuels electrosynthesis. *Nat. Catal.* 3 (2): 98–106.

7 Li, C.W. and Kanan, M.W. (2012). CO_2 reduction at low overpotential on Cu electrodes resulting from the reduction of thick Cu_2O films. *J. Am. Chem. Soc.* 134 (17): 7231–7234.

8 Liu, W., Zhai, P., Li, A. et al. (2022). Electrochemical CO_2 reduction to ethylene by ultrathin CuO nanoplate arrays. *Nat. Commun.* 13 (1): 1877.

9 Kim, J., Choi, W., Park, J.W. et al. (2019). Branched copper oxide nanoparticles induce highly selective ethylene production by electrochemical carbon dioxide reduction. *J. Am. Chem. Soc.* 141 (17): 6986–6994.

10 Chang, K., Zhang, H., Cheng, M.-j. et al. (2020). Application of ceria in CO_2 conversion catalysis. *ACS Catal.* 10 (1): 613–631.

11 Wang, W., Zhang, Y., Wang, Z. et al. (2016). Reverse water gas shift over In_2O_3–CeO_2 catalysts. *Catal. Today* 259: 402–408.

12 Aitbekova, A., Wu, L., Wrasman, C.J. et al. (2018). Low-temperature restructuring of CeO_2-supported Ru nanoparticles determines selectivity in CO_2 catalytic reduction. *J. Am. Chem. Soc.* 140 (42): 13736–13745.

13 Yan, X., Chen, C., Wu, Y. et al. (2021). Efficient electroreduction of CO_2 to C_{2+} products on CeO_2 modified CuO. *Chem. Sci.* 12 (19): 6638–6645.

14 De Luna, P., Quintero-Bermudez, R., Dinh, C.-T. et al. (2018). Catalyst electro-redeposition controls morphology and oxidation state for selective carbon dioxide reduction. *Nat. Catal.* 1 (2): 103–110.

15 Zhuang, T.-T., Liang, Z.-Q., Seifitokaldani, A. et al. (2018). Steering post-C–C coupling selectivity enables high efficiency electroreduction of carbon dioxide to multi-carbon alcohols. *Nat. Catal.* 1 (6): 421–428.

16 Xiao, Y., Wang, M., Yang, H. et al. (2024). Multi-shell copper catalysts for selective electroreduction of CO_2 to multicarbon chemicals. *Adv. Energy Mater.* 14 (1): 2302556.

17 Ren, Q., Zhang, N., Dong, Z. et al. (2023). Structural evolution of Cu_2O nanocube electrocatalysts for the CO_2 reduction reaction. *Nano Energy* 106: 108080.

18 Fu, W., Liu, Z., Wang, T. et al. (2020). Promoting C_{2+} production from electrochemical CO_2 reduction on shape-controlled cuprous oxide nanocrystals with high-index facets. *ACS Sustain. Chem. Eng.* 8 (40): 15223–15229.

19 Li, H., Jiang, Y., Li, X. et al. (2023). C_{2+} selectivity for CO_2 electroreduction on oxidized Cu-based catalysts. *J. Am. Chem. Soc.* 145 (26): 14335–14344.

20 Wu, Q., Du, R., Wang, P. et al. (2023). Nanograin-boundary-abundant Cu_2O-Cu nanocubes with High C_{2+} selectivity and good stability during electrochemical CO_2 reduction at a current density of 500 mA/cm^2. *ACS Nano* 17 (13): 12884–12894.

21 Wang, P., Meng, S., Zhang, B. et al. (2023). Sub-1 nm Cu_2O nanosheets for the electrochemical CO_2 reduction and valence state-activity relationship. *J. Am. Chem. Soc.* 145 (48): 26133–26143.

22 Kibria, M.G., Edwards, J.P., Gabardo, C.M. et al. (2019). Electrochemical CO_2 reduction into chemical feedstocks: from mechanistic electrocatalysis models to system design. *Adv. Mater.* 31 (31): 1807166.

23 Landers, A.T., Fields, M., Torelli, D.A. et al. (2018). The predominance of hydrogen evolution on transition metal sulfides and phosphides under CO_2 reduction conditions: an experimental and theoretical study. *ACS Energy Lett.* 3 (6): 1450–1457.

24 Deng, Y., Huang, Y., Ren, D. et al. (2018). On the role of sulfur for the selective electrochemical reduction of CO_2 to formate on CuS_x catalysts. *ACS Appl. Mater. Interfaces* 10 (34): 28572–28581.

25 Shinagawa, T., Larrazábal, G.O., Martín, A.J. et al. (2018). Sulfur-modified copper catalysts for the electrochemical reduction of carbon dioxide to formate. *ACS Catal.* 8 (2): 837–844.

26 Li, S., Duan, H., Yu, J. et al. (2022). Cu vacancy induced product switching from formate to CO for CO_2 reduction on copper sulfide. *ACS Catal.* 12 (15): 9074–9082.

27 He, K., Yao, Z., Hwang, S. et al. (2017). Kinetically-driven phase transformation during lithiation in copper sulfide nanoflakes. *Nano Lett.* 17 (9): 5726–5733.

28 Mahé, L., Boughdiri, S.F., and Barthelat, J.-C. (1997). Electronic structures and energetics in the CuX and Cu_2X series (X = O, S, Se, Te, Po). *J. Phys. Chem. A* 101 (23): 4224–4230.

29 Peng, C., Luo, G., Zhang, J. et al. (2021). Double sulfur vacancies by lithium tuning enhance CO_2 electroreduction to n-propanol. *Nat. Commun.* 12 (1): 1580.

30 Jiao, J., Lin, R., Liu, S. et al. (2019). Copper atom-pair catalyst anchored on alloy nanowires for selective and efficient electrochemical reduction of CO_2. *Nat. Chem.* 11 (3): 222–228.

31 Liu, X., Xiao, J., Peng, H. et al. (2017). Understanding trends in electrochemical carbon dioxide reduction rates. *Nat. Commun.* 8 (1): 15438.

32 Guo, C., Guo, Y., Shi, Y. et al. (2022). Electrocatalytic reduction of CO_2 to ethanol at close to theoretical potential via engineering abundant electron-donating $Cu^{\delta+}$ species. *Angew. Chem. Int. Ed.* 61 (32): e202205909.

33 Dou, T., Du, J., He, J. et al. (2022). Sulfurization-derived Cu^0–Cu^+ sites for electrochemical CO_2 reduction to ethanol. *J. Power Sources* 533: 231393.

34 Liang, Z.-Q., Zhuang, T.-T., Seifitokaldani, A. et al. (2018). Copper-on-nitride enhances the stable electrosynthesis of multi-carbon products from CO_2. *Nat. Commun.* 9 (1): 3828.

35 Gu, J., Magagula, S., Zhao, J., and Chen, Z. (2019). Boosting ORR/OER activity of graphdiyne by simple heteroatom doping. *Small Methods* 3 (9): 1800550.

36 Ye, T.-N., Park, S.-W., Lu, Y. et al. (2020). Vacancy-enabled N_2 activation for ammonia synthesis on an Ni-loaded catalyst. *Nature* 583 (7816): 391–395.

37 Lee, D., Sun, S., Kwon, J. et al. (2020). Copper nitride nanowires printed Li with stable cycling for Li metal batteries in carbonate electrolytes. *Adv. Mater.* 32 (7): 1905573.

38 Pereira, N., Dupont, L., Tarascon, J.M. et al. (2003). Electrochemistry of Cu_3 N with lithium: a complex system with parallel processes. *J. Electrochem. Soc.* 150 (9): A1273.

39 Peng, C., Luo, G., Xu, Z. et al. (2021). Lithiation-enabled high-density nitrogen vacancies electrocatalyze CO_2 to C_2 products. *Adv. Mater.* 33 (40): 2103150.

Part VI

Advanced Functional Materials for Electrocatalytic Nitrogen Reduction Reaction

19

Metal-Based Nanomaterials for Electrocatalytic Nitrogen Reduction Reaction

Yaxiao Guo[1,*], *Chenhao Xiao*[1,*], *Shinuo Guo*[1,*], *Yi Liu*[1], *and Zhaoyang Yao*[2]

[1] *Tiangong University, School of Chemistry, State Key Laboratory of Separation Membranes and Membrane Processes, No.399 Binshuixi Road, Xiqing District, Tianjin, 300387, China*
[2] *Nankai University, College of Chemistry, Key Laboratory of Functional Polymer Materials, No.94 Weijin Road, Nankai Disrtict, Tianjin, 300071, China*

19.1 Introduction

Ammonia (NH_3) is an indispensable chemical that is widely employed as a carbon-free energy carrier, raw material for nitrogen-containing fertilizers, pharmaceuticals, etc. [1, 2]. Thus far, more than 90% of ammonia is produced through a highly energy-consuming Haber–Bosch process. During this process, iron- or ruthenium-based catalysts will be used under very high temperature and pressure, meanwhile, a lot of hydrogen gas is consumed [3, 4]. With the rapidly increasing demand for ammonia, the Haber–Bosch process worldwide has consumed approximately 2% of the world's energy (34 GJ ton^{-1} NH_3) and emitted more than 1% of the world's CO_2 (~2 ton CO_2 per ton NH_3) [5–7]. Therefore, it is crucially important to develop low-cost and environmentally benign methods for ammonia production. Currently, the main possible methods for green production of ammonia are nitrogenase ammonia synthesis [8, 9], photocatalytic ammonia synthesis [10, 11], plasma methods for ammonia synthesis [12], and electrocatalytic ammonia synthesis [3, 4, 13]. Among them, the complexity and instability of nitrogenase enzymes greatly limit the industrialization of nitrogenase ammonia synthesis process. Although photocatalytic ammonia synthesis requires only nitrogen, water, and solar energy, its chemical conversion efficiency is quite low presently, and the unpredictability of solar energy is another huge challenge. As regards the plasma methods for ammonia synthesis, constant high-energy inputs are usually essential. In contrast, electrocatalytic nitrogen to ammonia nitrogen reduction reaction (NRR) has been getting more and more attention in the past few years [2, 14, 15], mainly benefiting from its outstanding advantages, including mild reaction conditions, simple devices, and excellent operation process safety. In addition, multiple green

*These authors contributed equally to this work.

Functional Materials for Electrocatalytic Energy Conversion, First Edition.
Edited by Zhicheng Zhang, Meiting Zhao, and Yuchen Qin.

energies, such as solar energy, wind energy, tidal energy, and so on, could be applied as a source of electric energy instead of consuming traditional fossil fuels.

In spite of the sharply increasing interest, electrocatalytic ammonia synthesis is still confronted with several huge challenges: (i) the inertness of dinitrogen (N_2) itself. Unlike other diatomic molecules (dihydrogen and dioxygen), the binding energy of N≡N is very large (941 kJ mol^{-1}), which results in a kinetically complicated and energetically challenging process during NRR [16–19]. (ii) The competitive hydrogen evolution reaction (HER). Nitrogen fixation is involved with six protons and six electrons ($N_2 + 6H^+ + 6e^- \rightarrow 2NH_3$, $E^0 = 0.057–0.059 \times pH$ V vs. standard hydrogen electrode). In sharp contrast, HER occurs easily with a rapid kinetic process. In light of the similar reduction potentials of HER and NRR, a relatively low selectivity of NRR is usually observed [20–22]. Therefore, how to control the generation of by-products and inhibit HER is crucially important if both the high NH_3 yield rate and Faraday efficiency (FE) of NRR are expected. Thus far, plenty of metal-based materials have been explored as electrocatalysts for NRR [23–25], including precious metal-based catalysts, non-noble metal-based catalysts, metal–carbon-based catalysts, and central metal-coordinated catalysts. Herein, we summarized several typical catalysts above and provided an outlook on the development of electrocatalytic NRR at the end.

19.2 Precious Metal-Based Catalysts

Due to their unfilled d electron orbitals, the surface of several precious metals (Pt, Pd, Au, etc.) makes it easy to conduct molecular adsorption and desorption and further catalyze electrochemical reactions with excellent selectivity and catalytic activity [26–28]. In addition, some other desired characteristics, such as high temperature resistance, oxidation resistance, and corrosion resistance, can also be observed for precious metals, making them highly potential to work as excellent electrocatalysts for NRR. Herein, we will give a few typical examples below.

19.2.1 Pt- and Pd-Based Catalysts

Although Pt has been proven to demonstrate excellent activities in HER, some studies also showed its great potential for NRR [29]. By employing a wet chemical method, Huang and co-workers synthesized Pt_3Fe nanocrystals including nanoclusters, nanorods, and nanowires with different electronic structures [30]. As presented in Figure 19.1a, all the three nanocrystals of Pt_3Fe exhibit relatively good NRR activity, among which the nanowire of Pt_3Fe reaches the best NRR performance with an NH_3 yield rate of 18.3 μg h^{-1} mg^{-1} and also a good FE of 7.3% at the optimal testing potential of −0.05 V. A systematic investigation has unveiled the crucial role of d–d coupling in strengthening Pt–Fe interactions and further promoting NRR performance.

Extensive investigations have indicated that the NRR performance of Pd seems better than that of Pt. In 2018, Feng and co-workers delicately compared

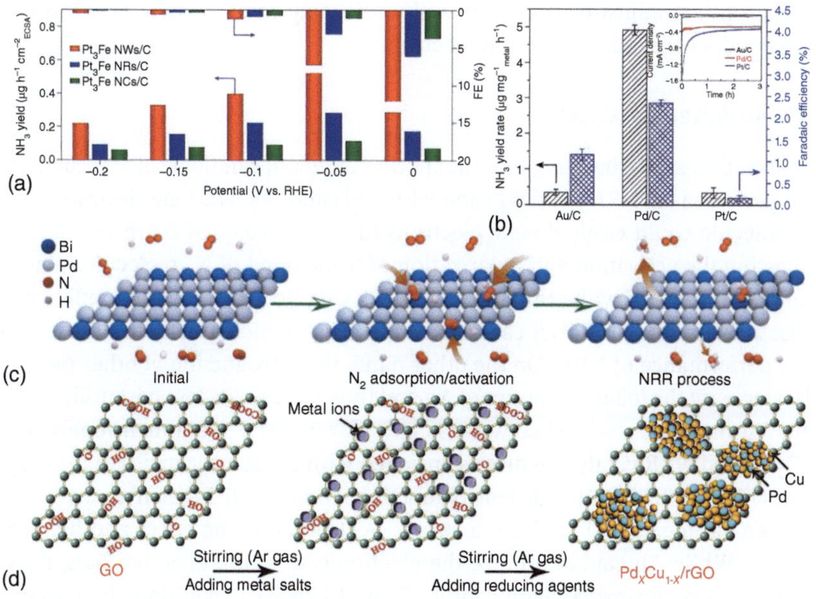

Figure 19.1 (a) NH_3 yield rate and FE of Pt_3Fe NCs/C, Pt_3Fe NRs/C, and Pt_3Fe NWs/C at different testing potentials. Source: Tong, et al. [30]/Oxford Univ Press/CC BY 4.0. (b) NH_3 yield rate and FE of Pd/C, Au/C, and Pt/C NRR catalysts at their individual optimal potentials. Source: Wang, et al. [31]/Spring Nature/CC BY 4.0. (c) The schematic illustration of the NRR process on np-Pd_3Bi. Source: Reproduced with permission of Wang et al. [32], Wiley-VCH. (d) The synthesis of NRR catalyst of Pd_xCu_{1-x}/rGO. Source: Reproduced with permission of Shi et al. [33], Wiley-VCH.

the NRR performance of carbon black-supported Pt (Pt/C), Pd (Pd/C), and Au (Au/C) nanoparticles, indicating that Pd/C could obtain a good NH_3 yield rate of 4.5 $\mu g\,h^{-1}\,mg_{Pd}^{-1}$ and FE of 8.6%, which are better than those of Pt/C and Au/C (Figure 19.1b) [31]. Given the high cost and fierce HER competition, coupling precious metals with some other metals is becoming a feasible approach to improve NRR performance while circumventing high cost. For example, the adsorption of N_2 molecules is the most important step in the initial activation of inert N_2. While the Bi atom could build a very strong interaction with N_2 molecules, more importantly, as well as Pd. Therefore, the strong coupling between Pd-Bi sites constructs an electron transfer channel between different metals, which may facilitate N_2 adsorption and activation (Figure 19.1c). Tan and co-workers performed electrochemical etching to fabricate the nanoporous Pd_3Bi catalyst (np-Pd_3Bi) successfully [32], which effectively inhibited HER and showed a good selectivity for NRR. Finally, an NH_3 yield rate of 59.05 ± 2.27 $\mu g\,h^{-1}\,mg^{-1}$ and FE of 21.52 ± 0.71% were achieved. In addition, the double continuous nanopore structure formed in Pd_3Bi catalyst could accelerate electron transport during the NRR process, thus boosting the catalytic performance of NRR. In similar fashion, Jiang et al. anchored PdCu non-nanoclusters on reduced graphene oxide (rGO) as an NRR catalyst (Figure 19.1d) [33]. Note that the molar ratio of Pd and Cu is a quite important factor in tuning the selectivity of NRR and

HER. Appropriate adjustment of bimetallic ratio may produce new active sites, which is conducive to activating N_2 molecules.

19.2.2 Au-Based Catalysts

Au-based catalysts have been regarded as another excellent candidate electrocatalyst for high-performance NRR besides Pt and Pd-based catalysts. The lone electron pairs on N_2 molecule could easily donate electrons to the unoccupied d orbitals of Au, thus boosting the activation and polarization of triple bond of N_2 molecule. When Au is further combined with other metals (e.g. Pt, Cu, Mo), the synergistic effect we have discussed in Pd_3Bi catalyst can also be observed, which greatly improves the catalytic performance of NRR. On the other hand, the introduction of other metals can also decrease the loading amount of Au and thus increase its low-cost utilization [34–39]. For instance, Guo and co-workers have developed a two-dimensional (2D) Au/TiO_2 hybrid NRR catalyst, which is enriched with oxygen vacancies in its TiO_2 substrate (Figure 19.2a) [40]. The density functional theory (DFT) calculations show that oxygen vacancies in TiO_2 have a strong electron-donating effect and help to activate N_2. While Au could optimize the electronic structure of active sites, thus improving NRR performance significantly (Figure 19.2b–d). In addition, this oxygen vacancy-rich TiO_2 can stabilize Au nanoparticles well and improve the long-term stability of catalysts.

19.2.3 Ru- and Rh-Based Catalysts

In addition to Pt, Pd, and Au-based catalysts, Ru- and Rh-based catalysts can also be applied for NRR. An efficient NRR catalyst of carbon nanotubes loaded with ultrafine Rh nanoparticles (NPs) (Rh/CNT) was prepared by Jiang and co-workers which affords an NH_3 yield rate of 26.91 $\mu g\,h^{-1}\,mg^{-1}$ and a really excellent FE of 23.48% in PBS [41]. DFT results show that the high spin and adsorption density of Rh/graphene well accelerates the activation and adsorption of molecular N_2, reduces free energy adsorption, and thus improves the NRR electrocatalytic activity. Therefore, the noble metal hybrid structure can promote the hydrogenation formation of N_2, thereby increasing the yielding rate of NH_3.

By using a micelle-assisted displacement reaction method, a mesoporous Au_3Rh film was built on the Ni foam (mAu_3Rh/NF) to establish an efficient catalyst for electrocatalytic NRR [42]. Benefiting from its bimetallic structure, interconnected mesoporous morphology, and binder-free features, the mAu_3Rh/NF catalyst finally gives rise to an NH_3 yield rate of 26.29 $\mu g\,h^{-1}\,mg^{-1}$ and a FE of 23.84% in 0.1 M Na_2SO_4 solution. In similar fashion, a mesoporous RhRu film built on Ni foam (mRhRu/NF) had been further established [43]. As a result, the as-prepared bimetallic mRhRu/NF NRR catalyst reaches an FE of 28.33% and NH_3 rate yield of 30.28 $\mu g\,h^{-1}\,mg^{-1}$ in a neutral electrolyte of Na_2SO_4 solutions, which demonstrates the slightly improved catalytic performance toward NRR compared to that of mAu_3Rh/NF catalyst.

To sum up, although precious metal-based catalysts have many advantages, their extremely high costs caused by very limited resources will become an insurmountable mountain on the road to industrialization.

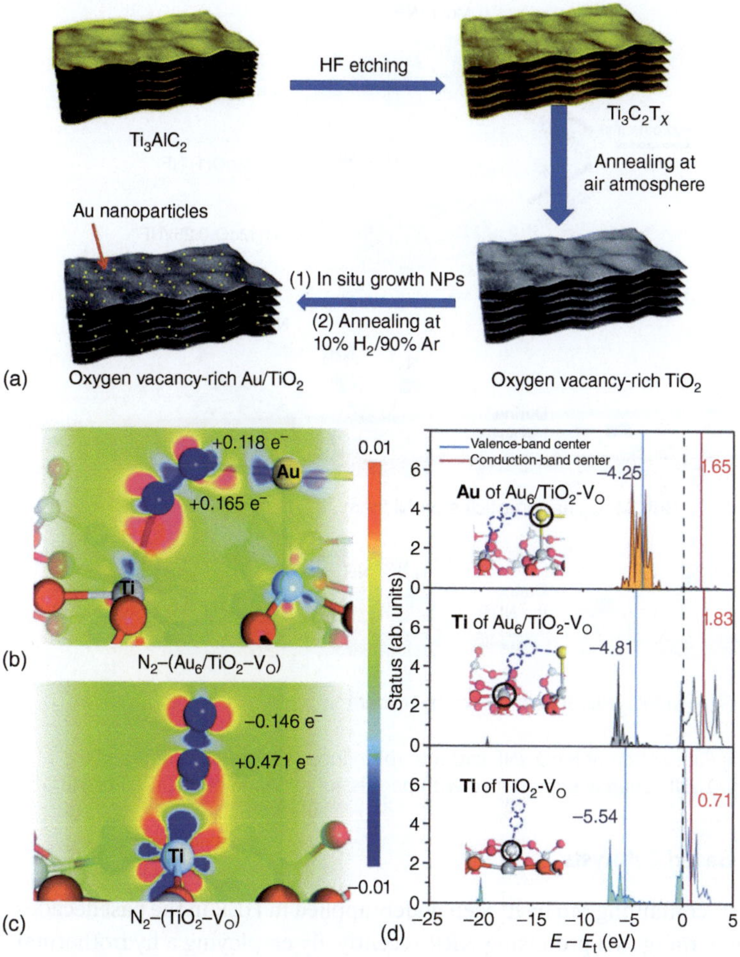

Figure 19.2 (a) Schematic illustration of the preparation process of the V_O-rich Au/TiO_2 nanosheet hybrid catalyst. Charge density difference for N_2 adsorption on (b) Au_6/TiO_2-V_O and (c) TiO_2-V_O. (d) The PDOS of the d band of selected adsorption sites without N_2. Source: Reproduced with permission of Zhao et al. [40], Royal Society of Chemistry.

19.3 Non-Noble Transition Metal-Based Catalysts

Non-noble metal-based catalysts, especially transition metal-based catalysts, are considered as the ideal alternative to noble metal-based catalysts because of their similar properties but dramatically reduced costs. Because of the unique d-band electronic structure of transition metal-based catalysts, N_2 molecules could adsorb at metal sites on the surface of catalysts. Thus far, Non-noble transition metal-based catalysts have exhibited excellent activity toward NRR characteristics of low reaction energy barriers and facilitated reaction kinetics [44–46].

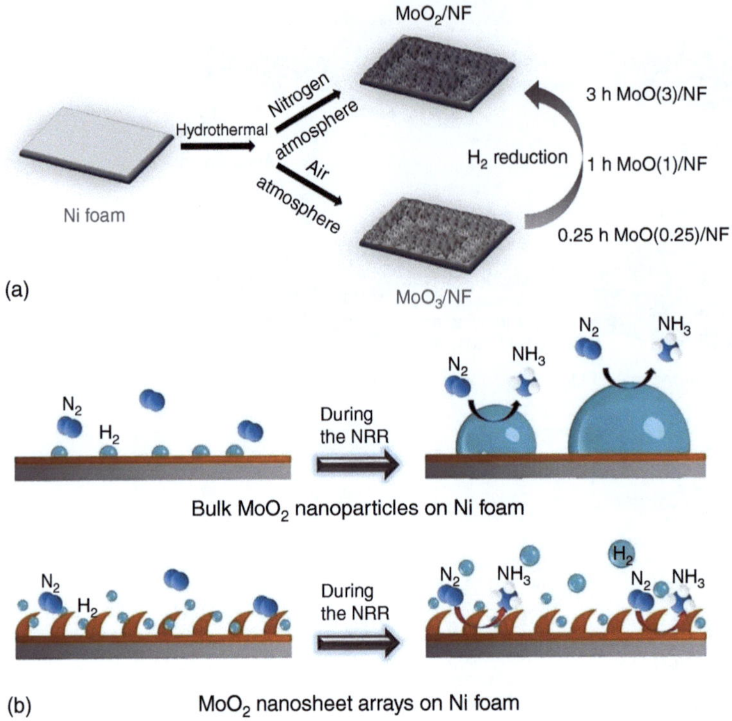

Figure 19.3 (a) Fabrication of MoO_2/NF catalyst. (b) Induced NRR on different surface structures of MoO_2/NF. Source: Reproduced with permission of Wang et al. [47], Elsevier.

19.3.1 Mo-Based Catalysts

Electrocatalysts containing Mo have been widely applied in HER in the past decades and are also emerging as a promising NRR recently. By employing a hydrothermal method (Figure 19.3a), MoO_2 and Ni foam (NF) were combined to afford MoO_2 nanosheets grown on NF (MoO_2/NF), which could work as an efficient NRR catalyst [47]. The porous structures of MnO_2 nanosheets and NF endow MoO_2/NF with a large surface area that provides plenty of catalytic active sites (Figure 19.3b). Finally, a good NH_3 yield rate of 9.56 μg h^{-1} cm^{-2} and FE of 5.14% can be afforded, demonstrating the great potential of Mo-containing materials for efficient electrocatalytic NRR.

The electronic structure of catalysts determines the adsorption of N_2 to a large extent. Only d-orbital elements having occupied d orbitals and appropriate symmetry could accept the lone electron pair at N_2 molecules. A series of investigations have disclosed that transition metal carbides possess a powerful capacity for the adsorption of electron-rich elements [48–50]. Therefore, molybdenum carbide (Mo_2C) was also selected as an NRR catalyst (Figure 19.4a) [51]. The unique electronic structure of Mo_2C has the potential to form a π bond with N_2 to polarize and activate the N≡N triple bond (Figure 19.4b). Unfortunately, the electrocatalytic process of NRR was accompanied by fierce HER by Mo_2C as the catalyst. In order

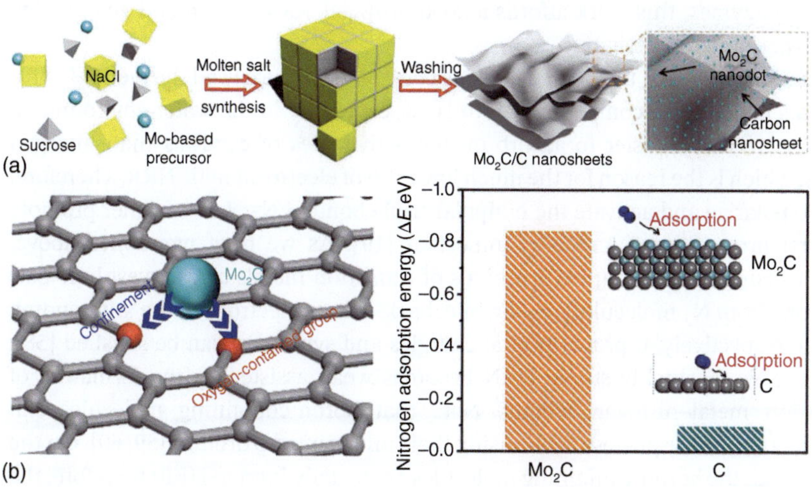

Figure 19.4 (a) Fabrication of Mo_2C catalyst. (b) Diagrams illustrating N_2 adsorption on surface of Mo_2C catalyst. Source: Reproduced with permission of Cheng et al. [51], Wiley-VCH.

to address this problem, a size-controlled synthesis strategy was further developed (Figure 19.4a). As a result, the HER process was suppressed successfully, because the uniform embedding of Mo_2C nanometers effectively buffers the bubble overflow range and resists the loss and deactivation of active sites caused by bubble overflow.

Given that biological N_2 fixation is usually proceeded by employing nitrogenase enzymes containing FeMo, FeFe, or FeV cofactors [52], further exploring artificial catalysts by complying with natural laws should be an effective method [53]. Thus far, several Fe-based catalysts like Fe_2O_3 [54] and FeS_2 [55] have worked as electrochemical NRR catalysts, despite the fact their morphologies are usually hard to control [55]. Note that the MoS_2 nanosheets are easy to synthesize and have exhibited promising catalytic performance for NRR with a moderate FE of 1.17% and NH_3 yield rate of 8.08×10^{-11} mol s^{-1} cm^{-2} in 0.1 M Na_2SO_4 solutions [56]. Therefore, Sun's group established a biomimetic catalytic electrode FeS@MoS_2/CFC consisting of FeS-dotted MoS_2 nanosheets covering carbon fiber cloth (CFC) [57]. Due to better control of the morphology of MoS_2 nanosheets, the resulting FeS@MoS_2/CFC catalytic electrode was also morphologically controllable. In addition, this hybrid catalyst of FeS@MoS_2 shows higher similarity to the elementary composition of FeMo-nitrogenases. Therefore, the possible synergistic effects between Fe and Mo may be expected, thus boosting the catalytic activity of NRR greatly. Consequently, the synergistic effect of uniformly distributed MoS nanosheets and FeS_2 nanoparticles makes the FeS@MoS_2/CFC catalytic electrode demonstrate efficient NRR with an NH_3 yield rate of 8.45 μg h^{-1} cm^{-2} and FE of 2.96% in a pH-neutral electrolyte (0.1 M Na_2SO_4 solutions). More interestingly, the FeS@MoS_2/CFC biomimetic catalytic electrode also displayed relatively good NRR activity in both acidic and alkaline electrolytes. By mimicking the FeMoS-nitrogen fixation enzymes, this work affords a good artificial NRR system and gives a clue for further material design.

In the NRR process, the main challenge confronted by nearly all the metal electrocatalysts is the fierce competition from HER, especially in the water environment. Protons are much easier to adsorb on the active sites of catalysts than nitrogen atoms, which is the reason for the much lower FE of electrocatalytic NRR. Therefore, how to polarize and activate the nonpolar triple bond of N≡N and further promote the first protonation is crucially important [19]. As we have mentioned above, the unoccupied and occupied d orbitals of transition metals can harvest lone-pair electrons from N_2 molecules, meanwhile, back-donate electrons to the antibonding orbital, respectively, if proper orbital energies and symmetry can be satisfied [58]. In this way, the highly strong N≡N becomes weak assisted by the formation of transition metal–nitrogen σ-bond. Note that boron-containing molecules can conduct the similar process above using their unoccupied p orbitals [59, 60]. On the other hand, the boron-containing molecules are usually inert to HER; therefore, the NRR process will be facilitated by combining transition metals and boron together. Herein, an efficient NRR catalytic electrode of B-MoS_2/CFC was established by Sun and co-workers [61]. Among that, the boron-doped MoS_2 nanosheets act as the NRR catalyst, and the CFC works as an excellent scaffold substrate. The two catalytic centers (e.g. boron and molybdenum) process different electron acceptance and back-donating abilities, which could synergistically improve the NRR activity. As a consequence, B-MoS_2/CFC catalytic electrode afforded an excellent NH_3 yield rate of 44.09 $\mu g\,h^{-1}\,cm^{-2}$ and a high FE of 21.72% in the acidic electrolyte (0.1 M HCl). A systematic investigation from theoretical calculations reveals that the boron and molybdenum with different σ-accepting and π-back-donating orbitals could construct a strong–weak electron polarization (SWEP) pair, which has been proven to polarize and activate N_2 initially, and further facilitate the first bond cleavage significantly. In addition, coupling with the solar cells and a bismuth vanadate $BiVO_4$ photo-anode with the catalytic cathode developed here, the authors further established a standalone photoelectrochemical cell (PEC) cell driven by visible light to overall produce ammonia, hydrogen, and oxygen from water and nitrogen for the first time. This demonstrates the great potential for large-scale electrocatalytic production of ammonia using multiple types of green energy.

19.3.2 Co-Based Catalysts

Recently, Co-based catalysts have also been proven to show good NRR activity [62]. For instance, Meng et al. fabricated an electrocatalyst for NRR with oxidized CoP nanoparticles encapsulated in carbon nanotube (O-CoP/CNT) [63]. The DFT calculations disclosed that the electron density at nitrogen atoms deviated to oxygen atoms, which helps weaken the N—N bond and further boost the formation of N—H bond. At the same time, the energy barrier for hydrogen production by O-CoP/CNT (0.257 eV) is much higher than that of O-CoP (0.137 eV), demonstrating a suppressed HER process. This mainly benefits from the hydrophobic CNT coating, which could hinder the proton accessibility and selectively adsorb N_2 molecules.

Atomic doping is a method to change the reaction intermediates/products, increase the number of active sites, and improve their quality by changing the internal microelectronic structure of catalysts [64, 65]. Three-dimensional Co_3O_4 nanospheres (Al-Co_3O_4/NF) rich in oxygen vacancies were developed by hydrothermal method and annealing [66]. The introduction of Al atoms improves the electronic environment of Co_3O_4, effectively enhances the activity of the catalyst surface active sites, and promotes the activation of nitrogen. Its unique sea urchin-like nanostructure demonstrates a good ability to limit bubble coalescence, prevent bubbles from covering the active sites, and reduce the reaction rate. Based on these synergistic advantages, the developed Al-Co_3O_4/NF affords a good electrochemical nitrogen fixation ability in alkaline electrolyte solutions, indicating the potential of these strategies in some other transition metal materials and providing new ideas for the rational design of highly efficient NRR catalysts.

19.3.3 Fe-Based Catalysts

Given that Fe is a component of the catalytic site of nitrogenase, Fe-based catalysts have also received widespread attention in research on electro-catalyzing NRR. Hao's research group fabricated the porous Fe_3O_4 nanosheets with abundant oxygen vacancies by employing a deep eutectic solvent (DES)-based regeneration strategy [67]. The resulting catalyst not only achieves the best FE of 34.38% among iron oxide-based catalysts and excellent NH_3 rate yield of 12.09 $\mu g\ h^{-1}\ mg^{-1}$ but also maintains high stability in Na_2SO_4 electrolyte solutions. The outstanding NRR activity is mainly ascribed to the porous structure of catalysts with a large surface area, plentiful oxygen vacancies, and graphitic carbon from the choline chloride/oxalic acid DES.

The FeS_2-embedded graphene oxide (FeS_2@GO) was developed for electrocatalysis of NRR by Gao et al. [68]. The NRR activities of FeS_2@GO electrode were fully tested in both acidic and alkaline media. Benefiting from a relatively strong interaction between carbonaceous material and FeS_2, a Mott–Schottky interface can be formed and further facilitate the reaction kinetics of NRR. The authors proposed that such a FeS_2-GO interaction could help the nanoparticles disperse uniformly and optimize the interfacial charge transfer; thereby, more active sites can be exposed for nitrogen adsorption at the initial step of NRR [69]. Theoretical investigation also showed that the large charge density at valence band edge and coordinatively unsaturated Fe sites are conductive to the adsorption and activation of $N\equiv N$.

19.4 Center Metal-Coordinated Catalysts

Central metal coordination materials, such as metal–organic framework (MOF), covalent–organic frameworks (COF), and other materials, are a new type of electrocatalyst for NRR. The organic frameworks are usually featured with high specific surface area and porosity. When embedding transition metal atoms as a catalytic active site to prepare new catalysts, a large amount of charge can be transferred from transition metals to the frames. These positively charged metals are expected to compete with H^+ for adsorption, thus greatly inhibiting HER and facilitating the continuous protonation process of NRR [70–72].

19.4.1 MOF- and COF-Based Catalysts

Luo et al. fabricated the C@NiO@Ni microtubules as electrocatalysts by employing MOF as a substrate, which could realize efficient NRR under alkaline environment [73]. The enriched O vacancies on NiO can facilitate proton adsorption at NiO/Ni interface, thus overcoming the difficulty of trapping protons in alkaline electrolytes.

Single-atom catalysts could combine the advantages of both homogeneous and heterogeneous catalysts, such as (i) high atom utilization and excellent natural activity; (ii) well-defined active sites and superb selectivity for different types of catalytic reactions; and (iii) excellent cycling stability [74–76]. Therefore, Tan et al. prepared an Au_{SA}/np-$MoSe_2$ electrocatalyst featuring single-atom Au isolated to nanoporous $MoSe_2$ [77]. The introduction of single Au atoms optimized the adsorption of NRR intermediates while suppressing the competing HER, affording an NH_3 yield rate of 30.83 $\mu g\ h^{-1}\ mg^{-1}$ and FE of 37.82%. Additionally, a new idea of diatomic catalyst was also put forward to be applied in NRR. The diatomic catalyst and 2D organic framework could work together to greatly improve the activity of catalyst [78–81]. For example, Zhao et al. designed a new type of NRR catalyst (M_1M_2-COFs) by embedding the transition metal diatoms into covalent–organic frameworks [82]. By performing theoretical studies, four NRR diatom-organic framework catalysts (MoRu-, MoMo-, MoCo-, and WRu-COFs) are figured out and expected to realize efficient NRR with the low limiting potentials and high FE efficiencies of 76–100% through enzymatic or distal mechanisms. This work points out great potential of diatom-organic framework materials for efficient NRR.

19.4.2 Metal Complex-Based Catalysts

As we have mentioned above, the process of electrocatalytic nitrogen reduction to produce ammonia involves the transfer of six electrons and six protons. There are often multiple reaction pathways and intermediate catalytic species on the surface of nanomaterial catalysts. Therefore, the theoretical modeling of catalytic process and the prediction of catalyst performance are very difficult, which limits the precise design of high-performance catalyst systems to a great extent. Compared with nanomaterial catalysts, molecular catalysts can provide more clues for mechanism research because the transition state can be separated and their chemical structure can be accurately identified [83, 84], which is also of great significance for unveiling the catalytic mechanism of nanomaterial catalysts.

The design of molecular nitrogen reduction catalysts began with biomimetic research on biological nitrogenase. After the pioneering research work on the dinitrogen metal complex (M1) [85] was reported in 1965 (Figure 19.5), many metal complexes that simulated the molybdenum/iron cofactor of nitrogenase were successively developed; however, their catalytic activity was not observed

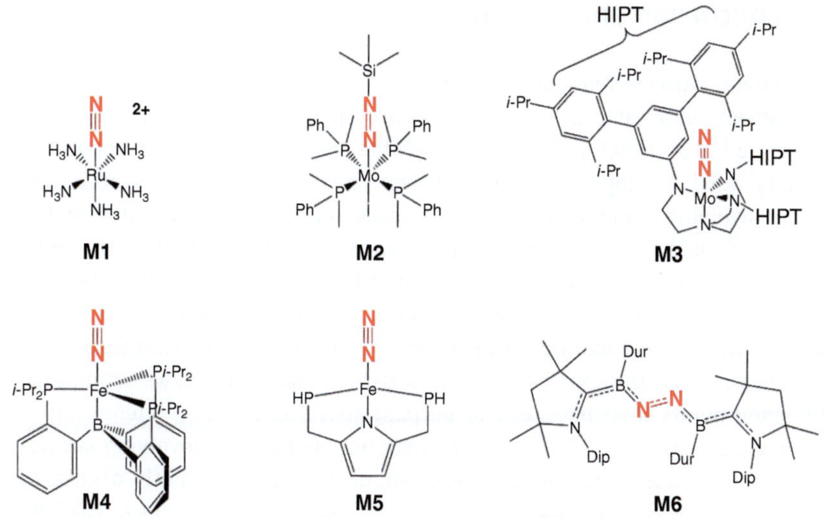

Figure 19.5 Chemical structures of classic nitrogen-containing metal complexes and molecular nitrogen reduction catalysts.

[86, 87]. In 1989, Hidai's group successfully converted dinitrogen into silylamine using molybdenum complex (M2) [88] as a nitrogen reduction catalyst. In 2003, Schrock and co-workers discovered that the transition metal dinitrogen complex (M3) [89] can directly convert dinitrogen into ammonia, meanwhile, revealing the reaction path through both the identification of reaction intermediates and theoretical calculations, which is the well-known Schrock ring. Then, Peters and co-workers found that the iron dinitrogen complex (M4) [89] with triphosphine borane and triphosphine alkyl groups can convert nitrogen into ammonia at a very low temperature (−78 °C). Subsequently, Nishibayashi and co-workers prepared a series of iron complexes (M5) with anionic PNP-clamping ligands [90] and discussed the nitrogen reduction catalytic reaction pathways in detail. In recent years, Braunschweig's group reported the first case of nonmetallic nitrogen reduction molecular catalyst (M6) [60], which uses B as a catalytic site to adsorb nitrogen molecules and catalyze the synthesis of ammonia.

Molecular catalysts for NRR display much larger turnover numbers and can reveal more clues for mechanism study because of the separable transition states, as compared with that of nanomaterial catalysts. However, much harsh conditions are demanded by molecular catalysts in order to maintain the good selectivity of NRR. Excessive soluble organic acids provide protons and highly strong reductants as sacrificial agents to afford electrons during the reaction. In order to depress their dramatic annihilation, the reaction generally has to settle for a very low temperature, such as −20 °C or −78 °C [91]. Moreover, the industrially feasible electrocatalysis in an aqueous electrolyte by molecular catalysts remains a great challenge due to lack of stability and the unclear electron and proton transfer pathways [92]. All the above hinder the further application of molecular catalysts.

19.5 Conclusion and Outlook

Electrocatalytic nitrogen reduction to synthesize ammonia has the great potential to realize green ammonia production and is currently the most promising strategy to replace the Haber–Bosch process. Although the electrocatalytic NRR has undergone a very rapid development in the last few years, there are still several huge challenges involving reducing the cost of catalysts, improving the NH_3 production selectivity and yield rate, etc. For example, the main problem faced by the most metal-based catalysts is the fierce competition of HER. Protons are much easier to adsorb on active sites of catalysts than nitrogen atoms, therefore, the further design of efficient NRR catalysts requires rational arrangement of the number and location of active sites and also the appropriate catalyst particle sizes. Herein, we summarize the following points that may need to be paid attention to when attempting to design new catalysts for NRR: (i) Lots of material engineering involving surface hydrophobicity, lattice matching, and heteroatom doping are urgently needed to suppress the fierce HER competition with NRR on the metal surface. (ii) The porous structures are highly desired in the view of exposing more active sites for NRR. (iii) The catalytic pathways for NRR are very complicated and still unclear because the processes involve six protons and six electrons. Given that the transition state of molecular catalysts can be separated and accurately identified, maybe the molecular catalysts could give some clues regarding the exploration of catalytic mechanisms. (iv) The initial polarization of nonpolar N≡N is a great challenge but quite important. Constructing a SWEP pair in catalysts may be an effective way to weaken N≡N and facilitate the first bond cleavage.

In conclusion, the current electrocatalytic nitrogen reduction ammonia synthesis is still a long way from industrial production; however, a key step forward has been made. We believe that the electrochemical ammonia synthesis technology even driven by renewable and clean energy will be widely promoted and replace the Haber–Bosch process for industrial ammonia synthesis.

Acknowledgment

This work was supported by the National Natural Science Foundation of China (22204119).

References

1 Marnellos, G. and Stoukides, M. (1998). Ammonia synthesis at atmospheric pressure. *Science* 282 (5386): 98.

2 Canfield, D.E., Glazer, A.N., and Falkowski, P.G. (2010). The evolution and future of earth's nitrogen cycle. *Science* 330 (6001): 192.

3 Kitano, M., Inoue, Y., Yamazaki, Y. et al. (2012). Ammonia synthesis using a stable electride as an electron donor and reversible hydrogen store. *Nat. Chem.* 4 (11): 934.

4 van der Ham, C.J.M., Koper, M.T.M., and Hetterscheid, D.G.H. (2014). Challenges in reduction of dinitrogen by proton and electron transfer. *Chem. Soc. Rev.* 43 (15): 5183.

5 Rodriguez, M.M., Bill, E., Brennessel, W.W., and Holland, P.L. (2011). N_2 reduction and hydrogenation to ammonia by a molecular iron-potassium complex. *Science* 334 (6057): 780.

6 Andersson, J. and Lundgren, J. (2014). Techno-economic analysis of ammonia production via integrated biomass gasification. *Appl. Energy* 130: 484.

7 Lai, F., Zong, W., He, G. et al. (2020). N_2 electroreduction to NH_3 by selenium vacancy-rich $ReSe_2$ catalysis at an abrupt interface. *Angew. Chem. Int. Ed.* 59 (32): 13320.

8 Seefeldt, L.C., Hoffman, B.M., and Dean, D.R. (2012). Electron transfer in nitrogenase catalysis. *Curr. Opin. Chem. Biol.* 16 (1): 19.

9 Lancaster, K.M., Roemelt, M., Ettenhuber, P. et al. (2011). X-ray emission spectroscopy evidences a central carbon in the nitrogenase iron-molybdenum cofactor. *Science* 334 (6058): 974.

10 Hirakawa, H., Hashimoto, M., Shiraishi, Y., and Hirai, T. (2017). Photocatalytic conversion of nitrogen to ammonia with water on surface oxygen vacancies of titanium dioxide. *J. Am. Chem. Soc.* 139 (31): 10929.

11 Medford, A.J. and Hatzell, M.C. (2017). Photon-driven nitrogen fixation: current progress, thermodynamic considerations, and future outlook. *ACS Catal.* 7 (4): 2624.

12 Winter, L.R. and Chen, J.G. (2021). N_2 fixation by plasma-activated processes. *Joule* 5 (2): 300.

13 Milton, R.D., Abdellaoui, S., Khadka, N. et al. (2016). Nitrogenase bioelectrocatalysis: heterogeneous ammonia and hydrogen production by MoFe protein. *Energy Environ. Sci.* 9 (8): 2550.

14 Liu, H. (2014). Ammonia synthesis catalyst 100 years: practice, enlightenment and challenge. *Chin. J. Catal.* 35 (10): 1619.

15 Nørskov, J., Chen, J., Miranda, R. et al. (2016). *Sustainable ammonia synthesis – exploring the scientific challenges associated with discovering alternative, sustainable processes for ammonia production.* https://doi.org/10.2172/1283146.

16 Weare, W.W., Dai, X., Byrnes, M.J. et al. (2006). Catalytic reduction of dinitrogen to ammonia at a single molybdenum center. *Proc. Natl. Acad. Sci.* 103 (46): 17099.

17 Pickett, C.J. and Talarmin, J. (1985). Electrosynthesis of ammonia. *Nature* 317 (6038): 652.

18 Moon, Y.H., Kim, N.Y., Kim, S.M., and Jang, Y.J. (2022). Recent advances in electrochemical nitrogen reduction reaction to ammonia from the catalyst to the system. *Catalysts* 12 (9): 1015.

19 Suryanto, B.H.R., Du, H.-L., Wang, D. et al. (2019). Challenges and prospects in the catalysis of electroreduction of nitrogen to ammonia. *Nat. Catal.* 2 (4): 290.

20 Wang, L., Xia, M., Wang, H. et al. (2018). Greening ammonia toward the solar ammonia refinery. *Joule* 2 (6): 1055.

21 Erisman, J.W., Sutton, M.A., Galloway, J. et al. (2008). How a century of ammonia synthesis changed the world. *Nat. Geosci.* 1 (10): 636.

22 Yao, Y., Zhu, S., Wang, H. et al. (2018). A spectroscopic study on the nitrogen electrochemical reduction reaction on gold and platinum surfaces. *J. Am. Chem. Soc.* 140 (4): 1496.

23 Dunwell, M., Lu, Q., Heyes, J.M. et al. (2017). The central role of bicarbonate in the electrochemical reduction of carbon dioxide on gold. *J. Am. Chem. Soc.* 139 (10): 3774.

24 Badea, G.E. (2009). Electrocatalytic reduction of nitrate on copper electrode in alkaline solution. *Electrochim. Acta* 54 (3): 996.

25 Hao, Y.-C., Guo, Y., Chen, L.-W. et al. (2019). Promoting nitrogen electroreduction to ammonia with bismuth nanocrystals and potassium cations in water. *Nat. Catal.* 2 (5): 448.

26 Ling, C., Zhang, Y., Li, Q. et al. (2019). New mechanism for N_2 reduction: the essential role of surface hydrogenation. *J. Am. Chem. Soc.* 141 (45): 18264.

27 Li, H.L., Robertson, D.H., Chambers, J.Q., and Hobbs, D.T. (1988). Electrochemical reduction of nitrate and nitrite in concentrated sodium hydroxide at platinum and nickel electrodes. *J. Electrochem. Soc.* 135 (5): 1154.

28 Zhang, H., Osmieri, L., Park, J.H. et al. (2022). Standardized protocols for evaluating platinum group metal-free oxygen reduction reaction electrocatalysts in polymer electrolyte fuel cells. *Nat. Catal.* 5 (5): 455.

29 Seh, Z.W., Kibsgaard, J., Dickens, C.F. et al. (2017). Combining theory and experiment in electrocatalysis: insights into materials design. *Science* 355 (6321): eaad4998.

30 Tong, W., Huang, B., Wang, P. et al. (2020). Exposed facet-controlled N_2 electroreduction on distinct Pt_3Fe nanostructures of nanocubes, nanorods and nanowires. *Natl. Sci. Rev.* 8 (1): nwaa088.

31 Wang, J., Yu, L., Hu, L. et al. (2018). Ambient ammonia synthesis via palladium-catalyzed electrohydrogenation of dinitrogen at low overpotential. *Nat. Commun.* 9 (1): 1795.

32 Wang, X., Luo, M., Lan, J. et al. (2021). Nanoporous intermetallic Pd_3Bi for efficient electrochemical nitrogen reduction. *Adv. Mater.* 33 (18): 2007733.

33 Shi, M.-M., Bao, D., Li, S.-J. et al. (2018). Anchoring PdCu amorphous nanocluster on graphene for electrochemical reduction of N_2 to NH_3 under ambient conditions in aqueous solution. *Adv. Energy Mater.* 8 (21): 1800124.

34 Shi, M.-M., Bao, D., Wulan, B.-R. et al. (2017). Au sub-nanoclusters on TiO_2 toward highly efficient and selective electrocatalyst for N_2 conversion to NH_3 at ambient conditions. *Adv. Mater.* 29 (17): 1606550.

35 Kim, H.S., Choi, J., Kong, J. et al. (2021). Regenerative electrocatalytic redox cycle of copper sulfide for sustainable NH_3 production under ambient conditions. *ACS Catal.* 11 (1): 435.

36 Zhao, M., Guo, C., Gao, L. et al. (2021). Vanadium-doped NiS_2 porous nanospheres with high selectivity and stability for the electroreduction of nitrogen to ammonia. *Inorg. Chem. Front.* 8 (13): 3266.

37 Li, W., Ye, Y., Jin, M. et al. (2023). Ambient electrochemical nitrogen fixation with aqueous V_2O_5 nanodots in a fluidized electrocatalysis system. *Chem. Eng. J.* 452: 139494.

38 Ren, J.-T., Chen, L., Wang, H.-Y., and Yuan, Z.-Y. (2021). Aqueous rechargeable Zn-N_2 battery assembled by bifunctional cobalt phosphate nanocrystals-loaded carbon nanosheets for simultaneous NH_3 production and power generation. *ACS Appl. Mater. Interfaces* 13 (10): 12106.

39 Monica, F.D. and Kleij, A.W. (2021). Synthesis and characterization of biobased polyesters with tunable T_g by ROCOP of beta-elemene oxides and phthalic anhydride. *ACS Sustain. Chem. Eng.* 9 (7): 2619.

40 Zhao, S., Liu, H.-X., Qiu, Y. et al. (2020). An oxygen vacancy-rich two-dimensional Au/TiO_2 hybrid for synergistically enhanced electrochemical N_2 activation and reduction. *J. Mater. Chem. A* 8 (14): 6586.

41 Zhang, Y., Zhang, Q., Liu, D.-X. et al. (2021). High spin polarization ultrafine Rh nanoparticles on CNT for efficient electrochemical N_2 fixation to ammonia. *Appl. Catal. B Environ.* 298: 120592.

42 Wang, H., Mao, Q., Yu, H. et al. (2021). Enhanced electrocatalytic performance of mesoporous Au-Rh bimetallic films for ammonia synthesis. *Chem. Eng. J.* 418: 129493.

43 Wang, Z., Tian, W., Dai, Z. et al. (2021). Bimetallic mesoporous rhru film for electrocatalytic nitrogen reduction to ammonia. *Inorg. Chem. Front.* 8 (18): 4276.

44 Zhang, M., Zhang, L., Huang, S. et al. (2021). 2D gallium molybdenum selenide grown on a hollow carbon nanofibrous aerogel for high-efficiency electroreduction of nitrogen: optimized basal plane activity via selenium vacancy modulation. *Appl. Catal. B Environ.* 292: 120175.

45 Nishibayashi, Y. (2015). Recent progress in transition-metal-catalyzed reduction of molecular dinitrogen under ambient reaction conditions. *Inorg. Chem.* 54 (19): 9234.

46 Rittle, J. and Peters, J.C. (2016). An Fe-N_2 complex that generates hydrazine and ammonia via Fe=NNH_2: demonstrating a hybrid distal-to-alternating pathway for N_2 reduction. *J. Am. Chem. Soc.* 138 (12): 4243.

47 Wang, H.-Y., Ren, J.-T., Weng, C.-C. et al. (2021). Insight into the valence state of sisal-like MoO_2 nanosheet arrays for N_2 electrolysis. *Chem. Eng. J.* 426: 130761.

48 Azofra, L.M., Li, N., MacFarlane, D.R., and Sun, C. (2016). Promising prospects for 2d d^2–d^4 M_3C_2 transition metal carbides (mxenes) in N_2 capture and conversion into ammonia. *Energy Environ. Sci.* 9 (8): 2545.

49 Naguib, M., Mashtalir, O., Carle, J. et al. (2012). Two-dimensional transition metal carbides. *ACS Nano* 6 (2): 1322.

50 Qin, B., Li, Y., Zhang, Q. et al. (2020). Understanding of nitrogen fixation electro catalyzed by molybdenum–iron carbide through the experiment and theory. *Nano Energy* 68: 104374.

51 Cheng, H., Ding, L.-X., Chen, G.-F. et al. (2018). Molybdenum carbide nanodots enable efficient electrocatalytic nitrogen fixation under ambient conditions. *Adv. Mater.* 30 (46): 1803694.

52 Eady, R.R. (1996). Structure–function relationships of alternative nitrogenases. *Chem. Rev.* 96 (7): 3013.

53 Cui, X., Tang, C., and Zhang, Q. (2018). A review of electrocatalytic reduction of dinitrogen to ammonia under ambient conditions. *Adv. Energy Mater.* 8 (22): 1800369.

54 Kong, J., Lim, A., Yoon, C. et al. (2017). Electrochemical synthesis of NH_3 at low temperature and atmospheric pressure using a γ-Fe_2O_3 catalyst. *ACS Sustain. Chem. Eng.* 5 (11): 10986.

55 Lashgari, M. and Zeinalkhani, P. (2018). Ammonia photosynthesis under ambient conditions using an efficient nanostructured FeS_2/CNT solar-energy-material with water feedstock and nitrogen gas. *Nano Energy* 48: 361.

56 Zhang, L., Ji, X., Ren, X. et al. (2018). Electrochemical ammonia synthesis via nitrogen reduction reaction on a MoS_2 catalyst: theoretical and experimental studies. *Adv. Mater.* 30 (28): 1800191.

57 Guo, Y., Yao, Z., Timmer, B.J.J. et al. (2019). Boosting nitrogen reduction reaction by bio-inspired FeMoS containing hybrid electrocatalyst over a wide pH range. *Nano Energy* 62: 282.

58 Lü, F., Zhao, S., Guo, R. et al. (2019). Nitrogen-coordinated single Fe sites for efficient electrocatalytic N_2 fixation in neutral media. *Nano Energy* 61: 420.

59 Qiu, W., Xie, X.-Y., Qiu, J. et al. (2018). High-performance artificial nitrogen fixation at ambient conditions using a metal-free electrocatalyst. *Nat. Commun.* 9 (1): 3485.

60 Légaré, M.-A., Bélanger-Chabot, G., Dewhurst, R.D. et al. (2018). Nitrogen fixation and reduction at boron. *Science* 359 (6378): 896.

61 Guo, Y., Yao, Z., Zhan, S. et al. (2020). Molybdenum and boron synergistically boosting efficient electrochemical nitrogen fixation. *Nano Energy* 78: 105391.

62 Castellano-Varona, B., Harb, M., Araña, J. et al. (2020). In silico design of novel NRR electrocatalysts: cobalt–molybdenum alloys. *Chem. Commun.* 56 (87): 13343.

63 Meng, Q., Hou, Y., Yang, F. et al. (2022). Modulation of surface properties on cobalt phosphide for high-performance ambient ammonia electrosynthesis. *Appl. Catal. B Environ.* 303: 120874.

64 Li, X., Li, T., Ma, Y. et al. (2018). Boosted electrocatalytic N_2 reduction to NH_3 by defect-rich MoS_2 nanoflower. *Adv. Energy Mater.* 8 (30): 1801357.

65 Chung, S., Ju, H., Choi, M. et al. (2022). Local proton source enhanced nitrogen reduction on a combined cobalt-molybdenum catalyst for electrochemical ammonia synthesis. *Angew. Chem. Int. Ed.* 61 (47): e202212676.

66 Lv, X.-W., Liu, Y., Hao, R. et al. (2020). Urchin-like Al-doped Co_3O_4 nanospheres rich in surface oxygen vacancies enable efficient ammonia electrosynthesis. *ACS Appl. Mater. Interfaces* 12 (15): 17502.

67 Ying, H., Chen, T., Zhang, C. et al. (2021). Regeneration of porous Fe_3O_4 nanosheets from deep eutectic solvent for high-performance electrocatalytic nitrogen reduction. *J. Colloid Interface Sci.* 602: 64.

68 Gao, L., Guo, C., Zhao, M. et al. (2021). Electrocatalytic N_2 reduction on FeS_2 nanoparticles embedded in graphene oxide in acid and neutral conditions. *ACS Appl. Mater. Interfaces* 13 (42): 50027.

69 Gao, L., Guo, C., Liu, X. et al. (2020). Co-doped FeS_2 with a porous structure for efficient electrocatalytic overall water splitting. *New J. Chem.* 44 (5): 1711.

70 Gu, Y., Xi, B., Tian, W. et al. (2021). Boosting selective nitrogen reduction via geometric coordination engineering on single-tungsten-atom catalysts. *Adv. Mater.* 33 (25): 2100429.

71 Yang, M., Wang, X., Gómez-García, C.J. et al. (2023). Efficient electron transfer from an electron-reservoir polyoxometalate to dual-metal-site metal-organic frameworks for highly efficient electroreduction of nitrogen. *Adv. Funct. Mater.* 33 (28): 2214495.

72 He, C., Xu, C., and Zhang, W. (2022). Shortening the screening process towards high-performance 2d-MOF NRR electrocatalysts with $\Delta(\mu_{B\_TM}\text{-}\mu_{B\_X})$ as the descriptor of N_2 activation capability. *Appl. Surf. Sci.* 606: 154904.

73 Luo, S., Li, X., Gao, W. et al. (2020). An MOF-derived C@NiO@Ni electrocatalyst for N_2 conversion to NH_3 in alkaline electrolytes. *Sustain. Energy Fuels* 4 (1): 164.

74 Cheng, N., Zhang, L., Doyle-Davis, K., and Sun, X. (2019). Single-atom catalysts: from design to application. *Electrochem. Energy Rev.* 2 (4): 539.

75 Lang, R., Du, X., Huang, Y. et al. (2020). Single-atom catalysts based on the metal–oxide interaction. *Chem. Rev.* 120 (21): 11986.

76 Chen, Y., Ji, S., Chen, C. et al. (2018). Single-atom catalysts: synthetic strategies and electrochemical applications. *Joule* 2 (7): 1242.

77 Chen, D., Luo, M., Ning, S. et al. (2022). Single-atom gold isolated onto nanoporous $MoSe_2$ for boosting electrochemical nitrogen reduction. *Small* 18 (4): 2104043.

78 Jiao, L. and Guo, L. (2022). Embedding double transition metal atoms in B-modified two-dimensional carbon-rich conjugated frameworks for efficient ammonia synthesis. *Inorg. Chem.* 61 (46): 18574.

79 Wei, X., Cao, S., Xu, H. et al. (2022). Novel two-dimensional metal organic frameworks: high-performance bifunctional electrocatalysts for OER/ORR. *ACS Mater. Lett.* 4 (10): 1991.

80 Huo, J., Wei, H., Zhang, K. et al. (2023). Nitrogen reduction reaction catalyzed by diatomic metals supported by N-doped graphite. *Catalysts* 13 (1): 49.

81 Li, W.-H., Yang, J., and Wang, D. (2022). Long-range interactions in diatomic catalysts boosting electrocatalysis. *Angew. Chem. Int. Ed.* 61 (52): e202213318.

82 Zhang, Z., Qi, S., Wang, J., and Zhao, M. (2021). Bi-atom active sites embedded in a two-dimensional covalent organic framework for efficient nitrogen reduction reaction. *Appl. Surf. Sci.* 563: 150352.

83 Foster, S.L., Bakovic, S.I.P., Duda, R.D. et al. (2018). Catalysts for nitrogen reduction to ammonia. *Nat. Catal.* 1 (7): 490.

84 Chalkley, M.J., Drover, M.W., and Peters, J.C. (2020). Catalytic N_2-to-NH_3 (or $-N_2H_4$) conversion by well-defined molecular coordination complexes. *Chem. Rev.* 120 (12): 5582.

85 Allen, A.D. and Senoff, C.V. (1965). Nitrogenopentammineruthenium(II) complexes. *Chem. Commun.* 24: 621.

86 Sippel, D., Rohde, M., Netzer, J. et al. (2018). A bound reaction intermediate sheds light on the mechanism of nitrogenase. *Science* 359 (6383): 1484.

87 Van Stappen, C., Decamps, L., Cutsail, G.E. et al. (2020). The spectroscopy of nitrogenases. *Chem. Rev.* 120 (12): 5005.

88 Komori, K., Oshita, H., Mizobe, Y. et al. (1989). Preparation and properties of molybdenum and tungsten dinitrogen complexes. 25. Catalytic conversion of molecular nitrogen into silylamines using molybdenum and tungsten dinitrogen complexes. *J. Am. Chem. Soc.* 111 (5): 1939.

89 Yandulov, D.V. and Schrock, R.R. (2003). Catalytic reduction of dinitrogen to ammonia at a single molybdenum center. *Science* 301 (5629): 76.

90 Kuriyama, S., Arashiba, K., Nakajima, K. et al. (2016). Catalytic transformation of dinitrogen into ammonia and hydrazine by iron-dinitrogen complexes bearing pincer ligand. *Nat. Commun.* 7 (1): 12181.

91 Arashiba, K., Miyake, Y., and Nishibayashi, Y. (2011). A molybdenum complex bearing PNP-type pincer ligands leads to the catalytic reduction of dinitrogen into ammonia. *Nat. Chem.* 3 (2): 120.

92 Chalkley, M.J., Del Castillo, T.J., Matson, B.D., and Peters, J.C. (2018). Fe-mediated nitrogen fixation with a metallocene mediator: exploring pK_a effects and demonstrating electrocatalysis. *J. Am. Chem. Soc.* 140 (19): 6122.

20

Carbon-Based Materials for Electrocatalytic N_2 Reduction Reaction

Wuyong Zhang[1], Xinsheng Liu[2], and Lei Dai[2]

[1] *University of Chinese Academy of Sciences, Ningbo Institute of Materials Technology and Engineering, Qianwan Institute of CNITECH, Zhongguan West Road, Ningbo, 315201, China*
[2] *Henan University, Academy for Advanced Interdisciplinary Studies, School of Materials Science and Engineering, Key Laboratory for Special Functional Materials of Ministry of Education, Dongjing Road, Kaifeng, 475004, China*

20.1 Introduction

The pursuit of artificial nitrogen (N_2) fixation has persisted until today without waning [1, 2]. While the Haber–Bosch process for nitrogen reduction has been notably successful and has greatly contributed to the progress of human civilization, it is also energy-intensive and environmentally impactful. This process consumes nearly 1% of the world's energy supply and generates 1.87 tons of CO_2 per ton of NH_3 produced, representing almost 1% of the total emissions of greenhouse gases [3]. Nature has taught us that nitrogen fixation under atmospheric conditions is achievable through sophisticated forms of the chemistry of living matter, particularly in certain nitrogenase bacteria [4, 5], but we are still struggling to catch up to the efficiency of nature until today. Thankfully, many alternatives, like electrocatalytic process, have been developed in recent decades, which shows the determination and ambition of scientific community for nitrogen-fix revolution [6]. This way, we can produce ammonia fully based on electrochemical reduction of dinitrogen as nitrogen source and water as hydrogen source with renewable energy [7]. From the US Department of Energy (DoE)'s REFUEL program, the practical performance targets for electrochemical nitrogen reduction reaction (NRR) are reaching a current density of 300 mA cm^{-2} with Faradaic efficiency (FE) of 90% for ammonia and 60% energy efficiency [8].

However, these given indicators in the present NRR system with lows still have a long way to go to meet the demand. The underlying reason can be attributed to the inert nature of N_2 molecule; the reader will find N_2 familiar yet enigmatic if it is scrutinized thoroughly (Figure 20.1) [9]. N_2 is linear and centrosymmetric with N≡N triple bond and sp hybridization formed by the two N atoms. The N≡N triple bond has a short bond length of 109.8 pm with high bond energy of 941 kJ mol^{-1}, but it cannot totally explain the chemically inert reason of N_2. In contrast, the bond energies of acetylene (HC≡CH, 962 kJ mol^{-1}) and carbon monoxide (CO, 1070 kJ mol^{-1})

Functional Materials for Electrocatalytic Energy Conversion, First Edition.
Edited by Zhicheng Zhang, Meiting Zhao, and Yuchen Qin.

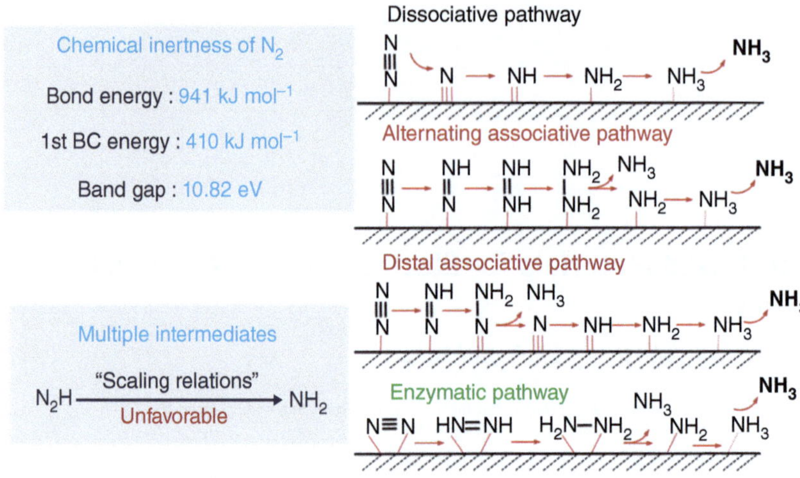

Figure 20.1 The physical–chemical properties and reduction pathway of N_2.

are even higher, but as we know, their chemical properties are significantly more active than N_2. The meaningful difference is the first bond cleavage energy for N_2 is 410 kJ mol^{-1}, which is much higher than that of HC≡CH (222 kJ mol^{-1}). There are also other underlying reasons that can explain the inertness of N_2 like the lack of permanent dipole, low proton affinity (493.8 kJ mol^{-1}), negative electron affinity (−1.90 eV), high ionization potential (15.84 eV), and the large energy gap (10.82 eV) between HOMO and LUMO [10, 11]. Therefore, the direct pronation of N_2 is hard under normal conditions even with strong acid. Activating N_2 through electrochemistry seems to be thermodynamically favorable with a low theoretical potential of 0.092 V vs. reversible hydrogen electrode (RHE). Theoretically, if a sufficiently negative potential is applied, the given electrochemical system can drive nitrogen reduction under mild reaction. However, there is a competing hydrogen evolution reaction (HER) present in aqueous system, which only requires two electrons per H_2 produced [12]. And for NH_3 produced from NRR, whether based on dissociative, associative, or enzymatic pathway, it involves six-electron transfer with multiple intermediates, creating an unfavorable scaling relation [13]. It is noteworthy that the first electron transfer of N_2 to form N_2^- has a redox potential of −2.78 vs. RHE with an extremely high-energy barrier [14, 15]. The energy barrier with the multiple intermediates must be circumvented by catalyst design [16]. Therefore, the top priority for practical NRR systems with space–time yield and energy efficiency is the development of novel catalytic nanomaterials with high-energy adsorption sites [17].

Among the materials investigated for NRR, noble metal-based materials seem to be good candidates as their versatility in nearly all electrocatalytic reactions if the scarcity and high cost can be ignored [18–20]. Transition metal-based materials have also been studied and undoubtedly show decent activity, but their usually poor conductivity, limited chemical stability, corrosion susceptibility, and potential environmental risk have severely restricted their large-scale application [21, 22]. Conversely, cost-free and abundant carbon-based materials (CBMs) have higher electrical conductivity, which is crucial for efficient electron, leading to better overall performance, and their excellent chemical stability contributes to their longevity and durability during harsh chemical conditions [23]. Moreover, the nature of large surface area can provide more active sites for nitrogen adsorption and activation, and the properties can often be tailored by doping different heteroatoms [24, 25]. Further, the heteroatom-doped carbon can anchor metal sites to construct metal–carbon composite material, which extensively further expands its scope of application [26].

In this chapter, we provide a summary of the state-of-the-art research trajectory in CBMs for electrochemical NRR under ambient conditions. We highlight advancements in CBM fabrication strategies and explore the underlying activation mechanisms. Additionally, the challenges and future opportunities of utilizing CBMs for NRR are presented.

20.2 Heteroatom-Doping Carbon-Based Materials (HDCBMs)

Heteroatom doping involves chemically attaching or incorporating non-carbon elements (heteroatoms) into the backbone of the respective CBMs, which are commonly employed to modify and enhance the properties of CBMs [25, 27, 28]. A commonly accepted viewpoint is that the proximity of heteroatoms with varying electronegativity or electron affinity can perturb the electron distribution of the ordered sp^2 hybridized carbon. This disturbance leads to increased charge delocalization, electron transfer rates, metal properties, and catalytic sites in CBMs. These factors are deemed strategic pathways for activating nitrogen by enhancing the chemisorption of reactants. Undoubtedly, chemisorption can be synergistically enhanced through the incorporation of a porous structure with high physisorption [29]. For HDCBMs, the first candidate that comes to mind is nitrogen-doped carbon, which has been developed over 100 years with number of works published [23, 30, 31]. Many studies also show that nitrogen doping can be extended to electrocatalytic NRR. Zhao and co-workers reported an N-doped porous carbon (NPC) synthesized by pyrolyzing zeolite imidazolate framework (ZIF) under different temperatures to regulate the nitrogen content and species (Figure 20.2a) [32]. As the pyrolysis temperature increased, the nitrogen content of NPCs decreased and a highly ordered carbon structure formed. The resulting NPC-750 with low temperature showed a robust activity with NH_3 yield rate of $1.40\ \text{mmol}\ g_{cat}^{-1}\ h^{-1}$ at −0.9 V vs. RHE (Figure 20.2b). Combining theoretical and experimental analysis, the authors claim that the high pyrrolic and pyridinic N content of NPCs are the active centers for NH_3 production, and the promising route for NH_3 formation is $^*N{\equiv}N \rightarrow {}^*NH{=}NH \rightarrow {}^*NH_2{-}NH_2 \rightarrow 2NH_3$ (Figure 20.2c). Coincidentally, Wu and co-workers also employed ZIF-8 as precursor to obtain N-doped carbon nanoporous carbon (C-ZIF) under different pyrolysis temperatures [33]. Interestingly, different from the former one, in this case, the best-performing one is C-ZIF-1100 with high temperature, which exhibits maximum NH_3 production

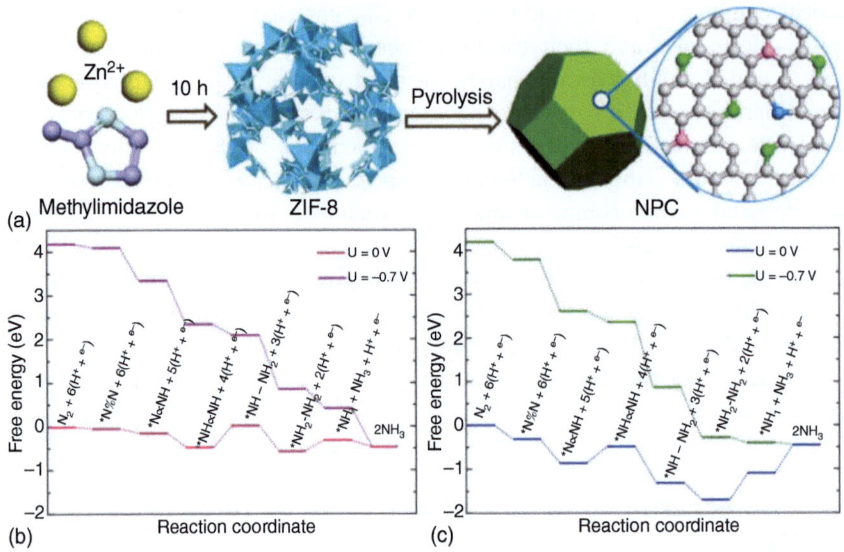

Figure 20.2 (a) The preparation illustration of NPC. Free energy diagram for ammonia synthesis on NPC with (b) pyridinic and (c) pyrrolic N. Source: Reproduced with permission from Liu et al. [32], American Chemical Society.

rate of 3.4×10^{-6} mol cm^{-2} h^{-1} and FE of 10.2% at −0.3 V vs. RHE. DFT calculations suggested that pyridinic N_3 moieties consist of one protonated pyridinic nitrogen and one adjacent vacancy in the graphitic layer, which are active sites for nitrogen reduction. With the pyridinic nitrogen, the adjacent vacancy can effortlessly absorb and initiate N_2. Although NPCs have been proven to be promising NRR catalysts with different mechanisms for N_2 activation proposed, the complexity associated with nitrogen doping variations makes it hard for researchers to figure out the real active sites. The controllable synthesis of NPCs with well-defined mechanisms is imperative to uncover the NRR mechanism. Oschatz and co-workers employed a C_2N-type NPC of HAT-700 with high concentration of pyrazinic nitrogen and terminal nitrile groups by pyrolyzing organic precursor at given temperature [34]. HAT-700 can go through an unexampled electrochemical activation process with terminal nitrile groups etched to expose carbon radical sites. The resulted C_2N exhibits excellent NRR performance with the highest FE of 11.4% and a remarkable NH_3 yield of 5.86 µg mg$^{-1}{}_{cat}$ h^{-1}. Benefiting from the well-defined active sites, the corresponding NRR mechanism based on a synergistic effect between carbon radicals and adjacent pyrazinic nitrogen atoms is proposed, which is supported by DFT calculations and experimental results.

Besides the widely used nitrogen doping, boron (B) has also been considered as promising candidate for active doping for nitrogen reduction [35]. B, as a p-block element with three valence electrons and empty p-orbitals [36, 37], has lower electronegativity than that of C, so B will be positively charged and play the role of Lewis acid, which can activate N_2 by elongating N≡N bond (Figure 20.3a) [41]. Moreover, after hybridization, the formed unoccupied orbitals of B can incorporate the lone pair electrons of nitrogen, and the electrons on the filled orbitals can donate back to the antibonding orbitals of N_2, this effect is called "π Back-Donation" (Figure 20.3b) [15, 42]. Many B-doping carbons have emerged recently and show decent NRR activity and selectivity [43, 44]. Zheng and co-workers reported a boron-doped graphene prepared by thermally reducing graphene oxide (GO) and boric acid in H_2/Ar atmosphere (Figure 20.3c–e) [38]. The local structure of BC_3 enables the lowest energy barrier for NRR, which achieves an NH_3 production rate of 9.8 μg h^{-1} cm^{-2} and FE of 10.8% at −0.5 V vs. RHE. Sun and co-workers reported boron carbide (B_4C) NSs prepared by liquid exfoliation of bulk B_4C (Figure 20.3f–h), which displayed a NH_3 synthesis rate of 26.57 μg $mg^{-1}{}_{cat.}$ h^{-1} and maximum FE of 15.95% at −0.75 V vs. RHE [39]. To further maximize the active sites of boron-doped carbon, Tan and co-workers utilize metallurgical alloy and chemical etching methods to fabricate nanoporous $B_{13}C_2$ (Figure 20.3i), which highly improves the NRR performance to achieve NH_3 yield of 91.28 μg $mg^{-1}{}_{cat.}$ h^{-1} and maximum FE of 35.53% at −0.05 V vs. RHE [40].

Other elements such as sulfur (S), fluorine (F), oxygen (O), and co-doping have also been utilized as active doping agents for carbon, aiming to enhance the NRR activity [45, 46]. Sun and co-workers reported a sulfur-doped carbon sphere as an efficient and stable NRR catalyst with the active sites of carbon atoms close to substituted sulfur proposed by DFT calculations (Figure 20.4a,b) [47]. The variance in electronegativity between F and C instigates a repulsive interaction at the Lewis acid site, discouraging the binding of H^+ and conversely dynamically enhancing selectivity toward N_2, fostering a robust affinity between N_2 and the Lewis acid C site. Zeng and co-workers reported a porous F-doped carbon derived from MOF precursor and HF etching (Figure 20.4c), which reaches the highest FE of 54.8% and maximum NH_3 yield of 197.7 μg $mg^{-1}{}_{cat.}$ h^{-1} (Figure 20.4d) [48]. The underlying mechanism can be attributed to F-doping enhanced binding strength of N_2 and the facilitated dissociation of N_2 into $*N_2H$. Similarly, oxygen doping can change the hybridization state of carbon from sp^2 to sp^3, which will cause the disordered carbon structure to undergo charge redistribution. Kong and co-workers employed in situ electroreduction on GO to controllably reduce the oxygenated functional groups on the surface. DFT and experimental results uncovered that the C=O group is pivotal to nitrogen fixation following the acceptance–donation mechanism and also affords a weak adsorption for the NH_2 intermediate. Elements co-doping can in turn improve the NRR performance by the synergetic effect between dopants [49]. Qin et al. reported that N and B co-doped porous carbon materials with abundant unpaired frustrated Lewis pairs, which results in desirable performance for electrochemical NRR [50]. S and B co-doped carbon nanofibers were prepared by electrospinning method from Du and co-workers. The S doping-induced changing of center position of the p_z orbital of B facilitates the adsorption of N_2 on S–C–B sites and reduces the energy barriers for rate-determining steps, which result in the highest FE of 22.4% and NH_3 yield of 0.223 μmol h^{-1} cm^{-2} at −0.7 V vs. RHE [51].

In conclusion, heteroatom doping can improve the NRR performance of pristine carbon by means of local charge redistribution, Lewis acid sites, π Back-Donation, and synergetic effects, depending on the chemical nature of given doping elements. However, there have been limited reports on the multiple elements doping carbon

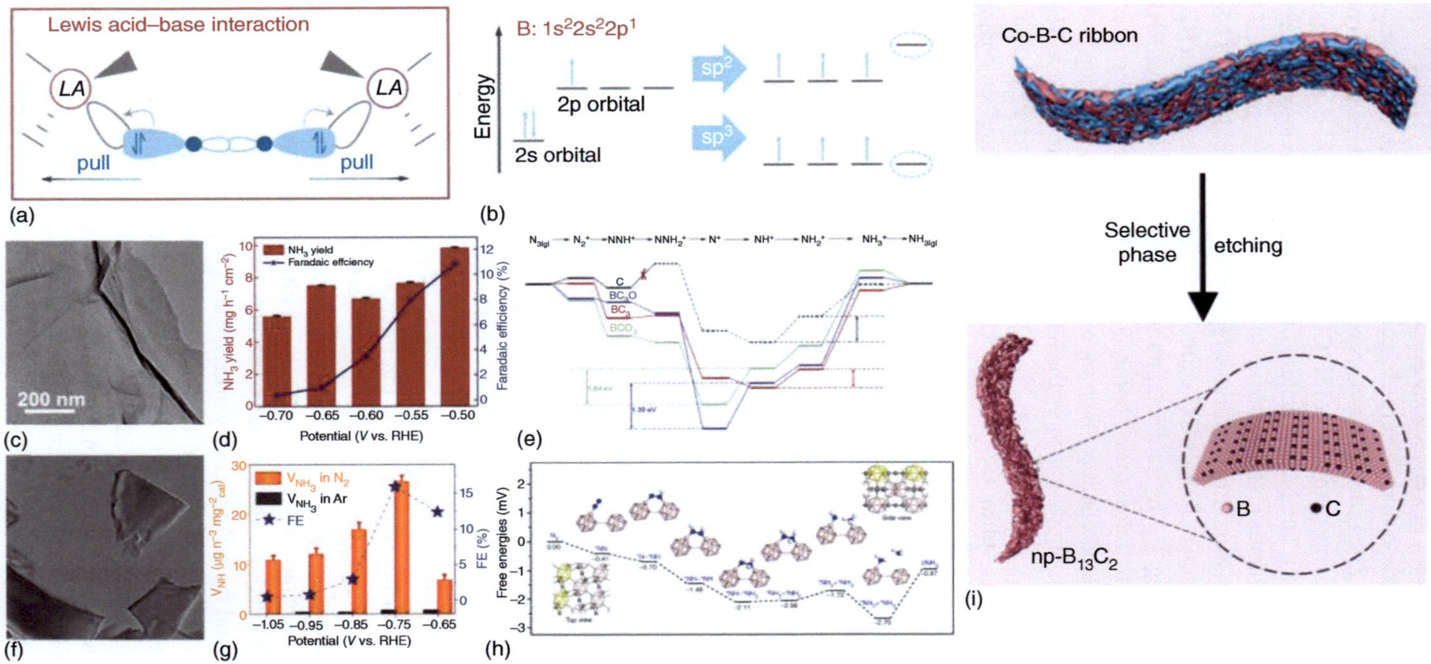

Figure 20.3 (a) Illustration of nitrogen activation by Lewis acid sites; (b) hybridization diagrams of boron. Source: Reproduced with permission from Li et al. [15], Elsevier. (c) TEM image and (d) NRR performance of boron-doped graphene. (e) Free energy diagrams of NRR on BC_3. Source: Reproduced with permission from Yu et al. [38], Elsevier. (f) TEM image and (g) NRR performance of B_4C NSs. (h) Free energy diagrams of NRR on B_4C (110) surface. Source: Qiu et al., [43]/Springer Nature/CC BY 4.0. (i) Schematic illustration of the fabrication process of 3D nanoporous $B_{13}C_2$. Source: Reproduced with permission from Lan et al. [40], Wiley-VCH.

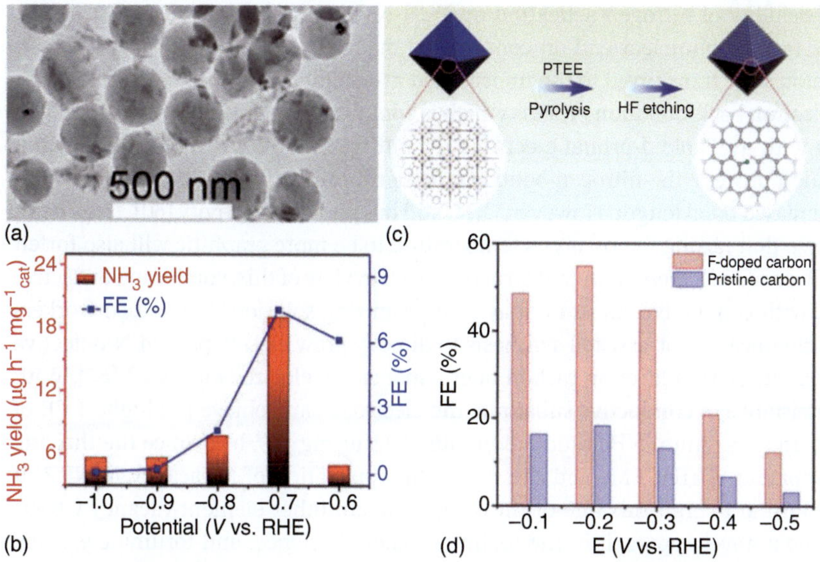

Figure 20.4 (a) TEM image and (b) NRR performance of sulfur-doped carbon sphere. Source: Reproduced with permission from Xia et al. [47], Wiley-VCH. (c) Scheme of the synthetic procedure for F-doped carbon. (d) FE comparison between pristine carbon and F-doped carbon. Source: Reproduced with permission from Liu et al. [48], Wiley-VCH.

in this field. With the introduction of the concept of "high-entropy doping," there is promising potential to create unexpected synergistic effects between elements and carbon [52]. The key questions revolve around understanding how the local structure and electronic state of carbon change when subjected to diverse effects from neighboring doping elements. Additionally, there is a need to explore strategies for oriented doping to enhance the efficiency of the NRR system.

20.3 Vacancy-Abundant Carbon Nitride Materials

Carbon nitride (C_3N_4), maybe the most famous one in the carbon family, which possesses the graphite-like layer structure composed of tris-s triazine units [53, 54]. Many publications including experimental and theoretical results have proved the high content of nitrogen can positively charge the adjacent carbon to increase electron donor-acceptor capabilities [55]. Moreover, some nitrogen in carbon skeleton will evaporate to from a series of N-vacancy during the pyrolysis process, which can provide abundant localized electrons for boosting N_2 activation [56–58]. Yu and co-workers reported the application of polymeric carbon nitride with tunable NVs (PCNs-NVs) in electrochemical NRR with different NV concentrations [59]. Due to the abundant nitrogen vacancies, PCN-NVs can produce ammonia with an NH_3 yield of 8.09 $\mu g\ h^{-1}\ mg^{-1}{}_{cat.}$ and a FE of 11.59% at −0.2 V vs. RHE whereas the ammonia yield of primary PCN is only 0.74 $\mu g\ h^{-1} mg^{-1}{}_{cat.}$. The calculation results confirm the feasibility of nitrogen activation on PCN-NVs, as N_2 can adsorb on PCN-NVs in the form of binuclear end-on coordinated mode. The electrons on adjacent carbon atoms are transferred to the unoccupied π^*-orbital of adsorbed N_2 similar with the electron back-donation process which is commonly observed in transition metal cases with available d-orbital electrons for nitrogen activation. With the electron donation process, the nitrogen bond can be significantly weakened as expressed in an increased bond length. However, C_3N_4 still has poor conductivity [60], it is a compromise that raising the pyrolysis temperature to be more graphitic will also forfeit the advanced nitrogen-abundant structure [61]. In view of this, combining the active C_3N_4 with conductive substrate may be a promising solution. Liu and co-workers used chemical vapor-assisted synthesis to directly grow closely packed N-defective carbon nitride thin layer on carbon fiber paper as the electrode for NRR [62]. With the assistance of conductive substrate, the electrode can achieve the highest FE of 62.1% and maximum NH_3 production rate of 2.9 $\mu g\ mg_{cat.}^{-1}\ h^{-1}$. Since the thriving development of HDCBMs and the successful application of N-vacancy in NRR, it is promising to graft this line of thinking to obtain other element-vacancy CBMs, forming a novel approach in the realm of material science, and fortunately, some pioneers have already endeavored on this path.

20.4 Metal–Carbon Composite Materials

Given that heteroatom doping, and abundant vacancies typically create unsaturated sites, these can serve as anchoring points for metals, leading to the formation of metal–carbon composite structures [63–65]. The robust interactions between the metal and the support facilitate the uniform dispersion of metal sites, allowing for the tailoring of geometric structures from nanoparticles to single atoms [66]. This capability further enables the regulation of the electronic configuration of active sites for specific reactions. Moreover, with metal sources, especially noble metals, becoming increasingly scarce, there is a heightened emphasis on the atomic efficiency of catalysts. The active sites of nanoparticles are perceived as confined to their surfaces, including planes, corners, and edges. This suggests that numerous internal atoms lack catalytic function and are, therefore, redundant. Atomizing metal nanoparticles to create single-atom catalysts (SACs) stands as a promising methodology for advancing materials in terms of heightened activity and enhanced utilization efficiency [67]. Moreover, SACs serve as a robust platform for scrutinizing the intricate mechanisms of NRR, owing to the precision of their well-defined active sites [68–70]. SACs have exhibited excellence in various fields, particularly in electrochemical conversion [71, 72]. So, it is not challenging to contemplate grafting them onto suitable supports for utilization in the NRR. To date, both theoretical and experimental results have confirmed the feasibility of NRR with SACs [73–76]. Qiao and co-workers conducted an extensive investigation into transition metal SACs, examining activity trends, electronic origins, and design strategies using DFT calculations [77]. In this study, a scaling relationship for nitrogen activation based on SACs was proposed, considering the influence of the metal center and its coordination environment. Consequently, a color contour plot of the limiting potential as a function of the Gibbs free energy for NNH (ΔG_{NNH^*}) and NH_2 ($\Delta G_{NH_2^*}$) was constructed to screen optimal SACs for NRR. The ideal catalyst should independently regulate ΔG_{NNH^*} and $\Delta G_{NH_2^*}$ to fall within the red region, surrounded by black lines, to achieve maximum NRR activity. With theoretical guidance, an increasing number of SACs have been developed and have demonstrated exceptional performance in the NRR field. With the guidance of theory, more and more SACs are developed and sparkle in NRR fields with different active center modes, like AuSAs-NDPCs [78], Ru SAs/N-C [79], Ru@ZrO_2/NC [80], Fe_{SA}-N-C [81], MoSAs-Mo_2C [82], Ti_1/rGO [83], Fe-SAs/LCC (Fe-$(O-C_2)_4$) [84], and many others. Oschatz and co-workers reported Au single atoms dispersed on N-doped porous carbons (AuSAs-NDPCs, Figure 20.5a,b) which exhibited NRR performance with FE of 12.3% and NH_3 yield

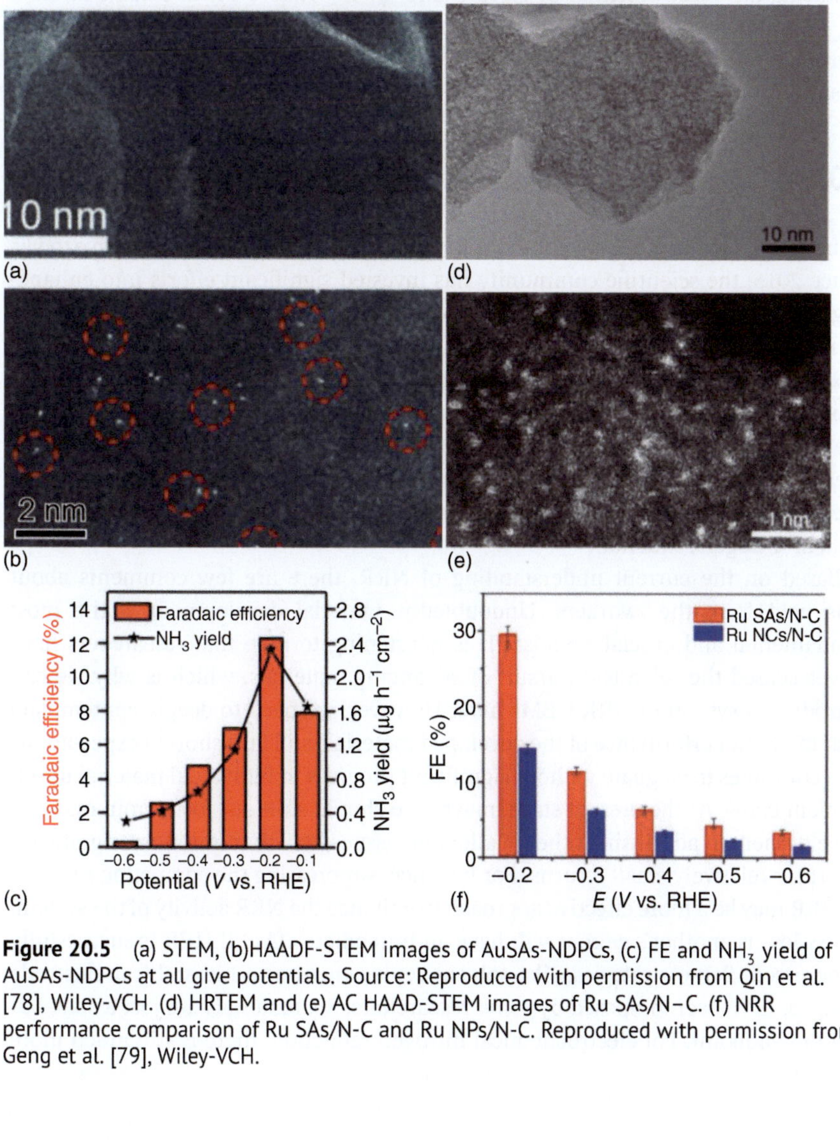

Figure 20.5 (a) STEM, (b)HAADF-STEM images of AuSAs-NDPCs, (c) FE and NH_3 yield of AuSAs-NDPCs at all give potentials. Source: Reproduced with permission from Qin et al. [78], Wiley-VCH. (d) HRTEM and (e) AC HAAD-STEM images of Ru SAs/N-C. (f) NRR performance comparison of Ru SAs/N-C and Ru NPs/N-C. Reproduced with permission from Geng et al. [79], Wiley-VCH.

of 2.32 μg h^{-1} cm^{-2} at −0.2 V vs. RHE (Figure 20.5c) [78]. In this case, the introduction of single gold sites enables the activation of nitrogen, and the open porous structure of carbon support is also pivotal to mass transport and enriching the local concentration of nitrogen in close proximity to the active sites. Zeng and co-workers reported a Ru single site decorated on nitrogen-doped carbon (Ru SAs/N–C) by pyrolyzing Ru-containing zeolitic imidazolate frameworks (Figure 20.5d,e) [79]. Due to the higher chemical affinity for N_2, Ru SAs/N–C exhibits a remarkable FE toward ammonia of 29.6% at the time with a yield of 120.9 μg h^{-1} $mg_{cat.}^{-1}$ at −0.2 V vs. RHE (Figure 20.5f). In contrast, Ru NPs/N–C only shows low yield rate of 5.2 μg h^{-1} $mg_{cat.}^{-1}$ and FE of 0.4%. Yan and co-workers employed the salt-templating method to synthesize isolated iron atoms on nitrogen-doped carbon (Fe_{SA}-N-C) [81]. The resulting catalyst exhibited notable efficacy in catalyzing nitrogen reduction, featuring an onset potential of 0.193 V and an impressive FE of 56.55% at 0.0 V vs. RHE. Supported by molecular dynamics simulations, the researchers asserted that N_2 could access single-atom iron with a minimal energy barrier, favoring N_2 adsorption over H adsorption through a robust exothermic process.

20.5 Conclusion and Outlook

Nitrogen reduction is a long-standing area of research that continues to thrive today. Since 2015, the scientific community has invested significant efforts into enhancing the performance of electrochemical NRR systems, resulting in the development of numerous materials. In this chapter, we provide a brief overview of CBMs for NRR, including mechanisms and performance improvement strategies such as heteroatom doping, vacancy engineering, and metal–carbon composites. Additionally, we propose some strategies that may not be fully developed, including high-entropy element strategies and multielement vacancy strategies. It is inspiring to note that the scientific community is exploring new possibilities for carbon materials in the field of nitrogen reduction.

Based on the current understanding of NRR, there are few comments about this field from the -workers. Undoubtedly, material development is the most fundamental and crucial premise. It is heartening to note that researchers have never ceased the relentless pursuit of advancing materials, which is why we can introduce a system of NRR CBMs here. However, we need to deeply contemplate that the NRR performance of the developed materials still falls short of expectations and continues to stagnate without significant progress in individual material development cases. At the present stage, maybe we should shift our focus from material development to addressing other challenging issues, which may require significant effort for relatively small returns. For instance, suppressing the competing reaction of HER may be a more effective approach to enhance the NRR activity of the system. One's loss is another's gain; it will have an immediate effect if HER is successfully suppressed. Some strategies, like solvent replacement and hierarchical electrode designs, have been proposed by some researchers. So, materials are just a start but not the endpoint. For electrochemical nitrogen reduction, there is still much more we can do.

References

1 Kyriakou, V., Garagounis, I., Vourros, A. et al. (2020). An electrochemical Haber-Bosch process. *Joule* 4: 142–158.

2 Rosca, V., Duca, M., de Groot, M.T., and Koper, M.T. (2009). Nitrogen cycle electrocatalysis. *Chem. Rev.* 109: 2209–2244.

3 Liu, H. (2013). *Ammonia Synthesis Catalysts: Innovation and Practice*. World Scientific.

4 Hoffman, B.M., Lukoyanov, D., Yang, Z.Y. et al. (2014). Mechanism of nitrogen fixation by nitrogenase: the next stage. *Chem. Rev.* 114: 4041–4062.

5 Fourmond, V. and Leger, C. (2017). Dinitrogen reduction: interfacing the enzyme nitrogenase with electrodes. *Angew. Chem. Int. Ed.* 56: 4388–4390.

6 MacFarlane, D.R., Cherepanov, P.V., Choi, J. et al. (2020). A roadmap to the ammonia economy. *Joule* 4: 1186–1205.

7 Qing, G., Ghazfar, R., Jackowski, S.T. et al. (2020). Recent advances and challenges of electrocatalytic N_2 reduction to ammonia. *Chem. Rev.* 120: 5437–5516.

8 Soloveichik, G. (2016). *Renewable energy to fuels through utilization of energy dense liquids (REFUEL) Program Overview.*

9 Cui, X.Y., Tang, C., and Zhang, Q. (2018). A review of electrocatalytic reduction of dinitrogen to ammonia under ambient conditions. *Adv. Energy Mater.* 8: 1800369.

10 Shilov, A.E. (2003). Catalytic reduction of molecular nitrogen in solutions. *Russ. Chem. Bull.* 52: 2555–2562.

11 Zhan, C.-G., Nichols, J.A., and Dixon, D.A. (2003). Ionization potential, electron affinity, electronegativity, hardness, and electron excitation energy: molecular properties from density functional theory orbital energies. *J. Phys. Chem. A* 107: 4184–4195.

12 Ren, Y.W., Yu, C., Tan, X.Y. et al. (2021). Strategies to suppress hydrogen evolution for highly selective electrocatalytic nitrogen reduction: challenges and perspectives. *Energy Environ. Sci.* 14: 1176–1193.

13 Bauer, N. (2002). Theoretical pathways for the reduction of N_2 molecules in aqueous media: thermodynamics of N_2H_n[1]. *J. Phys. Chem.* 64: 833–837.

14 Matus, M.H., Arduengo, A.J. 3rd, and Dixon, D.A. (2006). The heats of formation of diazene, hydrazine, $N_2H_3^+$, $N_2H_5^+$, N_2H, and N_2H_3 and the methyl derivatives CH_3NNH, CH_3NNCH_3, and $CH_3HNNHCH_3$. *J. Phys. Chem. A* 110: 10116–10121.

15 Li, L.Q., Tang, C., Jin, H.Y. et al. (2021). Main-group elements boost electrochemical nitrogen fixation. *Chem* 7: 3232–3255.

16 Koper, M.T.M. (2013). Theory of multiple proton-electron transfer reactions and its implications for electrocatalysis. *Chem. Sci.* 4: 2710–2723.

17 Wang, S.Y., Ichihara, F., Pang, H. et al. (2018). Nitrogen fixation reaction derived from nanostructured catalytic materials. *Adv. Funct. Mater.* 28: 1803309.

18 Shi, M.M., Bao, D., Wulan, B.R. et al. (2017). Au sub-nanoclusters on TiO_2 toward highly efficient and selective electrocatalyst for N_2 conversion to NH_3 at ambient conditions. *Adv. Mater.* 29: 1606550.

19 Zhang, W., Lin, W., Ren, J. et al. (2022). Electrochemical reduction of nitrogen to ammonia by Pd–S–Mo nanosheets on a hydrophobic hierarchical graphene support. *ChemElectroChem* 9: e202100052.

20 Wang, X., Luo, M., Lan, J. et al. (2021). Nanoporous intermetallic Pd_3Bi for efficient electrochemical nitrogen reduction. *Adv. Mater.* 33: e2007733.

21 Xiong, L., Qiu, Y., Peng, X. et al. (2022). Electronic structural engineering of transition metal-based electrocatalysts for the hydrogen evolution reaction. *Nano Energy* 104: 107882.

22 Liu, F., Zhang, L., Wang, L., and Cheng, F. (2021). The electrochemical tuning of transition metal-based materials for electrocatalysis. *Electrochem. Energy Rev.* 4: 146–168.

23 Liu, X. and Dai, L. (2016). Carbon-based metal-free catalysts. *Nat. Rev. Mater.* 1: 16064.

24 Lee, J., Kim, J., and Hyeon, T. (2006). Recent progress in the synthesis of porous carbon materials. *Adv. Mater.* 18: 2073–2094.

25 Miao, S., Liang, K., Zhu, J. et al. (2020). Hetero-atom-doped carbon dots: doping strategies, properties and applications. *Nano Today* 33: 100879.

26 Gerber, I.C. and Serp, P. (2020). A theory/experience description of support effects in carbon-supported catalysts. *Chem. Rev.* 120: 1250–1349.

27 Yang, Y., Zhang, L., Hu, Z. et al. (2020). The crucial role of charge accumulation and spin polarization in activating carbon-based catalysts for electrocatalytic nitrogen reduction. *Angew. Chem. Int. Ed.* 59: 4525–4531.

28 Asefa, T. and Huang, X. (2017). Heteroatom-doped carbon materials for electrocatalysis. *Chem. Eur. J.* 23: 10703–10713.

29 Yang, X., Sun, K., Ma, M. et al. (2020). Achieving strong chemical adsorption ability for efficient carbon dioxide electrolysis. *Appl. Catal. B Environ.* 272: 118968.

30 Jeon, I.-Y., Noh, H.-J., and Baek, J.-B. (2020). Nitrogen-doped carbon nanomaterials: synthesis, characteristics and applications. *Chem. Asian J.* 15: 2282–2293.

31 Thakur, A.K., Kurtyka, K., Majumder, M. et al. (2022). Recent advances in boron- and nitrogen-doped carbon-based materials and their various applications. *Adv. Mater. Interfaces* 9: 2101964.

32 Liu, Y., Su, Y., Quan, X. et al. (2018). Facile ammonia synthesis from electrocatalytic N_2 reduction under ambient conditions on N-doped porous carbon. *ACS Catal.* 8: 1186–1191.

33 Mukherjee, S., Cullen, D.A., Karakalos, S. et al. (2018). Metal-organic framework-derived nitrogen-doped highly disordered carbon for electrochemical ammonia synthesis using N_2 and H_2O in alkaline electrolytes. *Nano Energy* 48: 217–226.

34 Zhang, W., Zhan, S., Qin, Q. et al. (2022). Electrochemical generation of catalytically active edge sites in C_2N-type carbon materials for artificial nitrogen fixation. *Small* 18: e2204116.

35 Liu, C., Li, Q., Wu, C. et al. (2019). Single-boron catalysts for nitrogen reduction reaction. *J. Am. Chem. Soc.* 141: 2884–2888.

36 Wang, C., Huang, F., Liang, H. et al. (2023). d- and p-Block single-atom catalysts supported by BN nanocages toward electrochemical reactions of N_2 and O_2. *Phys. Chem. Chem. Phys.* 25: 25761–25771.

37 Légaré, M.-A., Bélanger-Chabot, G., Dewhurst, R.D. et al. (2018). Nitrogen fixation and reduction at boron. *Science* 359: 896–900.

38 Yu, X., Han, P., Wei, Z. et al. (2018). Boron-doped graphene for electrocatalytic N_2 reduction. *Joule* 2: 1610–1622.

39 Qiu, W., Xie, X.Y., Qiu, J. et al. (2018). High-performance artificial nitrogen fixation at ambient conditions using a metal-free electrocatalyst. *Nat. Commun.* 9: 3485.

40 Lan, J., Luo, M., Han, J. et al. (2021). Nanoporous $B_{13}C_2$ towards highly efficient electrochemical nitrogen fixation. *Small* 17: 2102814.

41 Biswas, A., Kapse, S., Ghosh, B. et al. (2022). Lewis acid dominated aqueous electrolyte acting as co-catalyst and overcoming N_2 activation issues on catalyst surface. *Proc. Natl. Acad. Sci.* 119: e2204638119.

42 Lv, C., Zhong, L., Yao, Y. et al. (2020). Boosting electrocatalytic ammonia production through mimicking "π Back-Donation". *Chem* 6: 2690–2702.

43 Li, N., Tong, Y., Li, H. et al. (2021). Boron-doped carbon nanospheres for efficient and stable electrochemical nitrogen reduction. *Carbon* 182: 233–241.

44 Ma, C., Zhang, Y., Yan, S., and Liu, B. (2022). Carbon-doped boron nitride nanosheets: a high-efficient electrocatalyst for ambient nitrogen reduction. *Appl. Catal. B Environ.* 315: 121574.

45 Xia, L., Yang, J., Wang, H. et al. (2019). Sulfur-doped graphene for efficient electrocatalytic N_2-to-NH_3 fixation. *Chem. Commun.* 55: 3371–3374.

46 Paraknowitsch, J.P. and Thomas, A. (2013). Doping carbons beyond nitrogen: an overview of advanced heteroatom doped carbons with boron, sulphur and phosphorus for energy applications. *Energy Environ. Sci.* 6: 2839–2855.

47 Xia, L., Wu, X., Wang, Y. et al. (2019). S-doped carbon nanospheres: an efficient electrocatalyst toward artificial N_2 fixation to NH_3. *Small Methods* 3: 1800251.

48 Liu, Y., Li, Q., Guo, X. et al. (2020). A highly efficient metal-free electrocatalyst of F-doped porous carbon toward N_2 Electroreduction. *Adv. Mater.* 32: e1907690.

49 Wang, S., Zhu, J., Zhang, Y. et al. (2020). Identifying the active site on graphene oxide nanosheets for ambient electrocatalytic nitrogen reduction. *Inorg. Chem.* 59: 11108–11112.

50 Qin, Q., Heil, T., Schmidt, J. et al. (2019). Electrochemical fixation of nitrogen and its coupling with biomass valorization with a strongly adsorbing and defect optimized boron–carbon–nitrogen catalyst. *ACS Appl. Energy Mater.* 2: 8359–8365.

51 Wen, Y., Zhu, H., Hao, J. et al. (2021). Metal-free boron and sulphur co-doped carbon nanofibers with optimized p-band centers for highly efficient nitrogen electroreduction to ammonia. *Appl. Catal. B Environ.* 292: 120144.

52 Ma, Y., Ma, Y., Wang, Q. et al. (2021). High-entropy energy materials: challenges and new opportunities. *Energy Environ. Sci.* 14: 2883–2905.

53 Cao, S., Low, J., Yu, J., and Jaroniec, M. (2015). Polymeric photocatalysts based on graphitic carbon nitride. *Adv. Mater.* 27: 2150–2176.

54 Kessler, F.K., Zheng, Y., Schwarz, D. et al. (2017). Functional carbon nitride materials—design strategies for electrochemical devices. *Nat. Rev. Mater.* 2: 17030.

55 Gong, K., Du, F., Xia, Z. et al. (2009). Nitrogen-doped carbon nanotube arrays with high electrocatalytic activity for oxygen reduction. *Science* 323: 760–764.

56 Kumar, A., Raizada, P., Hosseini-Bandegharaei, A. et al. (2021). C-, N-Vacancy defect engineered polymeric carbon nitride towards photocatalysis: viewpoints and challenges. *J. Mater. Chem. A* 9: 111–153.

57 Lowther, J.E. (1998). Defective and amorphous structure of carbon nitride. *Phys. Rev. B* 57: 5724–5727.

58 Zhao, Z., Long, Y., Chen, Y. et al. (2022). Phosphorus doped carbon nitride with rich nitrogen vacancy to enhance the electrocatalytic activity for nitrogen reduction reaction. *Chem. Eng. J.* 430: 132682.

59 Lv, C., Qian, Y., Yan, C. et al. (2018). Defect engineering metal-free polymeric carbon nitride electrocatalyst for effective nitrogen fixation under ambient conditions. *Angew. Chem. Int. Ed.* 57: 10246–10250.

60 Zhang, Y., Thomas, A., Antonietti, M., and Wang, X. (2009). Activation of carbon nitride solids by protonation: morphology changes, enhanced ionic conductivity, and photoconduction experiments. *J. Am. Chem. Soc.* 131: 50–51.

61 Xu, J., Xu, F., Qian, M. et al. (2017). Conductive carbon nitride for excellent energy storage. *Adv. Mater.* 29: 1701674.

62 Peng, G., Wu, J., Wang, M. et al. (2020). Nitrogen-defective polymeric carbon nitride nanolayer enabled efficient electrocatalytic nitrogen reduction with high Faradaic efficiency. *Nano Lett.* 20: 2879–2885.

63 Wang, Q., Lei, Y., Wang, D., and Li, Y. (2019). Defect engineering in earth-abundant electrocatalysts for CO_2 and N_2 reduction. *Energy Environ. Sci.* 12: 1730–1750.

64 Zhang, L., Shi, Y., Li, L. et al. (2022). Metal single-atom-confined electrocatalysts to water oxidation: development, innovation, and challenges. *Electrochem. Sci. Adv.* 2: e202100102.

65 Shi, Z., Yang, W., Gu, Y. et al. (2020). Metal-nitrogen-doped carbon materials as highly efficient catalysts: progress and rational design. *Adv. Sci.* 7: 2001069.

66 Yang, X.F., Wang, A., Qiao, B. et al. (2013). Single-atom catalysts: a new frontier in heterogeneous catalysis. *Acc. Chem. Res.* 46: 1740–1748.

67 Zhang, L., Ren, Y., Liu, W. et al. (2018). Single-atom catalyst: a rising star for green synthesis of fine chemicals. *Natl. Sci. Rev.* 5: 653–672.

68 Choi, C., Back, S., Kim, N.-Y. et al. (2018). Suppression of hydrogen evolution reaction in electrochemical N_2 reduction using single-atom catalysts: a computational guideline. *ACS Catal.* 8: 7517–7525.

69 Guo, X., Gu, J., Lin, S. et al. (2020). Tackling the activity and selectivity challenges of electrocatalysts toward the nitrogen reduction reaction via atomically dispersed biatom catalysts. *J. Am. Chem. Soc.* 142: 5709–5721.

70 Zhai, Y.L., Zhu, Z.J., Zhu, C.Z. et al. (2020). Single-atom catalysts boost nitrogen electroreduction reaction. *Mater. Today* 38: 99–113.

71 Zhu, Y.Z., Peng, W.C., Li, Y. et al. (2019). Modulating the electronic structure of single-atom catalysts on 2D nanomaterials for enhanced electrocatalytic performance. *Small Methods* 3: 1800438.

72 Liu, C.H., Wu, Y., Sun, K.A. et al. (2021). Constructing FeN_4/graphitic nitrogen atomic interface for high-efficiency electrochemical CO_2 reduction over a broad potential window. *Chem* 7: 1297–1307.

73 Pang, Y., Su, C., Xu, L., and Shao, Z. (2023). When nitrogen reduction meets single-atom catalysts. *Prog. Mater. Sci.* 132: 101044.

74 Zhang, Q. and Guan, J. (2020). Single-atom catalysts for electrocatalytic applications. *Adv. Funct. Mater.* 30: 2000768.

75 Chen, Z., Zhao, J., Cabrera, C.R., and Chen, Z. (2019). Computational screening of efficient single-atom catalysts based on graphitic carbon nitride (g-C_3N_4) for nitrogen electroreduction. *Small Methods* 3: 1800368.

76 Chen, Y., Lin, J., Jia, B. et al. (2022). Isolating single and few atoms for enhanced catalysis. *Adv. Mater.* 34: 2201796.

77 Liu, X., Jiao, Y., Zheng, Y. et al. (2019). Building up a picture of the electrocatalytic nitrogen reduction activity of transition metal single-atom catalysts. *J. Am. Chem. Soc.* 141: 9664–9672.

78 Qin, Q., Heil, T., Antonietti, M., and Oschatz, M. (2018). Single-site gold catalysts on hierarchical N-doped porous noble carbon for enhanced electrochemical reduction of nitrogen. *Small Methods* 2: 1800202.

79 Geng, Z., Liu, Y., Kong, X. et al. (2018). Achieving a record-high yield rate of 120.9 $\mu gNH_3\ mg_{cat.}^{-1}\ h^{-1}$ for N_2 electrochemical reduction over Ru single-atom catalysts. *Adv. Mater.* 30: e1803498.

80 Tao, H.C., Choi, C., Ding, L.X. et al. (2019). Nitrogen fixation by Ru single-atom electrocatalytic reduction. *Chem* 5: 204–214.

81 Wang, M., Liu, S., Qian, T. et al. (2019). Over 56.55% Faradaic efficiency of ambient ammonia synthesis enabled by positively shifting the reaction potential. *Nat. Commun.* 10: 341.

82 Ma, Y., Yang, T., Zou, H. et al. (2020). Synergizing Mo single atoms and Mo_2C nanoparticles on CNTs synchronizes selectivity and activity of electrocatalytic N_2 reduction to ammonia. *Adv. Mater.* 32: e2002177.

83 Liang, S., Zhu, C., Zhang, N. et al. (2020). A novel single-atom electrocatalyst Ti_1/rGO for efficient cathodic reduction in hybrid photovoltaics. *Adv. Mater.* 32: e2000478.

84 Zhang, S., Jin, M., Shi, T. et al. (2020). Electrocatalytically active Fe-(O-C_2)$_4$ single-atom sites for efficient reduction of nitrogen to ammonia. *Angew. Chem. Int. Ed.* 59: 13423–13429.

21

Porous Materials for NRR

Shun Zhang and Meiting Zhao

Tianjin University, Tianjin Key Laboratory of Molecular Optoelectronic Sciences, Institute of Molecular Aggregation Science, Department of Chemistry, No. 92, Weijin Road, Nankai District, Tianjin, 30007, China

21.1 Introduction

Efficient electrocatalytic reactions usually occur at specific sites with appropriate adsorption properties, such as the corners and edges featured with the low coordination number and thereby behave the enhanced orbital coupling degree toward reactant or intermediate. Among the strategies to obtain the active site with high intrinsic activity, constructing porous nanocomposites with rich channels and coordination-unsaturated sites is a promising way [1]. As for NRR, considering the extremely low solubility of N_2 in the aqueous solution, constructing porous electrocatalysts from a standpoint of facilitating the N_2 mass transfer or enrichment, and intrinsic activity enhancement should be a feasible approach to making NRR more active. More recently, as the new types of porous materials, metal–organic frameworks (MOFs) and covalent organic frameworks (COFs) are flourishing in the field of catalyst research. Based on the topology design, metal site choice, organic ligands engineering, coordination regulation, and diverse post-modification techniques, these materials demonstrated multiple applicable prospects in NRR. In addition, theoretical calculations results support the fact that MOFs and COFs-based catalysts are operable to produce ammonia electroreduction of N_2 [2, 3]. However, the research on NRR in MOFs or COFs is still in its infancy and a lot of work is needed to further promote this area of study.

In Section 21.3.3, porous materials, e.g. metal, metal alloys, metal chalcogenides, MOFs, and COFs as well as their derivatives will be represented. Particularly, the design strategies and synergistic effect of transition metal-based MOFs, the regulating effect, and precise active site construction of COFs will be covered for NRR. Totally, there will be ~23 pages for this chapter.

Functional Materials for Electrocatalytic Energy Conversion, First Edition.
Edited by Zhicheng Zhang, Meiting Zhao, and Yuchen Qin.

21.2 Porous Metal-Based Materials for NRR

21.2.1 Porous Au-Based Electrocatalyst

Metal-based catalysts that possess intrinsic activities for NRR could be further strengthened by porosity enhancement. For instance, porous Au film on Ni foam fabricated by micelle-assisted electrodeposition strategy had been adopted for NRR (Figure 21.1a,b) [4]. The porosity of the Au film could be tuned by the electrodeposition time and the micelle copolymer composition. Nonporous and blocked Au films would be obtained if the electrodeposition time was reduced below 1000 seconds and longer than 4000 seconds, respectively. An optimum 2000 seconds deposition time with the assistance of PVP-co-PS micelle would result in interconnected porous architectonics. Owing to the spatially and locally nanoporous structure, pAu/NF exhibits a high NH_3 yield rate of 9.42 μg h^{-1} cm^{-2} along with a FE of 13.36% at −0.2 V vs. reversible hydrogen electrode (RHE) in 0.1 m Na_2SO_4, which is superior to that analog with less porous property (i.e. lower C_{dl}), or that Au film without porosity (Figure 21.1c,d).

Porous hollow nanocatalysts could drive the reaction to proceed both at the outer and inner surfaces and hence be efficient NRR electrocatalysts. El-Sayed displayed a galvanic replacement technique to synthesize nanocage nanostructure [5]. During the preparation process, template silver nanocubes would be replaced by $HAuCl_4$ with non-equal stoichiometric numbers due to their oxidation potential and valence difference. As such, a relatively lesser number of Au atoms would dwell on the pristine Ag nanocube skeleton and the nanopores would form (defined as AuHNCs). Electrochemical NRR experiments in 0.5 M $LiClO_4$ show that the FE and NH_3 yield rates reached 30.2% and 3.9 μg cm^{-2} h^{-1}, respectively, at −0.4 V vs. RHE. After heating up to 50 °C, the FE amounted to 40.5% due to the speeded mass transfer. This NRR performance also surpassed various Au nanoparticles with different shapes (i.e. rods, spheres, or cubes), illustrating the importance of the higher surface area and the confinement effects to NRR. Hereafter, more detailed research on the pore size, density, and localized surface plasmon resonance (LSPR) offered a higher NRR FE (35.9%), further proving the catalytic efficiency of AuHNCs to NRR [6]. Alloying strategy also works for Au-based NRR electrocatalysts. According to Liu [7], the Cu, B co-doping can increase the active sites as well as efficiently increase the nitrogen adsorption, making a high NH_3 yield (13.2 μg h^{-1} $mg_{cat}{}^{-1}$) and FE (12.78%) in 0.1 M Na_2SO_4 solution at −0.2 V vs. RHE.

21.2.2 Porous Pd-Based Electrocatalyst

As a typical kind of NRR catalyst, Pd-based porous catalysts have also been widely inquired. To reduce the free energy barrier of rate-limiting step of $^*N_2H$ formation, Cheng developed nanoporous palladium hydride (np-$PdH_{0.43}$) via dealloying and in situ hydrogen injection method (Figure 21.2a). High resolution transmission electron microscopy (HRTEM) and X-ray photoelectron (XPS) results displayed the expanded lattice fringes and modified electronic structure due to the hydrogen insertion into the Pd lattice. The isotopic labeling experiments proved that

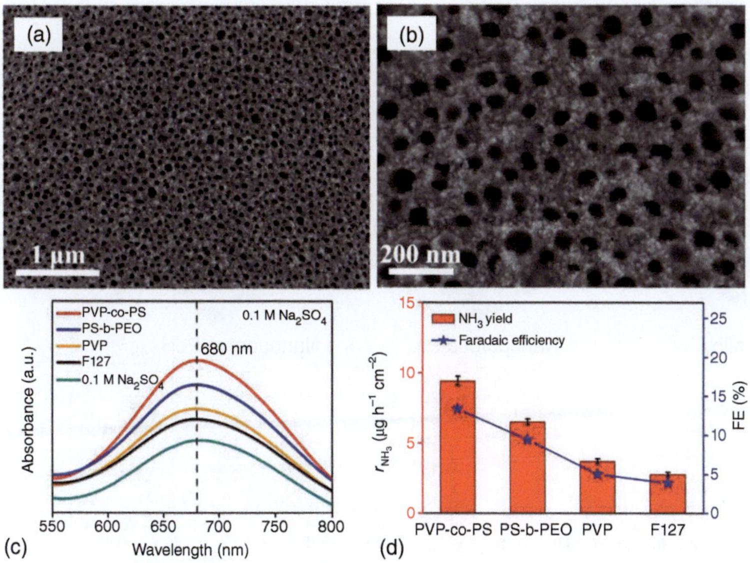

Figure 21.1 (a) Low- and (b) high-magnification top-surface SEM images of the pAu/NF. (c) UV–vis adsorption spectra of the electrolytes after electrolysis by different Au structures (d) NH_3 yield rate and FE at selected potentials. Source: Reproduced with permission of Wang et al. [4], Wiley-VCH.

the hydrogen dopants in np-$PdH_{0.43}$ could promote the activation of N_2 by the proton-coupled electron transfer process via lattice hydrogen atoms in the hydride, resulting in a relatively lower activation barrier (0.65 vs. 0.91 eV) of $^*N_2H$ formation than np-Pd [8] (Figure 21.2c). Henceforth, np-$PdH_{0.43}$ delivered a relatively high FE (43.6%) and a high yielding rate (20.4 mg h^{-1} mg^{-1} cat) at −0.15 V vs. RHE which are about three times value than np-Pd (Figure 21.2b).

Alloying two metals to form bimetallic structures is a promising way to improve the catalytic properties of metal materials [9]. For instance, taking advantage of the robust intrinsic activity of Ru-based materials [10], Wang performed NRR research on porous palladium–ruthenium nanosheets on Ni foam (PdRu NS-NF). Via the pluronic f127 micelle and KBr-assisted galvanic replacement reaction between Ni foam and $PdCl_4^{2-}/RuCl_3$ mixture, bimetallic nanosheets with interconnected porosity were grown on the Ni foam with uniformity. Derived from the abundant active sites, accessible channels, and its binder-free nature, improved NRR activity was obtained with 34.1 μg h^{-1} mg_{cat}^{-1}.

Compared to alloying with precious metals, combining noble metals with transition metals is a beneficial way to lower their dependence. As a representative sample, Zhang has demonstrated a two-step dealloying process to synthesize 3D hierarchical interconnected porous PdCu alloy electrocatalysts [11]. Using NaOH as the 1st step dealloying agent, majority of Al components in precursor $Pd_2Cu_{18}Al_{80}$ would be etched resulting in ordered porous Al–Cu–Pd intermetallic phase typical with about one hundred nanometer pores. After further etching by HNO_3, bimodal nanoporous Pd_3Cu_1 alloy could be finally obtained along with the abundant

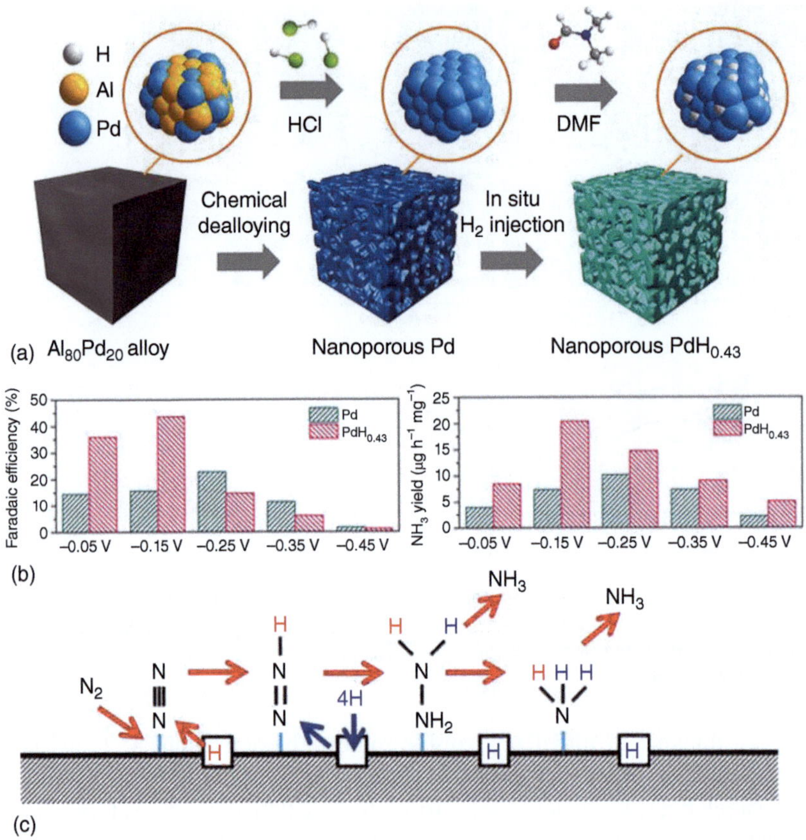

Figure 21.2 (a) The synthesis of nanoporous palladium hydride. (b) Faradaic efficiencies and NH_3 yield rates of np-Pd and np-$PdH_{0.43}$ at selected potentials. (c) Proposed lattice hydrogen-involved reaction pathway for nitrogen reduction on palladium hydride. Source: Reproduced with permission of Xu et al. [8], Wiley-VCH.

mesoporous pores below 10 nm. TEM and N_2 adsorption–desorption experiments proved the presence of the hierarchically porous structure. The appropriate component between Pd and Cu as well as its porosity profit jointly leads to the superior NH_3 yield rate of 39.9 $\mu g\,h^{-1}\,mg_{cat}{}^{-1}$ [11], 3.5 times higher than that of Pd_3Cu_1 nanoparticle and 4.2 and 4.6 times exceeding that monometallic nanoporous Pd and Cu, respectively.

Combining Pd with less HER-active metal is an efficient way to enhance its selectivity of NRR. Peng reported a nanoporous intermetallic Pd_3Bi electrocatalyst fabricated by chemical-electrochemical etching method [12]. Benefiting from the proper Bi doping, porosity adsorption capability for N_2, and the induced strong Pd–Bi coupling effect, np-Pd_3Bi facilitates the N_2 adsorption and activation and hence reduces the barrier for efficient NRR. As a result, an NH_3 yield rate of $59.05 \pm 2.27\,\mu g\,h^{-1}\,mg_{cat}{}^{-1}$ and a FE of $21.52 \pm 0.71\%$ were achieved at −0.2 V vs. RHE in 0.05 m H_2SO_4 electrolyte, much higher than those of np-$PdBi_2$ (6.50 $\mu g\,h^{-1}$ mg cat^{-1}, 18.56%) and Pd/C (8.70 $\mu g\,h^{-1}$ mg cat^{-1}, 0.14%).

Except for the strategy of alloying with metal, introducing non-metal components to the porous Pd catalyst is another choice to regulate its activity. In another study, wang devised a three-dimensional palladium–silver–sulfur porous nanosponges (Pd–Ag–S-PNSs) via chemical reduction-sulfuration method [13]. The porous structure should be derived from the H_2 bubble with $NaBH_4$ as reducing agent. After sulfuration, amorphous Pd–Ag–S PNSs amorphous material with rich "dangling bonds" occurred, which offered more active sites for the NRR. Besides, the sulfur doping could also help the hydrogenation of nitrogen and thus reduce the energy barrier of NRR. As such, Pd–Ag–S-PNSs show a NH_3 yield of 9.73 $\mu g\,h^{-1}$ $mg^{-1}{}_{cat}$ and FE of 18.41% at −0.2 V, superior to that of 3D Pd–Ag PNSs (6.08 $\mu g\,h^{-1}$ $mg^{-1}{}_{cat}$, 9.34%).

21.2.3 Porous Ru-Based Electrocatalyst

Since the lower overpotential on Ru than Pt or Pd to trigger NRR based on the theoretical evaluation [14], Ru-based electrocatalysts have been widely invested hitherto. Wang [15] synthesized mesoporous RhRu film on nickel foam (mRhRu/NF) by electrodeposition approach with the diblock copolymer as a soft template. The presence of Ru component shows enhanced adsorption ability toward N_2 according to the chemical adsorption experiment. The synergetic function between porosity and Ru significantly improved the NRR yield to 30.28 $\mu g\,h^{-1}$ $mg_{cat}{}^{-1}$ and FE of 28.33%, exceeding mRh/NF (13.97 $\mu g\,h^{-1}$ $mg_{cat}{}^{-1}$), mRu/NF (17.08 $\mu g\,h^{-1}$ $mg_{cat}{}^{-1}$) or other stoichiometric ratio bimetallic samples (mRh_4Ru_1/NF and mRh_1Ru_4/NF). Loading metal nanoparticles on the porous conductive support is also a suitable way to boost NRR. Using porous cellulose/CNT sponge as support, Jung synthesized Ru–Cu NPs with homogeneous mixing of Ru and Cu [16]. This catalyst delivered a robust FE of 30.95% and yield rate of 73.26$\mu g\,h^{-1}$ cm^{-2} at −0.1 V vs. RHE. Elaborate comparative experiments and DFT calculation identified that it is the homogeneous dispersion of Ru and Cu in Ru–Cu alloy NPs increased the adsorption of N_2 as well as suppressed the H binding and the high porous structure which ensured the mass transfer that jointly contributed to the superior performance.

21.2.4 Other Porous Noble-Based Electrocatalysts

In addition to the former kinds of noble metal-based electrocatalysts reports, noble metals such as Ir and Ag-based porous nanostructure have also been proven efficient for NRR. For example, Huang designed a new class of amorphous bullet-like M–Te (M = Ru, Rh, and Ir) glassy porous nanorods (PNRs) electrocatalysts (Figure 21.3a–c) [17]. Ir–Te PNRs stand out as the most active composition for NH_3 production with a yield rate of 51.1 $\mu g\,h^{-1}$ $mg_{cat}{}^{-1}$ and FE of 15.3% at −0.2 V vs. RHE in 0.1 M KOH (Figure 21.3d–f). This is closely related to the amorphous property and the component. Comparative experiments show that the crystalline $IrTe_4$ PNRs without porous structure in Ir–Te PNRs lack abundant dangling bonds which are short of active sites. However, merely increasing the active surface area would not further increase the NRR activity as the Rh–Te PNRs and Ru–Te PNRs, both of which possess large C_{dl} values. VB XPS measurements show that Ir–Te_4 PNRs present the highest d-band center (Figure 21.3g) relative to the Fermi level and thus could enhance the adsorption to N_2 which could be further supported by the N_2-TPD results (Figure 21.3h). Detailed DFT calculation demonstrated that the key reactants H_2O orbital matched well with the Ir 5d bands, implying good H_2O dissociation ability for consecutive NRR process. All the former factors in Ir–Te PNRs jointly result in the strongest adsorption ability to H_2O and N_2 and lowest energy barrier during the multistep NRR process. Similar work of amorphous IrTe material for NRR has also been reported by Yamauchi [18]. In their work, mesoporous amorphous IrTe metal alloy nanoparticle was fabricated by micelle-directed synthesis method. The amorphous nature as well as the interconnected porous scaffold induced a complementary effect that can simultaneously suppress HER and enhance N_2 absorption. Therefore, a high ammonia yield rate of 34.6 $\mu g\,mg^{-1}\,h^{-1}$ and FE of 11.2% was obtained at −0.15 V vs. RHE in 0.1 M HCl solution. As for Ag-based material, it is also working to design three-dimensional porous structures. Wang [19] synthesized boron-doped silver nanosponges (B–Ag NSs) utilizing a wet chemical reduction procedure with $NaBH_4$ as reductant and boron dopant in *N,N*-dimethylformamide (DMF). Due to the high boron content and their interconnected porous structures, a high yield of 26.5 $\mu g\,h^{-1}\,mg^{-1}{}_{cat}$ and FE of 8.9% could be obtained.

21.2.5 Porous Non-Noble-Based Electrocatalysts

Replace noble-metal-based electrocatalysts with low-cost electrocatalysts for ammonia synthesis study is a promising alternative. Among the transition-metal-based electrocatalysts such as Mo, Ni, Fe, Ti, and Mn-based electrocatalysts were well explored and demonstrated attractive prospects, and so were their porous derivatives [20]. For example, Lv devised a Mo_2C@C electrocatalyst with 3D hierarchical architectures that consisted of Mo_2C nanosheets supported on porous carbon matrix [21]. With the cubic NaCl crystals used as templates during the formation of Mo_2C@C, the high surface area reached 170 $m^2\,g^{-1}$, which is much higher than that of bulk sample Mo_2C@C. The 3D porous structures not only provide more exposed active sites but also speed the mass diffusion and electron transfer during NRR, resulting in a high yield of 12.55 $mg^{-1}\,h^{-1}\,mg^{-1}$at −0.20 V vs. RHE on Mo_2C@C-2/2. DFT calculations revealed that the Mo_2C content could balance the N_2 adsorption and the HER and the optimized Mo_2C@C-2/2 catalyst shows an end-on configuration while adsorbing N_2 and could further activate the molecule via a donation–backdonation mechanism.

Other typical Mo-based materials (MoN, MoS_2) have also been reported for NRR. In 2019, Wang reported Mo-vacancy-rich MoN nanocrystals loaded on N-doped hierarchical porous carbon (MoN@NC) via hard template method for NRR [22]. The Mo vacancies on MoN are derived from the SiO_2 template chemical etching process. Owing to the tuned electronic properties from Mo vacancies, the energy barrier of potential-determining step (PDS) for NRR shows an obvious decrease as well as a stronger hybridization between N_2 molecular orbitals and Mo d orbitals, and hence

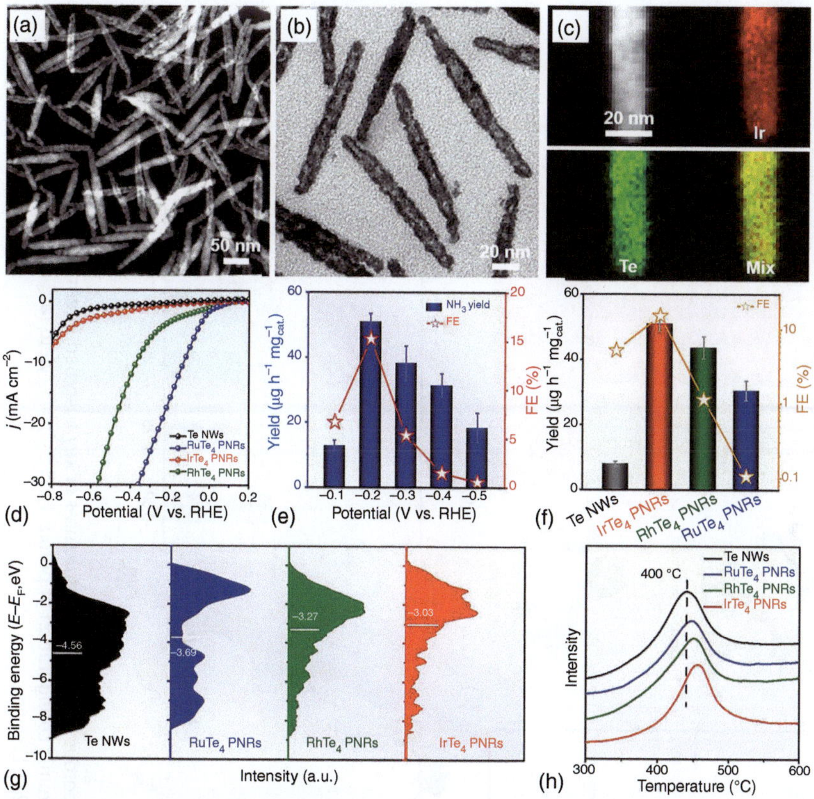

Figure 21.3 (a) HAAD-STEM image, (b) TEM image and (c) STEM-EDS elemental mappings of Ir-Te_4 PNRs. (d) Polarization curves of MTe_4 PNRs under ambient conditions. (e) Histograms of average NH_3 yield rate and corresponding FE at different applied potentials. (f) Histograms of average NH_3 yield rate and corresponding FE of various catalysts at −0.2 V vs. RHE. (g) Surface valence band photoemission spectra of MTe_4 PNRs. (h) The N_2-TPD profiles of MTe_4 PNRs. Source: Reproduced with permission of Wang et al. [17], Wiley-VCH.

an impressive NH_3 yield of 76.9 $\mu g\ h^{-1}\ mg^{-1}{}_{cat}$ could be obtained at −0.2 V vs. RHE in 0.1 M HCl solution. As for porous MoS_2, to boost its inert basal plane for NRR, Wang designed porous atomic layered 2H-MoS_2(PAL-MoS_2) for NRR (Figure 21.4c,d) [23]. During the preparation process, the hydrolysis of lithium compound in the MoS_2 lattice leads to the mesoporous structure and lots of adjacent Mo site pairs. DFT calculation results demonstrate that the N_2 is easily adsorbed on the adjacent Mo atom pairs with the laying-down manner on MoS_2-V_2 rather than the edge sites with the top-entry manner on MoS_2-V_1 (Figure 21.4a,b,e). As a result, PAL-MoS_2 delivered an excellent NH_3 yield of 3405.55 $\mu g\ h^{-1}\ mg^{-1}{}_{cat}$ and a high FE of 44.36% (Figure 21.4f).

Fe is an essential component of nitrogenase which is crucial for biological N_2 reduction. Therefore, Fe-based materials became the potential efficient electrocatalyst for NRR. For example, Li reported a DES-based regeneration strategy for constructing porous Fe_3O_4 nanosheets (Fe_3O_4-300) for NRR. Owing to the high

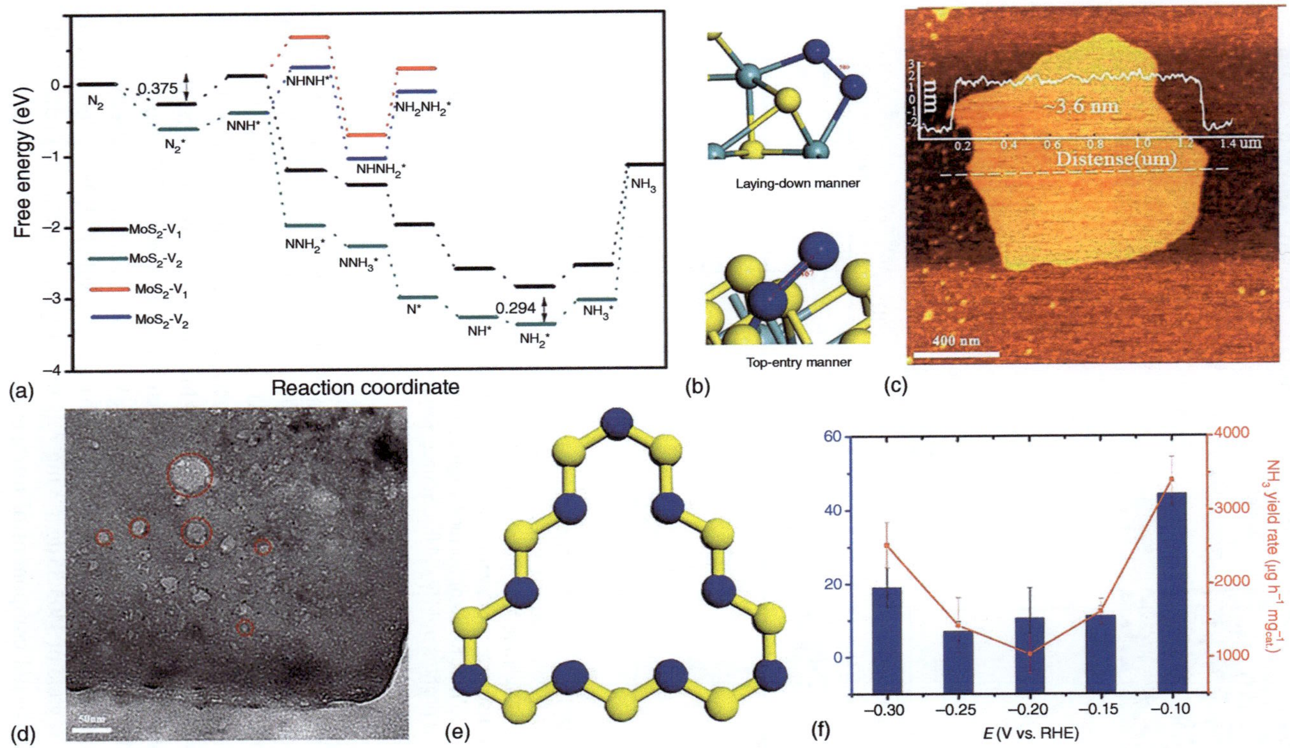

Figure 21.4 (a) Free energy diagram of NRR on defect-rich MoS_2 and porous MoS_2. (b) The structures of key adsorbed intermediates of N_2. (c) MoS AFM image and (d) HRTEM image shows porous structure in PAL-MoS_2. (e) Atomic configuration of porous MoS_2. (f) NH_3 yield rate and FEs of PAL-MoS_2 at various potentials. Source: Reproduced with permission of Chen et al. [23], Elsevier.

porosity, large active surface area, and rich oxygen vacancies, a notable yield of $12.09\,\mu g\,h^{-1}\,mg^{-1}{}_{cat}$ and FE of 34.8%. Mn-based material is also a potential NNR electrocatalyst due to its three-dimensional orbitals, which possess a vacant orbital to receive the electron of N_2 molecule and the occupied orbital could back-donate electrons to initiate N_2. Jiang synthesized a nanoporous CuMn alloy via dealloying method for NRR with the guidance of DFT calculation. As expected, the as-prepared np-CuMn alloy demonstrated a high ammonia yield rate of $28.9\,\mu g\,h^{-1}\,mg^{-1}{}_{cat}$ with a FE of 9.83% at −0.3 V vs. RHE in 0.1 M Na_2SO_4. Detailed experiments proved that, with the mass transfer promotion of porosity, there exists synergistic effect between Cu and Mn when their atom ratio is 1 : 1, which is the key to the high NRR performance.

Except for the noble-metal- and transition-metal-based electrocatalysts for NRR, main group metal-based materials such as Bi have also been studied due to the abundant reserves, low-cost, and tuned electronic structure. For instance, Bi is a suitable electrocatalyst for HER suppression but shows limited activity for the reductive protonation of N_2. Therefore, taking advantage of the chemical adsorption ability of Ni, Yu developed a three-dimensional amorphous BiNi alloy via a polymer gel-derived method [24]. As expected, BiNi alloy exhibits a high yield rate of $17.5\,\mu g\,h^{-1}\,mg^{-1}{}_{cat}$ and an FE of 13.8% at −0.6 V vs. RHE, which outperformed that of monometallic Bi. DFT results and comparative experiments confirmed that the synergistic effects of N_2 activation and adsorption, defect-rich surface, and enhanced mass/electro-transformation are the reasons for the high electrocatalytic activity.

21.3 Metal–Organic Frameworks for NRR

MOFs are a series of porous crystalline materials that consist of metal nodes and organic ligands through the principle of coordination chemistry. These kinds of materials can be designed with unique porosity and open structure which is beneficial for the active site exposure and charge mass transfer and thus to be the potential NRR electrocatalysts.

21.3.1 Pristine MOF Electrocatalyst

MOF materials without further transformation could be directly used as the NRR electrocatalysts. Up to now, several kinds of pristine MOF electrocatalysts have been proven to be efficient. In 2017, Zhao reported the first work of MOF electrocatalyst for NRR. After screening the NRR measurements on metal-based MOFs, including MIL-100(Fe), ZIF-67, and HKUST-1(Cu), MIL-100 turns out to be the optimal one with a yield rate of $2.12\times10^{-9}\,mol\,s\,cm^{-2}$ and FE of 1.43% at −1.2 V vs. Ag/AgCl at 90 °C. More importantly, an ammonia formation rate of $1.52\times10^{-9}\,mol\,s\,cm^{-2}$ could be achieved using air and water as raw materials, proving the catalytic efficiency of pristine MOF catalysts. In another report, Wei designed a self-supported Cu^{II}-MOF material on carbon cloth (JUC-1000/CC) for NRR [25]. Compared with the former reports, using as the bifunctional catalyst at cathodic and anode, JUC-1000/CC delivered a high NH_3 yield rate of 24.7 μg h^{-1} $mg^{-1}{}_{cat}$, yield of 11.9% and glucaric acid selectivity of 96.96% when coupled with for the anodic sodium gluconate (SG) oxidation reaction. This is superior to the condition that is coupled with oxygen evolution reaction. Post-characterization analysis confirmed the good preservation of the MOF structures, which are derived from the acidic (–OH) and basic (–NH–,1,3,5-trazine) groups in the ligand of MOF, proving the real active site is the pristine MOF.

Constructing conductive MOF is a feasible way to improve the current density during electrocatalytic reactions. Using hexahydroxytriphenylene (HHTP), which has been extensively used in the field of capacitors and sensors, as the MOF linker, Sun synthesized conductive MOF Co_3HHTP_2 nanoparticles for NRR [26]. The high-dispersed Co sites as well as the high charge transfer ability collectively result in a yield of 22.14 μg h^{-1} $mg^{-1}{}_{cat}$ and FE of 3.34%. Designing bimetallic MOF electrocatalyst is also an alternative way to tune the activity through the synergistic effect. Yan [27] shows an example of optimized composite of Co_3Fe MOF for NRR with a yield of 8.79 μg h^{-1} $mg^{-1}{}_{cat}$ and FE of 25.64% at −0.2 V vs. RHE in 0.1 M KOH, which can be attributed to the positive coupling effect between Co and Fe sites and the abundant active centers. In another work, the combination of nickel and iron in the bimetallic MOF also works for NRR . A yield of 9.3 μg h^{-1} $mg^{-1}{}_{cat}$ and FE of 11.5% was obtained on NiFe-MOF and it is superior to that of monometallic MOF. Density functional theory (DFT) calculations highlight the reduced energy barrier of the first ammonia release step induced by iron doping effect (3.1 wt% of Fe). Besides, the optimal electronic structures and the enlarged porosity aroused from the downsizing of the MOF would contribute to the reaction kinetics at the electrode/electrolyte interface. Main group elements such as Al, Bi and B-based MOFs are also effective for NRR. By taking advantage of the high selectivity for nitrogen capture, Ma reported an Al-MOF electrocatalyst for NRR. The MIL-100(Al) exhibits a yield of 10.69.3 μg h^{-1} $mg^{-1}{}_{cat}$ and FE of 22.6% [28]. Investigation indicates the activity is derived from the unsaturated metal sites (defects) which show high affinity to the N_2 and the strong interaction between N 2p orbitals and Al 3p. Defective strategy has also been applied on the UiO-66 MOF for NRR. Using formic acid and acetic acid as regulators, UiO-66-HAc and UiO-66-FA with defection have been investigated for NRR [29]. The presence of acid regulator leads to the missing of the organic linker in UiO-66-HAc and UiO-66-FA. Electrocatalytic experiments show that UiO-66-HAc with the highest amount of defects, exhibits the highest activity with a yield of 31.81 μg h^{-1} $mg^{-1}{}_{cat}$ and excellent FE of 48.06%. It is the high exposure of the Zr atom, which could enhance the adsorption of N_2 and suppress the HER at the same time.

Coordination microenvironment regulation is effective for the electronic structure tunning of the active sites and their catalytic performance. Recently, based on the theoretical calculation, Lee reported a symmetry-breaking and well-defined zinc metal-organic framework (Zn–N_2S_2-MOF) to be the best candidate for N_2 activation (Figure 21.5a) [30]. Using different ligands with varied S/N ratios, Zn-MOF with controlled N, S-coordination environment could be obtained (Zn-N_2S_2-MOF, Zn-N_3S-MOF, and Zn–N_4-MOF) (Figure 21.5a). After screening their electrochemical performance, Zn–N_2S_2-MOF stands out as the most robust one with an NH_3 yield of 25.07 μg h^{-1} $mg^{-1}{}_{cat}$ and FE of 44.57% (Figure 21.5c), which is well in

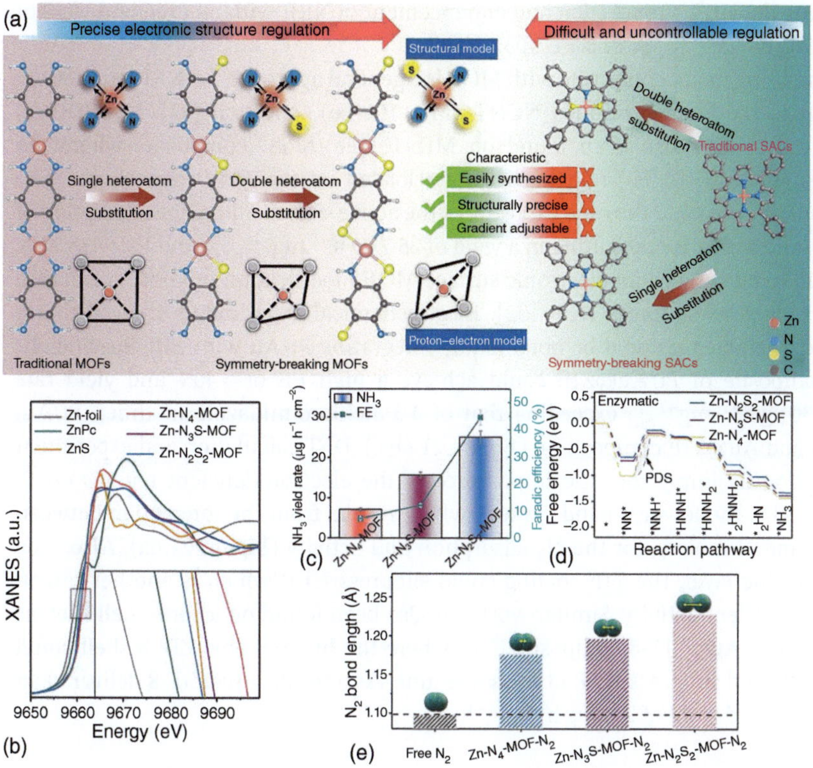

Figure 21.5 (a) Schematic illustration of tuning the electronic structure of a metal–organic framework. (b) Zn K-edge XANES spectra for Zn-N_2S_2-MOF, Zn-N_3S-MOF and Zn-N_4-MOF. (c) Comparison of the best NH_3 yield rates and FE of Zn-N_2S_2-MOF, Zn-N_3S-MOF and Zn-N_4-MOF. (d) Calculated free energy diagram through enzymatic pathway and (e) typical N≡N triple bond lengths after N_2 adsorption on Zn-N_2S_2-MOF, Zn-N_3S-MOF and Zn-N_4-MOF. Source: Reproduced with permission of Sun et al. [30], Wiley-VCH.

accord with the DFT results. XANES analysis shows that the distorted coordination environment could regulate the delocalized electrons around the Zn sites favoring the formation of low-valence $Zn^{\delta+}$, which could facilitate the N_2 adsorption and activation (Figure 21.5b,e). Besides, during NRR, theoretical calculations show that low-valence $Zn^{\delta+}$ in Zn-N_2S_2-MOF can promote the NRR kinetics by the reduced energy barrier of potential-determining step (Figure 21.5d).

21.3.2 MOF Composite Electrocatalyst

Combining MOF with other kinds of material is a feasible approach to tackle its disadvantages such as inferior conductivity, competitive reaction of HER, or poor NRR performance. For example, Ding developed a carbon nanotube insertion method to construct NCNT@MIL-101(Fe) material [31]. During NRR, the hydrophobicity in the MOF offers a N_2 enrichment environment near the active sites and hence could facilitate the HER suppression and enhancement of NRR with an excellent yield of 60,215 $\mu g\,h^{-1}\,mg^{-1}{}_{NCNT}$ and FE of 37.28%.

Combining metal composite with MOF is a general approach for NRR promotion. Similar to the aforementioned NCNT@MIL-101(Fe) condition [32], MIL-101(Fe) modification has also been found on MIL-101(Fe)/MoS_3 composite, where the ultra-small MIL-101(Fe) nanoparticles are loaded on the MoS_3 nanosheets. The high surface area and the high exposure of the active sites facilitate the adsorption of the N_2 and activation, resulting in a yield of 25.7 $\mu g\,h^{-1}\,mg^{-1}{}_{NCNT}$ and FE of 36.71%. In 2020, Ling showed an electronic surface MOF modification method to enhance NRR as well as suppress HER [33]. Electrochemical experiments showed that a 44-fold FE increase could be obtained by integrating Pt/Au with ZIF. Specifically, the composite of Pt/Au@ZIF could achieve a high FE of >40% and yield rate of >130 $\mu g\,h^{-1}\,mg^{-1}{}_{cat}$, exceeding that of 4,5-dichloroimidazole modified Pt/Au, Pt/Au and Au@ZIF composition (Figure 21.6b,c). DFT calculation and experiment results jointly imply that the importance of the electron-deficient regions on Pt catalyst (i.e. the lower d-band center), which results from the interaction between Pt/Au and N moiety, for the N_2 adsorption and activity (Figure 21.6a). Also, as a hydrophobic layer, the ZIF coating could suppress HER which is another reason for the excellent activity. Similar work has also been found on a core–shell such as nanoporous Au@ZIF-8 composite [34], where the hydrophobic ZIF-8 shell could enrich N_2 and suppress HER at the same time. As a result, Au@ZIF-8 delivered an FE of 44% and yield of 28.7 $\mu g\,h^{-1}\,mg^{-1}{}_{cat}$.

21.3.3 MOF-Derived Electrocatalyst

MOF materials can be further transformed into its derivation materials, such as metal composition, single metal site, and carbon-based electrocatalysts, for a specific application in NRR.

21.3.3.1 MOF-Derived Metal Composite Electrocatalyst

For the majority of MOF-derived metal composite electrocatalysts, metal–carbon materials obtained from the high-temperature pyrolysis could partially keep the original structure feature and enhance the conductivity, thus being the optimum material platform for NRR. In 2020, Liu et al. fabricated MOF-derived cobalt-doped carbon material Co/C-900 electrocatalyst by the calculation method using ZIF-67 as precursor [35]. After calculation, Co/C-900 retained the typical rhombohedral morphology. Owing to the porosity and the enhanced conductivity derived from carbonization, Co/C-900 catalyst exhibits a high yield of 4.66 $\mu mol\,cm^{-2}\,h^{-1}$ and FE of 11.53% at −0.3 V vs. RHE in 0.1 M KOH. Likewise, using ZIF-67 as precursor, Luo developed a MOF-derived core–shell such as nitrogen-doped carbon/Co_3O_4 nanocomposites (Co_3O_4@NCs) using a two-step annealing method . During the post-annealing and oxidation process under high temperatures, the Kirkendall effect caused the carbon atoms to migrate out more quickly than the metal atom hence leading to the formation of graphitized carbon shell and finally

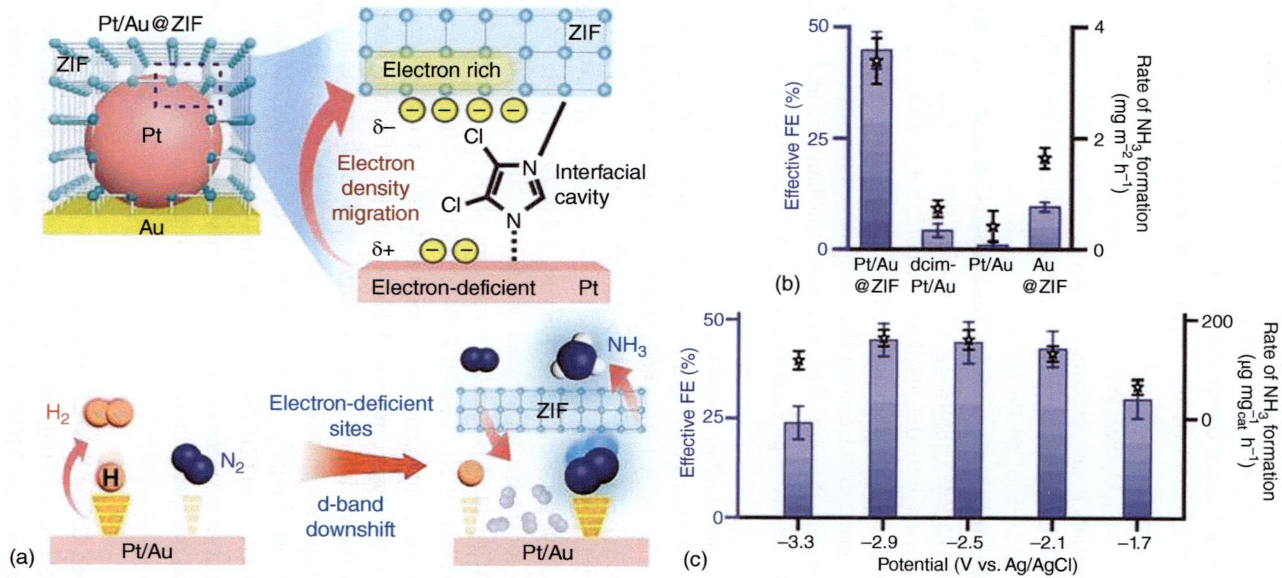

Figure 21.6 (a) Scheme illustration of Pt/Au@ZIF composition for NRR. (b) corresponding effective faradaic efficiency and rate of ammonia formation for Pt/Au@ZIF, (ii) dcim-Pt/Au, (iii) Pt/Au, and (iv) Au@ZIF. (c) NRR performance using 60 nm Pt/Au@ZIF platform at various electrochemical potentials. Source: Reproduced with permission of Sim et al. [33], Wiley-VCH.

Co_3O_4@NCs were obtained. appropriate annealing time of 5 min in air would generate suitable porosity and oxygen vacancies, which leads to an FE of 8.5% and yield of 42.58 μg h^{-1} mg^{-1}$_{cat}$. This is superior to that of Co_3O_4 nanoparticles, commercial Co_3O_4, and N-doped carbon, indicating the importance of the core–shell structure, oxygen vacancy, and then-doped carbon to NRR.

Bimetallic MOF-derived metal composite electrocatalysts have also been investigated for NRR. In 2022, inspired by the Mo-nitrogenase with FeMo cofactor in biological fixation of nitrogen, Tan fabricated bimetallic MOF-derived FeMo electrocatalysts ($Fe_{1.89}Mo_{4.11}O_7/FeS_2$@C, $FeMoO_4/FeS_2$@C) for NRR test using PMo12@MOF-100(Fe)@PVP as precursor (Figure 21.7a) [36]. Impressively, $Fe_{1.89}Mo_{4.11}O_7/FeS_2$@C delivered excellent activity with a high NH_3 yield of 105.3 μg h^{-1} mg^{-1}$_{cat}$ and FE of 54.7% at −0.4 V vs. RHE in acidic environment, which is higher than that $FeMoO_4/FeS_2$@C and FeS_2@C and are superior to the most reported electrocatalysts (Figure 21.7c). During the synthesis process, PVP layer would be carbonized into the graphitic carbon sheets which is beneficial for electron transformation and catalytic stability (Figure 21.7b). DFT calculation implies that the synergistic effect among the metallic composites could facilitate the NRR kinetics process through the enzymatic mechanism, and reduce the potential-determining step (the protonation of N_2 to form $*N_2H$ species) of NRR (Figure 21.7d,e).

MOF-derived single metal atom dispersed catalysts (SACs) have become intriguing in recent years due to the near 100% utility of the active sites. ZIF-8 is the typical precursor to construct SACs by taking advantage of the metal volatilization property of Zn during pyrolysis. In 2018, Zeng reported the first Ru-SACs for NRR using ZIF-8 as precursor [37]. Due to the low mass loading (0.18%), Ru SAs/N–C delivered a high yield of 120.9 μg h^{-1} mg^{-1}$_{cat}$ as long as a FE of 29.6% at −0.2 V vs. RHE, which paved the way for the design of effective SACs for NRR. After this work, Fe, Ni, and Co-based SACs had also been developed for NRR [38, 39]. For example, using the metal substitute of Ni to Zn in ZIF-8 ($Ni_xZn_{(1-x)}$BMOF), Ni-based SAC Ni_{single}-N–C(700-3 h) were obtained after pyrolysis at 700 °C for three hours [40]. Further increasing the pyrolysis time would destroy the pyridinic N (which is beneficial for NRR) and the vanish of single Ni atom. And the lower temperature at 600 °C could not induce enough defects. As a result, Ni_{single}-N-C(700-3 h) exhibits a high NH_3 yield of 115 μg h^{-1} mg^{-1}$_{cat}$ at −0.8 V vs. RHE and an FE of 22.9% at −0.21 V vs. RHE. Ni K-edge EXAFS and DFT results confirmed that Ni–N_3 sites are responsible for the high activity and selectivity through the associative alternating pathway and this configuration is superior to that of Ni–N_4 moiety for N_2 adsorption or side reaction suppression.

21.3.3.2 MOF-Derived Carbon-Based Electrocatalyst

As a kind of cost-effective material, porous carbon material is a rational choice for the electrocatalytic application. MOFs could be pyrolyzed into porous carbon materials which possess the pristine porosity or morphology nature. In 2018, Zhao gave an example of N-doped porous carbon for NRR [41]. Using ZIF as precursor, N-doped porous carbon with different N species and contents could be acquired by tuning the

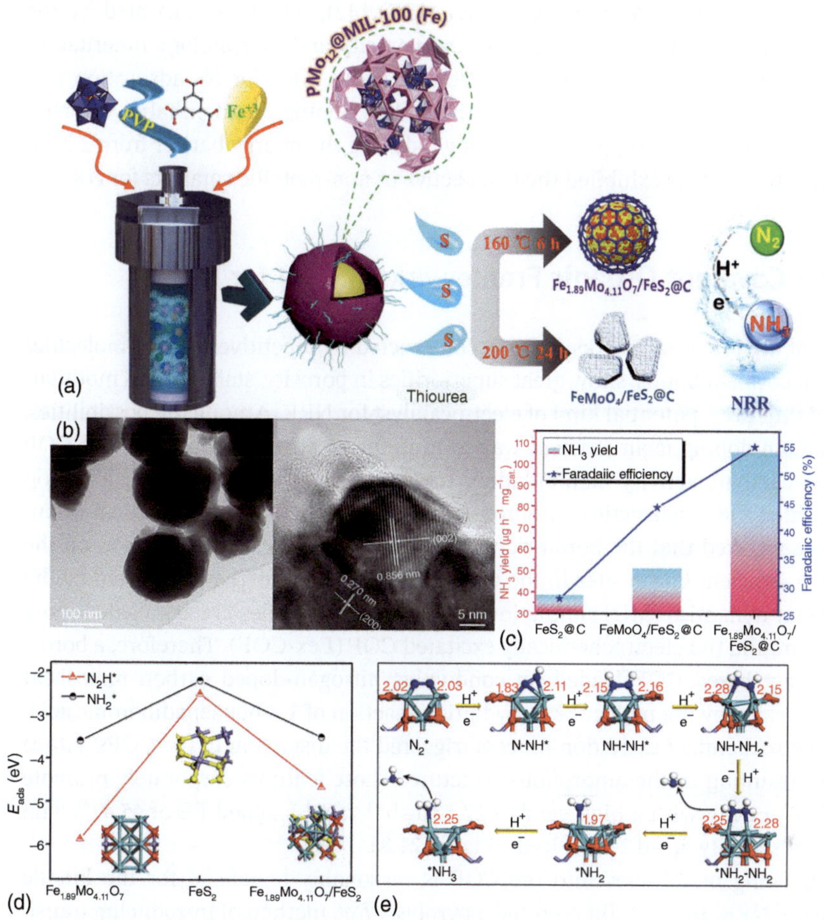

Figure 21.7 (a) Scheme illustration of the preparation of $Fe_{1.89}Mo_{4.11}O_7/FeS_2$@C. (b) TEM image and HRTEM image of $Fe_{1.89}Mo_{4.11}O_7/FeS_2$@C. (c) NH_3 yield rate and FE of FeS_2@C, $FeMoO_4$@FeS_2@C, and $Fe_{1.89}Mo_{4.11}O_7$@FeS_2@C (1.6 wt%) at −0.6 V, −0.5 V and −0.4 V vs. RHE. (d) Adsorption energies of $^*N_2H$ and $^*NH_2$ species on $Fe_{1.89}Mo_{4.11}O_7/FeS_2$, $Fe_{1.89}Mo_{4.11}O_7$ (002), and FeS_2 (200). (e) Structures corresponding to the minimum-energy enzymatic pathway for catalyzed NRR on $Fe_{1.89}Mo_{4.11}O_7/FeS_2$@C. Source: Reproduced with permission of Wang et al. [36], Royal Society of Chemistry.

pyrolysis temperature. Experiments show that NPC-750 with the highest N contents and pyrrolic N achieved the highest NH_3 yield of 1.40 mmol g^{-1} h^{-1} and FE of 1.42%. DFT calculation and experiments revealed pyridinic and pyrrolic N were active sites for ammonia synthesis and their contents were crucial for promoting ammonia production on NPC. Dual-atom doping could offer more selectivity opportunities for the activity regulation of carbon materials. Using MOF-5 as precursor, dicyandiamide (DCDA), and triphenylphosphine (TPP) as the source of N and P doping, porous N, P co-doping carbon (NP-C-MOF-5) could be obtained. The co-doping strategy facilitated the NRR reaction to proceed through the associative reaction mechanism and resulted in a yield of 1.08 μg h^{-1} $mg^{-1}{}_{cat}$. Similar study has also been observed on sulfur-modified N-MPC material(S/N-MPC) [42], which is fabricated by the pyrolysis of sulfur-doped MAF. Due to the porosity and morphology inheritance, S/N-MPC could provide abundant space and active sites for N_2 adsorption and activation. Besides, experiments show that sulfur doping could enhance the NH_3 yield from 18.65 to 45.51 $\mu g\,h^{-1}\,mg^{-1}{}_{cat}$ by reducing the energy barrier from 2.23 to 2.23 eV. These works exhibited the perspective of non-metallic catalysts for NRR.

21.4 Covalent Organic Frameworks (COFs) for NRR

Covalent organic frameworks (COFs), constructed by repetitive organic molecules through covalent bonds, show great superiorities in porosity, stability, and modularity and thus are a potential kind of electrocatalyst for NRR. Among the possibilities, heteroatom doping is an efficient way to tune the electronic property of the COF and for catalytic activity manipulation. In 2019, Yan reported a DFT calculation that guided the construction of boron-rich COF for NRR [43]. Theoretical calculations suggested that the boron-rich COFs could facilitate the N_2 fixation on the electron-deficient boron sites in the COF which could be further supported by the molecular dynamics (MD) simulations that the N_2 molecules are aggregated more closely around the electrochemically excitated COF (Eex-COF). Therefore, a boroxine linkage-based COF loaded on conductive nitrogen-doped carbon nanosheet was fabricated by the molecular dehydration reaction of 1,4-benzenediboronic acid. The electrochemical excitation process triggered the distortion of the COFs' lattice planes, resulting in the amorphous structure. These features corporately promote the NRR activity with a high yield of 12.53 $\mu g\,h^{-1}\,mg^{-1}{}_{cat}$, good FE of 45.43%, and excellent stability at −0.2 V vs. RHE (Figure 21.8).

Introducing metal sites into the COF is reasonably beneficial for the kinetic process of NRR. In 2022, Jin reported a pyrolysis-free method of introducing transitional metal into the COF via the unit of phthalocyanine block [44]. After screening the Ti, Cu, and Co elements with the $M–N_4–C$ center, Ti-COF turns out to be the best one due to its superior activity for activating inert N_2 molecules with a high NH_3 yield of 26.89 $\mu g\,h^{-1}\,mg^{-1}{}_{cat}$ and FE 34.62% in 0.05 M HCl. Theory calculation identified that the strong back-bonding between the d-orbital of Ti and the N_2 π^* orbital on Ti-COF is stronger than that of other transition metal catalysts (e.g. Cu and Co) and the NRR followed the alternate path instead of the distal pathway. This is in accord with the partial density of states (PDOS) simulation that Ti-COF could simultaneously adsorb and activate N_2 molecules, and efficiently inhibit HER side reactions. In another report, Feng designed a series of crystalline two-dimensional COF (2D c-COFs) with different $M–N_4–C$ centers (M = Fe, Co, Ni, Mn, Zn, and Cu) by the metal-phthalocyanine and pyrene units (Figure 21.9a,b) [45]. After screening the varied metal-based 2D c-COFs for NRR, $Fe–N_4–C$ center-based 2D c-COFs (FePc-pz) turn out to be the optimal one corresponding with a high NH_3 yield of 33.6 $\mu g\,h^{-1}\,mg^{-1}{}_{cat}$ and FE of 31.9% at −0.1 V vs. RHE (Figure 21.9c–e). Detailed experiments and characterization demonstrated that the $Fe–N_4–C$ center behaves in a localized electronic state at Fermi level (Figure 21.9f–h), which can bind with N_2 stronger than that of other metal-based $M–N_4–C$ centers.

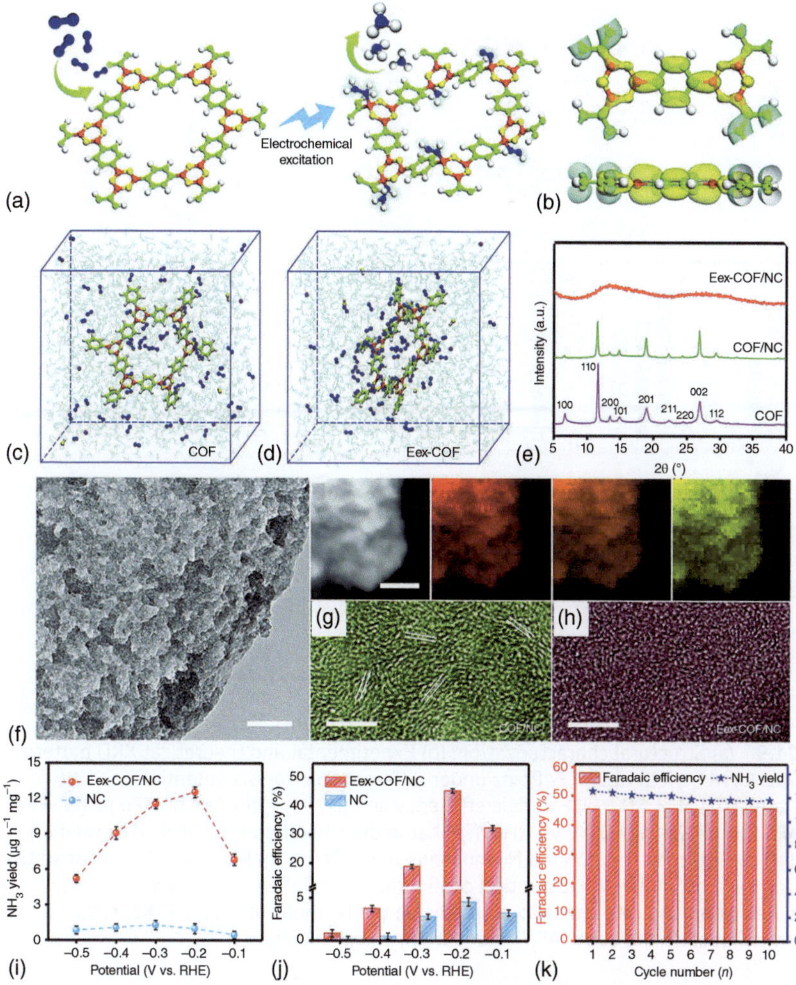

Figure 21.8 (a) Schematic illustration of the electrochemical excitation of COF. (b) The calculated charge distribution of COF. (c) Molecular dynamics simulation snapshots of e COF and (d) Eex-COF. (e) X-ray powder diffraction (XRD) patterns of Eex-COF, COF/NC and COF. (f) Transmission electron microscopy (TEM) image and element mappings of Eex-COF/NC. High-resolution TEM images of (g) COF/NC and (h) Eex-COF/NC. The red, green, blue, yellow, and gray spheres represent B, C, N, O, and H atoms, respectively. Electroreduction of N_2 to NH_3 at ambient conditions. (i) NH_3 yield rates, (j) corresponding faradaic efficiencies and the (k) durability test of Eex-COF/NC. Source: Liu et al., [43]/Springer Nature/CC BY 4.0.

Except for direct use as electrocatalysts, COFs that possess selective transport ability toward typical molecules could be used as the surface modifier to the NRR electrocatalysts. In 2021, Yan reported proton-filtering COFs with superior nitrogen penetration flux ability promoting ambient ammonia synthesis [46]. The COF was constructed by the 4,4′,4″-(1,3,5-triazine-2,4,6-triyl) trianiline and 1,4-phthalaldehyde vis Schiff base reaction. MD simulation demonstrates that the charge centers in the COF would prevent the active H^+ transfer, which

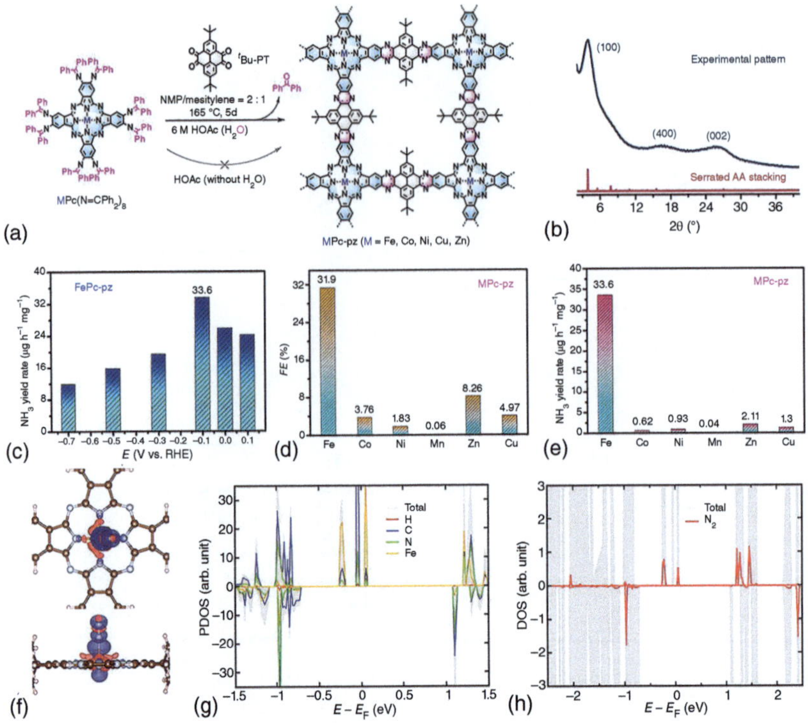

Figure 21.9 (a) Structural characterization. (b) Experimental and theoretical XRD pattern of FePc-pz. (c) NH_3 yield rate of FePc-pz under different electrolysis potentials in N_2-saturated 0.01 M H_2SO_4. (d) Faradaic efficiency and (f) NH_3 yield rate of MPc-pz (M = Fe, Co, Ni, Mn, Cu, and Zn) at −0.1 V vs. RHE. (f) Charge density differences of N_2 adsorbed on FePc-pz as compared to the isolated N_2 molecule and MPc-pz. (g) Projected density of states (PDOS) for the most stable N_2 adsorption configuration on FePc-pz with FeN_4. (c) DOS of FePc-pz. Red colors indicate partial projections on the N_2 molecule. Source: Reproduced with permission of Zhong et al. [45], American Chemical Society.

would be beneficial for the suppression of HER. More importantly, the strong van der Waals (vdW) interaction between COFs and N_2 could facilitate its mass transporting. As a result, the electrode of loading the exfoliated COF (ECOF) on the boron-doped-carbon paper (ECOF@BCP) can dramatically promote the NH_3 yield from about 45 to 287.2 $\mu g\,h^{-1}\,mg^{-1}{}_{cat}$ and FE from about 4% to 54.5% when compared with that PVDF (polyvinylidene fluoride, a long carbon chain but poor porosity polymer) modified electrode (ECOF@BCP) at −0.3 V vs. RHE, offering a feasible pathway for suppressing the compulsive electron-stealing HER reaction as well as promoting the adsorption of N_2 during NRR.

21.5 Conclusion and Outlook

Efficient electrocatalysts are the foundation for the future practical application of NRR. Among these, porous electrocatalyst is a promising alternative due to its abundant active sites and rich channels, which are beneficial for the activity enhancement and mass transfer of reactant, respectively. In this part, we have summarized the different types of porous electrocatalysts, including porous metal-based materials, MOFs, and COFs, providing a comprehensive and insightful investigation into the design of robust electrocatalysts. The MOF and COF materials are highlighted due to their ordered structure and distinct active sites, providing an ideal platform to explore the structure–activity relationship. Specifically, through defection, coordination, and synergistic effect strategy, the NRR activity could be optimized to a higher level on MOFs and COFs. Although these exciting developments have been achieved, there are still important challenges remaining for the application of porous materials in the NRR.

First, the mechanism of NRR is the foundation to interpret the structure–activity relationship and thus be the rational principle for the activity regulation study. So, persistent efforts on the mechanism research using the theoretical calculation and in-situ measurement technologies (such as Raman, FTIR, ESR, XPS, XANES, and EXAFS spectroscopy) on the porous materials with parodic structure and distinct active site should be emphasized during the NRR investigation.

Second, current performances of electrocatalysts are far from the industrial level with respect to selectivity, current density, and stability. Therefore, it is urgent to design efficient porous materials for NRR from the standpoint of intrinsic activity enhancement, mass transfer promotion, HER suppression, and conductivity improvement. For this purpose, combining the porous feature of ordered materials with other strategies such as heteroatomic doping, defect engineering, interface structure, and hydrophobic characteristics should be well studied.

Third, with the progressive development of catalytic materials and electrolyte systems, the NRR performance with respect to current density, selectivity, and stability has witnessed a continual increment. However, present works are mainly carried on the H type, single tank type, and polymer electrolyte membrane type electrolyzes, and little attention is paid to the design of suitable NRR electrolyzes, which is the key to the practical application of this infant technology. Future research can be focused on the appropriate reactor forms according to the specific system, and further optimized structural parameters.

Nowadays, despite the electrocatalytic NRR process is not as efficient as the Haber–Bosch process, and the ammonia production rate and Faraday efficiency remained to be improved, it is rational to speculate that the performance of this process could be significantly improved based on the cumulative NRR researches.

Acknowledgment

This work was supported by the National Natural Science Foundation of China (No. 21905195, 22473083), the Natural Science Foundation of Tianjin City (No. 20JCYBJC00800), and the PEIYANG Young Scholars Program of Tianjin University (No. 2020XRX-0023).

References

1 Yang, D.-H., Tao, Y., Ding, X., and Han, B.G. (2022). Porous organic polymers for electrocatalysis. *Chem. Soc. Rev.* 51 (2): 761–791.

2 Li, C., Qiu, Z., Sun, H. et al. (2022). Recent progress in covalent organic frameworks (COFs) for electrocatalysis. *Chin. J. Struct. Chem.* 41 (11): 2211084–2211099.

3 Ohashi, K., Iwase, K., Harada, T. et al. (2021). Rational design of electrocatalysts comprising single-atom-modified covalent organic frameworks for the N_2 reduction reaction: a first-principles study. *J. Phys. Chem. C* 125 (20): 10983–10990.

4 Wang, H., Yu, H., Wang, Z. et al. (2019). Electrochemical fabrication of porous Au film on Ni foam for nitrogen reduction to ammonia. *Small* 15 (6): 1804769.

5 Nazemi, M., Panikkanvalappil, S.R., El-Sayed, M.A. (2018). Enhancing the rate of electrochemical nitrogen reduction reaction for ammonia synthesis under ambient conditions using hollow gold nanocages. *Nano Energy* 49316–49323.

6 Nazemi, M., El-Sayed, M.A. (2018). Electrochemical synthesis of ammonia from N_2 and H_2O under ambient conditions using pore-size-controlled hollow gold nanocatalysts with tunable plasmonic properties. *J. Phys. Chem. Lett.* 9 (17): 5160–5166.

7 Wang, Z., Niu, J., Xu, Y. et al. (2020). Effects of AuCuB catalysts with porous nanostructures on electrosynthesis of ammonia. *ACS Sustain. Chem. Eng.* 8 (33): 12588–12594.

8 Xu, W., Fan, G., Chen, J. et al. (2020). Nanoporous palladium hydride for electrocatalytic N_2 reduction under ambient conditions. *Angew. Chem. Int. Ed.* 59 (9): 3511–3516.

9 Nakaya, Y., Furukawa, S. (2023). Catalysis of alloys: classification, principles, and design for a variety of materials and reactions. *Chem. Rev.* 123 (9): 5859–5947.

10 Li, Y., Yu, H., Wang, Z. et al. (2020). One-step synthesis of self-standing porous palladium-ruthenium nanosheet array on Ni foam for ambient electrosynthesis of ammonia. *Int. J. Hydrog. Energy* 45 (11): 5997–6005.

11 Pang, F., Wang, Z., Zhang, K. et al. (2019). Bimodal nanoporous Pd_3Cu_1 alloy with restrained hydrogen evolution for stable and high yield electrochemical nitrogen reduction. *Nano Energy* 58: 834–841.

12 Wang, X., Luo, M., Lan, J. et al. (2021). Nanoporous intermetallic Pd_3Bi for efficient electrochemical nitrogen reduction. *Adv. Mater.* 33 (18): 2007733.

13 Wang, H., Liu, S., Zhang, H. et al. (2020). Three-dimensional Pd–Ag–S porous nanosponges for electrocatalytic nitrogen reduction to ammonia. *Nanoscale* 12 (25): 13507–13512.

14 Skúlason, E., Bligaard, T., Gudmundsdóttir, S. et al. (2012). A theoretical evaluation of possible transition metal electro-catalysts for N_2 reduction. *Phys. Chem. Chem. Phys.* 14 (3): 1235–1245.

15 Wang, Z., Tian, W., Dai, Z. et al. (2021). Bimetallic mesoporous RhRu film for electrocatalytic nitrogen reduction to ammonia. *Inorg. Chem. Front.* 8 (18): 4276–4281.

16 Kim, C., Song, J.-Y., Choi, C. et al. (2022). Atomic-scale homogeneous Ru–Cu alloy nanoparticles for highly efficient electrocatalytic nitrogen reduction. *Adv. Mater.* 34 (40): 2205270.

17 Wang, J., Huang, B., Ji, Y. et al. (2020). A general strategy to glassy M-Te (M = Ru, Rh, Ir) porous nanorods for efficient electrochemical N_2 fixation. *Adv. Mater.* 32 (11): 1907112.

18 Jiang, B., Xue, H., Wang, P. et al. (2023). Noble-metal-metalloid alloy architectures: mesoporous amorphous iridium–tellurium alloy for electrochemical N_2 reduction. *J. Am. Chem. Soc.* 145 (11): 6079–6086.

19 Li, Y., Yu, H., Wang, Z. et al. (2019). Boron-doped silver nanosponges with enhanced performance towards electrocatalytic nitrogen reduction to ammonia. *Chem. Commun.* 55 (98): 14745–14748.

20 Huang, Z., Rafiq, M., Woldu, A.R. et al. (2023). Recent progress in electrocatalytic nitrogen reduction to ammonia (NRR). *Coord. Chem. Rev.* 478: 214981.

21 Wan, Y., Zheng, M., Lv, R. (2023). Rational design of Mo_2C nanosheets anchored on hierarchically porous carbon for boosting electrocatalytic N_2 reduction to NH_3. *Mater. Today Energy* 32: 101240.

22 Yang, X., Ling, F., Su, J. et al. (2020). Insights into the role of cation vacancy for significantly enhanced electrochemical nitrogen reduction. *Appl. Catal., B* 264: 118477.

23 Chen, J., Zhang, C., Huang, M. et al. (2021). The activation of porous atomic layered MoS_2 basal-plane to induce adjacent Mo atom pairs promoting high efficiency electrochemical N_2 fixation. *Appl. Catal., B* 285: 119810.

24 Fang, Z., Wu, P., Qian, Y., Yu, G. (2021). Gel-derived amorphous bismuth–nickel alloy promotes electrocatalytic nitrogen fixation via optimizing nitrogen adsorption and activation. *Angew. Chem. Int. Ed.* 60 (8): 4275–4281.

25 Zhao, L., Kuang, X., Chen, C. et al. (2019). Boosting electrocatalytic nitrogen fixation via energy-efficient anodic oxidation of sodium gluconate. *Chem. Commun.* 55 (68): 10170–10173.

26 Xiong, W., Cheng, X., Wang, T. et al. (2020). Co_3(hexahydroxytriphenylene)$_2$: a conductive metal–organic framework for ambient electrocatalytic N_2 reduction to NH_3. *Nano Res.* 13 (4): 1008–1012.

27 Li, W., Fang, W., Wu, C. et al. (2020). Bimetal–MOF nanosheets as efficient bifunctional electrocatalysts for oxygen evolution and nitrogen reduction reaction. *J. Mater. Chem. A* 8 (7): 3658–3666.

28 Fu, Y., Li, K., Batmunkh, M. et al. (2020). Unsaturated p-metal-based metal–organic frameworks for selective nitrogen reduction under ambient conditions. *ACS Appl. Mater. Interfaces* 12 (40): 44830–44839.

29 He, X., Liao, Y., Tan, J. et al. (2022). Defective UiO-66 toward boosted electrochemical nitrogen reduction to ammonia. *Electrochim. Acta* 409: 139988.

30 Sun, Y., Fan, W., Li, Y. et al. (2023). Tuning coordination structures of Zn sites through symmetry-breaking accelerates electrocatalysis. *Adv. Mater.* 36: e2306687.

31 Lv, Y., Wang, Y., Yang, M. et al. (2021). Nitrogen reduction through confined electro-catalysis with carbon nanotube inserted metal–organic frameworks. *J. Mater. Chem. A* 9 (3): 1480–1486.

32 Xu, W.-Y., Li, C., Li, F.-L. et al. (2022). A hybrid catalyst for efficient electrochemical N_2 fixation formed by decorating amorphous MoS_3 nanosheets with MIL-101(Fe) nanodots. *Sci. China Chem.* 65 (5): 885–891.

33 Sim, H.Y.F., Chen, J.R.T., Koh, C.S.L. et al. (2020). ZIF-induced d-band modification in a bimetallic nanocatalyst: achieving over 44% efficiency in the ambient nitrogen reduction reaction. *Angew. Chem. Int. Ed.* 59 (39): 16997–17003.

34 Yang, Y., Wang, S.-Q., Wen, H. et al. (2019). Nanoporous gold embedded ZIF composite for enhanced electrochemical nitrogen fixation. *Angew. Chem. Int. Ed.* 58 (43): 15362–15366.

35 Liu, A., Liang, X., Yang, Q. et al. (2020). Metal-organic-framework-derived cobalt-doped carbon material for electrochemical ammonia synthesis under ambient conditions. *ChemElectroChem* 7 (24): 4900–4905.

36 Wang, X., Feng, Z., Xiao, B. et al. (2020). Polyoxometalate-based metal-organic framework-derived bimetallic hybrid materials for upgraded electrochemical reduction of nitrogen. *Green Chem.* 22 (18): 6157–6169.

37 Geng, Z., Liu, Y., Kong, X. et al. (2018). Achieving a record-high yield rate of 120.9 for N_2 electrochemical reduction over Ru single-atom catalysts. *Adv. Mater.* 30 (40): 1803498.

38 Gao, Y., Han, Z., Hong, S. et al. (2019). ZIF-67-derived cobalt/nitrogen-doped carbon composites for efficient electrocatalytic N_2 reduction. *ACS Appl. Energy Mater.* 2 (8): 6071–6077.

39 Zhang, R., Jiao, L., Yang, W. et al. (2019). Single-atom catalysts templated by metal–organic frameworks for electrochemical nitrogen reduction. *J. Mater. Chem. A* 7 (46): 26371–26377.

40 Mukherjee, S., Yang, X., Shan, W. et al. (2020). Atomically dispersed single Ni site catalysts for nitrogen reduction toward electrochemical ammonia synthesis using N_2 and H_2O. *Small Methods* 4 (6): 1900821.

41 Liu, Y., Su, Y., Quan, X. et al. (2018). Facile ammonia synthesis from electrocatalytic N_2 reduction under ambient conditions on N-doped porous carbon. *ACS Catal.* 8 (2): 1186–1191.

42 Wang, J., Huang, H., Wang, P. et al. (2021). N, S synergistic effect in hierarchical porous carbon for enhanced NRR performance. *Carbon* 179358–179364.

43 Liu, S., Wang, M., Qian, T. et al. (2019). Facilitating nitrogen accessibility to boron-rich covalent organic frameworks via electrochemical excitation for efficient nitrogen fixation. *Nat. Commun.* 10 (1): 3898.

44 Jiang, M., Han, L., Peng, P. et al. (2022). Quasi-phthalocyanine conjugated covalent organic frameworks with nitrogen-coordinated transition metal centers for high-efficiency electrocatalytic ammonia synthesis. *Nano Lett.* 22 (1): 372–379.

45 Zhong, H., Wang, M., Ghorbani-Asl, M. et al. (2021). Boosting the electrocatalytic conversion of nitrogen to ammonia on metal-phthalocyanine-based two-dimensional conjugated covalent organic frameworks. *J. Am. Chem. Soc.* 143 (47): 19992–20000.

46 Liu, S., Qian, T., Wang, M. et al. (2021). Proton-filtering covalent organic frameworks with superior nitrogen penetration flux promote ambient ammonia synthesis. *Nat. Catal.* 4 (4): 322–331.

Part VII

Advanced Functional Materials for Liquid Fuel Oxidation

22

Metal-Based Materials for LFO

Yuchen Qin, Chao Gao, Ning Su, and Xia Sheng

Henan Agricultural University, College of Science, Department of Applied Chemistry Wenhua Road 95, Zhengzhou, 450002, China

22.1 Introduction

The excessive dependence and overuse of fossil energy leads to serious environmental pollution and energy shortage. To address the looming resource crisis and achieve a sustainable future, it is crucial to develop alternative pathways that can replace traditional petrochemical processes. Electrochemical liquid fuel oxidation (ELFO) has been regarded as a promising technology to meet the need of energy conservation and emission reduction. It can not only release a lot of energy but also be converted into abundant value-added chemicals via various energy conversion devices [1–10]. ELFO processes are particularly advantageous as they can be conveniently performed at room temperature and ambient pressure. This method is known for its simplicity, efficiency, and environmental friendliness.

Liquid fuels generally refer to small molecules of organics including formic acid, alcohols (methanol, ethanol, and glycerol), aldehydes (5-hydroxymethylfurfural (HMF), and furfural), and biopolymers (lignin and chitin). In recent decades, a series of organics (such as methanol, ethanol, and formic acid) have been investigated as the ideal fuels for direct fuel cells owing to their high mass-energy density, renewable sources, convenient transport, and storage (Figure 22.1a) [11–17]. Meanwhile, several liquid fuels (such as HMF, glycerol, and glucose) have been studied as promising anodic half-reactions to replace the oxygen evolution reaction (OER). These reactions can be coupled with cathodic reactions, including the hydrogen evolution reaction (HER), carbon dioxide reduction reaction (CO_2RR), and nitrogen reduction reaction (NRR) (Figure 22.1b) [1, 2, 5, 18–24]. It can effectively reduce the electricity consumption caused by the high potential of anodic OER and produce value-added chemicals instead of O_2 with negligible economic benefits. For instance, the anodic electrooxidation of glycerol can lower electricity consumption by up to 53% during electrochemical CO_2RR process [6].

Noble metal-based (NMB) nanostructures have been demonstrated to be the most efficient electrocatalysts for ELFO under different media. Numerous research efforts

Functional Materials for Electrocatalytic Energy Conversion, First Edition.
Edited by Zhicheng Zhang, Meiting Zhao, and Yuchen Qin.

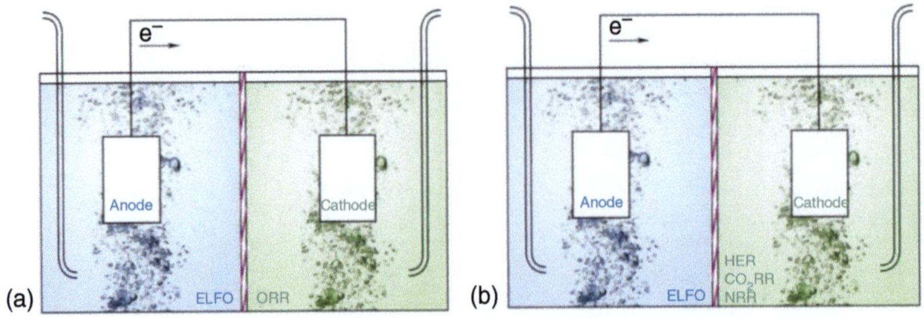

Figure 22.1 Schematic diagrams of (a) fuel cell and (b) coupled reactors.

have paid close attention to tuning the morphology, electronic structure, surface environment, and crystalline phase of NMB catalysts for the enhancement of electrochemical LFO performance [25–32]. Herein, this chapter will first summarize the reaction pathway of several common ELFO reactions. Then, the state-of-the-art advanced NMB electrocatalysts will be discussed by different elements. Finally, the future challenges and perspectives in the NMB design for ELFO reactions will be briefly proposed.

22.2 Reaction Pathway

22.2.1 Methanol Oxidation Reaction

Compared to hydrogen fuel cell, direct methanol fuel cell (DMFC) possesses higher theoretical energy density (6.13 kW h kg^{-1}) and volumetric energy density (17.28 MJ l^{-1}) [33]. Methanol oxidation reaction (MOR) takes place in the anode of DMFC and produces CO_2, H^+, and e^- ($CH_3OH + H_2O \rightarrow CO_2 + 6H^+ + 6e^-$). The thermodynamic equilibrium potential of MOR is 0.04 V, but the multistep reaction pathways result in sluggish kinetics and poor stability.

MOR typically involves a six-electron transfer process. It can proceed through two reaction pathways including CO or a non-CO pathway. In the CO pathway, known as the indirect pathway, CO* is produced through the deprotonation of COH* or CHO*. Afterward, CO* undergoes further oxidation to CO_2. In the non-CO pathway (direct pathway), possible intermediates, such as CHO* or COH*, are known to interact with OH* to form HCOOH*. These intermediates then undergo dehydrogenation to produce CO_2 (Figure 22.2).

The "CO" pathway is known to dominate in most traditional NMB electrocatalysts, leading to a strong interaction between activity sites and CO*. This interaction frequently causes the NMB catalysts, such as Pt-based catalysts, to be easily poisoned, resulting in poor stability. Therefore, improving the selectivity of the "non-CO" pathway on NMB catalysts is essential for preserving the active sites during the MOR process and enhancing practical application.

22.2.2 Ethanol Oxidation Reaction

Ethanol is an excellent fuel choice for alkaline fuel cells due to its high mass-energy density (8.01 KWh kg^{-1}), renewable sources, and convenient transportation and

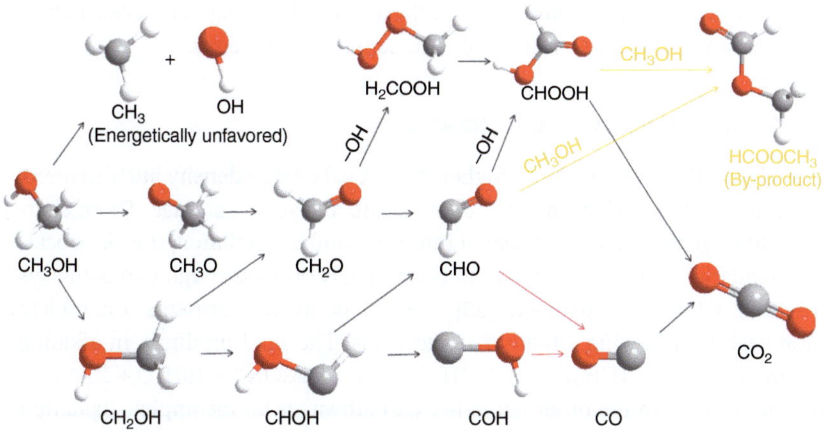

Figure 22.2 The pathway of electrocatalytic MOR.

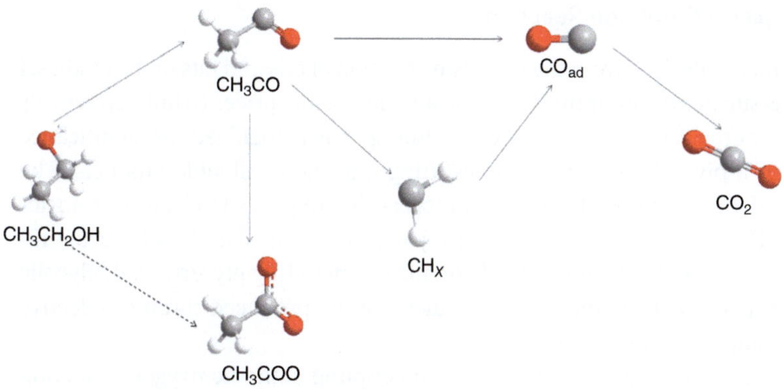

Figure 22.3 The pathway of electrocatalytic EOR.

storage. The electrocatalytic ethanol oxidation reaction (EOR) can be divided into two pathways (Figure 22.3): the incomplete oxidation pathway (C2 pathway) and the complete oxidation pathway (C1 pathway). In principle, ethanol remains bonded together (C—C bond) and is oxidized to acetic acid (acetate in alkaline media) through the C2 pathway, with the transfer of only four electrons. On the other hand, the C1 pathway involves the cleavage of the C—C bond, which releases a significant amount of energy stored within the bond and results in the production of CO_2 (carbonate in alkaline solution) [34]. Thus, the C1 pathway offers a higher energy conversion efficiency. However, even at the superior NMB electrocatalysts, C2 pathway is still predominant due to the weak C—C bond breaking capability.

Previous research has demonstrated the significant role of the CH_3CO intermediate (CH_3CO^*) in the reaction pathway of EOR. Studies have revealed that CH_3CO^* plays a pivotal role in two pathways: C1 pathway, where it undergoes C—C bond cleavage leading to carbonate formation, and C2 pathway, where it is directly oxidized to acetate. Therefore, it is crucial to enhance the C—C bond splitting of CH_3CO^* in order to promote the C1 pathway for improving cell performance, increasing ethanol conversion rates, and enhancing fuel efficiency.

22.2.3 Ethylene Glycol Oxidation Reaction

Ethylene glycol (EG) not only has a higher theoretical energy density but also offers the advantages of low toxicity, a high boiling point, and secure storage. These characteristics have generated a great deal of interest. Similar to ethanol, the electrocatalytic EG oxidation reaction (EGOR) can theoretically occur through two pathways (Figure 22.4): C1 and C2 pathway [35]. The C1 pathway represents a complete oxidation process, releasing a total of 10 electrons. The final products, in alkaline media, are carbonate ($CH_2OH–CH_2OH + 14OH^- \rightarrow 2CO_3^{2-} + 10H_2O + 10e^-$) or CO_2 in acid media. On the other hand, the C2 pathway is an incomplete oxidation process that occurs through a 4-electron mechanism, resulting in the generation of glycolate and oxalate in alkaline media or acetic acid in acid media.

22.2.4 Glycerol Oxidation Reaction

Glycerol, which is derived from the transesterification of oils and fats in the biodiesel industry, has substantial quantities and is available at a low price. To fully utilize this economic resource, the U.S. Department of Energy has recognized the significance of glycerol as a pivotal molecule for converting into more valuable products. The complete oxidation of glycerol to CO_2 is a process that requires 14 electrons transfer ($C_3H_8O_3 + 3H_2O - 14e^- \rightarrow 3CO_2 + 14H^+$) [36]. Due to this, it is unlikely to occur efficiently. Instead, a range of value-added chemicals, including glyceric acid, glycolic acid, hydroxypyruvic acid, and other organics, can be produced through selective glycerol oxidation reaction (GOR).

GOR can be carried out in both a fuel cell coupling cathodic oxygen reduction reaction (ORR) and electrolysis cell combing with cathodic HER, NRR, and CO_2RR.

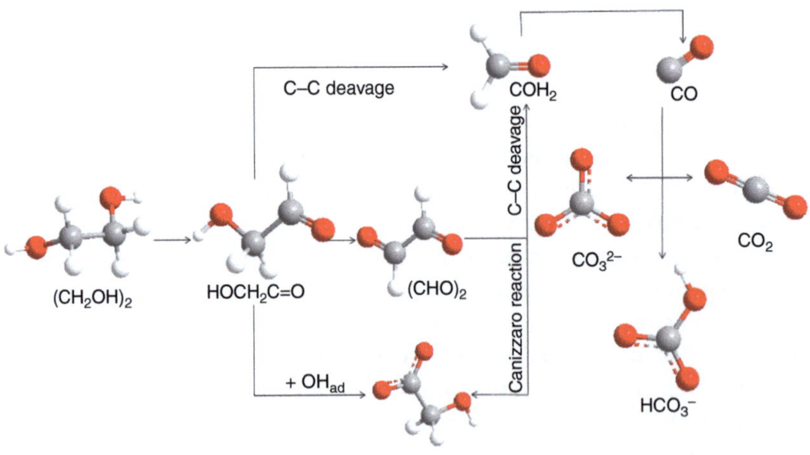

Figure 22.4 The pathway of electrocatalytic EGOR.

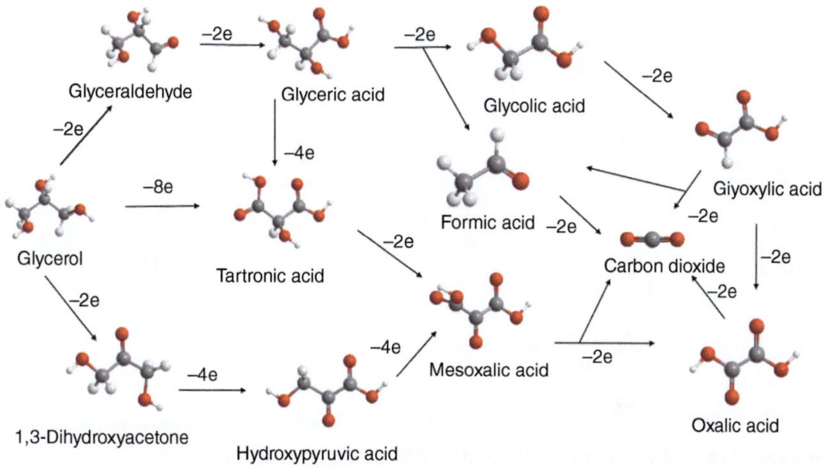

Figure 22.5 The pathway of electrocatalytic GOR.

Owing to the release of energy during GOR process, electricity can be generated at the anode in a fuel cell with the production of value-added chemicals. In electrolysis cells, the cell voltage can be effectively reduced when OER is replaced with GOR, as well as the target products can be obtained at both electrodes.

In general, glycerol electrooxidation is a complex process (Figure 22.5) that can be broken down into three main pathways. The first pathway involves the oxidation of the primary hydroxyl group, resulting in the formation of glyceraldehyde, which is further oxidized to glyceric acid. This process is followed by C—C bond cleavage and additional oxidation, ultimately leading to the production of CO_2. The second pathway involves the oxidation of the secondary hydroxyl group, resulting in the formation of 1,3-dihydroxyacetone. Subsequently, the terminal hydroxyl group undergoes oxidation, leading to the generation of hydroxypyruvic acid. Last, the third pathway involves the oxidation of both terminal hydroxyl groups, resulting in the production of tartronic acid.

22.2.5 Formic Acid Oxidation Reaction

Formic acid (HCOOH) has emerged as a highly promising option for a commercially viable fuel cell feedstock due to its advantageous oxidation properties under low temperatures, high theoretical cell potential, low toxicity, and minimal fuel crossover. It is widely acknowledged that the electrooxidation of HCOOH on Pt catalysts follows a dual-path mechanism, containing indirect and direct pathways [37]. The indirect pathway leads to the formation of the intermediate CO_{ad}, which can potentially degrade the catalyst. It is worth noting that the indirect pathway typically contributes to less than 5% of the overall rate of HCOOH oxidation. Based on the initial activation of either C—H or O—H bonds in HCOOH, the direct pathway can be bifurcated into two routes that involve a transient intermediate or a formate intermediate. This differentiation helps to comprehend the diverse reaction pathways and mechanisms involved in HCOOH electrooxidation.

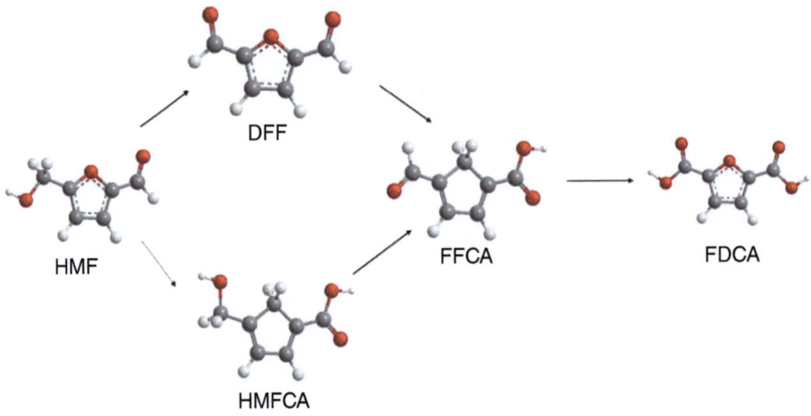

Figure 22.6 The pathway of electrocatalytic HMF oxidation reaction.

22.2.6 HMF Oxidation Reaction

HMF is a small molecule derived from non-cellulosic biomass. It contains active hydroxyl and aldehyde groups, which render it suitable for conversion into valuable chemical precursors utilized in the chemical industry. As an example of such a precursor, 2,5-furan dicarboxylic acid (FDCA) is obtained by oxidizing the two oxygen-containing groups of HMF to carboxyl groups [38, 39]. The U.S. Department of Energy recognizes FDCA as one of the top 12 platform chemicals derived from sugar.

As shown in Figure 22.6, the electrooxidation of HMF to FDCA possesses two possible pathways (pathway-1 and pathway-2), depending on the oxidation order of hydroxyl and aldehyde groups. Pathway-1 involves oxidizing the alcohol group of HMF, resulting in the formation of diformylfuran (DFF) as the initial intermediate. On the other hand, in pathway-2, the aldehyde group of HMF is oxidized, leading to the formation of 5-hydroxymethyl-2-furancarboxylic acid (HMFCA) as the first intermediate. Both DFF and HMFCA undergo further oxidation, ultimately yielding 5-formyl-2-furancarboxylic acid (FFCA), which eventually leads to the formation of FDCA.

Previous reports demonstrate that the dominance of each pathway is dependent on the reaction conditions. In strongly alkaline media with a pH of 13 or higher, pathway-2 becomes the primary route. Conversely, in

Conversely, in environments with a pH below 13 that are not strongly alkaline, pathway-1 is the preferred pathway.

22.3 Advanced Metal-Based Electrocatalysts

22.3.1 Pt-Based Electrocatalysts

Pt-based electrocatalysts have long been regarded as the most potent electrode materials utilized in energy conversion devices. Numerous modulation strategies have been successfully implemented to regulate the morphologies, electronic structures, components, and surface microenvironments of Pt-based nanostructures, achieving remarkable development in electrochemical LFO performance [40–47].

22.3.1.1 Methanol Oxidation Reaction (MOR)

The Pt–Ru system is widely recognized as the most popular bimetallic system, primarily due to its induced synergistic effect, enhanced CO tolerance, and superior catalytic activity for the MOR. Recently, Poerwoprajitno et al. presented a process that facilitates the growth and dispersion of Pt islands on Ru-branched nanoparticles, resulting in the formation of single-Pt-atom-on-Ru catalysts [48]. The single-Pt-atom-on-Ru catalyst exhibits remarkable performance in terms of high current density and mass activity for the MOR. The underlying reason for these exceptional characteristics lies in the unique capability of the single Pt atom structure to effectively adsorb CH_3OH, driving the MOR, while displaying weak binding affinity toward the CO intermediate, thereby minimizing CO poisoning. The enduring stability of the single-atom structure and its resistance to CO poisoning contribute significantly to the maintenance of this high MOR activity over time. Huang et al. have demonstrated the pioneering exploration of facet-dependent electrocatalytic activities in [49] catalysts by manipulating the exposed {111}/{100} facets on the surface of uniform nanocrystals. Interestingly, the {111}-terminated PtRu nanowires exhibit significantly higher stability and electrocatalytic mass activity toward methanol oxidation compared to the {100}-terminated PtRu nanocubes and commercial Pt/C catalyst. Specifically, the {111}-terminated PtRu nanowires demonstrate a performance improvement of 2.28 and 4.32 times over the {100}-terminated PtRu nanocubes and commercial Pt/C catalyst, respectively. This clearly indicates the superior methanol oxidation activity and resistance to CO poisoning of PtRu {111} facets in comparison to {100} facets.

Besides tuning the component of Pt-based electrocatalysts, the catalytic activity of platinum (Pt) is primarily governed by its electronic structure, specifically the position of its d-band center. According to d-band center theory, the adsorption strength can be modulated by electronic structure [50–52]. An increase in atom distance induces a tensile strain, causing an upward shift in the d-band center of the catalyst. Consequently, the antibonding states become less occupied, resulting in a stronger interaction between the adsorbates and the catalyst. Conversely, compressive strain reduces atomic distance and downwardly shifts the d-band center, leading to a weaker interaction between the adsorbates and the catalyst [53–59]. Therefore, the electronic structure is highly sensitive to the presence of lattice strain, which can significantly influence the catalytic performance of Pt-based catalysts. In a recent study by Jin et al. [56], ultrathin Pt shells were deposited onto Pd-based nanocubes. Through a phosphorization and dephosphorization process, a range of strain changes from −5.1% to 5.9% was achieved. The authors observed that the electrocatalytic MOR activity of the Pt shells exhibited an M-shaped curve in response to continuous strain changes. Importantly, the authors demonstrated that only moderate lattice strains, either tensile strains smaller than 4.7% or compressive strains smaller than 3.9%, enhanced the catalytic activity of Pt for MOR by promoting CO removal.

22.3.1.2 Ethanol Oxidation Reaction (EOR)

Pt is widely employed as a catalyst for alcohol oxidation, such as ethanol, despite its susceptibility to surface poisoning by strongly adsorbed CO and CH_x intermediates generated during the oxidation process. The EOR on a single Pt catalyst is highly influenced by its surface structure. Herrero et al. discovered that the (111) surfaces of Pt catalysts exhibit limited capability in breaking C—C bonds during alcohol oxidation, resulting in the predominant formation of acetic acid [60]. On the other hand, the (100) facets of Pt catalysts are highly active basal planes for C—C bond cleavage, leading to the dominant production of CO as a byproduct. Meanwhile, high-index planes typically exhibit significantly higher catalytic activity compared to particles with low Miller indices. The Sun group conducted a study where they synthesized Pt particles with a unique tetrahexahedral (THH) morphology through an electrochemical treatment of spherical Pt precursors on glassy carbon [61]. The resulting THH Pt particles are characterized by 24 high index facets (HIFs), which possess abundant atomic steps and dangling bonds, such as (730), (210), and (520) surfaces. These high-energy surfaces demonstrate significantly enhanced catalytic activity for EOR.

In addition to catalytic properties of EOR, the reaction pathways (C1 and C2 pathways) have also received extensive attention and research. Developing efficient electrocatalysts for the oxidation of ethanol to CO_2 (C1 pathway), particularly those capable of breaking the C—C bond, has posed a significant challenge in the field of electrocatalysis. In a previous study, Kowal et al. successfully synthesized a ternary $PtRhSnO_2/C$ electrocatalyst that demonstrates remarkable efficacy in splitting the C—C bond of ethanol at room temperature, leading to its predominant oxidation to CO_2 [62]. Detailed analyses identified the key factors contributing to the catalytic activity of this electrocatalyst. The synergy among the three constituents of the electrocatalyst is crucial. SnO_2, in the presence of H_2O, provides OH species that aid in the oxidation of dissociated CO at Rh sites (Figure 22.7). Pt, on the other hand, facilitates the dehydrogenation of ethanol. Additionally, Pt modifies the electronic structure of Rh, resulting in a moderate bonding between Rh and ethanol intermediates and products. This modification facilitates the breaking of the C—C bond in ethanol.

Recently, Chang et al. have demonstrated that the introduction of single-atomic Rh at O species on the metallic Pt(100) surface significantly enhances the cleavage of C—C bonds and promotes the oxidation of *CO, thereby achieving complete oxidation of ethanol to CO_2 [63]. The Rh at O-decorated Pt nanoclusters on carbon (Rh_{atO}-Pt NCs/C) exhibited an impressive 11.4-fold increase in current density compared to Pt/C at 0.75 V, along with a remarkable CO_2 selectivity exceeding 99.9% over a wide potential range of 0.35–0.75 V. The investigations have revealed that the presence of single-atomic Rh at O sites offers distinct advantages through the reversible configuration of Rh—O and C—Rh—O bonds during the EOR. This configuration facilitates the cleavage of C—C bonds, promoting the conversion

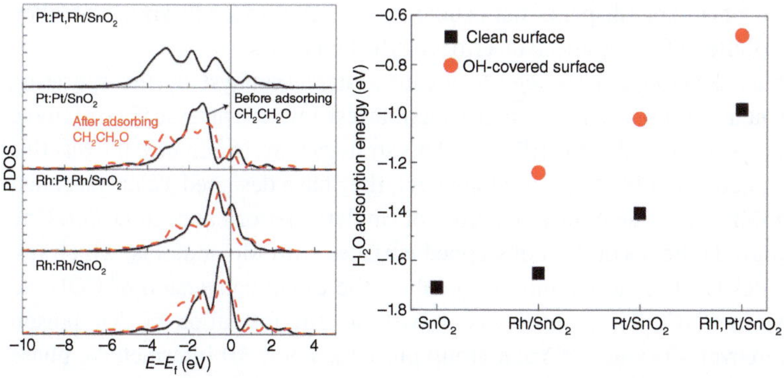

Figure 22.7 Calculated PDOS of the *d*-state of Pt or Rh on the $Rh/SnO_2(110)$, $Pt/SnO_2(110)$ and $RhPt/SnO_2(110)$ surfaces. Reproduced with permission of [62]. Springer Nature.

of ethanol, while also enhancing the formation of adsorbed −OH species that effectively remove the poisoning *CO intermediates. Liang et al. presented a novel ternary Au@PtIr core-shell catalyst for EOR in alkaline solutions [64]. The Au@PtIr catalyst exhibited a remarkable activity enhancement of six orders of magnitude compared to AuPtIr alloy catalysts. They investigated the catalytic mechanisms of Au@PtIr and its bimetallic subsets, namely Au@Pt and PtIr alloy. The findings indicate that monatomic steps and tensile strain induced by Au on PtIr facilitate the splitting of C—C bonds through ethanol dissociative adsorption, while Ir promotes dehydrogenation at low potentials. Notably, the presence of a CO band was only observed for the PtIr alloy, which exhibited low activity for ethanol dissociative adsorption. The outstanding performance of Au@PtIr for EOR can be attributed to the complementary effects of the Au core and surface Ir. This is evidenced by a low onset potential of 0.3 V and an all-metals peak current of $8.3\ A\ mg^{-1}$, with 57% of the currents generated via full ethanol oxidation.

Ethylene Glycol Oxidation Reaction (EGOR) Numerous efforts have been paid to tailoring and modulating the geometric and electronic structure of Pt-based electrocatalysts for the improvement of EGOR performance. Many bimetallic alloyed Pt—M (M = Ni, Co, Fe, Cu, and Au) have been synthesized, taking advantage of synergistic effect [12, 15, 16, 65–68]. Among the various transition metals, Rh has been studied due to its robustness as a co-element with unoccupied d-states. Rh has been found to enhance electrocatalytic behaviors, especially in terms of cleaving C—C bonds at a lower potential, providing better stabilization of intermediates, and improving the ligand effect. Du group reported a reliable method for synthesizing PtRh-Small (PtRh-S) and PtRh-Large (PtRh-L)NCs with different sizes of 5.5 and 10.5 nm, respectively. Rhodium acetate and rhodium chloride were introduced in the synthesis process. It is important to note that PtRh-S NCs are the smallest synthesized PtRh NCs reported thus far. The results have shown that PtRh NCs with {100} facet coverage exhibit superior catalytic activity and stability compared to pure Pt NCs terminated with {100} facets, specifically in EGOR. The unique combination of a monodisperse nanocubic structure, ultrasmall size, and alloy effects contribute to the enhanced electrocatalytic behaviors.

Wang et al. developed a strategy to create a stable catalyst by incorporating non-noble metals or nonmetals as "active auxiliaries" into the near-surface of noble metal nanocrystals bound with HIFs [69]. This strategy aims to enhance the reactive stability and activity of Pt. As proof-of-concept, they have designed a catalyst called 0.4%Mo/Pt3Mn, which demonstrates superior catalytic performance in EGOR. One notable feature is the decoration of stepped surfaces with Mo adatoms, which not only facilitates C—C cleavage but also enables the direct conversion of COH_x to CO_2 on Pt_3Mn CNCs. This process occurs without the formation of CO_L poison species. Moreover, the study of Yuan group presents a new 3D intermetallic phase of PtBi, doped with Au building blocks that are less than 7 nm in size. The inclusion of Au as an "[70]" significantly enhances the activity and protects against CO poisoning during EGOR on 3D PtBiAu [70]. This enhancement is achieved through the cleavage of C—C bonds and the attainment of a complete electrooxidation via a CO-free pathway that involves 10 electrons. The in situ Fourier transform infrared spectroscopy (FTIR) spectra confirm the successful implementation of a CO-free pathway for the complete EGOR using the 3D $Pt_{53.1}Bi_{43.4}Au_{3.5}$ intermetallic. When employed as an anode electrocatalyst in a practical direct ethylene glycol fuel cell (DEGFC), the 3D $Pt_{53.1}Bi_{43.4}Au_{3.5}$ intermetallic demonstrates a much higher power density (145/92 mW cm^{-2} in O_2/air, respectively) and long-term durability compared to commercial Pt/C.

Glycerol Oxidation Reaction (GOR) Pt nanostructures are extensively investigated as a glycerol electrooxidation catalyst because of their remarkable activity and ability to function across a broad pH range. In general, Pt has strong C—C bond breaking ability under high potential, resulting in the less C3 products. Alloying other metals to form binary or ternary catalysts can effectively modulate the catalytic activity and reaction pathway of GOR. Bismuth (Bi) is the most investigated element for GOR. Moreover, In the study of Souza et al., the addition of an optimal quantity of Bi ions to the solution was found to significantly enhance the activity of polycrystalline platinum (Pt_p) catalyst, resulting in a fivefold increase [71]. Additionally, the presence of adatoms on the catalyst surface exerts a strong influence on the reaction products. Specifically, the adatoms suppress pathways involving C—C bond breaking, thereby hindering the formation of CO and other unknown intermediates. Conversely, the production of glycerate is enhanced. Notably, the behavior of Pt_p–Bi systems in alkaline media differs from that in acid media. While Bi blocks the oxidation pathway through the primary carbon in acid media, glycerate becomes the main product in alkaline media. Dihydroxyacetone, on the other hand, is either produced in extremely low quantities or not produced at all.

In addition to Bi, Ru, and Sn are also usually employed to alloy with Pt for the enhancement of GOR performance. Sun's group conducted a comprehensive study on the GOR using various nanocatalysts including Pt/C, PtxRuy/C, and PtxRhy/C [72]. Through cyclic voltammetry characterization, it was observed that alloying Ru with Pt significantly enhanced the GOR activity, particularly at low potential.

However, alloying Rh did not have as much of an effect when compared to Pt/C. In order to investigate the GOR mechanism at a molecular level, electrochemical in situ FTIR and ^{13}C NMR spectroscopies were combined. The results revealed that the catalyst type and oxidation potential influenced the selectivity of products. While both PtRu/C and PtRh/C facilitated oxygen insertion reactions, resulting in a greater selectivity for carboxylic acids, the PtRh/C surface favored tartronic acid at higher potentials. Moreover, the increased ratio of the forward peak current to the reverse peak current (I_f/I_r) observed in the cyclic voltammograms for glycerol electrooxidation reaction (GEOR) on the binary and ternary catalysts suggests that the accumulation of oxidation products on these catalysts is significantly lower compared to Pt. Kim et al. conducted a study in which they synthesized PtRu/C and PtRuSn/C ternary catalysts [73]. The researchers discovered that the catalyst with an atomic ratio of Pt : Ru : Sn of 5 : 4 : 1 exhibited a lower onset potential (0.44 V) and a higher peak current density for GOR compared to Pt_5Ru_5/C and Pt/C catalysts. Additionally, the I_f/I_r ratio on $Pt_5Ru_4Sn_1$ was significantly higher than that of Pt and PtRu, indicating that there was less accumulation of intermediate residues on this catalyst during GEOR. Additionally, Falase et al. carried out a study on GEOR using Pt nanostructured catalysts prepared through spray pyrolysis. They analyzed the $Pt_{84}Ru_{16}$, $Pt_{96}Sn_4$, $Pt_{88}Ru_6Sn_6$, and Pt catalysts. The research findings indicated that the binary catalyst $Pt_{84}Ru_{16}$ had a higher oxidation current and better stability compared to the other catalysts. Interestingly, the addition of Ru and Sn to Pt did not contribute to maximizing current generation. This was evident from the lower current density observed on the ternary catalyst in comparison to the two binary catalysts. However, the coverage of reaction intermediates, which can impede electrode performance and reduce overall efficiency, was lower on the ternary catalyst compared to the binary PtRu, PtSn, and Pt nanocatalysts. This deduction was supported by the higher I_f/I_r ratio.

Formic Acid Oxidation Reaction (FAOR) Improving the FAOR performance has become a significant concern in past decades. In previous studies, alloying is a commonly used method to enhance the catalytic activity and CO resistance of Pt-based catalysts during FAOR process through indirect pathway. Perales-Rondón et al. reported that Bi adatom-modified Pt (111) electrodes could be utilized for HCOOH electrooxidation [74]. As the quantity of Bi coverage increased, the onset potential of HCOOH electrooxidation decreased, and the formate adsorption shifted toward higher potential values. Additionally, the activation energy for the C—H bond cleavage decreased. Peng et al. carried out a study on the Sb/Pt electrodes to examine their effectiveness as HCOOH electrooxidation catalysts [75]. They investigated the reaction mechanism by real-time electrochemical (EC)-surface-enhanced IR absorption spectroscopy (SEIRAS) in combination with DFT calculations. The results showed that at low potential, the CO band intensity on the Sb/Pt electrodes decreased significantly, while the formate band intensity was very weak. The increased anodic current primarily occurred through a non-formate direct transient pathway. DFT calculations revealed CO adsorption was reduced on Sb/Pt (111) surfaces. Importantly, the presence of Sb facilitated the adsorption of

HCOOH with a CH-down configuration, which is believed to be the precursor to CO_2 and the primary source of the oxidation current.

Due to the inevitable CO adsorption during indirect reaction pathway of FAOR, the direct dehydrogenation pathway of formic acid has received extensive attention recently. Huang's group has achieved a series of advanced Pt-based electrocatalysts in direct dehydrogenation of formic acid. They developed a new type of high-performance electrocatalyst, named PtBiPbNiCo hexagonal nanoplates (HEA HPs) [76]. The structure analysis reveals that the HEA HP consists of a PtBiPb medium-entropy core and a PtBiNiCo high-entropy shell (Figure 22.8a). Remarkably, these HEA HPs exhibit specific and mass activities of 27.2 mA cm^2 and 7.1 A $mgPt^{-1}$ for the FAOR. In addition, the HEA HPs achieve a power density of 321.2 mW cm^{-2} in a fuel cell with membrane electrode assembly (MEA). Both experimental and theoretical analyses provide evidence that the hexagonal intermetallic core/atomic layer shell structure and the synergistic effect of multiple elements greatly enhance the direct dehydrogenation pathway of formic acid molecules while suppressing the formation of CO. Furthermore, they fabricated surface-uneven PtPbBi/PtBi core/shell nanoplates (referred to as PtPbBi/PtBi NPs) as highly active and selective catalysts for FAOR (Figure 22.8b) [77]. Even in the complex MEA environment. These catalysts demonstrate unprecedented specific and mass activities of 25.1 mA cm^{-2} and 7.4 A $mgPt^{-1}$ for FAOR, respectively. Additionally, these catalysts exhibit weak CO adsorption and high selectivity for the dehydrogenation pathway in the FAOR test. These remarkable characteristics, combined with the ability to reach a power density of 161.5 mW cm^{-2} and maintain stable discharge performance (with only a 45.8% decay of power density at 0.4 V for 10 hours), demonstrate their great potential in single direct formic acid fuel cell devices. The in situ FTIR and X-ray absorption spectroscopy (XAS) results indicate a local electron interaction between PtPbBi and PtBi. Moreover, the PtBi shell exhibits excellent tolerance toward CO production/adsorption, allowing for the complete presence of the dehydrogenation pathway during FAOR. Besides Pt alloyed catalysts, they also report that the monoclinic phased platinum–tellurium nanotrepang (m-PtTe NT) can be utilized as an extremely active, selective, and stable catalyst for FAOR, following a direct reaction pathway [78]. The m-PtTe NT exhibits exceptional specific and mass activities, measuring at 6.78 mA cm^{-2} and 3.2 A $mgPt^{-1}$, respectively. These values are significantly higher than those of commercial Pt/C, rhombohedral-phased Pt_2Te_3 NT (r-Pt_2Te_3 NT), and trigonal-phased $PtTe_2$ NT (t-$PtTe_2$ NT), with increases of 35.7/22.9, 2.8/2.6, and 3.9/2.9 times, respectively. Furthermore, the m-PtTe NT demonstrates the greatest propensity for the direct FAOR pathway and exhibits superior tolerance to the poisonous CO intermediate. Notably, even in a single-cell medium, the m-PtTe NT achieves a much higher MEA power density of 171.4 mW cm^{-2} and a stability of 53.2% voltage loss after 5660 s, surpassing commercial Pt/C. In situ FTIR and XAS confirm that the unique nanostructure of m-PtTe NT effectively optimizes dehydrogenation steps, inhibits CO intermediate adsorption, and promotes the oxidation of the toxic CO intermediate. Density functional theory calculations further support

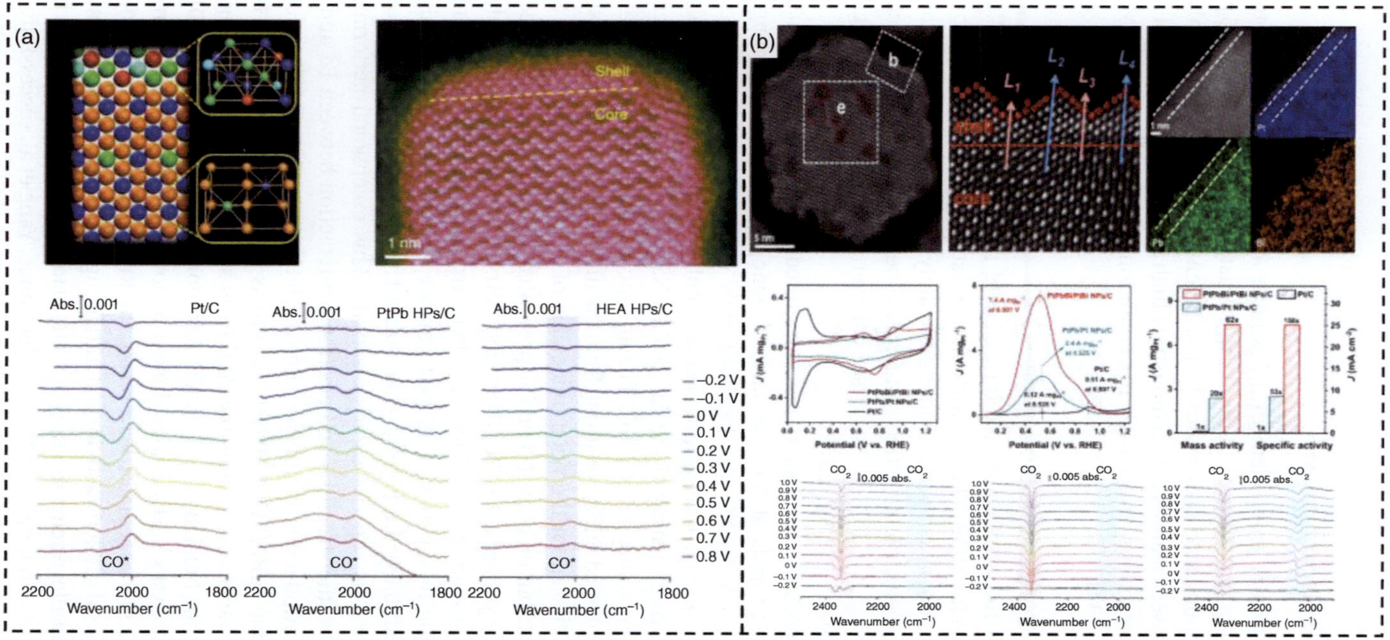

Figure 22.8 (a) Corresponding top view atomic model and AC-HAADF-STEM image of an individual HEA HP from the side view. In situ FTIR of voltage tracking from 0.2 to 0.8 V for Pt/C, PtPb HPs/C and HEA HPs/C. (b) AC-HAADF-STEM images of PtPbBi/PtBi NPs. FAOR performances and in situ FTIR of different catalysts. Source: (a) Reproduced with permission of [76]. Wiley-VCH. (b) Reproduced with permission of [77]. American Chemical Society.

these findings, revealing that the direct pathway is most favorable on m-PtTe NT compared to r-Pt_2Te_3 NT and t-$PtTe_2$ NT.

22.3.2 Pd-Based Electrocatalysts

As another important Pt group element, Pd is more abundant as compared to Pt, with a concentration of 0.015 ppm in the Earth's crust compared to Pt's 0.005 ppm by weight. This abundance contributes to the potential for lower-cost catalysts. Additionally, Pd-based catalysts also demonstrate high stability, which can facilitate anodic oxidation reactions under high pH conditions, making alkaline DMFCs practically feasible [34].

22.3.2.1 Methanol Oxidation Reaction (MOR)

Intensive research has been conducted on Pd-based monometallic, bimetallic, or trimetallic catalysts for MOR. This research aims to establish a solid foundation for the commercialization of next-generation Pt-free catalysts. Pd-based electrocatalysts generally exhibit excellent MOR activity under alkaline media (pH > 13). Especially, the adsorption strength of CO on Pd surface is lower than that of Pt, leading to a higher CO tolerance and better stability.

Huang et al. reported a novel supra-nanostructured Pd/Au composite (Au@Pd-SprNS) with Au nanorods at the core and Pd nanoparticles (NPs) forming the shell [79]. The Au@Pd SprNS was then uniformly decorated on graphene oxide (GO) nanosheets through electrostatic self-assembly for use in MOR catalysis. The MOR peak current density of the Au@Pd SprNSs/GO reached 1.74 mA cm^{-2}, a 25% improvement compared to Pd/C. The enhanced MOR performance can be attributed to the enlarged ECSAs resulting from the unique architecture: the discontinuous Pd shell allowed for thorough immersion of electrolytes into the Au core, while the even distribution of SprNS on GO prevented the aggregation of the active species. Kim et al. synthesized PdRu, PdSn, and PdIr binary nanoparticle catalysts, which were then dispersed on reduced graphene oxide (RGO) at a mass loading of 60 wt%, resulting in superior MOR performance [80]. Among these catalysts, the PdRu/RGO exhibited the highest peak current density and the lowest onset potential, indicating its superior catalytic activity. This exceptional performance can be attributed to the structural and electronic interaction between Pd and Ru. Specifically, the introduction of Ru can alter the lattice parameter and oxidation state of the resulting PdRu/RGO catalysts, as well as facilitate charge transfer from Ru to Pd. These changes can modify the Pd d-band states, ultimately lowering the activation barrier for breaking C—H bonds. Additionally, the formation of OH_{ads} species on the catalyst's surface is facilitated at lower potentials, aiding in the consumption of CO intermediates. As a result, the Ru species play a crucial role in enhancing the overall catalytic activity and mitigating the impact of CO poisoning.

Besides Pd-based alloyed electrocatalysts, metal oxides, and carbide materials have also been employed to improve the MOR performance. Yang et al. successfully created a loose network of Cu_2O-decorated palladium, referred to as the Cu_2O/Pd Network [81]. This network showed catalytic activity for MOR that was twice as high as both the commercial 20%Pd/C catalyst and the Pd black catalyst. Additionally, the Cu_2O/Pd Network demonstrated greater durability against CO poisoning. These results suggest that the appropriate amount of Cu_2O can cause the d-band center in Pd to shift upward, which promotes the capture of oxygen species and weakens the Pd–CO bond, thus aiding in the removal of CO intermediates. Furthermore, the open structure of the Cu_2O/Pd Network improved the exposure of Pd active sites, resulting in significantly enhanced MOR performance. Zhu et al. successfully fabricated nanoporous CuO/TiO_2/Pd–NiO–3 catalysts, demonstrating a significant improvement in MOR electrocatalytic performance and exceptional resistance to poisoning [82]. The highest anodic peak current density was nearly 4.40 times greater than that of the commercial Pd/C catalyst for MOR. The presence of Ni in the catalysts enhances the adsorption of oxygen-containing species on the catalyst surface, while the electron interaction between Pd and Ni supports charge transfer.

22.3.2.2 Ethanol Oxidation Reaction (EOR)

Pd has also received significant attention as a potential cost-effective alternative to the rare metal Pt for electrocatalytic EOR. Numerous efforts have been made to fabricate advanced Pd-based electrocatalysts for the enhancement of EOR performance and reveal the reaction mechanism. Su et al. showed that the morphological features of the nanowire arrays have a significant impact on operational durability in EORs, as opposed to the catalyst nanoparticles that are adsorbed on substrates (Figure 22.9) [83]. Chronoamperometric measurements revealed that sulfide-mediated Au@Pd nanowire arrays exhibited exceptionally long durability, with 86% of the initial current retained after one hour and an impressive 38% after 56 hours. A significant discovery is that the turn-off voltage in the cyclic voltammetry could be delayed up to 5.2 V, indicating delayed inhibition of the catalytic sites. The vertical arrays create open diffusion channels with a concentration gradient, allowing the active sites to gradually move downward as inhibition occurs, forming Pd–O–Pd. They propose that the inhibition depends on the coupling between two Pd–OH groups, which is more likely to occur in the Pd–Ox-rich regions at the top and less likely in the ethanol-rich regions at the bottom.

Zhang's group presents the synthesis of binary heterostructured nanoplates composed of palladium–phosphorus (Pd–P) using $Cu_{3-\chi}P$ nanoplates as templates through cation exchange [39]. The resulting nanoplates consist of a crystalline core and an amorphous shell with the same elemental components, which are denoted as c-Pd–P@a-Pd–P. Furthermore, these c-Pd–P@a-Pd–P nanoplates can be used as templates to create ternary heterostructured nanoplates by further alloying with nickel, resulting in Pd–Ni–P@c-Pd–Ni–P@a-Pd–Ni–P. The amount of nickel in the resulting ternary heterostructured nanoplates can be adjusted within a range of 9.47–38.61 at%. When employed as a catalyst, the c-Pd–Ni–P@a-Pd–Ni–P nanoplates containing 9.47 at% of nickel demonstrate exceptional electrocatalytic activity for ethanol oxidation, exhibiting a high mass current density of up to $3.05\ A\ mgPd^{-1}$, which is 4.5 times higher than that of the commercial Pd/C catalyst ($0.68\ A\ mgPd^{-1}$).

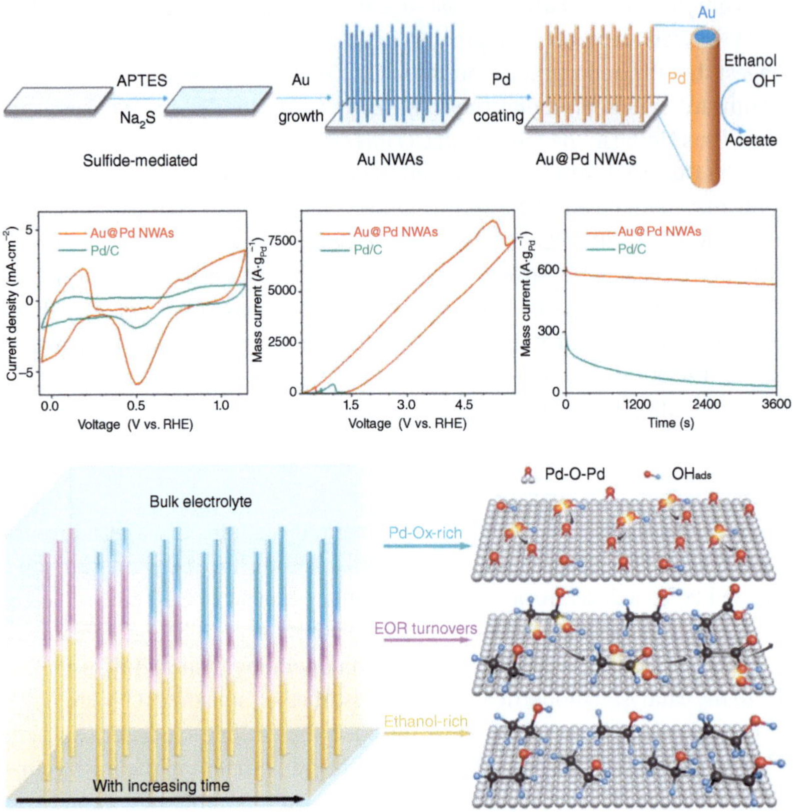

Figure 22.9 Schematic illustration and EOR performance of Au@Pd NWAs. Reproduced with permission of [83]. Elsevier.

Besides promoting the catalytic activity and stability of Pd-based catalysts, strengthening the ability of C—C bond cleavage and improving C1 pathway selectivity has also attracted wide attention. Jin's group presents findings on the effectiveness of amorphous PdCu (a-PdCu) nanomaterials as superior electrocatalysts for EOR [84]. Amorphization of the PdCu catalysts significantly facilitates the cleavage of C—C bonds, resulting in a Faradaic efficiency for the C1 path as high as 69.6%. By carefully controlling the size and shape of a-PdCu nanocatalysts, we can achieve a maximized mass activity of up to 15.25 A $mgPd^{-1}$, surpassing most reported catalysts. Importantly, accelerated durability tests demonstrate that both the isotropic structure and one-dimensional shape can dramatically enhance the catalytic durability of the catalysts. Qin et al. proposed the d-state-editing strategy to create more unoccupied d-states of Pd to enhance its interaction with CH_3CO^* and break C—C bonds [85]. As expected, the reduced number of eg electrons and the increased unoccupied d-states of Pd were successfully formed on the as-prepared porous RhAu–PdCu nanosheets (PNSs). Theoretical calculations demonstrate that the optimized d-states of RhAu–PdCu PNS can effectively enhance the adsorption of CH_3CO^* and significantly reduce the energy barrier of C—C bond cleavage, thereby boosting the complete oxidation of ethanol. The charge ratio of the C1 pathway on RhAu–PdCu PNSs is 51.5%, which is more than two times higher than that of Pd NSs. Xie's group introduces the "Metal-Edge-Driven" concept and proposes a one-step method for synthesizing ultrathin 2D PdPtCu trimetallic nanosheets and nanorings with increasing low-coordinated edge proportions of 11.74%, 23.11%, and 45.85% [86]. These materials function as advanced electrocatalysts for EOR. The enrichment of edges is achieved through a competitive surface capping and etching strategy, which allows for control over reaction kinetics. Electrochemical tests demonstrate that the edge-rich 9.0-nm-$Pd_{61}Pt_{22}Cu_{17}$ nanorings show exceptional activity and greatly improved durability in EOR. Molecular mechanistic studies confirm that the unsaturated edge sites on these 2D catalysts play a crucial role in triggering C—C bond scission and subsequent CO removal, facilitating the 12-electron-transferring EOR process.

22.3.2.3 Ethylene Glycol Oxidation Reaction (EGOR)

Alloying Pd with metals has been shown to be a highly effective catalyst for enhancing EGOR performance in alkaline environments, among various other catalysts.

Zhang et al. present a new method to generate significant lattice strains in a Pd-based catalyst through the radial growth of a Pd-rich phase on Au–Ag alloy nanowires with a thickness of no more than 1.5 nm [87]. This process creates tensile strains in the Pd-rich sheath due to the core–sheath lattice mismatch, as well as lattice distortion and twinning, resulting in nonhomogeneous local strains that act as hotspots for catalysis. When applied to the electrochemical EGOR, these highly strained nanowires outperformed versions with less strain, achieving up to 18.2 A $mgPd^{-1}$, respectively.

In addition to improving the catalytic activity, boosting the complete oxidation of EGOR is also an important approach to promote energy conversion efficiency. In the study of Qin's group, heterostructural nanosheets (HNSs) consisting of Pd and PdSe (Pd–PdSe HNSs) were successfully fabricated using a one-step approach [35]. These unique nanosheets possess unconventional electronic and geometrical structures, showcasing the coexistence of p-d hybridization interactions and tensile strain effect. Compared to commercial Pd/C and Pd NSs catalysts, Pd–PdSe HNSs demonstrate a 5.5-fold enhancement of specific activity and a 6.6-fold enhancement of mass activity for the EGOR. Moreover, the optimum C1 pathway selectivity of Pd–PdSe HNSs reaches 44.3%, demonstrating superior C—C bond cleavage ability. Electrochemical in situ FTIR spectroscopy and theoretical calculations show that the extraordinary p-d hybridization interaction and tensile strain effect effectively decrease the activation energy of C—C bond breaking and accelerate CO* oxidation, consequently enhancing the complete oxidation of EG and improving the catalytic performance.

22.3.2.4 Glycerol Oxidation Reaction (GOR)

Owing to the prominent performance of GOR in alkaline media, various advanced Pd-based electrocatalysts have been synthesized. In Zalineeva's study, the sacrificial support method is used to synthesize highly active self-supported Pd_xBi

catalysts [88]. These catalysts have a porous nanostructured morphology and high surface areas, ranging from 75 to 100 $m^2\ g^{-1}$, which makes Pd_xBi a cutting-edge catalyst. The Pd_4Bi catalyst exhibits the highest activity for GOR. In situ, FTIR demonstrates the unique catalytic behavior of self-supported Pd_xBi materials due to their specific structure and morphology. The confinement of reactants and intermediates in pores, which act as nanoreactors, is responsible for the high selectivity based on the electrode potential, yielding aldehyde and ketone at low potentials, hydroxypyruvate at moderate potentials, and CO_2 at high potentials. Simões et al. demonstrated that Pt-free $Pd_{0.9}Bi_{0.1}/C$ exhibited the same catalytic activity for GOR as Pt/C, while $Pd_{0.9}Bi_{0.1}/C$ displayed a GEOR onset potential that was 0.2 V more negative than Pt/C [89]. Results from the electrochemical experiments and in situ FTIR revealed that the modification of Pd and Pt surfaces by Bi did not alter the mechanism of GOR, with only the applied potential affecting the distribution of the final products. Chronoamperometric experiments, in combination with high performance liquid chromatography (HPLC) measurements, showed that the primary reaction products at low potentials were glycerate, dihydroxy acetone, and tartronate. As the electrode potential increased, mesoxalate was formed. When the potential exceeded 0.8 V, oxalate and formate resulting from C—C bond breaking were detected.

Besides Pd-based alloyed catalysts, adding metal oxides to Pd is a potential method for enhancing the efficiency of GOR. Xu et al. demonstrated that incorporating oxides such as CeO_2, NiO, Co_3O_4, and Mn_3O_4 into Pd/C greatly enhanced the catalytic activity and stability during the electrooxidation of alcohols [90]. Specifically, the addition of Pd–Co_3O_4 (2 : 1, w : w)/C resulted in the highest GOR activity, while Pd–Mn_3O_4/C exhibited superior performance stability compared to other oxides. Additionally, non-metallic components are also influential in the strategic development of catalysts.

22.3.2.5 Formic Acid Oxidation Reaction (FAOR)

Pd is one of the few metals capable of forming hydrides, and its catalytic properties depend on the elemental composition and spatial distribution of H atoms in the lattice. Xia's group presented a simple method for completely transforming Pd nanocubes into a stable phase composed of $PdH_{0.706}$ by treating them with aqueous hydrazine at a concentration as low as 9.2 mM [91]. To investigate the structure–catalytic property relationship of the resulting nanocubes with varying degrees of hydride formation, they used FAOR as a model reaction and the findings showed that the current density at 0.4 V increased fourfold when the nanocubes were fully converted from Pd to $PdH_{0.706}$. Additionally, using a set of slab models with PdH (100) overlayers on Pd (100), they performed density functional theory calculations to demonstrate that the degree of hydride formation can influence both the activity and selectivity toward FAOR by modulating the relative stability of HCOO and COOH intermediates. Yan et al. conducted a study on the synthesis of a series of catalysts, denoted as Au_xPd_{1-x}/NH_2–N–rGO, with varying ratios of Au and Pd atoms for the decomposition of formic acid [92]. Among all the prepared Au_xPd_{1-x}/NH_2–N–rGO composites, the $Au_{0.5}Pd_{0.5}/NH_2$–N–rGO catalyst exhibited the highest catalytic activity. The DFT calculations revealed that the $Au_{0.5}Pd_{0.5}$

catalyst had the most favorable energy barrier for its controlling reaction steps, including O—H bond cleavage, bi-HCOO* rearrangement, C—H bond cleavage, and H_2 desorption, in comparison to the other catalysts. This finding helps to explain its superior catalytic activity.

Moreover, in addition to metals, metal oxides can also serve as co-catalysts for effectively enhancing catalytic performance. Zhou et al. developed catalysts for the electrooxidation of formic acid, using Pd nanosheets decorated with SnO_2 nanoflakes (Pd@SnO_2–NSs/C) [23]. The in situ ATR-IR spectra indicated that the Pd@SnO_2–NSs/C catalysts could mitigate the accumulation of CO*. DFT calculations showed that, with SnO_2 modification, Pd(111) was preferable for converting HCOOB* to HCOOM* and for the C–H cleavage of HCOOM* to CO_2.

22.3.3 Other Noble Metal-Based Electrocatalysts

In addition to Pt-based and Pd-based catalysts, other noble metals such as Au- and Rh-based electrocatalysts have also shown unique electrocatalytic performance for various reactions [65, 93–95].

22.3.3.1 Advanced Au-Based Catalysts for LFO

Haldar et al. analyzed the impact of nanostructure and electrocatalytic activity on Au/Rh bimetallic composition [96]. A bio-assisted technique was utilized to create various ratios of Au : Rh nanoplates (25 : 75, 50 : 50, and 75 : 25). X-ray diffraction (XRD) and X-ray photoelectron spectroscopy (XPS) studies revealed the coexistence of Au and Rh phases within the bimetallic nanostructure, while electron microscopy confirmed the formation of triangle-shaped nanoplates. Notably, Au0.50Rh0.50 demonstrated increased activity for MOR in comparison to other compositions. Theoretical analyses suggest that the electrocatalytic improvement observed for MOR is attributed to the electronic modification effects of the surface, leading to optimization of the overall reaction energy profile through Au/Rh d-band mixing. Wang et al. represent the demonstration of the development of a nanoporous gold (NPG) with an irregularly shaped Au nanobranch structure prepared using a solid-phase reaction method [97]. Electrochemical testing results indicated improved catalytic activity and durability by the developed NPG electrocatalyst for the EOR as compared to other Au structures. These improvements are highly dependent upon the unique morphological features, high defect density, and presence of low-coordination atoms capable of participating in the EOR. These results actively expand the study of EOR using porous Au electrodes and promote further applications of the developed NPG as an electrocatalyst. Recently, Duan's group presented an electrocatalytic method for recycling glycerol (a biodiesel byproduct) to lactic acid and ethylene glycol (found in polyethylene terephthalate waste) to glycolic acid [98]. These products are valuable monomers for creating biodegradable polymers. By utilizing a nickel hydroxide-supported gold electrocatalyst (Au/Ni[OH]$_2$), we achieved high selectivities of lactic acid and glycolic acid (77% and 91%, respectively) with high current densities at moderate potentials (317.7 mA cm^{-2} at 0.95 V vs. RHE and 326.2 mA cm^{-2} at 1.15 V vs. RHE, respectively). We discovered that glycerol and ethylene glycol are concentrated at the Au/$Ni(OH)_2$ interface through their adjacent hydroxyl groups, significantly increasing local concentrations and, as a result, leading to high current densities. To demonstrate the concept, they used a membrane-free flow electrolyzer to recycle triglyceride and PET bottles, which yielded 11.2 g of lactic acid alongside 9.3 l of H_2 and 13.7 g of glycolic acid along with 9.4 l of H_2. This study highlights the potential for simultaneously producing valuable chemicals and H_2 fuel from waste in an environmentally friendly manner.

22.3.3.2 Advanced Rh-Based Catalysts for LFO

Wang et al. introduced a straightforward dual-template strategy for in situ fabrication of one-dimensional rhodium mesoporous nanotubes (Rh MNTs) [99]. In this method, the resulting Ni nanowires and di-block copolymer micelles play a key role in shaping the tubular interiors and mesoporous exteriors, respectively. The well-defined hollow mesoporous structure not only offers ample active centers and enhanced channels but also hinders catalyst aggregation. As a result, the produced Rh MNTs demonstrate exceptional activity, favorable kinetics, and high durability for the MOR, surpassing those of traditional commercial Pt black. In the study of Sung's group, a surfactant-free microwave-assisted method was used to easily synthesize Rh–Sn catalysts for the electrocatalytic oxidation of ethanol [100]. The electrocatalyst exhibits remarkable activity and durability at high fuel concentrations due to its bifunctional mechanism and electronic modification with C—C bond splitting. This allows for a significant reduction in the volume and weight of the fuel cell system. Chen et al. have developed a simple hydrothermal method to synthesize ultrathin nanoflakes nanoaggregates of bimetallic RhCu (RhCu-BUNNs) with atomic thickness [101]. In the context of the NRR, bimetallic RhCu-BUNNs demonstrate superior electrocatalytic activity compared to monometallic Rh ultrathin nanoflakes nanoaggregates (Rh-UNNs). This is due to the effective weakening of the competitive HER achieved by the introduction of Cu, resulting in a high NH_3 yield rate (95.06 $\mu g\, h^{-1}\, mg_{cat}{}^{-1}$) at a potential of −0.2 V. DFT calculations indicate that the addition of Cu can significantly reduce the energy demand in the *NN reductive process, thus enhancing NRR activity. Additionally, RhCu-BUNNs also display excellent electrocatalytic activity and selectivity for the GOR. In addition, they further used the GOR to assist the NRR with bifunctional RhCu-BUNN electrocatalysts to address the increased overall cell voltage caused by the sluggish anodic OER. Specifically, only a whole voltage of 1.2 V is required for electrochemical NH_3 production in the presence of glycerol, resulting in the generation of valuable glyceraldehyde at the anode.

22.4 Summary and Perspectives

In conclusion, ELFO finds wide applications not only in novel energy conversion devices but also in facilitating the coupling of multiple reactions, resulting in reduced cell voltage and increased production of high-value-added products. NMB nanomaterials have emerged as highly efficient catalysts for ELFO. Researchers have made significant progress in enhancing the intrinsic activity of these catalysts and reducing their usage by carefully controlling factors such as crystal structure, electronic structure, surface environment, composition, synergy effects, and carrier effects. These advancements have contributed to cost reduction and improved overall efficiency. However, despite these achievements, several challenges persist in this research field.

(1) **Strengthening the understanding of reaction mechanism**
The mechanisms of catalysts with diverse structures and compositions exhibit significant variations, and the precise reaction pathways and mechanisms of certain processes remain incompletely understood. Leveraging advanced modern spectroscopic characterization techniques in conjunction with theoretical calculations can provide deeper insights into reaction pathways and mechanisms. This research endeavor holds immense importance in the design and development of highly efficient catalysts, as well as in gaining a comprehensive understanding of the fundamental nature of catalytic reactions and enhancing catalyst performance.

(2) **Development of catalysts with low noble metal loading**
The development of low-loading noble metal catalysts plays a crucial role in advancing ELFO technology. Noble metal catalysts generally exhibit high catalytic activity, their limited reserves and high cost pose significant barriers to the widespread application of ELFO. The preparation of single-atom catalysts, incorporation of non-noble metals, or modulation of geometric/electronic structures are currently hot research topics aimed at reducing the atomic utilization of noble metals and enhancing their intrinsic catalytic activity and stability, also remaining great challenges.

(3) **Fabrication of multifunctional catalysts**
ELFO can be coupled with multiple cathodic reactions, effectively reducing cell voltage and increasing the production of high-value-added products. Therefore, the development of efficient multifunctional catalysts that can be applied to both the anodic and cathodic reactions in batteries is of great significance. In particular, the development of multifunctional catalysts suitable for different reaction environments to target various high-value-added products is currently a hot research topic and holds significant research value.

(4) **Design of novel electrolytic cell reactors**
Developing and designing advanced electrolyzer tailored to different coupling reactions is of paramount importance for enhancing the overall performance of the device. By considering the specific reaction medium, reaction voltage, and reaction conditions, the design of an efficient flow electrolyzer reactor can effectively improve reaction efficiency, minimize reactant losses, reduce the energy consumption of the device, and enhance the yield and selectivity of the desired products.

References

1 Li, S., Ma, R., Hu, J. et al. (2022). Coordination environment tuning of nickel sites by oxyanions to optimize methanol electro-oxidation activity. *Nat. Commun.* 13: 2916.

2 Xie, J., Zhang, Q., Gu, L. et al. (2016). Ruthenium–platinum core–shell nanocatalysts with substantially enhanced activity and durability towards methanol oxidation. *Nano Energy* 21: 247.

3 Tang, J.-X., Tian, N., Xiao, L.-P. et al. (2022). Helical PdPtAu nanowires bounded with high-index facets selectively switch the pathway of ethanol electrooxidation. *J. Mater. Chem. A* 10: 10902.

4 Li, S., Lai, J., Luque, R. et al. (2016). Designed multimetallic Pd nanosponges with enhanced electrocatalytic activity for ethylene glycol and glycerol oxidation. *Energy Environ. Sci.* 9: 3097.

5 Fan, L., Ji, Y., Wang, G. et al. (2022). High entropy alloy electrocatalytic electrode toward alkaline glycerol valorization coupling with acidic hydrogen production. *J. Am. Chem. Soc.* 144: 7224.

6 Verma, S., Lu, S., and Kenis, P.J.A. (2019). Co-electrolysis of CO_2 and glycerol as a pathway to carbon chemicals with improved technoeconomics due to low electricity consumption. *Nat. Energy* 4: 466.

7 Yang, Y. and Mu, T. (2021). Electrochemical oxidation of biomass derived 5-hydroxymethylfurfural (HMF): pathway, mechanism, catalysts and coupling reactions. *Green Chem.* 23: 4228.

8 Tian, Z., Da, Y., Wang, M. et al. (2023). Selective photoelectrochemical oxidation of glucose to glucaric acid by single atom Pt decorated defective TiO_2. *Nat. Commun.* 14: 142.

9 Thakur, N., Mehta, D., Chaturvedi, A. et al. (2023). Glucose oxidation assisted hydrogen and gluconic/glucaric acid production using NiVP/Pi bifunctional electrocatalyst. *J. Mater. Chem. A* 11: 15868.

10 Xu, X., Dong, X., Li, D. et al. (2023). Pt nanoflowers as a highly effective electrocatalyst for glucose oxidation in abiotic glucose fuel cells. *ACS Appl. Mater. Interfaces* 15: 17969.

11 Li, H., Zeng, R., Feng, X. et al. (2022). Oxidative stability matters: a case study of palladium hydride nanosheets for alkaline fuel cells. *J. Am. Chem. Soc.* 144: 8106.

12 Bu, L., Liang, J., Ning, F. et al. (2022). Low-coordination trimetallic ptfeco nanosaws for practical fuel cells. *Adv. Mater.* 35: 2208672.

13 Lv, J., Lang, Z., Fu, J. et al. (2022). Molecular iron oxide clusters boost the oxygen reduction reaction of platinum electrocatalysts at near-neutral pH. *Angew. Chem. Int. Ed.* 61: e202202650.

14 Feng, Q., Wang, X., Klingenhof, M. et al. (2022). Low-Pt NiNC-supported PtNi nanoalloy oxygen reduction reaction electrocatalysts—in situ tracking of the atomic alloying process. *Angew. Chem. Int. Ed.* 61: e202203728.

15 Xu, F., Cai, S., Lin, B. et al. (2022). Geometric engineering of porous PtCu nanotubes with ultrahigh methanol oxidation and oxygen reduction capability. *Small* 18: 2107387.

16 Song, T., Chen, M., Yin, P. et al. (2022). Intermetallic PtFe electrocatalysts for the oxygen reduction reaction: ordering degree-dependent performance. *Small* 18: 2202916.

17 Zhao, Y., Zhang, Z., Liu, L. et al. (2022). S and O co-coordinated Mo single sites in hierarchically porous tubes from sulfur–enamine copolymerization for oxygen reduction and evolution. *J. Am. Chem. Soc.* 144: 20571.

18 Wang, X., Liu, Y., Ma, X.-Y. et al. (2023). The role of bismuth in suppressing the CO poisoning in alkaline methanol electrooxidation: switching the reaction from the CO to formate pathway. *Nano Lett.* 23: 685.

19 Li, M., Wang, T., Zhao, W. et al. (2022). A pair-electrosynthesis for formate at ultra-low voltage via coupling of CO_2 reduction and formaldehyde oxidation. *Nano-Micro Lett.* 14: 211.

20 Pan, Y., Li, Y., Dong, C.-L. et al. (2023). Unveiling the synergistic effect of multi-valence Cu species to promote formaldehyde oxidation for anodic hydrogen production. *Chem* 9: 963.

21 Zhou, B., Li, Y., Zou, Y. et al. (2021). Platinum modulates redox properties and 5-hydroxymethylfurfural adsorption kinetics of $Ni(OH)_2$ for biomass upgrading. *Angew. Chem. Int. Ed.* 60: 22908.

22 Zhang, Y., Zhou, B., Wei, Z. et al. (2021). Coupling glucose-assisted Cu(I)/Cu(II) redox with electrochemical hydrogen production. *Adv. Mater.* 33: 2104791.

23 Zhou, Y.-W., Chen, Y.-F., Qin, X. et al. (2021). Boosting electrocatalytic oxidation of formic acid on SnO_2-decorated Pd nanosheets. *J. Catal.* 399: 8.

24 Ye, F., Zhang, S., Cheng, Q. et al. (2023). The role of oxygen-vacancy in bifunctional indium oxyhydroxide catalysts for electrochemical coupling of biomass valorization with CO_2 conversion. *Nat. Commun.* 14: 2040.

25 Li, Y., Zha, X., Zhao, Y. et al. (2022). Enhancing the spin–orbit torque efficiency by the insertion of a sub-nanometer β-W layer. *ACS Nano* 16: 11852.

26 Xu, B.-B., Fu, X.-B., You, X.-M. et al. (2022). Synergistic promotion of single-atom Co surrounding a PtCo alloy based on a g-C_3N_4 nanosheet for overall water splitting. *ACS Catal.* 12: 6958.

27 An, Z., Li, H., Zhang, X. et al. (2022). Structural evolution of a PtRh nanodendrite electrocatalyst and its ultrahigh durability toward oxygen reduction reaction. *ACS Catal.* 12: 3302.

28 Feng, Y., Zhao, Z., Li, F. et al. (2021). Highly surface-distorted Pt superstructures for multifunctional electrocatalysis. *Nano Lett.* 21: 5075.

29 Hong, Y., Dutta, S., Jang, S.W. et al. (2022). Crystal facet-manipulated 2D Pt nanodendrites to achieve an intimate heterointerface for hydrogen evolution reactions. *J. Am. Chem. Soc.* 144: 9033.

30 Lai, W., Yu, P., Gao, L. et al. (2022). Boosting the interfacial hydrogen migration for efficient alkaline hydrogen evolution on Pt-based nanowires. *J. Mater. Chem. A* 10: 16834.

31 Li, M., Duanmu, K., Wan, C. et al. (2019). Single-atom tailoring of platinum nanocatalysts for high-performance multifunctional electrocatalysis. *Nat. Catal.* 2: 495.

32 Huang, B., Bu, L., Ning, F. et al. (2022). Three-dimensional porous platinum-tellurium-rhodium surface/interface achieve remarkable practical fuel cell catalysis. *Energy Environ. Sci.* 15: 3877.

33 Tong, Y., Yan, X., Liang, J. et al. (2019). Metal-based electrocatalysts for methanol electro-oxidation: progress, opportunities, and challenges. *Small* 17: 1904126.

34 Bai, J., Liu, D., Yang, J. et al. (2019). Nanocatalysts for electrocatalytic oxidation of ethanol. *ChemSusChem* 12: 2117.

35 Qin, Y., Zhang, W., Wang, F. et al. (2022). Extraordinary p–d hybridization interaction in heterostructural Pd-PdSe nanosheets boosts C–C Bond cleavage of ethylene glycol electrooxidation. *Angew. Chem. Int. Ed.* 61: e202200899.

36 Ozoemena, K.I. (2016). Nanostructured platinum-free electrocatalysts in alkaline direct alcohol fuel cells: catalyst design, principles and applications. *RSC Adv.* 6: 89523.

37 Liu, X., Jacob, T., and Gao, W. (2022). Progress of fundamental mechanism of formic acid decomposition and electrooxidation. *J. Energy Chem.* 70: 292.

38 Chen, D., Ding, Y., Cao, X. et al. (2023). Highly efficient biomass upgrading by a Ni–Cu electrocatalyst featuring passivation of water oxidation activity. *Angew. Chem. Int. Ed.* 62: e202309478.

39 Zhang, Z., Liu, S., Wu, Z. et al. (2023). High efficiency coupled electrocatalytic CO_2 reduction to C_2H_4 with 5-hydroxymethylfurfural oxidation over Cu-based nanoflower electrocatalysts. *Green Chem.* 25: 5404.

40 Hu, Y., Guo, X., Shen, T. et al. (2022). Hollow porous carbon-confined atomically ordered $PtCo_3$ intermetallics for an efficient oxygen reduction reaction. *ACS Catal.* 12: 5380.

41 Chao, T., Luo, X., Chen, W. et al. (2017). Atomically dispersed copper–platinum dual sites alloyed with palladium nanorings catalyze the hydrogen evolution reaction. *Angew. Chem. Int. Ed.* 56: 16047.

42 Yang, L., Li, G., Ge, J. et al. (2018). TePbPt alloy nanotube as electrocatalyst with enhanced performance towards methanol oxidation reaction. *J. Mater. Chem. A* 6: 16798.

43 Luo, X., Liu, C., Wang, X. et al. (1967). Spin regulation on 2D Pd–Fe–Pt nanomeshes promotes fuel electrooxidations. *Nano Lett.* 2020: 20.

44 Li, H., Han, Y., Zhao, H. et al. (2020). Fast site-to-site electron transfer of high-entropy alloy nanocatalyst driving redox electrocatalysis. *Nat. Commun.* 11: 5437.

45 He, X., Zhang, Y., Gu, X. et al. (2023). Pt-induced atomic-level tailoring towards paracrystalline high-entropy alloy. *Nat. Commun.* 14: 775.

46 Haddadnezhad, M., Park, W., Jung, I. et al. (2022). Synthesis of Pt double-walled nanoframes with well-defined and controllable facets. *ACS Nano* 16: 21283.

47 Chen, Y., Pei, J., Chen, Z. et al. (2022). Pt atomic layers with tensile strain and rich defects boost ethanol electrooxidation. *Nano Lett.* 22: 7563.

48 Poerwoprajitno, A.R., Gloag, L., Watt, J. et al. (2022). A single-Pt-atom-on-Ru-nanoparticle electrocatalyst for CO-resilient methanol oxidation. *Nat. Catal.* 5: 231.

49 Huang, L., Zhang, X., Wang, Q. et al. (2018). Shape-control of Pt–Ru nanocrystals: tuning surface structure for enhanced electrocatalytic methanol oxidation. *J. Am. Chem. Soc.* 140: 1142.

50 Wu, J., Qi, L., You, H. et al. (2012). Icosahedral platinum alloy nanocrystals with enhanced electrocatalytic activities. *J. Am. Chem. Soc.* 134: 11880.

51 Jiao, S., Fu, X., and Huang, H. (2021). Descriptors for the evaluation of electrocatalytic reactions: d-band theory and beyond. *Adv. Funct. Mater.* 32: 2107651.

52 Huang, H., Jia, H., Liu, Z. et al. (2017). Understanding of strain effects in the electrochemical reduction of CO_2: using Pd nanostructures as an ideal platform. *Angew. Chem. Int. Ed.* 56: 3594.

53 Luo, M. and Guo, S. (2017). Strain-controlled electrocatalysis on multimetallic nanomaterials. *Nat. Rev. Mater.* 2: 17059.

54 Qin, Y., Yu, T., Deng, S. et al. (2022). RuO_2 electronic structure and lattice strain dual engineering for enhanced acidic oxygen evolution reaction performance. *Nat. Commun.* 13: 3784.

55 Wu, G., Han, X., Cai, J. et al. (2022). In-plane strain engineering in ultrathin noble metal nanosheets boosts the intrinsic electrocatalytic hydrogen evolution activity. *Nat. Commun.* 13: 4200.

56 He, T., Wang, W., Shi, F. et al. (2021). Mastering the surface strain of platinum catalysts for efficient electrocatalysis. *Nature* 598: 76.

57 Li, J., Sharma, S., Wei, K. et al. (2020). Anisotropic strain tuning of $L1_0$ ternary nanoparticles for oxygen reduction. *J. Am. Chem. Soc.* 142: 19209.

58 Han, S., Wang, C., Wang, Y. et al. (2021). Electrosynthesis of nitrate via the oxidation of nitrogen on tensile-strained palladium porous nanosheets. *Angew. Chem. Int. Ed.* 60: 4474.

59 Strasser, P., Koh, S., Anniyev, T. et al. (2010). Lattice-strain control of the activity in dealloyed core–shell fuel cell catalysts. *Nat. Chem.* 2: 454.

60 Paredes-Salazar, E.A., Herrero, E., and Varela, H. (2023). Mass transfer phenomena induced by surface gas flow rate in the hanging meniscus configuration: a case study of the methanol electro-oxidation reaction on Pt(100). *Electrochim. Acta* 464: 142917.

61 Tian, N., Zhou, Z.-Y., Sun, S.-G. et al. (2007). Synthesis of tetrahexahedral platinum nanocrystals with high-index facets and high electro-oxidation activity. *Science* 316: 732.

62 Kowal, A., Li, M., Shao, M. et al. (2009). Ternary Pt/Rh/SnO_2 electrocatalysts for oxidizing ethanol to CO_2. *Nat. Mater.* 8: 325.

63 Chang, Q., Hong, Y., Lee, H.J. et al. (2022). Achieving complete electrooxidation of ethanol by single atomic Rh decoration of Pt nanocubes. *Proc. Natl. Acad. Sci.* 119: e2112109119.

64 Liang, Z., Song, L., Deng, S. et al. (2019). Direct 12-electron oxidation of ethanol on a ternary Au(core)-PtIr(Shell) electrocatalyst. *J. Am. Chem. Soc.* 141: 9629.

65 Kong, F., Liu, X., Song, Y. et al. (2022). Selectively coupling Ru single atoms to PtNi concavities for high-performance methanol oxidation via *d*-band center regulation. *Angew. Chem. Int. Ed.* 61: e202207524.

66 Wu, J., Kong, Z., Li, Y. et al. (2022). Unveiling the adsorption behavior and redox properties of PtNi nanowire for biomass-derived molecules electrooxidation. *ACS Nano* 16: 21518.

67 Gao, J., Zhou, X., Wang, Y. et al. (2022). Exploiting the synergistic electronic interaction between Pt-skin wrapped intermetallic PtCo nanoparticles and Co-N-C support for efficient ORR/EOR electrocatalysis in a direct ethanol fuel cell. *Small* 18: 2202071.

68 Wei, M., Huang, L., Li, L. et al. (2022). Coordinatively unsaturated PtCo flowers assembled with ultrathin nanosheets for enhanced oxygen reduction. *ACS Catal.* 12: 6478.

69 Wang, Y., Zhuo, H., Sun, H. et al. (2018). Implanting Mo atoms into surface lattice of Pt_3Mn alloys enclosed by high-indexed facets: promoting highly active sites for ethylene glycol oxidation. *ACS Catal.* 9: 442.

70 Yang, X., Xin, Y.K., Ye, J.Y. et al. (2021). Interface-rich three-dimensional Au-doped PtBi intermetallics as highly effective anode catalysts for application in alkaline ethylene glycol fuel cells. *Adv. Funct. Mater.* 31: 202103671.

71 de Souza, M.B.C., Vicente, R.A., Yukuhiro, V.Y. et al. (2019). Bi-modified Pt electrodes toward glycerol electrooxidation in alkaline solution: effects on activity and selectivity. *ACS Catal.* 9: 5104.

72 Huang, L., Sun, J.-Y., Cao, S.-H. et al. (2016). Combined EC-NMR and in situ FTIR spectroscopic studies of glycerol electrooxidation on Pt/C, PtRu/C, and PtRh/C. *ACS Catal.* 6: 7686.

73 Kim, H.J., Choi, S.M., Green, S. et al. (2011). Highly active and stable PtRuSn/C catalyst for electrooxidations of ethylene glycol and glycerol. *Appl. Catal. B Environ.* 101: 366.

74 Perales-Rondón, J.V., Ferre-Vilaplana, A., Feliu, J.M. et al. (2014). Oxidation mechanism of formic acid on the bismuth adatom-modified Pt(111) surface. *J. Am. Chem. Soc.* 136: 13110.

75 Park, J.M., Cao, Y., Watanabe, K. et al. (2021). Tunable strongly coupled superconductivity in magic-angle twisted trilayer graphene. *Nature* 590: 249.

76 Zhan, C., Bu, L., Sun, H. et al. (2022). Medium/high-entropy amalgamated core/shell nanoplate achieves efficient formic acid catalysis for direct formic acid fuel cell. *Angew. Chem. Int. Ed.* 62: e202213783.

77 Hu, X., Xiao, Z., Wang, W. et al. (2023). Platinum–lead–bismuth/platinum–bismuth core/shell nanoplate achieves complete dehydrogenation pathway for direct formic acid oxidation catalysis. *J. Am. Chem. Soc.* 145: 15109.

78 Dong, C., Wang, X., Zhu, Z. et al. (2023). Highly selective synthesis of monoclinic-phased platinum–tellurium nanotrepang for direct formic acid oxidation catalysis. *J. Am. Chem. Soc.* 145: 15393.

79 Tao, Y., Dandapat, A., Chen, L. et al. (2016). Pd-on-Au supra-nanostructures decorated graphene oxide: an advanced electrocatalyst for fuel cell application. *Langmuir* 32: 8557.

80 Lim, E., Kim, Y., Choi, S. et al. (2015). Binary PdM catalysts (M = Ru, Sn, or Ir) over a reduced graphene oxide support for electro-oxidation of primary alcohols (methanol, ethanol, 1-propanol) under alkaline conditions. *J. Mater. Chem. A* 3: 549.

81 Ji, Y., Ying, Y., Pan, Y. et al. (2016). Palladium networks decorated by cuprous oxide for remarkably enhanced electrocatalytic activity of methanol oxidation reaction with high CO-tolerance. *J. Power Sources* 329: 115.

82 Niu, M., Xu, W., Zhu, S. et al. (2017). Synthesis of nanoporous CuO/TiO_2/Pd-NiO composite catalysts by chemical dealloying and their performance for methanol and ethanol electro-oxidation. *J. Power Sources* 362: 10.

83 Su, D., Lam, Z., Wang, Y. et al. (2023). Ultralong durability of ethanol oxidation reaction via morphological design. *Joule* 7: 2568.

84 Wang, W., Shi, X., He, T. et al. (2022). Tailoring amorphous PdCu nanostructures for efficient C–C cleavage in ethanol electrooxidation. *Nano Lett.* 22: 7028.

85 Qin, Y., Wang, F., Liu, P. et al. (2023). Enhancement of CH3CO* adsorption by editing d-orbital states of Pd to boost C–C bond cleavage of ethanol eletrooxidation. *SCIENCE CHINA Chem.* https://doi.org/10.1007/s11426-023-1756-8.

86 Wang, W., Zhang, X., Zhang, Y. et al. (2020). Edge enrichment of ultrathin 2D PdPtCu trimetallic nanostructures effectuates top-ranked ethanol electrooxidation. *Nano Lett.* 20: 5458.

87 Zhang, S., Liu, K., Liu, Z. et al. (2021). Highly strained Au–Ag–Pd alloy nanowires for boosted electrooxidation of biomass-derived alcohols. *Nano Lett.* 21: 1074.

88 Zalineeva, A., Serov, A., Padilla, M. et al. (2014). Self-supported Pd_xBi catalysts for the electrooxidation of glycerol in alkaline media. *J. Am. Chem. Soc.* 136: 3937.

89 Simões, M., Baranton, S., and Coutanceau, C. (2010). Electro-oxidation of glycerol at Pd based nano-catalysts for an application in alkaline fuel cells for chemicals and energy cogeneration. *Appl. Catal. B Environ.* 93: 354.

90 Xu, C., Tian, Z., Shen, P. et al. (2008). Oxide (CeO_2, NiO, Co_3O_4 and Mn_3O_4)-promoted Pd/C electrocatalysts for alcohol electrooxidation in alkaline media. *Electrochim. Acta* 53: 2610.

91 Shi, Y., Schimmenti, R., Zhu, S. et al. (2022). Solution-phase synthesis of $PdH_{0.706}$ nanocubes with enhanced stability and activity toward formic acid oxidation. *J. Am. Chem. Soc.* 144: 2556–2568.

92 Kim, Y., Kim, S., Ham, H.C. et al. (2020). Mechanistic insights on aqueous formic acid dehydrogenation over Pd/C catalyst for efficient hydrogen production. *J. Catal.* 389: 506.

93 Foucher, A.C., Yang, S., Rosen, D.J. et al. (2022). Synthesis and characterization of core-shell Cu-Ru, Cu-Rh, and Cu-Ir nanoparticles. *J. Am. Chem. Soc.* 144: 7919.
94 Li, T., Han, S., Cheng, C. et al. (2022). Sulfate-enabled nitrate synthesis from nitrogen electrooxidation on a rhodium electrocatalyst. *Angew. Chem. Int. Ed.* 61: e202204541.
95 Li, F., Jiang, M., Lai, C. et al. (2022). Yttrium- and cerium-codoped ultrathin metal–organic framework nanosheet arrays for high-efficiency electrocatalytic overall water splitting. *Nano Lett.* 22: 7238.
96 Biswas, R., Dastider, S.G., Ahmed, I. et al. (2023). Bio-assisted synthesis of Au/Rh nanostructure electrocatalysts for hydrogen evolution and methanol oxidation reactions: composition matters. *Energy Fuel* 37: 13231.
97 Zhang, A., Chen, Y. et al. (2020). Enhanced electrocatalytic activities toward the ethanol oxidation of nanoporous gold prepared via solid-phase reaction. *ACS Appl. Energy Mater.* 3: 336.
98 Yan, Y., Zhou, H. et al. (2023). Electrocatalytic upcycling of biomass and plastic wastes to biodegradable polymer monomers and hydrogen fuel at high current densities. *J. Am. Chem. Soc.* 145: 6144.
99 Wang, S., Liu, S. et al. (2021). Mesoporous Rh nanotubes for efficient electro-oxidation of methanol. *J. Mater. Chem. A* 9: 4744.
100 Ahn, M., Cha, I. et al. (2015). Heterogeneous rhodium–tin nanoparticles: highly active and durable electrocatalysts for the oxidation of ethanol. *J. Mater. Chem. A* 3: 17130.
101 Bai, J., Huang, H. et al. (2019). Glycerol oxidation assisted electrocatalytic nitrogen reduction: ammonia and glyceraldehyde co-production on bimetallic RhCu ultrathin nanoflake nanoaggregates. *J. Mater. Chem. A* 7: 21149.

23

Non-Noble Metal-Based Materials for Electrocatalytic Liquid Fuel Oxidation

Qi Zhou[1,2] *and Zhicheng Zhang*[1,2]

[1] *Tianjin University, Tianjin Key Laboratory of Molecular Optoelectronic Sciences, School of Science, Department of Chemistry, No. 92 Weijin Road, Nankai District, Tianjin, 300072, China*
[2] *Collaborative Innovation Center of Chemical Science and Engineering (Tianjin), No. 92 Weijin Road, Nankai District, Tianjin, 300072, China*

23.1 Introduction

Since the Industrial Revolution, the rapid development of science and technology has caused a large amount of traditional fossil energy consumption [1, 2]. At present, most of the oil fields and coal mines have faced resource depletion, and the exploitation has declined greatly, while causing a series of environmental problems, such as global warming, air pollution, water pollution, and so on [3–5]. These problems not only destroy the ecological balance but also have a serious impact on the survival and development of mankind. Therefore, the search for clean renewable energy that can replace fossil energy has become a research hotspot to alleviate the energy crisis [6, 7]. Among them, methanol, ethanol, hydrazine, and other organic small molecule liquid fuels have attracted much attention because of their wide sources, low cost, easy storage, and clean and pollution-free products [8–10]. So far, several methods have been developed for the liquid fuel oxidation reaction, including thermal catalysis, electrocatalysis, photocatalysis, direct combustion, and so on [11–14]. Electrocatalytic oxidation has become the focus of research because of its simplicity, clean oxidation products, and mild reaction conditions [15]. In addition, in the anodic oxidation process, the corresponding cathode reaction such as CO_2RR, hydrogen evolution reaction (HER) obtained products are also good clean energy [16]. Therefore, liquid fuel electrocatalytic oxidation can realize the efficient storage and utilization of renewable clean energy and has broad application prospects [17].

Methanol is not only one of the most commonly used commercial chemicals but also an excellent energy carrier [18, 19]. It is characterized by a high energy density (22.7 MJ l^{-1}), several times that of liquid H_2 (4.7 MJ l^{-1}). Methanol is liquid at ambient temperature and pressure, which is easy to store and transport and can be distributed using existing infrastructure. In addition, methanol can be regenerated from

Functional Materials for Electrocatalytic Energy Conversion, First Edition.
Edited by Zhicheng Zhang, Meiting Zhao, and Yuchen Qin.

biomass, organic waste, or CO_2 [20]. All of these advantages offer great potential for direct methanol fuel cells (DMFC) to power future electric vehicles and mobile applications [21]. However, the high consumption of DMFC is currently hampered by the lack of active, durable, and cost-effective catalysts, especially for methanol oxidation reactions (MOR). Ethanol is the most frequently used alcohol in daily life and is widely used in food, medicine, and chemical industry, mainly through biomass fermentation and synthesis gas catalysis [22]. Its oxidation products, such as CH_3CHO and CH_3COOH, also have high application value [23]. Hydrazine hydrate is an essential component of rocket fuel, and the reaction kinetics of hydrazine hydrate in the electrocatalytic oxidation process is far superior to oxygen evolution reaction (OER) [24, 25]. This makes the hydrazine hydrate oxidation reaction (HzOR) the best choice to replace OER.

Traditional electrocatalytic oxidation often relies on noble metal catalysts, despite their very low oxidation potential, which can save energy and reduce consumption [26]. But its high cost, low current density, poor stability, and the CO intermediates generated during the reaction process are easy to adsorb on the surface of the catalyst so the characteristics of the catalyst poisoning make researchers urgently need to find abundant reserves and cheap substitutes [27, 28]. Non-noble metal catalysts have been used by more and more researchers in the field of electrocatalytic oxidation because of their abundant reserves and easy availability on Earth, high current density, and high catalytic stability [29]. The catalytic mechanism of non-noble metal catalyst is relatively simple, in the electrocatalytic oxidation process, metal active center (M^{n+}) as the coordination of H_2O and to H^+ by oxidation state for active metal, and the price ($M^{(n+1)+}OH$, $M^{(n+1)+}OOH$). The organic matter with suitable reducing properties is dissolved in an aqueous solution, and the oxygen atoms in the water can be transferred to the added organic matter by hydroxylation reaction ($M^{[n+1]+}OH + RH \rightarrow M^+ + ROH + H^+$).

In the past few years, great efforts have been made to improve the current density of non-noble metal catalysts and reduce their reaction potential, proposing the subsequent strategies, by adjusting their size, increasing the specific surface area to expose more catalytic active sites, constructing heterostructures, exposing more crystal faces, changing and giving them a unique electronic structure through element doping, and improving the specific surface area [30–35]. Building bimetal or even more metal catalysts for collaborative catalysis and other strategies to achieve the aforementioned purpose [36, 37].

The main purpose of this chapter is to provide a comprehensive overview of the latest strategies and developments in non-noble metal catalysts for electrocatalytic oxidation of liquid fuels from both theoretical and experimental aspects, including mono-metal catalysts, bimetallic catalysts and metal oxides, hydroxides, phosphates, sulfides, and nitrides. Specifically, the synthesis strategies, the relationship between structure and properties, and the catalytic mechanism of the latest non-noble metal catalysts are discussed systematically. Finally, the opportunities and challenges of non-noble metal catalysts in electrocatalytic oxidation are presented.

23.2 Non-Noble Metal Catalysts are Used for Electrocatalytic Oxidation of Liquid Fuels

23.2.1 Single Metal Catalyst

In 1970s, the electrooxidation of methanol on nickel electrodes in basic media was reported, but it has not received more attention until the past 20 years [38]. In 2004. Abdel Rahim et al. carried out electrocatalytic oxidation of methanol on bulk Ni, but there was almost no activity [39]. Later, they loaded nickel on carbon black and showed some catalytic activity. In 2014, Azizi et al. prepared a catalyst (Ni/ANA) loaded with nickel on zeolite for methanol electrocatalytic oxidation under alkaline conditions and explored the effects of reaction conditions such as potential scanning rate, methanol concentration, and zeolite dosage on the reaction [40]. The rate constant of methanol catalytic reaction measured by chronoamperometer is $6 \times 10^3\ cm^3\ mol^{-1}\ s^{-1}$. The electrode significantly increased the electron transfer rate and decreased the overpotential of methanol oxidation. In recent years, two-dimensional metal–organic frameworks (MOF) with highly adjustable pore structure and function, as well as high density of active sites, have aroused wide interest as electrocatalysts. Due to the strong coordination ability of nickel, Ni-based MOF materials have gradually entered the field of view of scientists. In 2021, Liang et al. reported the synthesis of an atomically dispersed nickel two-dimensional molecular organic framework with rich carbonyl (C=O) functional groups in a nickel ion coordination environment [41]. Its atomic structure and excellent electrocatalytic activity for the electrochemical oxidation of methanol, ethanol, and benzyl alcohol were analyzed, which are related to the alcohol adsorption capacity of the carbonyl functional group and the high unsaturated coordination number of nickel ions. This kind of MOF coordination compound is formed by periodic bridge between metal and ligand, and has clear structural information, which is very useful for the study of the mechanism of catalytic reaction (Figure 23.1a). The structure can also be designed according to the specific requirements of catalytic reaction, and the selectivity can be adjusted by changing the ligand or the supporting group. In 2022, Sun et al. prepared the coordination compound Ni8-TET for MOR, whose formic acid selectivity and Faraday efficiency (FE) can reach 93.5% [42]. Through density functional theory (DFT) calculation, the electrocatalytic mechanism and the reasons for the improvement in activity were analyzed. The results showed that as long as an electric field was applied during the reaction, the electronic structure of the catalyst would change. The charge will be transferred from the Ni8 cluster to the ligand, thereby improving the adsorption capacity of the catalyst to the reaction intermediate and improving the catalytic activity (Figure 23.1b).

23.2.2 Transition Metal Oxides

Transition metal oxides are rich in crystal structure, with spinel and perovskite structures. They have long been used in the field of electrocatalytic oxidation and have

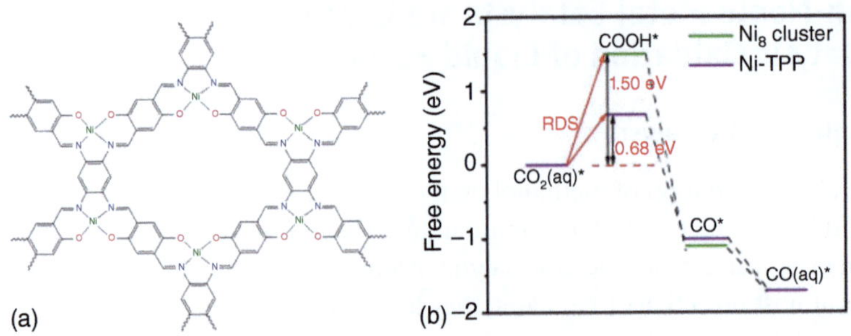

Figure 23.1 (a) Structure diagram of NI-2D-O-SA. Source: Reproduced with permission of Liang [41], Wiley-VCH. (b) Free energy of Ni8 cluster and Ni-TPP for MOR. Source: Reproduced with permission of Sun et al. [42], Wiley-VCH.

very high catalytic activity for small organic molecule electrooxidation. However, the current density of transition metal oxides in the catalytic process is relatively low, so it is necessary to change the synthesis strategy to improve the catalytic efficiency. Dai et al. obtained atomic-scale ultrathin Co_3O_4 nanosheets by ammonia-assisted hydrothermal method [43]. By exposing 111 crystal surfaces, the catalytic activity of electrooxidation of ethanol to ethyl acetate under alkaline conditions was improved, and the FE could reach 98%. Liu et al. synthesized highly dispersed ultrathin carbon-coated NiO nanoparticles on CC using an air-assisted transient thermal shock strategy as an efficient catalyst for alkaline fuel oxidation [44]. These nanoparticles are abundant in oxygen vacancies and high-valence Ni (Figure 23.2a). ATS-tech drives rapid heating and quenching processes that can provide an energy barrier for the movement and rearrangement of atoms at steady state, thereby inducing defects and high energy states of oxygen atoms. That is metastable. Combining with the advantages of NiO@C, which optimizes the electronic structure through the synergistic interaction between oxygen-rich vacancy, high concentration of Ni^{3+}, and ultrathin carbon layer, the prepared NiO@C has excellent electrocatalytic activity and stability for the electrooxidation of methanol and ethanol. Hao et al. prepared the Fe_2O_3/NiO heterogeneous catalyst on NF, and the current density reached 500 mA cm^{-2} at 1.645 V versus reversible hydrogen electrode (vs. RHE) [45]. In situ infrared spectroscopy and theoretical calculation showed that the heterogeneous structure regulated the electronic state of NiO through strong electron interaction (Figure 23.2b–d). It provides a unique synergistic active site for dynamic conversion of methanol to formate and inhibition of further oxidation. In addition, the interfacial constraint effect also stabilizes the metastable nickel active site and ensures the stability of the catalyst structure during the reversible redox cycle, thus achieving a stable and dynamically enhanced catalytic process.

23.2.3 Transition Metal Hydroxides

Transition metal hydroxides are the most common catalysts used for electrocatalytic oxidation reactions. Chen et al. investigated the catalytic mechanism of $Ni(OH)_2$ in MOR reaction and characterized the structural evolution of OER by X-ray

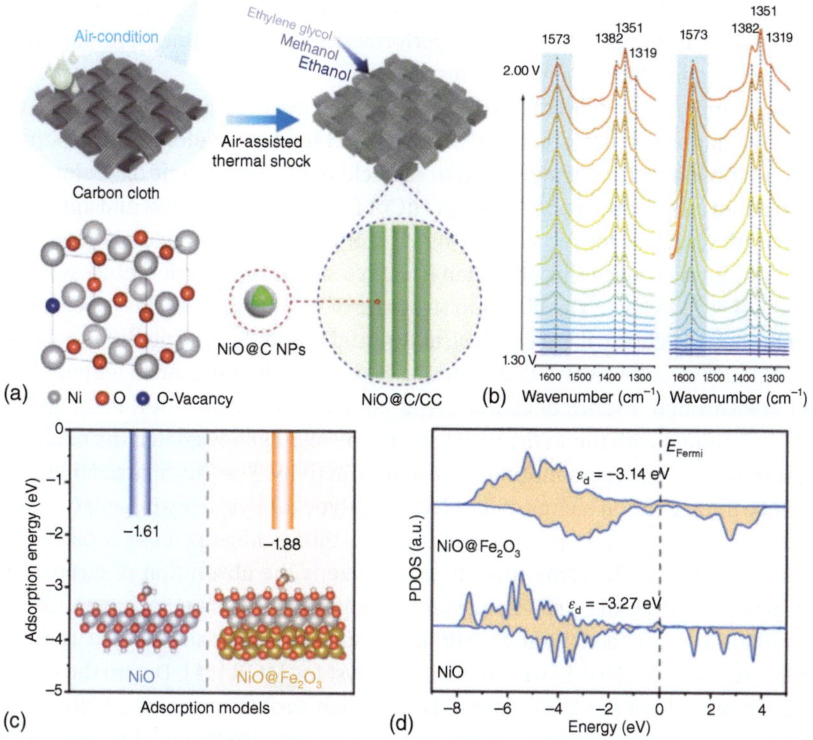

Figure 23.2 (a) Schematic of the synthesis process at NiO@C/CC Schematic illustration of the synthesis route of NiO@C/CC. Source: Reproduced with permission of Liu et al. [44], Wiley-VCH. (b) In situ IRRAS spectra in the wave number range from 1250 to 1650 cm^{-1}. (c) Calculated adsorption energy of MeOH on NiO (111) and Fe_2O_3/NiO. (d) Projected density of states (PDOS) of NiO and Fe-NF-500 regarding the Ni-3d orbitals. Source: Reproduced with permission of Hao et al. [45], Royal Society of Chemistry.

photoelectron spectroscopy (XPS) [46]. The mechanism of MOR was identified by electrochemical impedance spectroscopy (EIS), in situ Raman spectroscopy, and in situ X-ray absorption spectroscopy. Hydrogen isotopes labeled on nucleophiles during MOR were tracked by time-of-flight secondary ion mass spectrometry. Finally, the catalytic activity was improved by doping Co in $Ni(OH)_2$, which provided ideas for the design of catalyst. Li et al. prepared a $Co(OH)_2$@$Ni(OH)_2$ heterostructure catalyst for electrocatalytic oxidation of ethanol to acetate under alkaline conditions by using a one-pot hydrothermal method [47]. The heterostructure regulated the electronic structure of the catalytic active site of Ni. It was proved by in situ Raman spectroscopy and theoretical calculation that the generation of heterostructure reduced the adsorption energy of ethanol and increased the adsorption energy of acetic acid. The substrate is easy to adsorb, the product is easy to desorption, and the catalytic efficiency is improved. The catalyst has good ethanol oxidation reaction (EOR) activity (j_{10} = 1.30 V) and high stability. Zhang et al. prepared a three-dimensional ternary electrocatalyst for Pt/$Ni(OH)_2$ nitrogen-doped graphene on two-dimensional $Ni(OH)_2$ nanosheets [48]. The ternary structure of the catalyst promoted the oxidation of the reaction intermediate CO, prevented the catalyst poisoning, and improved the catalytic performance. It is 2.8 times and 4.8 times that of commercial PtRu/C and Pt/C, respectively.

Layered double hydroxide (LDH) is a unique metal hydroxide with a unique ultrathin layered nanostructure. Since its discovery, LDH has been widely used in various fields, and it has also been widely used in the field of electrocatalytic oxidation. Chi et al. prepared a self-supporting ultrathin NiCoLDH structure catalyst and applied it to MOR with excellent formic acid FE [49]. Through in situ Raman and theoretical calculation, it was proved that the catalytic active site in the reaction process is the oxidation of Ni^{2+} to Ni^{3+}. The ultrathin structure of LDH facilitates this process. Shi et al. modified the classical NiFeLDH structure and found that the modified catalyst showed higher ethanol electrooxidation activity by introducing a small amount of F atom [50]. Through a series of characterization tests such as in situ Raman and in situ EIS, combined with theoretical calculation, the results showed that the modified catalyst exhibited higher ethanol electrooxidation activity. The fluorinated NiFeLDH can reduce the activation energy of the electrocatalytic catalyst dehydrogenation and promote the deprotonation process of NiFeLDH, thus achieving a lower oxidation potential of EOR. At the same time, it also weakens the absorption of electrooxidation products of small organic molecules at high potentials and achieves higher EOR oxidation current density and acetic acid selectivity. Das et al. prepared a porous nanostructured CeO_2/NiV-LDH composite catalyst for MOR [51]. Due to the synergistic integration between CeO_2 nanorods with high surface area and 2D NiV-LDH nanosheets with short carrier diffusion distance, the electrochemical properties of the anode materials were effectively improved. CeO_2 has no catalytic effect, but the rare earth element's unique electron orbital regulates the electronic structure of the Ni active site, which improves the catalytic performance and catalyst stability.

23.2.4 Transition Metal Phosphide

In recent years, transition metal phosphide (TMP) has been widely studied as an excellent bifunctional catalyst for water electrolysis. In 2017, Zhu et al. reported a simple sugar-blowing technique and low-temperature phosphating for the preparation of three-dimensional self-supporting metallic carbon nanostructures called Co_2P@Co/N–C for MOR [52]. By utilizing high surface area, uniformly dispersed active sites, close interactions between active sites, and 3D porous nanostructures of 3D-N-doped carbon, Co_2P@Co/N–C exhibits satisfactory MOR properties. It is noteworthy that, compared with the large current density loss of RuO_2, Co_2P@Co/N–C showed significantly enhanced catalytic activity in the stability test, and the current density increased by 1.8 times after the stability test. Li et al. used Ni_2P nanosheet array with Cu_2O–Cu dendrites on NF (Cu_2O–Cu@Ni_2P/NF) as an efficient and stable electrocatalyst for MOR, reaching a low potential of 1.41 V vs. RHE at $100\,mA\,cm^{-2}$, which has excellent stability for at least 24 hours in 1.0 M KOH and 1.0 M methanol solutions [53]. In situ Raman analysis showed that Cu_2O–Cu promoted the formation of β-NiOOH active phase and was converted to Cu(II)O under electrolytic conditions. In addition, a methanol electrolytic cell using Cu_2O–Cu@Ni_2P/NF as the cathode and anode provides a current density of 10 mA cm^{-2} at a low cell voltage of 1.40 V, and a 210 mV reduction in the cell voltage compared to water electrolysis at a current density of 100 mA cm^{-2}. Liu and Wen et al. prepared Ni_xP nanowire arrays on NF by electrodeposition and one-step phosphating using hypophosphite as phosphorus source, respectively [35, 54]. The Ni_xP/NF catalyst shows high synergistic activity, excellent durability, and nearly 100% selectivity for the complete electrooxidation of hydrazine under alkaline conditions, which is the best performance among the reported hydrazine electrooxidation catalysts. The catalytic mechanism of Ni_2P on the electrooxidation of hydrazine was studied by first principle calculation. Li et al. regulated the crystal structure and electronic configuration of Ni_2P by introducing exogenous Mo element to enhance the catalytic performance of Ni_2P for OER and MOR [55]. Doping high-valence Mo as a catalytic site creates more structural defects and regulates the D-band structure of Ni to give it a higher oxidation state, thus better-adsorbing oxygen intermediates.

23.2.5 Transition Metal Sulfides

Li et al. prepared the most basic NiS catalyst for MOR and EOR [56]. Experimental studies showed that the prepared NiOOH had excellent electrocatalytic performance, high electrochemical surface area, and the presence of S^{2-} on the surface of NiOOH formed in alkaline media. At 1.6 V vs. RHE, the current density of ethanol and methanol oxidation are as high as 175.5 and 145.1 mA cm^{-2}, respectively. At these high current densities, the FE of methanol to formate is 98% and that of ethanol to acetate is 81%. DFT calculations show that the presence of the generated sulfate groups changes the electronic properties of the NiOOH surface, improving conductivity and electron transfer. In addition, the reaction energy barrier was determined by calculation, which revealed that the dehydrogenation of ethoxy group on the catalyst surface was more favorable than that of methoxy group. Inspired by the relatively recognized properties of transition metal sulfides (TMSs) in hydrazine oxidation, Sun et al. calculated and compared the catalytic properties of FeS_2 and Fe_3S_4 by DFT [57]. Due to the different Fe–S coordination numbers, the free energy of the dehydrogenation step on FeS_2 is much smaller than that of Fe_3S_4, which makes the catalytic performance of FeS_2 much better. With the continuous development of the field of electrocatalysis, researchers are no longer satisfied with catalysts that only catalyze a single reaction, and bifunctional catalysts have become the focus of research. However, mono-metallic sulfides often have poor HER performance on cathode, so researchers hope to improve it and make TMSs excellent bifunctional catalysts. Si et al. prepared spinel structure $NiCo_2S_4$ nanoflowers on CC by electrodeposition method [58]. By constructing array structure and adjusting surface magnetism, water/methanol co-electrolysis can achieve energy-saving hydrogen production at high current density. The magnetic properties of the CC@$NiCo_2S_4$ nanosheet arrays were optimized and fine-tuned so that the surface is generally exposed (011), has high catalytic activity of MOR, and has long-term stability at high current density. XPS combined with DFT confirmed the valence

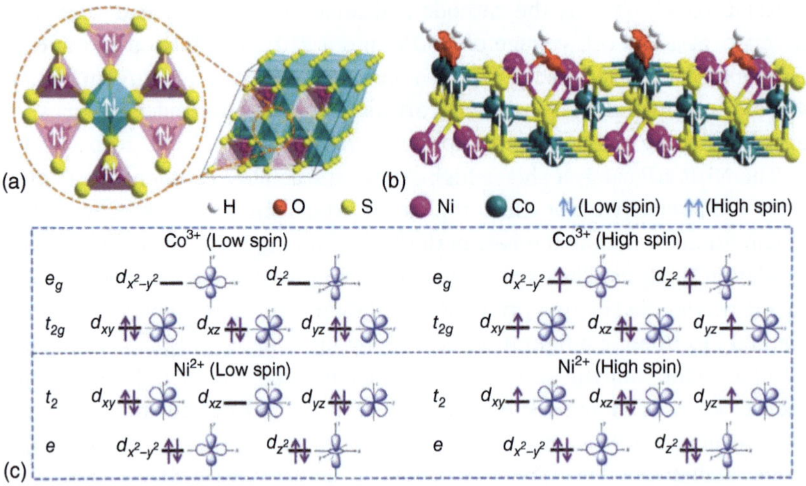

Figure 23.3 DFT calculation results for $NiCo_2S_4$. (a) Spinel structure of $NiCo_2S_4$. (b) Illustration of the spin for the (011) facet of $NiCo_2S_4$. (c) The occupation of d electrons for the high spin and low spin for Ni^{2+}, Co^{3+}. Source: Reproduced with permission of Si et al. [58], Wiley-VCH.

electron states and D-electron spins on the surface of $NiCo_2S_4$, thus enhancing the surface stability of the catalyst (Figure 23.3). The technique can be used to change the surface magnetism of nickel-based electrocatalytic materials and improve the stability of the materials. Heterogeneous interface engineering and defect engineering are also good strategies for catalyst design. Wang et al. prepared bifunctional electrocatalysts for ORR and EOR by anchoring Cu/NiS_2 nanoparticles with rich heterogeneous interfaces and sulfur vacancies on graphene (Cu_1Ni_2–S/G) [59]. It achieves an ultralow potential (1.37 V) at a current density of 10 mA cm^{-2}, while providing a high FE of 96% for the production of ethyl acetate, and can also generate value-added organic products and hydrogen under environmental conditions.

23.2.6 Transition Metal Nitrides

Transition metal nitrides (TMNs) have been widely used in electrocatalysis because of their crystal structure similar to metal. Zhang et al. prepared a highly porous Ni_3N nanosheet array supported by an NF substrate by a simple hydrothermal annealing with ammonia [60]. High-porosity nanosheets grown on NF (NSAs/NF) expose more active sites and allow for rapid charge/mass transfer, creating a synergistic effect between Ni_3N and NF. As a result, the strong interaction between Ni_3N and NF prevented leaching and kept the catalyst highly stable for more than 20 hours with a retention rate of 72.58%. Similar to TMPs, TMNs also have good bifunctional catalytic activity. Wang et al. report the rational construction of Mo-doped Co_4N nanoarrays with an open skeleton structure (Mo–Co_4N) as a powerful bifunctional electrocatalyst for the simultaneous electrolysis of high-purity hydrogen and added formate in both cathode and anode processes [61]. Thanks to the unique structural characteristics of Mo doping, more exposed active sites, and optimized electronic synergies, Mo–Co_4N exhibits interesting HER activity. An ultralow cell voltage of only 1.427 V vs. RHE is needed to achieve a current density of $10\,mA\,cm^{-2}$, with lower energy consumption compared to traditional holistic water splitting. In addition, in hydrogen evolution and value-added formate production, FE is close to 100% and has a long-term durability of 60 hours. After research, the addition of a small amount of noble metal elements has a great impact on the electrocatalytic performance of non-noble metal catalysts. Li et al. reported the synthesis of Pd/Co_3N–Ni_3N nanocomposite catalysts supported by CC by hydrothermal method, ammoniation treatment, and electrochemical deposition [62]. It is found that the addition of metallic Co_3N–Ni_3N can effectively improve the EOR durability of Pd catalyst under alkaline conditions while maintaining good electrical conductivity of the electrocatalyst. In addition, the high surface area of the nanostructure Co_3N–Ni_3N promises to expose a wealth of accessible active sites and favorable reactant/product mass transfer kinetics. Thanks to these favorable properties, Pd/Co_3N–Ni_3N/CC catalysts exhibit high activity and improved durability against EOR under alkaline conditions.

23.3 Conclusion and Outlook

So far, non-noble metal-based catalysts have made great progress in improving the electrocatalytic oxidation performance of liquid fuels. The enhancement of its performance is mainly achieved through the strategies of size adjustment, crystal surface engineering, element doping, and surface modification. In addition, the activity, selectivity, and stability of non-noble metal-based catalysts for electrocatalytic oxidation of liquid fuels can be effectively improved by constructing heterogeneous structures and forming hybrid tracks with doping elements to adjust the electronic structure.

Despite great efforts, the further development of electrocatalytic oxidation of liquid fuels still faces great challenges. On the one hand, since the main application direction of these liquid small molecule fuels is the field of fuel cells, often requires high temperature and concentrated alkaline reaction conditions, but the chemical stability of the catalyst is difficult to ensure good stability under such conditions, and for example, methanol and ethanol have a single reaction path, but the oxidation products also change with the change of reaction mechanism, which also involves changes in the number of electrons transferred during the reaction, there is still a need for a deeper and more comprehensive exploration of the key reaction intermediates in the oxidation process and their adsorption–desorption mechanisms. The high selectivity of a single product means that the product is more easily separated and helps to industrialize the reaction. Therefore, it is necessary to determine the reaction mechanism by means of surprise characterization to reveal the structure–performance relationship. In general, the identification of fine surface and interface structures requires a clever combination of various advanced characterization techniques. At the same time, in order to understand the activity, selectivity, and stability of each component under actual operating conditions, as well as the role of the reaction pathway, there is an urgent need to use operando/in situ characterization techniques, such as operando optical, X-ray and electron-based technologies. In addition, theoretical calculations combined with operational/in situ techniques allow for a more complete understanding of the underlying reaction mechanisms.

On the other hand, most non-noble metal-based catalysts are still a long way from practical application. The future research direction is to improve the stability of catalyst and increase the current density. Improve reaction efficiency. In addition, it is important to explore ways to improve selectivity without losing other key performance parameters such as cell voltage, current density, and energy efficiency.

Acknowledgment

This work was supported by the National Natural Science Foundation of China (22071172 and 22375142), 2022 Subsidized Project of Tianjin University Graduate Arts and Science Excellence Innovation Award Program (B2-2022-002).

References

1 Shi, R., Wang, Z., Zhao, Y. et al. (2021). Room-temperature electrochemical acetylene reduction to ethylene with high conversion and selectivity. *Nat. Catal.* 4 (7): 565.

2 Shi, X., Xiao, H., Azarabadi, H. et al. (2020). Sorbents for the direct capture of CO_2 from ambient air. *Angew. Chem. Int. Ed.* 59 (18): 6984.

3 Huang, X. and Tian, P. (2023). Polluting thy neighbor or benefiting thy neighbor: effects of the clean energy development on haze pollution in China. *Energy* 268: 126685.

4 Yun, J., Zhu, C., Wang, Q. et al. (2019). Catalytic conversions of atmospheric sulfur dioxide and formation of acid rain over mineral dusts: molecular oxygen as the oxygen source. *Chemosphere* 217: 18.

5 Franta, B. (2018). Early oil industry knowledge of CO_2 and global warming. *Nat. Clim. Chang.* 8 (12): 1024.

6 Gallezot, P. (2012). Conversion of biomass to selected chemical products. *Chem. Soc. Rev.* 41 (4): 1538.

7 Sun, X., Jiang, S., Huang, H. et al. (2022). Solar energy catalysis. *Angew. Chem. Int. Ed.* 61 (29): e202204880.

8 Kim, H.J., Huh, J., Kwon, Y.W. et al. (2019). Biological conversion of methane to methanol through genetic reassembly of native catalytic domains. *Nat. Catal.* 2 (4): 342.

9 Xu, K., Qin, L., Bai, W. et al. (2020). Multilevel defense system (MDS) relieves multiple stresses for economically boosting ethanol production of industrial saccharomyces cerevisiae. *ACS Energy Lett.* 5 (2): 572.

10 He, L., Liang, B., Huang, Y. et al. (2018). Design strategies of highly selective nickel catalysts for H_2 production via hydrous hydrazine decomposition: a review. *Natl. Sci. Rev.* 5 (3): 356.

11 Norsic, C., Tatibouët, J.-M., Batiot-Dupeyrat, C. et al. (2016). Non thermal plasma assisted catalysis of methanol oxidation on Mn, Ce and Cu oxides supported on γ-Al_2O_3. *Chem. Eng. J.* 304: 563.

12 Yang, F., Ye, J., Gao, L. et al. (2023). Ultrathin PtNiGaSnMoRe senary nanowires with partial amorphous structure enable remarkable methanol oxidation electrocatalysis. *Adv. Energy Mater.* 13 (34): 2301408.

13 Wang, L., Sun, Y., Zhang, F. et al. (2023). Precisely constructed metal sulfides with localized single-atom rhodium for photocatalytic C–H activation and direct methanol coupling to ethylene glycol. *Adv. Mater.* 35 (5): 2205782.

14 Gao, L.G., Zheng, J., Fernández-Ramos, A. et al. (2018). Kinetics of the methanol reaction with OH at interstellar, atmospheric, and combustion temperatures. *J. Am. Chem. Soc.* 140 (8): 2906.

15 Miao, B.-Q., Sun, B., Wang, T.-J. et al. (2023). Efficient promotion of ethanol complete electrooxidation by anti-poisoning rhodium-bismuth alloy nanodendrites. *Appl. Catal. B Environ.* 337: 122967.

16 Yang, Y. and Mu, T. (2021). Electrochemical oxidation of biomass derived 5-hydroxymethylfurfural (HMF): pathway, mechanism, catalysts and coupling reactions. *Green Chem.* 23 (12): 4228.

17 Wan, W., Tackett, B.M., and Chen, J.G. (2017). Reactions of water and C1 molecules on carbide and metal-modified carbide surfaces. *Chem. Soc. Rev.* 46 (7): 1807.

18 Wu, Y., Jiang, Z., Lu, X. et al. (2019). Domino electroreduction of CO_2 to methanol on a molecular catalyst. *Nature* 575 (7784): 639.

19 Kuld, S., Thorhauge, M., Falsig, H. et al. (2016). Quantifying the promotion of Cu catalysts by ZnO for methanol synthesis. *Science* 352 (6288): 969.

20 Navarro-Jaén, S., Virginie, M., Bonin, J. et al. (2021). Highlights and challenges in the selective reduction of carbon dioxide to methanol. *Nat. Sci. Chem.* 5 (8): 564.

21 Shi, Q., He, Y., Bai, X. et al. (2020). Methanol tolerance of atomically dispersed single metal site catalysts: mechanistic understanding and high-performance direct methanol fuel cells. *Energy Environ. Sci.* 13 (10): 3544.

22 Senne de Oliveira Lino, F., Bajic, D., Vila, J.C.C. et al. (2021). Complex yeast–bacteria interactions affect the yield of industrial ethanol fermentation. *Nat. Commun.* 12 (1): 1498.

23 Chen, L., Lu, L., Zhu, H. et al. (2017). Improved ethanol electrooxidation performance by shortening Pd–Ni active site distance in Pd–Ni–P nanocatalysts. *Nat. Commun.* 8 (1): 14136.

24 Zhai, Y., Jin, C., Xia, Q. et al. (2023). Atomically confined Ru sites in octahedral Co_3O_4 for high-efficiency hydrazine oxidation. *Adv. Funct. Mater.* **34** (13): 2311063.

25 Luo, F., Pan, S., Xie, Y. et al. (2023). Hydrazine-assisted acidic water splitting driven by iridium single atoms. *Adv. Sci.* 10 (32): 2305058.

26 Han, S.-H., Liu, H.-M., Chen, P. et al. (2018). Porous trimetallic PtRhCu cubic nanoboxes for ethanol electrooxidation. *Adv. Energy Mater.* 8 (24): 1801326.

27 Zhou, Y.-W., Chen, Y.-F., Jiang, K. et al. (2021). Probing the enhanced methanol electrooxidation mechanism on platinum-metal oxide catalyst. *Appl. Catal. B Environ.* 280: 119393.

28 Ji, L., Che, H., Qian, N. et al. (2023). Unconventional s-p-d hybridization in modulating frontier orbitals of carbonaceous radicals on PdBi nanosheets for efficient ethanol electrooxidation. *Appl. Catal. B Environ.* 328: 122521.

29 Rozmanowski, T. and Krawczyk, P. (2018). Influence of chemical exfoliation process on the activity of $NiCl_2$–$FeCl_3$–$PdCl_2$-graphite intercalation compound towards methanol electrooxidation. *Appl. Catal. B Environ.* 224: 53.

30 Vyas, A.N., Saratale, G.D., and Sartale, S.D. (2020). Recent developments in nickel based electrocatalysts for ethanol electrooxidation. *Int. J. Hydrog. Energy* 45 (10): 5928.

31 Song, J., Zhang, F., Hu, Q. et al. (2023). Dealloying fabrication of nickel oxyhydroxide/oxide-based composites boosting ethanol electrooxidation. *Appl. Surf. Sci.* 637: 157915.

32 Zhang, M., Zhu, J., Wan, R. et al. (2022). Synergistic effect of nickel oxyhydroxide and tungsten carbide in electrocatalytic alcohol oxidation. *Chem. Mater.* 34 (3): 959.

33 Chen, S., Yang, X., Tong, X. et al. (2020). Design of 3D hollow porous heterogeneous nickel–cobalt phosphides for synergistically enhancing catalytic performance for electrooxidation of methanol. *ACS Appl. Mater. Interfaces* 12 (31): 34971.

34 Catherin Sesu, D., Patil, I., Lokanathan, M. et al. (2018). Low density three-dimensional metal foams as significant electrocatalysts toward methanol oxidation reaction. *ACS Sustain. Chem. Eng.* 6 (2): 2062.

35 Liu, J.-F., Wen, H., Zhang, Z.-Y. et al. (2023). An amorphous/nanocrystalline Ni_xP/Ni heterojunction for electrooxidation of hydrazine. *J. Mater. Chem. A* 11 (26): 14213.

36 Hu, M., Jin, L., Zhu, Y. et al. (2020). Self-limiting growth of ligand-free ultrasmall bimetallic nanoparticles on carbon through under temperature reduction for highly efficient methanol electrooxidation and selective hydrogenation. *Appl. Catal. B Environ.* 264: 118553.

37 Wu, D., Zhang, W., and Cheng, D. (2017). Facile synthesis of Cu/NiCu electrocatalysts integrating alloy, core–shell, and one-dimensional structures for efficient methanol oxidation reaction. *ACS Appl. Mater. Interfaces* 9 (23): 19843.

38 Fleischmann, M., Korinek, K., and Pletcher, D. (1972). The kinetics and mechanism of the oxidation of amines and alcohols at oxide-covered nickel, silver, copper, and cobalt electrodes. *J. Chem. Soc.* 2 (10): 1396.

39 Abdel Rahim, M.A., Abdel Hameed, R.M., and Khalil, M.W. (2004). Nickel as a catalyst for the electro-oxidation of methanol in alkaline medium. *J. Power Sources* 134 (2): 160.

40 Azizi, S.N., Ghasemi, S., and Gilani, N.S. (2014). An electrode with Ni(II) loaded analcime zeolite catalyst for the electrooxidation of methanol. *Chin. J. Catal.* 35 (3): 383.

41 Liang, Z., Jiang, D., Wang, X. et al. (2021). Molecular engineering to tune the ligand environment of atomically dispersed nickel for efficient alcohol electrochemical oxidation. *Adv. Funct. Mater.* 31 (51): 2106349.

42 Sun, S.-N., Dong, L.-Z., Li, J.-R. et al. (2022). Redox-active crystalline coordination catalyst for hybrid electrocatalytic methanol oxidation and CO_2 reduction. *Angew. Chem. Int. Ed.* 61 (34): e202207282.

43 Dai, L., Qin, Q., Zhao, X. et al. (2016). Electrochemical partial reforming of ethanol into ethyl acetate using ultrathin Co_3O_4 nanosheets as a highly selective anode catalyst. *ACS Cent. Sci.* 2 (8): 538.

44 Liu, C., Zhou, W., Zhang, J. et al. (2020). Air-assisted transient synthesis of metastable nickel oxide boosting alkaline fuel oxidation reaction. *Adv. Energy Mater.* 10 (46): 2001397.

45 Hao, Y., Yu, D., Zhu, S. et al. (2023). Methanol upgrading coupled with hydrogen product at large current density promoted by strong interfacial interactions. *Energy Environ. Sci.* 16 (3): 1100.

46 Chen, W., Xie, C., Wang, Y. et al. (2020). Activity origins and design principles of nickel-based catalysts for nucleophile electrooxidation. *Chem* 6 (11): 2974.

47 Li, Z., Ning, S., Xu, J. et al. (2022). In situ electrochemical activation of $Co(OH)_2$@$Ni(OH)_2$ heterostructures for efficient ethanol electrooxidation reforming and innovative zinc–ethanol–air batteries. *Energy Environ. Sci.* 15 (12): 5300.

48 Zhang, K., Wang, H., Qiu, J. et al. (2021). Multi-dimensional Pt/$Ni(OH)_2$/nitrogen-doped graphene nanocomposites with low platinum content for methanol oxidation reaction with highly catalytic performance. *Chem. Eng. J.* 421: 127786.

49 Chi, H., Lin, J., Kuang, S. et al. (2023). Self-supported ultrathin NiCo layered double hydroxides nanosheets electrode for efficient electrosynthesis of formate. *J. Energy Chem.* 85: 267.

50 Shi, J., He, H., Guo, Y. et al. (2023). Enabling high-efficiency ethanol oxidation on NiFe-LDH via deprotonation promotion and absorption inhibition. *J. Energy Chem.* 85: 76.

51 Das, A.K., Pan, U.N., Sharma, V. et al. (2021). Nanostructured CeO_2/NiV–LDH composite for energy storage in asymmetric supercapacitor and as methanol oxidation electrocatalyst. *Chem. Eng. J.* 417: 128019.

52 Zhu, C., Fu, S., Xu, B.Z. et al. (2017). Sugar blowing-induced porous cobalt phosphide/nitrogen-doped carbon nanostructures with enhanced electrochemical oxidation performance toward water and other small molecules. *Small* 13 (33): 1700796.

53 Li, L., Zhang, L., Gou, L. et al. (2023). High-performance methanol electrolysis towards energy-saving hydrogen production: using Cu_2O-Cu decorated Ni_2P nanoarray as bifunctional monolithic catalyst. *Chem. Eng. J.* 454: 140292.

54 Wen, H., Gan, L.-Y., Dai, H.-B. et al. (2019). In situ grown Ni phosphide nanowire array on Ni foam as a high-performance catalyst for hydrazine electrooxidation. *Appl. Catal. B Environ.* 241: 292.

55 Li, L., Gao, W., Tang, K. et al. (2021). Structure engineering of Ni_2P by Mo doping for robust electrocatalytic water and methanol oxidation reactions. *Electrochim. Acta* 369: 137692.

56 Li, J., Tian, X., Wang, X. et al. (2022). Electrochemical conversion of alcohols into acidic commodities on nickel sulfide nanoparticles. *Inorg. Chem.* 61 (34): 13433.

57 Sun, J., Liu, C., Kong, W. et al. (2022). Rational design of FeS_2 microspheres as high-performance catalyst for electrooxidation of hydrazine. *J. Mater. Sci. Technol.* 110: 161.

58 Si, F., Liu, J., Zhang, Y. et al. (2023). Surface spin enhanced high stable $NiCo_2S_4$ for energy-saving production of H_2 from water/methanol coelectrolysis at high current density. *Small* 19 (2): e2205257.

59 Wang, Z., Liao, X., Zhou, M. et al. (2022). Interfacial and vacancies engineering of copper nickel sulfide for enhanced oxygen reduction and alcohols oxidation activity. *Energy Environ. Mater.* 6 (5): e12409.

60 Zhang, W., Rafiq, M., Lu, J. et al. (2023). Durable Ni_3N porous nanosheets array for non-noble metal methanol oxidation reaction. *APL Mater.* 11 (5): 051112.

61 Wang, T., Cao, X., Qin, H. et al. (2021). Integrating energy-saving hydrogen production with methanol electrooxidation over Mo modified Co_4N nanoarrays. *J. Mater. Chem. A* 9 (37): 21094.

62 Li, C., Tang, P., Gao, T.H. et al. (2020). Hierarchical nanostructured Pd/Co_3N-Ni_3N as an efficient catalyst for ethanol electrooxidation in alkaline media. *Adv. Mater. Interfaces* 7 (7): 1901875.

24

Nonmetal Materials for Electrocatalytic Liquid Fuel Oxidation

Qi Zhou[1,2] *and Zhicheng Zhang*[1,2]

[1] *Tianjin University, Tianjin Key Laboratory of Molecular Optoelectronic Sciences, School of Science, Department of Chemistry, No. 92 Weijin Road, Nankai District, Tianjin, 300072, China*
[2] *Collaborative Innovation Center of Chemical Science and Engineering (Tianjin), Tianjin Key Laboratory of Molecular Optoelectronic Sciences, No. 92 Weijin Road, Nankai District, Tianjin, 300072, China*

24.1 Introduction

The rapidly growing global energy demand still relies heavily on the consumption of traditional energy sources such as coal, oil, and gas [1]. However, in addition to causing serious climate change, burning these expendable conventional fuels to a greater extent is unsustainable [2]. Therefore, producing energy in a more environmentally friendly and sustainable way is of great scientific and public interest. Among the many renewable energy systems being developed to produce energy in a sustainable way (such as fuel cells, batteries, photovoltaic cells, and electrolyzers), fuel cells that can directly convert various compounds stored in molecular bonds (such as dihydrogen, alcohols, organic acids, and hydrazine) into electricity have attracted considerable interest [3, 4]. The key to designing efficient direct liquid fuel cells lies in the development of cost-effective electrocatalysts that can provide some solutions to society's grand challenges related to sustainability and an energy future [5, 6]. Among the many types of fuel cells, direct hydrazine fuel cells (DHFC) have received particular attention due to their high theoretical cell voltage and clean emissions. However, DHFC currently uses noble metal-based electrocatalysts, the scarcity and high cost of which has prevented the large-scale practical application of these fuel cells [7]. In order to replace noble metal-based electrocatalysts with sustainable electrocatalysts to help DHFC achieve widespread application, efforts are being made to develop stable heteroatom-doped carbon electrocatalysts such as B, N, O, P, and S, whose activity is comparable to or superior to that of noble metals.

Heteroatom-doped carbon nanomaterials have rapidly developed into the most important nonmetal catalysts for a wide range of chemical and electrochemical reactions [8, 9]. These carbon-based catalysts play an effective role in promoting oxygen evolution reaction, hydrogen evolution reaction, oxygen reduction reaction, and other reactions, and become electrocatalysts for clean energy conversion and

Functional Materials for Electrocatalytic Energy Conversion, First Edition.
Edited by Zhicheng Zhang, Meiting Zhao, and Yuchen Qin.

storage due to their extremely high flexibility, large surface area, good electrical and mechanical properties, and high stability in electrochemical environments [10, 11].

This chapter mainly introduces the research progress of heteroatom-doped carbon materials in recent years, the general properties of heteroatom-doped carbon materials, the electrocatalytic activity of heteroatom-doped carbon materials on hydrazine oxidation reaction (HzOR), and the electrocatalytic performance of carbon materials on HzOR. The chapter includes views on the different directions that future research efforts in this area need to take.

24.2 Synthetic Strategies for Heteroatom-Doped Carbon Materials

24.2.1 Hard-Templating Synthesis

The hard-templating synthesis method generally uses physically rigid nanostructured materials, such as nanoporous and colloidal SiO_2, zeolite, and $CaCO_3$ as hard templates to prepare heteroatomic-doped carbon materials with high porosity and large surface area [12]. As shown here in the case of SiO_2 (Figure 8.2.1a) [13]: (i) precursors containing carbon (and heteroatoms) are fixed in the pores of prefabricated nanoporous SiO_2 materials; (ii) mixing and polymerizing precursors containing carbon (and heteroatoms) with prefabricated colloidal SiO_2 templates or nanoporous SiO_2; (iii) and SiO_2 shells are deposited around precursors containing carbon (and heteroatoms), forming a core–shell structure. The resulting composites are then pyrolyzed at high temperatures. After the Si in the carbonized product is etched away, sometimes the product is further modified with a reagent containing heteroatoms, and nanoporous or nanostructured heteroatom-doped carbon materials can be obtained. The pores and morphology of the material produced using this method are usually the same as the template.

24.2.2 Soft-Templating Synthesis

There are significant problems in hard-template synthesis method, particularly the time and chemicals required to synthesize porous inorganic templates (typically SBA-15, KIT-6, and nano SiO_2) and ultimately remove/etch them. In addition, some chemicals, especially those used to etch templates, are corrosive and dangerous (for example, alkaline and HF solutions used to etch SiO_2 templates). As a result, soft-template synthesis is increasingly becoming an attractive alternative synthesis strategy for manufacturing a variety of heteroatom-doped porous carbon materials, especially since it typically involves fewer synthesis steps and does not require as many dangerous chemicals. For example, Wei et al. show a simple and controllable one-pot synthesis method to prepare N-doped ordered mesoporous carbon using sol, dicyandiamide, and commercial triblock copolymer F127 as C sources, N sources, and soft templates, respectively (Figure 24.1b) [14]. By changing the ratio of F127:dicyandiamide, the mesoporous structure, pore size, surface area, and

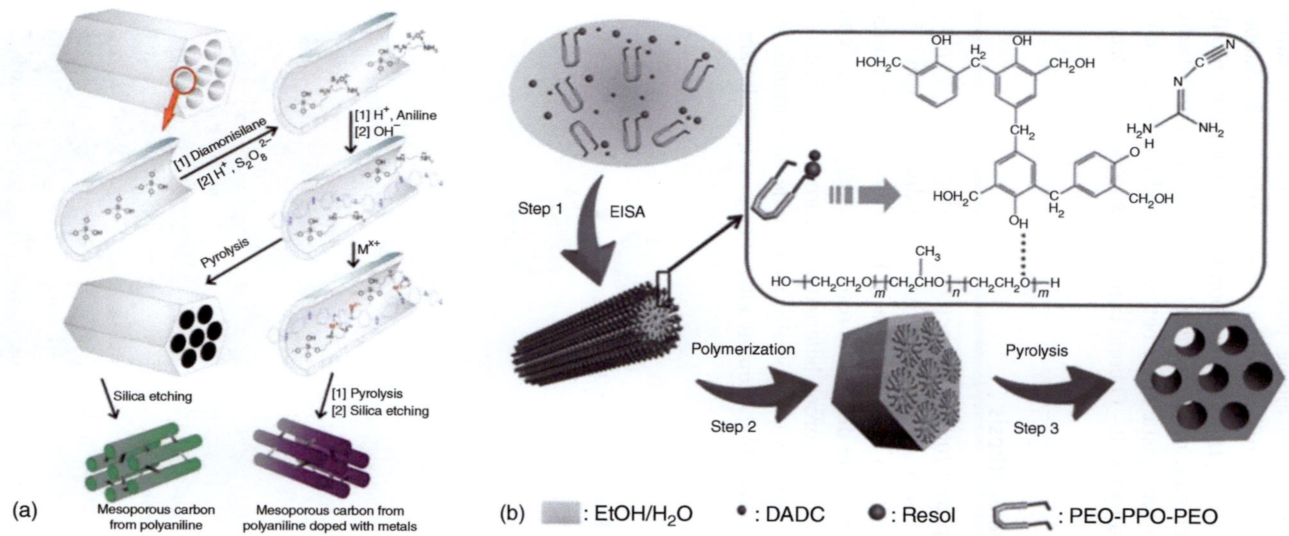

Figure 24.1 (a) Mesoporous carbon with N- and O-doped or undoped metals is synthesized by carbonizing polyaniline/SBA-15 and then etching SBA-15. Source: Adapted from Meng et al. [13]. (b) The formation process of ordered N-doped mesoporous carbon from a one-pot method using dicyandiamide as a nitrogen source. Source: Reproduced with permission of Wei et al. [14], Wiley-VCH.

nitrogen content of the obtained carbon material can be adjusted. The soft-template synthesis method also allows the synthesis of carbon materials codoped with various elements. For example, by applying an evaporation-induced self-assembly process on sol and triphenylphosphine precursors, followed by carbonization, electrocatalytically active, stratified, and ordered porous N, P codoped carbon materials can be synthesized [15]. The incorporation of P dopants into the carbon lattice increases the open edge of the material and the density and disorder of the defects. In addition, the synergistic effect between N and P doping and the charge delocalization enhanced by the P—O bond contributes to the high electrocatalytic activity.

24.2.3 Template-Free Synthesis

Although the soft-template synthesis of heteroatom-doped carbon materials has many advantages mentioned above, it also has some disadvantages. Most of the soft templates that have been used are relatively expensive surfactants or block copolymers, many of which are difficult to recover and recycle [16]. Template-free synthesis, which can produce porous heteroatomic carbon materials, is therefore becoming increasingly attractive, especially as various sustainable methods of material synthesis are increasingly sought after. For example, dopamine, which

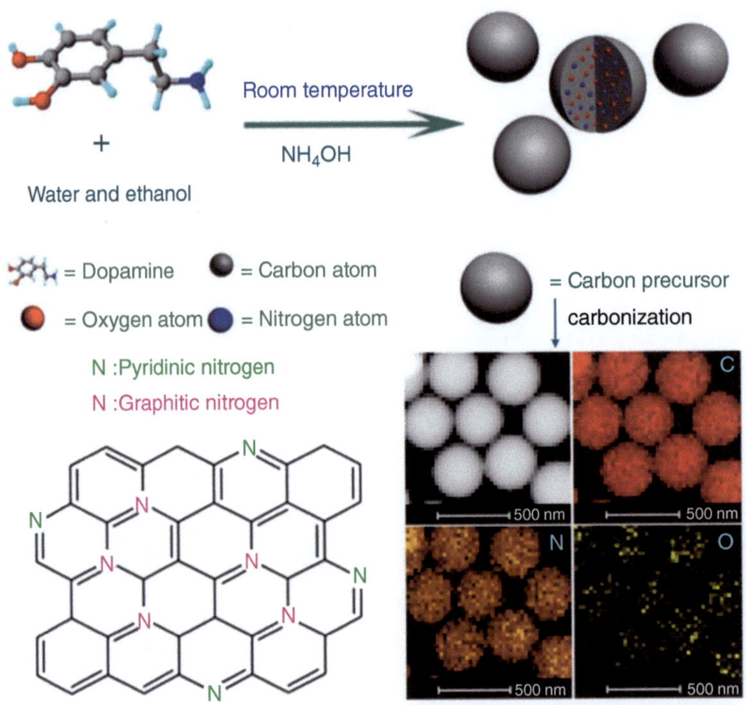

Figure 24.2 Synthesis of polydopamine-derived N-doped carbon SMSs. Source: Reproduced with permission of Ai et al. [17], Wiley-VCH.

is nontoxic and self-polymerized in alkaline solutions, has been successfully used as a precursor and self-template for the preparation of heteroatomically doped carbon submicron spheres (SMSs) with a high density of N-doped (Figure 24.2) [17]. Such precursors are more suitable for the manufacture of heteroatom-doped carbon materials than toxic precursors such as phenolic resins. In addition, due to its structure, dopamine produces more sp^2 C groups in the carbon matrix during carbonization than precursors such as phenolic resins. Therefore, carbon materials formed by dopamine exhibit higher electrical conductivity and better electrocatalytic activity. In another interesting example of self-template synthesis, Gao et al. used the N-containing polysaccharide in shrimp as a source of N and used the $CaCO_3$ in it as a template to make mesopore and macropore, thus producing N-doped carbon with layered ordered pores [11]. Furthermore, by adding an activation step involving a KOH solution to the synthesis process, the authors introduced micropores into the material. By changing the pyrolysis temperature, the porosity (or pore size) of the material can be further adjusted.

24.3 Heteroatom-Doped Carbon Materials for HzOR

Hydrazine can undergo electrooxidation under both acidic and alkaline conditions: $N_2H_5^+ \rightarrow N_2 + 5H^+ + 4e^-$ (acidic solution) and $N_2H_4 + 4OH^- \rightarrow N_2 + 4H_2O + 4e^-$ (alkaline solution) [18, 19]. In both cases, N_2 is produced. In addition to these electrochemical (Faraday) reaction pathways, competitive catalytic oxidative decomposition (non-Faraday) pathways for hydrazides have been reported, especially in the presence of noble and non-noble metal catalysts: $N_2H_4 \rightarrow N_2 + 2H_2$ and $3N_2H_4 \rightarrow N_2 + 4NH_3$. Hydrazine can also be hydrolyzed to $N_2H_4 + H_2O \rightarrow N_2H_5^+ + OH^-$.

Because the reaction produces OH^-, the pH of the hydrazine aqueous solution gradually becomes basic, and this change in pH will affect the performance of various electrocatalysts in the HzOR process [20]. Given these complex reaction pathways that hydrazine may undergo and the various intermediates and products it may produce (such as N_2, H_2, and NH_3), studying HzOR on a given catalyst may require careful examination of the total number of electrons involved in the reaction during electrocatalysis, and the products may need to be closely monitored using online mass spectrometry and other analytical techniques.

Silva et al. were the first to discover that heteroatom-doped carbon materials can electrocatalyze HzOR (Figure 24.3a) [21]. They prepared N-doped carbon nanoneedles (CNNs) by sol–gel method using cellulose nanowhiskers (CeNWs) material as a template. Then the CeNWs were carbonized and dehydrated by high-temperature treatment, and the SiO_2 shell was etched with NaOH aqueous solution to obtain N-doped CNNs. By XPS analysis, the material has a nanoneedle structure and high-density heteroatom doping (≈9% N), mainly in the form of pyridine. In addition, micro-Raman spectroscopic analysis shows that they have a high proportion of edge planes in the structure. The material was used as an electrocatalyst for HzOR and the initial potential of the 0.01 mM hydrazine solution

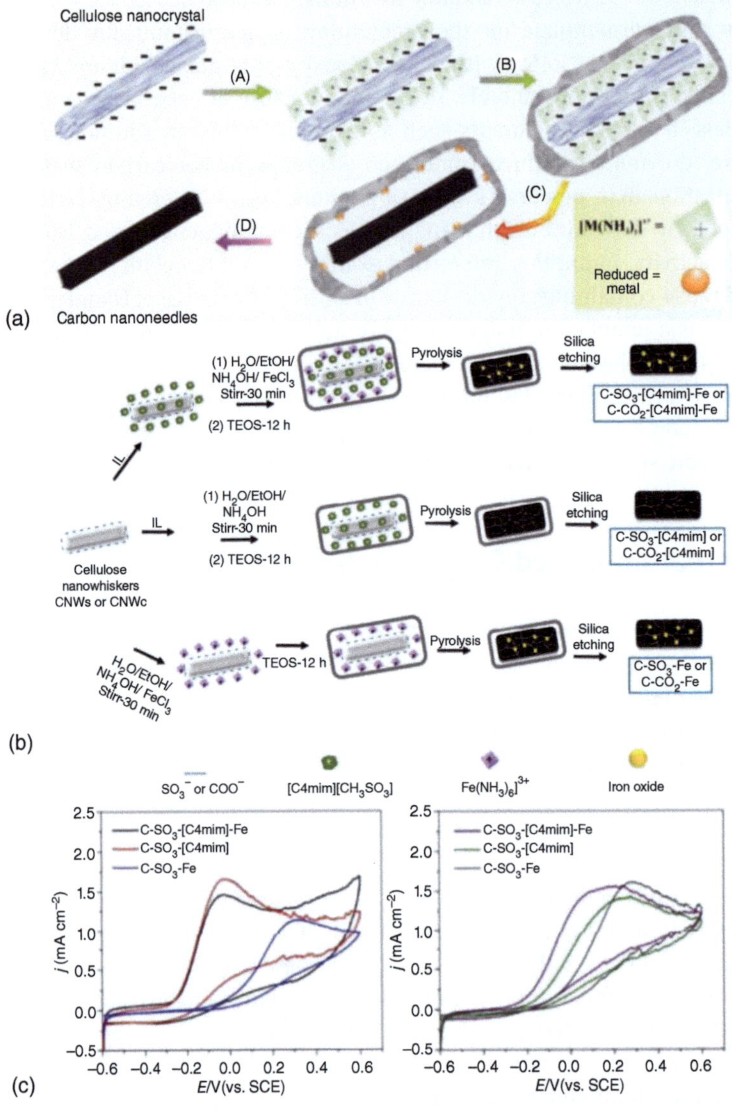

Figure 24.3 (a) The synthesis of cellulose-derived N-doped carbon nanoneedles via a core–shell strategy. Source: Reproduced with permission of Silva et al. [21], Wiley-VCH. (b) The synthetic strategy of N- and S-codoped nanostructured carbon materials derived from CeNWs. (c) Cyclic voltammetry in 0.1 M PBS at pH 7.4 for HOR of 50 mM of hydrazine over the different carbon materials derived from CNWs-SO_3^- and CNWs-CO_2^-. Source: Reproduced with permission of Fragal et al. [22], Royal Society of Chemistry.

at 0.15 V versus saturated calomel electrode (vs. SCE). This was indirectly verified by measuring the electrocatalytic activity of two different control materials, which modified the CNNs with Fe^{3+} or treated the CNN with HNO_3 to dissolve and remove any soluble Fe^{3+} impurities. There was hardly any difference in the electrocatalytic activity of these materials against HzOR compared to the original material (CNNs), suggesting that the CNN's electrocatalytic activity against HzOR is likely due entirely to the N-doped, rather than Fe^{3+} impurities. Further studies showed that heteroatomic doping of pyridines in its structure and the high proportion of edge planes appeared to be the main cause of its electrocatalytic activity.

While this work is the first to demonstrate that N-doped carbon can be a promising alternative material for the electrocatalysis of HzOR by conventional noble metals, these specific materials catalyze reactions at potentials well below the thermodynamic values of the reactions (−0.38 V vs. SCE). Therefore, other precursors and synthesis methods are explored to find more catalytic activated carbon materials for HzOR. Fragal et al. synthesized N and S codoped porous carbon materials by mixing CeNWs with heteroatom ionic liquid (IL) as a precursor (Figure 24.3b) [22]. Using SO_3^- or CO_2^- groups formed inherently in CeNWs after acid hydrolysis, the IL cations are anchored by electrostatic layer assembly, which later serve as sources of N and S-doped atoms. The subsequent coating of SiO_2 shells around the material, combined with the nonvolatile nature of IL, results in most of the heteroatoms being captured and evenly distributed in the carbon structure during pyrolysis. The material with the best electrocatalytic activity catalyzed HzOR at a starting potential of −0.29 V vs. RHE (Figure 24.3c). Furthermore, the electrocatalytic activity of these materials was found to be related to the specific type of dopant-associated chemical composition, rather than to the total density of dopants in the material. In other words, the highest density of N-doping does not necessarily make the material the most effective electrocatalyst. By comparing the atomic percentage of different N kinds in the materials and their corresponding electrocatalytic activities, we can indirectly conclude that N kinds of pyridine are the main reason for the electrocatalytic activity of these materials to HzOR.

Meng et al. obtained polypyridine-derived N and O codoped mesoporous carbon materials (PPY–NOMPCs) by hard-template pyrolysis (Figure 24.4a) [23]. Raman spectral analysis of PPY–NOMPCs at different pyrolysis temperatures showed that with the temperature increasing from 600 to 1000 °C, the material I_D/I_G (D band strength vs. G band strength) decreased from 1.00 to 0.93, indicating that higher pyrolysis temperature was conducive to the formation of graphite N groups (Figure 24.4b). It is not conducive to the formation of pyridine N group. Moreover, as the temperature increases from 600 to 900 °C, the electrocatalytic activity of the material to HzOR also increases (Figure 24.4c). In 0.1 M PBS, the materials with the highest electrocatalytic activity catalyze HzOR with an initial potential of −0.36 V and a peak potential of −0.06 V vs. SCE, which is comparable to Pt/C. Based on these results, it is found that the electrocatalytic activity of the material is directly related to the degree of its graphite structure. Due to the N-doped and the presence of N-doped-induced defect sites, hydrazine adsorbs well on the surface of PPY–NOMPCs and is dissociated/oxidized. At the same time, the conductive graphite network helps these electrons move well from electrode to electrode on the surface of the material. The presence of electronegative N and O doping on the material forms partially positively charged carbon atoms, promoting the oxidation of surface-adsorbed N_2H_4 to N_2 and H^+. The latter can react with OH^- to form water. The XPS results showed that the number of pyridine groups in the material

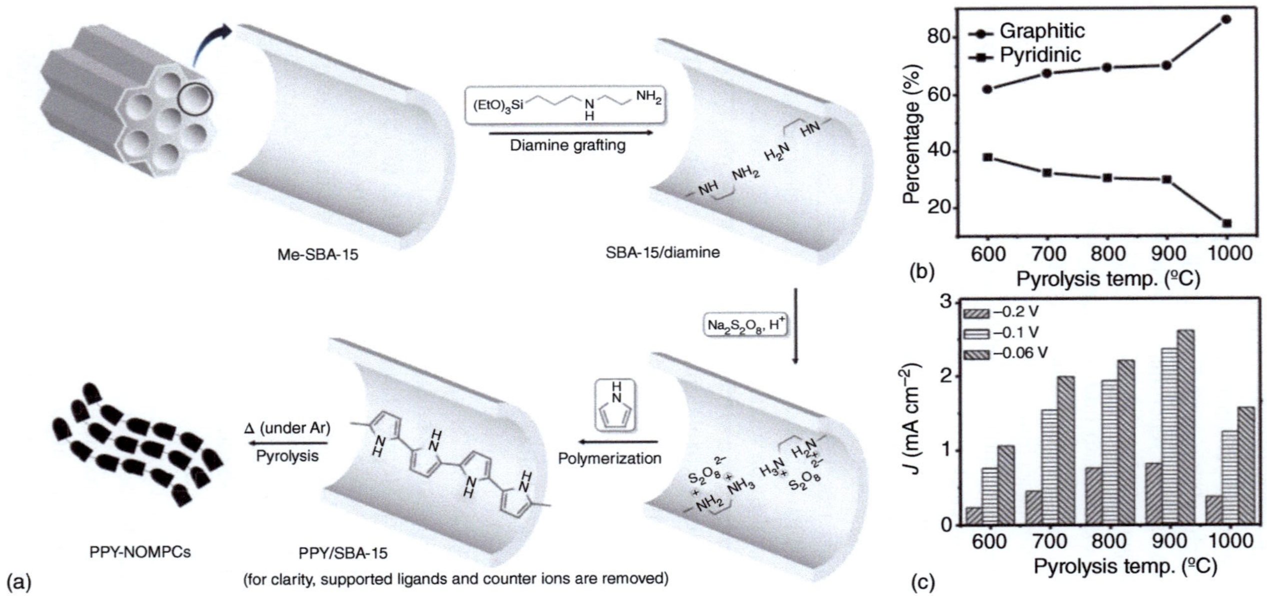

Figure 24.4 (a) Schematic synthesis of PPY-derived N and O codoped mesoporous carbon (PPY–NOMPCs). (b) The atomic content of different N kinds in different PPY–NOMPC synthesized at different temperatures. (c) Current density of the HzOR at different potentials in 0.05 M hydrazine solution in 0.1 M PBS over PPY–NOMPCs, which are synthesized using different pyrolysis temperatures. Source: Reproduced with permission of Meng et al. [23], Wiley-VCH.

gradually decreased with the increase of pyrolysis temperature from 600 to 900 °C. Therefore, it can be concluded that the graphite N group plays a major role in the electrocatalytic activity of HzOR, and the defect sites related to N-doped may be the main reason for the high electrocatalytic activity of these materials on HzOR.

Cazetta et al. used bone char as self-template and precursor of N and S atoms to synthesize N and S codoped nanoporous carbon materials by pyrolysis [24]. This material shows a higher HzOR electrocatalytic activity than N-doped carbon materials because S, which is much larger than N, causes a higher degree of structural defects in the material, changes the material band gap, leads to a reduction in the distance between the conductive and valence bands, and the material electronic work function, resulting in higher electron mobility at their solid/liquid interfaces. In addition, the 3p empty orbital of S can accommodate the lone pair electrons of many reactants, such as hydrazine, as well as their intermediates. Therefore, S-doped carbon materials generally exhibit better electrocatalytic activity for HzOR than N-doped carbon materials.

24.4 Conclusion and Outlook

At present, the research of heteroatom-doped carbon electrocatalysts is still in the initial stage, and the exact mechanism of these materials showing electrocatalytic activity to HzOR still needs to be further studied and determined. We highlight recent advances and results on heteroatom-doped carbon materials, including important synthetic strategies for synthesizing materials and controlling their dopant/structure-dependent electrocatalytic activity. It is emphasized that controlling and optimizing the type and density of heteroatom dopants (such as pyridine and graphite N) and designing the structure of materials (such as surface area and porosity) are key steps to obtaining heteroatomic-doped carbon materials with good electrocatalytic efficiency. The selection of suitable precursors and the development of simple and efficient synthesis routes for the production of heteroatomic-doped carbon materials with high density of catalytically active doping groups for HzOR require further research work. This includes finding sacrificial templating agents, especially those that are naturally available and more readily available, that can help produce high surface area, large porosity, and heteroatomic dopants in the material. Little is known about how multiple types of heteroelement doping work together to enhance the electrocatalysis of these materials, and how to optimize them to make the electrocatalytic activity of HzOR as high as possible. For example, research work related to B, P, and O-doped carbon materials is currently limited in HzOR. Even for N-doped carbon materials, there are relatively few theoretical studies that can provide the reaction mechanism of HzOR on these materials under different conditions.

Finally, HzOR is a multielectron transfer process involving various intermediates and products. The underlying reaction pathways and mechanisms involved in them are not fully understood, especially in electrocatalysis with different types of heteroatom-doped carbon catalysts. Although online electrochemical mass spectrometry measurements and some density functional theory (DFT) calculations have been performed on metal surfaces to gain some insights into the HzOR mechanism, little such research has been done for heteroatom-doped carbon materials electrocatalysts. In order to understand the reaction mechanism and active sites of heteroatom-doped carbon, in situ spectral analysis, and advanced theoretical calculations may be required. The results may in turn contribute to the development of more efficient or desirable HzOR electrocatalysts.

Acknowledgment

This work was supported by the National Natural Science Foundation of China (22071172 and 22375142), 2022 Subsidized Project of Tianjin University Graduate Arts and Science Excellence Innovation Award Program (B2-2022-002).

References

1 Küfeoğlu, S. and Özkuran, M. (2019). Bitcoin mining: a global review of energy and power demand. *Energy Res. Soc. Sci.* 58: 101273.

2 Semieniuk, G., Taylor, L., Rezai, A., and Foley, D.K. (2021). Plausible energy demand patterns in a growing global economy with climate policy. *Nat. Clim. Chang.* 11 (4): 313.

3 Fukuzumi, S. (2017). Production of liquid solar fuels and their use in fuel cells. *Joule* 1 (4): 689.

4 Jiang, C., Ma, J., Corre, G. et al. (2017). Challenges in developing direct carbon fuel cells. *Chem. Soc. Rev.* 46 (10): 2889.

5 Zhang, Y., Wang, J.-G., Yu, X. et al. (2019). Potential-dynamic surface chemistry controls the electrocatalytic processes of ethanol oxidation on gold surfaces. *ACS Energy Lett.* 4 (1): 215.

6 Wang, L., Nitopi, S., Wong, A.B. et al. (2019). Electrochemically converting carbon monoxide to liquid fuels by directing selectivity with electrode surface area. *Nat. Catal.* 2 (8): 702.

7 Wang, Y., Wang, Q., Wan, L.-Y. et al. (2020). KOH-doped polybenzimidazole membrane for direct hydrazine fuel cell. *J. Colloid Interface Sci.* 563: 27.

8 Li, D., Yang, Z., Yang, J. et al. (2022). Application of a catechol-polyamine Co-deposition method for synthesis of heteroatom-doped carbon nanomaterials. *Chem. Eng. J.* 429: 132363.

9 Li, L., Yang, H., Miao, J. et al. (2017). Unraveling oxygen evolution reaction on carbon-based electrocatalysts: effect of oxygen doping on adsorption of oxygenated intermediates. *ACS Energy Lett.* 2 (2): 294.

10 Ortiz-Medina, J., Wang, Z., Cruz-Silva, R. et al. (2019). Defect engineering and surface functionalization of nanocarbons for metal-free catalysis. *Adv. Mater.* 31 (13): e1805717.

11 Gao, K., Wang, B., Tao, L. et al. (2019). Efficient metal-free electrocatalysts from N-doped carbon nanomaterials: mono-doping and Co-doping. *Adv. Mater.* 31 (13): e1805121.

12 Xu, Y. and Zhang, B. (2014). Recent advances in porous Pt-based nanostructures: synthesis and electrochemical applications. *Chem. Soc. Rev.* 43 (8): 2439.

13 Meng, Y., Voiry, D., Goswami, A. et al. (2014). N-, O-, and S-tridoped nanoporous carbons as selective catalysts for oxygen reduction and alcohol oxidation reactions. *J. Am. Chem. Soc.* 136 (39): 13554.

14 Wei, J., Zhou, D., Sun, Z. et al. (2012). A controllable synthesis of rich nitrogen-doped ordered mesoporous carbon for CO_2 capture and supercapacitors. *Adv. Funct. Mater.* 23 (18): 2322.

15 Deng, C., Zhong, H., Li, X. et al. (2016). A highly efficient electrocatalyst for oxygen reduction reaction: phosphorus and nitrogen co-doped hierarchically ordered porous carbon derived from an iron-functionalized polymer. *Nanoscale* 8 (3): 1580.

16 Dutta, S., Bhaumik, A., and Wu, K.C.W. (2014). Hierarchically porous carbon derived from polymers and biomass: effect of interconnected pores on energy applications. *Energy Environ. Sci.* 7 (11): 3574.

17 Ai, K., Liu, Y., Ruan, C. et al. (2013). Sp2 C-dominant N-doped carbon sub-micrometer spheres with a tunable size: a versatile platform for highly efficient oxygen-reduction catalysts. *Adv. Mater.* 25 (7): 998.

18 Sanabria-Chinchilla, J., Asazawa, K., Sakamoto, T. et al. (2011). Noble metal-free hydrazine fuel cell catalysts: EPOC effect in competing chemical and electrochemical reaction pathways. *J. Am. Chem. Soc.* 133 (14): 5425.

19 Zhang, P.-X., Wang, Y.-G., Huang, Y.-Q. et al. (2011). Density functional theory investigations on the catalytic mechanisms of hydrazine decompositions on Ir(111). *Catal. Today* 165 (1): 80.

20 Dhiman, N., Pradhan, D., and Mohanty, P. (2022). Heteroatom (N and P) enriched nanoporous carbon as an efficient electrocatalyst for hydrazine oxidation reaction. *Fuel* 314: 122722.

21 Silva, R., Al-Sharab, J., and Asefa, T. (2012). Edge-plane-rich nitrogen-doped carbon nanoneedles and efficient metal-free electrocatalysts. *Angew. Chem. Int. Ed.* 51 (29): 7171.

22 Fragal, E.H., Fragal, V.H., Huang, X. et al. (2017). From ionic liquid-modified cellulose nanowhiskers to highly active metal-free nanostructured carbon catalysts for the hydrazine oxidation reaction. *J. Mater. Chem. A* 5 (3): 1066.

23 Meng, Y., Zou, X., Huang, X. et al. (2014). Polypyrrole-derived nitrogen and oxygen co-doped mesoporous carbons as efficient metal-free electrocatalyst for hydrazine oxidation. *Adv. Mater.* 26 (37): 6510.

24 Cazetta, A.L., Zhang, T., Silva, T.L. et al. (2018). Bone char-derived metal-free N- and S-co-doped nanoporous carbon and its efficient electrocatalytic activity for hydrazine oxidation. *Appl. Catal. B Environ.* 225: 30.

Part VIII

Advanced Functional Materials for Electrocatalytic Biomass Conversion

25

Metal-Based Materials for Electrocatalytic Biomass Conversion

Xu Li[1], Mengyi Xu[1], Yurong Su[2], Xianfu Zheng[1], Yuchen Qin[1], Fawen Zhang[3], Guangxin Ru[3], Xiuhong Zhu[3], Meirong Song[1], Shun Wang[1], and Xia Sheng[1]

[1] *Henan Agricultural University, College of Science, Department of Applied Chemistry, Nongye Road 63, Zhengzhou, 450002, China*
[2] *Zhengzhou University, Green Catalysis Center, and College of Chemistry, 100 Science Avenue, Zhengzhou, 450001, China*
[3] *Henan Agricultural University, College of Forestry, Department of environment, Nongye Road 63, Zhengzhou, 450002, China*

25.1 Introduction

With the rapid development of human society and the rapid depletion of fossil resources, energy shortages and environmental problems are becoming increasingly prominent. In response to the energy crisis and to reduce greenhouse gas emissions, there has been an international consensus to vigorously develop renewable energy. Biomass is recognized as a zero-carbon energy source and the largest reservoir of renewable hydrocarbon resources on the planet, making it an ideal raw material for replacing fossil resources in the production of a wide range of green fuels, chemicals, and materials [1].

Biomass conversion and utilization are of great strategic significance, and both domestic and foreign countries have listed bio-based chemicals and downstream materials as the core of bio-economic strategies. Investigating, developing, and utilizing biomass resources will help humanity reduce carbon emissions. According to documents from the National Energy Research Institute (NERI), the exploitation of biomass in China is set to reduce carbon emissions by more than 900 million tons by 2030 when combined with carbon capture and storage technologies. Currently, large-scale biomass energy utilization is mainly focused on power generation, heat and gas supply, and other fields to achieve substitution of fossil energy. In recent years, basic research and future carbon-negative technologies have paid more attention to the conversion and utilization of biomass resources, combined with the structural characteristics of biomass, to produce biomass energy and biobased products, and gradually developed in the direction of cheap and easy-to-obtain raw materials, green and low-carbon processes, and higher value-added products [1, 2].

As shown in Figure 25.1, wood biomass raw materials have a rich chemical structure and group, structural components are mainly cellulose (40–50%),

Functional Materials for Electrocatalytic Energy Conversion, First Edition.
Edited by Zhicheng Zhang, Meiting Zhao, and Yuchen Qin.

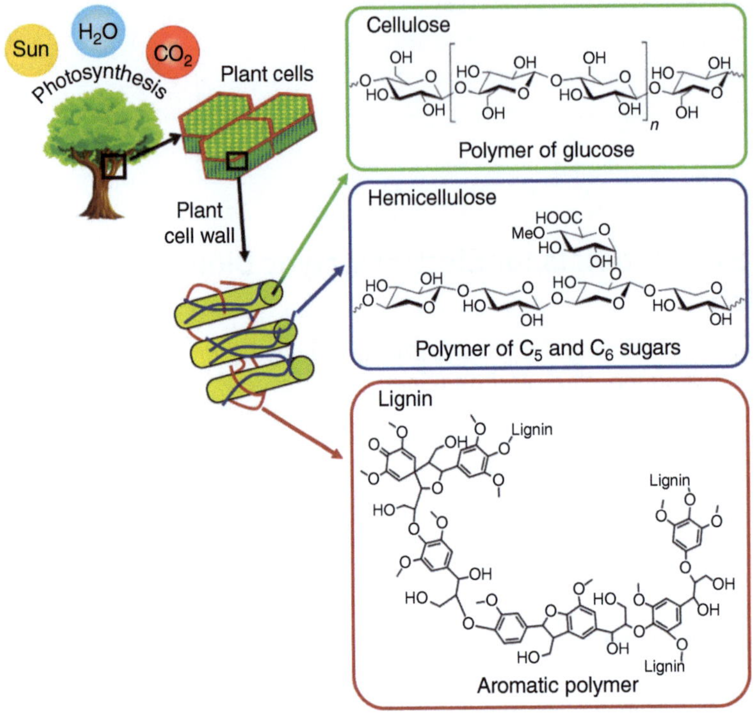

Figure 25.1 Composition of wood biomass.

hemicellulose (15–25%), and lignin (15–30%) [3–5]. Cellulose is a homogeneous polymer formed by the linear connection of glucose units through β-1,4 glycosidic bonds, cellulose fibers interact with each other through hydrogen bonding, can form crystalline and amorphous regions, and can be hydrolyzed or hydrogenated into glucose, HMF, LA, ethylene glycol, and hexanol [5, 6]. Hemicellulose is a heterogeneous polysaccharide consisting of different types of monosaccharides (including five-carbon sugars and six-carbon sugars) and is an important biomaterial for the production of furfural and lactic acid [7]. Lignin is an amorphous, aromatic polymer with a molecular structure rich in propyl alkoxyphenol or its derived structural units [8]. Lignin is dispersed between cellulose fibers and mainly plays the role of compression resistance, and hemicellulose runs between lignin and cellulose fibers and mainly plays the role of connectivity, promoting the formation of a very strong cellulose–hemicellulose–lignin network structure, which provides a barrier and protection for plant tissues [5].

According to the composition and characteristics of lignocellulose, the full utilization and selective conversion of its structure is an important way to achieve efficient utilization of biomass resources. The processing of lignin, cellulose, and hemicellulose can produce a huge variety and quantity of small molecule fine chemicals [7].

For lignin, the products are mainly aromatic phenols, aldehydes, and alcohols. Cellulose, glucose, and sorbitol are emphasized. The treatment of hemicellulose involves the generation of pectic acid, hemicellulosic alcohols, and organic acids. Overall, these compounds have potential applications in the chemical industry, pharmaceuticals, and food additives, highlighting the importance of these natural polymers in sustainable development and green chemistry [9, 10].

Electrocatalytic upgrading of biomass-derived intermediate compounds to value-added products has begun to attract increasing attention over the past decade [11–14]. Here, we highlight these representative examples and focus on candidate approaches with potential large-scale applications. Most of the reported electrocatalytic systems have focused on the upgrading conversion of HMF, glucose, methanol, ethanol, diacid, sorbitol, and glycerol.

Conventional thermochemical oxidation and catalytic reduction processes of woody biomass-derived compounds generally rely on reactions at high temperatures and pressures, and require the addition of hydrogen sources, oxidizing agents, and large amounts of organic solvents [15, 16]. For example, achieving more than 90% FDCA yield and HMF conversion in conventional HMF thermochemical oxidation typically requires precious metal catalysts (e.g. Pt, Au, Pd, or Ru), harsh conditions (high alkali concentrations, temperatures higher than 100 °C, and relatively high pressures), and toxic/expensive organic solvents. Under these reaction conditions, side reactions are difficult to control, and the catalysts are easily deactivated and difficult to recycle [17]. Therefore, the search for new catalytic technology means to solve the biomass conversion problem has received increasing attention.

Electrochemical catalysis is an environmentally friendly catalytic oxidation and reduction method that meets the requirements of green chemistry and plays an important role in improving the utilization of electric energy, developing a diversified energy situation, and solving the problem of high-value utilization of biomass. Currently, several studies have reported clean and efficient electrocatalytic conversion mechanisms and reaction technologies for biomass derivatives [3, 17].

Electrocatalytic reactions of biomass resources are usually carried out in a single cell or in cells separated by proton or anion exchange membranes [18, 19]. For single cells, both the anodic and cathodic reactions are carried out in the same cell chamber, whereas, for separated cells, the reactions are carried out in different cell chambers, as shown in Figure 25.2. The latter type of reaction cell is essential for the practical application of biomass electrochemical catalysis, as this type of reaction

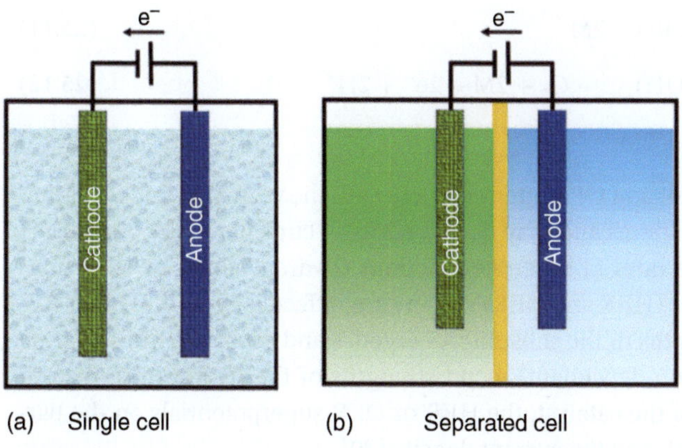

Figure 25.2 (a) Single cell and (b) Separated (H-shaped) electrolytic cell.

cell improves the reaction efficiency (Faraday efficiency [FE] and product yield) by preventing the oxidation/reduction of the target product at the opposite electrode. This type of cell is also conducive to product separation and simplifies operating procedures. However, the electrolytic cells need to be finely designed and constructed practically to achieve the goals of low cost, high performance, and ease of operation. The electrocatalytic performance could be evaluated by FE, which describes the transfer efficiency of charges in a system that facilitates the electrocatalytic reaction toward the target product.

Cathodic reactions such as electrocatalytic hydrogenation or deoxygenation (ECH) occur through multistep processes involving hydrogen chemisorption at the electrode surface [equations (25.1) and (25.2)], substrate reduction [(25.3)–(25.5)], and hydrogen evolution reaction (HER) [(25.6) and (25.7)].

$$H_2O + e^- + M \rightarrow (H)_{ads}M + OH^- \tag{25.1}$$

$$H_3O^+ + e^- + M \rightarrow (H)_{ads}\,M + H_2O \tag{25.2}$$

$$(RC = CR) + M \rightarrow (RC = CR)_{ads}M \tag{25.3}$$

$$(RC = CR)_{ads}\,M + 2(H)_{ads}M \rightarrow (RHC - CHR)_{ads}\,M \tag{25.4}$$

$$(RHC - CHR)_{ads}\,M \rightarrow RHC - CHR + M \tag{25.5}$$

$$(H)_{ads}\,M + e^- + H_2O \rightarrow H_2 + M + OH^- \quad \text{(Heyrovsky)} \tag{25.6}$$

$$(H)_{ads}\,M + (H)_{ads}\,M \rightarrow H_2 + 2M \quad \text{(Tafel)} \tag{25.7}$$

On the anode side, the electrocatalytic oxidation (ECO) reaction includes hydroxyl radical chemisorption and metal oxide formation [Eqs. (25.6) and (25.7)], substrate oxidation [Eqs. (25.8) and (25.9)], and oxygen evolution reaction (OER) [Eqs. (25.10) and (25.11)].

$$H_2O + M \rightarrow (\cdot OH)_{ads}M + e^- + H^+ \tag{25.8}$$

$$M(\cdot OH)_{ads} \rightarrow MO + e^- + H^+ \tag{25.9}$$

$$(R) + M \rightarrow (R)_{ads}M \tag{25.10}$$

$$(R)_{ads}M + MO \rightarrow RO + 2M \tag{25.11}$$

$$M(\cdot OH)_{ads} + M(\cdot OH)_{ads} \rightarrow O_2 + 2M + 2e^- + 2H^+ \tag{25.12}$$

$$MO + MO \rightarrow O_2 + 2M \tag{25.13}$$

In electrocatalysis, HER and OER often compete with the reduction and oxidation reactions of organic substrates and may lead to a loss of current efficiency. In ECH and ECO processes, the rates of substrate reactions (hydrogenation or oxidation) and competing processes (HER and OER) are, in turn, affected by several other factors, such as bond strengths in the substrate, energetics and probability of substrate adsorption (concentration-dependent), bond strengths of the hydrogen or oxygen species on the surface of the catalyst, the HER or OER superpotentials on the particular electrode material, and the current density [20].

Metals, metal oxides, or combinations of them are the most widely used multiphase catalysts in several important industrial applications. Metals exhibit a wide range of abilities to stabilize the oxidation state, especially during oxidation and hydrogenation [21]. Coupling of one metal with another metal or metal oxides can lead to improved redox properties due to synergistic metal–metal interactions excited by electron transfer interactions at the interfacial structure [22]. Precious metals have shown excellent performance and high stability as electrocatalysts for various reactions and are considered to be the best-performing electrocatalysts [23]. However, the scarcity and high cost of precious metals have limited their development. Recently, some low-cost and high-efficiency nonprecious metal catalysts have been developed, which mainly include transition metal oxides, including single-metal oxides and mixed-metal oxides, transition metal hydroxides, nonprecious metal carbon materials, which mainly include nanostructured and doped carbon materials, and other nonprecious metal catalysts [19].

Despite these advantages of metal catalysis, there are challenges to overcome, such as catalyst stability, selectivity control, waste management, and commercial scale-up [19]. Therefore, researchers continue to work on the development of more efficient metal catalysts and process conditions to maximize the potential of biomass resources, reduce dependence on traditional petrochemicals, and promote sustainable development.

In the past few years, considerable efforts have been made to study the electrowinning of biomass-derived intermediates [24]. HMF is an important building block with a simple molecular structure including a furan ring and groups such as –C–O and –OH, providing a promising model system for studying the upgrading of cyclic compounds [25, 26]. HMF is a hydrolysis product of cellulose, which is widely regarded as the "sleeping giant" of sustainable chemistry [26, 27]. Oxidation reaction of HMF (HMFOR) produces a range of important chemicals such as dicarbonylfuran (DFF) and 2,5-furandicarboxylic acid (FDCA), as shown in Figure 25.3. Among them, FDCA is the most important as it is often used as a precursor for the production of the biopolymer polyfurylate (PEF) [28]. Compared to petroleum-based terephthalic acid (TPA)-derived polyethylene terephthalate (PET), biobased PEF offers excellent gas barrier properties, recyclability, and extended mechanical properties [29]. Eerhart et al. believe that PEF has the potential to replace PET in the future [30]. However, companies producing FDCA (for PEF) have not been able to find a sufficient market for PEF at this time.

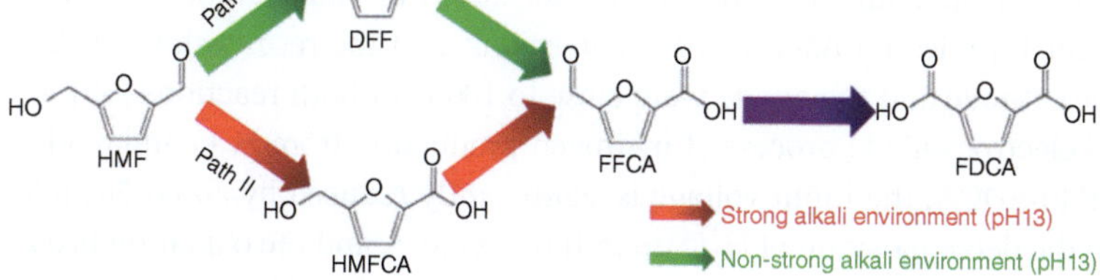

Figure 25.3 The oxidation pathway of HMF.

HMF molecules are composed of furan rings, aldehyde, and hydroxyl functional groups, and they can be used to synthesize a variety of high-value-added chemicals through processes such as aldehyde and hydroxyl oxidation. For example, the main electrooxidation product of HMF, FDCA, is a monomer for renewable polycondensation reaction, which can be widely used for the synthesis of furan polymers similar to PET. The electrocatalytic HMFOR mainly include noble (Au and Pd), nonprecious (Ni and Co) metals or bimetals, as well as transition metal-based oxides, sulfides, borides, and phosphides [31] In the process of electrooxidation of HMFs, HMFOR is carried out by a series of processes such as aldehyde and hydroxyl oxidation. During the electrooxidation reaction of HMF, the anode catalyst and electrode potential have an important influence on the distribution of its electrooxidation products.

In recent years, the electrocatalytic conversion of HMF to FDCA using nickel-based monometallics, metal oxides, or metal compounds (Ni_2P, Ni_xB, NiCoFe-LDHs, and $NiCo_2O_4$) as the anode has attracted a lot of attention [27, 32–34]. Barwe et al. used Ni_xB-modified foam Ni as anode to electrocatalyze the conversion of HMF to FDCA under alkaline conditions with FE of 100% and FDCA yield as high as 98.5% [35]. Zhang et al. reported a precious metal-free trimetallic layered double hydroxide (NiCoFe-LDHs) nanosheet to selectively oxidize HMF to FDCA under the optimal conditions, and the conversion of HMF reached 95.5% and the yield of FDCA reached 84.9% [36]. Co is a good co-catalyst for the electrooxidation of HMF. Kang et al. investigated the performance of Co^{3+} in the electrooxidation of HMF over the catalysts Co_3O_4 and $NiCo_2O_4$ spinel structures and found that the catalyst $NiCo_2O_4$ catalyzed the selectivity of HMFOR to FDCA as high as 90.8%, along with the good recovery performance of the catalyst [37]. At the same time, the catalyst showed good recycling performance, with the conversion rate still exceeding 80% after recycling three times.

Generally speaking, the oxidation reaction at the anode requires a low oxidation overpotential, and the electrooxidation of H_2O is a typical oxidation half-reaction at the anode, but the kinetics of its oxygen precipitation reaction is slow, and the value of the O_2 produced is not high. In recent years, the cathodic ECO of H_2O and the anodic oxidation of HMF in electrochemical cells, together with the production of clean H_2 fuel and oxidation products such as FDCA, is a promising refining strategy to maximize the economic benefits of the electrocatalytic process [29, 38]. You et al. reported a hierarchically structured porous Ni_3S_2/Ni foam electrocatalyst (Ni_3S_2/NF) for the anodic electrolysis of several representative biomass derivatives (ethanol, benzyl alcohol, furfural, furfuryl alcohol, and HMF) and cathodic electrolysis of H_2O for H_2 release under alkaline conditions, and the results showed that the FE of the cathode and the anode were close to 100% for both reactions [39]. In this coupled electrocatalytic process of hydrogen production from H_2O and oxidation of HMF to FDCA, the input voltage is significantly reduced by about 200 mV compared to the decomposition of H_2O up to 100 mA cm^{-2}, and the oxidation product (FDCA) at the anode is much more valuable than O_2. Considering the cost of precious metals and the unsatisfactory catalytic activity of HMF electrooxidation, the development of highly efficient nonprecious metal catalysts or nonmetallic catalysts has remained the mainstream so far. In particular, the catalytic activity of nonprecious metal catalysts, represented by nickel-based materials, far exceeds that of other metals, including precious metal alloys [40, 41].

Due to the abundant three-dimensional electron number and unique eg orbitals enhancing the covalency of the transition metal–oxygen bond, nickel-based catalysts can provide faster kinetics (or electron transfer coefficients) and are generally considered to be one of the most active nonprecious metal materials to date for the electrooxidation of HMF in terms of conversion, yield, and FE [42, 43].

Over the past decade, a great deal of research work has been carried out on the development of a variety of novel multiphase solid catalysts for the efficient conversion of biomass into sustainable advanced fuels and fine chemicals. These efforts have focused on fine-tuning the particle size, shape, surface structure, and porosity of solid catalysts with the aim of electronically and geometrically enriching the surface-active centers [44–46]. This will control the interaction of the catalytically active centers with biomass molecules and direct preferred reaction pathways to obtain the desired products. These advances have been made mainly in the hydrotreating, oxidation, hydrolysis, (trans)esterification, isomerization, and dehydration of various biomass feedstocks and their derived compounds. The most notable characteristics of solid catalysts that can play a beneficial role in catalytic reactions are particle size, shape, and porosity [47–49]. Compared to bulk catalysts, the nanocatalysts provide a higher surface area to volume ratio and their specific surface area can be greatly increased, thus making the surface-active sites more accessible to biomass molecules.

On the other hand, fine-tuning the particle morphology of the catalysts, especially for metals or metal oxides, will optimize the electronic and geometrical properties of the catalyst surface. Shape-controlled nanoscale metal oxides can preferentially expose a large number of surface-active crystal planes. As a result, the interactions between active sites and reactive species can be enhanced, which is a key driver for increasing the rate of biomass value-added reactions [19].

25.2 Morphology Control

Conformational control refers to a method of regulating the properties and applications of a material by changing its morphology, particle size, or pore size and distribution. The effect of morphology on performance comes from two main sources: (i) exposing specific crystalline planes that are highly active or selective, (ii) increasing the surface area, thereby exposing more active sites [19].

The modulation of material morphology is achieved by controlling various factors. These factors include the selection of raw materials, the preparation process, and the use of additives.

Different raw materials affect the structure and size of the material [50]. The preparation process is also an important factor that affects the morphology. Different preparation processes produce different physical and chemical reactions that affect the shape and size of the sample. For example, high-purity nanopowders can be obtained by sol–gel methods, while highly homogeneous nanopowders can be prepared by vapor phase deposition methods.

The use of additives is also important for morphology modulation. Additives can change the surface properties of the prepared material, thereby affecting its morphology and size [51, 52]. For example, surfactants can cause a thin film to form on their surface, thereby controlling morphology and size [53]. In addition, additives have the effect of altering the crystal structure of the material, thereby affecting its properties and applications [54].

The mechanism of morphology regulation is a complex process that requires comprehensive consideration of multiple factors. Through rational selection of raw materials, preparation process, and the use of additives, it is possible to realize the modulation of the morphology of samples to meet different application requirements. The construction of ordered nanostructures facilitates the diffusion of electrolytes and the exposure of more active sites, thus improving the activity of the catalyst.

The activity of catalysts strongly depends on the exposed crystal faces. In order to study the electrocatalytic properties of specific crystal faces, it is crucial to develop controlled synthesis methods for nanocrystals with well-defined structures and preferentially exposed crystal faces. Zheng and his coworkers reported an ultrathin Co_3O_4 nanosheet catalyst with exposed (111) facets, which was prepared by a one-pot template-free hydrothermal method assisted by NH_3 [55]. It was shown that NH_3 not only promoted the oxidation of the $Co(OH)_2$ nanosheet template to Co_3O_4 but also preferentially bound to Co_3O_4(111), ensuring the anisotropic growth of the nanosheets. X-ray photoelectron spectroscopy (XPS) analysis showed that the Co_3O_4 nanosheets exposed more Co^{3+} sites than the Co_3O_4 nanocubes. The Co_3O_4 nanosheets exhibited stronger EOR activity compared to the Co_3O_4 nanocubes with exposed (100) surface, which can be attributed to the fact that more Co^{3+} sites were exposed on the (111) surface than on the (100) surface.

In order to clearly understand the effect of exposed surface on the properties, Poerwoprajitno et al. prepared hexagonal close-packed (HCP) Ni branches on face-centered cubic (FCC) gold nanoparticles [56]. The FCC–Au nanoparticles provided the seeds for the growth of HCP–Ni branches. As shown in Figure 25.4a-e, the branched Ni nanoparticles were uniform in size and shape, with a length of ~70 nm and a width of ~20 nm. HCP–Ni branches were well-defined in shape, with flanks of [1010] and tips of [1011] and [0001]. The proportion of [1010] wafers increased with increasing branch length, which could be achieved by increasing the concentration ratio of $Ni(acac)_2$ to Au seeds during the synthesis process. The 70 nm-long HCP–Ni branch exhibited higher intrinsic spectroscopic activity than faceless Ni sphere nanoparticles as an electrocatalyst for HMFOR at 0.1 mol l^{-1} KOH. In addition, the tunable facet ratio allowed the identification of more active facets. As shown in Figure 25.4j, the catalyst exhibited higher HMFOR activity with increasing [1010] facet ratio, indicating that the [1010] facets were more active than the [0001] facets.

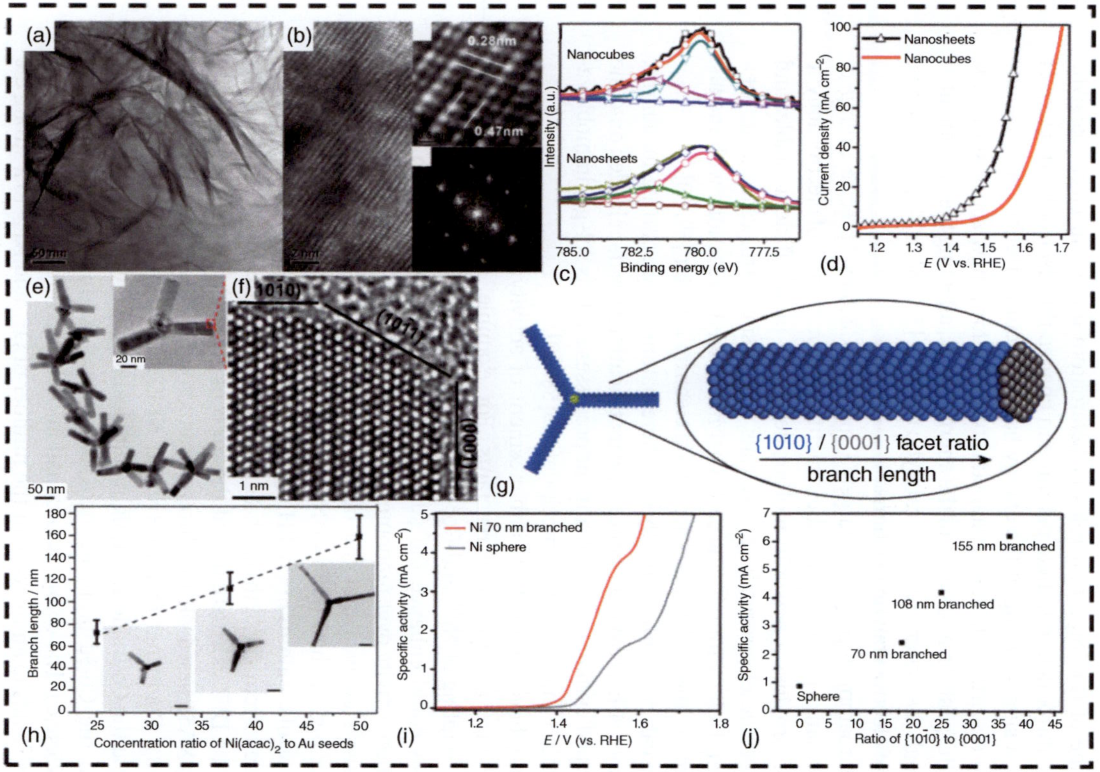

Figure 25.4 (a) TEM image, (b) HRTEM image, (c) Co 2p XPS spectra, and (d) EOR LSV curves of Co_3O_4 nanosheets. Source: Reproduced with permission of Dai et al. [55], American Chemical Society. (e) TEM image of branching Ni nanoparticles. (f) HRTEM image of the branch indicated by the red box in (e). (g) A model showing a surface with small facets along the branch. (h) Branch length (nm) vs. concentration ratio of $Ni(acac)_2$ to Au crystal species. Scale bar is 50 nm. (I) LSV curves of branched Ni nanoparticles and Ni spheres in 0.1 mol l^{-1} KOH and 10 mol l^{-1} HMF. (j) Specific activity of branched Ni nanoparticles versus [10−10] to [0001] crystal faces. Source: Reproduced with permission of Poerwoprajitno et al. [56], John Wiley and Sons.

25.3 Heteroatom Doping

Heteroatom (e.g. N, B, P, and S) doping is a method of introducing different kinds of atoms into electrocatalytic materials in order to change the electronic structure, surface properties, and catalytic activity of the materials [40]. The doping can affect the efficiency and selectivity of electrocatalytic reactions to a great extent. The following are some of the important roles of heteroatom doping in electrocatalytic materials – the introduction of heteroatoms can change the electronic structure of the material, causing a tuning of the energy band structure. Heteroatoms usually introduce local distortions in the crystal structure and form new active sites. This can modulate the position and width of the energy bands, affecting the movement of electrons through the material and the catalytic reaction [57, 58]. It helps to improve the efficiency of electrocatalytic reactions.

Overall, heteroatom doping is a targeted means to achieve precise tuning of catalytic performance by modulating the structure and properties of electrocatalytic materials. This provides a powerful tool for designing more efficient, stable, and functionally specific electrocatalytic materials. Many successful cases have been widely reported. Heteroatom doping can synergize catalytic reactions with metals, increase electronic interactions, and accelerate electron transfer [59, 60].

Heteroatoms can be obtained directly from heteroatom precursors (e.g. melamine, pyrrole monomer, 4-hydroxythiophene). The metal cations are then loaded onto the heteroatom-doped carbon by post-treatment processes such as sol–gel immobilization, impregnation, and deposition precipitation [58–60].

Recent papers have reported the enhancement of activity through heteroatom doping to enhance adsorption of reactants. For example, Zhai and coworkers reported that coupling of Ni^{2+} and Co^{2+} in perforated CoNi–hydroxide nanosheet (PHN) catalysts could enhance the adsorption of ethanol, resulting in superior ethanol oxidation activity over monometallic catalysts, i.e. Co–PHNs and Ni–PHNs [61]. XPS analysis showed that the binding energies (BEs) of Ni atoms in the Ni–PHNs were higher than those of Ni atoms, while the BEs of Co atoms were lower than those of Co atoms in Co–PHNs, suggesting that electrons in CoNi–PHNs were transferred from Ni^{2+} to Co^{2+} (Figure 25.4a,b). This result was further confirmed by the DFT calculations and the shifts of the peaks in soft X-ray absorption spectra (sXAS) based on synchrotron radiation of the Ni and Co L_3 edges (Figure 25.4c,d). Compared to Co-PHN and Ni-PHN, CoNi-PHN shows a lower peak intensity at the Co-L side and a slight enhancement at the Ni-L side, indicating an increase in the Co-3d occupied state and a decrease in the Ni-3d occupied state (Figure 25.4c,d). The complete occupation of the π-symmetric (t_{2g}) 3d orbital (t_{2g}-$e_g{}^2$) by N^{i2+} leads to an increase in the number of O^{2-} and Ni^{2+} to a strong e-e repulsion. In contrast, the unpaired electrons in the π-symmetric (t_{2g}) 3d orbitals of Co^{2+} interact with the bridging O^{2-} through the π-donor. Upon coupling of Ni^{2+} and Co^{2+} in Ni-PHN, the strong e^-–e^- repulsion between O^{2-} and Ni^{2+} leads to a partial transfer of charge from Ni^{2+} to Co^{2+}. Thus, the electron-rich Co^{2+} favors the adsorption of electrophilic alkyl groups, while the electron-deficient Ni^{2+} acts as an electrophilic reagent to adsorb nucleophilic hydroxyls (Figure 25.4e), which enhances the adsorption of ethanol molecules and thus improves ethanol oxidation activity.

In another case, Luo and coworkers demonstrated that doping nickel hydroxide with cobalt produces a higher proportion of Ni^{3+} sites with t_{2g}^6 e_g^1 configuration and e_g orbital occupancy close to 1 [62]. Considering that this structure facilitates the covalent metal–oxygen bonding, more Ni^{3+} sites promote the adsorption of HMF on the surface of the catalysts, and this may be the reason why the copper sulfide nanowires@NiCo-LDH ($Cu_xS@Ni_{0.75}Co_{0.25}O_mH_n$) is more active than $Cu_xS@NiO_mH_n$.

For electrocatalysts with excessive reactant adsorption, heteroatom doping is used to weaken the reactant adsorption and thus increase the activity. For example, Chen et al. used nonmetallic fluorine as a dopant to change the electronic structure of β-FeOOH [60]. The F dopant was introduced into β-FeOOH by solvent-thermal treatment of β-FeOOH in 1-butyl-3-methylimidazolium tetrafluoroborate. XPS analyses showed that the Fe 2p and O 1s spectra of the F-modified β-FeOOH shifted to higher BEs compared to β-FeOOH (Figure 25.4f,g), which can be attributed to the strong electronegativity of the F dopant. The effect of F doping on the electronic structure was further investigated by DFT calculations. Hirshfeld charge distributions (Figure 25.4h) showed that for β-FeOOH(110), C_2H_5O adsorption triggered the charge transfer from the Fe atoms of the Fe ooh to C_2H_5O, with a C_2H_5O adsorption energy of 4.71 eV. In contrast, F doping resulted in the Fe atoms carrying more positive charge, resulting in a less intense charge transfer from Fe atoms to C_2H_5O, and thus the adsorption energy of C_2H_5O was higher at 3.42 eV, indicating that F doping weakened the adsorption of C_2H_5O. As a catalyst for EOR, F-modified β-FeOOH provided a current density of 10 mA cm^{-2} at a potential of 1.207 V vs. RHE and a Tafel slope of 30.1 mV dec^{-1}. Both values were much lower than those of β-FeOOH, suggesting that F doping enhanced the activity and reaction kinetics of ethanol oxidation on β-FeOOH.

In addition, Liu and his coworkers found that Bi-doped Ni aerogel electrocatalysts exhibited better activity for MOR than Ni aerogels [63]. They proposed that one of the important reasons for the increased activity was that the Bi dopant optimized the d-band center of the catalyst by changing the electronic structure of Ni and inducing surface strain. As a result, the strong adsorption of cobalt-like intermediates on the surface of $Ni_{97}Bi_3$ catalysts was reduced to a low level, contributing to the activity enhancement. For $Ni(OH)_2$-based electrocatalysts following an indirect mechanism, the onset potential of catalytic reaction is limited by the autoxidation of $Ni(OH)_2$, which produces Ni(OH)O, and the electrophilic lattice oxygen as the active site.

Wang et al. found that doping Co^{2+} and Mn^{2+} into β-$Ni(OH)_2$ promoted catalytic activity by facilitating the conversion of β-$Ni(OH)_2$ to β-Ni(OH)O, whereas, Fe^{3+} and Al^{3+} dopants inhibited the autoxidation of β-$Ni(OH)_2$ and increased the onset potential [64]. As shown in Figure 25.5i, the onset potentials of ethanol oxidation over a series of $Ni(OH)_2$-based catalysts follow the order β-$Mn_{0.1}Ni_{0.9}(OH)_2$ < β-$Co_{0.1}Ni_{0.9}(OH)_2$ < β$Ni(OH)_2$ < β-$Fe_{0.1}Ni_{0.9}(OH)_2$ < β-$Al_{0.1}$ $Ni_{0.9}(OH)_2$, which is in agreement with the onset of catalyst autoxidation potential (Figure 25.5j). DFT calculations show that heteroatom doping can change the lattice oxygen ligand environment of β-$Ni(OH)_2$, and Co^{2+} and Mn^{2+} doping promotes the dehydrogenation of the lattice hydroxyl groups, whereas, Fe^{3+} and Al^{3+} increase the energy barrier of the dehydrogenation process (Figure 25.5k).

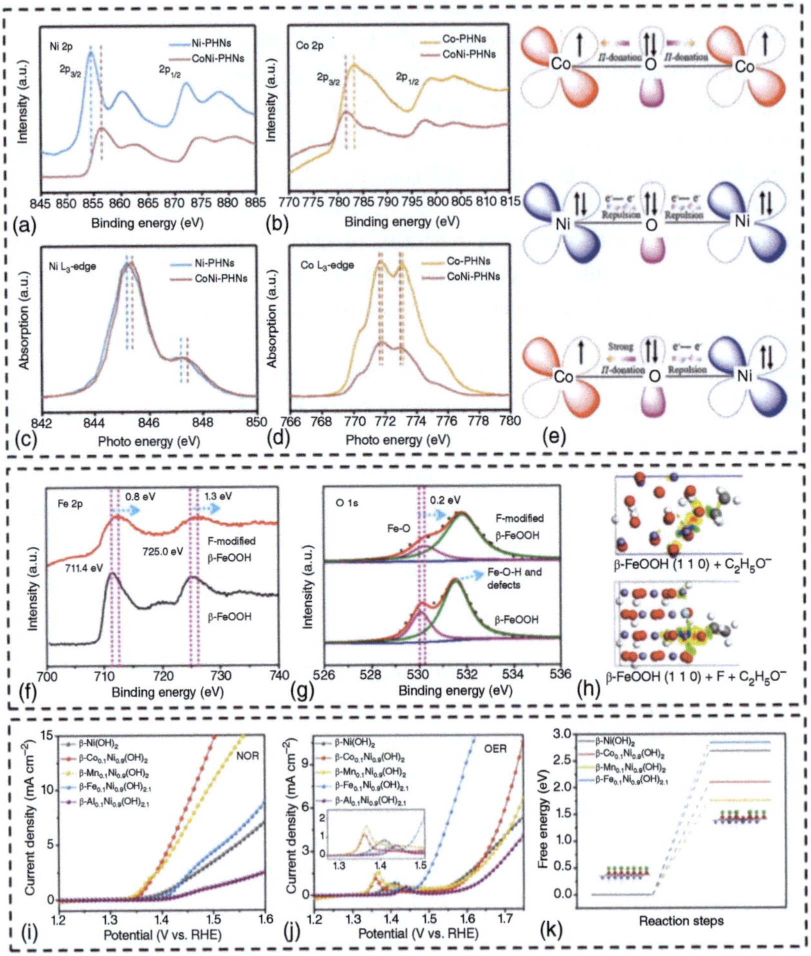

Figure 25.5 (a) XPS Ni 2p, and (b) Co 2p spectra of Co-PHNs and CoNi-PHNs. (c) Ni L_3-edge and (d) Co L_3-edge XAS spectra of Co-PHNs and CoNi-PHNs. (e) Schematic representation of the electronic coupling between Co and Ni in PHNs. Source: Reproduced with permission of Wang et al. [61], Wiley-VCH. (f) XPS Fe 2p and (g) O 1s spectra of F-modified β-FeOOH and pure β-FeOOH. Charge density difference plots (red contours correspond to charge accumulation and blue contours to charge depletion, red, white, gray, and light blue spheres represent O, H, Fe, and F atoms, respectively). β-$M_{0.1}Ni_{0.9}(OH)_x$ (M = Co, Mn, Fe, and Al) in (i) 1 mol l^{-1} KOH with 50 mmol l^{-1} ethanol and (j) 1 mol l^{-1} KOH LSV curves. (k) Free energy diagram of the dehydrogenation reaction from β-$M_{0.1}Ni_{0.9}(OH)_x$ to β-$M_{0.1}Ni_{0.9}(OH)_{x-1}O$. Source: Reproduced with permission of Chen et al. [64], Elsevier.

25.4 Defect Engineering

Defect engineering is potentially promising for electrocatalytic biomass utilization, where defect structures in electrocatalytic materials can be consciously introduced, modulated, or optimized to improve catalytic activity, selectivity, and stability. Defect engineering offers the possibility of designing more active and adaptable electrocatalytic materials, thus expanding the scope of electrocatalytic applications in biomass conversion [65–67].

To date, many defective nanomaterials have been prepared and investigated as electrocatalysts for electrocatalytic biomass utilization, and these defects have been shown to promote electrocatalytic performance by tailoring the adsorption of reactants, creating more active sites, and increasing conductivity [65–69].

Defects can enhance catalyst activity by increasing the adsorption strength of reactants. Shao and his coworkers reported oxygen-defect-rich LDH nanosheet arrays (E-CoAl-LDH-NSA), which were prepared by in situ electrochemical exfoliation (Figure 25.6a) of a bulk coal–LDH nanosheet array (B-CoAl-LDH-NSA)

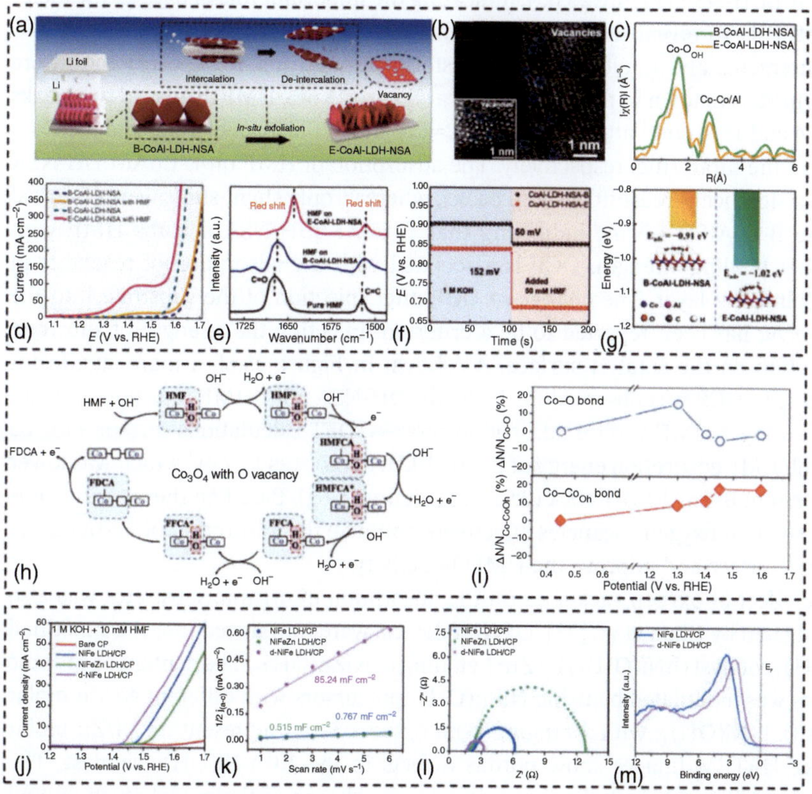

Figure 25.6 (a) Schematic of CoAl-LDH-NSA exfoliated in situ. (b) HADDF–STEM image of E-CoAl-LDH-NSA. (c) Co-edge FT–EXAFS spectra of B-CoAl-LDH-NSA and E-CoAl-LDH-NSA. (d) LSV curves of B-CoAl-LDH-NSA and E-CoAl-LDH-NSA with and without 10 mmol l^{-1} HMF in 1.0 mol l^{-1} KOH. (e) In situ FT–IR spectra of HMF adsorbed on B-CoAl-LDH-NSA. (f) OCP curves of B-CoAl-LDH-NSA and E-CoAl-LDH-NSA. (g) Optimized structures and adsorption energies of HMF on B-CoAl-LDH-NSA and E-CoAl-LDH-NSA. Source: Reproduced with permission of Song et al. [70], Elsevier. (h) Reaction mechanism of HMFOR on Vo-Co_3O_4. (i) Structural coherence changes in the EXAFS coordination number of Co–O and Co–CoOH bonds at an applied potential relative to the OCP state. Source: Reproduced with permission of Lu et al. [71], Wiley-VCH. (j) LSV curves, (k) double layer capacitance measurements, (l) EIS spectra of d-NiFe-LDH and NiFe-LDH, and (m) valence band spectra. Source: Reproduced with permission of Qi et al. [72], American Chemical Society.

[70]. High-angle annular dark-field scanning TEM (HADDF-STEM) images (Figure 25.6b) showed that defects were introduced into the catalyst after in situ stripping. XPS and electron paramagnetic resonance (EPR) measurements also confirmed the presence of oxygen defects in E-CoAl-LDH-NSA. In addition, Fourier transform extended X-ray absorption near-edge structure (FT–EXAFS) spectroscopy of Co K-edge (Figure 25.6c) showed that the coordination number of Co—O bonds in E-CoAl-LDH-NSA was 3.7, which was significantly lower than that of B-CoAl-LDH-NSA (4.2), suggesting that in situ electrochemical stripping through the introduction of oxygen vacancies altered the local structure of the Co sites. As an electrocatalyst for HMFOR, E-CoAl-LDH-NSA exhibited a low potential of 1.30 V vs. RHE at 10 mA cm^{-2}, which was 100 mV lower than that of B-CoAl-LDH-NSA (Figure 25.6d), indicating the important role of oxygen vacancies in HMFOR. To reveal the role of oxygen vacancies, open circuit potential (OCP) measurements and in situ Fourier transform infrared (FT-IR) spectroscopy were performed. As shown in the FT-IR spectra (Figure 25.6e), HMF displayed two peaks at 1674 and 1523 cm^{-1} attributed to the C=O bond of the formyl group and the C=C bond of the furan ring, respectively. The adsorption of HMF on E-CoAl-LDH-NSA led to a significant redshift of these peaks, whereas, only slight shifts were observed on the B-CoAlLDH-NSA, indicating that E CoAl-LDH-NSA adsorbs HMF with a higher binding strength. OCP is associated with the adsorption of reactants in the Helmholtz layer. The change in OCP after injection of the substrate into the electrolyte has been reported to be a criterion reflecting the strength of substrate adsorption on the catalyst surface. As shown in Figure 25.6f, a more significant decrease in OCP was observed for E-CoAl-LDH-NSA after HMF injection, indicating a stronger HMF surface adsorption process. DFT calculations further showed that the HMF adsorption energy of E-CoAl-LDH-NSA was 1.02 eV, which was lower than that of B-CoAl-LDH-NSA (0.91 eV) (Figure 25.6g). Based on these results, it is proposed that oxygen vacancies effectively optimize the adsorption of HMF on the catalyst surface and improve the HMFOR activity.

The effect of oxygen vacancies on the adsorption of organic substrates was also investigated by Chen et al. [73]. Oxygen defects were introduced into the sanctified $Ni(OH)_2$ catalyst (h$Ni(OH)_2$) by Zn^{2+} etching of $NiZn(OH)_2$. The content of cationic defects was modulated by using $NiZn(OH)_2$ precursors with different Ni/Zn molar ratios (x h-$Ni(OH)_2$, with x ranging from 0.5 to 1.5 and representing Ni/Zn molar ratios). HRTEM images show porous defects in the 1.0 h-$Ni(OH)_2$ catalyst. The oxidation of benzyl alcohol to benzoic acid was chosen as the model reaction, and 1.0 h-$Ni(OH)_2$ exhibited the best activity with a benzoic acid yield close to 100%. In addition, 1.0 h-$Ni(OH)_2$ was also effective for the electrooxidation of biomass derivatives such as HMF and furfural. DFT calculations showed that the adsorption energy of benzyl alcohol on defective $Ni(OH)_2$ was higher than that on nondefective $Ni(OH)_2$ and that oxygen atoms in benzyl alcohol gained more charge on defective $Ni(OH)_2$, suggesting that the oxygen vacancies enhanced the adsorption of benzyl alcohol.

The oxygen vacancies themselves can directly serve as adsorption sites for OH^-. Wang et al. found that the oxygen vacancies of Co_3O_4 can be filled with

OH^-, which binds to organic molecules to provide oxidation products and water molecules (Figure 25.6h) [71]. Thus, unlike the defect-free Co_3O_4 which is subject to competitive adsorption of OH^- and organic molecules on the metal sites, oxygen-vacancy-rich Co_3O_4 (Vo-Co_3O_4) can utilize the active sites more efficiently. As a HMFOR electrocatalyst, Vo-Co_3O_4 outputs 10 mA cm^{-2} at a potential of 1.35 V vs. RHE, which is superior to that of the defect-free Co_3O_4 (1.44 V vs. RHE). In situ EXAFS analysis showed that the coordination number of the Co—O bond increased until 1.30 V vs. RHE (the onset potential of HMFOR on Vo-Co_3O_4) and then decreased at 1.4–1.5 V vs. RHE, indicating that OH^- was filled into the oxygen vacancies and then depleted during the HMFOR process (Figure 25.6i). The Co K-edge's operando X-ray absorption near-edge structure (XANES) spectra also confirmed this process, which showed an increase in the valence state of Co on VoCo_3O_4 up to 1.30 V vs. RHE, whereas, there was no significant change in Co_3O_4 without defects.

These defects can expose more active sites and improve the conductivity. For example, Qi et al. prepared NiFe-LDH (d-NiFe-LDH) enriched with cationic defects by alkali etching treatment of NiFeZn-LDH precursor [72]. The HMFOR activity of d-NiFe-LDH was significantly enhanced compared with that of NiFe-LDH (Figure 25.6j). d-NiFe-LDH had a C_{dl} of 85.24 mF cm^{-2}, which was much higher than that of NiFe-LDH of 0.767 mF cm^{-2}, indicating that the cationic defects resulted in higher ECSA (Figure 25.6k). In addition, electrochemical impedance spectroscopy (EIS) measurements showed that d-NiFe-LDH has a much lower charge transfer resistance (R_{ct}) than NiFe-LDH, suggesting that dNiFe-LDH has faster HMFOR kinetics and better electrical conductivity (Figure 25.6l). Valence band spectroscopy further confirms the increased conductivity with an upward shift of the valence band maximum of d-NiFe-LDH (Figure 25.6m). Lei and co-workers reported an ammonia etching method for generating oxygen vacancies in $CuMn_2O_4$ catalysts. Defective $CuMn_2O_4$ catalysts exposed more Mn^{4+} sites to catalyze HMFOR via a Mn^{3+}/Mn^{4+} mediated indirect mechanism.

25.5 Heterostructuring

Heterostructures refer to structures composed of two or more different materials, where these materials differ significantly in crystal structure or electronic properties. The formation of such heterostructures is usually on the nanoscale and has been widely studied and utilized in electrocatalysis. Heterostructures have important application potential in electrocatalytic biomass utilization, which can be used to optimize the electrocatalytic performance and improve the efficiency of biomass conversion through rational design and construction of heterostructures. Heterogeneous structures have been shown to be an effective way to improve the efficiency of electrocatalytic biomass utilization [74]. Strongly coupled interfaces between two different compositions trigger band alignment and change the adsorption/desorption behavior of the catalyst [75, 76].

For example, Fu and coworkers prepared carbon-encapsulated MoO_2-FeP heterojunctions (MoO_2-FeP@C) by phosphatizing a polyoxymethylene metal–organic framework [77]. The HMF oxidation potential of MoO_2-FeP@C at $10\,mA\,cm^{-2}$ was lower at 1.359 V vs. RHE compared to MoO_2@C and FeP@C (Figure 25.7a). The increase in activity is attributed to the enhanced adsorption of HMF by the tailored electronic structure of MoO_2. As shown in the XPS spectra (Figure 25.7b,c), the BEs of Fe 2p and P 2p in MoO_2-FeP@C showed a negative shift compared to FeP@C. The BE of Mo 3d spectra in MoO_2-FeP@C is higher compared to MoO_2@C (Figure 25.7d), which indicates that the electrons are shifted from MoO_2 to FeP. DFT calculations also confirm the change in electron distribution. As shown in

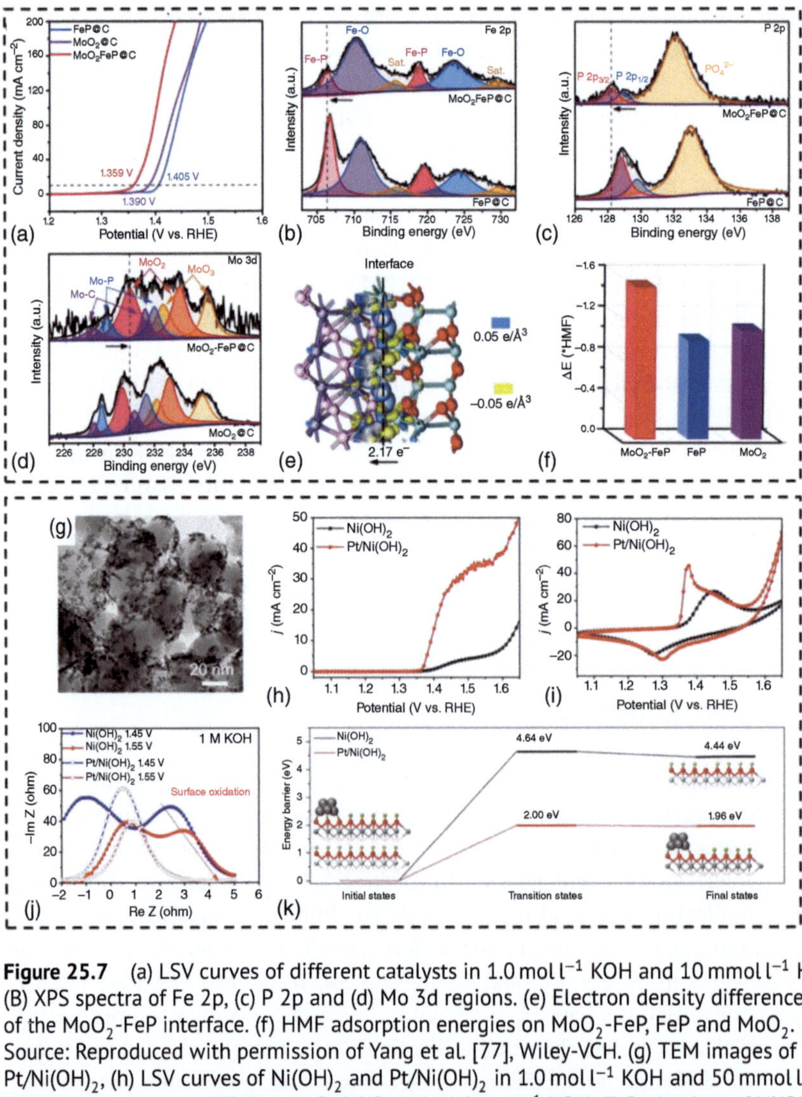

Figure 25.7 (a) LSV curves of different catalysts in $1.0\,mol\,l^{-1}$ KOH and $10\,mmol\,l^{-1}$ HMF. (B) XPS spectra of Fe 2p, (c) P 2p and (d) Mo 3d regions. (e) Electron density difference map of the MoO_2-FeP interface. (f) HMF adsorption energies on MoO_2-FeP, FeP and MoO_2. Source: Reproduced with permission of Yang et al. [77], Wiley-VCH. (g) TEM images of Pt/$Ni(OH)_2$, (h) LSV curves of $Ni(OH)_2$ and Pt/$Ni(OH)_2$ in $1.0\,mol\,l^{-1}$ KOH and $50\,mmol\,l^{-1}$ HMF. (i) CV curves of $Ni(OH)_2$ and Pt/$Ni(OH)_2$ in $1.0\,mol\,l^{-1}$ KOH. (j) Bode plots of $Ni(OH)_2$ and Pt/$Ni(OH)_2$ electrodes during OER. (k) Energy barrier distributions for Ni–OH dehydrogenation at $Ni(OH)_2$ and Pt/$Ni(OH)_2$ electrodes. Source: Reproduced with permission of Zhou et al. [78], John Wiley and Sons.

Figure 25.7e, the electrons are redistributed at the heterojunction interface of FeP and MoO_2, with electrons clustered on FeP and holes trapped by MoO_2. Further DFT calculations showed that the HMF adsorption energy of MoO_2-FeP@C was lower than that of FeP and MoO_2 (Figure 25.7f), suggesting that the heterojunction promotes HMFOR by enhancing HMF adsorption. In addition, density of states (DOS) analysis shows that MoO_2-FeP has better electrical conductivity, which favors the fast kinetics of HMFOR.

The optimal adsorption of the substrate on the heterogeneous catalyst surface should be neither too strong nor too weak. Liu and his coworkers found that the adsorption of glucose by metal Co was too strong with a low free energy of adsorption of −2.48 eV, whereas, the adsorption of glucose by CoP was weaker with a high free energy of adsorption of −1.20 eV [79]. In order to optimize the adsorption of glucose, they prepared Co/CoP heterojunctions with moderate free energy of adsorption of glucose of −1.41 eV. To optimize the glucose adsorption, they prepared Co/CoP heterojunctions with a moderate adsorption-free energy of −1.41 eV. Mott–Schottky measurements showed that the Mott–Schottky effect occurs at the interface between the semiconducting CoP and the metallic Co, resulting in the transfer of free electrons from the CoP to the Co. As an electrocatalyst for glucose oxidation, the Co/CoP heterojunctions efficiently catalyze the oxidation of glucose to gluconic acid, with the potential of 10 mA cm^{-2} being 10.5 mA cm^{-2} in an alkaline medium. mA cm^{-2} at a potential of 1.544 V vs. RHE. They concluded that the optimal adsorption of glucose molecules at the Co/CoP interface allows for a rapid desorption process that accelerates the regeneration of the active site, thus improving the accessibility of the active site. On the other hand, the charge redistribution at the Co/CoP interface increases the electrophilicity of CoP and promotes the activation of formyl C—H bonds in glucose molecules.

The heterogeneous structure can facilitate the indirect mechanism by modulating the redox properties of the $Mn^+/M^{(n+1)+}$ redox couple. Wang's group found that decorating $Ni(OH)_2$ with Pt nanoparticles (Figure 25.7g) reduced the dehydrogenation energy of $Ni(OH)_2$, thereby accelerating the formation of Ni(OH)O, which is an active species for oxidizing organic substrates [78]. As a HMFOR electrocatalyst, the current density of Pt/$Ni(OH)_2$ at 1.5 V vs. RHE was 37.31 mA cm^{-2}, which was 8.2 times higher than that of $Ni(OH)_2$ (Figure 25.7h). As shown in the CV curves recorded in 1.0 mol l^{-1} KOH without HMF (Figure 25.7i), the Ni^{2+} oxidation peak of Pt/$Ni(OH)_2$ shows a negative shift relative to that of $Ni(OH)_2$, suggesting that the dehydrogenation process of Pt/$Ni(OH)_2$ is more facile, which is further confirmed by the EIS measurements. As shown in the Bode plot (Figure 25.7j), the phase angle in the high-frequency region corresponds to the surface oxidation of the catalyst, which decreases with increasing potential, suggesting that the reaction rate of Ni^{2+} oxidation is higher at higher potentials. In addition, the phase angle of the catalyst was smaller than that of $Ni(OH)_2$ after loading Pt, suggesting a faster rate of Ni^{2+} oxidation for Pt/$Ni(OH)_2$. DFT calculations further showed that the free energy of the dehydrogenation reaction of Pt/$Ni(OH)_2$ (1.96 eV) was much lower than that of $Ni(OH)_2$ (4.44 eV), which suggests that Pt/$Ni(OH)_2$ is more favorable for the generation of Ni(OH)O reactive species (Figure 25.7k), which may be attributed to the Pt–O–Ni–OH coordination at the interface between Pt-$Ni(OH)_2$ and $Ni(OH)_2$.

25.6 Single-Atom Modification

The catalytic activity of a catalyst is highly dependent on the density and chemistry of the active sites, which can be improved by reducing the particle size. Single-atom catalysts (SACs) are the smallest-sized catalysts that can expose a larger surface area and tune the atomic configuration and electronic structure for optimal activity. Therefore, the development of SACs has become the frontier of electrochemical catalytic materials [80, 81]. SACs process well-dispersed and under-coordinated active centers in the form of single atoms, which can be coordinated, encapsulated, adsorbed, chemisorbed, or immobilized on high-surface-area solid carriers. The remarkable catalytic activity and selectivity of SACs are due to their unique geometrical and electronic properties, which include the absence of metal–metal bonding and positively charged individual metal atoms, which dramatically changes the adsorption properties and BEs of the intermediates to the metal centers and lowers the energy barriers [82, 83]. For these reasons, SACs are considered a transition state between homogeneous and inhomogeneous catalysts, with the added benefit of theoretical atom utilization efficiencies of up to 100% and the development of cost-effective metal-based catalysts [83]. However, SACs for electrocatalytic biomass upgrading utilization remain poorly exploited.

Wang's team found that modification of Co_3O_4 with single-atom Ir could modulate the adsorption of HMF and thus improve the efficiency of HMF ECO [84]. Single-atom Ir-loaded Co_3O_4 (Ir-Co_3O_4) was synthesized by Ir salt impregnation and subsequent roasting treatment (Figure 25.8a). The Ir atoms were atomically dispersed on Co_3O_4 as shown in the HAADF–STEM image (Figure 25.8b).

The FT–EXAFS spectrum of Ir-Co_3O_4 has a distinct peak at ~2 Å, corresponding to the Ir–O bond. No Ir–Ir bond was observed at ~2.6 Å, indicating that the Ir atom is isolated (Figure 25.8c). In addition, model-based fitting results indicated that the Ir atom is coordinated by eight O atoms, six of which are oxygen ligands and the other two from adsorbed hydroxyl groups or water. In alkaline medium, Ir-Co_3O_4, as an electrocatalyst for HMFOR, showed a significant increase in activity compared to Co_3O_4 with a low onset potential of 1.15 V vs. RHE, indicating that single-atom Ir strongly enhances the activity (Figure 25.8d). In order to reveal the reason for the enhanced activity, programmed temperature rise desorption (TPD) measurements were performed. As shown in Figure 25.8e, the desorption temperature of HMF on Ir-Co_3O_4 was much higher than that on Co_3O_4. A similar situation occurred for the desorption of ethylene (C=C) on Ir-Co_3O_4 and Co_3O_4 (Figure 25.8f).

These results suggest that monoatomic Ir can promote the adsorption of HMF by binding C=C groups on the furan ring of the HMF molecule. Density flooding calculations further confirmed that single-atom Ir preferentially adsorbed the C=C group of HMF with an adsorption energy of −3.77 eV (Figure 25.8g). Co_3O_4 (111) had a higher adsorption energy of 2.64 eV. Therefore, we conclude that the loading of single-atom-Ir enhanced the adsorption of HMF. This contributes to the activity of HMFOR.

A recent study has shown that single-atom Ru effectively enhances the HMFOR by promoting the water dissociation process ($H_2O \rightarrow OH^* + H^+ + e^-$) in a neutral

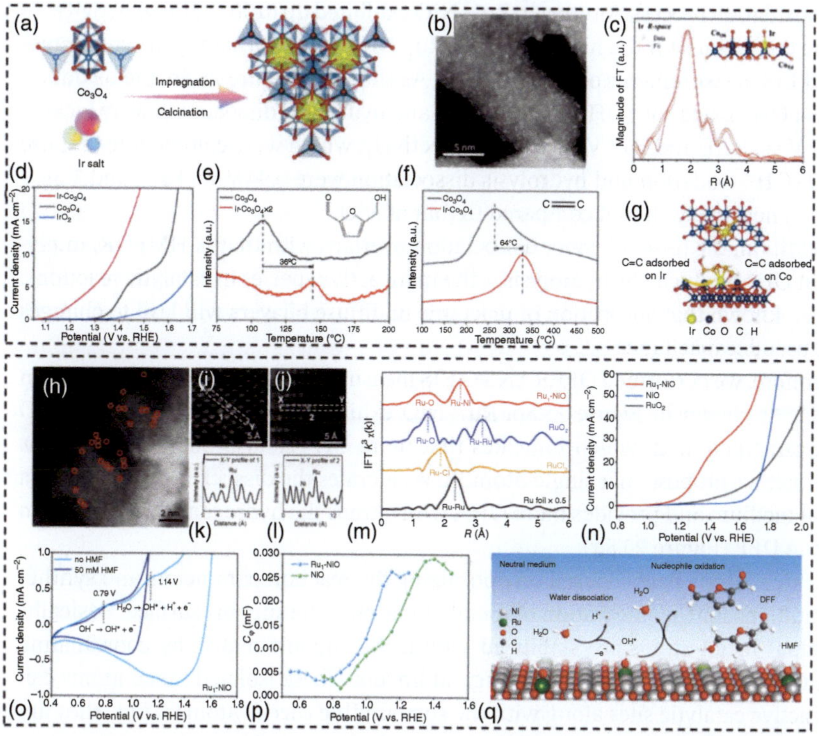

Figure 25.8 (a) Preparation scheme of Ir-Co_3O_4. (b) HAADF–STEM image of Ir-Co_3O_4. (c) FT–EXAFS spectrum of Ir-Co_3O_4. (d) TPD spectra of IrCo_3O_4, Co_3O_4, Ir-Co_3O_4, and Co_3O_4 under (e) HMF/He and (f) ethylene. (g) Modeling of adsorption of HMF molecules on Ir-Co_3O_4. Co_3O_4. Source: Reproduced with permission of Lu et al. [84], Wiley-VCH. (h) HAADF–STEM images of Ru_1-NiO. (i,j) HAADF–STEM images observed from the (i) [110] and (j) [211] directions of NiO. (k,l) Intensity distributions corresponding to the labeled lines in (i) and (j), (m) FT–EXAFS spectra of the Ru K-edge of Ru_1–NiO. (n) Solutions of NiO, Ru_1-NiO, and RuO_2 in 1.0 mol l^{-1} PBS containing 50 mmol l^{-1} HMF. (o) CV curves of Ru_1-NiO in 1.0 mol l^{-1} PBS with and without 50 mmol l^{-1} HMF. (p) $C_{(max)}$ vs. potential plots for catalysts in 1.0 mol l^{-1} PBS. (q) HMFOR mechanism of Ru_1-NiO in neutral medium. Source: Reproduced with permission of Ge et al. [85], Wiley-VCH.

medium [85]. NiO matrix (Ru_1NiO)-loaded single-atom Ru was prepared by impregnation of NiO with a Ru^{3+} solution followed by roasting in air. HAADF–STEM images show that the Ru sites are atomically dispersed on the NiO matrix by occupying the original lattice positions of the Ni atoms in NiO (Figure 25.8h-l). FT–EXAFS spectra indicate the presence of Ru–O and Ru–Ni scattering in Ru_1–NiO (Figure 25.8m).

Model-based fitting of the Ru R space spectrum further indicates that the coordination numbers for Ru–O and Ru–Ni scattering are 3.2 and 7.5, respectively, while the NiO matrix does not change significantly upon monoatomic Ru loading.

Interestingly, the low potential of Ru_1–NiO as an electrocatalyst for HMFOR in neutral medium was 1.283 V vs. RHE at 10 mA cm^{-2}, which was 433 mV lower than that of NiO (Figure 25.8n), suggesting that single-atom Ru plays a key role in the electrooxidation process. Moreover, Ru_1–NiO exhibited up to 90% DFF selectivity. To explore the reasons for the high activity of Ru_1-NiO, we performed cyclic voltammetry and EIS measurements of OPERANDO. As shown in Figure 25.8o, the oxidation peaks of OH^- oxidation ($OH^- \rightarrow OH^* + e^-$) and hydrolysis dissociation were located at 0.79 V vs. RHE and 1.14 V vs. RHE, respectively, which were comparable to those of NiO (OH^- oxidation and hydrolysis dissociation were 0.94 V vs. RHE and 1.36 V vs. RHE) negatively shifted compared to that of NiO.

In addition, the onset of water dissociation overlaps with that of HMFOR, implying that OH* produced by hydrolysis is the main active species in catalytic reactions. It is well known that adsorption of reactants in diffuse bilayers will lead to changes in the pseudocapacitance (C_φ).

Therefore, we performed OPERANDO EIS measurements to reflect the adsorption of OH*. As shown in Figure 25.8p, Ru_1–NiO exhibits higher C_φ values than NiO at the same potential, which indicates that more OH* is adsorbed onto Ru_1-NiO. Therefore, we propose that single-atom Ru accelerates the dissociation of water in a neutral medium and provides more OH*, which traps the hydrogen atom of HMF to generate DFF (Figure 25.8q).

Research efforts have focused on optimizing the reaction parameters and synthesis conditions for the integration of metal atoms onto the support surface. Basically, individual metal atoms are stabilized within the support matrix by coordinating unsaturated sites that act as surface ligand linkers. These isolated metal atoms can act as active catalytic sites along with the surrounding carrier atoms. Thus, they are localized at specific coordination sites, crystal faces, or surface defects (i.e. oxygen vacancies, metal vacancies, edges, steps, kinks, and corners) in order to maximize the metal–carrier interactions and obtain the best catalytic performance.

Various techniques have been used to characterize the structural properties of SACs at the support interface, such as HAADF–STEM, scanning tunneling microscopy (STM), diffuse reflectance infrared Fourier transform spectroscopy (DRIFTS), EXAFS, XANES, and XPS. Structural characterization of SACs involves determining the uniform distribution of SACs, information on metal–carrier interactions, the stability and dispersion of SACs on various carriers, the electronic state of SACs, and the local structure. The development of structural characterization techniques provides an ideal platform for the preparation and utilization of SACs.

25.7 Challenges and Prospects of Metal-Based Materials in Electrocatalytic Biomass Conversion

Facilities for the conversion of cheap, renewable biomass into high-value-added fine chemicals are of considerable importance in terms of energy use, economic efficiency, and environmental protection. Metal catalysts that can be used for biomass upgrading include but are not limited to, transition metals and their (hydro)oxides, metal interstitial compounds (nitrides, phosphides, sulfides, and selenides), and noble metals. We then delve into emerging strategies for building catalysts, mainly including morphology modification, heteroatom doping, defect engineering, heterostructures, and the introduction of monoatomic structures.

These methods can optimize the arrangement of electrons, enhance the adsorption or desorption of reactants, and introduce new active sites. Strategies for electronic structure optimization are varied and can be combined with the properties of the material itself and electronic modulation methods, depending on the desired effect. Various biomass derivatives, including monohydric alcohols/aldehydes, polyols, and biopolymers, have been converted into high-value-added chemicals by electrooxidation using metal electrocatalysts.

Considerable progress must be made before economically and environmentally beneficial commercial uses can be found. Current research on catalysts for electrocatalytic upgrading of biomass resources has the following limitations.

(1) Lack of studies on the catalytic mechanism before and after optimization of the electronic structure of the catalyst.
(2) The traditional characterization techniques used to study the changes in the electronic structure of catalysts are not intuitive enough, and in situ characterization techniques (e.g. in situ Raman spectroscopy, in situ FT-IR, in situ EXAFS) should be used to observe the transient changes in the electronic structure.
(3) A large number of studies have focused on electronic structure modulation methods for catalyst synthesis. However, the fundamental changes brought about by optimizing the electronic structure, such as changes in d orbitals (t_{2g} and e_g orbitals), electronic configuration, electronic DOS, and electronic energy band structure, have not yet been studied in depth.
(4) The current study does not cover all electronic structure modulation methods. Many other effective strategies exist, such as phase transitions, electron alignment, and spin polarization. However, they have received much less attention, as evidenced by the lack of reports on these strategies.
(5) Explore more biomass-derived substrates for catalytic processes. To date, extensively studied substrates are still limited to a few model molecules such as HMF, ethanol, and glycerol. The desire to explore more substrates is urgent.
(6) Development of acidic nonprecious metal electrocatalysts. Acidic media can simplify the purification and separation of carboxylic acid products. For example, the separation of FDCA from alkaline media requires an acidification process, which adds additional costs for acid/base consumption and salt waste disposal. The use of acidic electrolytes can solve this problem, but the challenge is the lack of efficient and durable nonprecious metal electrocatalysts.
(7) Therefore, there is a need to combine the oxidation reaction with cathodic HER or other reactions (CO_2RR, NRR, organic reduction) in order to increase the efficiency of the overall reaction [86–88]. This is very important for future commercial applications.

References

1 Tan, X. and Tan, T. (2022). Biofuels from biomass toward a net-zero carbon and sustainable world. *Joule* 6 (7): 1396.

2 Luo, H., Barrio, J., Sunny, N. et al. (2021). Progress and perspectives in photo- and electrochemical-oxidation of biomass for sustainable chemicals and hydrogen production. *Adv. Energy Mater.* 11 (43): 2101180.

3 Lucas, F.W.S., Grim, R.G., Tacey, S.A. et al. (2021). Electrochemical routes for the valorization of biomass-derived feedstocks: from chemistry to application. *ACS Energy Lett.* 6 (4): 1205.

4 Queneau, Y. and Han, B. (2022). Biomass: renewable carbon resource for chemical and energy industry. *The Innovation* 3 (1): 100184.

5 Park, J., Mushtaq, U., Sugiarto, J.R. et al. (2022). Total chemocatalytic cascade conversion of lignocellulosic biomass into biochemicals. *Appl. Catal. B-Environ.* 310: 121280.

6 Baniamerian, H., Høj, M., Beier, M.J., and Jensen, A.D. (2023). Catalytic conversion of sugars and polysaccharides to glycols: a review. *Appl. Catal. B-Environ.* 330: 122650.

7 Collard, F.-X. and Blin, J. (2014). A review on pyrolysis of biomass constituents: mechanisms and composition of the products obtained from the conversion of cellulose, hemicelluloses and lignin. *Renew. Sust. Energ. Rev.* 38: 594.

8 Sun, Z., Fridrich, B., de Santi, A. et al. (2018). Bright side of lignin depolymerization: toward new platform chemicals. *Chem. Rev.* 118 (2): 614.

9 Sharifzadeh, M., Sadeqzadeh, M., Guo, M. et al. (2019). The multi-scale challenges of biomass fast pyrolysis and bio-oil upgrading: review of the state of art and future research directions. *Prog. Energ. Combust.* 71: 1.

10 Gérardy, R., Debecker, D.P., Estager, J. et al. (2020). Continuous flow upgrading of selected C_2–C_6 platform chemicals derived from biomass. *Chem. Rev.* 120 (15): 7219.

11 Akhade, S.A., Singh, N., Gutiérrez, O.Y. et al. (2020). Electrocatalytic hydrogenation of biomass-derived organics: a review. *Chem. Rev.* 120 (20): 11370.

12 Xu, C., Paone, E., Rodríguez-Padrón, D. et al. (2020). Recent catalytic routes for the preparation and the upgrading of biomass derived furfural and 5-hydroxymethylfurfural. *Chem. Soc. Rev.* 49 (13): 4273.

13 Fang, Z., Li, F., Wang, M. et al. (2023). Selective electrocatalytic upgrading of lignin to aryl aldehydes and carboxylic acids over dodecyl sulfate-intercalated cos nanocones. *Appl. Catal. B-Environ.* 323: 122149.

14 Qi, Y., Liu, B., Qiu, X. et al. (2023). Simultaneous oxidative cleavage of lignin and reduction of furfural via efficient electrocatalysis by p-doped comoo4. *Adv. Mater.* 35 (14): 2208284.

15 Baldwin, R.M., Magrini-Bair, K.A., Nimlos, M.R. et al. (2012). Current research on thermochemical conversion of biomass at the national renewable energy laboratory. *Appl. Catal. B-Environ.* 115–116: 320.

16 Wang, S., Dai, G., Yang, H., and Zhongyang, L. (2017). Lignocellulosic biomass pyrolysis mechanism: a state-of-the-art review. *Prog. Energ. Combust.* 62: 33.

17 Ayub, R. and Raheel, A. (2022). High-value chemicals from electrocatalytic depolymerization of lignin: challenges and opportunities. *Int. J. Mol. Sci.* 23: 3767.

18 Feng, Y., Long, S., Tang, X. et al. (2021). Earth-abundant 3d-transition-metal catalysts for lignocellulosic biomass conversion. *Chem. Soc. Rev.* 50 (10): 6042.

19 Ge, R., Li, J., and Duan, H. (2022). Recent advances in non-noble electrocatalysts for oxidative valorization of biomass derivatives. *Sci. China Mater.* 65 (12): 3273.

20 Yang, C., Chen, H., Peng, T. et al. (2021). Lignin valorization toward value-added chemicals and fuels via electrocatalysis: a perspective. *Chin. J. Catal.* 42 (11): 1831.

21 Tian, X., Wang, Y., Zeng, Z. et al. (2022). Research progress on the role of common metal catalysts in biomass pyrolysis: a state-of-the-art review. *Green Chem.* 24 (10): 3922.

22 Tang, D., Huang, X., Tang, W., and Jin, Y. (2021). Lignin-to-chemicals: application of catalytic hydrogenolysis of lignin to produce phenols and terephthalic acid via metal-based catalysts. *Int. J. Biol. Macromol.* 190: 72.

23 Du, L., Shao, Y., Sun, J. et al. (2018). Electrocatalytic valorisation of biomass derived chemicals. *Catal. Sci. Technol.* 8 (13): 3216.

24 Sheng, H., Janes, A.N., Ross, R.D. et al. (2022). Linear paired electrochemical valorization of glycerol enabled by the electro-fenton process using a stable $NiSe_2$ cathode. *Nat. Catal.* 5 (8): 716.

25 Zhang, Y. and Shen, Y. (2024). Electrochemical hydrogenation of levulinic acid, furfural and 5-hydroxymethylfurfural. *Appl. Catal. B-Environ.* 343: 123576.

26 Guo, L., Zhang, X., Gan, L. et al. (2023). Advances in selective electrochemical oxidation of 5-hydroxymethylfurfural to produce high-value chemicals. *Adv. Sci.* 10 (4): 2205540.

27 Delparish, A., Uslu, A., Cao, Y. et al. (2022). Boosting the valorization of biomass and green electrons to chemical building blocks: a study on the kinetics and mass transfer during the electrochemical conversion of HMF to FDCA in a microreactor. *Chem. Eng. J.* 438: 135393.

28 Li, K. and Sun, Y. (2018). Electrocatalytic upgrading of biomass-derived intermediate compounds to value-added products. *Chem-Eur. J.* 24 (69): 18258.

29 Zhu, B., Chen, C., Huai, L. et al. (2021). 2,5-Bis(hydroxymethyl)furan: a new alternative to HMF for simultaneously electrocatalytic production of FDCA and H_2 over CoOOH/Ni electrodes. *Appl. Catal. B-Environ.* 297: 120396.

30 Eerhart, A.J.J.E., Faaij, A.P.C., and Patel, M.K. (2012). Replacing fossil based PET with biobased PEF; process analysis, energy and GHG balance. *Energ. Environ. Sci.* 5 (4): 6407.

31 Yang, Y. and Mu, T. (2021). Electrochemical oxidation of biomass derived 5-hydroxymethylfurfural (HMF): pathway, mechanism, catalysts and coupling reactions. *Green Chem.* 23 (12): 4228.

32 Carvajal, D., Arcas, R., Gouda, L. et al. (2024). Electrochemical valorization of HMF using Ni/graphite electrodes. *Mater. Chem. Phys.* 311: 128510.

33 Luo, R., Li, Y., Xing, L. et al. (2022). A dynamic $Ni(Oh)_2$-NiOOH/NiFeP heterojunction enabling high-performance E-upgrading of hydroxymethylfurfural. *Appl. Catal. B-Environ.* 311: 121357.

34 Zhong, Y., Ren, R.-Q., Qin, L. et al. (2021). Electrodeposition of hybrid nanosheet-structured $NiCo_2O_4$ on carbon fiber paper as a non-noble electrocatalyst for efficient electrooxidation of 5-hydroxymethylfurfural to 2,5-furandicarboxylic acid. *New J. Chem.* 45 (25): 11213.

35 Barwe, S., Weidner, J., Cychy, S. et al. (2018). Electrocatalytic oxidation of 5-(hydroxymethyl)furfural using high-surface-area nickel boride. *Angew. Chem. Int. Ed.* 57 (35): 11460.

36 Zhang, M., Liu, Y., Liu, B. et al. (2020). Trimetallic nicofe-layered double hydroxides nanosheets efficient for oxygen evolution and highly selective oxidation of biomass-derived 5-hydroxymethylfurfural. *ACS Catal.* 10 (9): 5179.

37 Kang, M.J., Park, H., Jegal, J. et al. (2019). Electrocatalysis of 5-hydroxymethylfurfural at cobalt based spinel catalysts with filamentous nanoarchitecture in alkaline media. *Appl. Catal. B-Environ.* 242: 85.

38 Gidi, L., Amalraj, J., Tenreiro, C. et al. (2023). Recent progress, trends, and new challenges in the electrochemical production of green hydrogen coupled to selective electrooxidation of 5-hydroxymethylfurfural (HMF). *RSC Adv.* 13 (40): 28307.

39 You, B., Liu, X., Jiang, N. et al. (2016). A general strategy for decoupled hydrogen production from water splitting by integrating oxidative biomass valorization. *J. Am. Chem. Soc.* 138 (41): 13639.

40 Xu, P., Bao, Z., Zhao, Y. et al. (2023). Anionic regulation and heteroatom doping of Ni-based electrocatalysts to boost biomass valorization coupled with hydrogen production. *Adv. Energy Mater.* https://doi.org/10.1002/aenm.202303557.

41 Xiao, D., Bao, X., Dai, D. et al. (2023). Boosting the electrochemical 5-hydroxymethylfurfural oxidation by balancing the competitive adsorption of organic and OH^- over controllable reconstructed Ni_3S_2/NiO_x. *Adv. Mater.* 35 (45): 2304133.

42 Formenti, D., Ferretti, F., Scharnagl, F.K. et al. (2019). Reduction of nitro compounds using 3d-non-noble metal catalysts. *Chem. Rev.* 119 (4): 2611.

43 Dutta, S., Yu, I.K.M., Tsang, D.C.W. et al. (2019). Green synthesis of gamma-valerolactone (GVL) through hydrogenation of biomass-derived levulinic acid using non-noble metal catalysts: a critical review. *Chem. Eng. J.* 372: 992.

44 Yang, G., Jiao, Y., Yan, H. et al. (2021). Electronic structure modulation of non-noble-metal-based catalysts for biomass electrooxidation reactions. *Small Struct.* 2 (10): 2100095.

45 Yu, Z., Lu, X., Wang, X. et al. (2020). Metal-catalyzed hydrogenation of biomass-derived furfural: particle size effects and regulation strategies. *ChemSusChem* 13 (19): 5185.

46 Yuan, C.-Z., Hui, K.S., Yin, H. et al. (2021). Regulating intrinsic electronic structures of transition-metal-based catalysts and the potential applications for electrocatalytic water splitting. *ACS Mater. Lett.* 3 (6): 752.

47 Zheng, L., Cui, X., Wang, X. et al. (2023). Pt/N co-doped porous carbon derived from bio-tar: a remarkable catalyst for efficient oxidation of 5-hydroxymethylfurfural under alkali-free condition: performance, mechanism and kinetics. *Chem. Eng. J.* 454: 140179.

48 Goetz, M.K., Usman, E., and Choi, K.-S. (2023). Understanding and suppressing C–C cleavage during glycerol oxidation for C_3 chemical production. *ACS Catal.* 13 (24): 15758.

49 Li, J., Qiu, K., Sun, D. et al. (2023). Cu/ZIF-8 derived Cu/ZnO/C catalysts for efficient conversion of glycerol to lactic acid. *Catal. Lett.* https://doi.org/10.1007/s10562-023-04393-0.

50 Li, J., Zhang, C., Li, Y. et al. (2023). Rational design and structural regulation of robust catalysts for electrocatalytic hydrodechlorination: from nanostructures to single atoms. *ACS Catal.* 13 (14): 9633.

51 Torres-Pacheco, L.J., Álvarez-Contreras, L., Lair, V. et al. (2019). Electrocatalytic evaluation of sorbitol oxidation as a promising fuel in energy conversion using Au/C, Pd/C and Au–Pd/C synthesized through ionic liquids. *Fuel* 250: 103.

52 Xi, B. and Sun, X. (2019). Single-atom catalysts electrostatically stabilized by ionic liquids. *Chem* 5 (12): 3012.

53 Mark, L.O., Zhu, C., Medlin, J.W., and Heinz, H. (2020). Understanding the surface reactivity of ligand-protected metal nanoparticles for biomass upgrading. *ACS Catal.* 10 (10): 5462.

54 Li, G., Marinkovic, N., Wang, B. et al. (2022). Manipulating the microenvironment of surfactant-encapsulated Pt nanoparticles to promote activity and selectivity. *ACS Catal.* 12 (22): 13930.

55 Dai, L., Qin, Q., Zhao, X. et al. (2016). Electrochemical partial reforming of ethanol into ethyl acetate using ultrathin Co_3O_4 nanosheets as a highly selective anode catalyst. *ACS Central Sci.* 2 (8): 538.

56 Poerwoprajitno, A.R., Gloag, L., Watt, J. et al. (2020). Faceted branched nickel nanoparticles with tunable branch length for high-activity electrocatalytic oxidation of biomass. *Angew. Chem. Int. Ed.* 59 (36): 15487.

57 Zhou, S., Dai, F., Dang, C. et al. (2019). Scale-up biopolymer-chelated fabrication of cobalt nanoparticles encapsulated in N-enriched graphene shells for biofuel upgrade with formic acid. *Green Chem.* 21 (17): 4732.

58 Wang, T., Cao, X., Qin, H. et al. (2021). Integrating energy-saving hydrogen production with methanol electrooxidation over mo modified Co_4N nanoarrays. *J. Mater. Chem. A* 9 (37): 21094.

59 Song, X., Liu, X., Wang, H. et al. (2020). Improved performance of nickel boride by phosphorus doping as an efficient electrocatalyst for the oxidation of 5-hydroxymethylfurfural to 2,5-furandicarboxylic acid. *Ind. Eng. Chem. Res.* 59 (39): 17348.

60 Chen, G.-F., Luo, Y., Ding, L.-X., and Wang, H. (2018). Low-voltage electrolytic hydrogen production derived from efficient water and ethanol oxidation on fluorine-modified feooh anode. *ACS Catal.* 8 (1): 526.

61 Wang, W., Zhu, Y.-B., Wen, Q. et al. (2019). Modulation of molecular spatial distribution and chemisorption with perforated nanosheets for ethanol electro-oxidation. *Adv. Mater.* 31 (28): 1900528.

62 Deng, X., Kang, X., Li, M. et al. (2020). Coupling efficient biomass upgrading with H_2 production via bifunctional Cu_xS@NiCo-LDH core–shell nanoarray electrocatalysts. *J. Mater. Chem. A* 8 (3): 1138.

63 Dubale, A.A., Zheng, Y., Wang, H. et al. (2020). High-performance bismuth-doped nickel aerogel electrocatalyst for the methanol oxidation reaction. *Angew. Chem. Int. Ed.* 59 (33): 13891.

64 Chen, W., Xie, C., Wang, Y. et al. (2020). Activity origins and design principles of nickel-based catalysts for nucleophile electrooxidation. *Chem* 6 (11): 2974.

65 Xu, G., Chen, C., Li, M. et al. (2022). W exsolution promotes the in situ reconstruction of a NiW electrode with rich active sites for the electrocatalytic oxidation of 5-hydroxymethylfurfural (HMF). *Catal. Sci. Technol.* 12 (10): 3363.

66 Lu, Y., Dong, C.-L., Huang, Y.-C. et al. (2020). Hierarchically nanostructured NiO-Co_3O_4 with rich interface defects for the electro-oxidation of 5-hydroxymethylfurfural. *Sci. China Chem.* 63 (7): 980.

67 Liu, J. and Tao, S. (2023). Laser promoting oxygen vacancies generation in alloy via Mo for HMF electrochemical oxidation. *Adv. Sci.* 10 (27): 2302641.

68 Velázquez-Hernández, I., Olivas, A., Bañuelos, J.A. et al. (2024). Defect engineering on metal oxides (Zn, Mo and Fe) and their impact on the crude glycerol biofuel oxidation. *Int. J. Hydrogen Energy* 52: 1033.

69 Yang, M., Wu, K., Sun, S., and Ren, Y. (2022). Regulating oxygen defects via atomically dispersed alumina on Pt/WO_x catalyst for enhanced hydrogenolysis of glycerol to 1,3-propanediol. *Appl. Catal. B-Environ.* 307: 121207.

70 Song, Y., Li, Z., Fan, K. et al. (2021). Ultrathin layered double hydroxides nanosheets array towards efficient electrooxidation of 5-hydroxymethylfurfural coupled with hydrogen generation. *Appl. Catal. B-Environ.* 299: 120669.

71 Lu, Y., Liu, T., Dong, C.-L. et al. (2022). Tailoring competitive adsorption sites by oxygen-vacancy on cobalt oxides to enhance the electrooxidation of biomass. *Adv. Mater.* 34 (2): 2107185.

72 Qi, Y.-F., Wang, K.-Y., Sun, Y. et al. (2022). Engineering the electronic structure of nife layered double hydroxide nanosheet array by implanting cationic vacancies for efficient electrochemical conversion of 5-hydroxymethylfurfural to 2,5-furandicarboxylic acid. *ACS Sustain. Chem. Eng.* 10 (1): 645.

73 Chen, X., Zhong, X., Yuan, B. et al. (2019). Defect engineering of nickel hydroxide nanosheets by ostwald ripening for enhanced selective electrocatalytic alcohol oxidation. *Green Chem.* 21 (3): 578.

74 Thapa, L., Bhaumik, A., Mondal, S., and Retna Raj, C. (2023). A heterostructured electrocatalyst for the electrochemical valorization of 5-hydroxymethylfurfural coupled with the hydrogen evolution reaction. *J. Mater. Chem. A* 11 (47): 26242.

75 Wang, J., Zhao, Z., Shen, C. et al. (2021). Ni/NiO heterostructures encapsulated in oxygen-doped graphene as multifunctional electrocatalysts for the her, uor and hmf oxidation reaction. *Catal. Sci. Technol.* 11 (7): 2480.

76 Zhang, Y., Xue, Z., Zhao, X. et al. (2022). Controllable and facile preparation of Co_9S_8–Ni_3S_2 heterostructures embedded with N,S,O-tri-doped carbon for electrocatalytic oxidation of 5-hydroxymethylfurfural. *Green Chem.* 24 (4): 1721.

77 Yang, G., Jiao, Y., Yan, H. et al. (2020). Interfacial engineering of MoO_2-FeP heterojunction for highly efficient hydrogen evolution coupled with biomass electrooxidation. *Adv. Mater.* 32 (17): 2000455.

78 Zhou, B., Li, Y., Zou, Y. et al. (2021). Platinum modulates redox properties and 5-hydroxymethylfurfural adsorption kinetics of $Ni(OH)_2$ for biomass upgrading. *Angew. Chem. Int. Ed.* 60 (42): 22908.

79 Zhang, Y., Qiu, Y., Ma, Z. et al. (2021). Core-corona Co/CoP clusters strung on carbon nanotubes as a schottky catalyst for glucose oxidation assisted H_2 production. *J. Mater. Chem. A* 9 (17): 10893.

80 Li, S., Dong, M., Yang, J. et al. (2021). Selective hydrogenation of 5-(hydroxymethyl)furfural to 5-methylfurfural over single atomic metals anchored on Nb_2O_5. *Nat. Commun.* 12 (1): 584.

81 Zhu, W., Meng, Y., Yang, C. et al. (2021). Effect of coordination environment surrounding a single Pt site on the liquid-phase aerobic oxidation of 5-hydroxymethylfurfural. *ACS Appl. Mater. Inter.* 13 (41): 48582.

82 Hu, Y., Li, B., Yu, C., and Li, Z. (2023). Mechanochemical preparation of single atom catalysts for versatile catalytic applications: a perspective review. *Mater. Today* 63: 288.

83 Yang, Q., Jiang, Y., Zhuo, H. et al. (2023). Recent progress of metal single-atom catalysts for energy applications. *Nano Energy* 111: 108404.

84 Lu, Y., Liu, T., Dong, C.-L. et al. (2021). Tuning the selective adsorption site of biomass on Co_3O_4 by Ir single atoms for electrosynthesis. *Adv. Mater.* 33 (8): 2007056.

85 Ge, R., Wang, Y., Li, Z. et al. (2022). Selective electrooxidation of biomass-derived alcohols to aldehydes in a neutral medium: promoted water dissociation over a nickel-oxide-supported ruthenium single-atom catalyst. *Angew. Chem. Int. Ed.* 61 (19): e202200211.

86 Xu, Y., Liu, M., Wang, M. et al. (2022). Methanol electroreforming coupled to green hydrogen production over bifunctional NiIr-based metal-organic framework nanosheet arrays. *Appl. Catal. B-Environ.* 300: 120753.

87 Wang, G., Chen, J., Li, K. et al. (2022). Cost-effective and durable electrocatalysts for CO-electrolysis of CO_2 conversion and glycerol upgrading. *Nano Energy* 92: 106751.

88 Liu, H., Lee, T.-H., Chen, Y. et al. (2021). Paired electrolysis of 5-(hydroxymethyl)furfural in flow cells with a high-performance oxide-derived silver cathode. *Green Chem.* 23 (14): 5056.

26

Porous Materials for Electrocatalytic Biomass Conversion

Shun Zhang and Meiting Zhao

Tianjin University, Tianjin Key Laboratory of Molecular Optoelectronic Sciences, Institute of Molecular Aggregation Science, Department of Chemistry, Tianjin, 30007, China

26.1 Introduction

Taking into account the sequential reaction steps of the electrocatalytic process, i.e. reactant diffusion, adsorption, electron transfer, product desorption, and product diffusion, the electrocatalytic performance is strongly dependent on both the intrinsic activity and the mass transfer capacity of the electrocatalyst [1]. In this context, constructing porous materials for biomass conversion is one of the optimal strategies. Among the various types of porous materials, pure porous metal-based electrocatalysts are widely explored for biomass electrochemical reduction and oxidation reactions due to their abundant active sites. However, most of the porous metal-based electrocatalyst is synthesized via the template method, which bears the disadvantage of wasting reactants and using strong acid or alkali. By contrast, thriving MOFs and emerging COFs are attracting accumulated research attention due to their unique structural characteristics, such as highly exposed active sites, programmable skeletons, modifiable compositions, and feasible functionalization. Current examples of reported MOF and COF-based electrocatalysts have demonstrated attractive performance in biomass transformations as a consequence of their unique structures, which are comparable with the performance of conventional porous materials or even surpass them [2].

In this chapter, porous metal composites, transitional metal-based MOFs, and COFs will be involved. Particularly, the composition and nanostructure design of MOFs as well as the structure variation during electrocatalysis will be emphasized for electrocatalytic biomass conversion. Overall, there will be ~12 pages for this chapter.

26.2 Porous Materials for Biomass Oxidation Reaction

Compared with the bulk material counterparts, porous metal composite materials can enhance the electrocatalytic biomass oxidation activity by facilitating the mass

Functional Materials for Electrocatalytic Energy Conversion, First Edition.
Edited by Zhicheng Zhang, Meiting Zhao, and Yuchen Qin.

transport or decreasing the diffusion range due to their large surface area, and porosity channels. In 2020, the Fu group reported a novel porous nanospindle composite for both hydrogen evolution reaction (HER) and biomass electrooxidation reaction (BEOR) (Figure 26.1a) [3]. This electrocatalyst is composed of a carbon layer coated on MoO_2-FeP heterojunction (MoO_2-FeP@C) with abundant active interfaces (Figure 26.1b–d). During the two-step hydrothermal-phosphorization procedure, the Mo element is introduced by the polyoxometalate (POMOF), P is derived from the phosphorization reagent (NaH_2PO_4) or precursor phosphomolybdic acid, and the carbon layer arises from the pyrolysis of 1,3,5-benzenetricarboxylic acid. The generated interfaces between MoO_2 and FeP triggered the electron redistribution to the electro-rich FeP and electron deficiency on MoO_2 (Figure 26.1e,f). Besides, such porous structures of the nanospindles contribute to the sufficient contact between MoO_2-FeP heterojunction and reactants. These features jointly resulted in the near 100% conversion and 98.6% yield of 2,5-furandicarboxylic acid (FDCA), as well as high stability, which is superior to that of FeP and MoO_2 (Figure 26.1g–i). Density functional theory (DFT) calculation revealed that the stronger adsorption ability of MoO_2-FeP to 5-hydroxymethylfurfural (HMF), which is derived from its electron-poor nature on MoO_2, is the key to its outstanding activity (Figure 26.1j).

Au-based porous nanocomposites are widely used for glucose electrooxidation and sensors [4]. For example, Ding reported a nonporous Au–Ag alloy electrode for glucose oxidation [5]. The porosity features could be tuned by the dealloying time using HNO_3 as etching agent. As expected, the nano-structured Au–Ag alloy exhibited higher current density response than the bulk sample. The presence of tiny Ag could promote the activity of glucose oxidation through the enhanced chemisorption of OH^-. In another report, Li displayed a PtBi bimetallic nanoparticle-decorated nonporous gold electrocatalyst (NPG-PtBi) for direct glucose fuels application [6]. As a result, NPG-PtBi delivered a power output of 8 mW cm^{-1}, which is 33 times higher than that of Pt/C and PtBi electrodes. Detailed experiments proved that the superior activity of NPG-PtBi toward glucose oxidation is derived from the accessible channels, which are beneficial for the mass transfer, and the enhanced OH^- adsorption due to the synergetic effect of Bi. Glycerol electrooxidation has also been carried out on nanoporous metal composites. Yang shows an efficient electrocatalyst of Au–Cu bimetallic composite for ethylene glycol and glycerol electrooxidation [7]. The self-supported porous 2D AuCu triangular nanoprisms were fabricated by the wet-chemical galvanic replacement reaction method that uses copper nanoprism as template with the assistance of KBr and polyvinyl pyrrolidone (PVP). Benefited from the inherent nanoporous morphology features as well as the electronic effects between Au and Cu, Au_1Cu_1 exhibits the highest current density and mass activity (2877 and 2263 mA mg^{-1}) response for EG and glycerol electrooxidation, almost 3.0 and 3.9-fold enhancement over monometallic Au electrocatalyst.

Metal–organic frameworks (MOFs) are excellent candidates for biomass platform molecules for electrocatalytic oxidation due to their high surface areas, modifiable compositions, and easy functionalization. Therefore, researchers developed different kinds of MOF or MOF-derived electrocatalysts for biomass oxidation.

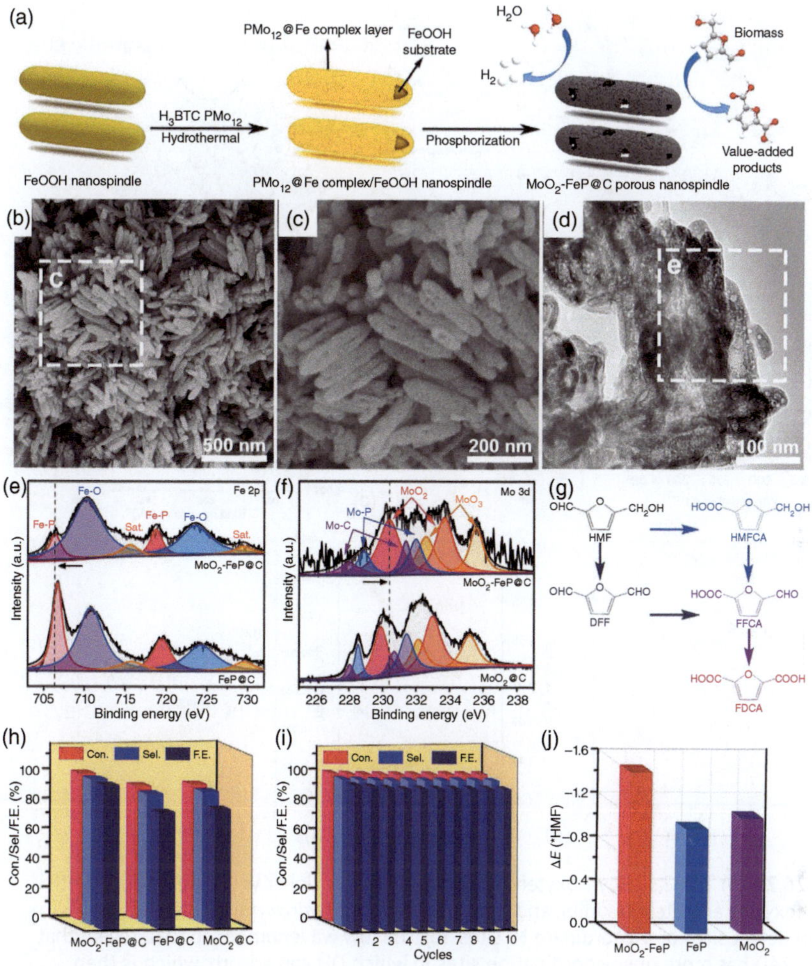

Figure 26.1 (a) Schematic illustration of the synthesis of the MoO_2-FeP@C hybrid. (b,c) SEM images, (d) TEM images. (e) The X-ray photoelectron spectroscopy (XPS) spectra of Fe 2p and (f) Mo 3d for MoO_2-FeP@C and FeP@C. (g) Two possible pathways of HMF oxidation to FDCA. (h) HMF conversion, FDCA selectivity, and FE of MoO_2-FeP@C, FeP@C, and MoO_2@C. (i) The HMF EOR of MoO_2-FeP@C under 10 successive cycles. (j) HMF adsorption energy on MoO2-FeP, FeP, and MoO_2 systems. Source: Reproduced with permission of Yang et al. [3], Wiley-VCH.

For example, taking the advantage of coordination strategy in MOF synthesis by partially replacing the 2-hydroxyterephthalic acid with 2,5-dihydroxyterephthalic acid, Nikolay developed a defective MOF (Ni-MOF-74D) electrocatalyst for HMF oxidation (Figure 26.2a) [8]. Experiments show that the Ni-MOF-74D is featured with a typical Ni-O_4 structure (Figure 26.2a). The unsaturated Ni sites and the induced electronic variation led to unique interactions with HMF and OH^- (Figure 26.2b,c) and therefore faster reaction kinetics on Ni-MOF-74D (Tafel slope, $160\,mV\,dec^{-1}$) than Ni-MOF-74 (Tafel slope, $300\,mV\,dec^{-1}$, Figure 26.2d,e), as

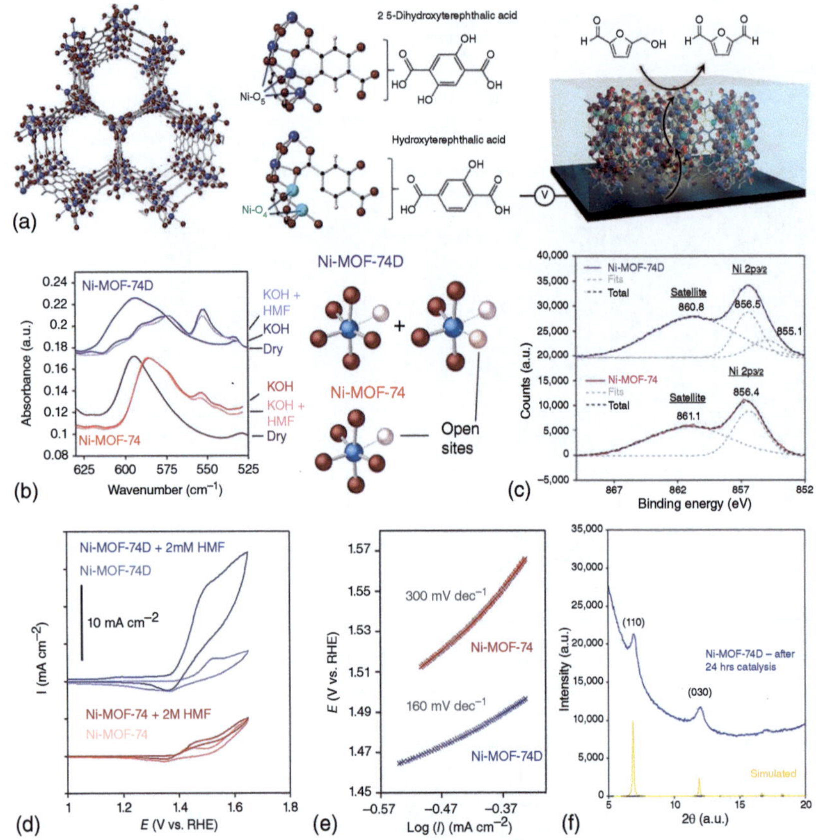

Figure 26.2 (a) The 2,5-dihydroxyterepthalic acid organic linker will bind through both the carboxylate and alcohol units, and replacing it with 2-hydroxyterepthalic acid will result in defect sites of 4-coordinate Ni sites. (b) The low wavenumber spectra reveal that Ni-MOF-74D has more open coordination sites in which OH can adsorb, which is then partially displaced by HMF. (c) XPS spectra of Ni-MOF-74D and Ni-MOF-74. (d) CV measurements in the presence of 2 mM HMF highlight the enhanced catalytic activity of Ni-MOF-74D. (e) Tafel slopes of Ni-MOF-74D and Ni-MOF-74. (f) PXRD pattern of Ni-MOF-74D after 24 hours of electrolysis shows that the principal (110) and (030) peaks remain. Source: Heidary et al., [8]/Royal Society of Chemistry/CC BY 4.0.

well as a high FDCA yield (95%). Further, the well-maintenance of the MOF XRD pattern shed light on the understanding of the structure–activity relationship during the reaction (Figure 26.2f). In another report, MOFs with similar M-O_4 coordination environments but varied metal centers (Ni, Co, Cu) were explored for HMFOR. During the electrooxidation experiments, the conductive ligand (2,3,6,7,10,11-hexahydroxytriphenylene hydrate) in the MOF helped the Co-CAT and Ni-CAT to be the active electrocatalyst for HMFOR with FE of 97.6% and 86.8%, respectively [9]. The relatively smaller FE of Ni-CAT might result from the occurrence of competitive oxygen evolution reaction (OER) reaction. Using MOF as active electrocatalyst is a practical method for catalyst design, which could be further supported by other reports. For example, Zhao designed a Ni-based MOF in situ grow onto $Ni(OH)_2$ nanosheet (S,N-MOFs@$Ni(OH)_2$-NSs/NF) using S,N-containing ligand (2,5-thiophenedicarboxylic acid, 4,4′-bipyridine) [10]. The presence of the S, N-containing ligand could promote the adsorption of HMF and suppress OER reaction at the same time, leading to a superior HMFOR activity of near 100% yield and FE at 1.4 V vs. reversible hydrogen electrode (RHE), which is higher than that of $Ni(OH)_2$ without the ligand.

MOF-derived electrocatalyst is also a typical kind for biomass conversion. The Luo group reported a "Nanoplatelet-on-Nanoarray" NiCo hydroxide-based electrocatalyst (t-NiCo-MOF) using bimetallic MOF as precursor (Figure 26.3a) [11]. The NiCo-layered double hydroxide possesses orderly nanoplatelets and rich surface-active sites, which were electrochemically transformed from the stacking of 2D nanosheet MOF on Ni foam (Figure 26.3b–g), among which the ligand is removed from the metal center and followed by the substitution of OH^-. Owing to the higher surface area and synergetic effect between Ni and Co, t-Ni_1Co_1-MOF delivered a rapid kinetic response, high current density (Figure 26.3h–i), high FE of 98%, as well as a high electrocatalytic stability performance (Figure 26.3j). Besides, t-Ni_1Co_1-MOF can be adopted for the electrochemical conversion of other biomass-derived platform molecules, such as glycerol and methanol, with high productive activity.

Similar work has also been observed on Ni-based two-dimensional MOFs reported by the Shalom group [13]. In their work, bimetallic Fe/Ni-MIL-53 MOF was electro-deposited on carbon cloth and further transformed into highly porous metal oxide catalyst (iron–nickel oxide, $NiFeO_x$). Experiments demonstrated that $NiFeO_x$ can efficiently oxide the methanol, HMF, and aniline into carboxylic acid and nitrile with high conversion and productivity. In another work, Zhang reported a bionic mineralization method for scalable MOF film fabrication for an HMF oxidation with ampere-level current density [12]. In their research, a universal strategy by combining the metal corrosion chemistry and bionic mineralization was proposed to synthesize 16 MOFs on 7 metal supports under ambient conditions (Figure 26.4a–d). For different metals, the preparation conditions varied according to their Pourbaix diagram (electrode potential–pH relationship) and could be classified into two categories: metal oxidation coupling with hydrogen evolution or oxygen reduction (Figure 26.4e). Since the interfacial metal cation concentration determined the surface corrosion and mineralization, droplet of the ligand solution (e.g. carboxylate-, N-heterocycle-, phenolic-, and phosphonate) was utilized to trigger the well-oriented and close-packed nanostructures according to the Kelvin equation, which is different from that homogeneous situation (Figure 26.4a–d). Among the fabricated MOFs, the Co-NiC_2O_4/NF electrode, which is derived from the droplet of $H_2C_2O_4$ and subsequent ion exchange, was investigated for the electrocatalytic oxidation of HMF. As expected, Co-NiC_2O_4/NF electrode showed a remarkably high conversion efficiency of 99%, FDCA yield of 98%, FE of 98% and high catalytic stability (Figure 26.4f,g). Operando attenuated total reflection-Fourier transform infrared (ATR-FTIR) spectroscopy identified the preferable aldehyde oxidation pathway (intermediate of HMFCA) for the product FDCA (Figure 26.4i).

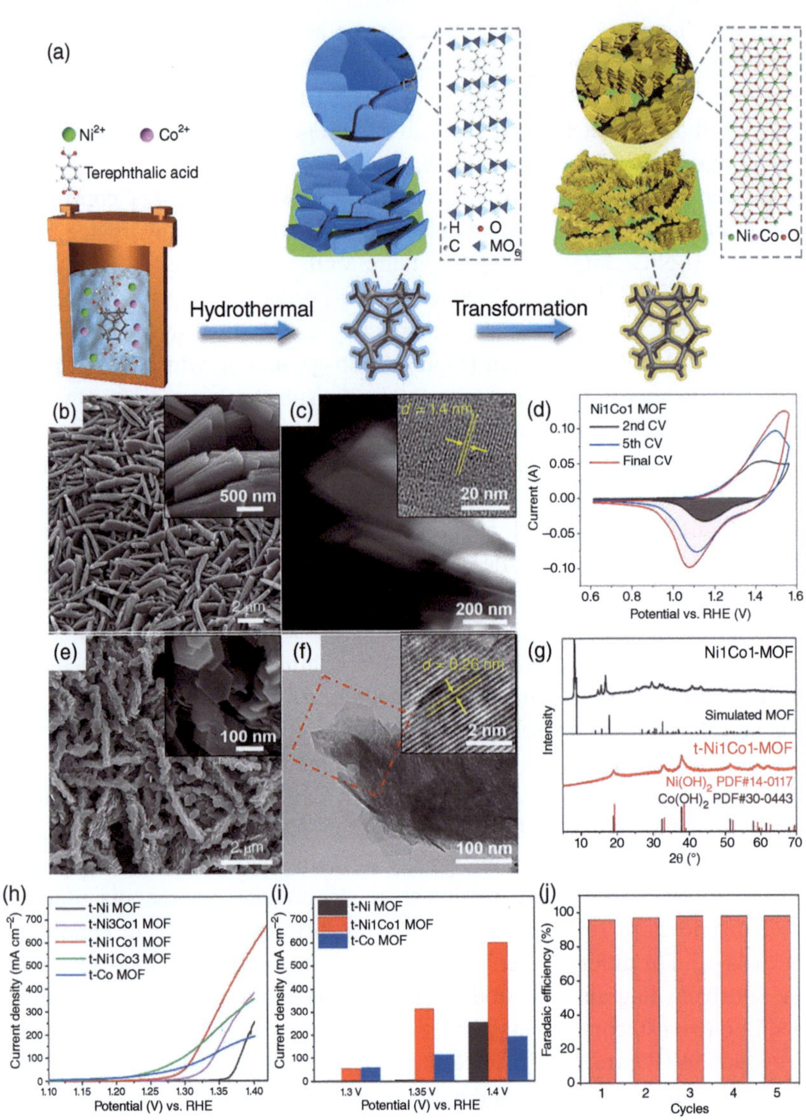

Figure 26.3 (a) Schematic illustration of the fabrication of 'nanoplatelet-on-nanoarray' electrocatalyst from transformation of hydrothermally prepared nickel–cobalt bimetallic MOF nanoarray on nickel foam. (b) SEM images, (c) TEM image of prepared Ni_1Co_1-MOF nanoarray. (d) SEM images, (e) TEM image of transformed Ni_1Co_1-MOF. (f) Development of CV curves and (g) XRD patterns of Ni_1Co_1-MOF and t-Ni_1Co_1-MOF with samples collected from multiple electrodes by sonication. (h) LSVs toward electrochemical HMF oxidation for t-MOFs with various Ni to Co ratios. (i) Comparison of current densities at 1.30, 1.35, and 1.40 V vs. RHE between t-Ni-MOF, t-Co-MOF, and t- Ni_1Co_1-MOF. (j) FEs toward FDCA in five successive runs. Source: Reproduced with permission of Deng et al. [11], Elsevier.

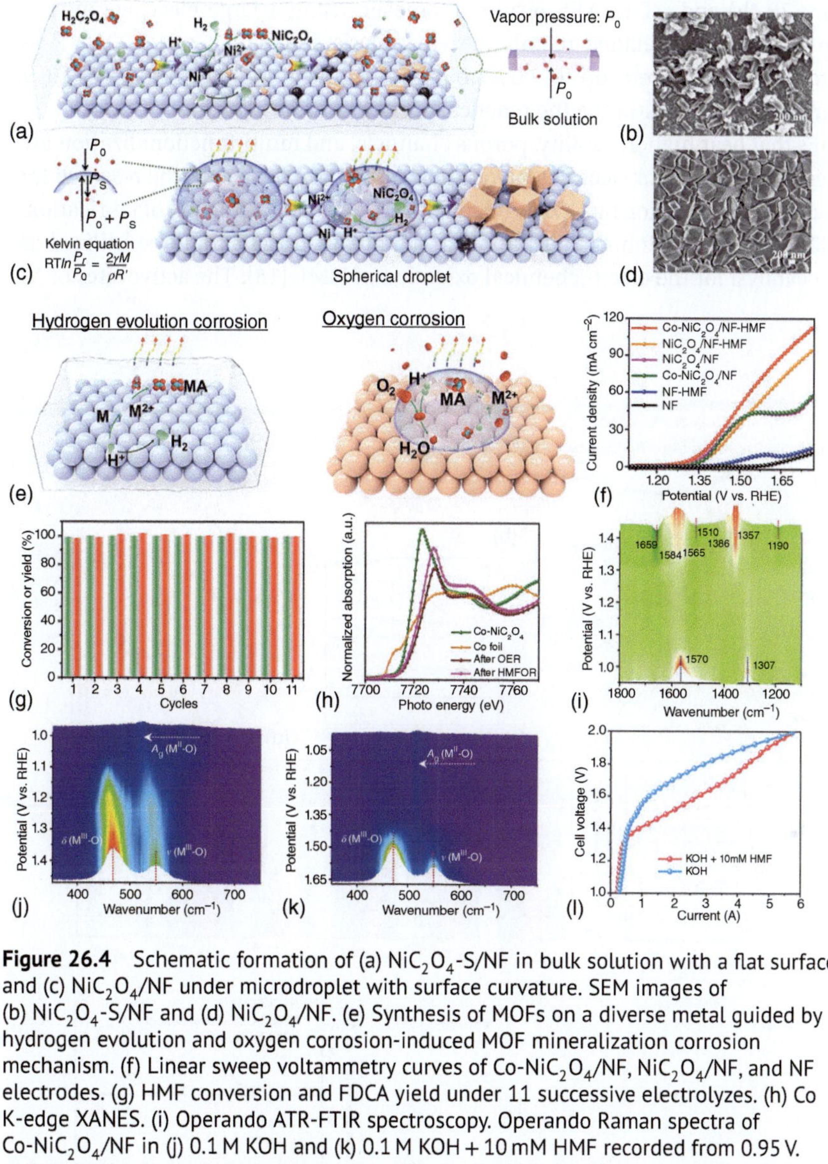

Figure 26.4 Schematic formation of (a) NiC_2O_4-S/NF in bulk solution with a flat surface and (c) NiC_2O_4/NF under microdroplet with surface curvature. SEM images of (b) NiC_2O_4-S/NF and (d) NiC_2O_4/NF. (e) Synthesis of MOFs on a diverse metal guided by hydrogen evolution and oxygen corrosion-induced MOF mineralization corrosion mechanism. (f) Linear sweep voltammetry curves of Co-NiC_2O_4/NF, NiC_2O_4/NF, and NF electrodes. (g) HMF conversion and FDCA yield under 11 successive electrolyzes. (h) Co K-edge XANES. (i) Operando ATR-FTIR spectroscopy. Operando Raman spectra of Co-NiC_2O_4/NF in (j) 0.1 M KOH and (k) 0.1 M KOH + 10 mM HMF recorded from 0.95 V. (l) Polarization curves of the electrolyzer with the Co-NiC_2O_4/NF electrocatalyst of 64 cm^2. Source: Reproduced with permission of Wang et al. [12], American Chemical Society.

To illustrate the real active sites during HMFOR, Operando Raman spectroscopy was utilized to reveal the structure transformation of Co-NiC_2O_4/NF during HMFOR. The metal oxalates would be changed into hydroxide under open-circuit potential in 0.1 M KOH. The typical bending vibration of Co-NiOOH under operational potential confirmed the formation of oxyhydroxide, which could be transformed back to hydroxide after injecting HMF and with the good maintenance of the morphology (Figure 26.4j,k), which could also be verified by the XANES

(Figure 26.4h) and FT-EXAFS results. Moreover, using a large electrode area of $64\,cm^{-2}$ in anion-exchange membrane electrolysis cell with $Co\text{-}NiC_2O_4/NF$ as electrocatalyst, a current up to 3.0 A could be obtained at a cell voltage of 1.6 V (Figure 26.4l), demonstrating the practical possibilities.

COFs that bear higher stability, porous channels, and multi-functionalization are promising platforms for electrocatalytic applications [14]. However, the research for biomass electrooxidation on COF is relatively rare due to the difficulty of fabrication. In 2020, Morris first found that the nickel(II)-modified COF could be utilized as electrocatalyst for the electrochemical oxidation of HMF [15]. The active sites of Ni

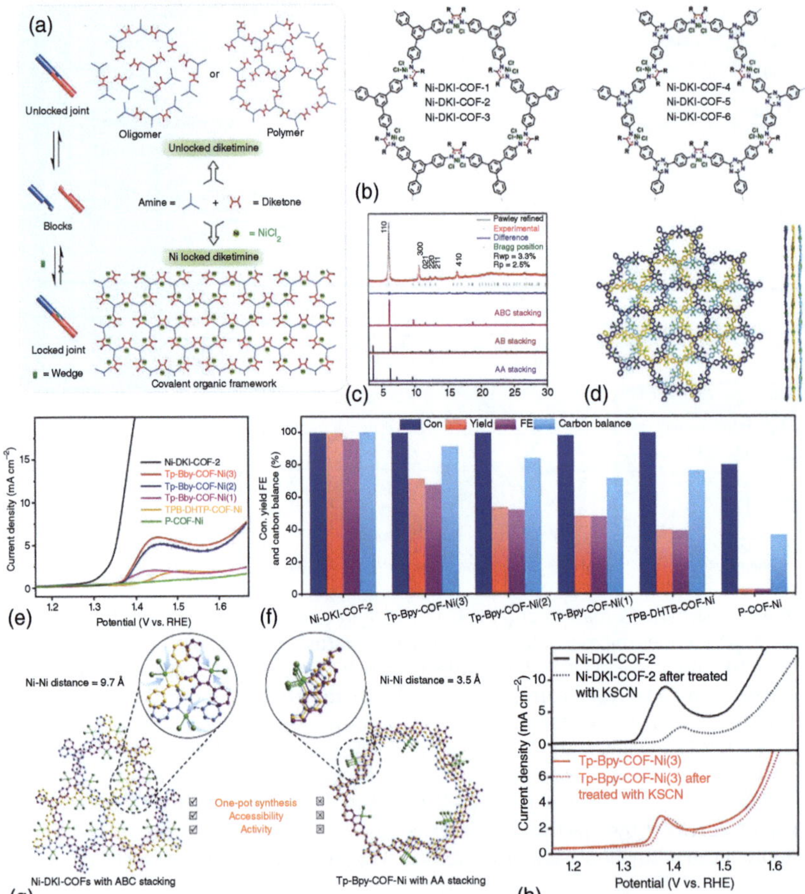

Figure 26.5 (a) Schematic illustration of the role of the wedge in locking the mortise and tenon blocks to form stable connection. (b) Structures of the Ni-DKI-COFs synthesized from corresponding diketones and amines. (c) Pawley refined, experimental, and simulated PXRD patterns of Ni-DKI-COF-1 with AA, AB, and ABC stacking models. (d) Top and side views of Ni-DKI-COF-1. (e) LSV curves and (f) the conversion, yield, FE, and carbon balance activity of HMFOR at various electrodes. (g) Schematic illustration of the different active site accessibility with ABC stacking and AA stacking. (h) LSV curves of COF electrodes before and after the treatment with KSCN. Source: Reproduced with permission of Zheng et al. [16], Wiley-VCH.

were introduced by the post-coordination method using TpBpy-COF as a platform. However, due to the lower metal loading and poor conductivity of the electrode, the COF composite TpBpy-Ni@FTO showed a conversion of 96% but an FDCA yield of 58%. Recently, the Zhao group reported nickel-diketimine-linked COFs (Ni-DKI-COFs) for the electrooxidation of biomass derivatives [16]. During the one-pot construction of the Ni-DKI-COFs, nickel site plays a key role in the lock of the highly flexible diketimine into the crystalline configuration with a typical ABC stacking mode (Figure 26.5a–d). Thanks to the high metal loading (up to 13.25 wt%) and the distinctive staggered stacking structure, Ni-DKI-COF-2 delivered a high HMFOR performance of fast kinetics procedure and stoichiometric production of FDCA with 99.9% yield and 99.5% FE, which is far beyond that of Ni-based COF with AA stacking model (Figure 26.5e,f). Detailed experiments demonstrated that the ABC stacking model endows the nickel sites with higher exposure and affinity to the reactant or OH^- than the AA stacking, resulting in the high activity (Figure 26.5g,h). Moreover, the electrochemical behavior of Ni-DKI-COF could be further found in the other biomass derivatives such as ethanol, glucose, furfuryl alcohol, and glycerol, indicating the potential multifunctional ability.

26.3 Porous Materials for Biomass Reduction Reaction

Due to the hydrogenation of biomass molecules mainly occurring on the metal-based catalyst, therefore, endowing the metal-based catalysts with the porous property is attractive for activity regulation. For instance, Ding reported a dealloying method to construct nanoporous Cu (NP-Cu) for electrocatalytic hydrogenation of furfural to furfuryl alcohol [17]. The porosity in the NP-Cu is derived from the alkali etching of the aluminum element from $Cu_{17}Al_{83}$ alloy ribbons, and the pore size could be tuned by the concentration of the alkali. Interestingly, the NP-Cu coated on the electrode shows relatively poor activity than fluidized NP-Cu due to the amounts of electrons that could pass through the catalyst and rapidly absorb reactant molecule, leading to the electrocondensation side reaction prevailing over the hydrogenation and resulting in a decline of the activity. Detailed experiments show that the fluidized NP-Cu-2 with a moderate active surface area and pore size delivered a higher performance of 96% selectivity and 95% FE on the electrochemical hydrogenation of furfural into alcohol in 0.1 M phosphate buffer solution than samples with larger or smaller pore size. Also, the activity is higher than that of commercial Cu sheet, Cu net, and fluidized Cu nanoparticles, proving the strategy efficiency of porous metal catalysts for biomass hydrogenation. Combining porous Cu with second metal is an applicable method to tune the electrocatalytic activity through the synergetic effect. For example, using 3D Cu open-cell foam, Benito developed a 3D Ag/Cu foam composite for electroreduction of HMF to 2,5-bis(hydroxymethyl)furan (BHMF) [18]. The layer of Ag particle assembly on the Cu skeleton is obtained by the elaborate pulse electrodeposition technic. Due to the large available surface area, enhanced mass, and electron transfer, 3D Ag/Cu foam delivered a high conversion (90%), selectivity (99%), and FE (90%) for the HMF electroreduction into BHMF, which is higher than that of Cu foam, Ag foam, and the foil samples (Cu, Ag, and Ag/Cu foil), illustrating the importance of the porosity structure for the electrocatalysis.

Except for the noble metal-based electrocatalyst, p-Block-element-based metal composite is also efficient for biomass derivative hydrogenation. For instance, Park reported an example of using porous dendritic BiSn electrocatalysts for HMF hydrogenation [19]. The highly porous dendritic BiSn nanostructure was fabricated through a dynamic hydrogen bubble templating method using electrodeposition technic on Cu foil. SEM images identified the morphology of BiSn as crater-like surfaces with vertically hierarchical pores with sizes ranging from ~10 to ~300 mm at the outermost layer, and the craters consisted of dendritic branches. This porous structure features guaranteed the sufficient mass transfer as well as the rich active sites in Bi and BiSn samples. Despite the higher FE (near 100%) happened on dendritic Bi for the lower HMF concentration situation (20 mM), BiSn sample delivered a higher FE (near 100%) and current density of $-144\ mA\ cm^{-2}$ for the high HMF concentration condition (2.0 M) at −1.8 V, which might be derived from the metallic states of Bi via electron transfer from Sn in BiSn. And, such a high current density surpassed most reported articles.

MOFs and their derivatives can also be used for electrochemical biomass reduction reactions. As for the derivatives, MOFs could be used as a sacrifice template through the pyrolysis method to construct metal- or carbon-based electrocatalysts with high conductivity, large surface area, and interconnected hierarchical porous channels with effective mass/electron transport. For example, Zhang shows an electrochemical hydrogenolysis reaction of furfural into 2-methylfuran (MF) using MOF-derived CuPdx catalysts (CuPdx/C) [20]. After the pyrolysis of HKUST-1, the octahedral morphology was preserved and also the partially porous property (Figure 26.6a,b). Among the various products of furfural hydrogenation, MF is the most attractive due to its high energy density. To promote the catalytic reaction pathway to the formation of MF, Zhang induced Pd elements into the Cu/C ($CuPd_x$/C) through the partial galvanic replacement reaction. The amount of Pd doping in $CuPd_x$/C could be tuned by the concentration of the Pd precursor, and consequently, the electron transfer from Pd to Cu happened for $CuPd_x$/C according to the XPS results (Figure 26.6c,d). Detailed experiments displayed that the current density (Figure 26.6e) with or without furfural would increase with the amount of Pd due to the increased HER and electrochemical hydrogenation. At the same time, the FE of MF shows a volcano-shaped curve, and the optimal one is $CuPd_{0.021}$/C corresponding to a value of 46% in 0.1 M acetate buffer solution (pH = 5, Figure 26.6f). The FE could be further enhanced to 75% when the pH of the electrolyte dropped to 2.9 (Figure 26.6g). In situ surface-enhanced Raman spectroscopy (SERS) study illustrated that the adsorption model of furfural on Cu or $CuPd_{0.021}$ is C=O bond, and the $CuPd_{0.021}$/C behaves with a higher H_{ads} coverage than Cu/C during the electrochemical hydrogenation process (Figure 26.6h,i), which is the reason for the higher performance. This could be further proved by the kinetic isotopic effects (KIE) experiment, which shows that KIE is more significant at Cu/C than $CuPd_{0.021}$. DFT calculation also confirmed that the incorporation of

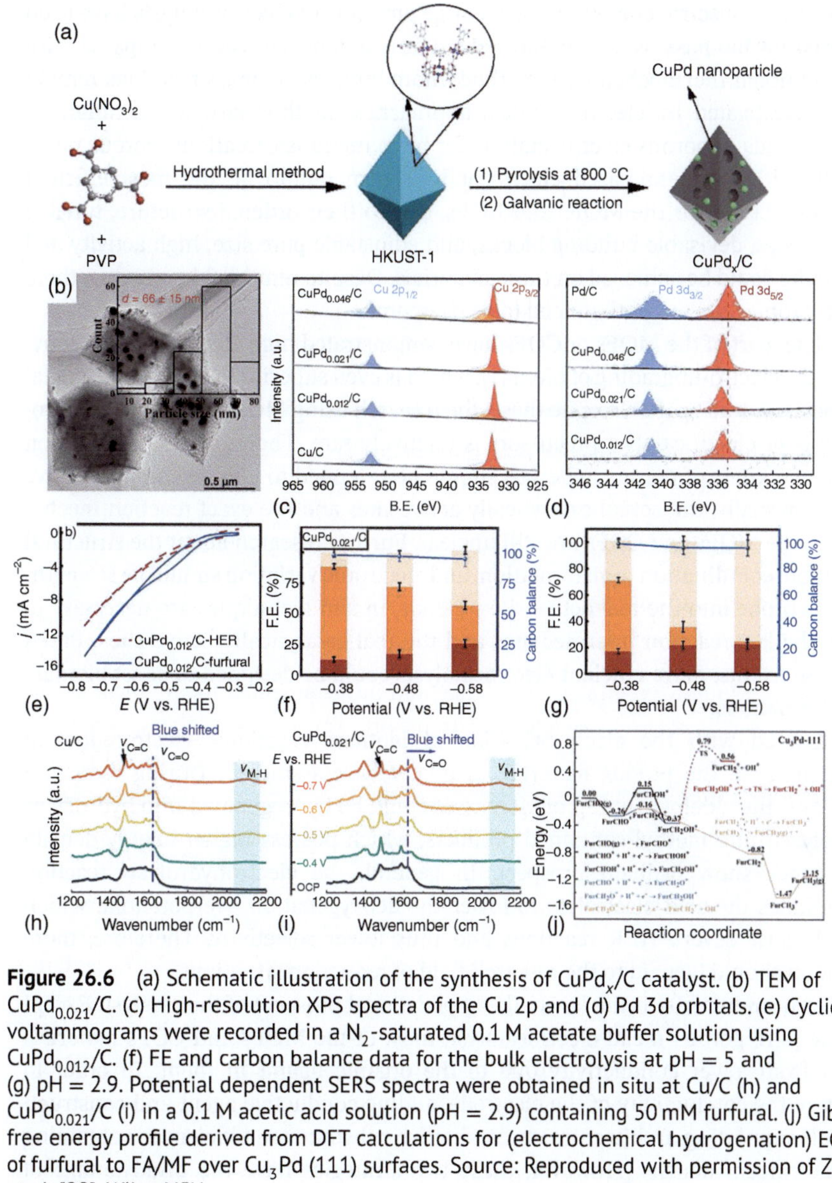

Figure 26.6 (a) Schematic illustration of the synthesis of $CuPd_x/C$ catalyst. (b) TEM of $CuPd_{0.021}/C$. (c) High-resolution XPS spectra of the Cu 2p and (d) Pd 3d orbitals. (e) Cyclic voltammograms were recorded in a N_2-saturated 0.1 M acetate buffer solution using $CuPd_{0.012}/C$. (f) FE and carbon balance data for the bulk electrolysis at pH = 5 and (g) pH = 2.9. Potential dependent SERS spectra were obtained in situ at Cu/C (h) and $CuPd_{0.021}/C$ (i) in a 0.1 M acetic acid solution (pH = 2.9) containing 50 mM furfural. (j) Gibbs free energy profile derived from DFT calculations for (electrochemical hydrogenation) ECH of furfural to FA/MF over Cu_3Pd (111) surfaces. Source: Reproduced with permission of Zhou et al. [20], Wiley-VCH.

Pd into Cu could reduce the key step of $FurCH_2$* to $FurCH_3$*, thereby facilitating the formation of MF (Figure 26.6j).

26.4 Conclusion and Outlook

Electrochemical upgrading of biomass coupled with clean energy (like solar, wind, or hydroenergy) can promote the sustainability of the biomass economy as well as the storage of electric energy. Hitherto, large amounts of electrocatalysts have been reported for biomass oxidation and reduction reactions. However, compared with the traditional thermochemistry method, more kinds of biomass reactions remain to be investigated by electrochemical approaches. In this part, we summarized various kinds of porous electrocatalysts for biomass transformations. Porous noble and transitional metal-based electrocatalysts turn out to be the most efficient ones. Especially, for the MOFs and COFs, due to their ordered structure, tunable metal nodes, devisable building blocks, and adjustable pore size, high activity and selectivity could be achieved on these materials. Despite remarkable progress, there are still some challenges that need to be overcome.

To date, part of the MOFs or COFs have demonstrated high electrocatalytic activity for the electro-upgrading of biomass, which is even superior to their porous metal composite counterparts. Despite these, the research using MOFs or COFs as electrocatalysts or electrocatalyst precursors is relatively rare. Therefore, the application of the two kinds of porous materials should be extended to a more comprehensive level. Especially, the actual catalytically active sites and the exact reaction mechanism of the MOFs and COFs are still unclear. Further research about the structural evolution, coordination transformation, and electronic variation should be strengthened to probe into the real active sites. Hence, in situ techniques are necessary to study the key reaction intermediates and the real catalytically active sites. Based on these results, more efficient electrocatalysts could be designed to boost biomass transformations.

Compared with the electrooxidation of biomass reactions, electroreduction of biomass on the porous materials is relatively less studied. During reduction reactions, the aldehyde group of biomasses could be hydrogenated into hydroxy or hydrolyzed into methyl-contained products, which possess higher energy density and hence show brilliant prospect. In general, the electrohydrolysis reaction necessitates the acid electrolyte to lower the deoxygenation overpotential, which would cause severe HER reactions and thus lower selectivity. Therefore, more attention should be paid to the research field of reduction situations.

Electroconductivity is the key to the applicational prospect of the catalyst. Except for the pure porous metal electrocatalyst, most of the MOFs and COFs materials suffer from lower conductivity due to the organic ligand in them. In order to improve the conductivity of the electrode, adding conductive agent and constructing self-supporting architecture on conductive substrate are efficient for obtaining higher current density. However, these two strategies have the disadvantage of using costly binder agents and wasting catalyst material. In this context, designing conductive MOFs and COFs via conjugated organic ligands is a promising way.

Acknowledgment

This work was supported by the National Natural Science Foundation of China (No.21905195, 22473083), Natural Science Foundation of Tianjin City (No. 20JCYBJC00800), PEIYANG Young Scholars Program of Tianjin University (No. 2020XRX-0023)...

References

1 Du, C., Li, P., Zhuang, Z. et al. (2022). Highly porous nanostructures: rational fabrication and promising application in energy electrocatalysis. *Coord. Chem. Rev.* 466: 214604.

2 Fang, R., Dhakshinamoorthy, A., Li, Y., and Garcia, H. (2020). Metal organic frameworks for biomass conversion. *Chem. Soc. Rev.* 49 (11): 3638.

3 Yang, G., Jiao, Y., Yan, H. et al. (2020). Interfacial engineering of MoO_2-FeP heterojunction for highly efficient hydrogen evolution coupled with biomass electrooxidation. *Adv. Mater.* 32 (17): 2000455.

4 Wang, J., Gao, H., Sun, F., and Xu, C. (2014). Nanoporous PtAu alloy as an electrochemical sensor for glucose and hydrogen peroxide. *Sensors Actuators B Chem.* 191: 612.

5 Liu, Z., Huang, L., Zhang, L. et al. (2009). Electrocatalytic oxidation of D-glucose at nanoporous Au and Au–Ag alloy electrodes in alkaline aqueous solutions. *Electrochim. Acta* 54 (28): 7286–7293.

6 Guo, H., Yin, H., Yan, X. et al. (2016). Pt-Bi decorated nanoporous gold for high performance direct glucose fuel cell. *Sci. Rep.* 6 (1): 39162.

7 Xu, H., Yan, B., Wang, J. et al. (2017). Self-supported porous 2D AuCu triangular nanoprisms as model electrocatalysts for ethylene glycol and glycerol oxidation. *J. Mater. Chem. A* 5 (30): 15932.

8 Heidary, N., Chartrand, D., Guiet, A., and Kornienko, N. (2021). Rational incorporation of defects within metal–organic frameworks generates highly active electrocatalytic sites. *Chem. Sci.* 12 (21): 7324.

9 Zhang, Y. and Kornienko, N. (2022). Conductive metal-organic frameworks bearing $M{-}O_4$ active sites as highly active biomass valorization electrocatalysts. *ChemSusChem* 15 (13): e202101587.

10 Wang, D., Gong, W., Zhang, J. et al. (2021). In situ growth of MOFs on $Ni(OH)_2$ for efficient electrocatalytic oxidation of 5-hydroxymethylfurfural. *Chem. Commun.* 57 (86): 11358.

11 Deng, X., Li, M., Fan, Y. et al. (2020). Constructing multifunctional 'Nanoplatelet-on-Nanoarray' electrocatalyst with unprecedented activity towards novel selective organic oxidation reactions to boost hydrogen production. *Appl. Catal. B Environ.* 278: 119339.

12 Wang, Y., Ma, J., Cao, X. et al. (2023). Bionic mineralization toward scalable MOF films for ampere-level biomass upgrading. *J. Am. Chem. Soc.* 145 (37): 20624.

13 Mondal, B., Karjule, N., Singh, C. et al. (2021). Unraveling the mechanisms of electrocatalytic oxygenation and dehydrogenation of organic molecules to value-added chemicals over a Ni–Fe oxide catalyst. *Adv. Energy Mater.* 11 (37): 2101858.

14 Kong, L., Zhong, M., Shuang, W. et al. (2020). Electrochemically active sites inside crystalline porous materials for energy storage and conversion. *Chem. Soc. Rev.* 49 (8): 2378.

15 Cai, M., Ding, S., Gibbons, B. et al. (2020). Nickel(II)-modified covalent-organic framework film for electrocatalytic oxidation of 5-hydroxymethylfurfural (HMF). *Chem. Commun.* 56 (92): 14361.

16 Zheng, C., Zhang, S., Li, Z. et al. (2023). Single site coordination enabled construction of metal-diketimine-linked covalent organic frameworks for boosted electrooxidation of biomass derivative. *Small* 19 (37): e2301331.

17 Yang, Z., Chou, X., Kan, H. et al. (2022). Nanoporous copper catalysts for the fluidized electrocatalytic hydrogenation of furfural to furfuryl alcohol. *ACS Sustain. Chem. Eng.* 10 (22): 7418.

18 Sanghez De Luna, G., Ho, P.H., Lolli, A. et al. (2020). Ag electrodeposited on Cu open-cell foams for the selective electroreduction of 5-Hydroxymethylfurfural. *ChemElectroChem* 7 (5): 1238.

19 Piao, G., Yoon, S.H., Cha, H.G. et al. (2022). Porous dendritic BiSn electrocatalysts for hydrogenation of 5-hydroxymethylfurfural. *J. Mater. Chem. A* 10 (45): 24006.

20 Zhou, P., Li, L., Mosali, V.S.S. et al. (2022). Electrochemical hydrogenation of furfural in aqueous acetic acid media with enhanced 2-methylfuran selectivity using CuPd bimetallic catalysts. *Angew. Chem. Int. Ed.* 61 (13): e202117809.

27

Summary and Perspective

Ruoqing Xu and Jingjie Ge

The Hong Kong Polytechnic University, Department of Applied Biology and Chemical Technology, 11 YuK Choi Road, Hung Hom, Kowloon, Hong Kong

The development of viable and sustainable new sources of energy has become a significant challenge for today's society, driven by increasing environmental concerns and the rapid decline of traditional fossil fuels. To fully utilize renewable energy sources, research on energy storage and conversion has attracted considerable attention. The design and optimization of catalysts are regarded as key strategies for achieving high performance in electrochemical energy conversion and storage devices. In recent years, functional materials have made remarkable progress in the field of electrocatalysis and have become essential for electrocatalytic energy conversion. These materials offer a more attractive solution for sustainable energy conversion, with lower costs and wider resource availability than traditional precious metal electrocatalytic materials. By studying various types of functional materials and gaining an in-depth understanding of their catalytic mechanism, active sites, electron transport properties, and other key parameters, we can provide valuable guidance for the design and preparation of highly efficient electrocatalytic devices and systems, ultimately enhancing the efficiency and stability of energy conversion. The design and synthesis of functional materials with excellent catalytic properties can facilitate key energy conversion processes, such as efficient water electrolysis for HER, ORR, carbon dioxide reduction CO_2RR, etc., thereby promoting the development and utilization of clean energy. Additionally, optimizing the structure, composition, and surface properties of functional materials can improve electrocatalytic activity and selectivity, reduce energy loss and side reactions, and increase energy conversion efficiency. While ongoing research on functional materials has led to a preliminary understanding of the design and mechanism of electrocatalysts, there remains a shortage of relevant studies and emerging research frontiers in the field, which presents obstacles to researchers and beginners.

This book aims to address all these aspects and discusses research advancements in various functional materials for electrocatalytic energy conversion. It is organized into four sections. The first part provides a brief overview of the origin, current

Functional Materials for Electrocatalytic Energy Conversion, First Edition.
Edited by Zhicheng Zhang, Meiting Zhao, and Yuchen Qin.

status, prospects, as well as challenges and opportunities associated with functional materials in electrocatalytic energy conversion. The second part focuses on electrocatalytic energy conversion from a theoretical perspective, offering insights into the mechanism and nature of electrocatalytic reactions. The third part reviews various electrocatalytic reactions on advanced functional materials and their properties in detail. Finally, the last section presents the challenges and opportunities for the rational design and controlled synthesis of advanced functional materials for electrocatalytic energy conversion.

Functional materials and electrocatalysis play a crucial role in various fields, such as energy conversion, environmental remediation, and chemical synthesis. Despite significant progress, the development of functional materials for electrocatalytic energy conversion remains in its early stages, with several challenges that need to be solved in both theoretical research and practical applications. For example, in practical applications, functional materials often encounter issues such as unclear reaction mechanisms and uncertain active sites. These challenges can lead to insufficient activity and dynamic oxidation of the materials during reactions, which in turn significantly affects their catalytic activity. Therefore, maintaining and enhancing the activity of functional materials is a crucial task for the future. To solve this problem, functional materials can be further regulated and optimized in terms of morphology, size, phase state, and atomic microstructure. Functional materials with high surface area and porous structure can be fabricated by controlling the shape, surface structure, and porosity of the materials to provide more active sites for the reaction and enhance the reaction interface, thereby enhancing catalytic performance. In addition, the electronic structure and chemical reaction properties of functional materials can also be fine-tuned by carefully selecting their crystalline structure and composition. Introducing suitable doping elements or modifying the crystal lattice can alter the oxidation state or electron transport properties, thereby enhancing the activity and stability of the functional materials. This leads to more efficient and reliable energy conversion and catalytic reactions. At the same time, ensuring the long-term stability and durability of functional materials is also vital. Under harsh operating conditions, such as high temperatures, high pressures, and corrosive environments, materials may suffer from structural degradation, surface corrosion, and inactivation of active sites. Developing strategies to improve their stability and durability is critical to their practical application. This may involve using protective coatings, optimizing material composition, and understanding of degradation mechanisms to mitigate their effects. These improvements will facilitate the development and application of functional materials, promoting scientific progress and technological innovation in related fields.

Furthermore, the design and development of electrocatalysts for efficient, economical, and sustainable energy conversion will be a key focus in the future. This includes the search for new catalyst materials, the exploration of suitable structures and compositions, and the optimization of the surface-active sites of the catalysts to enhance their catalytic performance. Understanding the mechanistic and kinetic behavior of electrocatalytic reactions, and the interactions between the catalyst and electrode interfaces, can guide catalyst design and performance optimization.

However, similar to the challenges faced by functional materials, there are pressing issues in the field of electrocatalysis that require attention. Currently, there still exist unclear reaction paths in catalytic reactions. For instance, in N_2RR, avoiding side reactions is difficult, leading to an increase in the number of product species and consequently reducing the selectivity and conversion of the desired products. In addition to the catalytic mechanism, achieving the industrialization of electrocatalysts is also a difficult task. Despite the remarkable results achieved in the laboratory, scaling up electrocatalytic technology to an industrial scale presents obstacles related to cost, sustainability, integration, and reactor design. Overall, the field of electrocatalysis has great prospects and opportunities, but it also faces challenges such as improving catalyst performance, addressing unclear reaction pathways and side reactions, achieving sustainability, and realizing the industrial application of electrocatalysts. By overcoming these challenges, electrocatalytic technology is expected to make significant contributions to energy conversion, environmental protection, and sustainable development.

Index

Functional Materials for Electrocatalytic Energy Conversion, First Edition.
Edited by Zhicheng Zhang, Meiting Zhao, and Yuchen Qin.

d

e

p